GEOLOGY AND LANDSCAPES OF AMERICA'S NATIONAL PARKS

GEOLOGY AND LANDSCAPES

OF AMERICA'S NATIONAL PARKS

David Osleger

UNIVERSITY OF CALIFORNIA, DAVIS

NEW YORK OXFORD

OXFORD UNIVERSITY PRESS

Oxford University Press is a department of the University of Oxford.
It furthers the University's objective of excellence in research, scholarship,
and education by publishing worldwide. Oxford is a registered trade mark of
Oxford University Press in the UK and certain other countries.

Published in the United States of America by Oxford University Press
198 Madison Avenue, New York, NY 10016, United States of America.

For titles covered by Section 112 of the US Higher Education
Opportunity Act, please visit www.oup.com/us/he for the latest
information about pricing and alternate formats.

Library of Congress Cataloging-in-Publication Data
CIP data is on file at the Library of Congress
978-0-19-930120-1

Printing number: 9 8 7 6 5 4 3 2 1
Printed by LSC Communications, Inc., United States of America

Brief Table of Contents

Contents

Preface

Americans love their national parks. The iconic image of the American family piling into an overstuffed minivan for their annual summer pilgrimage to visit national parks has been portrayed in movies, television shows, and countless books. Everyone appreciates the clockwork timing of Old Faithful, the sublime beauty of Half Dome, and the majesty of the Grand Canyon at dawn (Fig. P.1).

The aesthetic appeal of these landscapes is greatly enhanced by an understanding of the geologic framework in which they evolved. Being able to imagine the complex plumbing system through which superheated water explosively bursts to the surface adds a deeper pleasure to the geysers of Yellowstone. The recognition that the granitic rocks of Yosemite were once roiling masses of molten rock kilometers beneath the surface a hundred million years ago creates a more profound awareness as you gaze over the terrain from an overlook. Being able to visualize the colorful layers of rock at the Grand Canyon as an ever-changing tableau of shallow seas, river systems, and arid deserts makes the landscape more than just a treat for the eyes, but rather reaches deeper into the human psyche. The intent of this book is to enable the reader to transcend the experience of simply witnessing the grandeur of a national park by developing a deeper understanding of how those magnificent landscapes came to be.

The intended audience for this book is the curious nongeologist who might be taking an introductory class in geology or someone who might be planning a trip to a few national parks and wants to understand the landscapes they'll be seeing. My goal is to convey the geologic underpinnings of select national parks through clear images and graphics with just enough explanation for the reader to appreciate the landscapes of national parks at a level beyond the purely aesthetic.

Figure P.1   Colorado River within the Marble Canyon section of the Grand Canyon. View from the Nankoweap granaries. (NPS)

As the narrative builds from chapter to chapter, the reader will learn the fundamentals of tectonics, sedimentation, mountain-building, volcanism, glaciation, and other basic concepts as they are interwoven throughout the book in their appropriate context. The reader who wants greater detail can take a look at the numerous books, articles, and websites listed at the end of each chapter.

The parks and monuments described in the book are mainly the big, popular ones that are visited by most nature-lovers (Fig. P.2). In my attempt to keep the book from becoming unnecessarily encyclopedic, I had to reluctantly exclude a few noteworthy parks such as Big Bend, Guadalupe Mountains, Carlsbad Caverns, Great Basin, Isle Royale, Badlands, Cuyahoga Valley, and many others. In time, I'll compile chapters on each of these parks and include them in later editions or an online addendum of the digital version of this book. The vast and panoramic national parks of Alaska are a book unto themselves, and so they are not addressed in this book.

Structure and Themes

Geologists define discrete geologic provinces throughout North America that are characterized by similar landscapes and tectonic histories. Most of the nine Parts that make up this book use the geologic province to provide the broader context for the specific national parks that reside within that province. At the beginning of each Part, an introductory chapter provides a broad description of the characteristics and origin of the geologic province. These opening chapters establish a regional framework for each national park within that province, permitting a more focused view of individual parks in subsequent chapters.

For instance, Part 3, which is devoted to parks of the Sierra Nevada Province, begins with a description of the range, its drainage patterns, and overall topography. This information is followed by a short explanation of the mechanics of how tilted fault-block mountains such as the Sierra grow and evolve along active fault systems. Next, the origin of Sierran granitic rock is explained, introducing the concepts of plutonism and isostasy along the way. The final part of this chapter on the Sierra Nevada Province addresses convergent margin tectonism and the origin of magma. This expository chapter establishes the broader framework for a more detailed exploration of Yosemite and Sequoia–Kings Canyon national parks in the following chapters of Part 3.

A few overarching themes are woven into the texture of the book. One of these recurring themes is the role of plate tectonics in controlling the geologic origin and evolution of our national parks. The continual, imperceptible motion of large slabs of the Earth's outer shell, occurring over huge spans of geologic time, ultimately is responsible

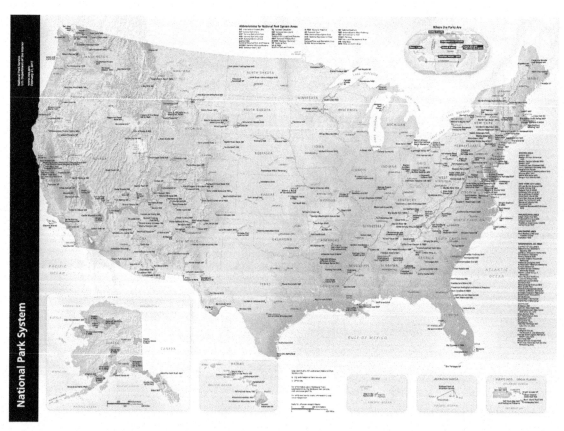

Figure P.2　Map of units within the National Park System. (NPS)

for the magnificent landscapes that we see at this moment in geologic time. Each of the national parks is a reflection of its particular tectonic history.

The integration of the fourth dimension of geologic time forms the second major theme of the book. An important aspect of understanding the geology of national parks is learning to incorporate the temporal chronology of events with the spatial elements of the landscape. In essence, I'd like to build an awareness of "what happened when" in addition to an understanding of how a particular geologic feature may have originated. This effort requires an understanding of the concept of what geologists call "deep time," the vast span of Earth history stretching back an incomprehensible 4.6 billion years. Abstract thinking in four dimensions can be difficult for the typical nongeologist, so I provide an overview of the geologic time scale and the concept of deep time in Chapter 1. Individual chapters are organized to maintain a simple chronologic framework that will keep the reader oriented. Chapters 33 and 35, which discuss the geologic evolution of the American West and the Appalachians, respectively, include a series of timelines that provide a visual history of events that influenced the western national parks. Comparable timelines for the geologic evolution of the Appalachians are integrated into Chapter 35.

A third theme of the book uses the national parks as "touchstones" to develop the broader story of the geologic origin and evolution of the North American continent. Anyone who's looked at a physiographic map of North America wonders how the rugged mountainous topography of the American West came to be. Why does it stretch from Denver to San Francisco and from Central America to Alaska? And why does the eastern flank of North America appear so subdued relative to the western flank? The structure of the first 8 Parts of the book methodically builds the larger story of the American West province by province, using individual national parks as touchstones to illustrate specific geologic features and events. In a similar manner, Chapters 35–39 use four national parks of the eastern United States to reveal greater detail about the larger narrative of the geologic evolution of the Appalachians. The intent is to have the reader view the parks not as isolated entities, but rather as related elements within a continuum of geologic events that affected all of the parks contemporaneously.

Excluding Alaska, the vast majority of America's national parks and monuments reside in the American West, so the balance of the book leans to the left side of the continent. Exceptional national parks do exist in the midwestern and eastern United States, of course, so Part 9 focuses on a few of the most popular of these parks. The Epilogue touches on some of the environmental threats to our national parks. It seeks to illustrate that our public lands shouldn't be thought of as pristine, unadulterated wilderness, but rather as communal property that deserves our thoughtful stewardship.

Suggestions for Using the Book

Chapter 1 provides a broad introduction to geologic concepts and principles such as the interior of Earth, plate tectonics, rock types, and geologic time. I would recommend *not* beginning with this chapter, but rather using it as a mini-textbook to refer back to as needed. In individual chapters on specific parks, I try to connect certain topics to the relevant parts of Chapter 1 to fill in details. For instance, when discussing the origin of the sedimentary rocks in the upper Grand Canyon, I refer the reader back to the section on sedimentary processes and products in Chapter 1. When addressing the relationship between rock and time in the Grand Canyon, I refer the reader back to the discussion of geologic time in Chapter 1. Try not to get bogged down by reading Chapter 1 in its entirety. Rather, use it strategically as a reference as you step through individual geologic provinces and parks.

Each chapter builds on information addressed in prior chapters to methodically construct the larger narrative. I would suggest to instructors teaching a class using this book that the most efficient organization is to begin with Part 1 on the Colorado Plateau and then move forward, referring back to specific sections of the broad introductory Chapter 1 as needed. *It's important to understand that each chapter is not autonomous, but rather works best as part of the integrated whole.* This was intentional to avoid repetition and to keep the narrative flowing from chapter to chapter. For instance, the effect of glaciation on the landscape is introduced in Part 2 on national parks of the Cascades and is expanded upon in Part 3 on parks of the Sierra Nevada. When glaciation is broached in subsequent chapters on national parks of the Rocky Mountains and Yellowstone, I don't repeat descriptions of glacial features and processes, but instead direct the reader to specific examples of U-shaped valleys, elongate glacial lakes, or cirques in those national parks.

Another example is volcanism and magmatism, a set of geologic concepts and features common to many national parks. The ideas are introduced incrementally, building through the body of the book. Part 2 on national parks of the Cascades describes the fundamentals of volcanic processes and products, including the role of silica. Part 3 on national parks of the Sierra Nevada addresses the mechanics of magma chambers. Chapter 15 on Crater Lake describes the ferocity of caldera eruptions. Chapters 21 and 23 discuss the importance of the "ignimbrite flare-up" in the American West. So by Chapter 30 on Yellowstone, I assume the reader has read at least a few of the earlier chapters where the concepts were introduced and can then plow ahead on volcanic events specific to Yellowstone. All of the principles and concepts collectively add up, chapter by chapter, into a coherent whole that culminates in summary chapters on the origin and evolution of the American West and the Appalachian Mountains.

I used mostly secondary sources for the content of the book, resorting to primary sources to clarify important points. These secondary sources include geologic guidebooks, textbooks, articles, and websites from the U.S. Geological Survey and National Park Service, as well as popular books written for a general audience.

Regarding the proliferation of terminology within the geosciences, I tried to walk a middle ground—not too much jargon so as to overwhelm the reader, but enough to communicate the larger concepts. Remember that science is not just a body of facts with a long list of terms to memorize, but rather a holistic system for understanding the natural world. The terminology is just part of the language for comprehending the broader concepts and principles. That said, terms in **bold** are those being introduced for the first time. Some important formations (e.g., Navajo Sandstone) and geologic events (e.g., Laramide orogeny) are highlighted in bold as well. These terms, formations, and events may be *italicized* where they are referred to later in the book, indicating to the reader that they can use the index to return to earlier chapters for clarification.

For simplicity, the ages of geologic events and features in national parks are described at the broadest of time units—the Precambrian, Paleozoic, Mesozoic, and Cenozoic. Shorter units such as the Neoproterozoic, the Devonian, and the Paleogene are avoided to reduce unnecessary complexity. Instead, qualifiers such as "early," "middle," and "late" are used. For instance, the seas transgressed across ancient North America during the early Paleozoic, rather than during the Cambro-Ordovician. Shorter-term time units such as the Quaternary, Pleistocene, and Holocene are used as necessary because so many of the landscape features of the national parks were formed over the last 2.6 million years.

As I write these pages, I'm torn between wanting to share my excitement about the geology and landscapes of the parks with you the reader and the conflicting desire to drop it all and take a trip over to Yosemite. My hope is that this book inspires you to gather your friends or family, throw your camping gear into the car, and take a roadtrip to visit those special places that we collectively own. Armed with a little geologic knowledge, your next visit will lead to a deeper, richer appreciation of America's national parks and monuments.

"I do not believe that any man can adequately appreciate the world of today unless he has some knowledge of . . . the history of the world of the past."

— THEODORE ROOSEVELT

Acknowledgments

The ideas for this book began bubbling up in my head after several solo visits to parks where I would camp in remote places and wander trails with the intent of understanding the geology of the landscapes I was seeing. Those random thoughts started to crystallize after a chat with Dan Kaveney, my first editor at Oxford University Press, who encouraged me to write a proposal for peer review. That's when things got serious. I'd like to thank him for triggering the transition from a vague array of underdeveloped ideas to the design, review, and refinement of the first several chapters.

My recent editors at Oxford, Dan Sayre, and Joan Kalkut, have ably guided me through the sometimes baffling process of seeing the ungainly collection of text and images become an actual book. The friendly and resourceful support team at Oxford that I've worked with through the years—Megan Carlson, Cailen Swain, Alex Foley, Christine Mahon, Arthur Pero, Patricia Berube, and Betty Pessagno—have made the process relatively smooth and efficient. I appreciate their affable and responsive contributions.

Of course, nobody really writes a book entirely on their own. They gain inspiration from casual conversations or incidental comments from friends and colleagues that lead down the path to a hint of an outline and eventually to the writing itself. Rob Zierenberg was the first to suggest a course on the geology of national parks long ago, while Jordy Margid, Jeff Mount, and Nicholas Pinter were instrumental in taking me on raft trips through the Grand Canyon where I could see the rocks from river level and talk about them around campfires at night. Long before the ideas in this book took shape, I was mentored by Fred Read, my PhD advisor at Virginia Tech, on how to properly do fieldwork as well as how to translate the data to publication. Fred's friendship and scientific guidance were invaluable to my entire career as well as to this book.

Although they don't realize it, the thousands of students who've taken my Geology of National Parks course at the University of California, Davis, have added immeasurably to the content of this book. Their curiosity and aptitude for asking questions that I can't immediately answer has led me to refine my understanding of the national parks and monuments far beyond my initial knowledge. At the professional level, the dozens of reviewers who spent hours evaluating drafts of the manuscript and sending along detailed comments have improved the presentation and content considerably. I've thanked a few of them individually, but I'm very appreciative of the time and effort they've all put into the task. Of course, I'm solely responsible for the accuracy of the science.

Kelly Bringhurst, *Dixie State University*
Ozeas S. Costa, Jr., *Ohio State University at Mansfield*
Christena Cox, *Ohio State University*
Patricia Deen, *Palomar College*
Christopher Fedo, *University of Tennessee*
Sue Finstick, *Southern Utah University*
Julie Fosdick, *University of Connecticut*
David Foster, *University of Florida*
Deborah Freile, *New Jersey City University*
Brian Frink, *Lakeland University*
Elizabeth Gierlowski-Kordesch, *Ohio University*
Priscilla C. Grew, *University of Nebraska-Lincoln*
Melissa Hage, *University of Wisconsin-Baraboo/Sauk County*
William H. Hirt, *College of the Siskiyous*
Paul Hudak, *University of North Texas*
Jacqueline Huntoon, *Michigan Technological University*
Gary Jacobson, *Grossmont College*
Charles E. Jones, *University of Pittsburgh*
Tor Lacy, *Cerritos College*
Lawrence L. Malinconico, *Lafayette College*
Keith O. Mann, *Ohio Wesleyan University*
Joy Nystrom Mast, *Carthage College*
David Mickelson, *University of Wisconsin-Madison*
Gustavo A. Morales, *Valencia College*
Stephen O. Moshier, *Wheaton College*
Alberto E. Patiño Douce, *University of Georgia*
Linda Pickett, *University of Nebraska-Lincoln*
Henry P. Scott, *Indiana University South Bend*
Cynthia Shroba, *College of Southern Nevada*
Mark R. Sweeney, *University of South Dakota*
Kathryn Thorbjarnarson, *San Diego State University*
Dylan Ward, *University of Cincinnati*

I'm particularly grateful to my family of enthusiastic adventurers (and a succession of faithful superdogs) who've shared in so many exceptional experiences in the outdoors, many in our national parks, others just out in the wilderness on trails, campsites, and mountaintops. My sons, Dillon and Alex, have taught me to look beyond the purely scientific aspects of natural landscapes to appreciate the fun to be had by simply being outside together. And I've done so much fieldwork with my wife, Isabel Montañez, in so many special and remote places that I've lost count. Her companionship and support over the years has been invaluable, and I'm not quite sure this project would have happened without her unwavering encouragement.

1 Inner Earth, Plate Tectonics, Rocks, and Deep Time

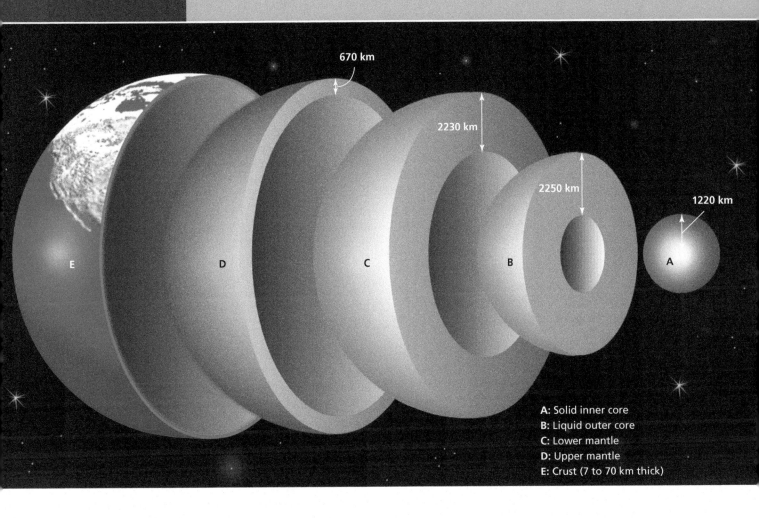

670 km

2230 km

2250 km

1220 km

E

D

C

B

A

A: Solid inner core
B: Liquid outer core
C: Lower mantle
D: Upper mantle
E: Crust (7 to 70 km thick)

Figure 1.1 Interior architecture of the Earth. The crust and mantle are composed of silicate rock, extending from the surface down to about 2900 km (1800 mi). The liquid outer core and solid inner core are composed of very dense, metallic iron alloy. Temperature, pressure, and density all increase toward the center of the Earth.

This first chapter provides the broader context for understanding some of the fundamental features and processes that characterize our planet. In order to comfortably comprehend the dimensions and time scales of geologic events that affected our national parks, it helps to understand the larger concepts of the structure of the Earth, the basic tenets of plate tectonics, the long, strange history of rocks, and the vast span of geologic time over which Earth's history unfolded. This chapter is intended to be a reference for students and instructors to consult as necessary. As various concepts and principles are introduced within each of the chapters on provinces and parks, the text repeatedly refers the reader back to the relevant sections of Chapter 1 for a more expansive discussion.

Earth's Inner Architecture

Earth's interior is composed of concentric spheres that differ in composition, density, and temperature (Fig. 1.1). The two primary materials that compose our planet are rock and metal, with the denser metallic component making up the **core** and the less dense rock making up the overlying **mantle** and **crust**. Pressure, density, and temperature all increase toward the center of our planet. At sea level on the surface of the Earth, the weight of the air in the atmosphere exerts almost 7 kg (~15 lb) on each square inch of the planet. With increasing depth inside the Earth, the weight of overlying rock causes the pressure to rise dramatically so that at 100 km (60 mi) beneath the surface, the pressures are greater than 230 metric tons per square inch. At the center of Earth, pressures are estimated to be about 3.6 million times the pressure of air at sea level. As you would expect, the increase in pressure with depth changes the atomic structure of materials, resulting in a proportional increase in density of the rock and metal composing Earth's interior. The inherently greater density of metallic iron in the core of the Earth relative to the overlying rock also contributes to the increase in density with depth. Temperature increases with depth in the Earth as well, although the rate of change varies with depth. Mathematical models suggest a temperature of ~4700°C near the center of our planet.

Beginning at the very center of the Earth, 6371 km (3959 mi) from the surface, a solid **metallic iron inner core** occupies a volume a little smaller than the size of our Moon. Because of the tremendous weight of the overlying rock and metal, atoms in the inner core are packed together very tightly and have a density of about 10–13 g/cm³. (Remember that density is measured in grams per cubic centimeter and that water has a density of 1 g/cm³. So the solid inner core has 10 to 13 times more mass per cubic centimeter than does water.) The solid inner core is an alloy, dominated by iron but containing trace amounts of nickel, sulfur, and other elements.

The solid inner core is surrounded by a **liquid outer core**—a dense, roiling ocean of liquid iron alloy about 2250 km (1400 mi) thick. Taken together, the inner and outer cores are approximately the size of Mars and contain about a third of our planetary mass. As opposed to the solid inner core, the liquid outer core remains a fluid because the temperatures at those depths are high enough to melt iron. The temperatures in the inner core are even higher, but the intense pressure in the inner core keeps the molecules of iron alloy packed so tightly together that melting is not possible. It's important to remember that the phase of a material (solid, liquid, or gas) is dictated by both temperature and pressure. The mobile liquid outer core, capable of conducting electricity and flowing slowly inside a rotating planet, is the source of Earth's surrounding magnetic field. Along with our atmosphere, the magnetic field protects life on Earth from the constant barrage of high-energy particles composing the solar wind.

We'll never visit the inner or outer core, of course, because the conditions are prohibitively beyond our technological know-how. But if we could somehow hold onto a piece of Earth's core, it would probably be a glowing silvery gray, the color of hot iron in both solid and liquid forms. As odd as it seems, we can touch pieces of the cores of asteroids in our solar system. They're called **iron meteorites**, and they fall to Earth free of charge, a result of catastrophic collisions in the asteroid belt that occurred long ago (Fig. 1.2). Debris from those impacts wandered through interplanetary space before accidentally intersecting the orbit of Earth. If they survived the intense frictional heat of Earth's atmosphere, they ended up as chunks of metal on the surface. The common occurrence of iron meteorites and the sheer abundance of iron in the universe as a whole provide compelling evidence that Earth's core is composed mostly of iron.

About 84 percent of Earth's volume and about 68 percent of Earth's mass are contained in the **mantle**, a sphere of very dense rock that extends from the core–mantle boundary at 2900 km (1800 mi) up to the base of the crust. The rock comprising the mantle is mostly composed of **silicate minerals**, dominated by the molecule SiO_2, with large amounts of iron and magnesium mixed in. Mantle rocks are very dense, ranging from about 3.5 to 5.8 g/cm³, but they are much less dense than the underlying metallic core. The mantle is commonly divided into upper and lower layers, separated by an abrupt increase in density at a depth of 670 km (415 mi).

It seems counterintuitive that such dense rock can flow, but in parts of the mantle high temperatures and pressures cause the silicate rock to behave like a **plastic** substance. ("Plastic" refers to a material that slowly flows and deforms under stress and never returns to its original form. A warm wax candle,

Figure 1.2 Iron meteorites. These fragments of long-ago collisions in the asteroid belt are interpreted to be chunks of the iron core of asteroids that were ejected outward toward the inner solar system where they happened to intersect the trajectory of Earth's orbit. The irregular, pitted outer surface is a result of melting and vaporization by frictional heat as it passed through Earth's atmosphere on its way to the surface.

for example, will behave plastically when squeezed or pulled.) Where the conditions are right, hot rock expands slightly, giving it a buoyancy that enables masses of rock to flow, ever so slowly, up through denser, "colder" rock of the mantle. You might imagine these masses of hot rock as acting like very dense taffy, moving at rates of a few centimeters per year—about the rate at which your fingernails grow. It's important to understand that this flowing rock of the mantle is a dense plastic solid and is not molten. The vast majority of the mantle is solid silicate rock, with only a small percentage being actual magma—molten rock. The visual of the "hot, molten mantle" is a myth perpetuated in bad movies and books and is better forgotten. Mantle rocks, sometimes coughed up through volcanoes, are commonly a dull green owing to the presence of the common mineral olivine. Earth may be the blue planet from space, but it is mostly green on the inside.

Relative to the voluminous mantle and core, Earth's **crust** forms a very thin

Figure 1.3 Map of global crustal thickness in kilometers. The crust is thinnest beneath the oceans and thickest beneath continental mountain ranges.

veneer on the surface of the planet, ranging from about 7 to 70 km (4 to 40 mi) in thickness. Civilization resides on the crust, most earthquakes occur in the crust, and all of our mineral and energy resources are found in the crust. The solid silicate rock composing our crust is different in composition than the solid silicate rock of the mantle. Crustal rocks are dominated by SiO_2, similar to mantle rocks, but the other elements composing crustal rocks tend to have lower atomic mass than those in mantle rocks, making rocks of the crust less dense (2.5–3.5 g/cm^3) than mantle rocks. Hence, the crust can be thought of as "floating" on the denser mantle. Furthermore, rocks of the crust are subject to much less overlying weight than those of the mantle, and thus they act, in general, as a **brittle** material, preferring to break rather than flow in response to stress.

The composition of Earth's crust is highly variable but is broadly divisible into **oceanic crust** and **continental crust**. Silicate rock composing the oceanic crust is dominated by **basalt**, a relatively dense rock that will be explained in greater detail below. Basaltic crust beneath the oceans varies in thickness from 7 to 10 km (4 to 6 mi) (Fig. 1.3). In contrast, silicate rock composing the continental crust is dominated by **granite**, which is less dense than basalt. Granitic continental crust averages 35–40 km (20–25 mi) in thickness but may reach 70 km (40 mi) beneath large mountain ranges. These differences in composition, density, and thickness between oceanic and continental crust are related to their tectonic origins and are addressed in this chapter.

A simple analogy with a hard-boiled egg makes the dimensions of each component of Earth's interior a bit easier

to visualize. The solid yolk of the egg would represent the combined inner and outer core of the Earth. The white of the egg could be viewed as the mantle, comprising the greatest volume of each object. The shell of the egg is proportional to the thickness of Earth's crust. Relative to the planetary radius, Earth's crust is only 0.1–1.0 percent of the total.

The deepest we've drilled is a little over 12 km (7 mi), and the deepest mines are less than 4 km (2.5 mi) in depth—tiny pinpricks into Earth's surface. So how can geoscientists possibly know the depth of the core–mantle boundary and the thickness of the crust so precisely? We know these dimensions because geoscientists listen for the echoes of earthquake waves that pass through the body of the planet. The seismic waves produced during earthquakes bounce off of interior surfaces (like the core–mantle boundary) and then rebound back up to seismometers spread across the globe. The study of earthquakes and the movement of seismic energy through Earth's interior enables the creation of three-dimensional images of the architecture of the inner Earth through sophisticated computational techniques and massive computer firepower. The ability of humans to "see" inside the Earth using mathematics, reason, and ingenuity is a truly remarkable achievement.

The landscapes that we see all around us are the uppermost surface of the crust and exhibit a wide range of elevations and topographic features. The Himalayan Mountains reach almost 9 km (>5 mi) above sea level, whereas the Challenger Deep in the western Pacific extends down to 11 km (~7 mi) below sea level. Thus, the maximum **topographic relief** of Earth's surface is

Figure 1.4 "Blue Marble" image of Earth from Apollo 17 showing the smoothly spherical outline of our planet.

planets like Earth tend to grow bigger and bigger until all of the loose debris in their orbit is swept up and incorporated into the planetary mass. The innumerable impacts generated enormous amounts of heat energy, and the accreted rocky and metallic debris contributed heat-producing radioactive elements to the growing planet. So planets like Earth are born as hot bodies, enveloped in a deep ocean of magma due to the intense heat keeping rock in a molten form. The hot, "soft" early Earth evolved into its spherical shape under the force of gravity.

Only after a few hundreds of millions of years, when enough of the planetary "heat of formation" had been given off to space, did the magma ocean cool sufficiently to form a solid outer shell. It was during this early super-hot phase of Earth history that the planet separated into its concentric spheres, driven by heat energy and gravity. Each layer of Earth's interior formed as denser elements like iron collapsed inward toward the center of the planet, while lighter, less dense elements (like silicon and oxygen, which together compose silicate rock) migrated outward. The end result is a planetary interior layered according to the density of its constituent materials: an interior dense metallic core, a surrounding less dense rocky mantle, and a thin, least dense surficial crust.

around 20 km (12 mi). On a human scale, Mount Everest is enormously high and the deepest ocean floor is unfathomably deep. Relative to Earth's radius, however, the surface topography forms almost imperceptible irregularities on a very smooth outer surface of our planetary sphere. Images from space illustrate the marble-like geometry of Earth (Fig. 1.4).

WHY IS EARTH'S INTERIOR LAYERED?

From the preceding text, you learned that Earth's layers get progressively denser toward the center. This is an inherent characteristic of all planets and is a consequence of planetary formation and evolution, which itself is a consequence of star formation. Over four and a half billion years ago, our Sun coalesced from a spinning cloud of gas and dust. Some of the remaining material self-organized into orbiting belts of metal, rock, and ice around the growing young star. Planets like Earth originally formed as a result of countless collisions between peanut- to mountain-sized bodies of rock and metal occupying the same orbital belt (Fig. 1.5). The constant addition of debris from the early solar system progressively increased the mass and thus the gravity of the young, infant planet. As the planetary gravity increased, more solid debris could be attracted. Thus,

These density trends continue to the layers above the solid Earth. The rocky crust is overlain by the **hydrosphere**, consisting of the world ocean, surface water like lakes and rivers, and groundwater. The low-density gases composing the **atmosphere** form an outer sheath, separating our planet from the emptiness of space. The **biosphere** is suffused through the hydrosphere and continental surface of the crust, while the **cryosphere** is composed of the continental ice sheets of Greenland and Antarctica as well as smaller volumes of glacial ice in high latitudes and high elevations. The intersection of Earth's crust with the hydrosphere, atmosphere, cryosphere, and biosphere is the focus of attention in our discussions of America's national parks. The juxtaposition of a mountain lake against a jagged rocky peak, perhaps with a glacier flowing through an alpine valley, is what we visit the parks to see. Add some vegetation and a bighorn sheep or two and you are witnessing the overlap of several of Earth's outermost spheres.

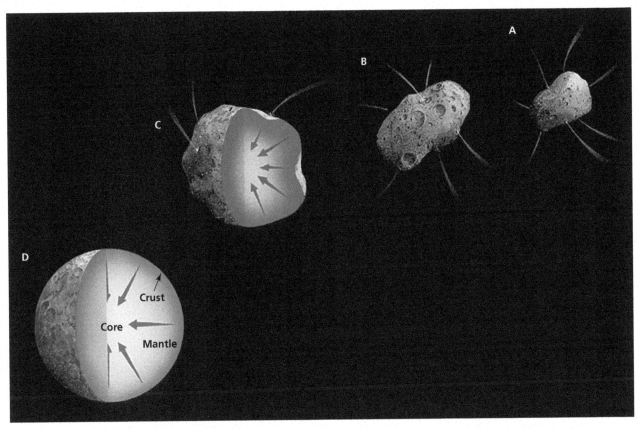

Figure 1.5 Creating a layered Earth. A) Over four and a half billion years ago, particles of gas and dust begin to coalesce by constant collisions while in orbit around the young Sun. B) As mass accumulated within the growing planetesimal, it attracted even more matter via its increasing gravity. C) Frictional heat generated by impacts of rock and metal, in concert with heat generated by radioactive decay, raised the temperature of the growing protoplanet. Denser metals flowed toward the center of the hot, "soft" planet, while less dense rocky material rose upward. D) With cooling, the dense iron core was surrounded by a less dense rocky mantle and crust.

Tectonics: Slow-Motion Evolution of Earth's Solid Surface

Plate tectonics is a unified set of ideas that describe the continual motion, creation, and destruction of Earth's active outer shell. The theory explains the uplift of mountain chains, the opening of ocean basins, and the occurrence of earthquakes and volcanoes. People often misuse the word "theory" to mean a tentative idea, as in the statement "It's *just* a theory," as if it's an unproven guess. But the definition is actually much more precise. A theory is a broad explanation of a natural phenomenon that has been tested and confirmed by evidence to such a degree that scientists have developed great confidence in the theory's ability to explain and predict a range of phenomena. Because the theory of plate tectonics has been tested and refined over several decades, with almost all geoscientists using it as a cornerstone of geologic thought, plate tectonics could justifiably be thought of as a set of natural facts.

Plate tectonics, although it still has several details to be worked out, is a truly powerful paradigm for understanding how the Earth works. There is no better way to gain a deeper awareness of the origins of America's national parks than via plate tectonics; indeed, it provides the guiding theme of this book. The fundamental precepts of plate tectonics are described in this section, but later chapters on geologic provinces and their national parks will dig deeper into specific tectonic models for their origin and evolution.

The term **plate** refers to the tabular shape of the mobile slabs of rock composing Earth's outer shell. Depending on how you count, there are 7 to 12 major plates and several smaller plates (Fig. 1.6). A single plate usually includes both granitic continental rock and basaltic oceanic rock, but some plates may consist of just continental or oceanic rock. The boundaries of tectonic plates are seldom directly coincident with the geographic boundaries of continents or oceans. Coastlines mark the geographic edges of continents, determined by the intersection of land and sea. The locations of plate boundaries, however, are typically marked by a linear pattern of earthquake epicenters and volcanoes due to the inexorable grinding of one plate against another. Plate boundaries may be located in the

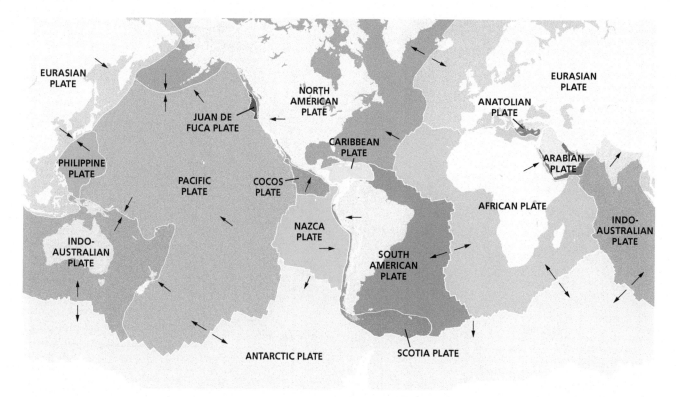

Figure 1.6   Global plate tectonic map. Only rarely do plate boundaries coincide with edges of continents. Plates may be composed of only oceanic rock (e.g., Pacific plate) or mostly continental rock (e.g., Arabian plate). Typically, tectonic plates are composed of both continental and oceanic rock (e.g., African plate, South American plate). Arrows indicate directions of plate motion.

middle of ocean basins or may cut through continents. It's been said many times that plate boundaries are "where geology happens."

Tectonics refers to the large-scale, long-term movement and deformation of Earth's outer shell. From the Greek "tekton"—meaning to build—tectonic motion explains the growth of linear mountain ranges such as the Sierra Nevada, the Cascades, the Rocky Mountains, and the Appalachians. Tectonics also explains the growth and demise of entire ocean basins. From the perspective of a human lifespan, the map of the world's continents and oceans seems permanent and eternal. But today's world map is a temporary arrangement when viewed over the enormous span of geologic time, with continents amalgamating and breaking apart and ocean basins opening and closing, all occurring at excruciatingly slow rates. As we explore the national parks in later chapters, you'll see how Earth's geography changes continually through time as plates migrate across the surface, as sea level rises and falls, and as global climate evolves.

How thick are tectonic plates? Earlier you read that the interior of the Earth is divided into concentric spheres, each defined by differences in composition and density, among other factors. To understand plate tectonics, we need to differentiate two distinct outer layers of the Earth defined on the basis of their strength: the **lithosphere** and the underlying **asthenosphere** (Fig. 1.7). You can think of "strength" as the potential to flow under certain conditions

of pressure and temperature. Strong materials are highly resistant to flow and tend to be rigid and brittle; weaker materials have a higher potential to flow and tend to act as plastic substances. The strength of a material is influenced not only by its composition, but also by the surrounding pressure and temperature conditions.

The lithosphere consists of the crust and upper mantle down to about 100–150 km (60–90 mi) beneath Earth's surface, with the lithosphere beneath continents being thicker than beneath oceans. The lithosphere is a "cool and strong" layer that behaves like a rigid body, meaning that it will flex, bend, and break but will not flow. Plates are stiff slabs of lithosphere; the commonly used phrase "crustal plate" is wrong since the crust is only the upper part of the lithosphere.

The temperature at which rock under pressure begins to slowly flow when exposed to a force defines the upper boundary of the underlying asthenosphere. Put another way, the lithosphere–asthenosphere transition occurs where a small portion of upper mantle rocks first begin to melt, commonly around 100 km (60 mi) beneath the oceans and about 150 km (90 mi) beneath the continents. At these depths and associated pressures, this temperature is around 1300°C and affects less than 1 percent of the rock. This **partial melting** of a small percentage of rock is enough, however, to weaken the rock and permit it to flow ever so slowly as a plastic material. As opposed to the "cool and strong" lithosphere ("cool" being a relative

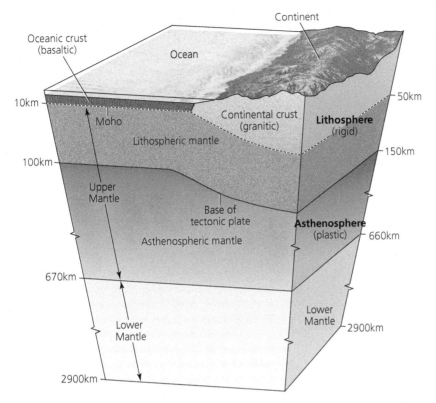

Figure 1.7 Cross section of outer shell of Earth illustrating the thickness of the lithosphere above the asthenosphere. The rigid lithosphere includes the crust and the uppermost part of the mantle. Note the thinness of the oceanic crust relative to continental crust. Broadly speaking, continental crust is composed of granitic rock, whereas oceanic crust is composed of denser basaltic rock.

term, of course), the asthenosphere is described as "hot and weak."

So the essence of plate tectonics involves the slow movement of stiff slabs of lithosphere above a mobile, weak asthenosphere, somewhat analogous to a cracker sliding above a slice of warm cheese. When you push down on the cracker with your finger, it will break; the cheese beneath, however, will flow in response to the stress. The lower boundary of the asthenosphere is not clearly defined because it imperceptibly merges into the rest of the mantle, which itself is capable of plastic flow as described earlier. The theory that explains the driving forces behind plate motion will be explained in a separate section later in this chapter.

Tectonic plates interact along their margins in three basic ways: They can pull away from one another, they can push against one another, or they can grind laterally past one another (Fig. 1.8). For the most part, interactions along plate boundaries affect only the outer margins of plates, with the interiors remaining largely intact. The three relative motions generate specific orientations of tectonic stress that result in characteristic topographic features and geologic processes at each of the three plate boundaries. We'll begin with a discussion of divergent plate boundaries, the ultimate source of the origin and growth of ocean basins.

DIVERGENT PLATE BOUNDARIES

Divergent boundaries are those where adjacent tectonic plates move away from each other. The tectonic forces that drive divergence result in **extension**, the stretching of lithosphere along the plate boundary. Divergent plate boundaries occurring on the seafloor contribute to the growth of ocean basins, whereas divergent boundaries on land mark sites where continents break apart.

Divergent plate boundaries in the ocean basins are marked by a 60,000 km (37,000 mi) elongate welt on the seafloor known as the **mid-ocean ridge system** that winds sinuously around the planet like a vast surgical scar (Fig. 1.9). The global mid-ocean ridge system rises imperceptibly from the deepest seafloor to a height about 2–2.5 km (1.2-1.6 mi) beneath sea level, forming a broad, linear swell. A series of parallel faults cut perpendicular to the trend of the ridge, breaking it into segments.

Mid-ocean ridges tend to be roughly symmetrical, with a deep, narrow valley extending along the crest of the ridge (Fig. 1.10). This **axial rift valley** marks the actual divergent boundary between the two plates and is the birthplace of new seafloor and underlying oceanic lithosphere. The slow-motion growth of ocean basins occurs within the axial rift valleys of the mid-ocean ridges through the process of **seafloor spreading**.

Here's how seafloor spreading works. As the two plates diverge away from one another due to extensional tectonic stress, the underlying asthenospheric rock plastically rises beneath the ridge axis and partially melts due to the decreased pressure at the shallower depths. It used to be thought that the molten rock, **magma**, accumulated in massive reservoirs a few kilometers beneath the crest of the mid-ocean ridge. But decades of research suggest that tiny droplets of melted rock migrate slowly toward the surface from the underlying asthenosphere. The droplets of melted rock coalesce to create larger streams of magma that rise upward toward the ridge axis. Extensional stress creates vertical fractures within the thin, brittle lithospheric rock along the axial rift valley, providing pathways for magma to buoyantly rise upward to the surface and flow outward onto the seafloor. Magma that finds its way to the surface is called **lava**; magma and lava are the same stuff, with the molten rock below the surface called magma and the molten rock above the surface called lava. As the oozing hot lava spreads out away from its elongate

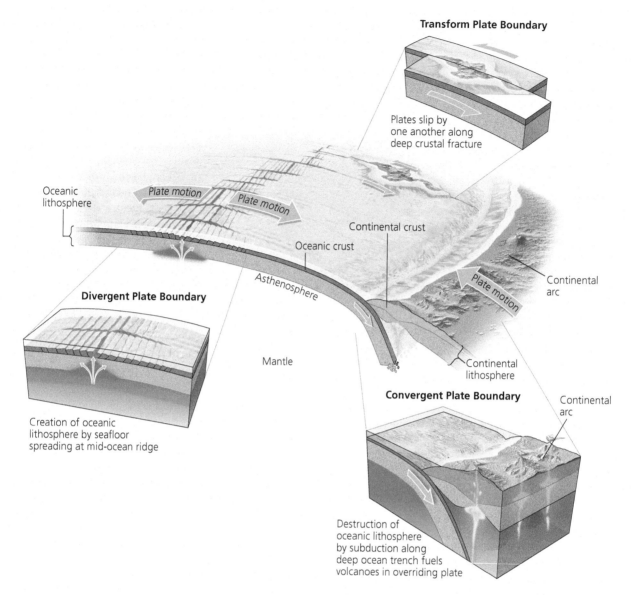

Figure 1.8   Three types of plate boundaries.

fracture along the rift valley, it becomes quenched by the cold waters of the deep sea and solidifies in place, creating an irregular, pillow-like pavement of brand-new volcanic seafloor.

Magma that never reaches the surface slowly cools and solidifies at depth within the fracture, becoming newly formed oceanic lithosphere. With continued extensional stress separating the plates bordering the mid-ocean ridge, the young, brittle, oceanic lithospheric rock fractures, permitting new molten material to rise along the axis of the rift valley. As time passes, the rock along the ridge axis gets displaced to either side, adding new real estate to the diverging margins of each plate. The tectonic and volcanic processes of seafloor spreading are currently active at a variety of sites along the Mid-Atlantic Ridge, the East Pacific Rise, and several other locales in the Indian, Arctic, and smaller ocean basins (Fig. 1.9).

Seafloor spreading is the process by which all seafloor and underlying oceanic lithosphere is created. The vast majority of the world's volcanism occurs in eternal darkness deep beneath the ocean depths along the crest of the mid-ocean ridge system, the most continuous physical structure on Earth. The end result is that the bottom of the entire world ocean is composed of volcanic rock, overlain by a variable thickness of fine sediment and organic debris drifting down from the water column above. Ocean basins grow outward laterally from the axis of the mid-ocean ridge system that bisects the world's oceans. A commonly used visual for seafloor spreading is a pair of conveyor belts imperceptibly moving in opposite directions away from the crest of the mid-ocean ridge system. Seafloor spreading along the East Pacific Rise occurs at ~10 cm/yr, whereas the Mid-Atlantic Ridge averages ~2.5 cm/yr. Although terrifically slow on human time spans,

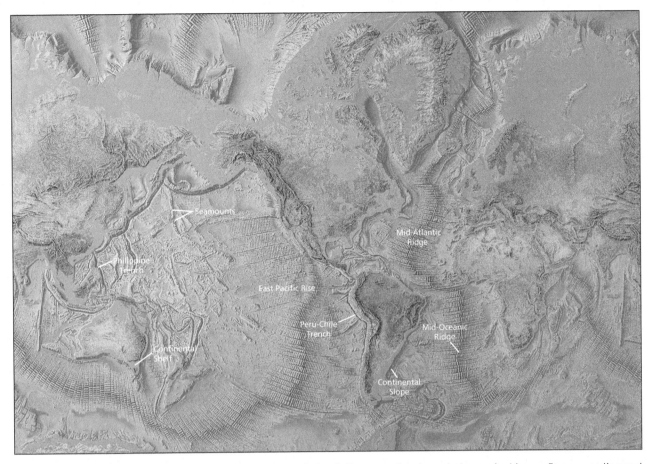

Figure 1.9 Global mid-ocean ridge system, a continuous chain of divergent plate boundaries marked by seafloor spreading and the growth of ocean basins.

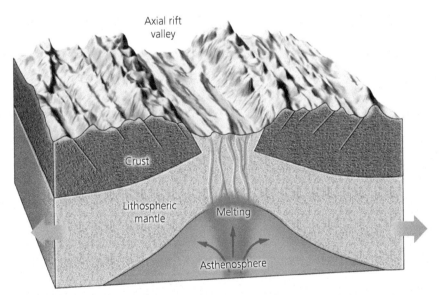

Figure 1.10 Features of a mid-ocean ridge system, illustrating extensional tectonic forces, upwelling of plastic asthenospheric rock, and volcanism occurring along fractures within the axial rift valley.

given a few hundreds of millions of years, ocean basins can grow thousands of kilometers wide.

Divergent plate boundaries that occur on land are called **continental rifts**. Tectonic extensional forces result in the breakup of continents, and continental rifting is the first step in the birth of ocean basins (Fig. 1.11). Seafloor spreading is not the driving process in continental rifts because by definition there is no seafloor on a continent. Instead, continental rifts are linear topographic features where continental lithosphere is actively stretched and broken by the upwelling of hot, buoyant asthenosphere trapped beneath the thick, insulating continent. As the continental rock rises and flexes above the heat source, it breaks along **faults**—planar fractures in rock along which movement takes place. Faulting along continental rifts typically results in linear rift valleys, analogous to the rift valleys along the axial crests of mid-ocean ridges (Fig. 1.11A).

As continental lithosphere stretches and breaks, it gets thinner, decreasing the lithospheric load and permitting the underlying asthenosphere to rise up beneath. Continental rifts can be geologically dynamic, marked by active volcanism and the occasional large earthquake. If rifting continues to the point where the continent actually splits apart, a **linear sea** may develop if the rift valley floor drops below sea level, permitting ocean water to flood the depression (Fig. 1.11B). Oceanic lithosphere will begin to form in the linear sea through seafloor spreading along the newly formed mid-ocean ridge axis, and a nascent ocean basin is born.

The classic modern example of a continental rift system is East Africa, where a sliver of eastern Africa is pulling away from the rest of the continent (Fig. 1.12A). A linear chain of rift valleys mark the youthful **divergent plate boundary** and are commonly filled with freshwater to form deep, elongate lakes like Tanganyika and Malawi. The nearby Red Sea is a young linear sea where continental rifting between the African and Arabian plates has progressed enough so that seafloor spreading is actively widening the seaway. Continental rifting is expressed today in national parks of the Basin and Range Province where the North American plate is currently experiencing tectonic extension. The evolution of continental rifting and the birth of ocean basins are particularly useful for understanding the synthesis chapters on the origin of the American West (Chapter 33) and the Appalachians (Chapter 35).

CONVERGENT PLATE BOUNDARIES

Convergent plate margins are characterized by **compression**, a style of tectonic stress where the interacting plates move toward each other, squeezing and contracting the lithosphere on either side (Fig. 1.8). As opposed to divergent plate boundaries where new lithosphere is created, convergent plate boundaries are marked by the destruction of lithospheric rock via burial in the mantle. This tectonic setting and its associated compression result in an array of topographic features such as linear mountain chains and deep ocean trenches. The highest mountains and the deepest ocean trenches on the planet are associated with convergent plate boundaries, as are the most violent volcanoes and largest magnitude earthquakes. As you might

suspect, the broad mountain belt of the American West as well as the narrower Appalachians were formed along convergent plate boundaries. Convergent plate boundaries are classified into three categories based on the composition of the interacting plates: continental-oceanic, oceanic-oceanic, and continent-continent.

Continental-oceanic convergent boundaries occur where basaltic oceanic lithosphere compresses against granitic continental lithosphere (Fig. 1.13). The denser oceanic plate bends and dives downward into the mantle beneath the less dense continental plate in a process called **subduction**. If you visualize the **subduction zone** in three dimensions, it will be easier to imagine the colossal amount of frictional grinding of the massive, descending slab of oceanic rock against the overriding slab of continental rock. Think of the slippery wet film of saturated seafloor sediment being smeared out along the broad contact between the two plates. At depths of about 100–150 km (60–90 mi), where the pressure and temperature conditions are just right, the water in the oceanic sediment

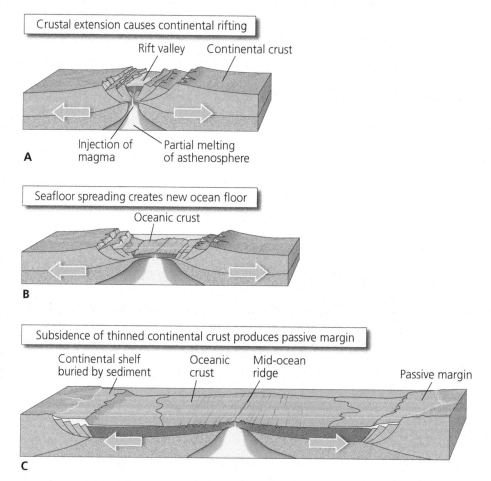

Figure 1.11 Growth of ocean basins via continental rifting and seafloor spreading. A) Hot rock and magma accumulates beneath a thick, insulating slab of continental rock. With continued extension, a fault-bounded continental rift may develop. B) If extension proceeds long enough, the rifted continent may sag beneath sea level, accompanied by the onset of seafloor spreading. C) As the lithospheric plates diverge away from the ridge axis, an ocean basin grows between the migrating continents. The time span from A to C is tens to hundreds of millions of years. Orange arrows indicate extensional tectonic stress.

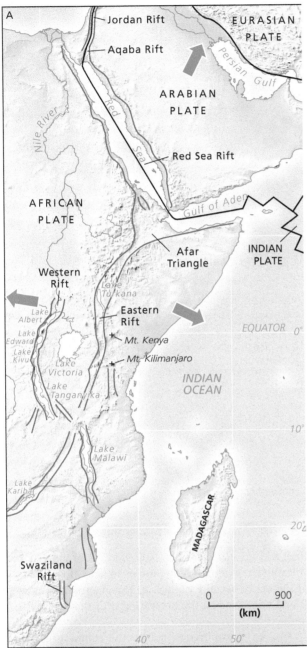

Figure 1.12 A) Map of East African continental rift illustrating relative directions of plate motion. Deep elongate lakes occupy some of the rift valleys. B) Erta Ale is a very active rift-related volcano located in the Afar region of Ethiopia.

is released into the overlying rock where it has the effect of lowering the melting point of the rock. The infusion of this water into the overlying mantle and the consequent melting trigger the generation of vast quantities of magma. These blobs of molten rock are less dense than the surrounding solid rock and are thus buoyant. The magma migrates slowly toward the surface, melting rock as it moves, growing in volume, and injecting itself into whatever fractures or zones of weakness may exist. Over millions of years, the plumes of magma ultimately find their way into the overlying continental lithosphere where they form reservoirs of molten rock a few kilometers beneath the surface called **magma chambers**.

The magma in a chamber slowly releases its contained heat to the surrounding rock, cooling and crystallizing into solid granitic rock. Granite formed in this manner adds mass and volume to the continental lithosphere. Some of the magma, however, buoyantly leaks upward to volcanically erupt on the surface. The end result is the formation of a linear chain of volcanoes along the edge of the continent aligned roughly parallel to the convergent margin—a feature called a **continental volcanic arc** (Fig. 1.13).

At the same time as these magmatic and volcanic processes are occurring, the preexisting rock of the continental lithosphere is being squeezed and deformed by convergence with the adjacent oceanic plate. The compressional squeezing of the continental rock causes it to narrow horizontally and thicken vertically, resulting in uplift of the rock to high elevations and the formation of a deep "root" beneath the mountain chain. The rising plumes of magma pierce through this massive body of deformed continental rock, while the volcanic arc forms a chain of isolated high peaks jutting above the mountain range. The Andes of South America and the Cascades of the Pacific Northwest are classic modern examples of elongate mountain belts capped by a linear chain of active volcanoes (Fig. 1.14). The ancestral Sierra Nevada of ~100 million years ago is an ancient example of continental–oceanic convergence.

In the adjacent ocean basin, a **deep ocean trench** forms along the linear contact between the subducting oceanic plate and overriding continental plate (Fig. 1.13). These narrow gashes in the seafloor may reach 8–11 km (5–7 mi) beneath sea level and are the bathymetric expression of the uppermost subduction zone. In essence, the deep ocean trench marks the actual line of contact between the two converging plates. Unless filled with sediment from a nearby river, a deep ocean trench characterizes all subduction zones. For instance, the Peru–Chile trench marks the subduction zone between the oceanic Nazca plate and the western edge of the South American continental plate (Fig. 1.14A). Whereas ocean basins *grow* at mid-ocean ridges via the creation of oceanic lithosphere, ocean basins *shrink* due to the consumption of oceanic lithosphere along subduction zones. On a global scale and averaged over time, the two processes cancel each other out and the surface area of Earth remains constant.

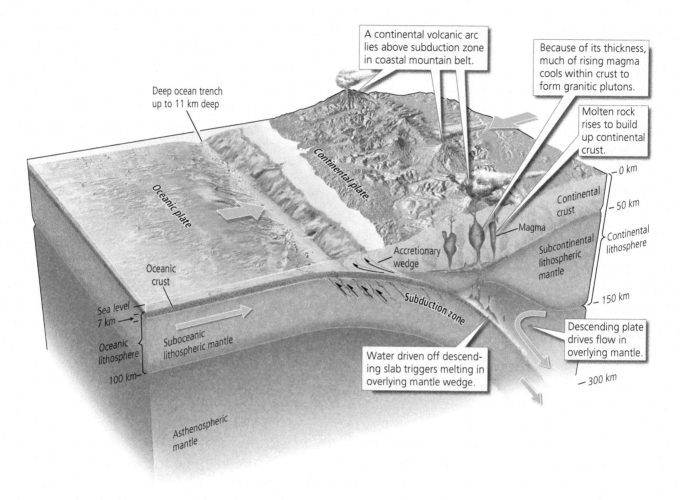

A continental volcanic arc lies above subduction zone in coastal mountain belt.

Because of its thickness, much of rising magma cools within crust to form granitic plutons.

Molten rock rises to build up continental crust.

Deep ocean trench up to 11 km deep

Continental plate

Oceanic plate

0 km

Continental crust

50 km

Magma

Continental lithosphere

Accretionary wedge

Subcontinental lithospheric mantle

Oceanic crust

Subduction zone

150 km

Sea level
7 km

Oceanic lithosphere

Suboceanic lithospheric mantle

Descending plate drives flow in overlying mantle.

Water driven off descending slab triggers melting in overlying mantle wedge.

100 km

300 km

Asthenospheric mantle

Figure 1.13   Subduction zone associated with a continental-oceanic convergent boundary.

The second type of convergent plate boundary occurs where an oceanic plate compresses against another oceanic plate. **Oceanic-oceanic convergent boundaries** have many of the same characteristics as continental–oceanic convergent margins, including the formation of a subduction zone and its associated topographic expression (Fig. 1.15). Where two oceanic plates converge, the older of the two plates is inherently denser than the other and will thus subduct. (As oceanic lithosphere is conveyed away from its formative mid-ocean ridge over time, the rock cools and contracts, becoming denser with age. Thus, older oceanic lithosphere is denser than younger oceanic lithosphere.) In a similar fashion to continental-oceanic convergence, the descending plate and its veneer of wet seafloor sediment release water, in the process generating large volumes of magma. The buoyant magma rises in plumes up into the overlying oceanic lithosphere. The magma accumulates in chambers a few kilometers beneath the seafloor, with most of the magma solidifying in place over time. Molten rock that escapes upward erupts onto the seafloor, forming submarine volcanoes. With

time and the eruption of greater volumes of magma, the volcanoes may grow high enough to breach the surface of the sea to become islands. In this manner, a **volcanic island arc** forms on the overriding oceanic plate parallel to the convergent boundary.

A deep ocean trench marks the submerged linear contact between the two oceanic plates. The deepest point in the world ocean is the Challenger Deep, almost 11 km (7 mi) beneath sea level, located along the Mariana Trench. (For comparison, the average depth of the world ocean is about 4 km [2.5 mi].) The Mariana Trench marks the subduction of old portions of the Pacific plate beneath the relatively younger Philippine plate to the west (Fig. 1.16). The Mariana Island volcanic arc overlies the subduction zone along the Philippine plate. Other modern examples of oceanic–oceanic convergent margin island arcs are the Japanese, Philippine, and Indonesian islands. A deep ocean trench parallels the linear archipelago of volcanic islands in each of these examples. Many rocks composing the mountains of the American West as well as the Appalachians are former

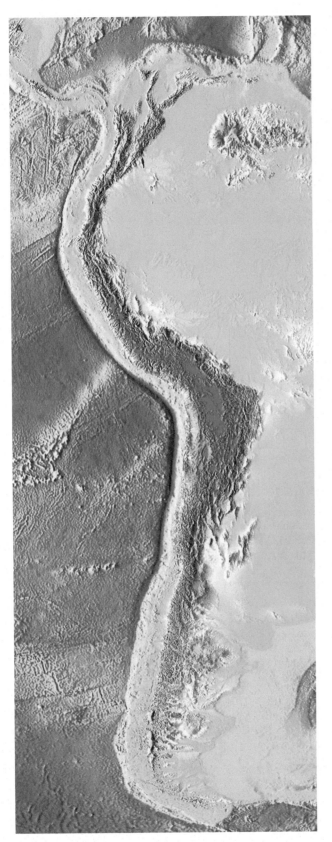

Figure 1.14 A) West coast of South America illustrating the linear Andean mountain chain (reds) and adjacent deep ocean trench (dark blue). B) Cotopaxi volcano, Ecuador—one of hundreds of volcanoes along the Andean volcanic arc.

volcanic island arcs created along this type of convergent boundary. Subsequent tectonic motion brought the once-distant island arcs into contact with a continent where they were connected to the outer margin of the continent along subduction zones. Examples of this fundamental process of continental growth are described and illustrated in Chapters 19, 21, 26, and 30.

The third type of convergent plate boundary occurs where two continental plates tectonically compress against one another. **Continent-continent convergent boundaries** are marked by the process of **continental collision** rather than subduction (Fig. 1.17). Since all continental lithosphere is composed of relatively lower density granitic rock, regardless of age, neither plate will subduct beneath the other. Instead, the continental plates jam against one another, with the formerly subducting plate wedging laterally beneath the other. This stacking of continental lithosphere results in very high elevations and a very thick lithospheric mass. Volcanism is absent during continental collision because no magma is generated by the interaction of the two low-density slabs.

The best modern example of continental-continental collision occurs along the Himalaya Mountains and Tibetan Plateau where rocks are lifted up to elevations approaching 9 km (>5 mi) above sea level. This linear mountain belt is created by the collision of the Indian part of the Indo-Australian plate with the southern Eurasian plate. Currently, the plates are jammed against one another, permitting the accumulation of tremendous amounts of tectonic stress. The abrupt release of this stress is expressed as an earthquake, such as the devastating Kashmir earthquake of 2005 that killed ~76,000 people. The Appalachian Mountains are the deeply eroded remnants of an ancient continental-continental collision between eastern North America and parts of Africa and Eurasia.

TRANSFORM PLATE BOUNDARIES

The plates on each side of a **transform plate boundary** move laterally in opposite directions past each other, resulting in the formation of large **transform faults** or networks of near-parallel faults. The tectonic stress associated with transform plate boundaries is **shear**, where the orientation of stress is parallel to the plate boundary rather than perpendicular as in extension and compression (Fig. 1.8). The classic example of a transform plate boundary breaking the continental lithosphere is the San Andreas fault of California, where the Pacific plate grinds laterally against the North American plate (Fig. 1.18). Other modern examples include the Dead Sea fault of the Middle East, the North Anatolian fault of Turkey, and the Alpine fault of New Zealand.

Transform plate boundaries are notable for their active **seismicity**, the frequency and distribution of **earthquakes** over time. Earthquakes occur along large transform faults because fault planes are not smooth features, but rather have many places along the fault plane where the rock on

Figure 1.14 (*continued*)

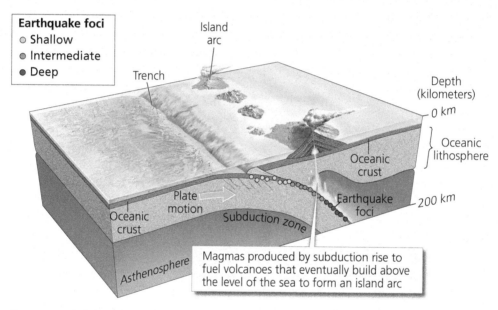

Figure 1.15   Subduction zone associated with an oceanic-oceanic convergent boundary. The contact between the two plates occurs within the deep ocean trench.

either side can "stick." The fault plane beneath the surface may stick due either to the frictional strength between the massive blocks of rock on either side or to protuberances or bends along the fault plane itself. This "sticking" behavior permits the accumulation of enormous amounts of tectonic shear stress, which seismologists refer to as a "locked" fault. Over some unpredictable time period, the accumulated tectonic stress eventually exceeds the frictional strength holding the two massive blocks of rock in place, resulting in abrupt slip along the fault plane and the violent release of that pent-up energy as an earthquake. The topographic expression of major transform faults and the **stick-slip** behavior that results in earthquakes will be addressed in greater detail in the chapters (27, 28, and 29, respectively) dealing with Joshua Tree and Pinnacles national parks and Point Reyes National Seashore, each of which is intimately associated with the San Andreas.

Transform plate boundaries not only occur on continents, but also bisect the world's mid-ocean ridge system

(Fig. 1.9). Oceanic transform faults play an active role in the origin of some continental transform faults, including the San Andreas. Chapter 26, which focuses on the tectonically active Pacific margin of North America, addresses the role of oceanic transform faults in the origin and evolution of the San Andreas fault and the national parks that reside along its length.

ENERGY FOR PLATE MOTION

The ultimate energy source for tectonic motion is the enormous reservoir of heat inside the Earth. As briefly discussed, planets are born hot and spend their entire lives transferring that heat outward into the coldness of space. It's that constant flow of heat from the interior, originally generated by the countless impacts that formed the planet plus the constant decay of radioactive elements, which supplies the energy for many geologic processes, including plate tectonics. Heat from the interior of the Earth is transferred toward the surface via **convection currents**,

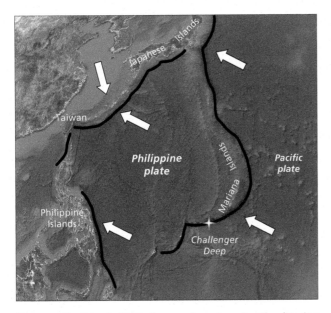

Figure 1.16 Google Earth image of western Pacific showing the Japanese, Philippine, and Mariana volcanic island arcs and associated deep ocean trenches (dark lines). Arrows show the relative motion of tectonic plates in the region.

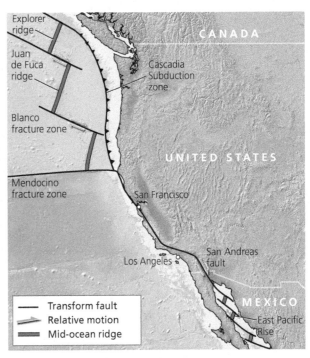

Figure 1.18 The San Andreas transform fault marks the plate boundary between the North American and Pacific plates along the continental margin of California. The San Andreas fault connects a divergent boundary in the Gulf of California with the convergent Cascadia subduction zone to the north.

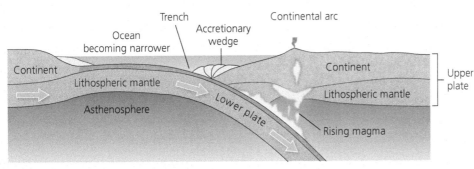

A Continent-oceanic subduction

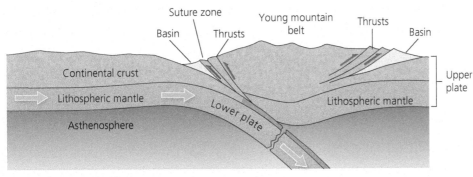

B Continent-continent collision

Figure 1.17   A) Oceanic lithosphere (and the overlying ocean basin) are consumed by subduction as two continents move toward one another. B) During continent-continent collision, subduction ends as one buoyant continental plate wedges beneath the other, thickening the crust and generating uplift of high mountains like the modern Himalayas or the ancestral Appalachians.

complex, plastic flows of rock in the mantle that move due to density contrasts (Fig. 1.19). As the metallic core cools through time, its heat is conducted outward into the dense rock of the mantle. The addition of thermal energy causes atoms in part of the lower mantle to vibrate a bit faster, permitting the slight expansion of a volume of rock. The larger volume of hot rock gives it a buoyant lift relative to adjacent regions that may be slightly less hot and thus slightly more dense, allowing for the plastic flow of hot rock upward.

As the plume of hot rock ascends, the pressure of overlying rock decreases, which has the net effect of increasing plume buoyancy even more. As the convective plume of

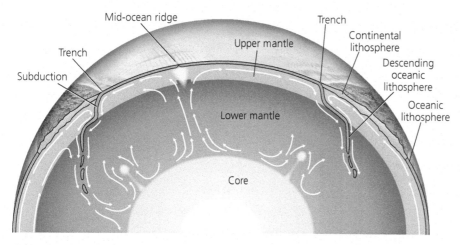

Figure 1.19   One model for the heat-driven circulation of convection currents within the mantle. The gravitational pull of subducting oceanic plates is an important factor in tectonic motion.

a tremendous pull on the rest of the plate behind it. This "slab pull" and other forces at the lithospheric level may ultimately be proven to be the direct drivers of plate tectonic movement and the resulting slow-motion rearrangement of the planetary map through time.

Rocks and the Rock Cycle

Armed with an understanding of Earth's interior and the dynamic nature of its outer shell, we can begin to address the origin and evolution of rocks, the solid matter from which the landscapes of our national parks are formed. We can more easily understand the origin of rocks if we also have an awareness of the processes by which they form. For instance, basalt is a common rock: fine-grained and black when you hold a chunk in your hand. But visualizing basalt as once-fiery lava with a temperature of 1000°C brings the otherwise unremarkable black rock to life, giving it a memorable origin. Moreover, when the process by which a rock originates is combined with its tectonic setting, rocks take on even greater significance and begin to reveal their long-held secrets. This section illustrates the general characteristics of the three main rock types and describes the processes that form each type. Many of the details are missing, but that's intentional since the specifics of various rock types will be addressed in each chapter where they are relevant to individual national parks.

hot plastic rock near the surface, some of the heat is lost to the atmosphere by venting through volcanism and escapes within hot water and gases. Most of the contained heat, however, is released to the surrounding rock beneath the surface, cooling the mass of rock and causing it to contract and increase in density. As its buoyancy decreases, the mass of rock slowly sinks back into the mantle, likely dragged downward along subduction zones. This entire process occurs at rates of a few centimeters per year. Given long spans of geologic time, convection is a very effective mechanism for transferring heat through the interior of the planet.

The trajectory of convective flow in the mantle is difficult to predict and is likely more chaotic than orderly. A common aid that will help you visualize the complexities of convective flow is the lava lamp you may have in your home. The heating element in the base would be equivalent to Earth's core; the blobs of colored wax would be the hot, expanded plumes of plastic rock; the clear liquid would be the cooler, denser mantle through which the plumes flow; and the upper tip of the lamp would be Earth's surface. As the warm, buoyant blob of wax rises toward the top, its heat dissipates to the surrounding liquid. The blobs of wax cool, contract, and become denser, and then sink back toward the base to start the convective motion all over again. Each convective cycle in the lava lamp is unique and enables, to a first approximation, the visualization of the complexities of actual mantle flow.

Heat, transferred through the planet via convection, is the energy source that makes Earth a dynamic entity. But the slowly churning mantle is only indirectly responsible for the tectonic motion of plates around the planetary surface. It may be that convection currents in the asthenosphere and underlying mantle exert a drag force along the base of the overlying stiff and cold lithosphere, but this is unlikely to be the primary driver of plate motion. Geoscientists today favor mechanical and gravitational forces acting along plate boundaries as the more direct factors influencing plate motion and velocity. For instance, the sheer weight of a downgoing oceanic slab within a subduction zone imparts

Rocks are composed internally of **minerals**, which themselves are composed of **elements** (Fig. 1.20). Of the 94 naturally occurring elements, only 4 dominate the mass of the Earth: iron, oxygen, silicon, and magnesium. You know from earlier in this chapter that much of the iron composing Earth is found in its core and that the mantle and crust are composed of rock. Most rocks are dominated by silica, SiO_2, which is a stable molecule formed by the chemical affinity that oxygen and silicon have for one another. Thus, most rocks composing the mantle and crust are **silicates**, named for the dominance of the silica molecule. Smaller amounts of iron, magnesium, aluminum, calcium, sodium, and potassium are incorporated into the silicate structure of rock-forming minerals.

There are about 4000 naturally occurring minerals on Earth, but only about 30 are common within rocks of the mantle and crust. Minerals are, by definition, solid, with the atoms composing each mineral arranged in a systematic, predictable fashion. This orderly atomic arrangement is reflected in the crystal structure of minerals. Not all crystals are the angular, beautifully geometric shapes exhibited in gemstones. Rather, most mineral crystals in

Granite (rock)

Quartz (mineral)

Feldspar (mineral)

Biotite (mineral)

Figure 1.20 Granite is an abundant rock of the continents and is commonly composed of three minerals: feldspar (pink and white grains), quartz (gray, glassy grains), and biotite (flake-shaped black grains). In turn, the minerals are dominantly composed of silicon and oxygen, with lesser amounts of aluminum, potassium, and other elements.

rocks are irregularly shaped grains arranged as an interlocking mosaic (Fig. 1.20). The individual minerals maintain a crystalline internal structure, but to the naked eye or through the microscope, the crystals commonly appear to have a random assortment of shapes.

Rocks are solid, naturally occurring aggregates of minerals. Common rock types include granite, basalt, sandstone, and marble. They are composed of common minerals such as quartz, feldspar, olivine, and calcite. A rock may be made up of several minerals, all combined into a single solid mass, or a rock may be composed of multiple grains of only one mineral. The processes that form the three main categories of rocks–igneous, sedimentary, and metamorphic—are described in the following sections.

IGNEOUS ROCKS

Magma and its surface equivalent, lava, are molten fluids composed of a variety of elements whose atoms bounce around chaotically within the molten mass. The intense temperatures necessary to keep rock in a molten state cause the individual atoms to vibrate at very high rates, with atoms colliding with one another and rebounding away to immediately collide with other atoms, precluding the formation of atomic bonds. Over time, heat is eventually released to the surrounding environment, and as the temperature falls, the atoms vibrate a little more slowly. With time and cooling, the atoms eventually link up via chemical bonds, forming the orderly atomic arrangement of mineral crystals. **Igneous rocks** form from the cooling and solidification of minerals from a molten fluid (Fig. 1.21).

Crystallization of specific minerals occurs systematically as the molten fluid progressively loses heat energy and the temperature decreases from perhaps 1250°C to less than ~600°C over time. As individual minerals solidify from the fluid, they ultimately form an interlocking mosaic of tightly welded crystals. Because mineral crystals growing from a molten fluid compete with each other for space, sharply defined crystal boundaries are rare, resulting in a homogeneous mass of irregular crystals (Fig. 1.20). The process is analogous to the freezing of ice from water as heat energy is lost, but the "freezing" of solid rock from molten rock occurs at much higher temperatures.

The solidification of igneous rocks may occur rapidly on the surface as lava is erupted from a volcano, or slowly a few kilometers beneath the surface within magma chambers (Fig. 1.22). Granite, a common igneous rock of the continents, is formed in the shallow crust as magma chambers slowly lose heat to the surrounding rock. The slow cooling within the magma chamber provides lots of time for individual crystals to grow; thus, the resulting texture of the granite is commonly composed of relatively large crystals visible to the naked eye. Over time, perhaps several hundred thousand to millions of years, the magma chamber will cool completely into a mass of granite called a **pluton.** Thus, coarse-grained igneous rocks such as granite are commonly called **plutonic igneous rocks** (named for Pluto, the mythical Greek god of the underworld). The majestic landscape of Yosemite National Park, including such iconic features as Half Dome and El Capitan, is formed in plutonic rock.

Molten lava erupted from a volcano comes in contact with the cold waters of the ocean or the cold air of the atmosphere, causing heat to be rapidly lost from the lava. Mineral crystals don't have much time to grow in these volcanic settings, so the resulting textures tend to be fine-grained, with individual crystals difficult to discern without a microscope. **Volcanic igneous rocks** thus differ from plutonic igneous rocks in the size of their constituent mineral crystals, due solely to the rate of cooling of the molten fluid. Volcanic and plutonic igneous rocks may have similar mineral assemblages, however, if they were derived from magmas with similar chemical compositions. For instance, a volcanic igneous rock called rhyolite (common at Yellowstone National Park) has the same mineral composition as granite, with the only difference being the size of the constituent crystals. Basalt, a very common volcanic rock in Hawaii Volcanoes National Park (Fig. 1.21), is chemically and mineralogically identical to its plutonic equivalent, gabbro, with both basalt and gabbro dominated by the minerals olivine, pyroxene, and feldspar. The only difference between the two is the size of the crystals composing each of them, with basalt being very fine-grained and gabbro coarse-grained, each a product of the cooling rate.

Figure 1.21   Lava slowly gives off its heat to the atmosphere, eventually solidifying into basalt, a very common igneous rock. Kilauea Volcano, Hawaii.

SEDIMENTARY ROCKS

Earlier in this chapter, you learned that many geologic processes are driven by the inexorable loss of heat from Earth's interior. But the Earth has another source of heat energy—the Sun—which drives the weather and climate systems of Earth's exterior. Weather, ranging from a sprinkling of rain to torrential downpours, acts to wear away solid rock exposed at the surface, producing smaller particles of loose sediment like sand and clay. Gravity, acting through the downhill flow of water and glacial ice, transports the loose particles from high regions such as mountains to low areas on Earth's surface, such as the world's ocean basins. Thus, the formation of **sedimentary rocks** from loose sediment is ultimately driven by solar energy, as opposed to the formation of igneous rocks, which requires the internal heat of the Earth.

The most obvious characteristic of all sedimentary rocks is their invariable arrangement as stacked layers of variable thickness (Fig. 1.23). The horizontally striped walls of the upper Grand Canyon form an arresting image of multicolored layers of sedimentary rock. Many landscapes of our national parks are characterized by the distinctive striped layering of sedimentary rock, so understanding their origin will provide a heightened awareness of Earth history and the unbelievable ancient geographies that predated today's world. There are two broad categories of sedimentary rock: clastic and biochemical.

Clastic sedimentary rocks are those produced by the weathering and erosion of preexisting rock to form smaller particles, followed by the transportation, deposition, and burial of those sediments to eventually form solid rock. The word "clastic" is derived from the Greek word "klastos," meaning to break or fragment. The loose particles of sediment may range in size successively from the largest boulders to fist-sized **cobbles**, **pebble**s, sand, silt, and finally, to the smallest flakes of **clay**. The genesis of clastic sedimentary rock can be broken down into four primary phases, as follows (Fig. 1.24).

First, preexisting rock residing in highlands must be weathered and eroded to form smaller particles of sediment. **Weathering** is the in-place breakdown of minerals in rock via chemical and physical activity. **Chemical weathering** typically involves water, either from rain, snowmelt, or groundwater, that acts to decompose the rock so that its constituent mineral grains are released from the solid mass. For instance, clay, the plate-like flakes used to make pottery as well as mud pies, is derived from the chemical disintegration of preexisting minerals such as feldspar and biotite. **Physical weathering** commonly occurs along natural fractures in rock where water can infiltrate and freeze, mechanically pushing the rock apart along the fractures. Roots reaching downward into fractures in search of water also act to physically push rock apart, enabling water to penetrate even deeper. Both chemical and physical weathering result in the in-place disintegration of solid rock.

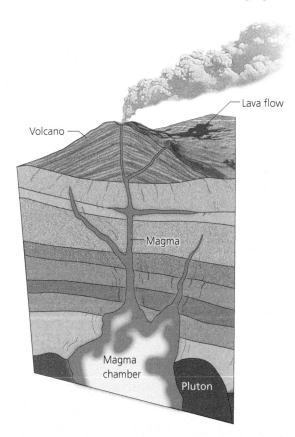

Figure 1.22   Igneous rocks may form by solidification of molten rock beneath the surface within magma chambers or above the surface by eruption from volcanoes.

Figure 1.23 Horizontally layered sedimentary rocks of the Grand Canyon, exposed through erosion by the Colorado River.

Erosion involves the removal of the newly weathered grains of sediment away from their source, perhaps within a rivulet of snowmelt or maybe as part of a landslide along a steep slope. The process of erosion might best be viewed as the transitional phase between the in-place weathering of rock and its eventual transportation downslope as loose particles of sediment. Grains of sand on a beach or in a desert are usually composed of quartz, a common,

resistant mineral found in many rocks. The quartz grains are the remnant loose sediment derived from the weathering and erosion of preexisting rock. From the point of view of national parks, weathering and erosion are the primary processes involved in the creation of scenic landscapes since rock exposed at Earth's surface is by nature subject to the effects of weather and its larger-scale cousin, climate.

The second phase in forming sedimentary rock involves the **transportation** of loose sediment downslope under the force of gravity. Rivers, wind, and glacial ice are the typical transporting agents. Networks of smaller streams feeding into larger rivers can move an enormous volume of sediment from highlands to lowlands. Winds can be strong enough to transport sand, silt, and clay, but they commonly leave the larger pebbles and cobbles behind. Glaciers are capable of moving everything from house-size boulders to minute particles of clay, usually by incorporating the sedimentary debris into the plastically flowing body of the glacier.

Phase three, **deposition**, begins when the energy of the transporting agent decreases to the point where it can no longer move particles along in the flow, such as where a fast-flowing river meets an open body of water like a lake or ocean. **Deposition** involves the physical settling of loose particles

Figure 1.24 This image of mudflats near the Copper River Delta in Alaska illustrates three of the four stages in the formation of sedimentary rock. The mountains in the background supply the raw sediment by weathering and erosion of the rocks. The Copper River transports that gravel, sand, silt, and clay downslope under the influence of gravity until it reaches the standing water of the Gulf of Alaska where the velocity of the river slows dramatically. The sediment is then deposited within the mudflats of the delta.

within a low region on Earth's surface. The site of deposition is typically a low-energy setting such as a river floodplain, a delta, a lake, or the margin of an ocean. Topographic depressions like these examples are called **depositional environments**. Because a particle of sediment will settle out to its lowest potential energy due to gravity, the countless particles will form a relatively smooth horizontal layer on the floor of a lake or seabed. As more sediment accumulates through time, the horizontal depositional surfaces build into thicker layers called **beds**. Organisms live in profusion in these usually watery depositional environments, and their presence may be preserved as **fossils** within the layers of sediment. Fossils provide crucial physical evidence for interpreting the depositional environment of a layer of sedimentary rock and for determining the relative age of the rock in which the fossil was found.

The fourth phase of creating a clastic sedimentary rock involves the **burial**, **compaction,** and **cementation** of the loose sediment into solid rock (Fig. 1.25). With time, more and more sediment may accumulate until older sediment is deeply buried beneath younger sediment. In active depositional settings such as the Mississippi Delta where the river empties into the Gulf of Mexico, sediment may build up over tens of millions of years to a pile as much as 20 km (12 mi) thick. As the loose sediment is buried deeper and deeper, water trapped in pore spaces between grains is squeezed outward, allowing the particles to pack closer together. However, pore space inevitably remains between the tightly compacted grains, providing circuitous pathways for groundwater charged with ions to percolate

Figure 1.26 Conglomerate composed of naturally cemented cobbles, pebbles, and sand grains derived from a variety of source rocks, Point Reyes National Seashore.

through the sediment pile. As the saturated groundwater flows laboriously through connected pores, minerals crystallize within the tiny spaces between grains, producing a natural cement that binds together the loose grains into a solid mass of sedimentary rock. The cement may be composed of iron oxide, calcite, silica, or other minerals precipitated from ions in the groundwater solution.

Natural cementation is the final step in creating solid sedimentary rock from a thick pile of loose sediment. Once cemented, sandy sediment becomes sandstone, silty sediment becomes **siltstone**, and clay-rich sediment becomes **shale** or mudstone. If the cemented sediment consists of a range of sizes, including rounded cobbles or pebbles, the rock is called a **conglomerate** (Fig. 1.26). In order to see these deeply buried sedimentary rocks at the surface, mountain-building processes must eventually uplift the entire mass, exposing them to a renewed cycle of weathering and erosion.

The second broad type of sedimentary rock is produced by biochemical processes that originate in living organisms. The accumulated skeletal debris from countless calcareous organisms that once lived in warm, shallow, tropical seas produces **biochemical sedimentary rocks** like limestone. Many marine organisms, including coral, clams, oysters, snails, sea stars, many algae, and some sponges, have evolved the ability to secrete hard skeletal frameworks of the mineral calcite (and its relative, aragonite). The calcium, carbon, and oxygen that compose calcite ($CaCO_3$) were originally derived from the surrounding seawater by the organism. If you've ever snorkeled along a coral reef, you know that these calcareous organisms may live in profuse abundance (Fig. 1.27).

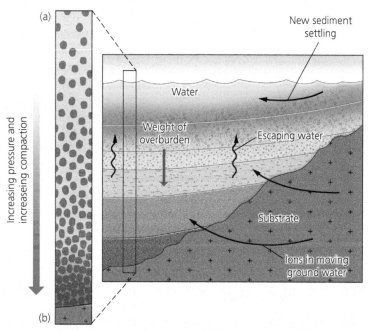

Figure 1.25 Burial, compaction, and cementation of sediment to become sedimentary rock. Deposited sediment is buried by younger sediment through time. With progressive burial, water is squeezed outward through pores between grains, compacting them into a smaller volume. The grains eventually become cemented together as saturated groundwaters filter through the remaining pore spaces, precipitating minerals.

Figure 1.27 A variety of corals and other calcite-secreting organisms, Florida Keys Marine Sanctuary. As the skeletal framework of these organisms breaks down due to natural causes through time, the calcite debris accumulates on the seafloor to form the raw material for limestone.

a wide variety of depositional environments, and the architecture and composition of accumulated sediment are affected by such factors as climate, tectonic setting, and changes in sea level. Examples of modern basins include the Gulf of Mexico, the Central Valley of California, Death Valley, the Gulf of California, and the continental shelf of the eastern United States, which is submerged beneath the relatively shallow waters of the western margin of the Atlantic Ocean. As these examples suggest, basins occur wherever the crust is at lower elevations than the surrounding land surface, commonly but not always below sea level.

Rivers invariably flow downslope toward sedimentary basins, transporting sediment that eventually will be

When the organisms die, their soft internal material is consumed by other animals or bacteria, leaving behind the hard debris of their skeletons (Fig. 1.28A). The calcareous fragments settle to the seafloor as "lime" sediment, where they may accumulate to tremendous thicknesses over broad areas of the seafloor through time. With burial and compaction of the limy seafloor sediment through time, groundwater will dissolve some of the calcite from the remnant shell material and re-precipitate it in pore spaces as a natural cement (Fig. 1.28B). Thus, the squishy, white lime sediment becomes solid, gray limestone. In a sense, limestones are the "rocks that were once alive."

There are many active limestone-forming regions today, most of them shallow tropical platforms like the Bahamas or broad, shallow continental shelves such as Florida Bay, which forms the southern third of Everglades National Park. Thick layers of limestone underlie the entire South Florida region, including the Everglades. Much older layers of limestone are exposed in high-elevation regions like the Grand Canyon, where they were raised far above sea level during a regional episode of uplift, then exposed to view through weathering and erosion. Limestones play an important role in forming landscapes in many of our national parks; understanding their origin is key to deciphering the ancient depositional environments that existed throughout Earth history.

There are other types of sedimentary rocks not discussed in this section such as natural salts (like gypsum and halite), dolostone, chert, travertine, and coal. Some of these rocks play an important role in the landscapes of our national parks and will be addressed in the text where relevant.

The larger-scale topographic depression in which beds of clastic and biochemical sediment accumulate through time is called a **sedimentary basin**. Basins may consist of

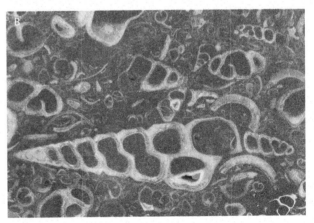

Figure 1.28 A) Calcareous skeletal debris derived from calcite-secreting marine invertebrates. This is the raw material that comprises limestone upon burial, compaction, and cementation. B) Photomicrograph of thin, near-transparent sliver of limestone showing the numerous fragments of calcareous organisms that compose the rock. The homogeneous dark material filling the space between grains is a natural calcite cement.

deposited and buried with the basin. Individual beds of sediment are deposited contemporaneously within a variety of depositional environments across the basin, adding mass to the pile of sedimentary debris. The weight of sediment exerts a load on the underlying lithosphere, causing it to sink in a process called **subsidence**. The weaker asthenosphere accommodates the sagging lithosphere by plastically migrating outward away from the sedimentary load. The continually subsiding depositional surface makes space for even more sediment to accumulate on top.

We are belaboring the tale of sedimentary rocks because they are inherently very important. First, stacked beds of sedimentary rock—**strata**—are commonly compared to pages in a history book. It's a book with several pages missing, but one that tells the story of ancient changes in climate, sea level, and environmental setting on the evolving surface of Earth. Second, sedimentary rocks contain fossils, the physical record of biological evolution and the history of life on Earth. Finally, our modern civilization exists due to the occurrence of energy resources such as coal, oil, and natural gas within sedimentary rock. We may gaze admiringly at colorful sedimentary rocks at places like the Grand Canyon, but we are dependent on them more than we might guess.

METAMORPHIC ROCKS

Deep in the cores of many mountain ranges, pressures and temperatures may increase to such a point that preexisting rocks physically transform into "new" rocks. A common example is **marble**, which is the end product of limestones that have been subjected to high enough pressures and temperatures that the minute fragments of fossils and the calcite cement within the limestone are entirely transformed into a dense interlocking mosaic of calcite crystals. Marble is a **metamorphic rock**, derived from the solid-state transformation of preexisting rock (Fig. 1.29).

It sounds somewhat mysterious, but metamorphism really does happen in the solid state, without the involvement of any melting or liquid phase. Under high pressures and temperatures, atoms physically diffuse through the crystal structure of preexisting minerals, in the process rearranging themselves into a different set of minerals. In essence, the elements just reshuffle themselves into a different, more compact atomic arrangement that is compatible with the ambient pressure and temperature conditions. Typically, the chemical composition of the metamorphic rock remains comparable to the precursor rock from which it was derived, but the mineral composition of the metamorphic rock is normally different from that of its progenitor.

The process of metamorphism has been reproduced in the laboratory as rocks are subjected to a range of pressure and temperature conditions. As an example, let's follow the path that a simple sedimentary rock might take as it is subjected to progressively higher temperatures and pressures (Fig. 1.30). Shale is a common sedimentary rock composed of a variety of tabular clay minerals that are stable at surface conditions. As the pressure and temperature are increased, the elements composing the clay minerals rearrange themselves into a more stable configuration, resulting in the transformation of those clay minerals into minerals of the mica group. Micas are platy minerals and include chlorite, biotite, and muscovite. The result of the mineralogical transformation is that the original shale

Figure 1.29 Banded marble derived from the metamorphism of alternating layers of limestone and shale.

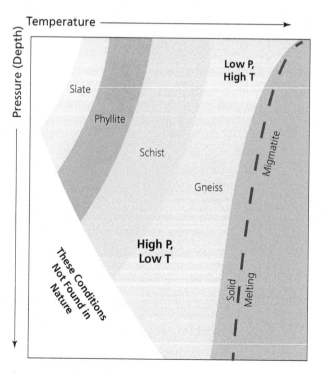

Figure 1.30 Gradational changes in mineral composition as sedimentary shale is subjected to progressively increasing temperature and pressure. The original shale transforms into various types of metamorphic rocks with increasing metamorphic grade, accompanied by incremental changes in the mineral composition.

metamorphoses into **slate**, a rock commonly used to make roofing shingles for homes in the eastern United States due to its tendency to break into flat, tabular slabs. The platy sheets of slate reflect the parallel alignment of remnant clay minerals and newly formed micas as they preferentially grow in orientations perpendicular to the direction of applied stress.

As pressure and temperature are ratcheted up, slate will evolve into **phyllite**, then **schist**, with the platy mica grains growing parallel to one another and creating a finely laminated texture to the rock. The metamorphic mineral garnet may begin to grow among the mica minerals since its atomic structure is stable at the higher pressures and temperatures. As the schist is subject to even higher pressures and temperatures, it will metamorphose into **gneiss**, a common metamorphic rock composed of mica minerals, quartz, feldspar, and perhaps a few others. All of these metamorphic transitions occur slowly over tens of thousands to millions of years.

If the temperatures get high enough that the minerals within gneiss begin to get soft prior to melting, the rock will begin to flow as a weak plastic, creating a beautiful rock called a **migmatite** that exhibits loopy banding and stretched mineral grains that look like pulled taffy (Fig. 1.31).

So, that homely pile of mud deposited at the bottom of a lagoon long ago may, with time and tectonic stress, become the dense, solid roots of a great mountain range or the cornerstone of a skyscraper. Metamorphic rocks are common throughout our national parks. We'll encounter them in the bottom of the Grand Canyon, adjacent to granite in the Sierra Nevada, along the stark mountainfronts of Death Valley, and at the highest elevations in the Appalachians.

Where does metamorphism occur? In order to generate the requisite high pressures and temperatures necessary for metamorphism to occur, the original rock must be subjected to either tectonic stress or hot fluids. Metamorphism is common within subduction zones where cold oceanic lithosphere is subjected to the intense pressures associated with its descent into the mantle. The minerals composing the metamorphic rocks created along subduction zones are stable at the high pressures and the relatively low temperatures of that regime. Another common metamorphic environment occurs within the deep roots of mountain ranges undergoing compressional tectonism. The "guts" of mountain ranges are commonly subject to both high pressures and high temperatures, so the suite of metamorphic minerals that form will reflect those conditions. A third metamorphic locale occurs near the margins of rock intruded by magma where the heat escaping from the molten rock is transferred into the surrounding colder rock. The metamorphic minerals that grow in this environment tend to reflect the high temperatures and lower pressures associated with regions intruded by magma.

Any existing rock may be transformed via metamorphism into a different rock, with a new set of minerals composing it. Sandstone may metamorphose into quartzite, granite may metamorphose into gneiss, or limestone may become marble. The intense conditions associated with metamorphism typically occur deep beneath the ground. In order for the rocks to be exposed at the surface, they need to be raised upward from their burial deep in the crust. This process of **exhumation** commonly happens where deeply rooted metamorphic rocks have been uplifted by mountain-building or where the combined effects of regional uplift and erosion have incised deeply enough to expose the rocks at the surface. Metamorphic rocks have been exhumed to the highest peaks in Grand Teton and Rocky Mountain national parks, and the process of exhumation plays an important role in the evolution of landscapes in many of our national parks.

THE ROCK CYCLE

The preceding sections illustrate how existing rock may be eroded and redeposited to become a sedimentary rock or, alternatively, may be melted and incorporated into a magma chamber beneath a volcano, or even slowly transformed by high pressures and temperatures into an entirely new rock. Clearly, the matter that composes rock may be recycled throughout its history, taking on different configurations as the conditions change. The transformation of matter into different forms that occurs continuously within the Earth and on its surface is called the **rock cycle**. The recycling of

Figure 1.31 Highly metamorphosed migmatite, Phantom Ranch area, Grand Canyon National Park. Migmatites form under very high pressures and temperatures that are close to the melting temperature of rock. Under these extreme conditions, the rocks behave as a thick plastic and tend to flow. Migmatites form in that netherworld between igneous and metamorphic rocks.

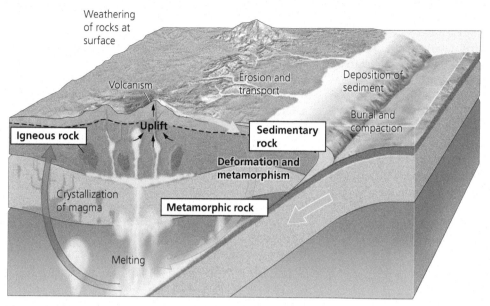

Figure 1.32 A view of the processes and products involved in the rock cycle that occur along subduction zones. Rock composing Earth is continually recycled through tectonism, driven by energy from the Earth's interior as well as energy from the Sun.

matter via the rock cycle requires energy, which is ultimately supplied by Earth's internal heat and by the Sun. The recycling process is facilitated by tectonic activity that moves rock around the face of the planet, creating a wide variety of conditions from the highest mountains to the deep ocean to the interior of subduction zones.

Matter can take numerous pathways within the Earth system (Fig. 1.32). Igneous rock formed at a mid-ocean ridge by seafloor spreading will be conveyed laterally before being dragged downward within a subduction zone. There it may be reanimated as a metamorphic rock due to the intense pressures associated with shear along the subduction front. If transported deep enough, that metamorphic rock may eventually melt to become part of a magma plume that slowly rises to the surface, perhaps to erupt from a volcano and once again become an igneous rock. That volcanic rock will eventually weather and erode, with the decomposed particles moved downslope and buried to eventually become part of a sedimentary rock. The recycling of rock within the Earth can go on and on as long as the energy supply continues to drive the system. Of course, it's the atoms composing the rock that are being recycled, rather than the rocks themselves. An individual oxygen atom may find itself within a quartz grain in sandstone at one point in its "life" or as part of an olivine crystal at another or within a garnet in yet another incarnation through time. That single oxygen atom potentially will cycle through the hydrosphere, atmosphere, and biosphere as well.

When you look at a landscape in our national parks, think about the dominant form of matter comprising that view: rock, made of minerals, which in turn are composed of a few common elements. The rock may be overlain in places by a thin veneer of **soil**, which is composed of

particles of rock mixed with organic matter. The soil may support a variety of plants, imparting the familiar green patina that is so pleasing to the eye (Fig. 1.33). The overwhelming volume of that landscape is rock, however, which is mostly unseen beneath the surface layer of soil and vegetation, but it is dominant nonetheless. Always try to project your view into what it may look like beneath the ground where the rocks of the crust and underlying mantle extend downward for literally a few thousand kilometers. Geologists always try to think in terms of four dimensions, so consider as well the long evolutionary history of that rock just beneath the soil: Perhaps it was once grains of sand on a beach, or part of a churning mass of molten rock within a magma chamber deep beneath a volcano, or maybe it was squeezed and baked within the bowels of a growing mountain range. Push your imagination beyond the simply aesthetic to see beneath the surface and into the past when you are enjoying vistas in our national parks.

Deep Time

Geologists have an innate understanding that Earth's history (i.e., time) is recorded within the rocks of Earth's crust. Many geologists spend their careers trying to decipher the chronology of specific events in Earth history by "reading the rocks." This section addresses the somewhat abstract

Figure 1.33 Reading the landscape, Glacier National Park.

concept of **geologic time**, commonly referred to as **deep time**. Perhaps the most important concept in geology that people should understand is a sense of the vast expanse of geologic time. This is especially true for a topic such as the geology of national parks where knowing what happened when is critical for a four-dimensional understanding of a particular landscape or geologic feature. To contemplate the 4.6 billion year age of the Earth is a sublime experience, incomprehensible to the human mind and even somewhat discomfiting since it's so far beyond the typical scale of human thinking. Developing a reasonable understanding of the vastness of geologic time is our goal here, which is effected primarily by visualizing time's arrow.

Humans invariably think in terms of hours to days to years to decades—and at most, just a few generations. But geologists have been trained to take the long view. Although abstract and nonintuitive, learning to think in terms of scales of thousands to millions to billions of years can be accomplished, especially if the concept of deep time can be somehow visualized. There are many examples of deep time metaphors. A few include the length of a football field, the hours in a day, or the calendar year. These metaphors lead to statements such as "The Cambrian Explosion of life occurred around Thanksgiving, the extinction of the dinosaurs occurred around Christmas, and the Ice Ages began around 4 hours before the end of the year." John McPhee, in his discussion of deep time in *Basin and Range* (1981), used the visual metaphor of the outstretched arms of a human, stating that "in a single stroke with a medium-grained nail file you could eradicate human history."

An inelegant visual aid that I use in many of my classes involves a tattered toilet paper roll, 46 sheets long, each sheet representing 100 million years, stretched across the front of the classroom and held by students at each end. (It's not really my idea. I borrowed it from a friend back when I began teaching.) At one end is the 4.6 billion year old origin of the Earth, while "today" anchors the other end. Certain key events in Earth history are marked at appropriate places along the ribbon, which I'll point out as I walk along. This permits me to make statements later in the course such as "The oldest rocks at the bottom of the Grand Canyon (~1.8 billion years) are about 18 toilet paper sheets from today," or "The age of the granite in Yosemite is only about 1 sheet (~100 million years) from today." Perhaps the most important idea in this simple visual exercise is that the outermost filament of toilet paper represents the entire history of civilization. The realization that humans and their constructs manifest only a tiny sliver of time in Earth's history is a profound moment for many nongeologists and puts the day-to-day ephemera of our lives in context.

RELATIVE AGE DATING

How do geologists know the age of the granite in Yosemite National Park or the age of the most recent eruption at Yellowstone? How do they know the age of Earth, for that matter, or the age of the most recent global glaciation? To answer these questions requires dates, values attributed to rocks that represent the time in Earth history when they were formed. Geologists acquire dates in two ways, one qualitative, one quantitative. **Relative age dating** qualitatively describes the age of one rock or geologic feature relative to another. Phrases such as "younger than" or "older than" are typically used in relative age dating. For example, rocks at river level in the Grand Canyon are older than rocks exposed along the upper rim (Fig. 1.23). Such statements are possible due to a set of logical principles defined by geologists over the last few centuries. Discussion of a few of these principles follows.

Superposition is an intuitive concept—in a sequence of layered rocks, those at the bottom of the stack are older than overlying layers. Think of how a chef makes a layer cake: first she places a layer of chocolate cake at the base, then she lays down other layers of white, then yellow, then another layer of chocolate, each with thin films of filling in between. The cake layer beneath had to have been there prior to slathering on the filling and placing another layer on top. This culinary sequence is analogous to the superposition of layers of bedded rock such as those exposed in the Grand Canyon.

Faunal succession relies on the fact that fossils—remnants of previously living organisms now preserved in rock—record the history of life on Earth. Species live for a duration of time and then invariably go extinct, never to return. This evolutionary reality means that the fossils of an organism preserved in a layer of rock will never reappear in younger, overlying layers after it has gone extinct. This results in an orderly, nonrepetitive succession of fossils from one layer to another that permits geologists to determine the relative ages of specific fossils and the rock in which they occur, regardless of where in the world the fossils are found (Fig. 1.34).

Uniformitarianism is a principle initially promoted by James Hutton, a renowned 18th-century geologist. The fundamental idea behind uniformitarianism is that physical processes occurring in today's world also likely occurred in worlds past. For example, the processes that we observe in a modern river are likely identical to those of ancient rivers. This principle is commonly simplified to "The present is the key to the past." However, the assumption of slow, gradual geologic change implicit in uniformitarianism does not describe the full range of rates of geologic processes, especially abrupt, catastrophic events that punctuate the geologic record such as volcanic eruptions and asteroid impacts. Therefore, uniformitarianism has been modified into a more realistic model of geologic change called **actualism** that takes into account the wide range of rates at which events occur through time. As we visit various national parks in the book, we'll use actualism as a powerful technique to understand the ancient world by comparison with the modern.

Original horizontality reflects the fact that sediment will be deposited in horizontal layers (i.e., "beds") because of the influence of gravity on the settling of grains. For

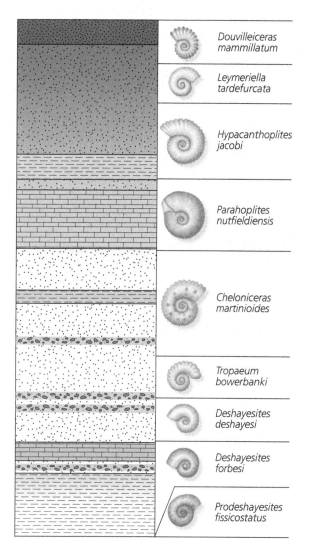

Douvilleiceras mammillatum

Leymeriella tardefurcata

Hypacanthoplites jacobi

Parahoplites nutfieldiensis

Cheloniceras martinioides

Tropaeum bowerbanki

Deshayesites deshayesi

Deshayesites forbesi

Prodeshayesites fissicostatus

Figure 1.34 Fossil species of ammonites, extinct marine invertebrates related to squid and octopuses, exhibit an orderly, predictable faunal succession within sedimentary strata worldwide. The systematic arrangement permits geologists to assign a relative age to layers of rock. For example, rocks containing fossils of *Cheloniceras* are the same age everywhere the rock is found.

instance, the velocity of a river will decrease dramatically when it encounters a lake or an ocean, slowing the current so that it is no longer able to transport its sediment load. The sediment will then accumulate on the lakebed or seafloor as a horizontal layer because each grain settles to its lowest possible position. The horizontal strata of the Grand Canyon exemplify this principle (Fig. 1.23). So when geologists see a layered sequence of rock tilted at an incline or folded into arcuate shapes, they know that the originally horizontal orientations were later deformed by tectonic stress.

Cross-cutting relations is easiest to explain with an example or two. If a series of rock layers are cut by a body of igneous rock, then the intrusion of magma must be a younger event since the layered rocks had to have been there to be intruded (Fig. 1.35A). Another example is a fault cutting through sedimentary layers (Fig. 1.35B).

Faulting had to have happened some time after the sedimentary layers were deposited. Many geologic settings can be highly complicated by cross-cutting relations, and unraveling the relative chronology of events is a methodical and challenging process. Cross-cutting relations is a particularly powerful dating technique when combined with the quantitative rigor of numerical dating.

NUMERICAL AGE DATING

Formerly known as "absolute" age dating, the terminology has evolved to its current "numerical" descriptor because the dating technology keeps getting better and better; previous absolute dates keep changing due to the increasing resolution of the instrumentation. As opposed to relative dating, **numerical age dating** places specific ages on individual rocks. For instance, the numerical age of the oldest rock yet found on Earth is 4.031 ±0.003 billion years, derived from rocks found in northwestern Canada. The ±0.003 billion years (±3 million years) expresses the range of uncertainty in the measurement. The age of the Earth is generally accepted as 4.56 billion years (± a few million years), based on the ages of certain types of rare meteorites that are assumed to have formed at around the same time as the rest of the solar system. The oldest age of Moon rocks returned by the Apollo missions is 4.48 billion years, suggesting that the Moon was formed soon after the formation of Earth. These values are determined by **radiometric dating**, a systematic procedure that relies on high-precision instrumentation as well as the fundamental laws of physics and chemistry.

Radiometric dating (informally referred to as "clocks in rocks") is based on the measurable natural decay of radioactive elements such as uranium, thorium, and potassium incorporated within the mineral structure of certain rocks. Radiometric dating works best on igneous rocks—those that solidify from a molten fluid. As minerals crystallize from the cooling magma, they may incorporate trace amounts of radioactive elements into their atomic structure. Once the mineral is fully crystallized, the radioactive elements within begin to spontaneously decay and the atomic clock is set in motion. (By the way, radioactive decay also releases prodigious amounts of heat. Heat production from the rocks of the mantle contributes to the overall heat budget of the Earth, and thus to convective circulation and the geologic dynamism of the planet.) As long as the mineral crystal remains a closed system, without the addition or loss of critical elements, it stays viable for radiometric dating. The workhorse mineral of radiometric age dating is **zircon**, a mineral formed in igneous rocks that preferentially incorporates uranium and thorium into its atomic structure. Zircons remain remarkably stable through time, resisting alteration from high temperatures and pressures as well as from erosion.

As the radioactive element decays with its tick-tock efficiency, it produces new elements. For instance, a form of uranium known as U-238 will decay to a form of lead known as Pb-206. In this case, U-238 is the **parent element**

Figure 1.35 A) Igneous dike intruding red sedimentary rock layers, illustrating the principle of cross-cutting relations. Lower Grand Canyon. B) Multiple faults in sedimentary rock layers illustrating principles of superposition and cross-cutting relations. Near Moab, Utah.

according to their atomic masses. The individual atoms are then counted within the instrument, providing raw values of both parent and daughter elements. The parent:daughter ratio is integrated mathematically with the known half-life of the parent to produce a numerical age of the original mineral crystal. The mineral is the specific object that is dated, thus determining the time at which the mineral first crystallized from the magma or lava. Radiometric dating techniques involve much more detail than discussed, but understanding the essence of the method is enough to move forward with our discussion of deep time and the **geologic time scale**.

INTEGRATING RELATIVE AND NUMERICAL TECHNIQUES

Most relative ages come from sedimentary rock, primarily because most fossils are found in sedimentary rock. In contrast, most numerical ages are derived from igneous rocks, as described above. The key to placing numerical ages on sedimentary (and metamorphic) rocks is to find exposures where the igneous rocks cross-cut layers of fossiliferous sedimentary rock. Geologists search specifically for exposures of solidified lava or volcanic ash interbedded with sedimentary rocks, thus enabling radiometric dates to bracket a sequence of fossiliferous layers.

In the hypothetical example of Figure 1.36, multiple volcanic ash layers are interbedded with fossil-bearing sedimentary strata. This implies that at some point in the past, nearby volcanoes spewed ash into the atmosphere, with the fine particles eventually falling onto the sea where the organisms lived. The ash settled through the water column and formed a layer on the seabed, entombing the organisms beneath. Much later in time, after burial and cementation, the stack of rocks of the ancient seabed was raised above sea level by tectonic activity, exposing the layers to view.

On Figure 1.36, radiometric dating of specific minerals within the volcanic ash layers above and below the trilobite fossil constrain the age of that species to between 440 and 465 million years. Likewise, the bivalve fossil higher in the sequence is bracketed by age dates to have lived between 400 and 415 million years. We can infer that the trilobite species lived between 440 and 465 million years ago everywhere they're found around the planet. The numerical dates surrounding the trilobite fossils can simply be transferred to other localities to estimate the ages of sedimentary layers containing the same fossils, even if no datable volcanic layers exist at the other locality. If the same trilobite species is found in Australia or Siberia or Nevada, it must have the same age range, based on the

and Pb-206 is the **daughter element**. The time over which half of the parent atoms decay to daughter atoms is a constant known as a **half-life**, a value unique to each radioactive element that can be determined experimentally in a laboratory. (Some sedimentary rocks can be dated using **radiocarbon dating**, but the short half-life of radioactive carbon limits the technique to rocks younger than about 50,000 years.)

Knowing the theoretical background, a geologist first collects a fresh rock sample by hammering or drilling it out from a larger exposure of rock, perhaps from a mountainside. Back in the lab, the rock is crushed so as to separate out the minerals most likely to contain radioactive elements. The select mineral crystals are then placed in a sensitive instrument called a mass spectrometer where they are vaporized to release the parent and daughter elements from the rest of the mineral. The now free atoms are magnetically separated within the spectrometer

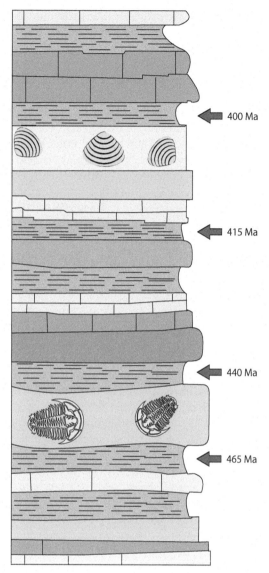

Figure 1.36 In a sequence of fossiliferous layers interbedded with volcanic ash deposits, the ages of the fossilized organisms can be bracketed by radiometric dates of minerals within the ashes (gray layers).

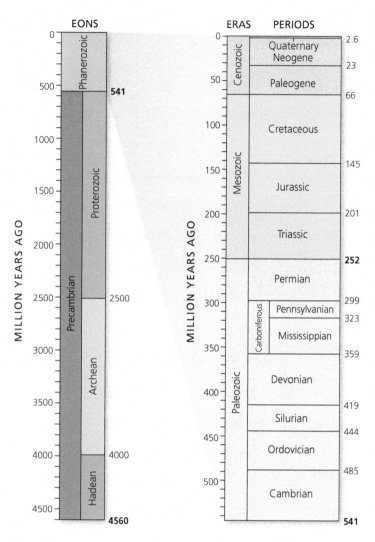

Figure 1.37 Geologic time scale, determined by integrating relative age dates with numerical age dates. Long eons of time are subdivided into eras, which in turn are subdivided into periods.

radiometric dates from our one hypothetical locality. In sum, the techniques of radiometric dating are integrated with the principles of superposition, faunal succession, and cross-cutting relations to bracket the range of ages for past species and the rocks in which they're found.

GEOLOGIC TIME SCALE

By careful observation and collection from thousands of rock exposures around the world, geologists have integrated relative ages of rock with numerical dates to produce the geologic time scale (Fig. 1.37). Merging evolutionary theory with the logical principles of relative age determination, further combined with the physics- and chemistry-based techniques of radiometric age dating, produces a powerful method for dating events through Earth history. Over the last century or two, careful study

of rocks from numerous localities around the world, publication of the data, and methodical correlation of rock strata have enabled the creation of the geologic time scale that calibrates the history of our planet and the evolutionary record of life. Most people take the geologic time scale for granted, but it is truly a monumental achievement. As the process of science dictates, the geologic time scale is constantly being modified, updated, and refined as new data are uncovered and as the instrumentation evolves to higher and higher degrees of precision.

Various subdivisions of time characterize the geologic time scale (Fig. 1.37). The longest scale is the **eon**, with the Hadean, Archean, and Proterozoic eons composing the **Precambrian** "supereon" of time, spanning almost 8/9ths of deep time, from 4.6 Ga to 541 Ma. (Ga is shorthand for "billions of years" ago and Ma stands for "millions of years" ago. A fuller explanation is provided below.) This long phase of Earth history is preserved primarily as complex metamorphic and igneous rocks of the continents.

Fossilized remnants of microbes represent the dominant life form through the Precambrian. Many of our national parks exhibit Precambrian rock at the tops of the highest peaks, as well as within the deepest river gorges.

Although fossil evidence of multicellular life is found in the youngest Precambrian rocks, it's the advent and evolution of complex life forms that define the succeeding **Phanerozoic Eon**. Eons are subdivided into smaller units called **eras**; the Phanerozoic is composed of the **Paleozoic**, **Mesozoic** and **Cenozoic** eras. Eras are further subdivided into **periods** such as the Devonian, the Triassic, and the Paleogene, which are themselves divided into even smaller increments of time called **epochs**. Each of these time units is defined by numerical ages at their upper and lower boundaries.

The four primary time units around which this book is organized are the Precambrian, the Paleozoic, the Mesozoic, and the Cenozoic. In certain instances, shorter time units may be used out of necessity, such as the Quaternary Period comprising the Ice Ages of the past 2.6 million years. But most of the text refers to the four major units both for the sake of simplicity and to avoid saturation by geologic terminology. With some rounding, the Precambrian ranges from 4.6 Ga to 541 Ma, the Paleozoic from 541 Ma to 252 Ma, the Mesozoic from 252 Ma to 66 Ma, and the Cenozoic spans the latest 66 m.y. of Earth history.

Geologists use specific abbreviations for time units. The absolute usage is somewhat in flux, but typically the units **ka**, **Ma**, and **Ga** are used to denote specific *points* in geologic time: "ka" (kilo-annum) refers to thousands of years, Ma (mega-annum) to millions, and Ga (giga-annum) to billions of years. For instance, the Phanerozoic Era begins at 541 Ma, a point in time. The units k.y., m.y., and b.y. are commonly used to define *durations* of geologic time; k.y. equates to thousands of years, m.y. to millions of years, and b.y. to billions of years. So one might say that the sedimentary rocks of the upper Grand Canyon span about 250 m.y. of Earth history. Or, combining the abbreviations, one might say that the majority of the igneous rock in the Sierra Nevada spans a range of about 40 m.y. between 120 Ma and 80 Ma. These abbreviations are occasionally redefined within individual chapters of the book.

Summary

This brief review of the fundamentals of geology is intended to supplement the more in-depth discussions of individual geologic provinces and national parks in the rest of the book. Many more fundamental concepts will be addressed in subsequent chapters, such as faulting, isostasy, seismicity, styles of volcanism, mountain-building, and glaciation as they relate to specific national parks. As you explore the geology and landscapes of the parks in the book, you'll be directed to return to certain parts of this first chapter to gain a deeper comprehension of the relevant subject matter.

Key Terms

actualism
alluvium
asthenosphere
atmosphere
axial rift valley
basalt
bed
biochemical sedimentary rock
biosphere
brittle
burial
calcite
cementation
Cenozoic
chemical weathering
clastic sedimentary rock
clay
cobble
compaction
compression
conglomerate
continent-continent convergent boundary
continental collision
continental crust
continental rift
continental volcanic arc
continental-oceanic convergent boundary
convection currents
convergent plate boundary
core
cross-cutting relations
crust
cryosphere
daughter element
deep ocean trench
deep time
deposition
depositional environment
divergent plate boundary
earthquake
element
eon
epoch
era
erosion
exhumation
extension
fault
faunal succession
fossil
Ga
geologic time
geologic time scale
gneiss
granite
half-life
hydrosphere
igneous rocks
iron inner core

iron meteorites
ka
lava
limestone
linear sea
liquid outer core
lithosphere
Ma
magma
magma chamber
mantle
marble
Mesozoic
metamorphic rock
mid-ocean ridge system
migmatite
mineral
numerical age dating
oceanic crust
oceanic-oceanic convergent boundary
original horizontality
Paleozoic
parent element
partial melting
pebble
period
Phanerozoic Eon
phyllite
physical weathering
plastic
plate
plate tectonics
pluton
plutonic igneous rocks
Precambrian
radiocarbon dating
radiometric dating

relative age dating
rock
rock cycle
sandstone
schist
seafloor spreading
sedimentary basin
sedimentary rock
seismicity
shale
shear
silicate minerals
silicate rocks
siltstone
slate
soil
stick-slip
strata
subduction
subduction zone
subsidence
superposition
tectonics
topographic relief
transform fault
transform plate boundary
transportation
uniformitarianism
volcanic igneous rocks
volcanic island arc
weathering
zircon

Related Sources

McPhee, John. (1981). *Basin and Range*. New York: Farrar, Straus, Giroux. p.126.

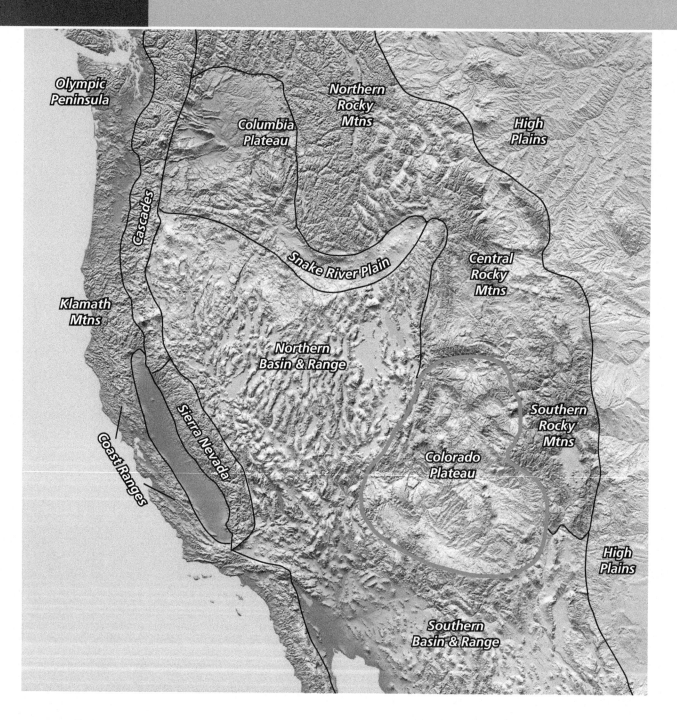

Figure 2.1 Physiographic provinces of the American West, with Colorado Plateau highlighted.

Geologists have subdivided the American West into several distinct physiographic provinces based on similarities in geology, topography, and landscape (Fig. 2.1). The focus of Part 1 is on seven national parks and three national monuments of the *Colorado Plateau*, a broad, high-elevation region of multicolored rock, flat-topped mesas, deeply

Figure 2.2 Canyonlands National Park illustrating tabular sedimentary layering and elongate escarpments typical of the Colorado Plateau. The Colorado River meanders through an incised canyon in the foreground. View from Dead Horse Point State Park.

rugged. Rather, it's a rocky landscape cut by deeply eroded canyons, giving it a stark, naked appearance. Open forests of juniper-piñon and ponderosa pine partly cover the higher elevations. Elongate cliff faces extend for tens of kilometers, defining the margins of individual plateaus composing the larger plateau. These **escarpments** mark the sharp, near-vertical boundaries separating the high tops of plateaus from the lowlands below. Escarpments may be hundreds of meters high, exposing horizontal layers of colorful sedimentary rock to the intrepid geologist or curious tourist. The Colorado Plateau is commonly referred to as "redrock country" due to the multihued varieties of red that color its rocks (Fig. 2.2).

The margins of the Colorado Plateau are defined on the north by the east–west-trending Uinta Mountains (Fig. 2.3). The plateau is bound along its western, southern, and southeastern flanks by the extensive Basin and Range Province. The east margin of the plateau in Colorado is a complex region bordered by the rugged topography of the Rocky Mountains.

The Colorado Plateau is named for the Colorado River system that crosses the landscape, incising deeply into the redrock as it flows to its ultimate destination in the Gulf of California (Fig. 2.4). Fed by snowmelt in the Colorado Rockies, the river flows westward toward Utah where it meets the Green River in Canyonlands National Park. The Green River begins in the Wind River Range of Wyoming and flows southward to its confluence with the Colorado. The combined flow of the rivers cuts southwest through southern Utah into Arizona where its path abruptly shifts westward on a circuitous route through the majestic gorges of Grand Canyon National Park. Near Las Vegas, the Colorado jags southward where it forms the border between California and Arizona before entering Mexico and emptying into the Gulf of California. Along the way, tributaries such as the Gunnison, the San Juan, the Little Colorado, the Paria, and the Salt rivers contribute to the flow in the Colorado.

The geographic region encompassing all of these larger tributaries as well as the thousands of smaller streams and rivulets that feed into the Colorado River define the **drainage basin** (aka **watershed**). A drop of rain or a droplet of snowmelt that finds itself within the boundaries of the Colorado River drainage basin will ultimately end up in the Gulf of California, an elongate embayment of the Pacific Ocean. At least it would have if the Colorado wasn't so heavily impacted by human modification. Today, the river system is highly stressed, with much of the water dammed and diverted for

incised canyons, and sweeping vistas (Fig. 2.2). Our journey through America's national parks begins with the Colorado Plateau because the geology of the region is extremely well exposed due to the long-term effects of erosion as well as the semi-arid climate. Soils are thin, vegetation is sparse, and rivers cut deeply into the surface, allowing for a full view of the rocks comprising the province. Furthermore, the horizontally layered rocks of the region tend to be relatively undisturbed by the complicating effects of deformation, permitting the reconstruction of a vibrant picture of the environmental, climatic, and geographic conditions of the past. Among the many concepts explored in the chapters in Part 1 are sedimentation, weathering and erosion, the expression of geologic time in rock, and landscape evolution.

The Colorado Plateau is centered over the Four Corners region where Colorado, Utah, Arizona, and New Mexico converge (Fig. 2.3). **Plateaus** are broad, high-elevation regions characterized by relatively level upper surfaces. The Colorado Plateau comprises several smaller, flat-topped plateaus whose elevations range between 1500 and over 3000 m (~5000–10,000 ft). The Plateau is not really characterized by "mountains," even though it's high in elevation and

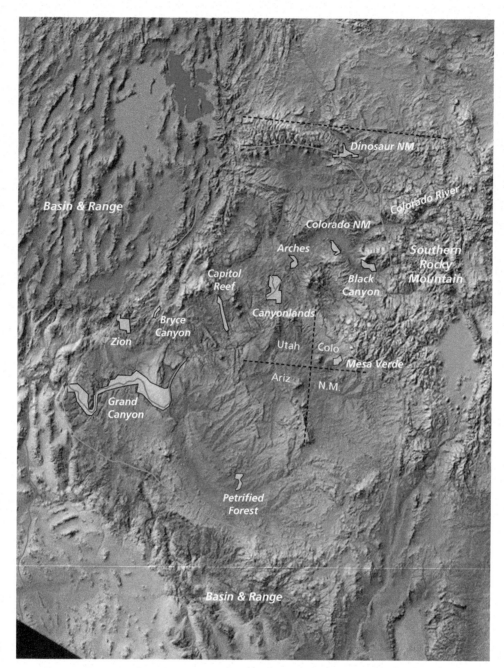

Figure 2.3 Locations of national parks of the Colorado Plateau addressed in Part 1. Crossed lines mark the Four Corners; blue lines mark the Colorado and Green rivers.

irrigation and urban use in the seven states within the drainage basin. The Colorado seldom has any water left by the time it reaches its mouth at the northern end of the Gulf of California. In the millions of years prior to the massive engineering of the modern river, however, the Colorado and its network of tributaries carved some of the most spectacular scenery in the American West.

This brief summary of the Colorado Plateau provides the physiographic background for more specific investigation of the national parks and monuments that reside here. We'll return to the geologic origins of the Colorado Plateau in future chapters, once you've had a chance to understand the characteristics of the geology and landscapes within each park and monument.

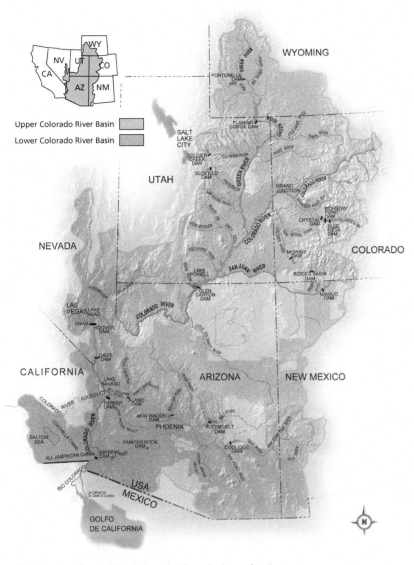

Figure 2.4   Map of the Colorado River drainage basin.

Key Terms

drainage basin
escarpment
geologic province
plateau
watershed

Related Sources

Baars, Donald L. (2000). *The Colorado Plateau: A Geologic History,* revised ed. University of New Mexico Press.
Blakey, R., and Ranney, W. (2008). *Ancient Landscapes of the Colorado Plateau,* 1st ed. Grand Canyon Association.
Chronic, H., and Chronic, L. (2004). *Pages of Stone: Geology of the Grand Canyon and Plateau Country National Parks and Monuments,* 2nd ed. Mountaineers Books.

3 Grand Canyon National Park

Figure 3.1 Grand Canyon panorama.

Only the worst of cynics would not be awed by the sublime panorama of the Grand Canyon: the countless shades of rock, the stark banding of the sedimentary layers, the alternating cliffs and slopes appearing like a poorly made giant's stairway, all extending downward toward the narrow ribbon of the Colorado River at the very bottom (Fig. 3.1). The planar upper surface of the rim on the far side of the Canyon expresses the park's location on the Colorado Plateau of northwestern Arizona (Fig. 2.3). The national park boundaries broadly follow the sinuous path of the Colorado River as it flows westward for 450 km (277 mi) through the canyon, including two distinctive, looping bends (Fig. 3.2). The river separates the South Rim of the Canyon at ~2100 m (6800 ft) in elevation from the higher North Rim at ~2500 m (8300 ft). The majority of the park's six million visitors per year head for Grand Canyon Village on the South Rim.

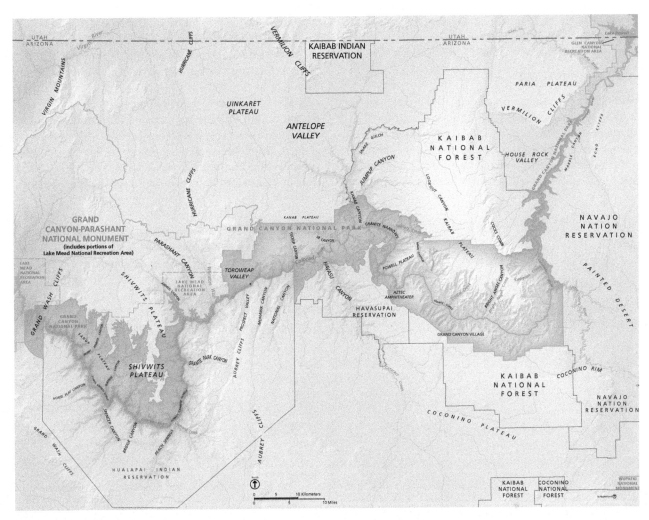

Figure 3.2   Map of Grand Canyon National Park.

The Colorado River in the Grand Canyon is constrained by dams at each end. At its upstream end near the Utah–Arizona border, Glen Canyon Dam has regulated flow through the Grand Canyon since 1966. The dam acts like a tourniquet on the river, trapping sediment in Lake Powell, the reservoir impounded by the dam. Like many dams, Glen Canyon generates power, modulates flow downstream, and provides an aquamarine waterpark in the otherwise dry redrock country of southern Utah. But the dam also flooded a uniquely beautiful canyon upstream and profoundly affected the riparian corridor downstream through Grand Canyon. Since water from Lake Powell exits Glen Canyon Dam through an outlet about 80 m (260 ft) above the floor of the reservoir, virtually all of the sediment transported by the Colorado is trapped below the outlet behind the dam. Thus, very little sediment remains to replenish sandbars and build beaches downstream, degrading riverine habitat in the Canyon. In turn, the plant ecology of the river corridor has changed since the dam was built, with native desert vegetation like mesquite, acacia, and ocotillo being replaced by tamarisk and willow.

The only real input of sediment into the main channel of the river occurs through **debris flows**, turbulent slurries of sediment and water flushed through tributaries as flash floods. Almost all of the 161 rapids in the Grand Canyon were created by debris flows and their bouldery deposits dropped in the main channel near the mouths of side canyons. Ongoing experimental high-flow releases from Glen Canyon Dam by the U.S. Geological Survey are aimed at discovering the optimal timing and amount of release that helps to rearrange sediment along the channel and thus aid habitat restoration.

At the western, downstream end of Grand Canyon, Hoover Dam has backed up the Colorado River into Lake Mead since the dam was completed in the 1930s. Lake Mead holds the largest volume of water of any reservoir in America, but some projections suggest that Lake Mead may be effectively depleted in the next few decades due to human water demand and future long-term droughts associated with climate change. The millions of people who depend on water from Lake Mead as well as hydroelectricity from Hoover Dam may need to turn to alternative sources to maintain the current population density in the American Southwest.

Sculpting Grand Canyon

Standing on the rim or beginning a hike down one of the trails, visitors are struck by the sheer enormity of the canyon relative to the thin string of the Colorado River far below (Fig. 3.3). The details of the geologic story of the formation of Grand Canyon are not completely resolved, but there's no doubt that the Colorado River and its tributaries were the main agents of erosion. Over the last several million years, the river system inexorably cut downward into the layered stack of rocks, resulting in an average depth of 1.6 km (1 mi) and an average width of 16 km (10 mi) from rim to rim. But the Colorado River within the park only averages about 90 m (300 ft) in width. How could that narrow strand of water have carved the grandest of canyons? The answer consists of four parts.

First, the Colorado River didn't do it alone. Numerous **tributary streams** flow into the Colorado within the Grand Canyon and countless smaller tributaries drain into the large ones, creating the feathery texture of the Canyon as seen from above (Fig. 3.4). Every droplet of water within the drainage basin of the Colorado near the Grand Canyon ultimately finds its way to the main channel at the bottom, driven relentlessly by gravity. As the main Colorado River incised deeper into the rock through time, its tributaries kept pace, forming deep side canyons that extended down to river level. The combined flow of these tributaries and their subsidiary streams, rivulets, and gullies erosionally widened the canyon and supplied sediment and water to the main river.

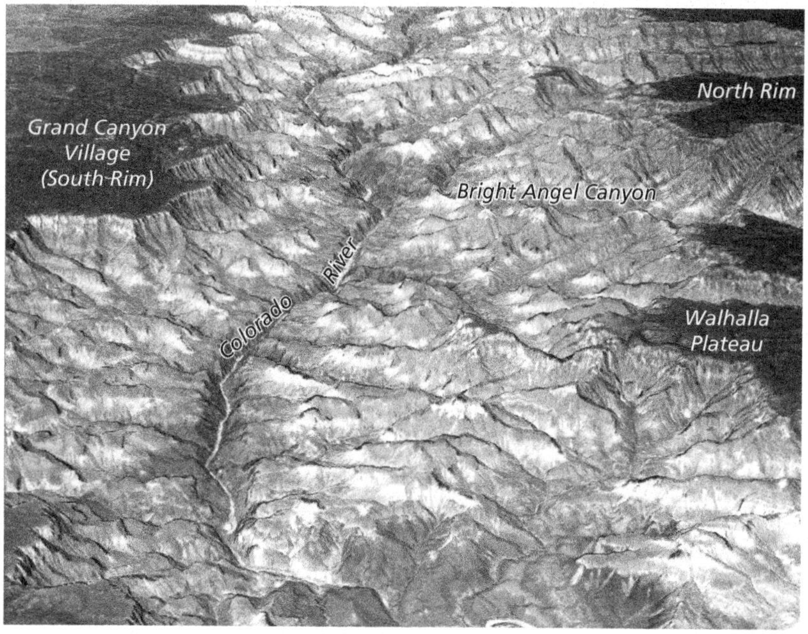

The second factor is the tremendous erosive power of rivers in flood. It's not the water within the rivers and streams that erodes the rock, but rather the sedimentary debris carried along in the flow. During **flood stage**, when the velocity of the river is high, an enormous, turbulent volume of sand, gravel, and even house-sized boulders are transported downstream. The impact of rapidly flowing debris against the rock walls of the canyon is highly efficient at breaking down the intact bedrock. The main canyon and its tributary canyons have deepened and widened during innumerable floods occurring over millions of years.

A third physical factor that contributes to the evolution of Grand Canyon is the regular occurrence of innumerable landslides, **rockfalls**, and debris flows. The decomposition and weakening of the rock that lead to its eventual abrupt collapse is influenced by the steepness of the terrain, the countless fractures that provide pathways for snowmelt to seep into the rock, and the intense seasonal storms. Mass movements of rock are a significant landscape-forming process on the Colorado Plateau and will be discussed in greater detail in later chapters.

Figure 3.3 Perspective view looking westward along a portion of Grand Canyon. The Colorado River winds through the deepest gorge, flanked by colorful, eroded rock alongside canyons. The forested plateau of the South Rim is to the left; the North Rim is to the right. The image was created by combining satellite photos with an elevation map.

Figure 3.4 Satellite photo of eastern Grand Canyon showing the Colorado River and the numerous tributaries that drain into it. The feathery texture lining the river is the canyon itself, carved by the unrelenting flow of water in concert with episodic mass movements. The green swath is the forested Kaibah Plateau.

The fourth factor is the immense amount of time that these erosive processes have had to carve the intricate texture of the Grand Canyon. The Colorado drainage system has been incising the Canyon for the past 5 to 6 million years—ample time for countless floods from countless drainages. Season after season of snow, rain, and flash floods have relentlessly broken down the rock, with the river system carrying the eroded particles far downstream, leaving behind the stupendous landscape of the Canyon. The uncompromising erosional effects of running water and mass movements, integrated over millions of years, have resulted in the natural drama that is the Grand Canyon. The rate of erosion has abruptly decreased since the construction of Glen Canyon Dam, however, with the volume and velocity of the Colorado River currently managed within a relatively narrow range.

Upper Grand Canyon: Ancient Coastal and Marine Environments

The most striking aspect of the Grand Canyon landscape is the bold, multicolored, horizontal striping that forms the walls of the canyon. Geologists have subdivided individual layers into **formations** based on distinct features that characterize each layer (Fig. 3.5). Formations may be tens to hundreds of meters thick and are composed internally of **beds** of variable thickness. The very fact that we can view these sedimentary rocks along the canyon walls expresses the principle of *cross-cutting relations* described in Chapter 1. That is, the rocks had to have been there prior to their incision and exposure by the Colorado River and its tributaries. Younger river erosion cross-cuts the stack of old, layered sedimentary rocks, exposing them to view.

Figure 3.5 Formations of the Grand Canyon.

A simple analogy is that of a multitiered wedding cake that has to be completely made, one layer at a time, prior to the ceremonial cutting by the bride and groom. As the knife slices downward and the wedge of cake is removed, the individual layers are revealed. The basic concept we need to understand here and regarding most vistas that you encounter in our national parks is that *the modern landscape that we see (e.g., deeply cut canyons, rugged mountains, dry desert valleys) is typically much younger than the rocks that compose the landscape.* In other words, geologic processes that formed the rocks occurred prior to the relatively recent weathering and erosion that created the various landforms visible on the surface. Regarding the Grand Canyon, visualize the sedimentary layers as continuous horizontal sheets stacked one on top of the other like a layercake. The principle of *superposition* tells us that the bottom layers are older than the upper layers (Chapter 1). The canyon itself is a relatively young feature, superimposed on the layercake by the relatively recent incision of the Colorado River and its innumerable tributaries.

With your awareness of the relative timing of the older rocks versus the younger landscape, we can now address the origin of the sedimentary layers and assess their meaning. It's important to be aware that the region that would become the Colorado Plateau lay near sea level during the Paleozoic and Mesozoic. All of the horizontal, multicolored layers represent a quarter-billion years of changing depositional environments, such as shallow tropical seas, meandering rivers and floodplains, and windswept deserts covered with migrating sand dunes. Through this vast span of time, a pile of sediment up to 1.5 km (~1 mi) thick accumulated in the low-elevation region that would much later be raised upward into the high-elevation Colorado Plateau. With burial and time, the sediment was transformed into sedimentary rock, then later uplifted and exposed to view in two episodes during the latest Mesozoic and Cenozoic. The evidence and timing of the **uplift** of the Colorado Plateau and the subsequent incision of the Colorado River to create the Grand Canyon is addressed in detail in Chapter 34.

Geologists decipher the environmental setting of sedimentary rocks through observation of grain size and composition, various structures in the rocks formed during deposition, and the types of fossils found in the rocks. These descriptive data are combined with fundamental techniques for dating rocks as well as the integration of modern analogs based on the principle of *actualism* (Chapter 1). Review the appropriate pages in Chapter 1 to refresh your understanding of how the formation of *clastic* sedimentary rocks is driven by solar energy, which acts to weather and erode previously existing rock to create new particles that

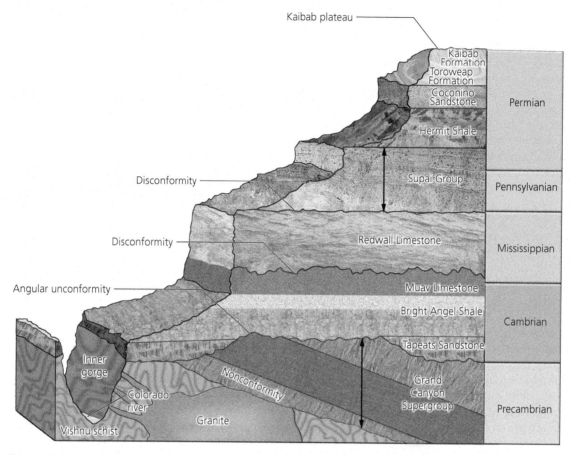

Figure 3.6   Graphic column of the main sets of rocks composing the Grand Canyon, with geologic ages. Precambrian crystalline basement occurs at river level in this figure and is overlain by the Grand Canyon Supergroup. The upper Grand Canyon is composed of Paleozoic sedimentary rocks with variable expression of erosion.

recombine to form sedimentary rocks such as sandstone and shale. Also reexamine how *biochemical* sedimentary rocks such as limestones form from the accumulated skeletal debris of hard-shelled marine organisms. Both types of sedimentary rocks are present in the "stairstep" horizontal layers of the upper Grand Canyon and can be interpreted to reveal their environmental and climatic setting as well as the ancient geographic conditions that existed at the time they were deposited. Rather than methodically describe each individual formation in the upper Grand Canyon, we'll look in detail at just a few formations to illustrate how sedimentary rocks act as an archive of events in the history of Earth. A graphic column of the rocks composing the Grand Canyon will keep you organized as you read the following sections (Fig. 3.6).

EARLIEST PALEOZOIC ROCKS OF THE GRAND CANYON

The lowermost Paleozoic formations in the Grand Canyon form a distinct triplet composed of the cliff-forming **Tapeats Sandstone**, the overlying broad slopes of the **Bright Angel Shale**, and the uppermost stepped ledges of the **Muav Limestone** (Fig. 3.7). Collectively, they are a

few hundred meters thick in the Grand Canyon region. These formations are Early Paleozoic (Cambrian) in age and formed over a few tens of millions of years around 500 million years ago. (To place this and other dates in perspective, refer to the section on "Deep Time" in Chapter 1 for a geologic time scale and a discussion of how geologic ages are determined.) The quartz sand grains composing the Tapeats are relatively equal in size and smoothly rounded, evidence for continual abrasion and reworking by waves on a broad beach. Burrows and traces of soft-bodied, worm-like animals hint at biologic activity on the ever-shifting sands of the Tapeats shoreline.

The overlying Bright Angel Shale consists of solidified silt and clay that accumulated on a quiet, low-energy seafloor below the depth of reworking by waves. Fossils of trilobites, hard-shelled arthropods that scavenged for food in the Paleozoic seas, are found within the shales and siltstones of the Bright Angel (Fig. 3.8). Trilobites were ocean-dwelling organisms and thus provide direct evidence for a marine origin of the mud composing the Bright Angel. The species of trilobites found in the Bright Angel pinpoint the relative age of the formation to the Middle Cambrian period.

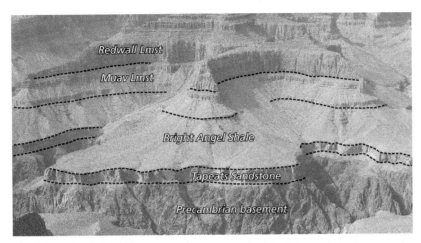

Figure 3.7 The oldest sedimentary rocks of the upper Grand Canyon are the cliff-forming Tapeats Sandstone, the broad slopes of the Bright Angel Shale, and the ledges and slopes of the Muav Limetone.

Figure 3.8 Trilobite fossil from the Bright Angel Shale—evidence for deposition in a shallow marine setting during the Cambrian Period.

The limestone of the overlying Muav is composed of the fragmental and intact remains of marine, calcite-shelled organisms that lived and died on a shallow seafloor that was slightly farther from shore than that of the muddy Bright Angel. Tracks and burrows preserved in the limestone suggest an active ecosystem. These three formations reveal a vision of the Grand Canyon region of northern Arizona around 500 Ma as a coastal setting adjacent to a shallow, tropical sea hosting a variety of invertebrates

hunting, reproducing, and dying over millions of years.

The geographic distribution of the sedimentary layers comprising the Tapeats–Bright Angel–Muav triplet ranges over several thousand square kilometers of the American West. How did the layers become so broadly spread, and how did they become stacked on top of each other? The answer lies in the continuous, slow-motion rise and fall of **sea level**. The world's shorelines are ephemeral, transitory features that mark the intersection of sea level with the gently inclined coastal plains of continents. But sea level can rise and fall up to several tens of meters to even a few hundred meters, occurring over a range of rates and time scales. Thus, the positions of shorelines are ever changing. When sea level rises, the shoreline migrates landward, flooding river valleys and coastal plains along the continental margins. As the shoreline shifts inland, the associated coastal and marine environments migrate along with the shoreline in a process called **transgression** (Fig. 3.9a, b). So as sea level rose during the early Paleozoic, the Tapeats beach sands migrated slowly landward over the exposed rocks of the continent. (Remember, the region that would eventually become the elevated Colorado Plateau was originally near sea level through the Paleozoic.)

Over a few million years, as sea level continued to rise over the future Grand Canyon region, the quiet, muddy, offshore environment of the Bright Angel migrated over and above the older Tapeats beach sand. As transgression slowly continued over millions of years, the limy sediments of the Muav environment migrated over the Bright Angel muds, depositing a thick blanket of calcitic skeletal debris across future northern Arizona. Even though we're focusing on what was happening in the future Grand Canyon region during the early Paleozoic transgression, it's important to understand that the entire coastline of North America was experiencing the same rise in sea level and landward migration of coastal environments.

The Tapeats–Bright Angel–Muav transgression is shown in the series of **paleogeographic maps** on Figure 3.10.*

*The ancient geographic locations of landmasses like western North America through time are determined by measuring the magnetic properties of certain rocks, which in turn preserve their latitudinal position at the time the rock was created. Using specialized instruments in a lab, the magnetic signature of rocks collected from the field can be measured, marking the changing latitude of a continent as it migrates across the planet. Paleogeographic maps like those of Figure 3.10 are partly based on magnetic measurements of rocks dated to specific geologic ages. The maps also rely on the distribution and composition of sedimentary rocks, as well as their assemblage of fossils. Visualizations of the ancient geographic appearance of a region are based on enormous amounts of data, mixed with a healthy dose of speculation.

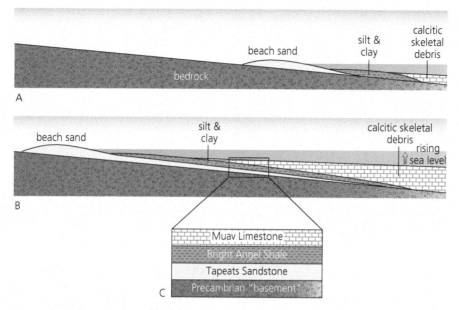

Figure 3.9 Stages in the Tapeats–Bright Angel–Muav transgressive sequence of deposition. A) The beach sand grades seaward into silt and clay deposited in quieter, deeper water. Further seaward is a zone of hard-shelled marine invertebrates living and dying on the seafloor. Each of the environments is laterally contemporaneous with one another. (B) As sea level rises, the shoreline migrates inland as much as hundreds or even thousands of kilometers over the bedrock of the continent. The sandy beach tracks the transgressing shoreline, followed by the muddy deeper water environment. The limy sediments migrate in tune with the other environments, eventually overlapping and burying the underlying layers. C) After burial and cementation, the beach sand became the broad layer of the Tapeats Sandstone. The silt and clay of the slightly deeper water setting became beds of Bright Angel siltstone and shale, and the calcitic skeletal debris formed the Muav Limestone.

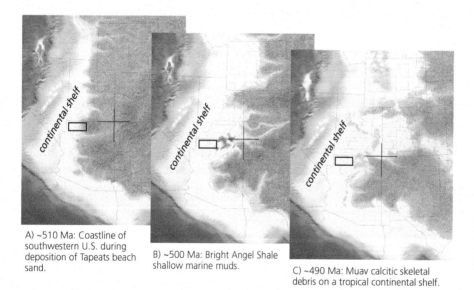

A) ~510 Ma: Coastline of southwestern U.S. during deposition of Tapeats beach sand.

B) ~500 Ma: Bright Angel Shale shallow marine muds.

C) ~490 Ma: Muav calcitic skeletal debris on a tropical continental shelf.

Figure 3.10 Paleogeographic maps such as these illustrate an interpretation of the prevailing environmental and geographic conditions during any particular phase of time. This set of maps showing the Tapeats–Bright Angel–Muav transgression is based on careful description of the rocks and fossils comprising the three formations at several locations. The rectangle shows the location of the future Grand Canyon.

The shoreline around 510 Ma extended through central Arizona and Utah, with sand accumulating on coastal beaches near the future Grand Canyon. The muddy and calcitic environments of the Bright Angel and Muav were located in slightly deeper water to the west. By ~500 Ma, sea level had risen, with the coastline transgressing farther onto the continent. The Tapeats beach sands were now being deposited farther inland and muddy sediments of the Bright Angel were being deposited in the Grand Canyon region. By ~490 Ma, the Tapeats shoreline reached far into Colorado, Wyoming, and New Mexico, with the muds of the Bright Angel deposited in slightly deeper water to the west. In the Grand Canyon region, the calcitic skeletal debris of the Muav was accumulating. The end result of this transgressing shoreline and associated depositional environments was the sequential development of the layered sediments of the Tapeats, Bright Angel, and Muav formations in the Grand Canyon region and throughout the entire American West. Through time, the sediments would become buried, compacted, and cemented to form the sedimentary rocks now exposed in the walls of the Grand Canyon.

Stepping outward to the continental scale, much of North America was submerged beneath shallow seas during the Cambrian (Fig. 3.11). Most of the land surface was centered over Canada and Greenland, which was attached to North America at the time. A peninsula extended from Minnesota to the American Southwest, and small islands rose above the sea in other parts of the continent. Most of California, Oregon, and Washington didn't yet exist, other than as a few offshore islands. Furthermore, North America was tectonically rotated about 90° clockwise with Canada to the east, Arizona to the west, and the equator extending along what would today be the Pacific shoreline. The coastal marine environments of the future Grand Canyon region were swept by warm trade winds and bathed in shallow, tropical seas. Land plants hadn't evolved yet, so the early Paleozoic continents were brown and gray

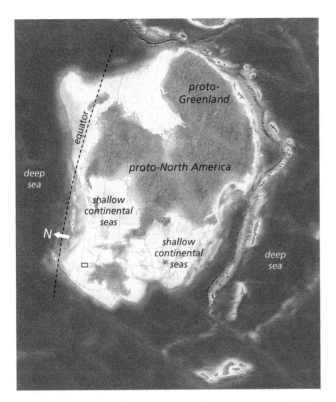

Figure 3.11 Paleogeographic map of North America during Cambrian time (~500 Ma). The extensive transgression of the shoreline deep onto the continent resulted in the deposition of the Tapeats–Bright Angel–Muav and equivalent formations across broad swaths of North America. The equator extended across what is today western North America, with the continent tectonically rotated ~90° clockwise from its current orientation. The small rectangle shows the future location of the Grand Canyon.

expanses of exposed Precambrian rock covered randomly by thin veneers of sediment. Without the stabilizing influence of plants, particles of sand, silt, and clay were readily swept into sediment-choked rivers that flowed the color of chocolate milk to the coastlines. These were the sediments that were deposited and then eventually solidified into the sandstone of the Tapeats and the siltstone and shale of the Bright Angel.

This bizarre image of western North America during the Cambrian was the geographic setting during the deposition of the Tapeats, Bright Angel, and Muav formations. Visualize this triplet of sedimentary layers in the Grand Canyon as once being laterally continuous over a broad region of the American West. Today, the layered rocks are broken up and discontinuous due to postdepositional faulting, mountain-building, and dissection by erosion. The widespread extent of these rocks, pieced together from isolated exposures in canyon walls and mountain ranges, is typical for most sedimentary units on the Colorado Plateau as well as elsewhere around the planet.

Of course, sea level can't rise forever, so when it inevitably falls, the coastal and marine environments of deposition migrate seaward with the shoreline in a process called **regression**. If the sea-level fall is rapid, the abrupt exposure of the underlying transgressive deposits may result in erosional beveling down to a broad, gently inclined land surface. But if the fall in sea level is gradual, sediment may accumulate as the various depositional environments migrate seaward along with the shoreline. Even though sea level may be falling, space is created for sediment to build up by *subsidence*, the sagging of the lithosphere imposed by the load of sediment (Chapter 1). We don't know what happened following the Tapeats–Bright Angel–Muav transgression because the top of the Muav is marked by a widespread surface along which an indeterminate thickness of overlying sedimentary rock was eroded and redeposited elsewhere during an extensive phase of exposure. (This surface is called an **unconformity** and will be discussed below in a broader discussion of how rocks preserve an archive of time.)

Geoscientists interpret the rise and fall of sea level through time by careful examination of sedimentary rocks and their preserved history of transgression and regression. Most sedimentary rocks in the Grand Canyon, as well as throughout the rest of the Colorado Plateau, were deposited in response to transgressions and regressions of the shoreline through time.

YOUNGER PALEOZOIC ROCKS OF THE GRAND CANYON

With this transgressive-regressive model in mind, we can traverse upward through the rest of the Grand Canyon and decipher the origins of the rocks. The >200 m (~660 ft) thick **Redwall Limestone** (Figs. 3.5, 3.6) records the transgression of a long-lived, shallow tropical sea that covered vast amounts of the American West around 340 Ma. Much of western North America was a mosaic of shallow shoals and banks populated by prodigious numbers of lime-secreting invertebrates such as crinoids, solitary corals, bivalve-like brachiopods, and sponges (Fig. 3.12). Primitive fish, including sharks, cruised the pristine blue waters looking for a meal. Deposition of the Redwall Limestone occurred within water shallower than about 10 or 20 m (30 to 60 ft), depths where the penetration of sunlight is greatest. The limestone composing the Redwall is actually bluish gray in color; only the surface is stained red by iron-rich rainwater that bleeds downward from the reddish rocks of the overlying Supai Group.

The ledges and slopes of the **Supai Group** (Fig. 3.13) are reddish sandstones, siltstones, shales, and less common limestones deposited within a variety of coastal settings that existed in the Grand Canyon region over a 30 million year interval around 300 Ma. Many of the clastic sediments accumulated in the river systems and adjacent floodplains that drained a broad coastal plain, which may have looked somewhat similar to the modern Gulf Coast of Texas. Footprints of amphibians and reptiles, as well as the remnants of terrestrial plants, attest to the evolutionary transformation of continental ecosystems that was occurring around this time. Thin limestones represent episodic incursions of the sea onto the continent, accompanied by the transgression of associated marine environments. The

Figure 3.12 Diorama of Mississippian seafloor (~340 Ma) showing "meadows" of flower-like crinoids and other invertebrates, along with primitive fish, including sharks. The Redwall Limestone is composed of countless particles of calcite debris shed by crinoids and other marine invertebrates.

clastic sediment composing the Supai was derived from erosion of the ancestral Rocky Mountains that were rising in Colorado at this time, a mountain-building event that will be addressed in later chapters.

To complete this brief summary of the rocks of the upper Grand Canyon and their respective depositional environments, the **Hermit Shale**, **Coconino Sandstone**, and **Toroweap** and **Kaibab** limestones (Fig. 3.13) represent a time when the continents were converging into a single supercontinent accompanied by extensive mountain-building and climatic aridity. Abundant plant fossils in the Hermit provide evidence of terrestrial deposition in temperate coastal forests around

280 Ma. A dramatic transition to a coastal desert system around 275 Ma is preserved in the overlying Coconino by wind-sorted sands deposited as dunes. Tracks of reptiles that wandered the margins of the desert in search of prey are imprinted on the surface of sandstone layers.

The Toroweap and Kaibab record the final Paleozoic incursions of the sea onto the North American continent. The Kaibab was deposited around 270 Ma and marks the uppermost layer in the canyon and thus the surface on which most visitors walk. The observant tourist can find numerous types of marine fossils like corals and mollusks, many of which became extinct soon after the Kaibab deposition in the greatest mass extinction of all time. The **end-Permian mass extinction** marks the Paleozoic–Mesozoic boundary worldwide and is characterized by the dramatic disappearance of >90 percent of marine species from the fossil record. The jury is still out on the ultimate cause of the mass extinction, but several lines of evidence point toward a global environmental catastrophe triggered by a prolonged phase of volcanism.

The Paleozoic rocks of the Grand Canyon were buried by kilometers of overlying Mesozoic and Cenozoic sedimentary rocks, almost all of which were subsequently eroded away from the area. Evidence for this "missing" long-term phase of deposition and eventual erosional removal will be addressed when we visit other parks on the plateau, but the key point is that these long-buried rocks

Figure 3.13 Formations of Pennsylvanian and Permian age that compose the highest canyon walls of Grand Canyon.

of the upper Grand Canyon had to have been uplifted to higher elevations for weathering and erosion to expose them to view. The uplift of the Colorado Plateau and the incision of the Grand Canyon occurred during two phases of the latest Mesozoic and Cenozoic that we'll discuss in detail in Chapter 34. For now, it's enough to know that these rocks, originally deposited in various depositional environments near a fluctuating shoreline, were relatively recently uplifted to their current elevations where they were subject to attack by river erosion and other effects of weather and climate.

Why do the sedimentary layers of the upper Grand Canyon appear as "stairstep" topography, with alternating cliffs and slopes (Fig. 3.5)? Each of the sedimentary rocks of the upper Grand Canyon exhibits a differing response to the constant assault by erosive processes. This **differential erosion** is typically related to grain size and the type of cementation that makes some rock types more resistant to erosion than others. Upon uplift and exposure, the rocks are attacked by rainfall and snowmelt that dribble downslope along the walls of the canyon, infiltrating pores and fractures in the rock and dissolving the natural cement holding grains together. Side canyons and tributary streams may look placid today, but they can flow as violent, sediment-choked torrents after summer thunderstorms. These churning *debris flows* have high erosive power and can remove large masses of the underlying rock in relatively short bursts of time. Countless flash floods, occurring in innumerable drainages throughout the canyon over vast amounts of geologic time, have removed immense volumes of rock.

The end result is that, in the semi-arid climate, sandstones and limestones tend to remain resistant to erosion and form near-vertical cliffs. Weaker, more poorly cemented siltstones and shales tend to decompose more easily, resulting in slopes mantled with a thin veneer of soil and desert vegetation. Near-vertical cliffs of sandstone and limestone are maintained by rockfalls that break off along vertical fractures that cut through the layers (Fig. 3.14). Differential erosion is a fundamental factor in producing the stairstep landscapes of the Colorado Plateau and will become easier to understand as we visit the other national parks in the province.

Lower Grand Canyon: Chaos beneath the Layercake

In the previous discussion of the sedimentary rocks of the upper Grand Canyon you may have wondered what lay beneath. The foundation of rock upon which the layered sedimentary rock was deposited consists of a

Figure 3.14 View of Colorado River from Toroweap Point, showing steep cliff face of Redwall Limestone where rock has broken away along vertical fractures.

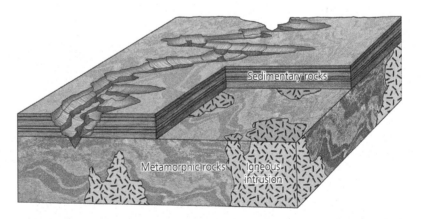

Figure 3.15 In the Grand Canyon, Paleozoic sedimentary rocks were deposited above a Precambrian "basement" of crystalline metamorphic and igneous rocks.

complex assemblage of metamorphic and igneous rock loosely called the **basement** (Fig. 3.15). These dense, **crystalline** rocks are the remnants of ancient Precambrian mountain-building events and form the crustal bedrock of North America and all other continents. The term *crystalline* refers to the interlocking mosaic of crystals within igneous and metamorphic rock. The crystalline basement is occasionally exposed where the combined effects of uplift and erosion exhume these rocks to the surface, providing a glimpse into events that occurred far back in time. The Inner Gorge within the deepest reaches of the Grand Canyon is one of those places where we can get a good look at what lies beneath the veneer of younger rock (Fig. 3.16). The basement rocks of the Inner Gorge are best accessed by boat, but they can also be seen where trails reach down to river level. To better understand the following description of events, it might be useful to review the processes involved in the formation of igneous and metamorphic rocks described in Chapter 1.

The oldest rocks of the crystalline basement of the Grand Canyon are about 1.8 billion years in age (aka 1.8 Ga). At the time these rocks were created, the geography of the region would have been unrecognizable relative to today (Fig. 3.17). A broad peninsula of stark brown and gray rock, devoid of vegetation, extended southwestward from Wyoming across Utah toward Southern California. A chain of volcanic islands arced across future Arizona and Colorado, marking the location of a northeasterly trending *subduction zone* (Chapter 1). The entire region was equatorial, with intense tropical rains washing enormous volumes of sediment off both the continental edge and the volcanic islands into an intervening, narrow ocean basin. As the island chain migrated tectonically toward the continental margin, the sedimentary and volcanic rock in between was crushed, deformed, buried to great depths, and then metamorphosed. These now-metamorphosed rocks were tectonically welded to the southern margin of Precambrian North America and formed a high mountain range in what would eventually become the American Southwest.

This is one of the ways that continents grow—by the tectonic addition of masses of rock along their margins, accompanied by the compressional uplift of mountain ranges. The process is occurring today in ultra-slow motion where the volcanic Indonesian islands are converging against the northwest margin of Australia. The intervening oceanic sediments between the islands and the mainland are

inexorably being squeezed and deformed, eventually over millions of years to be tectonically attached to the edge of the Australian continent. In similar fashion, Precambrian crystalline rocks formed large amounts of new land that were added to the southwest corner of the evolving North American continent. These rocks form much of the crustal basement of North America and lie hidden beneath younger rocks and sediments throughout the continent. We'll return to this story of the Precambrian assembly of North America when we visit other national parks with exposures of 1.8 Ga crystalline basement.

The remnants of this Precambrian convergent tectonic event are the metamorphic and igneous rocks of the basement in the Inner Gorge of the Grand Canyon (Fig. 3.18). Metamorphosed sedimentary and volcanic rocks are

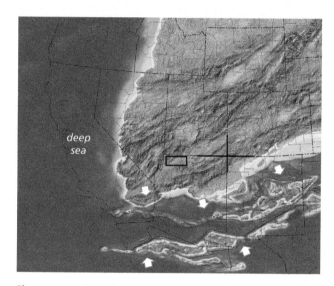

Figure 3.17 Speculative reconstruction of the southwestern United States during the Precambrian (~1.7 Ga). Note Four Corners region on the map as a point of reference. Rectangle shows location of future Grand Canyon. Arrows show possible directions of subduction along deep ocean trenches.

Figure 3.16 Inner Gorge of the Grand Canyon is cut into highly resistant crystalline rocks of the Precambrian basement. Dashed line separates overlying Paleozoic sedimentary rocks (Tapeats Sandstone) from significantly older crystalline rocks beneath

Figure 3.18 River-level view of Precambrian crystalline basement rocks. Dark gray schists are intruded by dikes of reddish granite.

grouped within the massive body of rock known as the **Vishnu Schist**. These schists are characterized by a platy, finely layered texture created as micas and other minerals grew in directions perpendicular to the orientation of tectonic stress during metamorphism (Chapter 1). Analysis of the minerals composing these schists suggest that they formed at temperatures between 500 and 700°C and at depths of about 20 km (~12 mi). The intense pressures at that depth precluded the melting of rock but facilitated the slow diffusion of elements within the rock to create new, more stable mineral assemblages now composing the schists.

Molten rock was subsequently injected along fractures into the dark-colored schists later in Precambrian time, solidifying into reddish granitic igneous rock (Fig. 3.18). This intrusive igneous rock used to be called the **Zoroaster Granite**, a name that is now being slowly discontinued after it was recognized that not all of the granite is of similar age and origin. The pink hues of the granitic rock contrast with the somber tones of the schists, creating a psychedelic tableau along the walls of the Inner Gorge. We see these formerly deep-seated rocks today at river level in the Grand Canyon due to the combined effects of *exhumation* during uplift of the Colorado Plateau and downcutting by the Colorado River system.

A second phase of Precambrian mountain-building is represented by a thick, tilted set of mostly sedimentary and lesser volcanic rocks known as the **Grand Canyon Supergroup** (Figs. 3.6, 3.19). This complicated interval of rock spans a thickness of 3700 m (~12,000 ft), about 2.5 times as thick as the Paleozoic sedimentary rocks composing the upper Grand Canyon. Ranging in age from ~1200 to 740 Ma, these rocks record a prolonged phase of deposition and volcanism that occurred in elongate basins formed as late Precambrian North America was being pulled apart by extensional tectonic forces. The great thickness of sediment accumulated in basins that were undergoing active faulting during deposition, a process that is likely related to *continental rifting* (Chapter 1).

The sedimentary rocks in the Supergroup represent a variety of nearshore and shallow marine environmental settings. It is considerably more difficult to determine the depositional environment in these rocks than in the overlying Paleozoic sedimentary rocks, however, since no hard-shelled, multicellular organisms existed during this phase of Earth history. Whereas fossils provide hard evidence of environmental settings in Paleozoic rocks, no such fossils are found in the Grand Canyon Supergroup. The dominant organisms at the time were microbial, single-celled bacteria and algae that formed laminated structures called **stromatolites**. We'll dig deeper into stromatolite morphology and genesis later in the book when we encounter them in Precambrian rocks in Glacier National Park.

The lack of fossils makes age determinations difficult in the Grand Canyon Supergroup, but fortunately volcanic ash deposits, cross-cutting igneous dikes (Fig. 1.35A), and interbedded lava flows can be dated radiometrically, bringing some chronologic coherence to this great thickness of rock. Further complicating the interpretation of these rocks is that exposures of the Supergroup are discontinuous and isolated by Precambrian faults, making the reconstruction of the total thickness a painstaking process. Even with these constraints, geologists have been able to determine that part of the tilting of these rocks occurred synchronously with deposition and part occurred after deposition ended ~740 Ma. As these rocks were uplifted and exposed to the elements, an unknown thickness of rock was eroded away, accompanied by the beveling of the upper surface to a broad, flat erosional plain (Fig. 3.19). The significance of this erosional surface will be addressed in the following section on the relationship of rocks to time.

The main point to take away from this discussion of the Precambrian rocks of the Grand Canyon is that they represent two major phases of mountain-building and subsequent erosion. The first phase began ~1.8 Ga during the tectonic convergence of a volcanic island chain with the North American continental margin and the consequent formation of the metamorphic and igneous rock of the crystalline basement. Around 800 million years ago, the Precambrian continent of future North America began to break apart, with sediment of the Grand Canyon Supergroup accumulating to tremendous thicknesses within fault-bounded rift basins. Later, the rocks of the Supergroup and underlying basement were uplifted and tilted in a second phase of mountain building to emerge as "fault-block" mountains

Figure 3.19 Gently tilted beds of Grand Canyon Supergroup rocks were erosionally beveled to a horizontal surface prior to the deposition of the Cambrian Tapeats Sandstone (light-colored cliff). The contact between the Supergroup rocks and the Tapeats is called an angular unconformity (marked by the dashed line).

on the late Precambrian landscape where they were exposed to the relentless effects of erosion. We'll encounter both sets of Precambrian rocks in other national parks later in the book. The complex rocks of the crystalline basement will be revisited at Colorado National Monument, Black Canyon of the Gunnison National Park, and Rocky Mountain National Park. Rocks equivalent to the Grand Canyon Supergroup will be seen at Dinosaur National Monument, Glacier National Park, and Death Valley National Park.

The Precambrian basement rocks of the Grand Canyon exert a distinct influence on canyon morphology due to their interlocking, crystalline texture and consequent resistance to erosion. In contrast to the stairstep topography formed by differential erosion of the Paleozoic sedimentary rocks of the upper Grand Canyon, the resistant basement rocks exposed in the Inner Gorge form deep, narrow, forbidding chasms (Fig. 3.16). The dense rocks of the basement are exposed along several kilometers of the river in narrow gorges, characterized by steep, near-vertical canyon walls.

Rocks of the Grand Canyon and Deep Time

You may have noticed from the previous sections that certain long spans of time weren't recorded by rocks in the Grand Canyon, both in the lowermost Precambrian rocks and in the overlying Paleozoic sedimentary strata. As you'll see, more geologic time is "missing" in the rocks of the Grand Canyon than is actually represented physically by the rocks. And stranger yet is the fact that the "missing" time is represented by near-horizontal surfaces separating rocks of vastly different ages. So, oddly enough, something (time) is recorded by nothing (a near-horizontal surface separating masses of rock).

Using the fundamental techniques for determining the ages of rock described in Chapter 1, geoscientists have dated many of the primary rock bodies in the Grand Canyon. As the chronology of events represented by rocks of the Grand Canyon was compiled, it became evident that long intervals in the record of Earth history were not preserved as rock (Fig. 3.20). It's somewhat analogous to reading a book where a mischievous toddler has ripped out random chapters, leaving behind sections of the book that have little continuity with each other. These gaps in the geologic record are called **unconformities** and are represented in rock as relatively planar surfaces that signify a missing span of time.

There are a variety of types of unconformities, but all of them share one distinguishing characteristic: their unconformable surfaces manifest an extended episode of exposure and erosion, a prolonged phase of nondeposition, or both (Fig. 3.21). It may be that entire mountain ranges were raised and then eroded. that sea level rose and fell

multiple times; or that climate changed during the "missing" time. But the physical evidence for those events was gradually washed away along the extent of the unconformity, leaving behind a beveled surface that represents the landscape just prior to its burial by overlying sediment. Unconformities can sometimes be difficult to recognize but may be identified by the abrupt juxtaposition of differing rock types, by a pronounced gap in the fossil record, or by differing angles of bedding above and below the unconformable surface.

Let's look at a few of the major unconformities in the Grand Canyon, beginning with the surface that separates the crystalline basement from the overlying Tapeats Sandstone (Figs. 3.16, 3.22). The majority of the crystalline rocks of the Precambrian basement are about 1800 Ma in age, whereas the Tapeats beach sands were deposited around 500 Ma. Thus, the time "missing" along the contact between the two rock units is a staggering 1.3 billion years. This widespread surface is known as the **Great Unconformity**. What events could we infer to have happened during that inconceivably long span of time? Remember that the basement rocks were formed several kilometers beneath the surface about 1.8 b.y. ago, so they must have been later uplifted within a growing mountain range, exposing the rock to weathering and erosion. Moreover, we can infer that the highlands were gradually worn down to a relatively level rocky plain (the unconformity itself). The rocky surface was eventually inundated by rising sea level and the transgressing Tapeats shoreline around 500 m.y. ago. An unconformable surface like the Great Unconformity, where sedimentary rock directly overlies igneous and metamorphic crystalline rocks, is called a **nonconformity** (Fig. 3.21A). The Great Unconformity is widespread across the American West; we'll encounter this same surface at other national parks later in the book.

The unconformable surface separating the inclined layers of the Grand Canyon Supergroup from the overlying Tapeats is known as an **angular unconformity** and represents over 200 m.y. of time (Figs. 3.19, 3.21B). The originally horizontal layers of the Supergroup rocks were tilted during a late Precambrian phase of uplift. Those mountains were beveled by erosion, then buried beneath the transgressing Tapeats beach. Many significant evolutionary events are lost along that single surface. The only evidence of life in the rocks beneath the unconformity are photosynthesizing bacterial colonies (stromatolites), whereas complex, multicellular marine invertebrates are abundant above the unconformity. Many other events came and went over the unimaginably vast span of time represented by the unconformity, but evidence for those events isn't found in the vicinity of the Grand Canyon.

Multiple unconformities occur within the Paleozoic rocks of the upper Grand Canyon, spanning durations of up to 70 million years (Fig. 3.20). An erosional surface that separates sedimentary units of vastly different

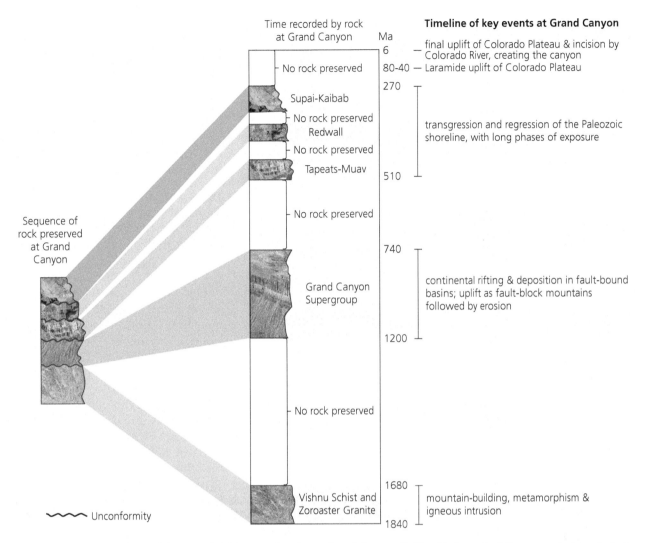

Time recorded by rock
at Grand Canyon

Timeline of key events at Grand Canyon

Ma

6 — final uplift of Colorado Plateau & incision by
Colorado River, creating the canyon

80-40 — Laramide uplift of Colorado Plateau

270

No rock preserved

Supai-Kaibab

No rock preserved
Redwall

transgression and regression of the Paleozoic
shoreline, with long phases of exposure

No rock preserved

Tapeats-Muav

510

Sequence of
rock preserved
at Grand
Canyon

No rock preserved

740

Grand Canyon
Supergroup

continental rifting & deposition in fault-bound
basins; uplift as fault-block mountains
followed by erosion

1200

No rock preserved

1680

Vishnu Schist and
Zoroaster Granite

mountain-building, metamorphism &
igneous intrusion

1840

〰〰〰 Unconformity

Figure 3.20 Comparison of the actual thickness of rocks at Grand Canyon (left) with the actual time recorded (center). It should be apparent that more time is "missing" than is preserved in rock. The intervals of missing time are recorded as single surfaces (unconformities) separating vastly different ages of rock.

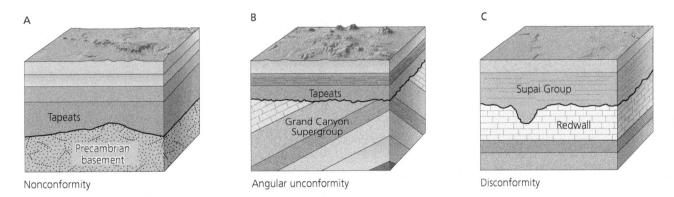

A

Tapeats

Precambrian
basement

Nonconformity

B

Tapeats

Grand Canyon
Supergroup

Angular unconformity

C

Supai Group

Redwall

Disconformity

Figure 3.21 Three main types of unconformities, each represented at Grand Canyon. Unconformable surface noted by irregular dark line. A) A nonconformity, representing about 1.3 billion years of time, separates Precambrian crystalline rocks of the basement from the overlying Tapeats. B) An angular unconformity separates the Grand Canyon Supergroup from the overlying Tapeats. C) A disconformity separates sedimentary rock of the Redwall from the overlying Supai Group.

Figure 3.22 Pinkish Precambrian crystalline rock directly overlain by bedded Tapeats Sandstone. Dark line of the Great Unconformity represents ~1.3 billion years of missing time in the rock record. Cobbles on the floor of Blacktail Canyon for scale.

ages is called a **disconformity** (Fig. 3.21C). It's likely that sediment was deposited during the time represented by a disconformity but was eroded away during regressive phases when sea level was low and the coastal margins of the continent were exposed. A disconformity occurs at the top of the Redwall Limestone where long-term weathering and erosion resulted in the dissolution of the upper layers of limestone and the creation of a complex network of caves and caverns within the formation. In time, coastal river systems of the Supai Group meandered across the exposed limestone surface, filling in low areas, smoothing the ancient landscape, and burying the disconformable surface.

One of the superlatives frequently used by thoughtful visitors to the Grand Canyon is "timeless." Looking out at the vast expanse of colorful, layered rock, one can't help but be struck by the ineffable depth of time represented by the panorama. It's a bit ironic, however, to learn that the rocks of the Canyon don't preserve a continuous archive of Earth history over the past 1.8 billion years. In fact, the rocks of the Grand Canyon only record about 40 percent of that span of time, with the remaining 60 percent lost to the ravages of erosion. Rather than being "timeless," the rocks of the Grand Canyon might better be characterized as "episodic phases of deposition fortuitously preserved between prolonged interludes of exposure and erosion," although that description doesn't roll off the tongue quite as easily. Of course we wouldn't be able to decipher any of this long and complicated geologic history if the Colorado

River hadn't carved deeply into these rocks in relatively recent geologic time, exposing them to view.

Formation of the Grand Canyon: No Easy Answer

The Grand Canyon is the premier locale for "reading the rocks" to gain an understanding of the Precambrian and Paleozoic history in the region. But the subsequent story detailing the uplift of the Colorado Plateau and the intertwined tale of the origin of the Grand Canyon has been one of the most difficult puzzles to decipher for geologists. The models are continually evolving as more and more data accumulate, but the timing and mechanics of the origin of the Grand Canyon are not completely resolved at this point. The quandary facing geologists who work to determine the formation of the Grand Canyon is that there's nothing there. That is to say, the rock that used to occupy the vast gap between the North and South Rims has been eroded and carried away by the Colorado River, leaving empty space and spectacular vistas. The mass of rock that used to exist prior to the incision of the canyon now consists of countless tons of sand, silt, and clay deposited at the mouth of the Colorado River at the northern margin of the Gulf of California. So geologists must search the nearby region and use ever more sophisticated techniques to find evidence to support a robust and verifiable model for when and how the Grand Canyon was created.

Be patient—the narrative of the uplift of the Colorado Plateau and the origin of Grand Canyon will be more fully appreciated once you've grasped important concepts such as tectonism, volcanism, mountain-building, and the effects of the recent Ice Ages on the modern landscape. After you've been exposed to the geologic history of several other national parks, a comprehensive explanation for the origin and evolution of Grand Canyon will be provided in Chapter 34. For now it's enough to know that Grand Canyon is the magnificent product of millions of years of erosional downcutting by the Colorado River and its tributaries.

Key Terms

angular unconformity
basement
crystalline rock

debris flow
differential erosion
disconformity
end-Permian mass extinction
flood stage
formation
Great Unconformity
nonconformity
paleogeographic map
regression
rockfall
sea level
stromatolite
transgression
tributary stream
unconformity
uplift

Related Sources

Beus, S. S., and Morales, M. (Eds.). (2003). *Grand Canyon Geology*. New York: Oxford University Press.

Powell, J. L. (2005). *Grand Canyon—Solving Earth's Grandest Puzzle*. New York: Pi Press.

Ranney, W. (2012). *Carving Grand Canyon: Evidence, Theories, and Mystery*, 2nd ed. Grand Canyon Association.

Geologic formations of Grand Canyon National Park. Accessed January 2021 at http://www.nps.gov/grca/naturescience/geologicformations.htm.

Grand Canyon fly-through animation. Accessed January 2021 at https://www.nps.gov/grca/learn/photosmultimedia/fly-through.htm.

National Park Service—Geology of the Grand Canyon. Accessed January 2021 at https://www.nps.gov/grca/learn/nature/grca-geology.htm.

4 Zion National Park

Figure 4.1 View into Zion Canyon from Angels Landing. The North Fork of the Virgin River meanders through the valley.

Zion National Park in southwestern Utah exhibits classic examples of Colorado Plateau physiography and geology: a high-elevation, flat-topped plateau, deeply incised canyons, steep cliffs of horizontal sedimentary redrock, and a stunning, stark beauty (Fig. 4.1). The North Fork of the Virgin River flows southward through the narrow green valley between the reddish, near-vertical canyon walls. The entire red and green landscape is commonly enveloped by an azure sky, creating soul-stirring vistas. In contrast to the immense scale of the Grand Canyon, Zion is much more compact and easier for tourists to explore over the course of a few days. Unfortunately, its proximity to Las Vegas makes Zion uncomfortably crowded

between May and September; October and April are the best times to go, months when the air is cooler and the people are fewer. Visitors to the park should be aware of the effects of intense heat, very steep dropoffs, rockfalls, lightning at higher elevations, and the random flash flood.

In the previous chapter on Grand Canyon National Park, we explored geologic concepts such as sedimentation and sea-level change, the way time is represented in rock, and the creation of topography through river incision, weathering, and erosion. We also learned about how a few fundamental principles of geology such as superposition and cross-cutting relations are expressed in the rock record. Armed with that understanding, we'll expand on those concepts in this chapter and add a few more as we work our way through the rocks and landscapes of Zion (Fig. 4.2). Remember, all of the processes and features that characterize national parks of the

Figure 4.2   Map of Zion National Park showing park boundaries and the path of Virgin River. (NPS)

Colorado Plateau will be tied together in later chapters on the uplift of the plateau and formation of the modern landscape.

Origin and Age of Rocks at Zion

The rocks of Zion are mostly Mesozoic in age and record the geologic history of the region subsequent to the deposition of the Paleozoic rocks of Grand Canyon (Fig. 4.3). The uppermost Paleozoic layer at Grand Canyon, the Kaibab Limestone, is the lowermost layer exposed at Zion National Park. The Mesozoic rocks that once covered the Grand Canyon region have long since been eroded away, but their geologic history is archived in the rocks at Zion. In terms of depositional setting, the future Colorado Plateau region was located near sea level during the Mesozoic, similar to the setting during the Paleozoic. The depositional environments recorded by the horizontally

Figure 4.4   Cross-bedding in the Navajo Sandstone, Zion National Park.

bedded Mesozoic rocks of Zion range from lazy river systems draining a gentle coastal plain, to vast coastal deserts, to shallow tropical seas. These Mesozoic redrocks were deposited over about 150 million years of changing sea level, mostly arid climate, and generally mild tectonic activity.

The dominant formation in Zion is the **Navajo Sandstone**, which forms the sheer cliff faces and monoliths that draw our attention. The Navajo is highly resistant to erosion due to the natural calcite and iron oxide cements that bind together the quartz sand grains that compose the rock. The particles of sand are well sorted; that is, they are relatively uniform in size and well rounded, meaning that they've been smoothed by abrasion prior to deposition. **Sorting** and **rounding** are attributes of populations of sand grains that reveal the degree of erosional reworking that occurred during transport of the grains.

The most obvious physical characteristic of the Navajo is that much of its internal bedding occurs as tilted layers that occur at a variety of angles and orientations (Fig. 4.4). This feature is known as **cross-bedding**—the physical preservation of sand deposition on inclined surfaces under the influence of a moving fluid like water or wind. In the case of the Navajo, the uniform grain size, the high degree of rounding of quartz sand grains, and the ubiquitous cross-bedding are evidence of deposition by wind across a vast coastal desert that extended across the American West about 185 m.y. ago (Jurassic). The overlapping sets of cross-beds are the preserved remnants of huge sand dunes that migrated across the sea of sand at the time.

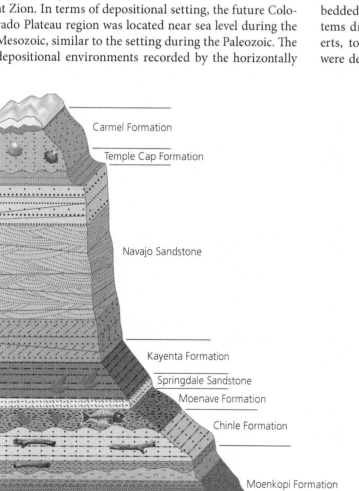

Carmel Formation

Temple Cap Formation

Navajo Sandstone

Kayenta Formation

Springdale Sandstone

Moenave Formation

Chinle Formation

Moenkopi Formation

Kaibab Formation

Figure 4.3   Graphic column of rocks exposed at Zion National Park

Cross-beds form as wind blows particles of loose sediment across a surface, eventually building upward to create a sand dune (Fig. 4.5). Wind is highly efficient at sorting grains by their size and density. The larger sand grains typically bounce as they move along, whereas the finer silts and clays are picked up by the wind to be redeposited far downwind in quieter settings such as a lake or ocean basin. If the wind blows strong enough, sand grains bounce up the gently inclined windward face of the dune, accumulating at the crest of the dune. There, the mass of sand grains builds up to a threshold when it abruptly cascades down the steep-walled "slip-face" as a mini-avalanche, building a single inclined cross-bed. Each individual cross-bed is composed of a lower layer of slightly coarser sand grains overlain by a slightly finer layer of sand grains, a product of self-sorting as the grains tumble down the steep front of the dune.

Younger dunes may cannibalize the sand from older dunes as they migrate, forming a sharp erosional surface separating individual sets of cross-beds. Some of the sharp surfaces between sets of cross-beds may manifest an ancient water table where the sand grains were held together by water in pores, making them more resistant to wind erosion. Cross-bedding can be observed forming today in desert settings where geoscientists apply the principle of actualism to interpret similar ancient settings.

Based on the sheer size of some sets of cross-beds in the Navajo, sand dunes in the Navajo desert of 185 m.y. ago had crests tens of meters high, comparable to huge migrating dunes in many modern deserts such as the Namib along the southwest coast of Africa (Fig. 4.6). The orientation of cross-beds represents the wind direction at the time of deposition, so differences in orientation of Navajo cross-beds manifest changing wind directions through time across the vast Navajo sand sea. Measurements of thousands of cross-bed orientations in the Navajo Sandstone across the American West suggest that Mesozoic winds blew mostly from the northwest, although there is considerable variability, reflecting seasonal and monsoon-related shifts in wind direction through time.

Based on the regional distribution of exposed rock of the formation, the Navajo desert system extended from southern Nevada to central Wyoming (Fig. 4.7). The Navajo Sandstone is over 760 m (~2500 ft) thick near Zion, but it changes thickness across its areal extent. In Wyoming, where the Navajo is known as the Nugget Sandstone, thicknesses range from 500 m (~1600 ft) in the west to tens of meters in the east. The thickness of the Navajo thins dramatically in western Colorado but thickens westward toward the ancestral Mesozoic coastline. Sedimentary layers like the Navajo may be laterally extensive, but they're neither tabular nor equally thick in all directions. They are more typically wedge-shaped, tapering to a thin edge in the landward direction.

The Navajo sand sea stretched along the western margin of Mesozoic North America at about 10–20°

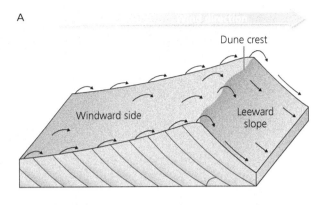

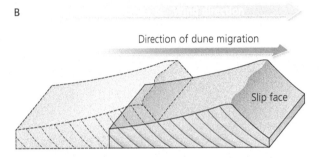

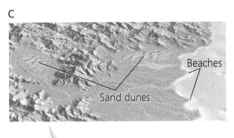

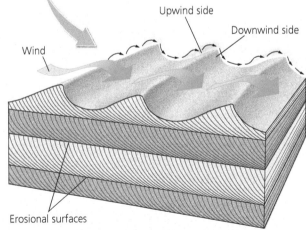

Figure 4.5 Formation of cross-bedding within sand dunes by wind. Dunes commonly migrate across the back of older dunes, cannibalizing sand from the upwind side of the dune to be redeposited on the downwind face.

north of the equator (Fig. 4.8). To the west, the rocks that would eventually become California were a partly submerged chain of volcanoes and isolated islands beginning to coalesce along the margin of the continent. Ancestral

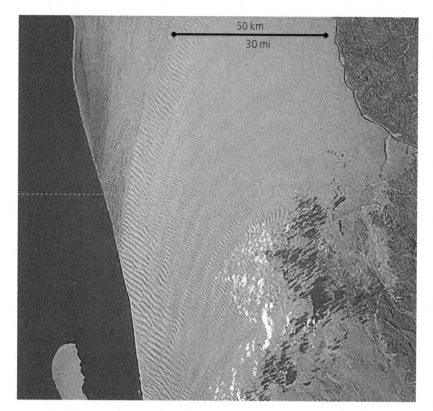

Figure 4.6 Namib coastal desert complex, southwest coast of Africa. This modern setting provides a good visual analog for the Mesozoic Navajo desert of the western United States. Note the subparallel orientation of dune systems due to prevailing winds.

mountains arced across central Arizona into Nevada, while low, remnant highlands lay to the east in Colorado. The quartz sand that accumulated in the Navajo desert was partly derived from sediment shed from these ancient mountains, but some of the sand may have been derived via immense continent-crossing rivers from the ancient Appalachians!

The geochemistry and age of highly resistant grains of minerals known as *zircons* found within the Navajo indicate that they had to have come from a specific body of rock in the Appalachians. The zircons and associated quartz grains probably began their life by crystallizing from a granitic magma deep beneath the eastern margin of North America. With uplift of the Appalachians toward the end of the Paleozoic, these grains of quartz and zircon were eroded out of the exposed granite, then carried along within a long-lost river that drained westward from the highlands. There was no Mississippi River, of course, nor any Rocky Mountains to break the westward flow. The river may have dumped its load of sandy sediment in the seaway to the west (Fig. 4.8) where the rounded grains of quartz sand and many fewer particles of zircon were picked up by the prevailing northwesterly winds and blown back onto the vast Navajo sand sea. With burial and cementation, these quartz sand grains now compose the massive Navajo Sandstone. Their destiny is to erode from the cliffs bounding Zion Canyon to once again become loose sand bouncing along the bed of the Virgin River. At some point

in the future, they may once again become part of a thick bed of sandstone to start the cycle anew.

Underlying the Navajo in Zion is a slope-forming, partly vegetated unit called the **Kayenta Formation** (Fig. 4.9). The thin beds of siltstone, shale, and sandstone comprising the Kayenta record deposition within Mesozoic streams flanked by broad floodplains. The streams meandered back and forth through time, depositing laterally extensive beds of channel sands. When the stream flow was high enough to overtop the banks, broad layers of clays and silts were deposited on the floodplains surrounding the main channel. Measurements of water-lain cross-beds within the stream deposits suggest that the flow was generally toward the west or northwest. Mesozoic dinosaurs left behind three-toed tracks in the wet, squishy sediments that were quickly filled and preserved by finer particles (Fig. 4.10A). With time, uplift, and the constant effects of erosion, the tracks were re-exposed to view, providing greater insight into the environmental conditions of this Mesozoic world.

Landscape Evolution at Zion

The main canyon at Zion is dominated by reddish and white Navajo cliffs that stand erect above green vegetated Kayenta slopes near the floor of the canyon (Fig. 4.9). The variable colors of the Navajo are a result of their original environment of deposition, as well as later events that occurred when the rock was buried deep beneath the surface. During Mesozoic deposition of the Navajo desert, iron-bearing minerals "rusted" by interaction with atmospheric oxygen, imparting the reddish hues to the quartz sand grains. After the layer of reddish sand was buried, groundwater percolating through the porous pile of sediment bleached the iron-rich minerals, creating shades of orange and white. It's likely that the groundwater was charged with natural gas, enhancing its ability to chemically react with the rock. When you gaze at the array of reddish and white Navajo cliffs you are seeing the effects of fluid migration through the rock when it was buried a kilometer or two beneath the surface.

The North Fork of the Virgin River meanders along the valley floor, its normally gentle flow masking the role of raging floodwaters in creating the Zion landscape. The North Fork originates in the highest reaches of the plateau country north of Zion. Countless drainages add their snowmelt runoff into the Virgin, which flows southward through the main canyon of Zion National Park before it emerges and combines its water with the East Fork Virgin

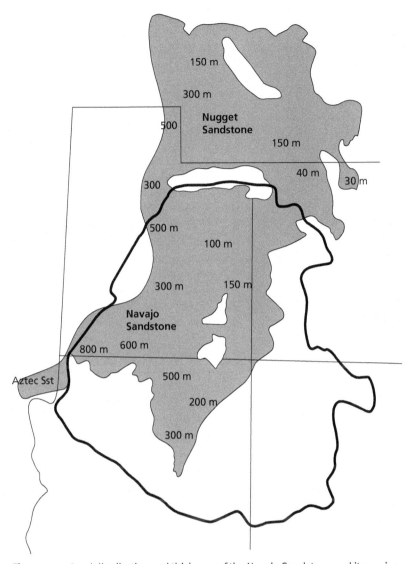

Figure 4.7 Areal distribution and thickness of the Navajo Sandstone and its equivalents. Note the wedge shape of the unit, thinner on the east and thicker on the west.

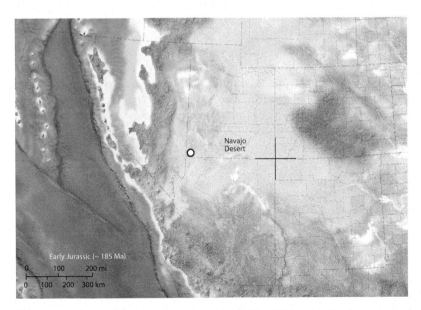

Figure 4.8 Paleogeography of southwestern United States showing the extent of the sand sea during deposition of the Navajo Sandstone (~185 Ma). Circle shows the location of the future Zion National Park.

River to flow westerly. After dropping off the southwestern edge of the Colorado Plateau, the Virgin River empties into Lake Mead where its waters merge with those of the Colorado River.

Similar to the Colorado River in the Grand Canyon, the Virgin River doesn't erode very much on most days of the year when visitors wade in the cool waters. The real action occurs during episodic **flash floods**, with the primary tools of erosion being the boulders, cobbles, and sand that crash and grind away at the underlying bedrock. A typical gentle summer flow of the Virgin through Zion may be around 200 cubic feet per second. (Cubic feet per second—cfs—is the standard unit of **discharge**, the volume of water passing by any one cross-sectional area on the stream at any second. Think of 200 cfs of water as roughly equal to the volume of 7 refrigerators passing by each second.) During a flash flood event in 1998, the discharge on the Virgin rose to 4500 cfs, causing significant damage to streambanks, campgrounds, and roads. (Think of 4500 cfs of water as equal to the volume of about 150 refrigerators passing by each second.) The huge mass of solid debris carried along in the rushing floodwaters acts like a giant scraper on the underlying bedrock, grinding away during those few hours of extreme flow. It is during these random, violent events that most of the downcutting occurs.

Flash floods can also be hazardous to people caught exploring narrow canyons in the park. In September 2015, seven people died when a flash flood swept through Keyhole Canyon after a half-inch of rain fell in under an hour. Rainfall runoff drains quickly across the rocky surface of the plateau, dropping into narrow canyons where it builds into a powerful *debris flow* that sweeps away anything in its path.

The width of Zion Canyon varies as a function of the differential resistance of the Navajo and Kayenta formations to the effects of stream erosion. For most of its course through Zion, the Virgin River cuts into the easily eroded rocks of the Kayenta Formation, creating a relatively broad valley. But upstream toward the north end of the main canyon, the Virgin slices a tight gorge through the resistant Navajo in a place appropriately called The Narrows (Fig. 4.11). Armed

Figure 4.9 The Towers of the Virgin are composed of Navajo Sandstone, which is underlain by the slope-forming, vegetated Kayenta Formation. The light-colored cliff within the Kayenta is a sandstone tongue of the Navajo that interfingers with the Kayenta, reflecting variations in climate and other environmental conditions during deposition.

with a long stick for probing, sensible footwear, a healthy sense of adventure, and the latest weather report, the intrepid visitor can explore deep into the canyon, wading in shallow waters of the Virgin while surrounded by Navajo cliffs about 600 m (~2000 ft) high.

You now know that the canyon *deepens* during countless, unpredictable flash floods that have been occurring for millions of years. And you know that *widening* occurs partly due to the relative resistance of the Kayenta and Navajo to erosion. But why are the canyon walls at Zion so steep? Part of the answer lies in the vertical fractures that penetrate downward through the Navajo Sandstone. Most bodies of rock are fractured, commonly occurring during uplift and flexure of the stiff layer of rock. These fractures are called **joints** and typically occur as multiple near-parallel fractures called **joint sets**. The Zion area has a well-developed joint set with a predominant north–northwest orientation (Fig. 4.12).

Many drainages in Zion are oriented in a north–northwest direction because water opportunistically seeps into the joints, widening them through time until they become pathways for streams. We'll dig deeper into the role of jointing and the various processes of weathering and decomposition of rock when we visit other parks on the plateau, but for now just accept that the joints permit rainwater or snowmelt to penetrate deeply into the body of rock. These waters are mildly acidic and are capable of slowly dissolving the calcite and iron oxide cements between quartz grains, weakening the rock as individual grains separate from the main body of rock. This process preferentially occurs along the sides of joints, in effect separating great slabs of rock along the vertical cracks. With time and perhaps triggered by monsoonal rains or the random earthquake, huge slabs of rock may dislodge from the Navajo cliffs and may free-fall or clatter downward as a *rockfall*. The countless angular blocks of rock mantling the base of the canyon walls (the pile of loose debris is called **talus**) are the remains of rockfalls. Innumerable rockfalls leave behind the sheer walls of the Navajo cliffs, which are in effect the exposed face of a joint plane.

Using the joint system as conduits, rainwater and snowmelt may reach all the way down to the horizontal

Figure 4.10 A) Three-toed track of a dinosaurian, likely *Megapnosaurus*. B) *Megapnosaurus* was a bipedal carnivore about 3 m (10 ft) long that hunted along the river edge during deposition of the Kayenta.

Figure 4.11 The Narrows at Zion National Park. Canyon width decreases dramatically where the Virgin River flows through the resistant Navajo Sandstone. Compare with Figure 4-1 where the Virgin flows through the less-resistant Kayenta. In the foreground, note the debris deposited by flash floods that rage through the narrow canyon.

slope. Alcoves are common at Zion, and the shade created by the overhanging Navajo has provided a welcome respite from the sun for humans, both modern and prehistoric. Both Weeping Rock and Hanging Gardens are alcoves formed near the Kayenta–Navajo contact, festooned by plants that take advantage of the springwaters. In later chapters, you'll see how alcoves are a key feature in the formation of natural arches.

Landslides and rockfalls witnessed by humans today in Zion National Park are miniscule compared with the magnitude of some ancient landslides. Geoscientists have recently mapped and computer-modeled the effects of a gigantic **rock avalanche** that broke off the east face of The Sentinel near the south end of the canyon about 4800 years ago (Fig. 4.13). A rock avalanche begins as a massive rockfall but evolves upon impact into a chaotic flow of shattered rock that spreads out laterally across the surface. The pile of debris from the Sentinel rock avalanche was about 1.5 by 3 km (1 by 2 mi) in extent, covering the valley floor to an average thickness of 95 m (310 ft). The ancestral Virgin River was dammed by the pile of rock, creating a lake that extended upstream almost to The Narrows. Today's Zion Lodge would have been under 115 m (380 ft) of water. After about 700 years, the lake filled with sediment and the Virgin River began to cut downward into the level surface of the lake sediments. As you walk among the cottonwoods that grow along the flat valley floor on either side of the Virgin River, think about that surface being the floor of a sediment-filled lake created by a long-ago catastrophe.

Remember, too, that the area that would become the Colorado Plateau was located just above or just below sea level during the Paleozoic and Mesozoic, long before the landscape we now see was created. In order for rivers like the Colorado and Virgin to cut downward into solid rock, the entire region had to have risen upward toward its current elevations of 2–3 km (1–2 mi). Incision of the main canyon at Zion by the Virgin River occurred during two phases of uplift of the plateau, with the first phase beginning in the latest Mesozoic. Similar to the earlier section on the Grand Canyon, we'll save the full story of the origin of Zion Canyon for a later chapter, after we've visited other parks and you've been introduced to other important events related to

boundary with the underlying finer-grained Kayenta Formation, which is less fractured than the Navajo and relatively impermeable to the flow of water. The water has nowhere to go except laterally along the horizontal interface between the two formations, eventually emerging as a linear array of seeps and **springs** along the base of the Navajo cliffs. The exposed rock of the Kayenta just beneath its contact with the Navajo becomes saturated with this water, slowly decomposing as natural cements are dissolved and grains separate. The weakened rock loses its integrity and over time becomes less able to support the weight of the massive Navajo beds above. Rigid slabs of Navajo sandstone may break away as rockfalls where they are left unsupported by the decomposing Kayenta beneath, creating an **alcove** beneath the overhanging cliff. Over time, alcoves may grow upward as more rock breaks away and crashes downward onto the Kayenta

Figure 4.12 Oblique view of Zion Canyon looking northward. Note the north–northwest orientation of the vertical joints that cut through the rock. Also note the variable width of Zion Canyon, related to differential resistance of the Navajo and Kayenta formations to erosion by the Virgin River.

the evolution of the Colorado Plateau. Most of the erosional landscapes of the national parks of the Colorado Plateau occurred in response to the same tectonic events, so be patient as we work our way through the rest of the narrative.

Key Terms

alcove
cross-bedding
discharge
flash flood
joint
rock avalanche
rounding
sorting
spring
talus

Figure 4.13 Google Earth image illustrating the original source of the Sentinel rock avalanche (dashed white line) and the lateral extent of the debris (blue outline). The lake that formed behind the debris dam was over 100 m (330 ft) deep and reached upstream beyond the Temple of Sinawava.

Related Sources

Biek, R. F., Willis, G. C., Hylland, M. D, and Doelling, H. H. (2010). *Geology of Zion National Park, Utah.* In P. B. Anderson, T. C. Chidsey Jr., and D. A. Sprinkel (Eds.), *Geology of Utah's Parks and Monuments*, 3rd ed. Salt Lake City, Utah Geological Association.

Eaves, R. (2005). *Water, Rock and Time: The Geologic Story of Zion National Park.* Springdale, UT, Zion Natural History Association.

Graham, J. (2006). *Zion National Park Geologic Resource Evaluation Report.* Natural Resource Report, NPS/NRPC/GRD/NRR—2006/014. Denver, CO: National Park Service.

National Park Service—Geology of Zion National Park. Accessed January 2021 at https://www.nps.gov/zion/learn/nature/geology.htm

To see a computer animation of the Sentinel rock avalanche, go to: http://geohazards.earth.utah.edu/zion_anim.html (accessed January 2021)

5 Bryce Canyon National Park

Figure 5.1 Bryce Amphitheater, Bryce Canyon National Park.

The bizarre architecture of the landscape at Bryce Canyon National Park is significantly different from the deeply incised gorges of Grand Canyon and Zion national parks (Fig. 5.1). Bryce does not feature a main canyon with a river flowing along the bottom as do the other two parks. Instead, visitors peer downward from the rim of a high plateau into multicolored amphitheaters of ornately sculpted pinnacles and crenulated walls reminiscent of a psychedelic cityscape. The park itself is oriented more or less north–south (Fig. 5.2), outlining the sharp dropoff along the eastern edge of the Paunsaugunt Plateau. Weathering and erosion by running water have worn away the sedimentary rocks along the plateau margin to create a colorful escarpment known as the Pink Cliffs. The erosional landscape of Bryce Canyon National Park is formed within the Pink Cliffs that extend along the eastern and southern margin of the Paunsaugunt Plateau.

Particles of sediment eroded from the rocks exposed along the margin of the plateau are carried downslope through countless tiny rivulets and gullies that cut through the sculptured terrain of the park. The sediment eventually is transported into larger channels and streams east of the park that merge into the Paria River. The Paria flows south, ultimately emptying into the Colorado River near the eastern edge of Grand Canyon National Park.

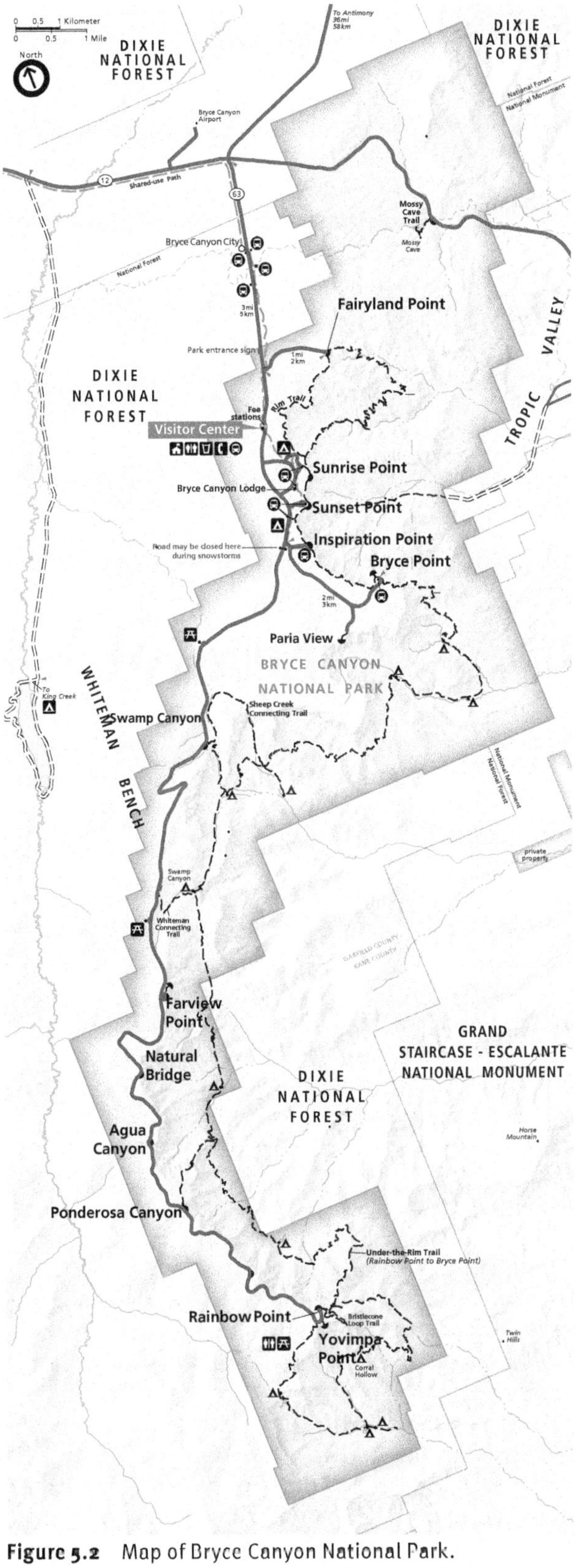

In this section, we'll expand upon many of the concepts that you learned in the earlier sections on Grand Canyon and Zion, but our focus will be on the details of weathering and erosion and the creation of the fantastic Bryce landscape. Incidentally, this astounding scenery is even more inspirational along one of the many trails that wind among the rocks. Never be content with just snapping a picture from an overlook on the rim, but rather hike down into the rocks and wander within the landscape to gain the full experience.

Origin and Age of Rocks at BryceCanyon

The erosional landscape at Bryce is cut into the horizontally bedded **Claron Formation**, a Cenozoic-age unit over 400 m (1300 ft) thick. In the Bryce Canyon region, the Claron is mostly Eocene in age (56–34 Ma). In lower-elevation parts of the park toward the Paria Valley, Upper Mesozoic formations are exposed beneath the Claron. Just outside the park, rocks as old as the Mesozoic Navajo Sandstone are exposed. So the Cenozoic rocks at Bryce Canyon enable us to continue the story of deposition that ended toward the top of the strata at Zion National Park. The continental depositional and paleogeographic settings of the Cenozoic world of the Claron Formation are considerably different from the near-sea-level settings of the Paleozoic and Mesozoic encountered earlier at Grand Canyon and Zion. The Eocene continental environments exposed in the Claron Formation at Bryce were established in response to an episode of tectonic uplift during the late Mesozoic and early Cenozoic that raised the Colorado Plateau by a kilometer or more. This important event created highlands throughout the American West and will be more fully addressed in later chapters.

The Claron Formation is composed of a range of clastic sedimentary rocks—from fine-grained shales, mudstones and siltstones to coarser-grained sandstones and conglomerates. Limestones are relatively common, as are volcanic ashes. These alternating rock types collectively form a thick stack of beds, each ranging from about a half-meter to a few meters in thickness. The sediment composing these rocks was laid down in a variety of depositional settings associated with a series of large **intermontane** lakes that extended from Wyoming to southern Utah (Fig. 5.3). Sediment was fed into the lake system via streams originating in nearby mountains in western Utah and Nevada, as well as from the ancestral Rocky Mountains to the east.

As the shoreline of the lakes transgressed and regressed through time in response to climatic or tectonic influences, different types of sediments were deposited one on top of the other. A sandstone may have accumulated as a lakeside beach, with waves constantly reworking and sorting the sandy sediment. A mudstone (composed of

Figure 5.2 Map of Bryce Canyon National Park.

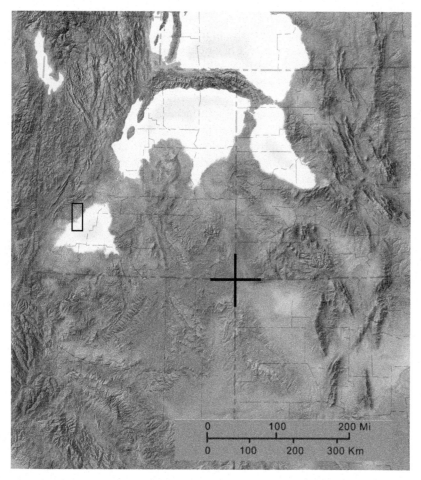

Figure 5.3 Paleogeography of the Colorado Plateau region during the early Cenozoic (Eocene, ~50 Ma). The depositional setting of the Claron Formation near Bryce Canyon National Park (rectangle) was a large lake fed by streams originating in nearby mountains. Four Corners marked by cross.

Claron time and reveal information about native vegetation and climate conditions in the region during the Cenozoic.

These continental stream and lake settings were well oxygenated, being in direct contact with the atmosphere. Oxygen combined with iron in the muds and sands, forming iron oxides and imparting a reddish-orange cast to the sediments and soils. These multiple hues of red, pink, apricot, ocher, and tan color the eye-popping scenery that dominate today's landscape at Bryce. The Claron also forms the scenery at **Cedar Breaks National Monument** to the west, where the formation is thicker and sandier than in the Bryce region.

The loose sediments of the Claron, deposited in a variety of environments associated with a long-lived, high-elevation lake system, were eventually buried by younger sediments, then compacted and cemented to become solid rock. During the late Cenozoic, a phase of broad, slow uplift affected the Colorado Plateau, raising the long-buried rocks up to higher elevations where they were exposed to the inexorable effects of weathering and erosion. That's when the current pinnacled landscape of Bryce began to develop.

clay and silt particles, as the name implies) may represent a quiet, deeper part of the lake where muddy sediment settled out of suspension after being carried in by streams. A limestone may record a phase when the chemistry of the lake was conducive to precipitating tiny crystals of calcium carbonate (calcite) that accumulated on the lakebed. Volcanoes in the surrounding mountains occasionally belched plumes of ash that blanketed the region downwind, with the ash slowly settling out through the water onto the floors of lakes.

Meandering networks of streams bordered by broad floodplains fed the Claron freshwater lakes, with the deposits of the stream network becoming interbedded with those deposited within the lake. Beds of conglomerate within the Claron record fast-rushing streams choked with cobbles and pebbles derived from nearby highlands. When the streams overtopped their banks during seasonal floods, fine clays and silts were deposited on the adjacent floodplains, eventually forming mudstones and siltstones after burial. The floodplain clays and silts developed into vegetated soils during the long spans of time between floods, eventually becoming preserved as **paleosols**. These ancient soils are records of the landscape during

Landscape Evolution at Bryce Canyon

The surreal array of pinnacles, spires, and turrets of rock carved into the Claron Formation are known as **hoodoos**, a general term that describes any ornately sculpted column of rock. Hoodoos are found in many places on the Colorado Plateau as well as worldwide; they are commonly carved into bedded volcanic rocks as well as sedimentary layers. At Bryce, the primary processes that form the scenery are *weathering* and *erosion* acting on the bedded and fractured Claron Formation where it is exposed along the eastern edge of the Paunsaugunt Plateau. Although the terms are sometimes used interchangeably, weathering and erosion are distinctly different processes (Chapter 1). Weathering involves the in-place decomposition of rock due to direct contact with the atmosphere, typically via interaction with water. Erosion is the next step after weathering and involves the movement of the newly weathered grains downslope, perhaps within runoff after a hard rain.

At Bryce and many other localities, rainwater and snowmelt preferentially infiltrate into rock along *joints*—planar fractures that penetrate deeply into masses of rock. As you may have read in Chapter 4 on Zion, joints commonly occur in near-parallel sets that cut vertically across horizontal bedding. Multiple joint sets may form in differing orientations as brittle slabs of rock are repeatedly flexed by large-scale tectonic stresses. Two near-vertical joint sets intersect in the Bryce region, one trending northwest and the other northeast, setting the stage for the creation of hoodoos. Mildly acidic rainwater infiltrates into joints, dissolving the natural cements holding grains together and causing the rock to disintegrate along the joint plane. As water erosively flushes the loose grains outward, the joint plane will widen, permitting the deeper infiltration of stormwater runoff and spring snowmelt. Focusing along joint planes, weathering and erosion work together to create deeper and wider gaps within the body of rock.

The high elevations at Bryce (~2100 m/~8000 ft) promote the freezing of water within widened joint planes during cold nights, with the ice expanding and pushing the rock apart along the joint plane ever so slightly. This highly effective process of physical weathering is called **frost wedging** (Fig. 5.4). As the ice melts during the warmth of the day, the meltwater washes away a few of the loose particles of rock before repeating the process the next cold night. This freeze–thaw cycling, repeated nightly and seasonally over long periods of time, acts to progressively widen the joint plane and create a physical separation of the remaining rock walls on either side of the fracture. Plant roots may work their way downward into widened joints in search of water, physically pushing the rock apart as well as chemically decomposing the rock with their natural organic acids. As joints slowly widen into narrow slots, rainwater and snowmelt use these slots as channels, funneling water downslope from the edge of the high plateau toward the lower elevations of the Paria Valley to the east. Through time, a youthful drainage network of gullies

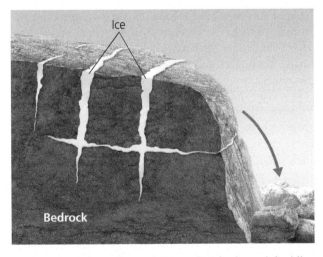

Figure 5.4 Vertical joints intersecting horizontal bedding planes provide a network of conduits for water to penetrate deeply into rock. When the water freezes and expands during cold winter nights, the rock is physically broken apart into angular blocks. This process of frost wedging is fundamental to the formation of hoodoos at Bryce Canyon National Park.

and rivulets was established along the intersecting joint planes cutting through the Claron Formation along the top of the Paunsaugunt Plateau.

At the very edge of the Paunsaugunt Plateau, the angular network of widened joint planes focused erosion that enabled stream channels to incise in an upslope direction in a process called **headward erosion** (Fig. 5.5). The **head** of a stream is its upslope point of origin, whereas the downstream end where the stream empties into a larger stream or a standing body of water is called its **mouth**. As storm runoff or snow meltwater flows off the edge of the plateau into the head of a joint-controlled gully, it gains velocity due to the abrupt increase in slope. The faster flow increases the erosive power of the water so that the rock at the highest tip of the channel is preferentially attacked. As this rock breaks down, the loose sediment is flushed through the head of the channel and transported downslope. The net effect of this concentrated erosion is that the heads of miniscule rivulets and small streams migrate upslope (i.e., headward), extending deeper into the edge of the plateau and exposing greater amounts of rock. Along the Pink Cliffs at Bryce Canyon, headward erosion incrementally wears away at the margin of the Paunsaugunt Plateau, causing it to recede through time.

Weathering and headward erosion along joint planes interact to create the array of hoodoos at Bryce National Park. Beneath the flat, forested top of the Paunsaugunt Plateau, the horizontal beds of the Claron Formation are intact, other than being crosscut by the two distinct joint sets (Fig. 5.6A). As weathering and headward erosion gradually carve away at the margin of the plateau along joint planes, the remnant rock between the widening joints is exposed to the influence of rain, wind, and gravity (Fig. 5.6B). Elongate walls and columns of rock become isolated as planar joints widen through time, eventually evolving into aligned rows of hoodoos. A close look at the labyrinth of walls and aligned hoodoos at Bryce reveals preferred patterns, reflecting the northwest and northeast orientations of joint sets (Figs. 5.7).

As newly formed hoodoos evolve from the breakdown of the intact Claron Formation along the edge of the Paunsaugunt Plateau, older hoodoos continually break down due to the long-term effects of weathering and erosion (Fig. 5.6C). In time, these older hoodoos will decompose entirely, and their constituent grains will become the gravel and sand of the Paria River system. The eastern edge of the Paunsaugunt Plateau progressively recedes due to headward erosion along the head of the Paria drainage network, accompanied by the ongoing formation and destruction of hoodoos.

Why do individual hoodoos have such ornate architecture? Many hoodoos look like poorly made chimneys or elaborate totem poles. In some instances, blocks of rock perch precariously as "capstones" above a narrow neck of rock (Fig. 5.8). Each of the different rock types composing the Claron Formation has variable resistance

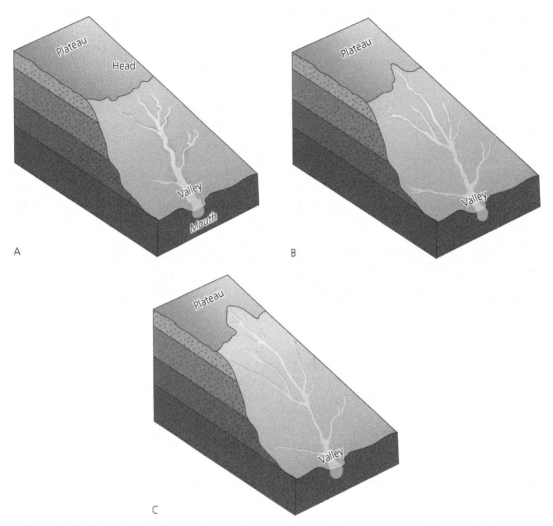

Figure 5.5 Stages in the headward erosion of a drainage network. The heavier line marking the edge of the plateau progressively recedes through time as erosion focuses along the highest tips of streams and rivulets.

to weathering and erosion. Within individual hoodoos, most sandstones and limestones protrude outward, reflecting their inherent strength. In contrast, less resistant beds of shales, siltstones, and volcanic ashes are more easily broken down and commonly pinch inward relative to underlying and overlying more resistant layers. This differential erosion of alternating layers of rock is what imparts the baroque appearance of many hoodoos.

In sum, the formation of the spectacular landscape at Bryce is an ever-evolving process. It begins with a 400 m (1300 ft) thick slab of colorful Claron rock, internally composed of stacked beds of highly variable rock types, resting above a thick pile of Mesozoic rocks. The unremitting effects of weathering and erosion act to remove material from the intact rock of the Claron, preferentially along vertical joint planes, creating walls and columns of rock left behind as the joint planes erode in between. With time, the walls and columns evolve into hoodoos,

the remnant erosional landforms that delight the eye. Natural changes in climate through time varied the rate of landscape formation, with the pace quickening during wet phases and slowing during drier phases. The enormous amount of erosional debris removed from the Claron over millions of years was transported through the Paria River into the Colorado River, which flushed it through the Grand Canyon and eventually into the northern margin of the Gulf of California where it resides today.

The Grand Staircase

How do the rocks at Zion and Bryce Canyon national parks relate to the rocks of Grand Canyon? Over a distance of about 150 km (90 mi), a series of east–west-oriented escarpments step northward from the rim of the Grand Canyon to Bryce Canyon in a topographic

A

B

C

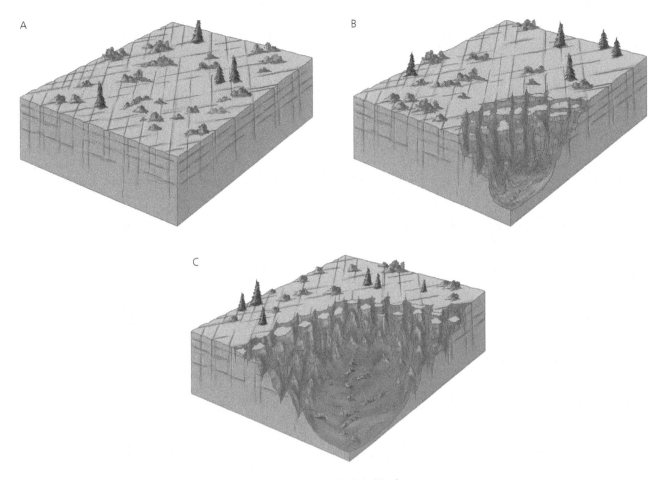

Figure 5.6 Stages in the formation of hoodoos at Bryce Canyon National Park.

Figure 5.7 Aligned walls of hoodoos eroding out of intact layers of Claron Formation underlying the forested Paunsaugunt Plateau. The broad gaps between the walls are the widened spaces of former joint planes. Compare this photo with the graphic of Figure 5.6B.

feature called the **Grand Staircase** (Fig. 5.9). Each escarpment extends for several tens of kilometers and is named for the dominant color of the rock exposed along the steep cliff face (e.g., Chocolate Cliffs, Vermillion Cliffs, White Cliffs, Pink Cliffs). Recognized since the 1870s, the rocks exposed in each escarpment become progressively younger from south to north. A geologic cross section extending across the Grand Staircase reveals the relationships between the rocks of all three parks (Fig. 5.10). The Precambrian and Paleozoic rocks that dominate the Grand Canyon extend beneath the surface toward the north, forming the deep bedrock beneath Zion and Bryce national parks. The Mesozoic rocks that characterize the White Cliffs of Zion Canyon (Navajo

Figure 5.8 The ornate architecture of the hoodoo known as Thor's Hammer is due to the variable resistance of individual beds of rock within the stack comprising the Claron Formation.

Pink Cliffs

White Cliffs

Vermillion Cliffs

Figure 5.9 Stepped erosional escarpments of Mesozoic and Cenozoic sedimentary rock of the Grand Staircase.

Formation) extend beneath the surface toward the Pink Cliffs of Bryce Canyon (Claron Formation).

It's very likely that the Mesozoic rocks exposed at Zion once extended southward across the Grand Canyon region but were worn away during later uplift and erosional downcutting. And it's likely that Cenozoic rocks today exposed at Bryce Canyon once extended laterally across the Grand Canyon region as well but were subsequently removed during uplift and erosion. The escarpments of the Grand Staircase manifest the current position of erosional retreat of those Mesozoic and Cenozoic sedimentary layers. The natural architect for the erosional landscape of the Grand Staircase was the Colorado River network of streams whose collective erosive power gradually removed thick layers of rock. Very few places in the world reveal

such a complete span of strata ranging from Precambrian through Cenozoic over such a small area.

This remarkable geologic wonderland is preserved within **Grand Staircase-Escalante National Monument**, a huge, undeveloped region popular among those hardy adventurers who like their redrock country empty and remote (Fig. 5.11). This is four-wheel drive territory, abounding in narrow slot canyons ripe for exploration. Paleontologists flock here for the dinosaur-rich fossil beds found in the Mesozoic rocks that dominate the monument. The spectacular array of dinosaur fossils includes a huge, near-complete skeleton of a duck-billed herbivore, a bizarre ceratopsid with 15 horns creating a broad frill around its skull, and a snub-faced tyrannosaur with an appropriately terrifying set of teeth. A lost world is preserved in the Mesozoic rocks of the Grand Staircase-Escalante; time, funding, and curiosity will eventually expose the secrets of this solidified Cretaceous Park.

By the way, since this is the first mention of a national monument in the book, it seems appropriate to discuss the differences between a monument and a park. **National monuments** can be declared by executive order of the president of the United States, without congressional approval as required for a national park. Monuments typically preserve a single unique feature of either cultural or natural significance, whereas national parks preserve scenic and natural regions of global and national significance. Monuments typically receive less funding than parks and are afforded less protection from hunting, grazing, and mining. The Bureau of Land Management administers Grand Staircase-Escalante National Monument, whereas most other national monuments are under the supervision of the National Park Service of the Department of the Interior. Many national monuments have "graduated" to become national parks, and perhaps that may happen to Grand Staircase-Escalante in the future. The National Park System as a whole comprises 423 parks, monuments, preserves, recreation areas, seashores, memorials, historic sites, and battlefields.

Origin of Bryce and Zion National Parks along Uplifted Plateaus

The north–northeast trending Paunsaugunt Plateau extends subparallel to the Markagunt Plateau to the west (Fig. 5.12A). Both plateaus are bound by *faults* on each side. Faults are simply planar features within rock along

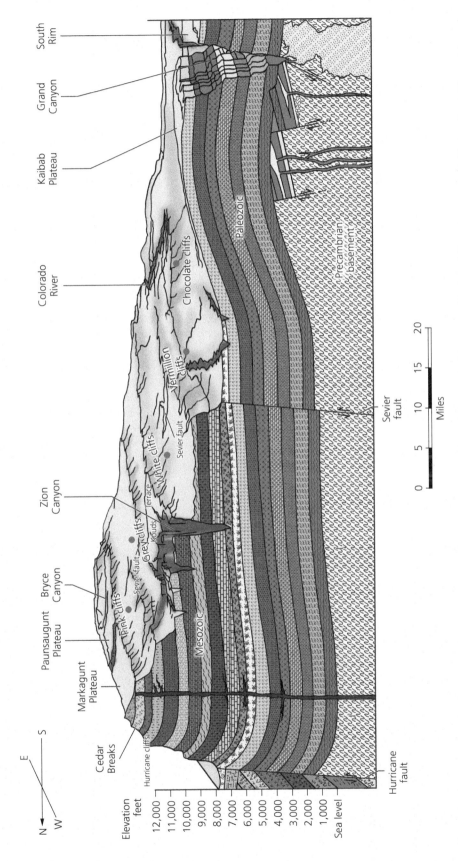

Figure 5.10 Cross section through the Grand Staircase relating Precambrian and Paleozoic rocks exposed in Grand Canyon to younger Mesozoic and Cenozoic rocks exposed in Zion (White Cliffs) and Bryce (Pink Cliffs). Red dots mark the five main cliff-forming "steps" in the Grand Staircase.

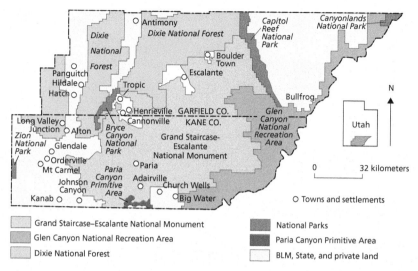

Figure 5.11 Map of Grand Staircase-Escalante National Monument showing its huge extent, ranging from the Arizona state line in the southwest to Capitol Reef National Park to the northeast.

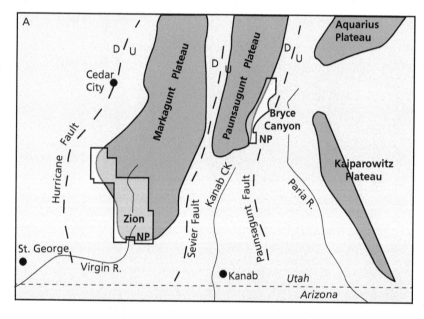

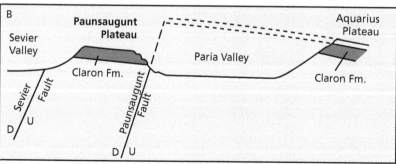

Figure 5.12 A) Map of the Zion–Bryce region showing locations of normal faults responsible for uplift of north–south-oriented plateaus. B) Simple cross section extending across Paunsaugunt Plateau, illustrating the sense of motion along faults bounding the plateau. D = down, U = up.

which displacement has occurred (Chapter 1). That is, a massive block of rock on one side of a fault has moved relative to a massive block of rock on the other side. Movement commonly happens abruptly during earthquakes, with displacement of tens or hundreds of meters accumulating over hundreds or thousands of earthquakes through time.

The faults bounding the Paunsaugunt and Markagunt plateaus are **normal faults** that form due to extensional stress (Fig. 5.12B). As a part of the crust is stretched, a large mass of rock on one side of the normal fault slides downward relative to the mass of rock on the other side of the fault. The tectonic origin and age of this faulting is related to the Cenozoic origin of the Basin and Range geologic province to the west that influenced the uplift of the Colorado Plateau. The details will be addressed in Chapter 33 on the origin of the modern landscape of the American West.

Three major north–south trending normal faults developed over the past several million years that resulted in the uplift of the Markagunt and Paunsaugunt plateaus. To the east of the Paunsaugunt Plateau is the Paunsaugunt fault, while to the west is the Sevier fault (Fig. 5.13), both with hundreds of meters of down-on-the-west vertical offset. A similar down-on-the-west normal fault exists on the west side of the Markagunt Plateau, the Hurricane fault, which marks the boundary between the Colorado Plateau and the Basin and Range geologic province to the west. (The Grand Wash fault to the southwest defines the boundary with the Basin and Range near the Grand Canyon.) As the plateaus were irregularly shifted during sporadic earthquakes on the faults, the horizontal strata were tilted slightly, influencing the flow of water along the top of the plateau. The steep, exposed margins of the plateaus became the focus of weathering and headward erosion because the flow of storm runoff and

Figure 5.13 Exposure of the Sevier fault in Red Canyon on the west flank of the Paunsagunt Plateau. Dark volcanic rocks have been dropped downward along the normal fault relative to the orange–pink sedimentary rocks of the Claron Formation.

snowmelt accelerated as it dropped off the steep escarpment toward the lower elevation valleys below.

On the southwest margin of the Markagunt Plateau, the Mesozoic rocks at Zion were deeply incised by the Virgin River drainage as it flowed southward off the plateau. On the west side of the Markagunt Plateau, the soft rocks of the Claron lakebeds were carved into Cedar Breaks National Monument. Along the east side of the Paunsaugunt Plateau, the fluted cliffs, turreted castles, and totem-like hoodoos were cut to form the strange landscape of Bryce Canyon National Park. The erosion rates at Bryce are very rapid, geologically speaking, with the plateau edge retreating westward about a meter or so every century. Occasional intense rainfall that produces flash floods contributes to the rapid erosion rate, as does the lack of a veneer of stabilizing soil and vegetation on the steep slopes of the escarpment. So the geographic locations of Zion and Bryce are not simply due to chance, but rather reflect the sequential interaction of deposition, faulting, jointing, weathering, and erosion acting over unimaginably long spans of time.

Key Terms

frost wedging
head
headward erosion
hoodoo
intermontane
mouth
national monument
national park
normal fault
paleosol

Related Sources

Davis, G. H., and Pollock, G. L. (2010). *Geology of Bryce Canyon National Park, Utah.* In P. B. Anderson, T. C. Chidsey Jr., and D. A. Sprinkel (Eds.), *Geology of Utah's Parks and Monuments*, 3rd ed. Salt Lake City, Utah Geological Association.

DeCourten, F. (1994). *Shadows of Time—the Geology of Bryce Canyon National Park.* Bryce Canyon, Utah: Bryce Canyon Natural History Association, 128 p.

Doelling, H. H., Blackett, R. E., Hamblin, A. H., Powell, J. D., and Pollock, G. L. (2010). *Geology of Grand Staircase-Escalante National Monument, Utah.* In P. B. Anderson, T. C. Chidsey Jr., and D. A. Sprinkel (Eds.), *Geology of Utah's Parks and Monuments*, 3rd ed. Salt Lake City, Utah Geological Association.

ThornBerry-Ehrlich, T. (2005). *Bryce Canyon National Park Geologic Resource Evaluation Report.* NPS/NRPC/GRD/NRR—2005/002. Denver, CO: National Park Service.

National Park Service—Geology of Bryce Canyon National Park (and links within) Accessed January 2021 at https://www.nps.gov/brca/learn/nature/geologicformations.htm and https://www.brycecanyon.com/bryce-canyon-geology

National Park Service—Grand Staircase Accessed January 2021 at http://www.nps.gov/brca/naturescience/grandstaircase.htm

6 Capitol Reef National Park

Figure 6.1 Landscape along the Waterpocket Fold, Capitol Reef National Park.

If you're searching for solitude to accompany sweeping desert vistas, you've found the right place at Capitol Reef National Park (Fig. 6.1). The summer crowds at Zion and Bryce are nowhere to be found at this nearby gem, likely because only a few roads pass through the park, with many of them being unpaved or requiring a high-clearance vehicle. Furthermore, the elongate north–south shape of the park with lots of terrain that are reachable only by foot discourages many visitors who prefer to stay on the relatively short paved roads in the park (Fig. 6.2). Then there's that name. What kind of park would be called "Capitol Reef" in the desert Southwest?

The park has a north–south orientation because of the **Waterpocket Fold**, a 160 km (100 mi) long wrinkle in the Earth's crust that created a daunting barrier to early settlers on their way west. They called it a "reef" to represent the jagged obstacle it presented, comparable to that of a biologic reef in impeding the progress of a ship at sea. The white, rounded, domal shapes of the upper Navajo Sandstone common in the park reminded easterners of the rotundas common to government buildings in the nation's capital (Fig. 6.3). "Capitol Reef" stuck as one of the more colorful park names in the country.

As is typical for all the parks of the Colorado Plateau, the erosional landscape of elongate ridges, isolated buttes and monoliths, and steep-walled gorges at Capitol

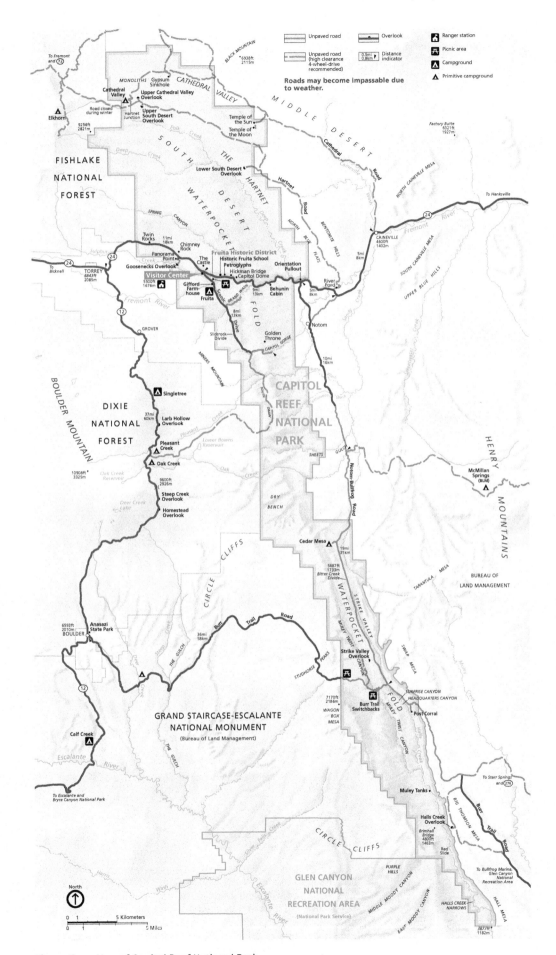

Figure 6.2 Map of Capitol Reef National Park.

Figure 6.3 Navajo Dome, formed in Navajo Sandstone.

Reef is primarily the result of flowing water. The park's main drainage is the Fremont River, which flows easterly across the north part of the park before it empties into the Dirty Devil River, which flows south into the Colorado River near the north end of Lake Powell (Fig. 2.4). Several smaller ephemeral streams cut narrow gorges directly across the Waterpocket Fold. The stream channels are normally dry, but like many drainages on the Colorado Plateau, they can rapidly transform into powerful, sediment-choked torrents. Traveling along Route 24 with my family in a spitting rain one day, we watched dumbstruck as the normally tranquil Fremont River raged the color of a chocolate milkshake, rumbling along with the click-clack sound of cobbles banging into one another along the base of the channel.

Many of the processes and features discussed in this chapter are familiar from previous sections, such as Mesozoic-age sedimentary rocks, jointing, and differential weathering and erosion. New concepts will be introduced in this chapter that relate to the way rocks deform (monoclines) and the mechanism by which flowing water erodes narrow gorges into solid rock (stream incision).

Origin and Age of Rocks at Capitol Reef

The rock record at Capitol Reef spans over 3000 m (~10,000 ft) of uppermost Paleozoic through lowermost Cenozoic strata. You might think of the strata as beginning with the uppermost rock of the Grand Canyon, duplicating the Mesozoic rock of Zion and continuing upward almost to the Cenozoic rock at Bryce. The dominant scenery-forming units are Mesozoic redrocks that preserve tens of millions of years of coastal deposition, arid climates, and the shifting sands of enormous deserts. We'll focus on a few of the most prominent formations in the park—the Moenkopi, Chinle, Wingate, Kayenta, and Navajo (Fig. 6.4).

The distinctive brownish-orange rocks of the **Moenkopi Formation** commonly mark the base of escarpments in the area near the park visitor center. Uniform beds of siltstone, shale, and sandstone impart a banded appearance to the formation. The sediments were deposited in shallow marine and tidal flat environments flanking a shallow seaway about 240 Ma. Tracks of reptiles and amphibians are preserved along bedding planes, left behind on the muddy nearshore landscape. The Moenkopi is relatively resistant to erosion and forms the Chocolate Cliffs of the Grand Staircase (discussed in Chapter 5 on Bryce Canyon National Park).

The purplish-gray and reddish-orange slopes of the **Chinle Formation** are widespread across the Colorado Plateau, but they are especially prominent here at Capitol Reef. A wide variety of sedimentary rock types compose the Chinle, but clay-rich mudstones and siltstones are dominant. The wide variety of colors mainly reflects the concentration of oxygen in the coastal depositional environments in which the Chinle sediments accumulated. Reddish hues are derived from iron minerals deposited in well-oxygenated conditions like high-velocity streams. Muted, purplish-gray tones reflect deposition in poorly oxygenated marshes or stagnant lakes.

The prominent reddish-orange cliffs of the **Wingate Sandstone** tower above the less resistant, slope-forming Chinle. The Wingate is characterized by large, high-angle cross-bedding, a depositional structure that was introduced

Figure 6.4 Oblique aerial view of Mesozoic redrock, with white domes of Navajo Sandstone at the top of the escarpment. The red roof of the visitor's center is in the foreground, with the green corridor of Sulphur Creek in the valley below. The Navajo overlies the well-bedded Kayenta Formation, which in turn overlies the reddish-orange vertical face of the Wingate Sandstone. The Chinle and Moenkopi formations underlie the Wingate all the way down to the valley floor. "The Castle" is perched as an isolated monolith of Wingate resting on the slope-forming Chinle.

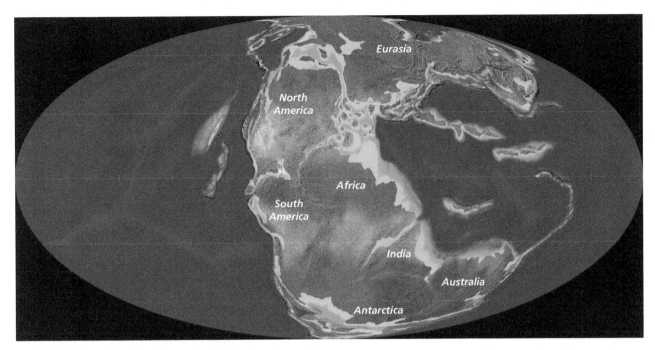

Figure 6.5 Global paleogeographic map showing the initial breakup of the supercontinent Pangea (~200 Ma). The vast Wingate sand sea is shown along the western flank of Mesozoic North America. The very early continental rifting between North America, Africa, and South America marks the birth of the Atlantic Ocean as a linear seaway.

in the section on Zion National Park. Stacked layers of cross-beds composed of quartz sand manifest the interiors of huge sand dunes that migrated with the Mesozoic wind. The Wingate represents a phase of Earth history when the American West was as dry as any modern sand sea like the Sahara or Arabian deserts. To the south and southwest was an intermittent drainage network that transported sand and gravel from ancestral highlands. These braided, sediment-choked streams fed the Wingate desert, maintaining the supply of resistant quartz sand over millions of years.

Cross-beds within the Wingate are commonly obscured by **desert varnish**, a micron-thin, dark coating of iron and manganese oxides mixed with clay and organic matter. Dark vertical streaks of desert varnish adorn many cliff faces on the Colorado Plateau, initiated by water dribbling downward over the rock. The thin film of water allows for microbes and windblown clay particles to take hold on the cliff face, setting the conditions for formation of the surficial crust. Over thousands of years, microbes concentrate the iron and manganese oxides on the wet rock surface, darkening the rock in the process. Native Americans scratched petroglyphs into desert varnish to reveal the lighter color rock beneath, animating their artwork. Desert varnish is a relatively recent feature of the rock related to long-term exposure in the arid climate and should not be mistaken as a primary depositional characteristic of the rock.

With time, the environment gradually shifted toward the meandering streams and broad floodplains recorded by the Kayenta siltstones, shales, and sandstones, a formation discussed earlier in the chapter on Zion. The overlying Navajo Sandstone, with its huge cross-beds and widespread extent, indicates a return to arid, desert conditions (Fig. 4.8). The Navajo at Capitol Reef is about half as thick as it is at Zion, but

the white, dome-forming character of the Navajo is prominent enough to give the park its first name. The white color of the Navajo at Capitol Reef (similar to the upper Navajo at Zion) may be the result of "bleaching" by methane gas that migrated through pores between sand grains before being expelled into the ancient atmosphere a few million years ago.

During the deposition of the vast Wingate and Navajo sand seas around 200 Ma, Earth was experiencing the initial tectonic breakup of **Pangea**, the single global continent that existed for the previous 100 million years or so (Fig. 6.5). Of course, this was the time of dinosaurs, but small tree-dwelling and burrowing mammals were also present, as were the very first birds. A single global ocean populated by giant marine reptiles bearing enormous sets of teeth surrounded Pangea. The North Atlantic was in its first stage of birth as a linear sea, created as continental rifting separated North America from the conjoined African and South American landmasses. There were no polar ice sheets, and the disintegrating supercontinent had generally warm and equable climates at all latitudes. The Wingate and Navajo deserts extended across much of the future Colorado Plateau, reflecting the widespread warmth and aridity of the American West at the time.

Waterpocket Fold

Armed with enough information about the key formations at Capitol Reef, we're now in a position to better understand the genesis of the main feature of the park—the Waterpocket Fold. From the air, the Waterpocket Fold looks like a row of giant shark teeth (Figs. 6.1, 6.6). The broad white band is readily identifiable as the rounded upper surface of the Navajo Sandstone. The sawtooth pattern of the redrocks is the erosional edge of a younger unit known as the Carmel

Figure 6.6 Oblique aerial photograph of the sawtooth pattern of the steeply inclined, tan Entrada and red Carmel formations stacked against the rounded white outcrop of the underlying Navajo along the Waterpocket monocline. Note the dry drainages that cut down through the highly angled beds of redrock that merge into the meandering dry streambed in the vegetated valley.

(synclines). These bends in the rock are called **folds**. The Waterpocket Fold is a specific type of fold called a **monocline**, a step-like flexure in the crust where a horizontally layered stack of rocks abruptly bends downward at a steep, near-vertical angle before returning to horizontality a short distance away (Fig. 6.7). Monoclines form in response to compressional tectonic stress and commonly relate to buried fault movement deep beneath the surface.

The trend of the Waterpocket monocline is oriented slightly northwest—southeast, which indicates that compressional stress was from the west–southwest, perpendicular to the axis of the fold. Brittle older rocks at depth likely ruptured along a fault or perhaps along a set of near-parallel faults that never broke through to the surface. The younger rocks above were pushed upward on the west and draped like a throw rug across the buried fault that lay beneath. The step-like draping of the layered sedimentary rocks is the monocline. For a sense of perspective, layers of Mesozoic rock on the west side of the monocline are pushed upward over 2 km (1.2 mi) above their equivalent layers on the east side, reflecting the displacement of rocks at depth along the buried fault (Fig. 6.7).

Formation juxtaposed against the even younger tan rocks of the Entrada Formation. How could these rocks, which are normally arranged in layercake fashion, have possibly formed this strange, angular landscape?

When subjected to compressional stress (Chapter 1), rocks can respond in a variety of ways, depending on the inherent strength of the rocks as well as the orientation, intensity, and rate of the applied tectonic stress. Some rocks will break, resulting in faults that record the offset along the plane of weakness where rupture occurred (Fig. 5.13). Other rocks will bend, some with a convex-upward shape (anticlines) and some with a concave-upward shape

The Waterpocket monocline formed sometime during the latest Mesozoic through early Cenozoic during a major episode of compression and mountain-building called the **Laramide orogeny**. This tectonic event played a significant role in the uplift of the Colorado Plateau and indeed of the entire American West. We'll address this profound mountain-building event in much greater detail in later chapters as we add more geologic evidence.

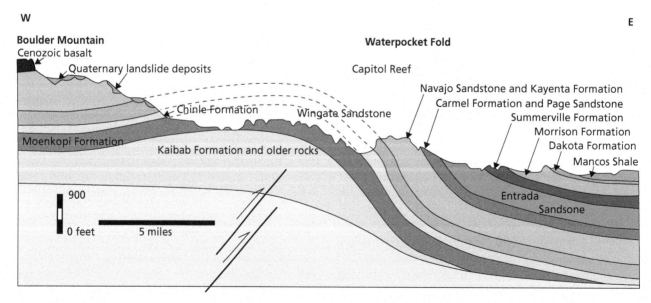

Figure 6.7 Cross section from Boulder Mountain on the west across the Waterpocket monocline, illustrating the steeply inclined bedding along the flexure of the fold. The thick lines below the cross section show the probable location of buried faults beneath the monocline, and the arrows indicate the likely sense of offset. The upper profile of the figure shows the differential erosion of the various formations across the modern landscape, accentuated by an expanded vertical scale. Dashed lines show the former position of rock layers that were removed by erosion.

Landscape Evolution at Capitol Reef

Why does the Waterpocket monocline have such a strange appearance in outcrop, with its angular sharktooth pattern (Fig. 6.6)? Weathering and erosion have acted continuously on the rocks, preferentially wearing away the highlands to the west and redepositing the debris downslope in the lower elevations to the east via a drainage system that is dry most of the time. Along the most steeply inclined layers of the monocline, weathering and erosion by running water have worn downward into the tilted, near-vertical rocks, leaving behind the jagged, eroded margins of each formation angling upward into the air as a steep escarpment. Each individual "tooth" of red Carmel and tan Entrada rock is bound on both sides by dry streambeds draining downslope to the east. During torrential rains when flash floods occur, the rushing water and its entrained rocky debris incise deeply into the angled layers of rock, likely along joint planes, leaving behind the serrated edges of the two formations where they abut the underlying rounded slopes of white Navajo. The "teeth" along the eroded monocline point upward and back toward the higher part of the fold. Go back to the photograph that introduces this chapter for a renewed understanding of the angular landscape created by the interaction of erosion with monoclinal folding of multicolored rock (Fig. 6.1).

Seasonal storms and the resulting flush of watery debris flowing through normally dry streambeds are the primary agents of erosion in Capitol Reef, as they are everywhere on the Colorado Plateau. Summer downpours can drop so much water over a short period of time that the water can't penetrate downward into the rock, instead flowing across the slickrock surface toward lower areas. Particles of sand as well as larger pebbles and cobbles transported by the flow may become trapped in small basins where they swirl around, abrading deeper into the rock. These shallow bedrock puddles are called "waterpockets" and can hold water long after the storm has passed, nourishing the surprisingly abundant wildlife of the desert.

Why do narrow gorges cut directly through the bedrock perpendicular to the trend of the Waterpocket monocline? Examples include Grand Wash, Capitol Gorge, Pleasant Creek, Sheets Gulch, and Oak Creek that cut east–west through the steep slopes of the monocline (Figs. 6.2, 6.8). If you were a droplet of water, you would take the path of least resistance on your gravity-induced flow downslope. That's what some of the drainages in the park do, flowing through relatively soft shale and siltstone parallel to the steep slopes of the monocline. Then why do other streams cut directly across the resistant rocks of the Wingate and Navajo that form the escarpment rather than taking the easy route and flow south parallel to the trend of the monoclinal cliffs?

The reason is that the streams were there first and the land beneath rose upward *into* the streams during a later uplift event. The Capitol Reef area (and likely much of the Colorado Plateau) was buried beneath a broad blanket of erosional debris several tens of millions of years ago in the early Cenozoic (Fig. 6.9). When the land beneath was subsequently uplifted in late Cenozoic time, the gradient of the streams increased, accompanied by an increase in the velocity of the water in the channels. Rather than change course, the streams incised downward through the carapace of loose sediment and then into the slowly rising solid rock.

Incision, strictly speaking, means that the streams cut downward into the underlying rock without changing their course. In the case of the incised gorges that crosscut the trend of the Waterpocket monocline, as the land was uplifted beneath the previously existing streams, younger rock that had buried the Laramide-age monocline was eroded away. As the monocline was gradually exposed, the overlying streams maintained their west–east paths and cut gorges into the emerging structure without regard for the generally north–south trend of the fold. These types of incised stream channels are called **superposed streams** in that they were superimposed upon the underlying structural grain by the synchronous processes of uplift and erosion.

Superposed streams and their deeply incised canyons are ubiquitous on the Colorado Plateau. The Grand Canyon is the most obvious example, formed by the superposition of the Colorado River and its tributaries (Chapter 34). The incised gorge of Zion Canyon was formed by superimposition of the North Fork of the Virgin River. At smaller scales, narrow and deep **slot canyons** may be only

Figure 6.8 Incised gorge of Grand Wash in Capitol Reef National Park. Steep walls are composed of Navajo Sandstone.

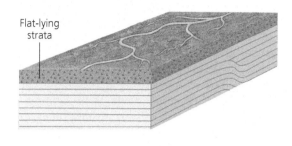

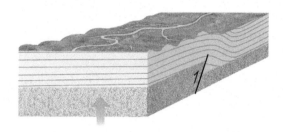

Figure 6.9 Development of a superposed stream as uplift and erosion expose buried, preexisting structures. The stream continues incising downward and across the trend of the underlying structure.

Figure 6.10 Rounded black boulders of volcanic basalt rest atop eroded slopes of redrock.

as wide as your outspread arms. Slot canyons are found in all national parks of the Colorado Plateau, created by the focused flow of water along joint planes. Regardless of size, all these incised canyons have a common origin related to uplift of the Colorado Plateau, initiated during the late Mesozoic to mid-Cenozoic Laramide orogeny and then rejuvenated during the late Cenozoic (explored in detail in Chapter 33).

Another distinctive and odd feature of the landscape at Capitol Reef National Park is the occurrence of rounded black boulders directly on outcrops of redrock (Fig. 6.10). The rock is volcanic basalt and its nearest source is Boulder Mountain about 10–20 km (6–12 mi) to the west. The rounded shape of the boulders suggests that they were abraded during transport, perhaps by banging against other rocks in a fast-flowing stream in flood. The boulders were likely part of a huge load of sediment transported through stream valleys toward the Capitol Reef area from the west. Boulders of this size could only be moved via high-velocity debris flows and were deposited along with large amounts of finer-grained muds and sands also carried along in the flow. Through time, the smaller particles were washed away by water and wind, leaving behind the boulders perched incongruously on the redrock. It may be that streams and rivers flowed with much greater *discharge* during wetter phases of climate when they would be more capable of moving large boulders.

There are many more geologic features to see in Capitol Reef National Park than are described in this brief chapter. Ancient oyster reefs preserve a time when shallow seas covered the region. Petrified logs and dinosaur bones provide glimpses of Mesozoic ecosystems. A sponge-like network of holes pockmark the sandstone within gorges, reflecting the weathering of natural cements binding sand grains together. Enormous monoliths of rock isolated by erosion resemble medieval cathedrals. As in all national parks, the best way to appreciate the array of geologic phenomena in Capitol Reef is to explore the diversity of trails that lead away from the roads.

Key Terms

desert varnish
fold
Laramide orogeny
monocline
Pangea
slot canyon
superposed stream

Related Sources

Graham, J. (2006). *Capitol Reef National Park Geologic Resource Evaluation Report. NPS/NRPC/GRD/NRR—2006/005.* Denver, CO: National Park Service.

Morris, T., H., Manning, V. W., and Ritter, S. M. (2000). *Geology of Capitol Reef National Park, Utah.* In D. A. Sprinkel, T. C. Chidsey Jr., and P. B. Anderson (Eds.), *Geology of Utah's Parks and Monuments.* Utah Geological Association Publication 28, pp. 85–106.

National Park Service—Geology of Capitol Reef National Park. Accessed January 2021 at http://www.nps.gov/care/naturescience/geology.htm

7 Arches National Park

Figure 7.1 Delicate Arch, a landform so iconic that it graces the license plate of many a Utahn. The top of the arch is about 16 m (52 ft) above the base.

At this point in the narrative of national parks of the Colorado Plateau, you know what to expect in Arches National Park—horizontally bedded Mesozoic redrock that has been weathered and eroded along joints to form a wide array of glorious landforms. The distinctive feature of the park is the greatest concentration of natural arches on the planet, giving the park its very name (Fig. 7.1). Moreover, Arches differs from most other parks and monuments of the Colorado Plateau in that salt plays a large role. Yes, salt—the same stuff we sprinkle on food to enhance the taste. In this case, the salt records the desiccating, hyperarid conditions of a long-lost inland sea that once occupied this part of Utah.

All of the typically dry washes that cut through the landscape drain toward the Colorado River that marks the southern boundary of the park (Fig. 7.2). The annual rainfall of about 20 cm/yr (8 in/yr) is overwhelmed by an evaporation rate of about 100 cm/yr (40 in/yr). The rain that does fall occurs during the summer monsoons that sweep up from the south, commonly generating turbulent flash floods that flow through the dry washes or plummet off impermeable sandstone tablelands as short-lived waterfalls. As in the rest of the Colorado Plateau, these occasional

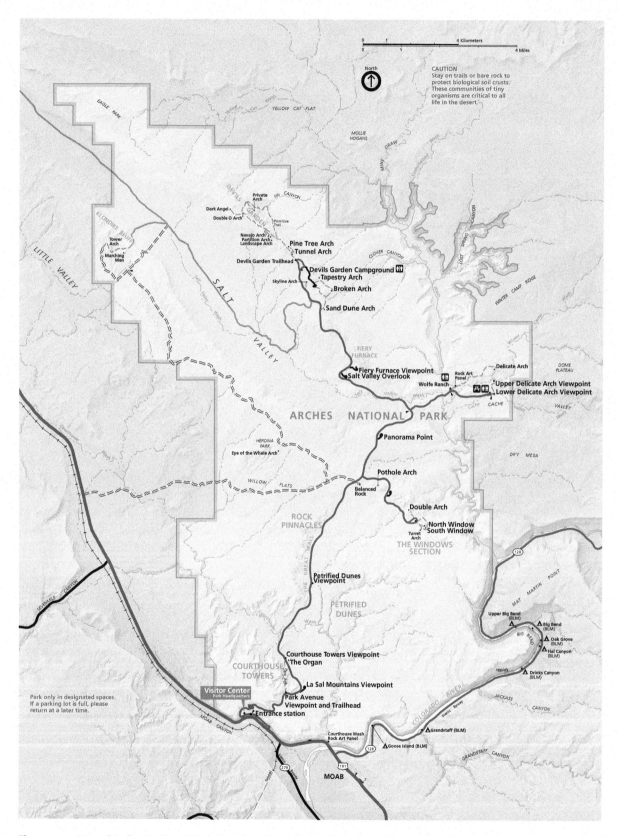

Figure 7.2   Map of Arches National Park. Note how the Colorado River forms the southern boundary.

bursts of running water laden with sedimentary debris are important sculptors of much of the desert landscape at Arches National Park. But it's the undetectable infiltration of storm runoff and snowmelt into the body of rock at Arches that works via weathering and erosion to create

its signature feature – the array of natural arches for which the park is named.

In line with the ongoing scheme of building upon concepts introduced in previous chapters, you have the requisite background to understand the origin of the dramatic erosional landscapes at Arches. In this chapter, you'll be introduced to one more Mesozoic formation and an important Paleozoic unit that will add to our growing stack of formations that characterize the Colorado Plateau. The new concepts that we'll address include the critical role of jointing in the creation of natural arches, as well as the curious behavior of sedimentary rock salt when subjected to pressure.

Origin and Age of Rocks at Arches

The majority of the natural arches and much of the landscape in the park are formed in the Mesozoic **Entrada Sandstone**, deposited about 170 million years ago. The salmon-colored Entrada overlies the tan shades of the Navajo Sandstone, a formation covered in Chapters 5 and 6, respectively, on Zion and Capitol Reef national parks. Much of the main road through the park runs along the rolling top of the cross-bedded sandstones of the Navajo. Pinnacles, arches, and elongate walls of rock formed in the Entrada rise upward above the mostly buried Navajo.

The Mesozoic geography of the region changed dramatically from the older, inhospitable Navajo desert, with sea levels rising to extend an arm of the ocean southward across central Utah. This shallow seaway, called the Sundance Sea by geologists, brought slightly less severe conditions to the region, as evidenced by dinosaur trackways preserved on bedding planes in the Entrada. The climate remained semiarid, however, with broad tidal flats separating the shoreline from a widespread coastal desert covered in shifting sand dunes.

The Entrada Formation is composed of three members, each distinguished by their characteristic appearance in outcrop. (Some geologists break these three members out into separate formations, but we'll stay with the member terminology for simplicity.) The lowermost Dewey Bridge Member is composed of wrinkly, thin beds of reddish-brown siltstone and muddy sandstone with a few thin shales (Fig. 7.3). This unit accumulated on broad tidal flats that extended across the southern Utah region during a phase of higher sea level when the shoreline environments of the Sundance Sea transgressed landward.

As sea level fell slightly over time and the shoreline regressed seaward to the west, a coastal desert system swept across the Arches area, preserved as the overlying Slick Rock Member of the Entrada. The Slick Rock forms massive layers of reddish-orange sandstone, with the well-sorted sand grains cemented together by calcite or iron oxides. Some *cross-bedding* is apparent in the Slick Rock, attesting to its deposition as migrating sand dunes on a widespread coastal desert. Many of the natural arches in the park are

Figure 7.3 Erosional remnant of lower two members of Entrada Sandstone, Park Avenue in Courthouse Towers area.

Figure 7.4 North Window Arch with Turret Arch in background, Arches National Park. Dashed line marks sharp contact between the Dewey Bridge Member below and the Slick Rock Member above.

formed along the sharp contact between the Dewey Bridge and Slick Rock members of the Entrada (Fig. 7.4).

The uppermost unit in the Entrada is called the Moab Member, which forms lighter colored beds capping the massive Slick Rock Member. The Moab forms a prominent rounded caprock on some of the landforms in Arches. For instance, counter to most of the arches in the park, Delicate Arch (Fig. 7.1) is formed from the Slick Rock Member along its base and lower pedestal, while the bridge at the top is formed in the Moab Member.

Landscape Evolution at Arches

So how do **natural arches** form? We'll start with the sequence of stages that occur at the surface to form arches, and later we'll dig deeper into the underlying controls.

The process begins with the broad slab of Entrada Sandstone and its three internal members sandwiched between other layers of Mesozoic redrock. Sometime in the geologic past, large-scale stresses acted upon this stiff layer of rock, flexing it to produce parallel vertical *joints* that penetrate through the horizontal beds of Entrada. (Jointing was introduced in the chapter on Zion.)

With tectonic uplift of the Colorado Plateau and erosion of overlying formations, the Entrada was exposed to the atmosphere and the unremitting effects of weathering and erosion. Water from rainfall or snowmelt infiltrated deeply into the rock via the parallel joints, slowly dissolving the natural cement holding sand grains together and decomposing the rock along the vertical fractures. Frost-wedging (as at Bryce National Park) forced apart the rock on either side of the joint plane, further weakening the rock. With time, the decomposing rock along the vertical joints was eroded away to create a narrow slot between adjacent remnant vertical slabs of Entrada. At Arches, slots developed along joint sets to form a distinct, parallel pattern with a northwest–southeast orientation (Fig. 7.5). The elongate, linear walls of rock between erosional slots are locally called "fins" and exhibit a variety of shapes and sizes across the Arches landscape.

Coincident with the preferential weathering and erosion along vertical joint planes, water also acts to break down the rock along the horizontal bedding contact between the Dewey Bridge and overlying Slick Rock members. The Slick Rock is a permeable sandstone, meaning that water can slowly percolate through tiny pores in the rock, as well as more rapidly along the vertical joint planes. In contrast, the thin clay- and silt-rich beds of the Dewey

Figure 7.6 Small alcoves forming along the contact between the Dewey Bridge Member and overlying Slick Rock Member. Contact marked by dark line. Outcrop along Route 313 near north entrance to nearby Canyonlands National Park.

Bridge are relatively impermeable and less jointed, so when rainwater or snowmelt migrating downward through the Slick Rock hits the uppermost Dewey Bridge, the water travels horizontally along the bedding plane, between the two members. The water eventually seeps outward onto the ground when it encounters the exposed face of the rock. Over time, seeps and springs of mildly acidic water decompose the rock along the contact between the two units, creating loose sand and silt grains and weakening the rock. The water seeping out along the surface tends to evaporate quickly in the intense aridity, but it is resupplied during the next rainstorm or spring snowmelt. (A similar scenario of water movement was described earlier for the Navajo–Kayenta contact in Zion National Park.)

As the rock of the Dewey Bridge member breaks down along the horizontal contact, the overlying massive sandstone of the Slick Rock loses its underlying support. Tabular slabs of rock may separate and episodically collapse as small rockfalls, enlarging the cavity along the contact and eventually widening to become a small *alcove* (Fig. 7.6). The alcove will become larger over time as blocks of Slick Rock detach along horizontal bedding planes or along fractures oriented subparallel to the upper arc of the alcove. The alcove may evolve into a freestanding arch as the vertical joint plane behind the alcove widens over time, isolating the wall of rock containing the arch from adjoining rock (Fig. 7.7).

This process is wonderfully illustrated by Landscape Arch, the longest arch in the park with an opening 93 m (306 ft) across (Fig. 7.8). This arch is formed in the Slick Rock Member, with the Dewey Bridge covered by soil and

Figure 7.5 Weathering and erosion along parallel joints in the Entrada Sandstone creates narrow slots separating elongate walls of rock oriented northwest–southeast. The lighter-colored jointed rock is the Moab Member, whereas the tan jointed rock to the southwest is the underlying Slick Rock Member. Field of view is about 3 x 4 km (2 x 2.5 mi).

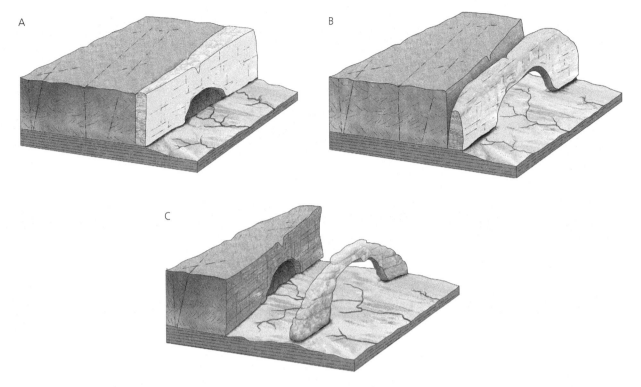

Figure 7.7 Sequential formation of natural arches. A) Creation of an alcove by localized weathering along a bedding plane. B) Separation of fin with alcove from adjoining wall of rock along widened joint plane. C) An arch is produced as the alcove enlarges and the adjacent rock face retreats via weathering and erosion, isolating the arch.

Figure 7.8 Landscape Arch, Arches National Park. Note the angular scars along the underside of the arch that mark the positions of slabs of rock that broke away. In time, more slabs of rock will break off, enlarging the arch. Eventually, the arch will collapse entirely. Landscape Arch is less than 2 m (6.6 ft) thick where it pinches toward the right side of the picture. The most recent rockfall from the underside of the arch happened in 1991.

collapse, reminding us of the continual change that characterizes the desert landscape at Arches.

Flowing Salt, Jointing, and Natural Arches

What is the ultimate control on the orientation of the joint system at Arches and why is there such a concentration of natural arches at the park? The answer begins about 300 million years ago during the late Paleozoic, a particularly dynamic phase of Earth history. The supercontinent Pangea was assembling, with huge mountains rising around the entire periphery of North America as plates converged along the margins. The ancient Appalachians rose to Himalayan heights during the late Paleozoic due to continental collision with northwest Africa. In the American West, the Ancestral Rockies formed rugged highlands composed of Precambrian crystalline basement. Massive continental ice sheets waxed and waned in the southern polar latitudes, causing sea level to fluctuate. During phases when the planet cooled, snow was supplied to the growing ice sheets by evaporation of water from the world ocean, causing sea level to fall by over 100 m (330 ft) or more. During phases of planetary warming, melting ice sheets sent the water back to the

talus along the ground surface. The fin in which Landscape Arch formed is a remnant of a larger parallel set of fins. Visualize this arch and the adjacent fins as once being part of a broad slab of Entrada that has been slowly dissected by the infiltration of water over perhaps a few tens of thousands of years to produce the landscape we enjoy today. The inescapable effects of erosion rule in the long run, however, resulting in the eventual collapse of arches to leave behind isolated columns and walls of rock. A particularly famous example is Balanced Rock, marked by a massive block of Slick Rock sandstone tenuously perched on a narrow neck of thin-bedded Dewey Bridge (Fig. 7.9). Sometime in the future, perhaps quietly at night or maybe in full view of a hundred tourists, Balanced Rock will

Figure 7.9 Balanced Rock, an erosional remnant of a collapsed arch or wall. A large block of the Slick Rock Member rests above a pedestal of thin-bedded Dewey Bridge Member.

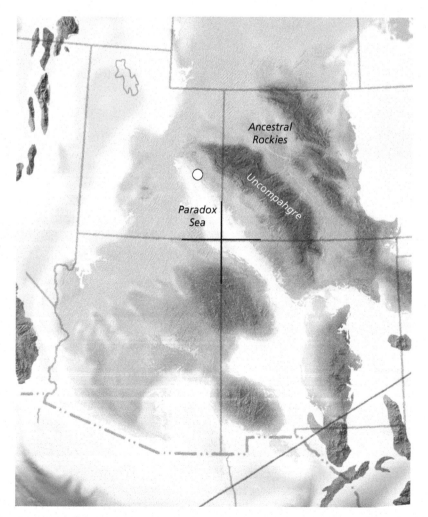

Figure 7.10 Paleogeographic map of future Colorado Plateau showing the restricted, evaporative Paradox Sea cutting northwest across the Four Corners region (308 Ma). Circle marks modern location of Arches National Park. Light blue shades represent shallow seas, whereas the shades of gray and brown represent exposed land of various elevations.

ocean basins, causing sea level to rise. These natural climatic variations and associated sea-level changes occurred over scales of hundreds of thousands of years, repeating in multiple cycles over a few tens of millions of years during the late Paleozoic.

To the southwest of the Ancestral Rockies, an elongate sag formed and was intermittently flooded by seawater leaking in from an extension of the ocean running through the middle of New Mexico (Fig. 7.10). This restricted basin is called the **Paradox Sea** and might best be visualized as comparable to the modern Caspian Sea and its semiarid climate. The Paradox Sea was located at about 10°N latitude and the regional climate during the late Paleozoic was hot and dry.

Glacially induced sea-level change caused the Paradox Sea to increase and decrease in areal extent repeatedly over a few million years. During phases of low sea level, the outlet to the open ocean would close. Intense evaporation in the semiarid climate caused the enclosed sea to shrink, resulting in the precipitation of thick layers of salt on the bed of the desiccating sea. The salt that crystallized was mostly **halite**, NaCl, along with appreciable amounts of sulfate and potassium salts.

During phases of high sea level, the Paradox Basin was replenished with water entering through the outlet to the open ocean on the southeast. A diversity of marine organisms proliferated in the shallow tropical sea. Over time, thin layers of limestone formed on the seabed from the accumulated remains of calcifying invertebrates. Millions of years of repeated sea-level fluctuations resulted in alternating salt deposits and limestones of the **Paradox Formation**, which built to a cumulative thickness of over a kilometer. The thickest mass of sedimentary deposits of the Paradox trended northwest–southeast beneath the future location of Arches. Through the rest of the Paleozoic, the Paradox Sea was gradually filled by clastic sediment derived from the eroding Ancestral Rockies. Finally, these late Paleozoic rocks of the Paradox Sea were buried by a broad blanket of Mesozoic and Cenozoic sedimentary rocks, including the Entrada (Fig. 7.11A).

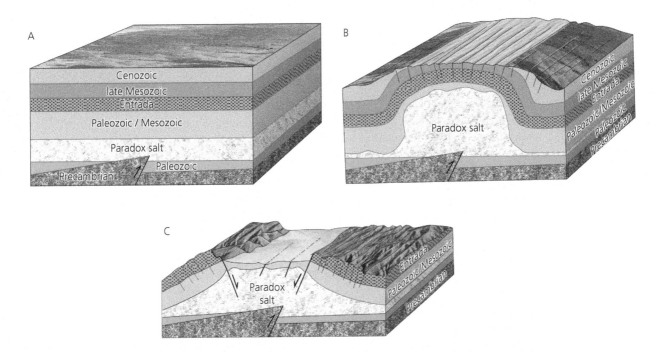

Figure 7.11 Sequence of events leading to concentration of natural arches at Arches National Park. A) Paradox salt deposits are buried by ~2 km (1.2 mi) of younger rock. B) The weight of overlying rock causes the lower density salt to flow like a thick plastic, forming broad walls of salt. Flexure of the stiff overlying rock creates northwest–southeast oriented joint sets. C) Dissolution of the uppermost salt by groundwater leaves the overlying rock unsupported, causing it to collapse along normal faults and creating elongate valleys.

This is where the story gets a little weird. When deeply buried beneath the weight of overlying sediment, salt will behave plastically and will slowly flow like thick taffy. The salt is less dense than the overlying rock and thus flows both laterally and vertically as a buoyant mass (Fig. 7.11B). The salt flows away from areas of high pressure where the overlying rock is thickest and toward lower pressure areas where the overlying rock is thinnest. The orientation of deeply buried faults may also have contributed to the direction of salt movement. This plastic flow results in bulges of salt pushing upward on overlying layers of rock, flexing the brittle slab into a broad, convex-upward fold. The bending of the rigid sedimentary rock results in joint sets that form in a preferred orientation. (Gently flex a chocolate-covered candy bar. The rigid chocolate crust will break as near-parallel cracks, whereas the gooey caramel filling will flow.)

Beneath the surface near Arches, salt has flowed to form **salt walls**, thick accumulations of mobile salt, oriented northwest–southeast, the predominant joint orientation in the overlying Entrada (Fig. 7.5). Exploratory drilling in the region by oil companies has determined that the salt wall beneath the Salt Valley in Arches is over 3 km thick, 5 km wide, and over 100 km long (2 x 3 x 60 mi). The salt layer on either side of the wall is very thin to absent, reflecting lateral flow to form the wall. Other salt walls occur beneath the nearby Moab and Castle valleys just outside the park.

Why do topographic valleys overlie the elongate salt walls? The uppermost surface of the salt wall is about 120 m (400 ft) beneath the ground, but freshwater from rainfall and snowmelt has penetrated downward through the joint system to dissolve the highest levels of salt near the top of the wall. The dissolution of salt and its removal within groundwater left the overlying sedimentary rock unsupported, causing it to collapse along normal faults (Chapter 5). The end result was an elongate valley directly above a salt wall (Fig. 7.11C). These elongate basins have been filled with late Cenozoic sediment, but the uplifted flanks of the valleys still preserve abundant exposures of jointed Entrada rock. So it's not surprising that some of the highest concentrations of natural arches occur along the uplifted margins of the fault valleys within Arches National Park. Famous arches such as Delicate Arch, Landscape Arch, and the dense array of fins and arches at Fiery Furnace, Devils Garden, and the Windows are aligned along the elevated edges of Salt and Cache valleys (Fig. 7.2).

Thus, the concentration of arches at the park is controlled by the plastic flow of Paleozoic salt beneath the surface that pushed up overlying layers of younger rock. The resulting flexural jointing in the Entrada exposed at the surface focused the relentless effects of weathering and erosion by water to create the array of over 2000 documented arches that we see in the modern Arches landscape. Isolated natural arches exist in nearby Canyonlands

National Park, as well as at Zion, Capitol Reef, and Bryce national parks where they are primarily controlled by weathering and erosion along joint planes. The localized thickening of salt beneath the Arches region, however, enhanced the development of closely spaced joints and the resulting cluster of natural arches.

Arches National Park is a beloved and special place, but it's not that big. Traffic can back up at the lone entrance, and parking spaces can be limited once one is in the park, frustrating the visitor who would rather commune with nature. The mobility that cars, RVs, and tourist buses permit allows everyone to access our national parks, which is in general a very good thing. But that easy mobility fuels overcrowding in some of our more popular national parks like Arches. The National Park Service is considering ways to alleviate the impact of too many people, such as a reservation system and increases in admissions fees. Until then, travel with friends or family, take a shuttle bus if one is available, or go during the off-season. (The busiest months are between May and October.) Once you're in Arches, get out of your car and wander the sandy trails, hike the slickrock, slip through the narrow slots between fins, and take in the superb panorama of the nearby La Sal Mountains, a resistant mass of igneous rock. Away from the roads and parking lots, the numbers of people decrease while the vistas become even more inspirational.

Key Terms

halite
natural arch
salt wall

Related Sources

Abbey, Edward. (1971). *Desert Solitaire: A Season in the Wilderness*. New York: Ballantine Books, 337 p.

Doelling, H. H. (2010). *Geology of Arches National Park, Grand County, Utah*. In P. B. Anderson, T. C. Chidsey Jr., and D. A. Sprinkel (Eds.), *Geology of Utah's Parks and Monuments*, 3rd ed. Salt Lake City, Utah Geological Association.

Graham, J. (2004). *Arches National Park Geologic Resource Evaluation Report*. NPS D-197. Denver, CO: National Park Service.

National Park Service—Geology of Arches National Park. Accessed January 2021 at https://www.nps.gov/arch/learn/nature/geologicformations.htm

8 Canyonlands National Park

Figure 8.1 Erosional landscape of Canyonlands National Park. Silhouette of the igneous La Sal Mountains in the background to the east.

The wilderness in Canyonlands National Park is remote, unworldly quiet, and packed with surreal scenery (Fig. 8.1). As opposed to the justifiable popularity of nearby Arches National Park, Canyonlands receives many fewer visitors per year. Some tourists are understandably content to simply gaze down at the sublime vistas from nearby Dead Horse Point State Park (Fig. 2.2). But if you have the proper spirit and common sense, drive the backroads and hike the trails to gain the full Canyonlands experience.

The **confluence** of the Colorado and Green rivers occurs in the heart of Canyonlands (Fig. 8.2). The Colorado meanders through a deep canyon in a generally south–southwest direction, while the Green flows southward from its headwaters in Wyoming (Fig. 2.4). After the rivers merge, their combined flow generates 23 km (14 mi) of world-class rapids through a tight gorge known as Cataract Canyon. The rapids are located where debris flows have dumped boulders of all shapes and sizes at the mouths of tributaries, enhancing turbulent flow as well as the pulse rate of whitewater rafters and kayakers. (*Debris flows* are dense slurries of watery sediment ranging in size from boulders to clay that are triggered by summer cloudbursts. See Chapter 3 on the Grand Canyon.)

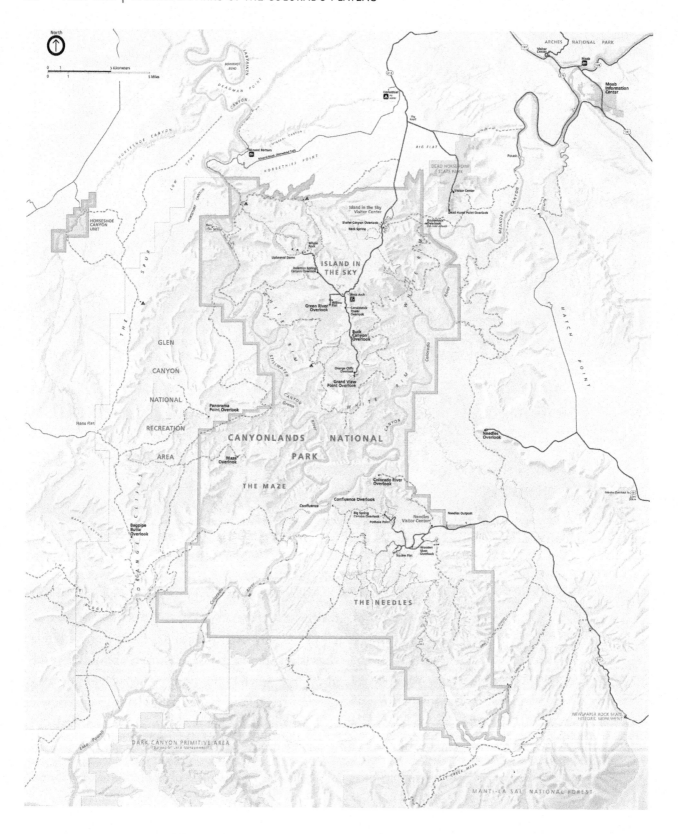

Figure 8.2   Map of Canyonlands National Park.

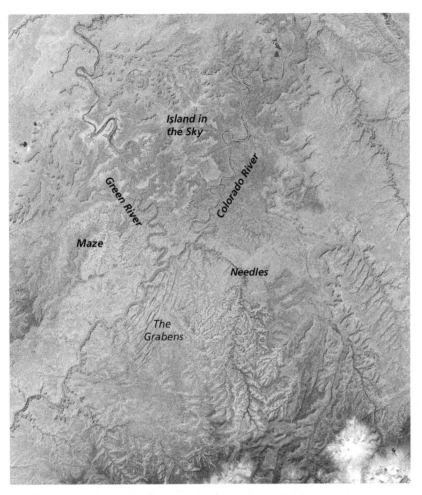

Figure 8.3 Satellite image of Canyonlands National Park illustrating the three main divisions of the park, each bound by the widely meandering channels of the Colorado and Green rivers.

art, the petroglyphs scratched into *rock varnish* at Newspaper Rock, are located outside the park boundary but are easily accessible via the road leading into the Needles District.

Closer to the present, humans have used the resources of the area for a very different purpose than did the Native Americans. Although the magnificent Canyonlands landscape appears to be worth nothing more than our aesthetic appreciation, riches hide within the rocks in the form of uranium ore. Many of the dirt roads that switchback down impossibly steep cliff faces or cross remote areas in the park were cut during the Cold War search for the raw material for atomic weaponry. The prospecting peaked in the 1970s, leaving behind leftover piles of mine waste at more than 30 abandoned mine openings inside the park boundaries. Gravelly mounds of tailings—the ground-up remnants of processing for uranium ore—are mildly radioactive and can potentially contaminate groundwater or be blown around by the wind. Remediation of these sites by the federal government is an ongoing process.

Similar to all the other parks of the Colorado Plateau, rock is the raw material on which time and water have created their magical artistry in Canyonlands. There are more lurid shades of red and orange than can possibly be imagined, all due to tiny amounts of iron oxides coating grains of quartz as well as cementing those grains together to create sedimentary rock. The interaction of differential weathering and erosion, headward erosion by streams, jointing, and salt movement deep beneath the surface have conspired to create a diverse spectrum of landforms at Canyonlands: deeply incised bedrock canyons featuring "gooseneck" meanders, the dissected remnants of plateaus, balanced rocks, linear fault valleys, and even an enigmatic circular structure whose origins are still being debated. The current landscape at Canyonlands reflects 300 million years of Earth history, culminating in the redrock wonderland that we enjoy today.

The converging rivers divide the park into three distinct sections. (1) The Island in the Sky is a triangular northern area bound by the two rivers (Fig. 8.3). This district is a deeply dissected tableland whose relatively flat upper surface resides at an elevation of ~2000 m (6600 ft), which is about 600 m (2000 ft) above river level. (2) The Needles occupies the area east of the Colorado River and is notable for a diverse array of colorful landforms, including The Grabens, a series of linear valleys and ridges formed by the migration of salt deposits beneath the surface. (3) The Maze forms the region west of the Green River and its confluence with the Colorado. As the name implies, this remote district is a jumble of bizarre erosional landforms, labyrinthine corridors, and slot canyons.

A fourth district in the park, Horseshoe Canyon, is physically separate from the rest of the park and is notable for its remarkable rock art. Bands of hunter-gatherers roamed the American West for thousands of years, recording stories of their exploits and belief system on cliff walls in Canyonlands. Some of the earliest American art is preserved in the Great Gallery of the Horseshoe Canyon district, ghostly images painted onto the rock over 3000 years ago (Fig. 8.4). Another remarkable example of rock

Origin and Age of Rocks in Canyonlands

The landscape at Canyonlands is composed of rock that ranges in age from late Paleozoic (Pennsylvanian Period) through middle Mesozoic (Jurassic Period). The future Colorado Plateau was near sea level through this interval,

Figure 8.4 A) Petroglyphs of Newspaper Rock. B) Rock art of the Great Gallery, Horseshoe Canyon district, painted by nomadic people over 3000 years ago.

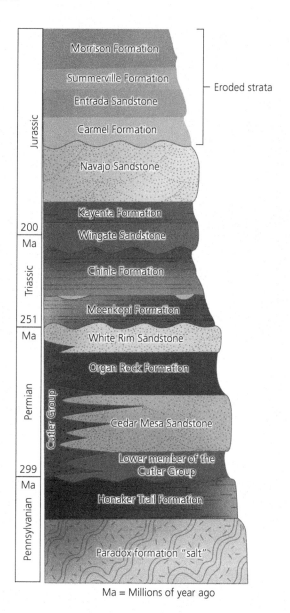

Ma = Millions of year ago

Figure 8.5 Graphic column of rocks at Canyonlands National Park.

recorded by the shallow water depositional environments of its rocks. The oldest unit in Canyonlands is the Paradox Formation, which played such an important role in the creation of arches at nearby Arches National Park (Fig. 8.5). Plastic movement of the Paradox salt has a less obvious influence in Canyonlands, but it does contribute to the creation of broad warps in the overlying rock as well as to the formation of the linear valleys and ridges known as The Grabens. A small plug of Paradox gypsum, a calcium sulfate mineral, is exposed at river level along the Colorado, but the Paradox remains deeply buried beneath younger rocks in the rest of the park. Remember that all of the older Paleozoic sedimentary rock and Precambrian basement forms the underlying crust beneath the Paradox deep below the surface.

The most widespread scenery-forming unit in Canyonlands is a complex mix of rock types called the **Cutler Group**. (A Group is simply a package of related formations.) This colorful late Paleozoic (Permian) unit is composed of interbedded shales, siltstones, sandstones, limestones, and conglomerates that in some places can be over 600 m (~2000 ft) thick. The Cutler Group is most easily identified by exposures of its resistant uppermost formation, the White Rim Sandstone, which tends to form a broad, flat, light-colored terrace viewable from many of the overlooks in the park. An older, resistant formation in the Cutler, the

Cedar Mesa Sandstone, commonly forms another broad terrace below that of the White Rim. The Cedar Mesa is widespread throughout the park, commonly forming picturesque landforms of banded reddish-orange to buff sandstone (Fig. 8.6).

The rock types in the Cutler are notable for their variability, both vertically and laterally across the park. This diversity reflects the broad array of closely spaced depositional environments in which the original sediments accumulated. During the late Paleozoic, North America was fully sutured to Africa, Eurasia, and South America along the young Appalachian mountain chain, collectively forming part of the Pangean supercontinent. Sea level fluctuated up and down on geologically rapid time scales of hundreds of thousands of years due to the waxing and waning of ice sheets located in the late Paleozoic Southern Hemisphere. In the Canyonlands area, depositional

Figure 8.6 Cedar Mesa Sandstone, showing interbedding of lighter-colored, cross-bedded units deposited as coastal sand dunes and reddish-orange beds deposited in streams that fed the Paleozoic Cedar Mesa desert.

Mesozoic redrocks above. The vibrant hues of the Moenkopi and Chinle formations that directly overlie the White Rim were deposited in coastal plain environments comparable to the other Mesozoic redrocks on the Colorado Plateau (Chapter 6). Above the slope-forming Chinle is the Wingate–Kayenta–Navajo triplet that should be familiar to you from Chapters 4 and 6, respectively, on Zion and Capitol Reef national parks. The Wingate is notable for forming steep cliffs covered by dark-brown desert varnish. The Navajo may form a thin caprock along the upper surface of many dissected plateaus in Canyonlands but is nowhere near as prominent as at Zion National Park. With Arches National Park located a short distance to the northeast, you'd think that the three members of the Entrada would dominate the surface of Canyonlands as well. But erosion has serendipitously stripped off the Entrada and younger formations in Canyonlands, leaving behind just a few isolated outcrops on the northern outskirts of the park.

environments migrated laterally in tune with the transgressing and regressing shoreline that extended from southern Nevada northeast across Utah.

A broad coastal desert system was the setting for the accumulation of windswept sand dunes that would become the Cedar Mesa and later the White Rim sandstones. Some of the Cedar Mesa sands were likely deposited as offshore barrier bars just seaward of the shoreline. Streams draining the nearby Ancestral Rockies to the east in Colorado fed sediment to the coastal plain, creating interbeds within the Cedar Mesa coastal sand dunes and offshore sandbars. The combination of rapidly changing sea level, global climatic shifts related to continent-scale glaciation, and the growth of massive mountain belts that supplied sediment created the variable conditions under which the Cutler Group was deposited.

These late Paleozoic rocks at Canyonlands are the lateral equivalents to the sandstones, shales, and limestones that form the uppermost cliffs at the Grand Canyon. Other than having unique formation names (a consequence of the original geologists who mapped these units), the only difference is that they represent distinct depositional environments spread out across the coastal region that would eventually become the Colorado Plateau. The Coconino, Toroweap, and Kaibab formations of the uppermost Grand Canyon were accumulating at essentially the same time as the sediments of the Cutler Group in contemporaneous environments a few hundred kilometers away. These formations grade laterally into one another in the intervening region between the two parks.

The broad, flat terrace formed by the White Rim Sandstone provides a convenient topographic boundary separating late Paleozoic rocks below from early to middle

Landscape Evolution in Canyonlands

DISSECTED PLATEAUS

The widespread extent of Chinle slopes and overlying Wingate cliffs throughout Canyonlands provides an excellent example of differential weathering and erosion, concepts introduced in earlier chapters that relate to the variable resistance of different rock types (Fig. 8.7). The relatively weak shales and siltstones of the Chinle weather easily, slowly decomposing to form a thin veneer of soil mantling a slope. In contrast, the massive, resistant Wingate Sandstone erodes episodically as weathering along vertical joints loosens slabs of rock that detach to form rockfalls. These abrupt, violent events are set up as the soft rocks of the underlying Chinle slope wear away beneath the cliff, leaving the thick slab of fractured rock above unsupported and prone to collapse. This process is called **cliff retreat** and is the fundamental reason for the ubiquitous stairstep topography that characterizes arid landscapes across the entire Colorado Plateau. The escarpments of the Grand Staircase, the cliff-slope alternations of the upper Grand Canyon, and the steep cliff faces of Navajo Sandstone at Zion are all examples of cliff retreat operating over long spans of time.

A related process that acts to dissect the tabular strata of plateaus is *headward erosion*, a process introduced in Chapter 5 on Bryce Canyon National Park. In Canyonlands, joints penetrating downward through the exposed

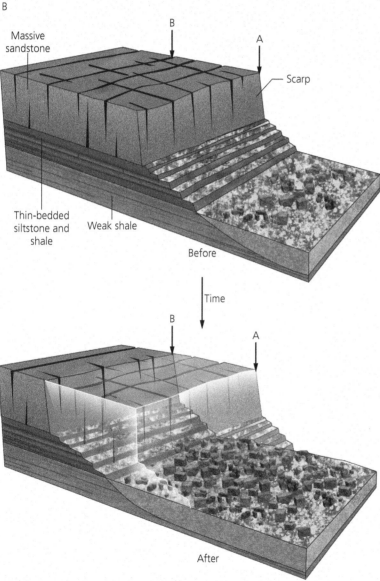

Figure 8.7   A) Vertical cliff of massive Wingate Sandstone overlain by bedded Kayenta Formation. The slope beneath the Wingate cliff is developed in the soft shales and siltstones of the Chinle Formation. B) Stages in the process of cliff retreat, a fundamental reason for the stairstep topography of cliffs and slopes that typifies arid-climate landscapes like those of the Colorado Plateau.

upper surface of the White Rim Sandstone provide conduits for water to slowly work its way into the rock. The subtle notches that develop by weathering along the upper parts of joint planes tend to focus flow across the relatively flat surface (Fig. 8.8A). As the joints deepen and widen through time, they funnel storm runoff toward the head of the canyon where the water may plunge over the edge as a temporary waterfall.

This focused erosion at the head of the canyon forms a **knickpoint**, an abrupt increase in slope that creates an acceleration in the flow of storm runoff. The concentration of weathering and erosion at the knickpoint causes the canyon to migrate headward back into the edge of the terrace (Fig. 8.8B). In this way, the plateaus are dissected into intricate patterns primarily determined by the original orientation of joint sets.

As time passes, the related processes of cliff retreat and headward erosion act to dismember the broad plateaus of horizontally bedded sedimentary rock. Canyons widen and plateaus reduce to **mesas**, which in turn evolve into smaller **buttes**, which themselves may narrow into **pinnacles** (Fig. 8.9). Ultimately, these narrow spires of remnant rock will collapse into a loose pile of rubble. Ironically, in the final stage, the remnant blocky debris is broken down to loose sediment, the original material that formed the rock long ago. In this incremental manner, rock is recycled back into sediment to be transported and redeposited, perhaps to become new rock in the future. It's likely that the sediments composing the rocks of the Colorado Plateau have been recycled many times throughout Earth history, with the current version simply being a temporary stage along the way.

INCISED MEANDERS

The Colorado and Green rivers crossing through the Canyonlands region are confined to narrow, meandering, incised canyons cut into bedrock (Fig. 8.3). The sinuosity of the channels is striking, including some that almost double back on themselves; this riverine feature is informally called a **gooseneck** (Fig. 8.10). A **meandering channel**

Figure 8.8 A) Shallow notch at top of joint plane that focuses runoff at the edge of a plateau. A knickpoint will form where the notch drops off into the canyon below, initiating headward erosion. B) Head of box canyon dissecting the edge of a terrace capped by White Rim Sandstone. The shallow washes draining the terrace into the canyon focus erosion at the knickpoint at the tip of the canyon, causing the canyon to grow into the terrace in a headward direction through time.

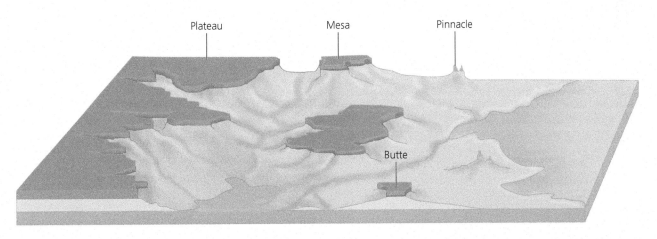

Figure 8.9 Arid-climate landforms produced by differential erosion, cliff retreat, and headward erosion acting over time.

is typical of lazy rivers that wander across a relatively gentle slope mantled by a veneer of soft, unconsolidated sediment, such as the Mississippi and similar rivers that drain the Gulf Coast of the United States. Rivers flowing along low **gradients** naturally form meanders as slightly faster-moving water on the outside of bends erodes the bank, extending the bend outward. Slower-moving water on the inside of the bend tends to deposit sediment, producing a growing gravel bar along the inner bank. These complementary processes—erosion along the outer bend and deposition along the inner bend—propagate the meandering sinuosity that is so common

to rivers flowing along low slopes. If meandering patterns are common to low-gradient coastal rivers, then why are they present on the highlands of the Colorado Plateau, elevated 2–3 km (1.2-1.8 mi) above sea level?

The formation of **incised meanders** at Canyonlands and many other places on the Colorado Plateau began during the late Mesozoic–early Cenozoic *Laramide orogeny* (introduced in Chapter 6). This major uplift event raised the Colorado Plateau, bringing rocks that were deposited near sea level up to a few kilometers in elevation. The highlands of the plateau formed a broad intermontane basin between surrounding mountains in Nevada

Figure 8.10 "Gooseneck" in the Green River formed during uplift of the plateau. The tightly sinuous meander loop was part of the original river pattern that was maintained as the land rose beneath.

and Colorado (Fig. 5.3). Rivers draining these mountains transported sediment onto the youthful Colorado Plateau where they were deposited as a broad layer of sand and gravel along floodplains and within lakes. The ancestors of the Colorado and Green river drainage systems migrated laterally across the relatively flat landscape, establishing meandering patterns on wide floodplains (Fig. 8.11).

Uplift of the entire plateau was rejuvenated in the late Cenozoic (~5-6 Ma), raising the region and its drainage system a kilometer or more above sea level. Instead of flowing into nearby lakes, the Colorado River and its

countless tributaries now had a new, lower elevation to empty into at the Gulf of California off to the southwest. The higher gradient triggered an increase in flow velocity, which in turn increased the capability of rivers and streams to incise downward through the overlying layers of unconsolidated sediment, then into the rock beneath. Once entrenched in the solid rock, however, the streams ceased their lateral motion, instead cutting vertically to create steep-walled canyons while maintaining their meandering pattern. The incised meanders of the Colorado and Green rivers in Canyonlands, along with their numerous tributaries, were thus formed by *superposed streams*. Recall from Chapter 6 on Capitol Reef National Park that superposed streams are those that were originally established on low-sloping landscapes and were later incised downward into the underlying rock beneath in response to tectonic uplift and the associated increase in gradient. The connection between superposed streams and incised meanders is widespread across the Colorado Plateau, related to the plateau's history of uplift and the Colorado River and its network of tributaries that flow across the province as the water drains toward the sea.

THE GRABENS

Other landscape features at Canyonlands are comparable to those of nearby Arches National Park in that they relate to the plastic migration of Paradox salt beneath the surface. Concentrations of near-parallel elongate fins of rock separated by slot canyons are common, and, as you might expect, 25 arches are found in Canyonlands, developed within the sandstones of the Cutler Group. In the Needles district, joint sets that intersect at sharp angles produce rectilinear blocks of rock that weather into dense fields of needles and pillars composed of colorfully banded Cedar Mesa Sandstone (Fig. 8.12).

In a region called The Grabens, elongate valleys and flat-topped ridges are aligned near-parallel to the Colorado River south of the confluence with the Green. This landscape is related to movement along *normal faults* (introduced in Chapter 5 on Bryce Canyon), with crustal blocks episodically collapsing downward between adjacent blocks along the fault plane. Downdropped blocks of rock are called **graben**, a German word referring to an elongate linear fault valley. Upraised blocks are called **horsts**, another German word that refers to the fault blocks that form elongate ridges between valleys.

Normal faulting is related to extensional stress, a stretching of the crust typically caused by tectonic forces. In the case of The Grabens, however, the extension is due to the lateral migration of Paradox salt about a half kilometer

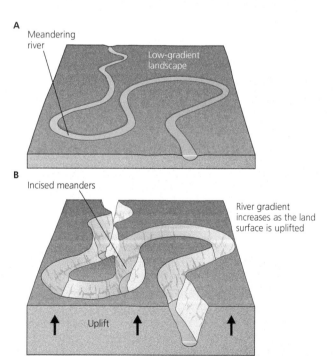

Figure 8.11 Development of an incised meander. A) Rivers tend to form sinuous channels as they flow across low gradient surfaces. B) With uplift, the gradient increases, triggering streams and rivers to incise downward into rock.

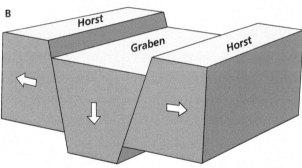

Figure 8.12 A) Dense concentration of blocks formed along intersecting joint sets in the Cedar Mesa Sandstone. The Grabens are a series of elongate, fault-bounded valleys and ridges formed in response to the movement of salt about a half-kilometer beneath the surface. The valleys are about 10–15 m (30–45 ft) deep. B) Graphic of horsts and graben formed by extension-related normal faulting.

beneath the surface (Fig. 8.13). The salt slowly flows toward the Colorado River canyon where the load from overlying rock was removed by river incision, alleviating the pressure bearing down on the salt layer. A gentle westward tilt of the rocks helps to direct the gravitational flow of salt toward the river corridor. It's somewhat analogous to peanut butter oozing out of your sandwich as you lift and squeeze it on the way to your mouth.

As the Colorado River incised progressively downward toward the salt layer, the overlying rigid slab of rock

slid westward toward the canyon due to the loss of lateral constraint. This extension of the half-kilometer thick slab of rock caused it to break along normal faults, resulting in the linear **horst-and-graben** topography. The dissolution of salt by groundwater aids in triggering fault motion by creating voids that the overlying rock can collapse into. The westward movement of the rock above the mobile salt layer constricts Cataract Canyon, contributing to the exhilarating set of rapids within the narrow gorge.

UPHEAVAL DOME

One of the most unique and enigmatic landscape features in all of the national parks is Upheaval Dome, located in the Island in the Sky district. On satellite images, the 5.5-km-wide (3.4 mi) bullseye clearly stands out against the intricate, water-sculpted textures of Canyonlands (Fig. 8.14). Up close, Upheaval Dome is composed of a chaotic nucleus of colorful Chinle Formation, surrounded by concentric rings of progressively younger Mesozoic rock. Whitish patches of Navajo Sandstone mark the youngest rock of the outermost rim. Each of these formations tilts radially away from the central core of deformed Chinle. There are two competing explanations for the origin of the feature.

Because of the presence of Paradox salt in the region and its inherent mobility, Upheaval Dome was initially interpreted as a salt-related structure. Under the right conditions, salt will migrate upward as a vertical column. Layers of overlying rock bow upward and outward as salt progressively pierces through each layer. When the salt nears the surface, the overlying rock will form a domal topographic bulge. Dissolution of the salt by fresh surface waters results in an eroded concentric circle of rock as found at Upheaval Dome. **Salt domes** are common features along the Gulf Coast of the United States where they are mined for table salt and, when hollowed out, are used as storage facilities for oil. A recent updating of this "salt piercement" model suggests that the salt dome itself and much of the overlying rock were entirely removed by erosion, leaving Upheaval Dome as the remnant of the thin salt stem where it pinched off from the domal bulge above.

The alternative explanation for Upheaval Dome was proposed in the post-Apollo years when it became evident that many circular features on Earth (as well as on the Moon and other terrestrial planets) were **impact craters**, the scars left behind by the ultraviolent collision of meteors. With the careful observation and modeling that characterizes all the geosciences, it was suggested that Upheaval Dome was the deeply eroded scar left behind by a meteor, perhaps 100–300 m (330–1000 ft) across, that hit sometime between 60 and 160 million years ago.

Most meteors that survive the fiery passage through Earth's atmosphere are rocky or iron-rich fragments of asteroids. They travel at velocities of tens of thousands of kilometers per hour, so that at impact their

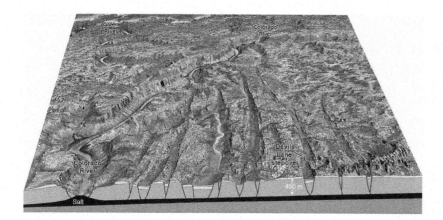

Figure 8.13 The lateral migration of Paradox salt westward toward the Colorado River corridor creates extensional stress in the overlying rock. The Grabens landscape developed as rigid blocks of rock dropped downward along normal faults as salt flowed beneath.

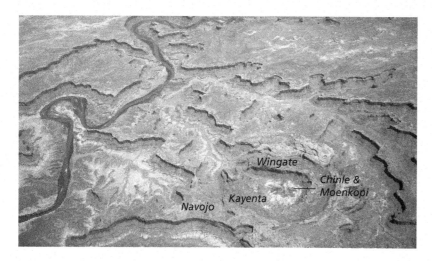

Figure 8.14 Oblique view of Upheaval Dome, with surrounding formations labeled.

outward catastrophically as fragmentary debris, leaving behind the bowl-shaped crater surrounded by an irregular ring of ejected rock.

In this impact model, Upheaval Dome is inferred to be the deep remnant of the rebounding shock wave exposed after the crater and most of the surface layers were eroded away. Some quartz grains show microscopic evidence of having been "shocked" by the intense pressures associated with meteor impact, but for now the final answer remains elusive. What we are sure of is that Upheaval Dome adds to the wide diversity of surface features that collectively create the soul-stirring landscape that is Canyonlands National Park.

Key Terms

butte
cliff retreat
confluence
gooseneck
gradient
horst-and-graben
impact crater
incised meander
knickpoint
meandering channel
mesa
pinnacle
salt dome

Related Sources

Baars, D. L. (2010). *Geology of Canyonlands National Park, Utah.* In P. B. Anderson, T. C. Chidsey Jr., and D. A. Sprinkel (Eds.), *Geology of Utah's Parks and Monuments*, 3rd ed. Salt Lake City, Utah Geological Association.

KellerLynn, K. (2005). *Canyonlands National Park Geologic Resource Evaluation Report.* Natural Resource Report NPS/NRPC/GRD/NRR—2005/003. Denver, CO: National Park Service.

Lohman, S. W. (1974). The Geologic Story of Canyonlands National Park. *U.S. Geological Survey Bulletin* 1327, 126 p.

National Park Service—Geology of Canyonlands National Park. Accessed January 2021 at https://www.nps.gov/cany/learn/nature/geologicformations.htm

kinetic energy is transferred into the planetary surface as tremendous amounts of heat and pressure. Rock at the surface is violently compressed and forms a dense wall of energy that penetrates downward into the Earth as a shock wave. This wave of highly compressed and deformed crustal rock eventually reaches a depth where the density of rock is high enough to cause the shock wave to reflect back toward the surface. This rebounding energy has the effect of bowing up deep layers of rock before blasting the surficial layers

9

Dinosaur National Monument

Figure 9.1 Green River flowing through Whirlpool Canyon in Dinosaur National Monument.

Dinosaur National Monument is far more than just a spectacular display of huge dinosaur bones spread out across a tilted slab of rock within the Quarry Visitor Center. The monument also has two large rivers entrenched in deep canyons, a stack of mostly sedimentary rocks that span over two billion years of time, and an austere desert landscape populated by a surprising diversity of wildlife (Fig. 9.1). Many people skip a visit ("oh, it's just a monument") to this monument located along the boundary of northeastern Utah and northwestern Colorado and instead spend time at more easily accessible national parks. Of course, the dinosaur exhibits are fascinating, but one should take the time to float the rivers, explore the backcountry, and learn the geology—it'll be time well spent.

Incidentally, there's very little difference between a national monument and a national park. As was discussed in Chapter 5 on Grand Staircase-Escalante National Monument, the difference is not, as many believe, that monuments have less inherent natural beauty than parks. Rather, the difference lies simply in the political process involved in their formation; as mentioned earlier, monuments can be established by presidential declaration without congressional approval, whereas a national park is created by an act of Congress. Presidents may designate regions with a unique beauty or natural feature as a national monument using the 1906 Antiquities Act, signed into law by President Theodore Roosevelt. The Bureau of Land Management administers some national monuments, but the National Park Service operates Dinosaur, one of 108 monuments in the U.S. National Park System. In order to protect the extraordinary fossil resources, Dinosaur was named a monument in 1915 and its area was expanded in 1938, but the surrounding landscape is full-on national park quality.

Dinosaur National Monument is located along the southeast corner of the Uinta Mountains, which extend across the northeastern border of Utah (Fig. 9.2). The Uintas are part of the Rocky Mountain geologic province and are notable for being the highest range in the contiguous United States with an east–west orientation (Fig. 2.3). Anyone who has looked at a physiographic map of North America recognizes that the predominant texture of mountainous topography in the West is slightly northwest–southeast, so the discordant trend of the Uintas readily stands out. The rise of the Uintas occurred during the Laramide phase of mountain-building (introduced in Chapters 7 and 8 on Capitol Reef and Canyonlands, respectively), and the uplift history of the range plays an important role in the development of landscape features at Dinosaur.

Dinosaur National Monument is situated at the transition zone where the northeastern boundary of the Colorado Plateau merges with the Rocky Mountain Province. The rocks that compose the landscape of the monument are essentially the same as those elsewhere on the Colorado Plateau, so Dinosaur is included within Part 1 of this book. The rocks are not horizontal, tabular layers as they are at other parks on the plateau, however, but are instead tilted, folded, and faulted. The deformation of the rocks is associated with uplift of the Uinta Mountains, so the landscape at Dinosaur reflects its location along the transition between the two provinces. Dinosaur could just as easily have been included within Part 4, which covers the national parks of the Rocky Mountains.

The Green and Yampa rivers merge within the national monument. The Green River originates in the Wind

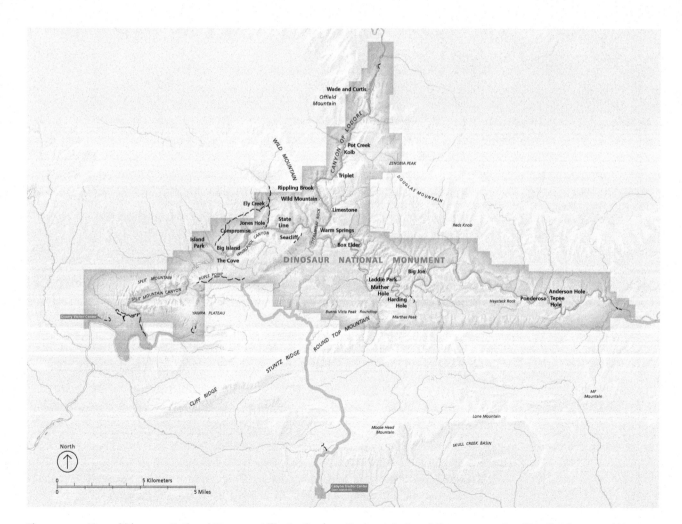

Figure 9.2 Map of Dinosaur National Monument illustrating how the boundaries of the monument outline the Green and Yampa rivers.

Figure 9.3 Green River watershed (light green) illustrating the broad bend in the river around the east nose of the Uinta Mountains and the confluence with the Yampa River in Dinosaur National Monument.

River Range of western Wyoming and then flows southerly across the vast, windswept Green River Basin before backing up in Flaming Gorge reservoir near the Utah state line (Fig. 9.3). The dam in northernmost Utah was completed in the early 1960s as a trade-off by conservationists who fought a dam proposed for Whirlpool Canyon within the monument boundaries (Fig. 9.1). Below the dam, the Green River swings easterly before abruptly cutting south across the eastern nose of the Uinta Mountains within the imposing Canyon of Lodore. In places, the canyon is almost a kilometer deep, with precipitous walls lining the narrow chasm.

The Green River flows relatively sediment-free through the gorge since most of the sand and silt is trapped within Flaming Gorge Reservoir upstream. With stretches of the river named Upper Disaster Falls and Hells Half Mile, the rapids within the tightly constrained canyon provide an adrenaline-charged experience for lovers of whitewater. The west-flowing Yampa River joins the south-flowing Green River inside the monument (Fig. 9.4), with the commingled waters tracing a sinuous path westward through Split Mountain Canyon. The

Yampa is the only undammed major river within the Colorado River watershed and thus flows brown, turbid with sediment.

Many of the same processes that were addressed in earlier sections in this chapter are expressed in Dinosaur National Monument. Weathering and erosion are dominant here, as they are in the other parks of the Colorado Plateau, and all of the familiar erosional landforms are on display. In keeping with the format used in previous chapters, we'll first introduce the rocks and the associated environments in which the sediments were originally deposited. We'll touch on the process of fossilization as well as the phenomenal record of dinosaur life preserved in the Morrison Formation. Finally, we'll focus on the creation of the landscape at Dinosaur, concentrating on the critical processes that control canyon incision.

Origin and Age of Rocks at Dinosaur

The span of sedimentary rocks exposed at Dinosaur National Monument rivals that of the Grand Canyon, ranging in age from Precambrian to middle Cenozoic. The Precambrian **Uinta Mountain Group** is as old as 800 Ma and form imposing reddish cliffs along the Canyon of Lodore. The rocks are mostly tightly cemented sandstones, with lesser amounts of pebble conglomerate and shale, which are

Figure 9.4 The relatively sediment-free Green River merges with the muddy Yampa River at the base of Steamboat Rock in Echo Park. The angular summit of Harpers Corner looms in the background.

likely deposited by sediment-choked rivers and streams. (If necessary, see Chapter 1 for a review of clastic sedimentary rocks.) The base of the Uinta Mountain Group is not exposed in the monument, but the total thickness of the unit is well over 4.5 km (15,000 ft).

The age of the Uinta Mountain Group in Dinosaur National Monument indicates that they are time-equivalent to the youngest rocks of the Grand Canyon Supergroup. Similar to the Precambrian sedimentary rocks of the Grand Canyon, the tremendous thickness and limited areal extent suggest that the Uinta Mountain Group was rapidly deposited within a narrow, elongate **rift basin** that extended landward into the continent from an ancient Precambrian ocean. These rapidly subsiding, fault-bounded basins are likely related to the earliest phases of *continental rifting* (Chapter 1) and mark the initial breakup of a late Precambrian supercontinent and the very beginnings of Paleozoic North America. This major paleotectonic event will be addressed in Chapter 22 on Glacier National Park.

Many of the Paleozoic rock units above the Uinta Mountain Group were addressed in previous chapters, reflecting the widespread extent of these sedimentary layers across the American West. The different formation names of rock units in Dinosaur simply record the whims of the geologists who originally mapped and named them in the early 20th century, unaware that they were the lateral equivalent of strata elsewhere in the American West. Many of the Paleozoic formations in Dinosaur National Monument are comparable in age and composition to those in the upper Grand Canyon. For example, the Cambrian Lodore Sandstone in Dinosaur represents the same transgressive shoreline deposit as the Tapeats Sandstone. The cliff-forming Madison Limestone in Dinosaur accumulated in the same widespread, shallow tropical sea as the Redwall Limestone of the Grand Canyon. The wind-deposited Weber Sandstone that forms massive sheer cliffs along the Yampa River (Fig. 9.5) was laid down

Figure 9.6 Outcrop of multicolored Morrison Formation, Dinosaur National Monument.

contemporaneously with the Supai Group in the Grand Canyon, though in different depositional environments.

In a similar fashion, most of the Mesozoic formations in Dinosaur are lateral and temporal equivalents to formations addressed in earlier chapters. As an example, the thick, cross-bedded Nugget Sandstone at Dinosaur is simply the combined Wingate, Kayenta, and Navajo formations described earlier from Zion, Capitol Reef, and Canyonlands national parks. The broad areal extent of these Paleozoic and Mesozoic formations across the American West illustrates the **principle of lateral continuity**, which posits that layers of sediment were originally deposited as widespread and continuous bodies across their areal extent. Sedimentary layers may be subsequently deformed during mountain-building, or they may be erosionally destroyed in places, but they were originally continuous in their depositional extent. We'll expand on this principle in later chapters.

The most important rock unit exposed in Dinosaur National Monument is the world-famous **Morrison Formation** because of its abundant and diverse array of dinosaur fossils (Fig. 9.6). The Morrison is also well known as a source of uranium ore, with much of U.S. production derived from the Morrison throughout the American West. The multicolored rocks composing the Morrison are dominated by thick intervals of gray, pink, and maroon **mudstones** interspersed with thinner beds of sandstone or conglomerate. A mudstone is similar to shale in that both are composed of clay- and silt-sized particles (i.e., "mud"). But mudstones tend to break apart as blocky fragments, as opposed to the flat chips and plates that characterize shale.

The Morrison was deposited between 155 and 148 Ma (late Jurassic) in a variety of depositional environments across a broad expanse of the American West, extending from Utah and Colorado up through Wyoming and Montana. The muddy sediments that dominate the Morrison accumulated in continental settings such as sluggish river

Figure 9.5 Exposures of whitish, cliff-forming Weber Sandstone along the Yampa River near Castle Park.

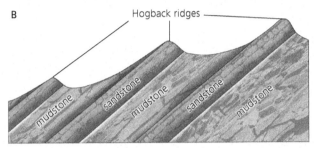

Figure 9.7 A) Highly tilted beds of the Morrison Formation near the Quarry Visitor Center. Less resistant mudstones form poorly vegetated slopes, whereas thinner, resistant sandstones stick out as hogback ridges. Tilt of beds is downward to the left. B) Diagram of hogback ridges formed in the Morrison.

channels and broad floodplains, as well as quiet lakes and ponds. Sandstones and conglomerates interbedded with muddy strata likely record episodes when rivers in flood spilled over their banks, depositing a lobe of coarser sediment on the floodplain. Current indicators preserved in the sandstones (e.g., small-scale cross-beds) suggest that streams flowed toward the northeast, likely from highlands in Nevada and Arizona. Volcanoes in those mountains erupted ash that fell out across the Morrison stream system; the glassy particles of ash later interacted with water to convert to a clay called bentonite, adding to the slick texture of the mudstones.

The Morrison is exposed primarily in the western margin of the monument where it has been tilted by Laramide deformation. Weathering and erosion of thick mudstones in the Morrison creates poorly vegetated slopes, but the more resistant sandstones jut out of the ground at steep angles in a landform called a **hogback ridge** (Fig. 9.7). One of these tilted slabs of sandstone encases thousands of dinosaur bones, which are now displayed within the Quarry Visitor Center. Over 400 individual skeletons of 11 different types of dinosaurs have been reconstructed from fossils delicately removed from the bone bed, including the spiky-tailed *Stegosaurus*, giant plant-eating sauropods like *Camarasaurus*, *Diplodocus*, and *Apatosaurus*, and the dominant predator of the Jurassic,

Allosaurus. Dinosaur National Monument is the original Jurassic Park.

Some of the dinosaur skeletons were preserved as intact specimens, as if their bodies were rapidly buried by sand deposited from raging floodwaters. Other bones not connected to the rest of a skeleton exhibit a preferential orientation, as if deposited within the flow of a stream in flood. **Fossilization** requires that sediment cover the corpse soon after death, allowing bacteria to break down the soft parts, but protecting the harder bones from scavengers and postmortem rearrangement. The scenario envisioned for the spectacular preservation of large-bodied dinosaurs in this bed of the Morrison is that they were drowned in a catastrophic flood, carried downstream, and then buried in sand as the floodwaters receded. Since the sandstone bed is encased in lake-deposited mudstone, it may be that the floodwaters carrying the corpses cut a new channel into an ancient lake and then dumped the bodies in a ghastly pile. In some cases, preservation is so good that braincases and even tiny ear bones can be recognized.

Many other fossils are found in the Morrison that reveal a surprisingly rich Mesozoic ecosystem. Mudstones deposited on the calm floors of lakes and ponds contain delicate fossils of frogs, salamanders, lizards, turtles, and crocodiles. A rodent-like mammal with large incisors has also been found in the Morrison at Dinosaur National Monument, attesting to the contemporaneity of dinosaurs and small mammals about 150 million years ago during the Jurassic. Fossil pollen and spores indicate that the Morrison ecosystem was warm and dominated by ferns as well as larger plants like *Araucaria*, a pine-like tree that could reach heights of 50 m (160 ft) or so. Ginkgos, cycads, horsetails, and tree ferns found in the Morrison are still with us today, falling under the category of "living fossils."

Landscape Evolution at Dinosaur

The sparsely vegetated highlands at Dinosaur are incised by the deep, meandering canyons of the Green and Yampa rivers, exposing the impressive panorama of rock reaching back to the Precambrian (Fig. 9.8).

The harsh landscape is softened somewhat by the green shades of the riparian corridor that flank the rivers deep in their gorges. The arrangement of the rocks as well as the terrain at Dinosaur is directly related to the uplift of the Uinta Mountains, a tale that expands upon the initial discussions of the Laramide orogeny introduced in Chapters 7 and 8 on Capitol Reef and Canyonlands. The Uinta Mountains are the elevated topographic expression of an elongate **structural arch**, a convex-upward flexure in the rock extending east–west for ~200 km (125 mi) across northern Utah into northeastern Colorado. The upfolding of the rock within the Uinta arch was facilitated by faults cutting the arch along its northern and southern boundaries, as if the mountains rose like

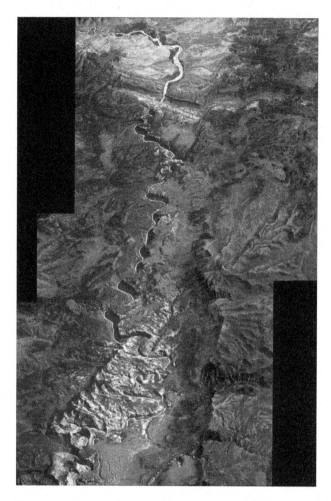

Figure 9.8 Entrenched, meandering canyon of the Yampa River, looking obliquely toward the east, Dinosaur National Monument. River flow is toward the bottom of the image.

an elongate mushroom pushed up from below along the faults (Fig. 9.9).

The Uinta arch formed from compressive forces generated during the late Mesozoic to mid-Cenozoic Laramide orogeny, a phase of mountain-building that raised the Colorado Plateau, as well as the Rocky Mountains from New Mexico to Montana. The sedimentary rock at Dinosaur National Monument was uplifted, folded, tilted, and faulted during the Laramide, and was subsequently exposed to view by the effects of erosion and river downcutting.

Oddly, the Green and Yampa rivers formed their deeply entrenched canyons directly across the hard rocks of the eastern Uintas. Why wouldn't the rivers follow the path of least resistance and cut their channels through the more easily erodible sediments surrounding the Uintas? The explanation is that, like other *superposed streams* of the Colorado Plateau (Fig. 6.9), the Green and Yampa rivers once flowed across a broad layer of unconsolidated sedimentary debris that buried the deformed rocks beneath. The path of the rivers was superimposed on the buried roots of the mountains as they were uplifted

and exhumed later in time, with the rivers incising deep, sinuous gorges through the underlying rock.

Recall from Chapters 7 and 8 that the creation of *incised meanders* within deep canyons on the Colorado Plateau is a two-step process. The first step in the region near Dinosaur National Monument began with late Mesozoic to mid-Cenozoic Laramide uplift that created the Uinta highlands. Later erosion wore down the mountains, creating sediment that accumulated in nearby lowland basins. The Uinta highlands were ultimately buried in their own erosional debris, with perhaps only the highest remnant peaks poking through. By mid-Cenozoic time, the ancestral Upper Green River established a meandering channel across the relatively level surface of sedimentary debris north of the modern Uintas, likely flowing east to the North Platte River (Fig. 9.10A). At the same time, a precursor of the Lower Green River flowed southward from the remnant highlands of the Uintas.

The second step in the development of incised meanders of the Green and Yampa rivers began in late Cenozoic time when uplift occurred again, this time as a broad regional upwarping that is still occurring today (Fig. 9.10B). The rejuvenated highlands increased the gradient of the ancestral Green and Yampa rivers, causing higher velocity flows and erosional downcutting of their meandering channels. The blanket of sedimentary debris that hid the mountains beneath was itself eroded, exhuming the underlying hard rock. With time, the ancestral Lower Green River eroded headward through the hard rock at the eastern end of the slowly rising Uintas, eventually breaking through to the north. (Browse Chapters 5 and 8 on Bryce Canyon and Canyonlands, respectively, for more on the process of headward erosion.)

As the head of the ancestral Lower Green River migrated northward, it eventually encountered the ancestral Upper Green River, stealing its flow in a process called **stream capture** (aka *stream piracy*). Headward erosion of the ancestral Lower Green River may also have captured the flow of the Yampa River during this same late Cenozoic phase. The complete Green River system was thus established, flowing from its head in the Wind River Range to its confluence with the Colorado River in Canyonlands National Park.

With the continued uplift and exhumation of the Uintas during the latest Cenozoic, the solid rock of the Uinta arch rose upward into the overlying rivers (Fig. 9.10C). The Green and Yampa rivers incised downward into the slowly rising mass of rock, maintaining their meandering pattern. Once entrenched within the resistant rock, lateral channel migration ceased, with vertical downcutting becoming the dominant process. The Canyon of Lodore, the Yampa River Canyon, and Split Mountain Canyon were thus incised into the hard rock of the eastern flank of the modern Uinta Mountains.

The interacting processes of uplift, burial by erosional debris, exhumation, headward erosion, and stream capture

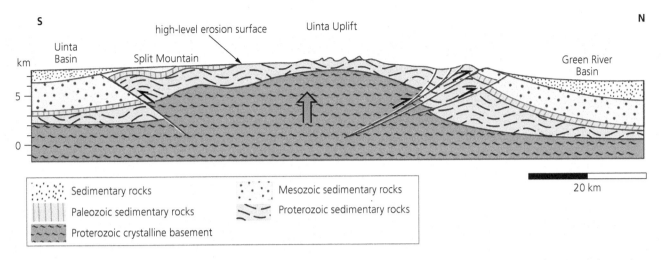

Figure 9.9 North–south geologic cross section across the Uinta Mountain structural arch. The deep-rooted Precambrian rocks of the Uinta Mountain Group pushed upward between high-angle faults along the northern and southern margins of the range.

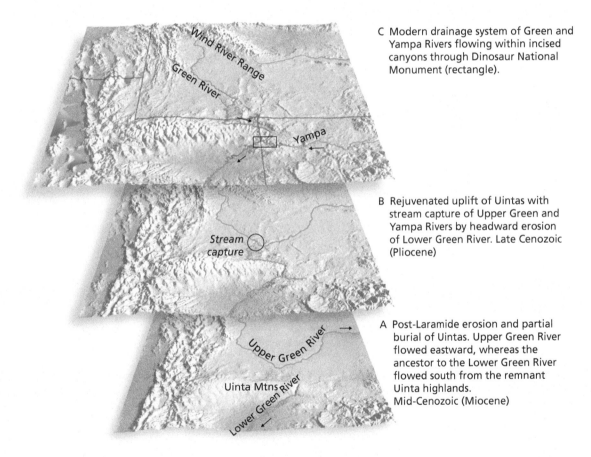

C Modern drainage system of Green and Yampa Rivers flowing within incised canyons through Dinosaur National Monument (rectangle).

B Rejuvenated uplift of Uintas with stream capture of Upper Green and Yampa Rivers by headward erosion of Lower Green River. Late Cenozoic (Pliocene)

A Post-Laramide erosion and partial burial of Uintas. Upper Green River flowed eastward, whereas the ancestor to the Lower Green River flowed south from the remnant Uinta highlands. Mid-Cenozoic (Miocene)

Figure 9.10 Mid- to late Cenozoic phases in the evolution of incised meanders of the Green and Yampa rivers through the eastern nose of the Uinta Mountains.

occurred not only at Dinosaur National Monument, but were widespread across the entire Colorado Plateau and Rocky Mountain provinces. We'll expand on the concepts of superposed streams and canyon incision in later chapters. As you'll see, the uplift of the Colorado Plateau, the evolution of the Colorado River drainage system, and the deeply incised topography of the national parks and monuments on the plateau all share a common history. This explanation of the narrow gorges that score the landscape of Dinosaur National Monument provides an introductory small-scale model for the broader account of incised canyons of the Colorado Plateau as a whole. That's a story that is more fully explored in Chapters 33 and 34. Until then, consider the spectacular geology and landscapes of Dinosaur National Monument as one more touchstone in our narrative of the origin of the American West.

Key Terms

fossilization
hogback ridge
lateral continuity
mudstone
rift basin
stream capture
structural arch

Related Sources

Graham, J. (2006). *Dinosaur National Monument Geologic Resource Evaluation Report. Natural Resource Report NPS/NRPC/GRD/NRR—2006/008*. Denver, CO: National Park Service.

Gregson, J. D., Chure, D. J., and Sprinkel, D. A. (2010). *Geology and Paleontology of Dinosaur National Monument, Utah-Colorado.* In P. B. Anderson, T. C. Chidsey Jr., and D. A. Sprinkel (Eds.), *Geology of Utah's Parks and Monuments*, 3rd ed. Salt Lake City, Utah Geological Association.

Hansen, W. R. (1975). *The Geologic Story of the Uinta Mountains*. Washington, DC: U.S. Geological Survey Bulletin 1291.

National Park Service—Geology of Dinosaur National Monument Accessed January 2021 at https://www.nps.gov/dino/learn/nature/geology.htm

National Park Service—Dinosaurs of Dinosaur National Monument Accessed January 2021 at https://www.nps.gov/dino/learn/nature/paleontology.htm

10

Figure 10.1 The monolith in the center of the image is Independence Monument, a 135 m (440 ft) tall erosional remnant of a wall of rock that once separated Monument Canyon in the foreground from Wedding Canyon on the far side.

Colorado National Monument is a compact park that can be visited in a day's tour along windy and picturesque Rim Rock Drive (Fig. 10.1). The locals in nearby Fruita and Grand Junction appreciate the easy accessibility, inherent beauty, and recreational potential of the monument, going so far as to push to have it redesignated as a national park. They have a very good argument since the natural features are quintessential Colorado Plateau, all of them expressed in a relatively small area of western Colorado (Fig. 2.3).

Many of the classic landscape features and processes that characterize other parks of the Colorado Plateau are also on display in Colorado National Monument (Fig. 10.2). Similar to Bryce Canyon National Park, the erosional scenery at the monument is cut into the northeastern escarpment of the Uncompahgre Plateau, which rises over 700 m (2300 ft) above the adjacent Grand Valley (Fig. 10.3). Many of the narrow, elongate walls and fins in the monument are the dissected remnants of the plateau margin and are reminiscent of those at Arches National Park. Similarities with Canyonlands National Park include incised canyons carved into the plateau edge by headward erosion.

Other comparable geologic features exhibited at Colorado National Monument include an elongate monocline along the eastern margin of the monument, similar

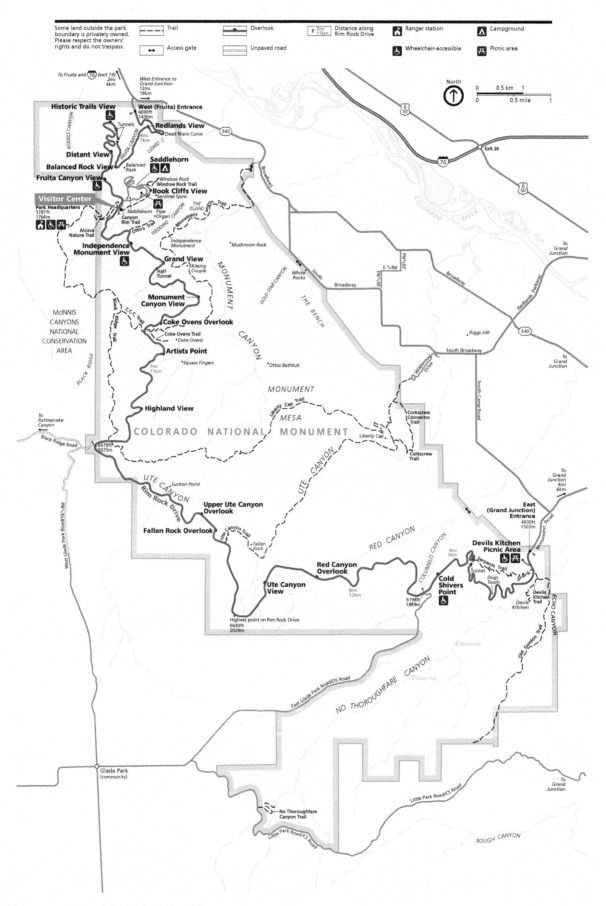

Figure 10.3 Map of Colorado National Monument.

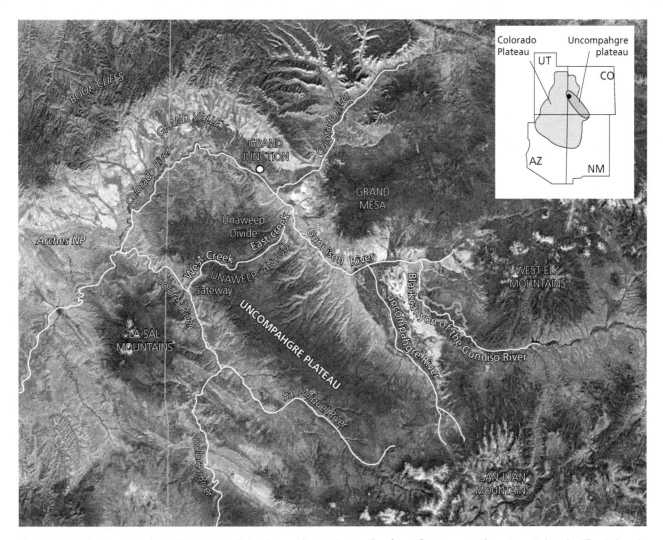

Figure 10.3   The Uncompahgre Plateau trends to the northwest for 150 km (90 mi) across southwestern Colorado. The Colorado River wraps around the north end of the plateau.

in origin to the Waterpocket monocline at Capitol Reef National Park. Within the stack of rocks comprising Colorado National Monument, a great unconformity exists that exceeds that of the Grand Canyon in the amount of time missing along the unconformable surface. Even the "monument" status is similar to Grand Staircase-Escalante and Dinosaur, other monuments with national park attributes. So there are a lot of familiar geologic features in such a small geographical area.

The origin of the modern landscape at Colorado National Monument is directly related to the geologic history of the Uncompahgre Plateau, the topographic expression of a block of Precambrian crystalline rock that stretches northwest for ~150 km (90 mi) along the northeast margin of the Colorado Plateau. Average elevations of the relatively flat upper surface of the Uncompahgre are almost 3000 m (~10,000 ft). Deep canyons incise the plateau, focusing their drainage toward the northeast where it adds to the flow of the Gunnison and Colorado rivers. These episodically flowing streams scour the eastern escarpment

of the Uncompahgre Plateau, creating the erosional landscape of Colorado National Monument.

This chapter will add to your understanding of now-familiar processes such as differential weathering and erosion, cliff retreat, stream incision, and headward erosion. We'll continue adding details to the story of one of the main mountain-building episodes in the American West, the Laramide orogeny, and you'll be introduced to an earlier phase of uplift that raised the Ancestral Rocky Mountains.

Origin and Age of Rocks at Colorado National Monument

The age of the rocks at Colorado National Monument are primarily Precambrian and Mesozoic; Paleozoic and Cenozoic rocks are absent. The crystalline rocks of the Precambrian basement at Colorado National Monument are very similar to those described in Chapter 3 on the Grand Canyon and will be expanded upon in the next chapter on nearby Black Canyon of the Gunnison National Park. And you are already familiar with most of the Mesozoic units in Colorado National Monument. Many of the soaring bluffs and monoliths at the monument are developed in the resistant Wingate Sandstone, discussed in earlier chapters on Capitol Reef and Canyonlands. The underlying mudstones of the Chinle Formation play an important role in the process of *cliff retreat* at Colorado National Monument.

The stream-deposited Kayenta Formation, above the Wingate, is tightly cemented and forms a protective cap atop Wingate cliffs. The Navajo Sandstone is entirely absent at Colorado National Monument, reduced by erosion to a feather edge just west of the park (Fig. 4.7). The Entrada Sandstone is less prominent at the monument than at nearby Arches where it has the starring role in the landscape. The Morrison Formation, so important at Dinosaur National Monument, preserves both fossils and trackways at Colorado National Monument but is commonly poorly exposed due to its low resistance to erosion and its susceptibility to landsliding.

The *lateral continuity* of each of these Mesozoic rock units across the Colorado Plateau attests to the extensive nature of their original depositional environments, whether as a coastal plain of meandering streams and floodplains, vast sandswept deserts, or broad tidal mudflats. Repeated transgressions and regressions of the Mesozoic shoreline across the ancient American West drove the widespread accumulation of sediments in a variety of depositional environments. After burial and solidification, these layered sedimentary rocks were raised toward the surface by uplift of the Colorado Plateau and fragmented by erosion into isolated exposures. These disparate outcrops may be exposed on mountainsides, along the edge of a plateau, or within a deep canyon, but we can reconstruct the original extent of each formation based on their position within the larger stack of strata, their fossil content, and their physical characteristics. The principle of lateral continuity may seem intuitive and somewhat obvious, but it provides the conceptual basis for reconstructing past environments and illustrating them as the paleogeographic maps you see throughout this book.

Uplift of Rocks at Colorado National Monument

The most startling aspect of the rocks composing Colorado National Monument is that 230 million-year-old Mesozoic sedimentary rock rests directly above 1.7 billion-year-old Precambrian crystalline rock, with Paleozoic strata completely missing. The origin of this great unconformity, representing 1.5 billion years of "missing" time, relates to a mountain-building event that raised the Ancestral Colorado Rockies in the late Paleozoic. Let's step back in time to tell the story chronologically.

The crystalline rocks of the Precambrian basement at Colorado National Monument were likely originally deposited as sedimentary rocks but were subsequently deformed and metamorphosed into the schists and gneisses we see today.

(Revisit Chapter 1 to review metamorphism and metamorphic rocks.) Metamorphism likely occurred during a Precambrian episode of mountain-building when the original sedimentary rocks were subjected to the intense pressures and temperatures within the deep core of a long-eroded mountain range. The 1.7 billion-year age of these rocks reflects the timing of metamorphism when the radioactive "clocks" within the rocks were reset by the high temperatures. The crystalline rocks were later intruded by magma that solidified to become granite, which itself was later metamorphosed to gneiss. This brief history of the basement rocks at Colorado National Monument should remind you of the similar age and sequence of events that formed the Precambrian crystalline basement in the Grand Canyon. That's because both sets of rocks were part of the same mountainous mass that was tectonically attached to the southwestern United States during the Precambrian (Fig. 3.17).

These Precambrian mountains were eventually eroded and submerged beneath the shallow seas of the Paleozoic that migrated back and forth across western North America between ~500 and ~330 Ma, leaving behind a variety of clastic sediments as well as limestones. These are the sedimentary rocks exposed in the upper Grand Canyon as well as at Dinosaur National Monument. These same Paleozoic rocks once likely formed a thick blanket above the Precambrian basement at Colorado National Monument, but they were subsequently eroded away. The reason for their absence lies with the birth of the Ancestral Rocky Mountains during a late Paleozoic phase of mountain-building that crested ~315 Ma. Compressional tectonics related to growth of the Pangean supercontinent caused elongate masses of basement and overlying sedimentary rock to rise upward along faults across Colorado. This **Ancestral Rocky Mountain orogeny** generated mountain ranges that are the precursors of today's Colorado Rockies (Fig. 7.10).

The Uncompahgre highlands, composed of Precambrian crystalline rocks, were uplifted as part of the Ancestral Rocky Mountains. They formed an elongate upland that extended from northern New Mexico across southwestern Colorado into eastern Utah. With uplift, running water and mountain glaciers attacked the carapace of early to mid-Paleozoic rocks, with the eroded particles transported downslope to the west and redeposited in adjoining lowland basins. Examples of these late Paleozoic recycled sediments are the Supai Group (Grand Canyon), the Cutler Group (Canyonlands), the Weber Sandstone (Dinosaur National Monument), and the Honaker Trail Formation that overlies the salt deposits of the Paradox Basin (Arches). By the end of the Paleozoic, the Uncompahgre highlands were beveled down to a remarkably planar surface that marks the unconformable top of Precambrian rocks in Colorado National Monument.

In early Mesozoic time, a depositional system of shallow, sinuous streams and broad, vegetated floodplains migrated over the exposed remnants of the Uncompahgre uplift, laying down the multicolored mudstones of the Chinle Formation on top of the Precambrian basement. The Great Unconformity that separates the 230 million-year-old Chinle from the underlying 1.7 billion-year-old crystalline rocks represents ~1.5 billion years of time, a span ~300 m.y. longer than that of the Great Unconformity in the Grand Canyon! Today, the Great Unconformity is expressed along the eastern escarpment of the monument where purplish-gray outcrops of schist and gneiss are overlain by reddish-orange slopes of Chinle (Fig. 10.4). The unconformable contact can be traced as a horizontal surface along the base of the Chinle slope within each of the canyons that cut perpendicular to the escarpment, with the Precambrian rocks forming the lower slopes and canyon floors. The unconformity is particularly well-exposed in nearby Unaweep Canyon, south of the national monument.

Through the rest of the Mesozoic, as sea level rose and fell, as climate fluctuated, and as flora and fauna evolved, sediments of the Wingate, Kayenta, Entrada, and younger formations were deposited above the Chinle, mostly in coastal continental settings. Toward the end of the Mesozoic, beginning about 80 million years ago, Laramide mountain-building once again raised the Uncompahgre region as part of the broader uplift of the Colorado Plateau. As at Capitol Reef National Park, long-dormant

faults cutting the Precambrian basement were reactivated, raising the crystalline rocks and the overlying Mesozoic sedimentary rocks. **Fault reactivation** occurs when an old fault, marking a plane of weakness in the crust, exhibits renewed motion after a long phase of inactivity. The rigid basement rocks rose as an intact block along the fault surface, but the overlying layers of Mesozoic redrock flexed and draped downward to the northeast as a *monoclinal fold* (Figs. 10.5, 10.6). The fault beneath the monocline that marks the edge of the escarpment is called the Redlands fault, and erosion has exposed it to view in certain locations.

Faults that have one block of rock pushed up and over the adjacent block along a high-angle fault plane are called **reverse faults** (Fig. 10.7A). These types of faults are produced by compressional stress, commonly induced by tectonic interaction along a plate boundary perhaps hundreds to thousands of kilometers away. Reverse faults show the opposite sense of motion to that along *normal faults*, which are produced by extensional stress (Fig. 10.7B). (Normal faults were discussed earlier in the chapters on Bryce Canyon and Canyonlands.) It's important to understand that motion along faults occurs episodically during

Figure 10.4 View toward west through mouth of Monument Canyon. Green line marks monument boundary. Wiggly black line is called a "contact" and marks the position of the unconformity above purplish Precambrian crystalline rocks (PC) along the front of the escarpment and along the lower portions of Monument Canyon. Orange-brown, rubble-strewn slope above the unconformity is the Chinle (Ch) and the overlying cliffs are Wingate (Wi) capped by a terrace of Kayenta (Ka). Think of the unconformable contact as a flat surface projecting back beneath the ground with the Chinle Formation resting on top. The overlapping triangular patterns in the lower right are the erosional remnants of tilted beds of the Entrada Formation (En) marking the eastern front of the monocline that dips toward the viewer.

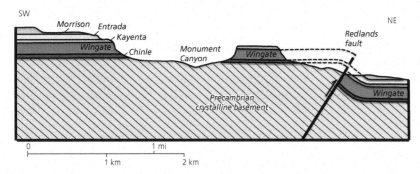

Figure 10.5 The Redlands reverse fault dictates the location of the monocline along the northeastern edge of the Uncompahgre Plateau (and Colorado National Monument). Precambrian basement rocks on the west were shoved up along the fault relative to the Precambrian rocks on the east. The overlying Mesozoic redrocks were partially broken along the Redlands fault, but for the most part drape over the edge of the fault block, with the lower part on the east about 700 m (2300 ft) lower than the equivalent rock to the west.

the same general orientation as the earlier highlands of the Ancestral Rocky Mountain and Laramide orogenies, extending toward the northwest with dimensions of 150 x 45 km (90 x 30 mi). The nature of late Cenozoic regional uplift is somewhat enigmatic and will be addressed in detail in Chapters 33 and 34 on the mechanisms controlling uplift of the Colorado Plateau. For now, it's enough to understand that the erosional terrain of Colorado National Monument is carved into the relatively young Uncompahgre Plateau, the current expression of a complicated history of repeated uplift and erosion.

Landscape Evolution at Colorado National Monument

Figure 10.6 In places, the Redlands fault is buried beneath the surface, and the Mesozoic redrock drapes over the fault edge as a monoclinal fold, marking the northeastern edge of the Uncompahgre Plateau. The rocks dip northeastward down toward the green Grand Valley. Lines highlight the monoclinal fold in the rocks.

Colorado National Monument is drained by a network of **ephemeral streams** that flow northeastward off the Uncompahgre Plateau toward the Gunnison and Colorado rivers that run northwest through the adjacent Grand Valley (Fig. 10.3). The streams are "ephemeral" in that they only flow seasonally when the amount of rainfall or snowmelt exceeds the rate of infiltration of water into the underlying water table. These types of streams are common in semiarid climates like those of the Colorado Plateau where precipitation is sparse and the rate of evaporation is high. The typically dry streambeds are deceptive, however, in that they can rage with flash floods after intense summer thunderstorms.

earthquakes. The nature of uplift along faults and the mechanics of earthquakes will be addressed in greater detail in later chapters, but for now you should be aware that the cumulative motion of thousands of earthquakes, occurring over millions of years, can add up to hundreds to thousands of meters of uplift.

To briefly recap, the Uncompahgre highlands were originally raised over a few tens of millions of years during the late Paleozoic as part of the Ancestral Rocky Mountain orogeny. During the late Mesozoic and early Cenozoic, Laramide mountain-building once again raised Precambrian through Mesozoic rocks upward along the Redlands fault and related faults. Yet again, a third phase of renewed regional uplift occurred in the late Cenozoic (~5-6 Ma) that exhumed the Uncompahgre Plateau to its current elevation over 2 km (1.2 mi) above sea level. The modern Uncompahgre Plateau is the topographic expression of this latest phase of uplift. The Plateau maintains essentially

The greatest degree of canyon modification occurs during flash floods and associated debris flows when large boulders are transported through the channels and deposited on flatter slopes downstream. The proximity of homes and roads to the northeastern margin of the monument raises the risk due to flash floods because the sheer volume of water and entrained debris can overwhelm developed areas where people live and work.

At Colorado National Monument, the overlapping processes of headward erosion, cliff retreat, differential weathering, and stream incision during flash floods conspire to create the delightful landscape of deep canyons and isolated monoliths. As at many other places on the Colorado Plateau, these processes tend to focus along the natural joint system that cuts vertically through the rigid layers of rock. Many of the cliff faces and monoliths in Colorado National Monument are formed in the Wingate Sandstone, commonly capped by resistant beds of the Kayenta Formation

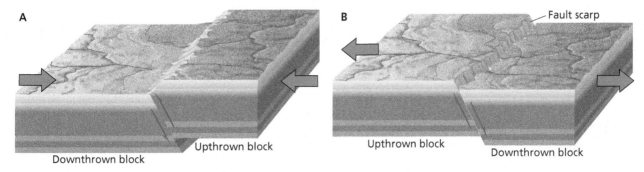

Figure 10.7 Compare the sense of motion between a reverse fault (A) and a normal fault (B). Weathering and erosion preferentially attack the edge of the uplifted block, smoothing the topographic transition across the fault.

(Fig. 10.6). Here in western Colorado, the stream-deposited sandy sediment of the Kayenta was cemented by silica-rich groundwaters when it was deeply buried and solidifying into sandstone. The silica cement makes the Kayenta highly resistant to erosion, protecting the vertical cliffs of Wingate below.

In many places, the Chinle Formation disintegrates by water seeping downward through vertical joints in the overlying Wingate as well as by surface water runoff after intense storms. The decomposition of the Chinle undermines the thick, jointed layer of Wingate above, allowing large slabs of sandstone to break away and collapse downward, either catastrophically as a rockfall or as an intact block that detaches along a joint surface. *Cliff retreat* is a common process on the Colorado Plateau due to the juxtaposition of easily eroded shales and mudstones against overlying resistant sandstones. In places where the protective cap of silica-cemented Kayenta is degraded, the calcite-cemented Wingate may form bizarre rounded shapes such as the Coke Ovens (Fig. 10.8).

The beauty of the landscape we see today at Colorado National Monument may appear stable and timeless as you gaze upon it from one of its trails. But the inexorable processes of weathering and erosion occur at two time scales. They happen slowly, grain-by-grain, over long spans of time, as well as rapidly during short, violent events such as flash floods, rockfalls, and landslides. A spectrum of geologic processes, occurring over a broad range of time scales, interact to create the moment in time that we see today in stunning landscapes such as that at Colorado National Monument.

Figure 10.8 The Coke Ovens are eroded into a broad wall of Wingate Sandstone, with the distinctive shapes formed as the protective Kayenta caprock slowly decays. Joints widened by weathering and erosion separate each of the "ovens" into discrete rounded masses. Contrast the Coke Ovens with the steep cliffs of Wingate and intact Kayenta in the distance.

fault reactivation
reverse fault

Related Sources

Baars, D. (1998). *The Mind-Boggling Scenario of the Colorado National Monument—A Visitor's Introduction.* Grand Junction, CO: Canyon Publishers, 14 p.

KellerLynn, K. (2006). *Colorado National Monument Geologic Resource Evaluation Report.* Natural Resource Report NPS/NRPC/GRD/NRR—2006/007. Denver, CO: National Park Service.

National Park Service—Geology of Colorado National Monument Accessed January 2021 at https://www.nps.gov/articles/nps-geodiversity-atlas-colorado-national-monument-colorado.htm

National Park Service videos of flash floods at Colorado National Monument Accessed January 2021 at https://www.nps.gov/colm/learn/education/exploring-the-evidence.htm

Key Terms

Ancestral Rocky Mountain orogeny
ephemeral stream

11 Black Canyon of the Gunnison National Park

Figure 11.1 The Gunnison River deep within the narrow gorge of Black Canyon of the Gunnison National Park.

The somber colors and perennial shadows that characterize Black Canyon of the Gunnison National Park create a moody atmosphere for the visitor (Fig. 11.1). As you gaze downward from one of the overlooks, you're struck by the thought that the thin little strand of the river far below was somehow able to cut the vertiginous walls of the gorge. The Gunnison River today is regulated by three dams located upstream in nearby Curecanti National Recreation Area (Fig. 11.2), but for the vast majority of its existence, the river raged through the canyon. As you'll see, the torrent of the Gunnison River combined with a variety of other geologic events and processes to cut the precipitous gash of the Black Canyon.

Black Canyon of the Gunnison National Park is located along the eastern edge of the Colorado Plateau in western Colorado, with the high peaks of the Rocky Mountain Province looming to the east and south (Fig. 2.3). It shares many characteristics with other parks of the plateau, such as the deeply incised gorge that is the centerpiece of the park and the relatively undeformed Mesozoic and Cenozoic sedimentary rock that overlies the crystalline Precambrian rock within

Figure 11.2   Map of Black Canyon of the Gunnison National Park.

the canyon. "The Black," as climbers call it, has some of the most extreme dimensions of any incised canyon on the Colorado Plateau. The canyon has a maximum depth of about 830 m (2700 ft) and yet is only 335 m (1100 ft) wide at its narrowest at the rim. The Painted Wall is the highest cliff in Colorado at 685 m (2250 ft), almost twice the height of the Empire State Building. At river level, the canyon is only 12 m (40 ft) across at its narrowest point. The dimensions are so steep and tight that sunlight rarely illuminates the dark rock in the depths of the gorge, giving the canyon its name. Wallace Hansen, a geologist with the U.S. Geological Survey, created a series of profiles of several popular canyons of the American West that enables comparison with the dimensions of the Black Canyon of the Gunnison (Fig. 11.3). Other canyons may be deeper, but none compare with the narrow vertical gorge that is the Black Canyon.

The Gunnison River flows for over 260 km (160 mi) across western Colorado from its origin in the Sawatch Range to its confluence with the Colorado River near Grand Junction (Fig. 10.3). The Black Canyon portion of the Gunnison extends for 85 km (50 mi), although only the most imposing 23 km (14 mi) are included within the

national park. The gradient of the river through the park is very steep, about 18 m/km (95 ft/mi) on average, with a maximum of 45 m/km (240 ft/mi). For comparison, the Colorado River in the Grand Canyon drops at an average gradient of 1.4 m/km (7.5 ft/mi), a fraction of the steepness of the Gunnison River through Black Canyon. Of course, today the flow of the Gunnison is modulated by the three dams upstream in the Curecanti, reducing the potential power of the river through Black Canyon. But in the distant past, especially during the Pleistocene Ice Ages, the Gunnison thundered through Black Canyon charged with sand, cobbles, and boulders that battered the rock of the canyon walls.

Now that the remarkable physical characteristics of Black Canyon and the Gunnison River have been established, let's investigate the nature of the rocks composing the park as well as the evolution of the landscape through time. In the following sections, we'll dig deeper into the Precambrian history of the Colorado Plateau, we'll examine an unconformity that represents 1.5 billion years of missing time, and we'll see how regional volcanic activity can influence the path of a river and its incision into the landscape.

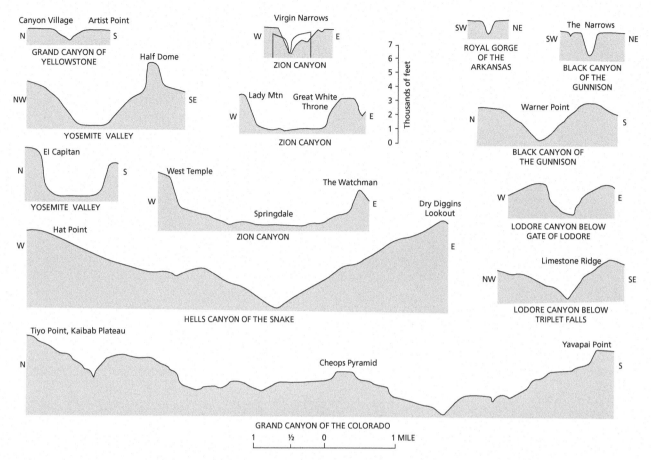

Figure 11.3 Profiles of well-known canyons of the American West.

Origin, Age, and Uplift of Rocks at Black Canyon of the Gunnison

As opposed to the Paleozoic, Mesozoic, and Cenozoic sedimentary rocks that dominate most of the other parks on the Colorado Plateau, the bulk of Black Canyon of the Gunnison is built from Precambrian crystalline rocks. At nearby Colorado and Dinosaur national monuments, you can see a few hundred meters of the uppermost basement. Deep in the Grand Canyon, you can see a comparable thickness of Precambrian basement at river level. But here at Black Canyon, you gaze at over 600 m (~2000 ft) of a continuous wall of the Precambrian crystalline crust of North America. This is the very same rock that is about a kilometer beneath the cornfields of Nebraska. Indeed, at Black Canyon, you view the crystalline rock that forms the Precambrian foundation of North America. To maintain perspective, it's important to recall from Chapter 1 that continental crust averages 35–40 km (20–25 mi) in thickness, so even at Black Canyon we're only glimpsing less than the uppermost kilometer. In your mind, project the crystalline rocks of Black Canyon downward for 35 or more kilometers to visualize the crust deep beneath your feet.

The metamorphic and igneous rocks that characterize the Precambrian crystalline basement at Black Canyon are similar in age, composition, and origin to those discussed previously in the chapters on the Grand Canyon National Park and Dinosaur and Colorado national monuments. The precursor rocks to the gneisses, schists, and quartzites that dominate the exposures at Black Canyon originated as clastic sediments and volcanic deposits that accumulated in an oceanic setting almost 2 billion years ago (Fig. 3.17). Tectonic convergence generated the necessary pressures and temperatures to metamorphose these precursor rocks into highly contorted gneisses and schists about 1.74 billion years ago, based on radiometric dating of the timing of metamorphism. In places, the metamorphic conditions were extreme enough to cause the rock to flow like a plastic, forming ribbon-like migmatites. (It may be useful to revisit the appropriate sections on metamorphic rocks in Chapter 1, as well as the more complete description of the tectonic origin of the Precambrian basement of the American West in Chapter 3 on the lower Grand Canyon.)

Several hundred million years after the main phases of metamorphism, beginning around 1.4 billion years ago, a succession of igneous intrusions penetrated into the existing basement. The volumetrically most important igneous rock is granitic in composition. These granitic rocks are notable for their huge feldspar crystals (up to 4–5 cm [up to 2 in] in length), which solidified early and then floated in the remaining magma. In places, the elongate crystals are aligned with one another, reflecting the direction of flow just prior to final solidification of the magma. The most distinguishing characteristic of these granitic rocks, however, is their extraordinary resistance to erosion. The most precipitous cliffs in Black Canyon are formed in places where the metamorphic country rock is strengthened by large masses of granite.

Perhaps the most dramatic igneous intrusions are the pinkish bands of **pegmatite** that slash across the metamorphic country rock, vividly expressed on the Painted Wall (Fig. 11.4). When viewed in three dimensions, the pegmatites form tabular-shaped bodies called **dikes**, created as water-rich magma was forcibly injected into fractures in the country rock deep beneath the surface. (The subparallel alignment of the dikes on the Painted Wall likely reflects the original orientation of fractures in the metamorphic rock.) The intense pressure associated with intrusion of the magma forced the rock apart along the fracture, with the intense heat of the magma partially melting the country rock along the margins. Pegmatites are notable for the enormous size of their interlocked mineral crystals, mostly quartz, feldspar, and platy muscovite mica. Some feldspars may exceed a meter or more in length. The reason for the huge crystals is the prolonged cooling of the magma within the dikes, perhaps a few kilometers beneath the surface, which allowed for the continual slow growth of crystals. Similar to the granites, these pegmatite dikes are highly resistant to erosion and add strength to the metamorphic country rock, contributing to the near-vertical character of cliffs like the Painted Wall.

These highly complex metamorphic and igneous rocks manifest the cores of ancient mountain ranges, similar to the Precambrian basement at other national parks. By the end of the Precambrian, these ancient mountains were beveled down to a planar lowland surface, ripe for repeated flooding during transgressions of the Paleozoic seas. Clastic and biochemical sedimentary rocks were laid down on top of the basement complex but were erosionally removed during late Paleozoic mountain-building. This Ancestral Rocky Mountain orogeny (Fig. 7.10) raised the Precambrian basement at Black Canyon as part of the Uncompahgre highlands (discussed in Chapter 10 on the Colorado National Monument). This uplift exposed the veneer of Paleozoic sedimentary rocks to the ravages of weathering and erosion, redepositing the debris in adjacent lowlands surrounding the Uncompahgre highlands. (We'll revisit these specific sedimentary rocks in Chapter 21 on Rocky Mountain National Park.) Erosion during this time likely planed off some of the Precambrian crystalline basement in the Black Canyon region.

This erosional surface remained exposed until the middle Mesozoic (~170 Ma) when coastal environments of the Entrada Sandstone migrated across the region. Thus, the planar unconformity at Black Canyon, expressed by the 170 Ma sedimentary rocks of the Entrada juxtaposed above the underlying 1.7 Ga basement, represents a span of over 1.5 billion years (Fig. 11.5). The "missing time" along

Figure 11.5 "Great Unconformity" at Black Canyon of the Gunnison National Park. The planar surface above the Precambrian basement (~1.7 Ga) is overlain by vegetated hills of Entrada Sandstone (~160 Ma).

Figure 11.4 Pink pegmatite dikes cross-cutting metamorphic gneiss, Painted Wall.

the Great Unconformity at Black Canyon is a few tens of millions of years longer than the Great Unconformity at nearby Colorado National Monument, which itself was about 300 m.y. longer than the Great Unconformity at the Grand Canyon. The Great Unconformity is a widespread surface, traceable across much of the American West.

The lowermost Entrada sandstones enclose eroded fragments of the underlying Precambrian rock, attesting to the migration of a Mesozoic sandy coastal system across the deeply weathered crystalline basement exposed along the shoreline. Deposition continued through the remainder of the Mesozoic as sea level rose and fell, inundating western Colorado multiple times. (These sedimentary rocks are best exposed in the Curecanti area, outside the park boundaries.) The accumulation of sediment imparted a load to the underlying crust, which subsided in response, creating more space for sediment to fill. These quiescent conditions lasted into late Mesozoic time when events occurred that would set the stage for the modern landscape that we see today.

Landscape Evolution at Black Canyon of the Gunnison

The Laramide Orogeny of late Mesozoic to mid-Cenozoic age was first introduced in Chapter 6 on Capitol Reef National Park and expanded upon in subsequent discussions on Canyonlands National Park and Dinosaur and Colorado national monuments. Clearly, this phase of mountain-building was a formative event in the evolution of landscapes of the American West. The tectonic origins of the Laramide will be discussed in Chapter 20, but for now let's focus on the direct effects in the Black Canyon region. This section addresses the question of how the Gunnison River carved the narrow chasm of the Black Canyon.

Laramide compressional stress in the Black Canyon area reactivated deeply rooted *reverse faults* in the Precambrian basement, arching these rocks and their overlying sedimentary carapace back toward the surface. The highland fault block formed during this event in the Black Canyon region is called the Gunnison Uplift, located to the east of the nearby Uncompahgre highlands (Fig. 11.6A). The high elevations of the Mesozoic sedimentary rocks and underlying Precambrian crystalline rocks were subjected to the inescapable processes of weathering and erosion. By the mid-Cenozoic, the Gunnison Uplift was beveled down to the Precambrian basement, with fragments of the Mesozoic rocks preserved around the margins (Fig. 11.6B). Rivers flowed from remnant Laramide highlands across a broad, flat surface. It was at this point that the events happened that set the course of the modern Gunnison River and the incision of the Black Canyon.

Beginning about 30 million years ago, volcanic eruptions occurred in the nearby West Elk Mountains to the northeast and the San Juan Mountains to the south (Fig. 11.6C). (This mid-Cenozoic volcanism is called the *ignimbrite flare-up* and was widespread across the American West. The tectonic trigger for the volcanism is explained in greater detail in Chapter 23.) Huge volumes of ash and other volcanic debris covered southern Colorado, including the rocks of the Gunnison Uplift. The ancestral precursor of the Gunnison River, draining westward from the Laramide-age Sawatch Range to the east, became confined to the now-buried Gunnison lowlands between the two volcanic ranges to the north and south. Between about 10 and 15 million years ago, the ancestral Gunnison River meandered back and forth across the blanket of volcanic rocks overlying the buried Gunnison Uplift, carving its path northwestward toward its convergence with the Colorado River.

By the latest Cenozoic, around 2–3 Ma, much of the Colorado Plateau and adjoining Rocky Mountains were broadly uplifted, increasing the gradient of rivers and triggering a rejuvenation of downcutting (Fig. 11.6D). As the crust beneath was exhumed, the Gunnison River incised downward through the thick veneer of Cenozoic volcanics and Mesozoic sedimentary rocks until it encountered the more resistant rocks of the Precambrian basement. Once the river began to incise into the exposed core of the Gunnison Uplift, it could no longer meander laterally but rather was constrained to the narrow notch within the resistant, crystalline rock. Thus, the Gunnison River, like most rivers on the Colorado Plateau, is a *superposed river* (Fig. 6.9), carved into the crystalline core of the underlying Gunnison Uplift. The incised gash of the Black Canyon was cut by the Gunnison River at a rate of around 6 cm per hundred years (2.5 in/100 yr).

The back and forth swings of Ice Age climate over the past 2 million years or so contributed significantly to the downcutting. During cold phases, glaciers in the Colorado Rockies ground away at the rock, creating enormous volumes of loose sediment. During the warm interglacial phases, torrents of glacial meltwater picked up the sedimentary debris and transported it through the Gunnison drainage network. The swollen Gunnison River and its entrained rocky debris pounded away at the crystalline rock of the Black Canyon, deepening it over time.

Northwesterly oriented faults and joints, generated during the multiple phases of uplift of the Gunnison region, provided planes of weakness for weathering and erosion to focus their effects. The Gunnison River opportunistically followed these patterns, dictating the orientation of the Black Canyon (Fig. 11.1). Vertical joints also focused the infiltration of snowmelt that methodically froze and remelted, slowly wedging apart great slabs of rock. Angular, house-sized blocks of gneiss and granite line the channel at the bottom of the canyon, attesting to episodic violent rockfalls that widened the gorge. The steep gradient and high-volume flows of the river maintained the rapid pace of downcutting, however, exceeding the rate of lateral widening and resulting in the extreme depth-to-width ratio of the Black Canyon.

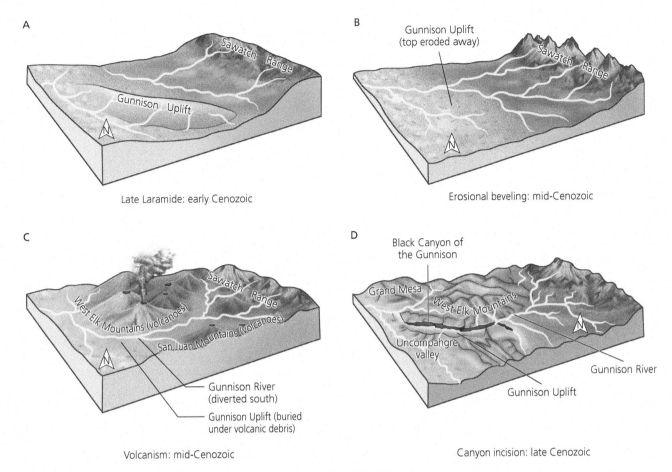

Figure 11.6 Evolution of the Gunnison River drainage and incision of the Black Canyon.

Many other canyons on the Colorado Plateau are much wider than they are deep; Grand Canyon and Zion Canyon are two notable examples (Fig. 11.3). Widening in these canyons is driven not only by mass wasting events like rockfalls and landslides, but also by rapid downcutting of tributary streams that keep pace with incision of the main river. In the Black Canyon, however, the tributary streams are no match for the rate of incision by the Gunnison River. The watersheds of these tributaries are relatively small, and the erosive power of these low-flow drainages is minimized by the sheer strength of the underlying rock. Many of the tributaries end as dry, V-shaped notches high on the canyon walls, whereas other tributaries have given up and have reoriented themselves away from the main canyon that runs along the crest of the Gunnison Uplift.

The sheer cliffs and narrow dimensions of the Black Canyon of the Gunnison are thus the result of a long series of geologic events and processes. Precambrian episodes of mountain-building and metamorphism formed the crystalline core of the canyon. Repeated phases of uplift and erosion alternately raised these ancient rocks to the surface and beveled them down again. Faulting, jointing, and regional volcanism superimposed the orientation of the Gunnison River directly above the Gunnison Uplift.

And finally, the erosional action of the river, charged with sedimentary debris, wore away at the crystalline rock to create the awe-inspiring gorge of the Black Canyon of the Gunnison.

Key Terms

dike
pegmatite

Related Sources

Hansen, W. R. (1965). The Black Canyon of the Gunnison Today and Yesterday. *U.S. Geological Survey Bulletin* 1191, 76 p. Accessed January 2021 at https://pubs.usgs.gov/bul/1191/report.pdf

Thornberry-Ehrlich, T. (2005). *Black Canyon of the Gunnison National Park and Curecanti National Recreation Area Geologic Resource Evaluation Report*. Natural Resource Report NPS/NRPC/GRD/NRR—2005/001. Denver, CO: National Park Service.

National Park Service—Geology of Black Canyon of the Gunnison National Park Accessed January 2021 at https://www.nps.gov/blca/learn/nature/precambrian.htm

PART 2
Volcanic National Parks of the Cascade Range

12 Cascades Geologic Province

Figure 12.1 Cascades Geologic Province. Triangles mark the locations of the three national parks and one national monument addressed in Part 2.

The Cascades Geologic Province consists of a linear chain of active volcanoes that extends for about 1100 km (700 mi) roughly parallel to the coastline of the Pacific Northwest (Fig. 12.1). Of the world's 1511 active volcanoes, 13 reside within the U.S. portion of the Cascade Range, 7 of which have erupted in historical times. The most recent eruption was in 1980 when Mount St. Helens blasted 1.3 cubic km (0.3 mi³) of volcanic debris skyward, accompanied by a series of volcanic hazards that resulted in 57 fatalities (Fig. 12.2). It's only a matter of time, perhaps within a few decades, before another of the Cascades volcanoes unleashes its pent-up fury. That said, the next few chapters include concepts related not only to volcanism in national parks and monuments of the Cascades, but also to the wide variety of hazards and risks associated with volcanic eruptions.

Figure 12.2   The May 18, 1980, eruption of Mount St. Helens in Washington State.

In northern California, the Cascade Range is about 50 km (30 mi) wide, bound on the east by the Basin and Range Province and on the west by the Klamath Mountains (Fig. 12.3). The Cascades progressively broaden northward, reaching about 200 km (125 mi) in width in northern Washington. Flat-lying volcanics of the Columbia Plateau bound the Cascades to the east in Oregon and Washington, whereas the Willamette Valley and Puget Lowland flank the range on the west. The highest peak in the Cascades is Mount Rainier in Washington at 4392 m (14,410 ft), and the second highest is Mount Shasta in California at 4317 m (14,162 ft). The other summits of Cascades volcanoes range between about 3000 and 3700 m (~10,000 and 12,000 ft).

The volcanic peaks are extremely conspicuous landforms in the Pacific Northwest, not only for their white crowns of snow and ice, but for the fact that many of them rise dramatically above the surrounding landscape. The only major break in the range is formed by the Columbia River that flows within the Columbia Gorge just north of Mount Hood in Oregon. Early nineteenth-century explorers called the rugged peaks on either side of the river "the mountains of the cascades" after a series of whitewater rapids that complicated boating through the Gorge. The Cascade Range takes its name from these long-lost rapids, now submerged beneath the waters backed up behind Bonneville Dam.

The high peaks of the Cascades act as a natural barrier to storm fronts moving eastward from the Pacific. Most precipitation falls as rain along the coastal lowlands west of the range and as snow on the highest peaks. This moisture supports dense conifer forests and a verdant undergrowth of ferns along the western slopes of the range.

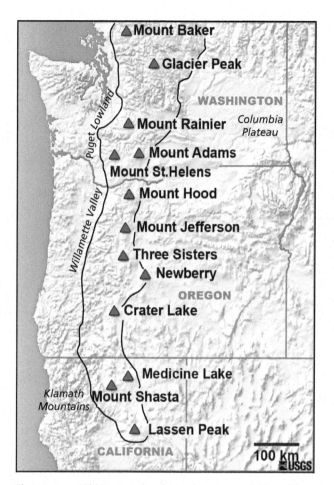

Figure 12.3 Thirteen volcanic peaks arranged in a linear chain characterize the Cascade Range in the Pacific Northwest. Blue line marks the diffuse margins of the Cascades Geologic Province.

In the rain shadow east of the Cascades, a sparse flora of pine, juniper, and sagebrush marks the dry, brown landscape. Among the towering volcanoes of the Cascades are three national parks: Mount Rainier, Crater Lake, and Lassen Volcanic. Mount St. Helens National Volcanic Monument was designated in the years following the 1980 eruption to commemorate the devastation and loss of life from that cataclysmic event.

Tectonic Setting of the Cascades Geologic Province

The linear array of Cascades volcanoes is related to a *continental-oceanic convergent boundary* (Fig. 1.13). It's a good idea to go back to Chapter 1 and refresh your understanding of important concepts such as compression, subduction, the generation of magma, and the tectonic origins of continental volcanic arcs and deep ocean trenches. Figure

1.8 is particularly useful to review since it shows all of the main types of plate boundaries in one composite diagram. In this section, we'll build on the fundamentals of tectonics to develop a deeper understanding of the geologic setting and active volcanism of the Cascades Province.

The tectonic setting along the Pacific Northwest is called the **Cascadia subduction zone**. On Figure 12.4, the line with the eastward-pointing triangles marks the convergent plate boundary where the North American continental plate to the east overrides the much smaller, denser Juan de Fuca, Gorda, and Explorer oceanic plates to the west. A deep ocean trench typically occurs along the boundary between the converging plates, but in this case the trench is mostly filled with prodigious amounts of sediment from the Columbia River that drains a large watershed in the Pacific Northwest. The Cascadia "trench," is located about 100–160 km (60–100 mi) from the coastline and reaches a maximum depth of over 2900 m (~9600 ft).

The western boundary of the Juan de Fuca plate and two smaller **microplates** is a *mid-ocean ridge divergent boundary* where *seafloor spreading* and associated volcanism create new lithospheric rock. With time and extensional motion, the newly formed oceanic lithosphere of the Pacific plate is translated laterally to the west. The lithosphere of the Juan de Fuca and smaller microplates migrates to the east where it eventually subducts beneath the North American plate. The third type of plate boundary in the Cascadia region is the *transform fault,* which offsets mid-ocean ridge segments. The very southernmost margin of the Cascadia convergence zone is the Mendocino fault, the offshore extension of the San Andreas transform fault that bends westward at Cape Mendocino. Note how Lassen Peak, the southernmost volcano in the Cascades, is located due east of the southern margin of the Cascadia subduction zone.

Figure 12.5 helps to visualize the upper few hundred kilometers beneath the surface of the Cascadia subduction zone. Notice the proximity of the subduction zone to the spreading ridges and the relatively small size of the Juan de Fuca plate. Lithosphere is being destroyed by subduction along the eastern boundary of the plate at the same time new rock is being created by seafloor spreading along its western margin. The net rate of convergence is about 4 cm/yr (1.6 in/yr), measured to a high degree of accuracy by the global positioning system of satellites.

You may wonder why the mid-ocean ridge separating the Juan de Fuca and smaller plates from the huge Pacific plate is not located in the middle of the Pacific Ocean. The reason is that the smaller oceanic plates are subducting beneath the North American plate faster than the rate of seafloor spreading along their divergent boundaries to the west. The rapid eastward subduction of the Juan de Fuca and smaller plates drags along the spreading ridges behind, with the net effect being the slow, progressive destruction of the smaller plates. The Juan de Fuca plate and the Gorda and Explorer microplates are small remnants of an ancient, gigantic plate called the Farallon that has been subducting beneath the North American plate for the past 200 m.y. or so. In time, the spreading ridges will likely be pulled under the North American plate, marking the demise of the Juan de Fuca and adjacent smaller plates. (As you'll read later in the book, the Farallon plate figured prominently in the formation of the American West and its array of national parks and monuments.)

Active convergent margins like Cascadia are notable not only for their explosive volcanism, but also for related hazards like earthquakes and tsunami. As the two lithospheric plates inexorably grind against each other deep along the subduction zone, enormous amounts of tectonic strain accumulate. The plates remain locked together, however, by the frictional resistance along the interface between the colossal lithospheric slabs. At some unpredictable future threshold, the pent-up tectonic strain will exceed the frictional resistance, and the overriding continental plate will suddenly lurch forward, releasing the accumulated energy as a massive earthquake.

Based on some clever geologic detective work by geoscientists from the U.S. Geological Survey, as well as by colleagues in Japan, the last major earthquake on the Cascadia subduction zone occurred on January 26, 1700. The quake had an estimated magnitude of 9.0 and was accompanied by a tsunami that inundated the coastal zone of the Pacific Northwest in a matter of minutes, reaching all the way to the western Pacific over a few hours. The Cascadia subduction zone has not experienced a quake of that magnitude in the years since, but it's just a matter of time. The recurrence interval of giant earthquakes along subduction zones typically ranges over a few hundred years. There's no reason to think that a major earthquake is imminent along the Cascadia margin, but then again it's likely that someday soon the cities of Vancouver, Seattle, and Portland will be violently rocked by seismic shaking with all the attendant effects on buildings, roads, railways, and other human-made structures. It's

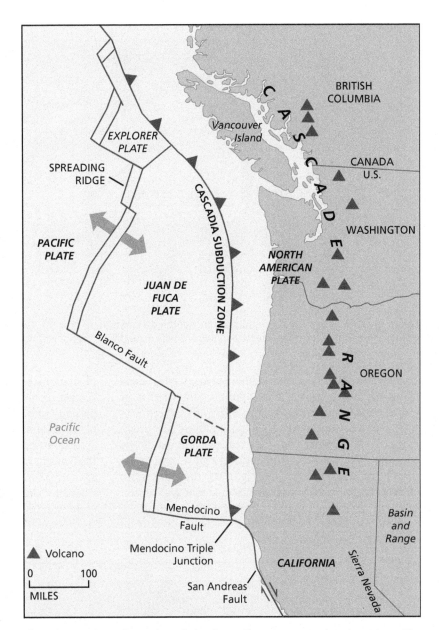

Figure 12.4 Tectonic setting of the Cascadia convergent plate boundary. The Cascadia trench is located along the line with the eastward-pointing triangles and marks the exact plate boundary.

highly probable that a tsunami will be triggered that will surge onto the shoreline, inundating coastal lowlands. The populace of the Pacific Northwest understands the magnitude of the risk, and contingency planning is an ongoing process.

Silica and Its Influence on Volcanic Eruptions

Many active volcanoes erupt violently, spewing gray ash and debris tens of kilometers into the atmosphere and causing widespread devastation to the surrounding

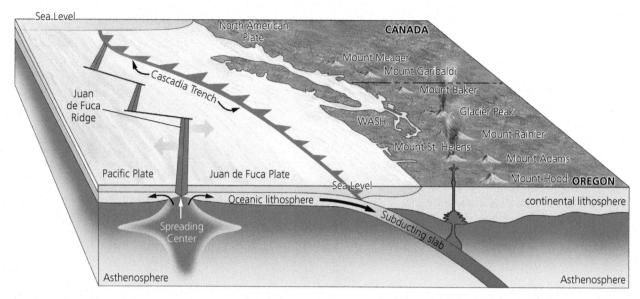

Figure 12.5   Cross section of Cascadia subduction zone near northern Oregon, Washington, and southern British Columbia. Dense oceanic lithosphere of the Juan de Fuca plate descends beneath the continental lithosphere of the North American plate.

countryside. The volcanoes of the Cascades and others along the "Ring of Fire" that circumscribes the Pacific Ocean erupt in this **explosive** style (Fig. 12.6). Other volcanoes, like those of Hawaii, erupt in a relatively nonthreatening manner, emitting flowing streams of lava that slowly cool and solidify, covering the landscape with an irregular veneer of black, volcanic rock (Fig. 1.21). These are called **effusive** eruptions. A common molecule that is abundant in Earth's mantle and crust, **silica (SiO_2)**, governs these two end-members of volcanism. The relative proportion

of silica within magma determines the amount of gases trapped in the magma, which in turn influences the explosivity of eruptions as well as the shape and appearance of the volcano.

Before we delve deeper into the style of volcanism in the Cascades as well as differences in the behavior of volcanoes, it's a good idea to review the relevant section of Chapter 1 on igneous rocks. You'll want to refresh your understanding of fundamental concepts such as (1) the progressive crystallization of minerals from magma or

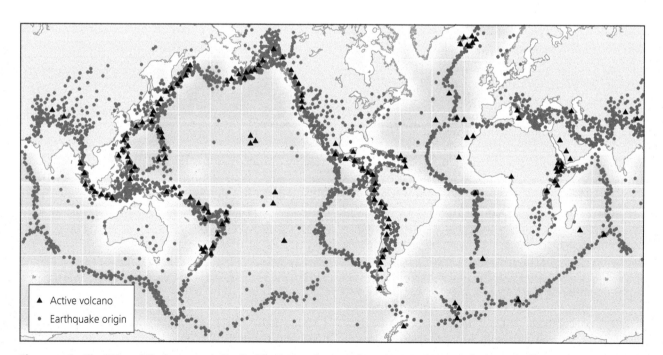

Figure 12.6   The "Ring of Fire" surrounds the Pacific plate to the east along the Americas, north along Alaska, and west along Asia and the southwest Pacific. The zone is defined by active volcanism and seismicity associated with convergent and transform plate boundaries.

lava as the molten fluid cools through time, and (2) the rapid crystallization of lava and ash on Earth's surface to form the fine crystals that comprise volcanic igneous rocks. (Remember that molten rock is called *magma* when it is located beneath the surface and *lava* when it has erupted onto the surface.) The integrated knowledge of volcanic processes from Chapter 1 with this chapter will set the stage for understanding not only the nature of Cascades volcanism, but also for future discussions of national parks like Hawaii Volcanoes and Yellowstone.

All magmas have a distinctive chemical composition governed by the particular tectonic setting as well as the evolution of the magma as it ascends toward the surface. Magma consists of three main components: (1) a liquid phase (called the **melt**), (2) any solid crystals that might have formed from the melt as it cooled, and (3) **volatile gases** such as water, carbon dioxide (CO_2), sulfur dioxide (SO_2) and hydrogen sulfide (H_2S). Think of magma as a mixture of solid crystals floating in a liquid melt, which is infused with common gases dissolved in the melt. Magma may also contain solid fragments of rock derived from the walls of the magma chamber or from the conduit to the surface. The gases, dominated by water vapor, may reach up to 5 percent of the magma's weight and are particularly important since gas pressure is the ultimate source of the explosivity of the magma. In general, the more gases trapped in the magma, the more powerful the potential eruption. The amount of dissolved gases in magma is controlled by the relative amount of silica comprising the magma. Ultimately, the amount of silica in magma is determined by the tectonic setting of the volcano.

At oceanic-continental convergent boundaries such as Cascadia, the generation of magma is triggered by the release of water from the descending slab of oceanic lithosphere at depths of 100–150 km (~60–90 mi) (Fig. 12.7). The water interferes with the chemical bonds between atoms in the overlying asthenospheric rock, lowering the melting point and permitting some of the rock to convert to magma (Chapter 1). But not all asthenospheric rock melts equally—minerals composing the rock that are dominated by lighter elements such as oxygen, silicon, and aluminum tend to melt first, enriching the magma in silica. The remaining

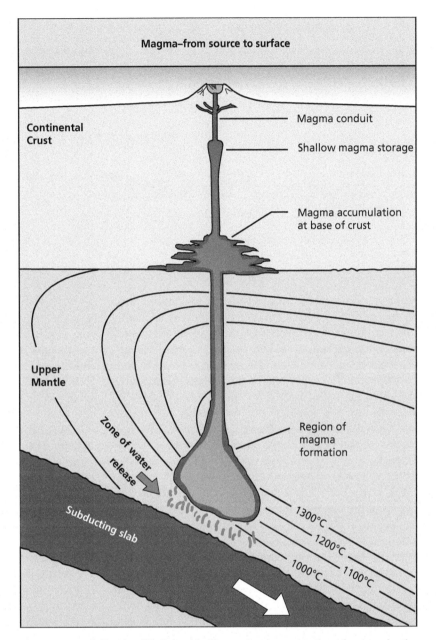

Figure 12.7 Highly simplified graphic illustrating the generation of magma by the release of water from a subducting oceanic plate. Magma is less dense than the surrounding rock and thus buoyantly rises toward the surface, evolving chemically along the way.

silica-poor minerals are left behind, unmelted. In addition, volatile gases like water and CO_2 are released from the subducting slab, ultimately permeating into the magma, similar to the way CO_2 exists in solution in carbonated soda or champagne. Since the silica- and gas-enriched magma is less dense than the surrounding rock at those depths, it buoyantly rises, chemically evolving as it melts rock of the upper mantle and crust along its path toward the surface.

The amount of silica in the magma influences the magma's resistance to flow, a characteristic called **viscosity**. (For instance, water flows relatively easily and thus has a low viscosity. Syrup is thicker than water and flows more slowly; thus syrup has a higher viscosity than

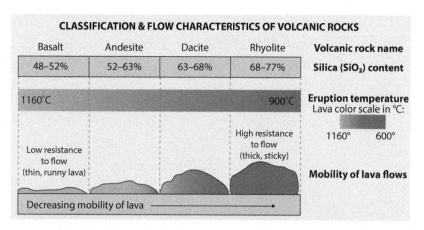

CLASSIFICATION & FLOW CHARACTERISTICS OF VOLCANIC ROCKS

Basalt	Andesite	Dacite	Rhyolite	Volcanic rock name
48–52%	52–63%	63–68%	68–77%	Silica (SiO₂) content

Figure 12.8 Relationship between silica concentration, the viscosity of lava, the temperature of eruption, and the four main types of volcanic rocks.

Mountains, a *continental volcanic arc* generated by oceanic–continental convergence very similar to that of the Cascades. **Dacite** has a slightly higher silica content than andesite, and **rhyolite** is the most silica-rich volcanic rock. Figure 12.8 illustrates how viscosity increases with silica concentration. You know that as viscosity increases, the amount of trapped gases increases. And the higher the gas pressure, the higher the probability of explosive eruptions. So rhyolitic magma tends to be the most viscous and gas-rich, leading to the most violent eruptions on the planet.

water. A solid substance is very resistant to flow and thus has a very high viscosity.) Silica molecules tend to chemically bind together, imparting a thick, sticky texture to silica-enriched magma. Gases become trapped within the highly viscous melt, enhancing the gas pressure and thus the potential explosivity of the magma. Conversely, the lower the silica content of a magma, the lower the viscosity. As you might guess, the thinner texture of low-silica magma permits the release of gases, lowering the gas pressure and thus the potential explosivity of the magma.

Silica content defines the four main types of volcanic rocks, and all of them are common components of volcanoes of the Cascades Range (Fig. 12.8). Other elements composing the minerals within volcanic rock, such as aluminum, calcium, iron, magnesium, potassium, and sodium, are present in variable amounts.

Basalt has the lowest silica content and is the most common product of volcanism associated with seafloor spreading at mid-ocean ridges—the entire global seafloor is composed of basalt (Chapter 1; Fig. 1.21). Basaltic lava is also dominant at "hotspot" volcanoes like Hawaii or Iceland. Basalt dominates the seafloor and hotspot volcanic islands primarily because the source magma is derived directly from the underlying low-silica mantle. Since the viscosity of basaltic lava is low, gases tend to leak out easily, traveling through fractures in the surrounding **country rock** to the ocean above or into the atmosphere. The low gas pressure of basaltic magma tends to create effusive eruptions of lava that ooze out of craters or fractures before spreading out laterally across the seafloor or down the sides of a volcano. As you'll see when we get to Hawaii Volcanoes National Park, runny basaltic lava flows tend to produce broad, low-profile edifices called shield volcanoes. (The **volcanic edifice** is its physical structure, composed of volcanic materials that accumulate through time.)

Andesite has an intermediate silica content and is named for the common volcanic rock of the Andes

Pyroclastic Eruptions and Stratovolcanoes

The magma residing within chambers beneath the volcanoes of the Cascades is always trying to buoyantly escape to the surface. As magma circuitously winds its way upward through whatever natural plumbing system exists, the pressure from the weight of overlying rock is reduced, allowing gases to bubble out of solution from the viscous magma. If a conduit to the surface suddenly opens that permits the magma to rush upward into the volcano, the rapid expansion of bubbles of gas instantaneously shatter the magma into countless solid particles of rock and glass called **pyroclastic debris** that explosively erupt from the volcano (Fig. 12.2). Driven by the pent-up energy of gas pressure, **pyroclastic eruptions** may form a vertical column that reaches several kilometers into the atmosphere. Prevailing winds typically transport the solid particles of rock and glass across great distances, with the debris eventually falling to the surface as a **pyroclastic fall** where it forms a widespread blanket of solid particles on the landscape.

Fragments of pyroclastic debris are given the general name **tephra** and range across a variety of sizes, from tiny particles of ash to boulder-sized blocks (Fig. 12.9). In the Cascades, tephra produced by pyroclastic eruptions typically has an andesitic, dacitic, or rhyolitic composition. **Volcanic ash** is composed of dust- to sand-sized particles of rock and microscopic shards of glass that are highly abrasive. Ash poses an acute hazard to airplanes as well as to plants and animals on the ground. The term *ash* is somewhat of a misnomer in that it is not the same as the soft, papery ash produced by burning wood, but rather consists of solid grains of rock and glass that pose a significant peril to humans and the structures we build, as well as to farmland, forests, grasslands, and the animals that live in those ecosystems.

Another abundant type of tephra deposit is **pumice**—low-density, porous fragments of glass spontaneously

Figure 12.9 A) Ash-sized tephra from Mount St. Helens 1980 eruption. B) The ash from the 1980 eruption formed a thick blanket across the landscape.

Figure 12.10 Pumice is a low-density, porous volcanic rock created during explosive, pyroclastic eruptions.

produced by the rapid degassing of magma during eruptions (Fig. 12.10). You might think of pumice as a substance similar to the gas-filled foam that forms the head of a beer as gases are released while the beer is being poured into a glass. Particles and chunks of pumice consist of a solid matrix of glass surrounding a network of frozen gas bubbles. Other than ash and pumice, pyroclastic debris may include large blocks of rock ripped from the walls of the throat of the volcano as the magma violently blasted outward.

Volcanoes created by intermediate- to high-silica magmas typically take the form of steep-sided, conical edifices called **stratovolcanoes**, produced by the accumulation of alternating layers of pyroclastic debris and viscous lava flows (Fig. 12.11). Most volcanoes of the Cascades, as well as most active volcanoes of the Ring of Fire, are stratovolcanoes. The individual layers composing a stratovolcano are the physical expression of hundreds of discrete eruptive events occurring over tens to hundreds of thousands of years. Stratovolcanoes typically have a summit crater where most eruptions originate, but many have multiple **vents** along their flanks and throughout the surrounding area, sourced from a fractured network of conduits beneath the surface. There are 13 active volcanoes in the Cascades, but there are over 3000 identified vents.

Eruptions in the Cascade Range over the Past 4000 Years

The imposing volcanic peaks of the Cascade Range were formed very recently in geologic time, primarily over the last 2+ m.y. of the Quaternary (Fig. 1.37). Several of the Cascades volcanoes are only a few thousand years old, extremely young by geologic standards. These fresh volcanoes are the latest manifestation of volcanism that's been occurring in the Cascades Province for the past 40 m.y. or so, as long as the Cascadia subduction zone has been active. Southward from central Washington, the active volcanoes of the modern Cascades rest on a platform of older, deeply eroded volcanic rocks from eruptions earlier in the Cenozoic. In the northern Cascades, two Quaternary volcanoes have pierced a foundation of much older nonvolcanic rocks.

USGS volcanologists, along with colleagues from academia and state geologists, have constructed a chronology of volcanic events in the Cascades over the past 4000 years (Fig. 12.12). This compilation represents a huge amount of work, both in the field and in the lab. Careful mapping and examination of the variety of volcanic rock units comprising individual volcanoes and the surrounding area permit volcanologists to identify unique eruptive events of the past. Sampling of individual volcanic units combined with geochemical analysis in the lab allows for the

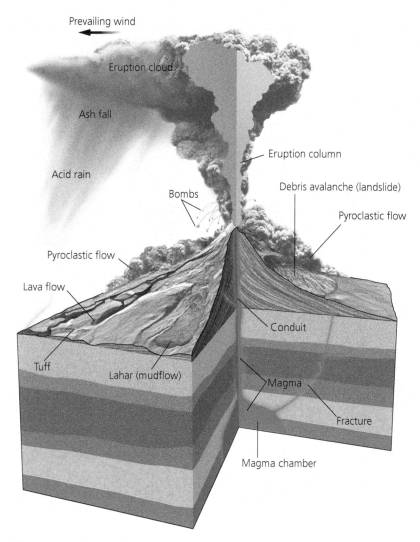

Figure 12.11 Stratovolcanoes are composed of irregularly alternating layers of pyroclastic debris and lava flows, built up over hundreds of eruptions.

Cascade Eruptions During the Last 4,000 Years

Silverthrone		▲
Franklin Glacier		
Meager		▲
Cayley		
Garibaldi		
Baker		▲
Glacier Pk.		▲ ▲ ▲ ▲ ▲ ▲
Rainier		▲ ▲ ▲ ▲ ▲
St. Helens		▲ ▲ ▲ ▲ ▲ ▲ ▲ ▲ ▲ ▲ ▲ ▲ ▲ ▲ ▲
Adams		▲
Hood		▲ ▲
Jefferson		
Three Sisters		▲ ▲ ▲ ▲
Newberry		▲ ▲
Crater Lake		
Medicine Lake		▲ ▲ ▲ ▲ ▲ ▲
Shasta		▲ ▲ ▲ ▲ ▲ ▲ ▲
Lassen		▲ ▲

4,000   3,000   2,000   1,000   200 0
Years Ago

Figure 12.12 Documented eruptions in the Cascade Range during the past 4000 years.

determination of a chemical "fingerprint" of specific eruptive units that permits correlation over a broad area. A variety of numerical dating techniques (discussed in Chapter 1) enable the age of specific volcanic units to be calculated.

Integrating all the fieldwork, lab analysis, and age-dating from each of the Cascades volcanoes results in the chronology presented in Figure 12.12. The eruptive episodes shown in the diagram are likely a minimum, however, because the evidence of many other events has been eroded beyond recognition or buried completely by younger events. It's important to remember that even though all of the Cascades volcanoes are born from the same tectonic system, each of them has its own unique magma chemistry and history of eruptive activity.

A few very interesting trends are revealed in Figure 12.12. (1) Of the 13 active volcanoes of the Cascades, 11 have erupted in the past 4000 years. Crater Lake shows no activity over the past 4000 years because it exploded catastrophically a little over 7000 years ago, a devastating eruption experienced by Native Americans in the region who suffered from the effects. (2) The duration between eruptions at any one volcano ranges widely from a few decades to a few thousand years. (3) Mount St. Helens in southern Washington has been the most consistently active volcano in the recent past, followed closely by Mount Shasta and Mount Rainier. (4) No eruptions appear to have occurred over the past 1000 years or so within the 450 km (280 mi) span between Mount Hood in northern Oregon and Mount Shasta in northern California. It's unlikely that the volcanoes in that stretch are extinct. Rather, their relative inactivity may be lulling us into a false sense of security.

Reaching closer toward the present, a total of 45 eruptive events from 7 Cascades volcanoes have been documented since 1750. Two national parks (Mount Rainier and Lassen Peak) are among the 7, while a third is a national monument (Mount St. Helens). Both Lassen Peak and Mount St. Helens have erupted within the 20th century. Furthermore, of these 7 historically active volcanoes, 6 are located relatively close to cities and towns, creating a significant and unpredictable risk.

Our human perception of volcanic eruptions naturally focuses on the hazards they pose to humans and the constructs of our modern civilization. But the products of volcanism through time have impacted the planet in many beneficial ways. Volcanic ash and weathered volcanic rocks release potassium and phosphorus, contributing to the fertility of volcanic soils and the resulting agricultural largesse. Many industries and our modern technological society depend on elements concentrated by volcanic processes. And countless eruptions through Earth history have transferred water vapor from the interior of the planet to the atmosphere where it condenses into the essential fluid for our hydrosphere and biosphere.

Differences between the Cascade Province and the Colorado Plateau

There are two features of the Cascades Geologic Province that differentiate it from the geology and landscapes of the Colorado Plateau discussed in Part 1. First, the age of the rocks and the age of the mountainous terrain of the Cascades are both geologically very young, related to the fact that both the rocks and the mountainscape form contemporaneously by the same youthful volcanic processes. This is in direct contrast to the Colorado Plateau where the rocks are very old but the erosional landscape is relatively young.

The second feature that distinguishes the national parks and monuments of the Cascades from those of the Colorado Plateau is the presence of glaciers and snow crowning the volcanic summits. The ponderous movement of glacial ice acts to sculpt volcanic landscapes of the Cascades, in contrast to the primary role of running water that carved the deeply incised canyons, broad mesas, and elongate escarpments of the Colorado Plateau. Moreover, the combination of explosive eruptions and the presence of glacial ice and snowy peaks make the Cascades volcanoes particularly hazardous to residents of the Pacific Northwest.

In the following chapters on national parks of the Cascades, we'll address many new geologic concepts, including the role of glaciers (Mount Rainier National Park), unimaginably huge eruptions of the past (Crater Lake National Park), historic eruptions in the Cascades (Lassen Peak National Park), and the threat of volcanic hazards to the rapidly growing population of the Pacific Northwest. We'll begin our journey, however, with Mount St. Helens National Volcanic Monument, the site of the most recent explosive eruption in the contiguous United States.

Key Terms

andesite
country rock
dacite
effusive eruption
explosive eruption
melt
microplate
pumice
pyroclastic debris
pyroclastic eruption
pyroclastic fall
rhyolite
silica
stratovolcano
tephra
vent
viscosity
volatile gases
volcanic ash
volcanic edifice

Related Sources

Harris, S. L. (2005). *Fire Mountains of the West*. Missoula, MT: Mountain Press.

Orr, W. N., and Orr, E. L. (2018). *Geology of the Pacific Northwest*, 3rd ed. Long Grove, IL: Waveland Press.

Tucker, D. (2015). *Geology Underfoot in Western Washington*. Missoula, MT: Mountain Press.

Cascades Range Volcanoes
    Accessed February 2021 at http://en.wikipedia.org/wiki/Cascade_Volcanoes

Cascades Volcano Observatory
    Accessed February 2021 at http://volcanoes.usgs.gov/observatories/cvo

USGS Volcanic Glossary and Images
    Accessed February 2021 at http://volcanoes.usgs.gov/images/pglossary/index.php

13

Mount St. Helens National Volcanic Monument

Figure 13.1 View to the south of the horseshoe-shaped amphitheater of Mount St. Helens today. The summit bowl is over 600 m (2000 ft) deep and 2 km (1.2 mi) wide.

If you're interested in witnessing the aftereffects of a recent pyroclastic eruption and don't have the resources to visit Indonesia or the Andes, then drive to Mount St. Helens National Volcanic Monument in southern Washington. There you'll see the regeneration of a volcanic landscape a few decades after the volcano exploded in May 1980, the most recent eruption in the contiguous United States (Fig. 13.1). The buildup and eventual eruption of the volcano mesmerized the American public for several months and was the most intensely documented and studied volcanic event ever at the time.

Today, the monument is a natural laboratory for a variety of scientific interests. Volcanologists monitor the mountain continuously with an array of instruments that sense the slightest of seismic vibrations beneath the surface and sniff the air for volcanic gases. They use satellites to assess changes in topography and heat flow, and they perform experiments to image the interior structure of the volcano. Landscape ecologists and biologists interested in the recovery of ecosystems impacted by cataclysmic events roam the area, recording changes in environmental rejuvenation. And tourists and adventurers visit the monument to hike, ski, mountain-bike, fish, watch wildlife, and learn about the immense natural forces associated with volcanoes.

Mount St. Helens National Volcanic Monument was established in 1982 to preserve the area around the volcano for education, scientific study, and recreation. Because it is located within the Gifford Pinchot National Forest, the U.S. Department of Agriculture's Forest Service administers the monument (Fig. 13.2). (From earlier chapters, you will recall that the only difference between a national park and a national monument is related to the political process; monuments can be established by presidential declaration without congressional approval, whereas a national park is created by an act of Congress.) Portland, Oregon, is the nearest major city, located 80 km (50 mi) to the southwest. Nearby volcanoes of the Cascade Range include Mount Adams, about 70 km (~40 mi) to the east, and Mount Rainier, 80 km (50 mi) to the northwest.

The Lewis River flows westward just south of the monument, whereas the Cowlitz River runs just north of the monument. Both rivers are tributaries within the Columbia River watershed. Most volcanoes, being somewhat conical edifices, exhibit **radial drainage** patterns; runoff of rain, snowmelt, and glacial meltwater flows radially away from the summit. Notice the way the smaller streams on Mount St. Helens (e.g., South Fork of Toutle River, Pine Creek) radiate outward from the main mass of the mountain (Fig. 13.2). The radial pattern changes once the flowing waterways reach the surrounding rolling terrain, with streams flowing westward in accord with the regional slope.

The goal of this chapter is not to describe each and every aspect of the monument, but rather to illustrate the wide variety of volcanic events that occur during explosive pyroclastic eruptions in the Cascades Range, based on the eruption of 1980. Refer back to Chapter 12 on the Cascades Geologic Province as necessary to refresh your memory on volcanic processes (e.g., how magma is generated, the role of silica) and products (e.g., volcanic rock types, pyroclastic debris).

Age and History of Eruptions at Mount St. Helens

Today Mount St. Helens is the decapitated remnant of a majestic stratovolcano that had been growing for only about 2200 years prior to the eruption of 1980. But the volcano has had a long and complicated history that can be traced back over the past 40,000–50,000 years. The volcano grew in volume during eruptive phases that lasted hundreds to thousands of years, only to erode and shrink during dormant phases of comparable duration. This is a typical story for most volcanoes: alternating episodes of growth and decay as activity in the magma chamber deep beneath the volcano waxes and wanes through time.

Ancestral versions of Mount St. Helens grew during four eruptive phases since ~50,000 ka (ka = thousands of years ago). Each of the four active episodes lasted thousands of years, and each is separated by dormant intervals between 35–28 ka, 18–16 ka, and 12.8–3.9 ka. Each new phase of eruptive activity partially destroyed or covered the evidence of previous eruptions, complicating the painstaking forensic geology necessary for determining the ancient chronology of events. The most recent phase of growth since 3.9 ka occurred in bursts of pyroclastic eruptions and lava flows of dacitic magma, with lesser amounts of andesitic and even basaltic composition. (Dacite has a silica content of around 65 percent.) The pre-1980 edifice of Mount St. Helens (Fig. 13.3) was primarily constructed over the last 2200 years following a dormant phase lasting a few hundred years during which the volcano was deeply weathered and eroded. Over the past 4000 years, Mount St. Helens has been the most active volcano in the Cascades, erupting with a recurrence interval ranging from a few decades to a few centuries (Fig. 12.12).

Precursor Events to the 1980 Eruption of Mount St. Helens

The story of the Mount St. Helens eruption in 1980 has been told and retold in professional journal articles from the U.S. Geological Survey (USGS) and academics, in popular books and videos, and in several websites. Recounted here are only the highlights of the events that preceded the eruption, the eruption itself, and the aftermath. Since the eruption was so well documented, it provides a model for understanding and potentially forecasting future eruptions of Mount St. Helens or other Cascades volcanoes, possibly reducing the risk to the populace of the Pacific Northwest. To help imagine the immense amount of energy involved in the eruption, compare the photographs of modern Mount St. Helens in 2014 (Fig. 13.1) with the pre-eruption image in 1979 (Fig. 13.3). The mountainscape

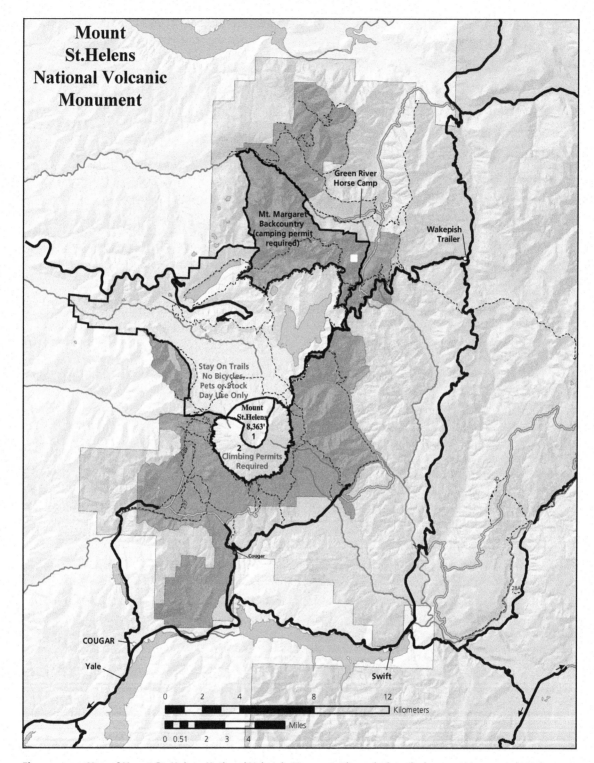

Figure 13.2 Map of Mount St. Helens National Volcanic Monument boundaries, drainage patterns, and roads.

of Mount St. Helens changed in about a minute and a half on that fateful day in May 1980.

After 123 years of dormancy, the first indication of imminent change at the volcano was a magnitude 4.1 earthquake that shook the mountain in March 1980. Seismometers installed by the USGS in the following weeks recorded a relatively continuous series of low-magnitude, rhythmic quakes known as **harmonic tremor**. These distinctive

quakes are commonly ascribed to the intermittent movement of magma through narrow fractures within the volcano, forcibly widening the cracks and creating a constant stream of small quakes. An alternative interpretation is that harmonic tremors are the noise created by the vibration of gas bubbles as they separate from liquid magma.

A small **phreatic eruption** of steam and ash occurred in late March, caused by the interaction of rising magma with

Figure 13.3 View of pre-eruption Mount St. Helens looking northeast toward Mount Adams in the distance, September 1979. Before the eruption, Mount St. Helens was 2950 m (9677 ft) in elevation; after the eruption, the elevation was 2549 m (8363 ft).

cool groundwater. As the two fluids of different temperature interacted, the groundwater abruptly flashed to vapor, forcing a column of steam and ash upward a few kilometers into the atmosphere. This event attracted curious sightseers to the mountain, complicating safety measures enacted by decision makers who acted on the warnings of USGS geologists of the potential for a catastrophic eruption. As a drumbeat of small earthquakes and several steam eruptions continued to rattle the mountain, a state of emergency was called in early April, with access to the immediate area cordoned off by manned roadblocks. This decision likely saved many lives since the main eruption was just a few weeks away.

From mid-April into May, a **topographic bulge** grew on the north side of the mountain at a rate of 2 m (6.5 ft) per day, pushed upward by the accumulation of magma just beneath the surface. The bulge grew to 1.5 km (~1 mi) across and over 100 m (330 ft) high. An arcuate fracture developed along the bulge near the summit, suggesting the possibility of a disastrous collapse of the north face of the volcano. The progressive buildup of precursor activity on the mountain pointed toward an imminent eruption.

Events Related to the Main Eruption of Mount St. Helens

The main eruption began at 8:32 a.m. on May 18 and consisted of five interrelated, sequential events. (1) An earthquake with a magnitude of 5.1 shook the mountain, abruptly releasing

accumulated strain along the arcuate fracture along the margin of the bulge near the summit. (2) Simultaneous with the earthquake, the volcano's north flank detached in an enormous, chaotic landslide called a **debris avalanche**, consisting of giant blocks of rock, glacial ice, tens of thousands of trees, and a loose slurry of volcanic ash and soil (Fig. 13.4). Traveling at more than 200 km/hr (125 mph) and with an estimated volume of 2.5 km³ (0.6 mi³), this was perhaps the largest landslide in recorded history. (3) About 20 to 30 seconds later, a **pyroclastic flow** of ash, superheated gases, and huge blocks of rock rocketed laterally to the north, derived from the release of pressurized magma residing beneath the bulge (Fig. 13.5). This **lateral blast** of pyroclastic debris traveled at speeds of over 500 km/hr (>300 mph) and rapidly accelerated past the debris avalanche. The temperature of the flow was estimated to be about 300°C, so every plant and animal in a 600 km² (230 mi²) area was scorched as the lateral blast swept by. The rapidly moving, turbulent flow scoured the ground surface, ripping up the soil and blowing down trees, stripping them of bark and branches.

(4) The lateral blast was followed immediately by the vertical explosion of gas-charged pyroclastic debris (Fig. 12.2). In 15 minutes, the **eruption column** reached 25 km (16 mi) into the atmosphere and lasted for 9 hours before tapering off. High-altitude winds carried the ash eastward, where it eventually fell as a thin, dense layer of *tephra* across eastern Washington, Idaho, and western Montana. As the skyward thrust of the eruption diminished over time, portions of the pyroclastic column collapsed onto the flanks of the volcano, hurtling downslope as isolated pyroclastic flows.

Figure 13.4 Debris avalanche of north flank of Mount St. Helens about 10 to 20 seconds after the beginning of the massive landslide.

Figure 13.5 About 40–50 seconds after the start of the debris avalanche, the northward-directed, light-colored pyroclastic flow was immediately followed by the skyward explosion of dark-colored pyroclastic debris.

(5) The intense heat of the eruption melted glacial ice and snow near the summit of the mountain that mixed with volcanic ash to form a thick slurry that flowed down the flanks of the volcano. These **volcanic mudflows** (aka **lahars**) surged into river valleys near the lower elevations around the mountain where they mixed with sediment and soil to form a rapidly moving mush with the consistency of runny, wet cement. The lahars ripped up tens of thousands of trees lining the banks of the streams, adding to the destructive power of the flows. Several bridges, roads, and homes were demolished as the flows swept past at speeds of about 40 km/hr (25 mph), but thankfully nobody died due to speedy evacuations from the lahar hazard zone along the river corridors. The mudflows eventually reached the Columbia River where they clogged its shipping channel. By the end of the day, the main events of the eruption had ended, leaving behind a barren moonscape, devoid of life.

Aftermath of the Eruption

The combined effects of the massive debris avalanche, the lateral blast, and the vertical explosive eruption removed the top 400 m (1300 ft) of Mount St. Helens. The mountain that had a conical symmetry before the eruption had been beheaded, leaving behind a gaping amphitheater, open to the north (Fig. 13.6). Fifty-seven people died, most of them from suffocation after inhaling the hot ash and gases of the main pyroclastic flow. A few hundred homes were destroyed, several bridges and railways were ruined, and volcanic deposits covered about 300 km (180 mi) of highway. Millions of trees were flattened across a huge area, many of which were harvested as lumber in the succeeding months. Wildlife was devastated, with millions of small and large animals incinerated or buried beneath ash and solidified lahar deposits (Fig. 13.7). Fallout of tephra from the eruption column extended across a broad region east of the volcano, choking engines and paralyzing the populace for weeks. Economic losses to the affected region were estimated at $1.1 billion.

The majority of the dacitic magma was erupted during the main event on May 18, but within a few months a **lava dome** of viscous, gas-depleted dacite had risen to plug the vent at the summit of the volcano (Fig. 13.8). Over the succeeding decades, the dome grew via a series of small eruptions of lava alternating with phases of decay by rockfalls

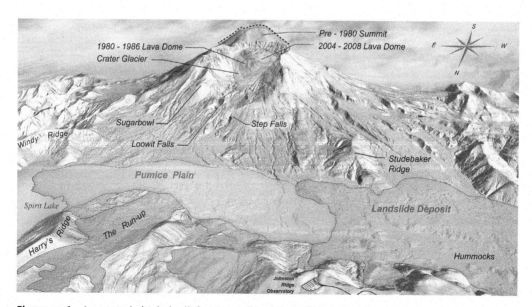

Figure 13.6 Annotated shaded relief map produced from digital elevation data showing the pre- and posteruption summits of Mount St. Helens as well as the distribution of volcanic deposits from 1980 to 2008. The north-facing summit amphitheater is almost 2 km (1.2 mi) across. The "Pumice Plain" was created by pyroclastic fallout during the eruption.

Figure 13.7 Lahar deposit left behind as the flows diminished, Muddy River. Over 200 km (120 mi) of river channels were affected by the mudflows. Note the peak height of the lahar marked by the mudline partway up the standing trees. Geoscientist in orange for scale.

Figure 13.8 Dacitic lava dome plugging the crater of Mount St. Helens, including a whaleback-shaped spine of lava, 2005. Cracked gray lobe to the left of the lava dome is an ash-covered glacier.

and glacial erosion. Ominously, the dome acts to trap gases in the magma reservoir deep beneath the surface, building pressure toward the next major eruption.

Geologists consider Mount St. Helens to be the most likely volcano in the Cascades to erupt in the near future. No one is sure when that event may occur, but the mountain is heavily instrumented and continuously monitored for rumblings that may presage the next major eruption. By human standards, the 1980 eruption of Mount St. Helens was catastrophic, but it was a relatively small event

by volcanic standards. You'll be forced to imagine the unimaginable when we visit Crater Lake and Yellowstone national parks in later chapters where you'll learn about the magnitude of colossal eruptions in the geologic past.

A final thought: Some call the decapitated remnant of Mount St. Helens "ugly," but many others consider the other-worldly landscape and surrounding national forest to be a wonderland of beauty and rebirth. The monument is very much worth a visit to witness the power of nature, both as an agent of destruction and a wellspring of environmental rejuvenation. In the next chapter, we'll visit another active Cascades volcano that poses an even greater risk to the Pacific Northwest: Mount Rainier National Park.

Key Terms

debris avalanche
eruption column
harmonic tremor
lahar
lateral blast
lava dome
phreatic eruption
pyroclastic flow
radial drainage
topographic bulge

Related Sources

Doukas, M. P. (1990). Road Guide to Volcanic Deposits of Mount St. Helens and Vicinity. *U.S. Geological Survey Bulletin* 1859, 53 p. Accessed February 2021 at http://pubs.er.usgs.gov/publication/b1859

Foxworthy, B. L., and Hill, M. (1982). Volcanic Eruptions of 1980 at Mount St. Helens: The First 100 Days. *U.S. Geological Survey Professional Paper* 1249, 125 p. Accessed February 2021 at http://pubs.er.usgs.gov/publication/pp1249

Mount St. Helens National Volcanic Monument. Accessed February 2021 at https://www.fs.usda.gov/recarea/giffordpinchot/recarea/?recid=341439)

Mount St. Helens Science and Learning Center. Accessed February 2021 at http://www.mshslc.org/gallery/1980-eruption

USGS videos of the 1980 eruption. Accessed February 2021 at https://volcanoes.usgs.gov/observatories/cvo/st_helens_videos.html

USGS Volcano Hazards Program: Mount St. Helens. Accessed February 2021 at http://volcanoes.usgs.gov/volcanoes/st_helens

14

Mount Rainier National Park

Figure 14.1 View of Mount Rainier from the west. Note the elongate white glaciers extending downslope from the cap of snow and ice at the crest. Narrow ribs of rock separate the glaciers. The long glacier to the right of center descending from the crest is the Tahoma, whereas the adjacent glacier to the left of Tahoma is the Puyallup. The steep concave cliff above the Puyallup Glacier near the crest is Sunset Amphitheater.

All of the Cascades volcanoes are unique in their appearance, but the most iconic has to be Mount Rainier, visible to millions of residents of the Seattle–Tacoma metropolitan area (Fig. 14.1). Rainier is sheathed in a radiating series of alpine glaciers that form the largest single-peak glacial system in the contiguous United States. Narrow ridges of dark volcanic rock separate each of the glaciers, creating a stark contrast with the white rivers of ice. Lower on the slopes, dense vegetation creates a green blanket juxtaposed against the monochromatic shades of the upper mountain.

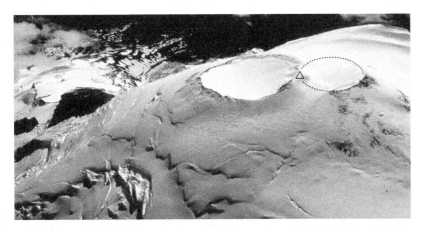

Figure 14.2 View of the snow- and ice-covered crest of Mount Rainier and the overlapping summit craters with their rims of exposed rock. The main crater is about 400 m (1300 ft) across, with the highest point on the rim forming the Columbia Crest, the true summit of the mountain (triangle). Dashed line marks snow-obscured margin of one summit crater.

Mount Rainier is far from a classic conical stratovolcano, but rather has a craggy asymmetry that changes depending on the viewer's perspective. The crest of the mountain appears decapitated, crowned with a rounded mantle of snow and ice that partially masks two overlapping summit craters (Fig. 14.2). The narrow rim of rock that defines the main crater reaches an elevation of 4392 m (14,411 ft) at the Columbia Crest, making Rainier the highest mountain in the Cascades and the second highest in the coterminous United States. The broad white summit of the mountain abruptly gives way to steep shoulders of dark rock that form precipitous cliffs gouged by glacial erosion. As the narrative of the history of Rainier unfolds in this chapter, you'll see how the heat of volcanism interacted with the bone-chilling cold of glacial ice to govern the construction of the mountain and its modern, asymmetric appearance.

Mount Rainier National Park is located about 90 km (55 mi) southeast of Seattle (Fig. 12.3) and is notable for its old-growth forests, spectacular wildflower meadows, and abundant wildlife. Consequently, the park accommodates around 2 million visitors per year who hike the outstanding trail system, attempt the oxygen-sapping climb to the summit, or simply gaze at the stunning vistas. Much of the park is inaccessible in the winter due to the abundant snowfall, which typically exceeds 7.5 m (>300 in) at Paradise just south of the mountain.

The copious amount of snow is generated by moisture-laden weather systems that move eastward from the Pacific. The air masses progressively expand and cool as they rise up the west flank of the mountain, with the moisture condensing to drop massive amounts of snow during much of the year. Snow high on Rainier's flanks feed the network of **alpine glaciers**, elongate "rivers" of ice that extend downslope from the crest. Meltwater from snow and glacial ice, as well as rainfall, drain into six major rivers and numerous streams that flow radially

away from the mountain. As they reach the surrounding lowlands, most of the river systems reorient westward and flow into Puget Sound. Along the way, many of the rivers are dammed for the generation of hydroelectric power.

This chapter has four main goals. The first is to build on the concepts introduced in Chapter 12 regarding the fundamentals of volcanism and the tectonic setting of Cascadia to explain the geology and landscape evolution of Mount Rainier National Park. A second goal is to address the critical role of the Ice Ages on the modification of mountainscapes. The Ice Ages of the past 2.6 m.y. had a profound impact not only on the terrain of the Cascades, but on almost all of our national parks. The cyclic changes in climate of the Ice Ages are briefly introduced in this chapter and will be built upon in subsequent chapters as necessary.

A third goal is to describe the behavior of the modern glaciers that drape Rainier's slopes (Fig. 14.3). Mount Rainier has a total volume of glacial ice greater than all other Cascades volcanoes combined and is thus the ideal place to understand how they work. Glaciers are not only important landscape features on their own, but they also create a variety of landforms as they expand and retreat through time. Many of the most famous vistas in our national parks are the result of glacial activity. This is particularly true at Mount Rainier National Park where glaciation and volcanism are not separate entities, but rather influence the behavior of each other through time to dictate the landscape evolution of the mountain.

The fourth goal is to address the volcanic hazards that pose a threat to the populace of western Washington. The natural beauty of Mount Rainier masks the potential for catastrophe. Because of its proximity to the Seattle–Tacoma metropolitan area, Mount Rainier is commonly known as "the most dangerous volcano in America." This notoriety is not solely related to the possibility of renewed volcanism, but rather to the inherent geologic conditions of the mountain that are conducive to generating *volcanic mudflows* (Chapter 13). This risk to the general public of the region around Rainier is addressed by comparing the eruptive history and volcanic hazards of Rainier to the variety of volcanic events described in the preceding chapter on Mount St. Helens.

Pleistocene Ice Ages

About 2.6 million years ago, the global climate became significantly colder than it was previously, likely triggered by changes in ocean circulation and the consequent shift in the transfer of heat energy around the planet. This time marks the beginning of what is known as the **Pleistocene**

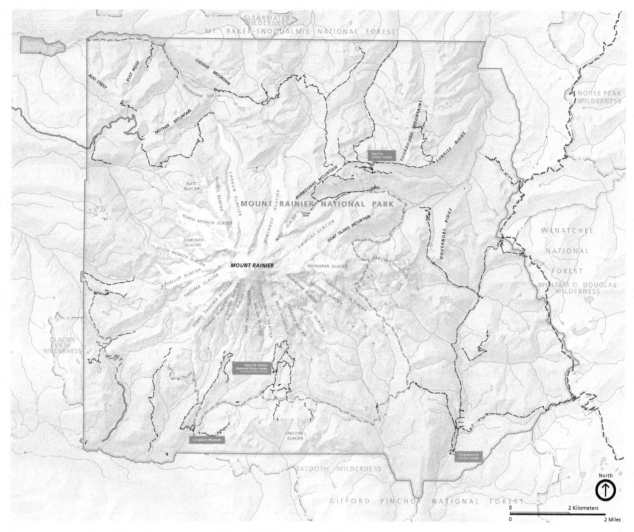

Figure 14.3 National Park Service map of Mount Rainier National Park. Twenty-five named alpine glaciers radiate downslope from the crest of the mountain.

Ice Ages, which lasted to about 12,000 years before present (12 ka). With colder temperatures and increased snowfall, thick sheets of ice began to grow in high northern latitudes and to extend southward across North America and Eurasia (Fig. 14.4). These **continental ice sheets** reached thicknesses of 3–4 km (2-2.5 mi) and are very different from the much smaller alpine glaciers that dominate Mount Rainier and other mountain ranges globally.

Over the course of the Pleistocene, continental ice sheets expanded and contracted 20 to 30 times, controlled by cyclic variations in Earth's orbit and the tilt of its axis. During glacial maxima in North America, vast white sheets of ice covered most of Canada and extended across New England and the upper Midwest. Glacial ice covered up to 32 percent of the land area on Earth during these episodes of peak glaciation. During glacial minima, the ice sheets would recede back toward higher latitudes and higher elevations. Swift-flowing streams and rivers emanated from the melting front of the massive sheet of ice. Earth is currently in a warm, interglacial climate phase,

with continental ice sheets constrained to Greenland and Antarctica.

The most recent phase of peak glaciation occurred only about 20,000 years ago and is known as the **Last Glacial Maximum** (LGM). During the LGM, continental ice sheets extended into northern Washington. A broad lobe of the ice sheet perhaps a thousand meters thick covered what is today the Puget Lowlands. Cold temperatures and the ready availability of moisture from the Pacific caused the formation of an interconnected system of ice caps and alpine glaciers crowning the peaks of the Cascades. The volcanic edifice of Mount Rainier would have been barely visible beneath its thick blanket of glacial ice during the LGM. (The details of Pleistocene glaciation in coastal Washington are more fully addressed in Chapter 30 on Olympic National Park.)

As global temperatures began to rise after the LGM, the continental ice sheets began to melt and retreat northward. The Pleistocene ended about 12,000 years ago as the planet entered a phase of globally equable climate known

Figure 14.4 Maximum extent of continental ice sheets during the most recent Ice Age about 20,000 years ago. Arrows show flow directions of ice.

Alpine glaciers consist of elongate lobes of ice that ponderously flow downslope under their own weight, driven by gravity (Fig. 14.5). Glaciers move in a process called "creep," with the ice literally flowing as a plastic material within the body of the glacier. ("Plastic" refers to a material that slowly flows and deforms under stress and never returns to its original form—Chapter 1.) Alpine glaciers may also move downslope due to sliding along the underlying soil or bedrock, sometimes with the aid of a basal layer of liquid water. The ice at the base of a glacier may melt due to the sheer weight of the ice above, providing a lubricant for the ice to glide on.

Alpine glaciers are divided into two general regions. At the highest elevations where the glacier originates (also known as the "head" of the glacier), the area is called the **zone of accumulation**. At these high elevations, snow falls and accumulates during a prolonged winter. If the summer is not too warm, the previous winter's snow survives and gets buried and compacted by the following winter's snows. Over several successive cold years, the older snow at the bottom of the pile becomes compressed into denser particles, with trapped air forced outward. With time and increased accumulation, the dense snow particles transform into solid crystals of ice, with most of the air squeezed out by the pressure of overlying ice and snow. This crystalline ice commonly takes on a bluish cast. At some point, the mass of the ice becomes great enough that the glacier begins to flow downhill under the force of gravity. To reach this threshold when the ice begins to flow, it might take a thousand years and 400 m (1300 ft) of excess snow to make 40 m (130 ft) of ice.

At the lowest end of the glacier (called the "toe" or the "terminus"), the temperatures are commonly warm enough that the ice begins to melt. This creates meltwater streams and ponds along the toe of the glacier. The ice at the terminus may also vaporize directly into the atmosphere in a process called **sublimation**. And, if the toe of the glacier ends in the ocean or a lake, huge chunks of ice may break off and float away as icebergs. The decay of glacial ice by melting, sublimation, and physical breaking occurs at the lower end of an alpine glacier in the **zone of ablation**.

If the rate of accumulation at the head exceeds the rate of ablation at the toe, the glacier advances downslope. This typically happens during phases of colder climates. Conversely, if the rate of ablation exceeds the rate of accumulation, the glacier retreats upslope by loss of ice at its toe. This typically happens during warmer climatic phases.

as the **Holocene**. The thick mass of glacial ice that mantled the Cascades disappeared as the global climate warmed. The alpine glaciers we see in high elevations today, including those of Rainier, are not remnants of the LGM, but rather were born during the **Little Ice Age** that spanned about 500 years from ~1350 to ~1850 CE (Common Era). We'll explore the evolution of glacial landscapes during the Pleistocene Ice Ages in greater depth in later chapters. The effects of Ice Age glaciers might arguably be the single most important factor in the formation of landscapes in many of our national parks.

How Alpine Glaciers Work

Twenty-five named glaciers and several smaller snowfields dominate the high elevations of Mount Rainier (Fig. 14.3). Collectively, the glaciers cover 90 km² (35 mi²), about 10 percent of the park area. The long tongues of glacial ice are **alpine glaciers**, which are sometimes referred to as "mountain" glaciers or even "valley" glaciers because they typically occupy and modify former stream valleys. In this chapter, we'll focus on the mechanics of how alpine glaciers work, while introducing a few of the landscape features that they create. We'll delve deeper into the erosional and depositional landforms left by Ice Age alpine glaciers when we visit the national parks of the Sierra Nevada in Part 3.

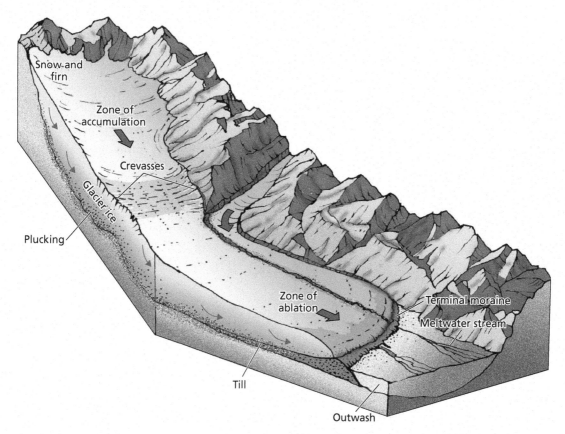

Figure 14.5 Main features of alpine glaciers. Curved arrows within the body of ice represent the internal plastic flow of ice.

(Incidentally, the glacier doesn't actually move upslope; rather, it just appears to move that way due to melting and loss of ice at the toe. It's really just the *position* of the toe of the glacier that retreats upslope, not the entire glacier itself.) Other than global and regional changes in temperature, the other key factor that influences the advance or retreat of alpine glaciers is the availability of moist air masses that control the supply of snow. The close proximity of the Pacific Ocean to Mount Rainier and the other Cascades volcanoes ensures a steady source of moisture.

For most of Rainier's lifetime of about a half-million years, massive alpine glaciers sculpted the morphology of the mountain and physically interacted with episodic pulses of volcanic activity. The modern alpine glaciers mantling the slopes of Mount Rainier are relatively puny compared to their ancestors that existed during the LGM of the Pleistocene about 20,000 years ago. Careful mapping of a variety of glacial landscape features in the region surrounding Rainier indicates that some alpine glaciers of the LGM reached over 100 km (60 mi) outward from the base of the mountain. The progressive climatic warming that marks the Holocene interglacial stage caused a dramatic retreat of Rainier's glacial system, with the modern glaciers regrowing during the Little Ice Age phase of global cooling. Observations of Rainier's glaciers since the mid-1800s indicate that they've

lost up to 35 percent of their surface area and are continuing to lose mass yearly to a combination of natural climatic warming, human-induced warming, and drier winters. The south-facing Nisqually Glacier has receded over 2 km (1.2 mi) from its Little Ice Age peak extent, leaving behind piles of sedimentary debris in its wake (Fig. 14.6).

Figure 14.6 Unconsolidated glacial debris left behind as the Nisqually Glacier (NG) recedes upslope.

Glacial Landforms on Mount Rainier

The exposed ice at the terminus of many glaciers on Rainier can be reached by trails where you may be surprised by how rocky and gritty the toes of glaciers can be. Alpine glaciers transport enormous amounts of loose rock and sediment within the body of the ice, on the surface of the ice, and at the front of the flowing mass of ice. As the toe of the glacier melts and recedes during warm phases, the entrained rocky debris is left behind as moraines and outwash plains (Fig. 14.7). **Moraines** are elongate ridges of unconsolidated mixtures of boulders, pebbles, sand, and clay. **Outwash plains** are formed as debris left behind by receding glaciers is eroded and redeposited by meltwater streams draining the glacial terminus. The moving water in the streams sorts these outwash deposits so that the coarse gravels are laid down near the moraines while finer particles are deposited farther away. Meltwater streams are commonly choked with glacial debris and switch their course repeatedly, commonly laying down a broad plain of outwash sediment through time.

Several alpine glaciers on Rainier are notable for their classic morphology and associated landforms. Tahoma Glacier, on the mountain's southwest face (Fig. 14.8), descends steeply from the summit to its current terminus at 1700 m (5577 ft). Narrow ridges of bedrock called **arêtes** bound the Tahoma on both sides. (At Rainier, rib-like arêtes are informally known as **cleavers**.) On most mountains, arêtes are erosional remnants created as glaciers on either side scour the rocky wall separating them. On Rainier, however, arêtes are constructional features as well in that they were built upward by the focusing of lava flows along the flanks of alpine glaciers. Ancient lava flows diverted by the heads of glaciers commonly streamed along

Figure 14.8 Tahoma Glacier (TG) descends from the summit ice cap of Rainier. The glacier is bound on both sides by narrow ribs of rock, locally called "cleavers." The steep headwall of the concave cliff labeled SA is Sunset Amphitheater, formed by glacial erosion as well as by catastrophic collapse. SM is South Mowich Glacier and PG is Puyallup Glacier.

the narrow crests of arêtes where they cooled and solidified, building the arêtes upward in the process.

Another spectacular glacial landform on Rainier is Sunset Amphitheater, the steep, concave cliff face located near the head of South Mowich glacier (Fig. 14-8). Known as a **cirque**, the amphitheater formed as the growing glacier "quarried" the adjacent rock, breaking it down and steepening the cliff by countless rockfalls. A catastrophic collapse of the steep cirque walls of Sunset Amphitheater around 2600 years ago was likely the trigger of a huge *debris avalanche* that evolved into a *volcanic mudflow* as it rumbled down the steep slopes of Rainier.

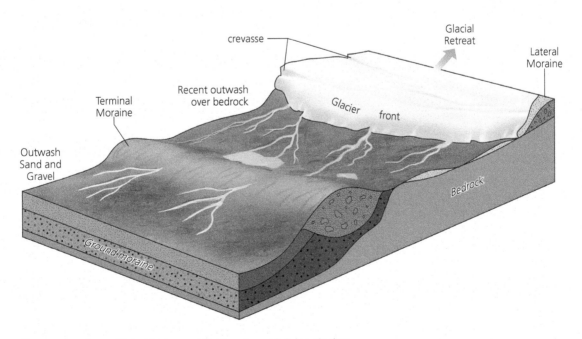

Figure 14.7 Depositional features at the terminus of alpine glaciers.

Figure 14.9 Google Earth image of Mount Rainier looking southeast toward Mount Adams in the distance. The Carbon Glacier flows northward (outlined). The steep cliff at the head of Carbon Glacier is the Willis Wall, a 1100-m-high cirque headwall. (The green line marks part of the national park boundary.)

One other Rainier glacier described by superlatives is the Carbon Glacier that flows northward from the mountain (Fig. 14.9). The Carbon is not only the largest glacier in the park, but the thickest and most massive glacier in the coterminous United States. Over 200 m (660 ft) thick and not quite a cubic kilometer in volume, the Carbon Glacier reaches over 9 km (5.6 mi) from head to toe—the longest glacier on Rainier. Its terminus ends at 1067 m (3500 ft) in elevation, the lowest in the lower 48 states. The Carbon Glacier originates at the base of the Willis Wall, an 1100 m (3600 ft) high cirque headwall composed of stacked layers of ancient lava flows. The Willis Wall cirque is likely the tallest and most precipitous of any Cascades volcano.

Pleistocene Eruptions at Mount Rainier

Mount Rainier has had a long and sporadic history of eruptions and growth. Multiple volcanic events, irregularly spaced over the last half-million years of the Pleistocene, occurred coincidentally with Ice Age glaciation. Like many volcanoes, eruptive phases of volcanic upbuilding alternated with phases when glacial erosion dominated; only here at Rainier the interaction of volcanism and glaciation is more pronounced due to the perennial crown of glaciers mantling its slopes. Lava flows occurred beneath glacial ice, along the margins of glaciers, and likely even on top of glacial ice as well. Glacial-interglacial cycles

Figure 14.10 Stacked layers of andesite lava flows, lower Tahoma Cleaver.

march to their own rhythms, dictated by the orbital variations that determine the periodicity of Ice Age climate. But volcanic activity is much less periodic and less predictable. When volcanic phases coincided with warmer interglacial stages, the volcano experienced net growth. When glacial phases dominated over volcanic activity, the volcano experienced net erosion and decay. The bottom line is that the history of eruptions that built Mount Rainier can't really be understood without integrating the synergistic role that glaciation played in shaping the mountain.

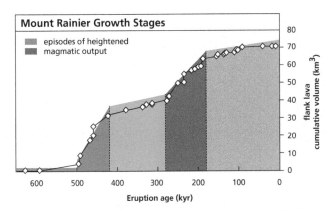

Figure 14.11 Plot of growth stages of Mount Rainier illustrating long-term phases of heightened magmatic output when Rainier would experience net growth and equally long phases of reduced output when Rainier would experience net decay.

The Mount Rainier stratovolcano is composed primarily of andesite and lesser amounts of dacite produced during hundreds of effusive lava flows over the last 500,000 years (Fig. 14.10). Pyroclastic debris composes a relatively small percentage of Rainier's volume, although explosive eruptions were relatively common based on multiple thin layers of tephra interbedded with solidified lava flows. Mapping suggests that most eruptions occurred from summit craters, but many other eruptions emanated from satellite vents and fissures located along the flanks of Rainier. Some lava flows reached up to 22 km (14 mi) away from the summit during active eruptive phases, but most flows (being moderately viscous andesitic composition) stayed within about 8 km (5 mi) of the summit, building the stratovolcano upward. Ninety percent of the mass of Rainier is composed of lava flows, with the remaining mass including pyroclastic deposits, lahar deposits, and glacial debris.

The chronology of events described below is not intended to be comprehensive, but rather is a generalized synthesis of the major phases of volcanic growth and decay at Rainier, complicated by the overlapping effects of glaciation. The sequence was constructed by careful geologic mapping combined with geochemical "fingerprinting" of various lava flows and pyroclastic deposits. Radiometric dates were determined for each of the volcanic units so that a numerical chronology of events could be defined (Fig. 14.11). Much of the evidence for past eruptions is now deeply buried beneath glaciers and younger lava flows, likely making the chronology somewhat incomplete. But more than enough volcanic material remains to tell the story of Rainier's evolution.

Modern Mount Rainier rests on an underlying foundation of 14- to 18-million-year-old granitic rock, exposed in outcrops surrounding the mountain. A precursor andesite volcano to Rainier pushed up through the granitic rock 1 to 2 m.y. ago but was almost entirely eroded by Pleistocene glaciers prior to the origin of the modern mountain. The Rainier that we know today originated about 500,000 years ago as a voluminous series of andesitic and lesser dacitic lava flows that pierced through the older rocks and erupted intermittently till about 420,000 years ago (Figs. 14.11, 14.12). Andesite within this age range is found over a broad area along the east, northeast, and west flanks of Rainier. Volcano growth via eruptions was modulated by ongoing glacial erosion whose intensity varied with the rhythms of Pleistocene climate change. Production of magma subsequently tailed off during a phase of reduced volcanic output that ranged from 420,000 to about 280,000 years ago. Glacial erosion dominated during this phase, and the mountain likely was diminished in volume, with the debris carried off by the ancient drainage system.

A renewed surge of volcanism and cone-building began around 280,000 years ago and lasted to about 180,000 years ago (Fig. 14.11). Many of these lava flows are evident as stacked layers composing narrow cleavers radiating from the crest. Active lava flows from the summit were diverted around the heads of alpine glaciers to pathways along their margins, which were composed of rib-like arêtes that separated adjacent glaciers. The lava flowed along the axes of

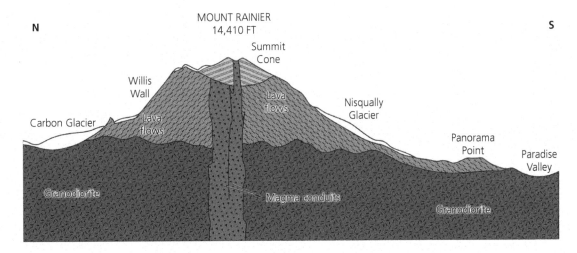

Figure 14.12 Schematic cross section of Mount Rainier oriented north–south showing the main bodies of rock composing the mountain.

Figure 14.13 View of Mount Rainier showing the horseshoe-shaped outline of the Osceola collapse from the northeast flank (red dashed line). Later lava flows have mostly filled in the collapsed crater, and the heads of the Winthrop and Emmons glaciers form an icy veneer.

the arêtes, building them upward into narrow ridges of rock as the lava cooled and solidified with time (Fig. 4.10).

Individual lava flows from this growth phase range in thickness from about 4 to 18 m (13 to 60 ft), with some of the more extensive flows thickening downslope to 60 m (200 ft) or more as the lava piled up on the lowlands surrounding the volcanic cone. Magma production leveled off after this pulse ended around 180,000 years ago. Even though glacial erosion was highly active during this latest phase, Rainier continued to build upward from small eruptions, perhaps reaching over 4600 m (>15,000 ft) in elevation by the end of the Pleistocene around 12,000 years ago.

Holocene Eruptions at Mount Rainier

The last 12,000 years of volcanic activity on Rainier is relatively well preserved because the effects of glacial erosion have diminished as alpine glaciers receded during the climatic warming of the Holocene. Some of these Holocene eruptions are recorded as lava flows, but many are preserved as lahars in the valleys surrounding the mountain. One of the larger eruptive events produced the **Osceola mudflow, which** occurred around 5600 years ago (~3600 BCE) when magma accumulated in the core of the volcano, eventually erupting laterally from the northeast flank of Rainier's summit. The eruption triggered a massive *debris avalanche* that roared downslope, leaving a horseshoe-shaped collapse crater in its wake (Fig. 14.13). The volume of the Osceola debris avalanche was about 1.5 times the size of the 1980 debris avalanche at Mount St. Helens.

Snow and glacial ice melted and transformed the chaotic jumble of rock into a *volcanic mudflow* (aka *lahar*) that plunged into tributaries of the White River near the base of Rainier's northeast margin (Fig. 14.14). The dense slurry of debris grew in volume as it mixed with river water, stream gravel, glacial debris, soil, and uprooted trees to flow northward as a tumultuous mass over 100 m (~330 ft) thick and moving at up to 60 km/hr (~40 mph). Turning west, the raging lahar traveled almost 120 km (75 mi) from its source, spreading out over a huge area of the Puget Lowlands before flowing into Puget Sound near what is today the port of Tacoma. Within a few hours, what had once been

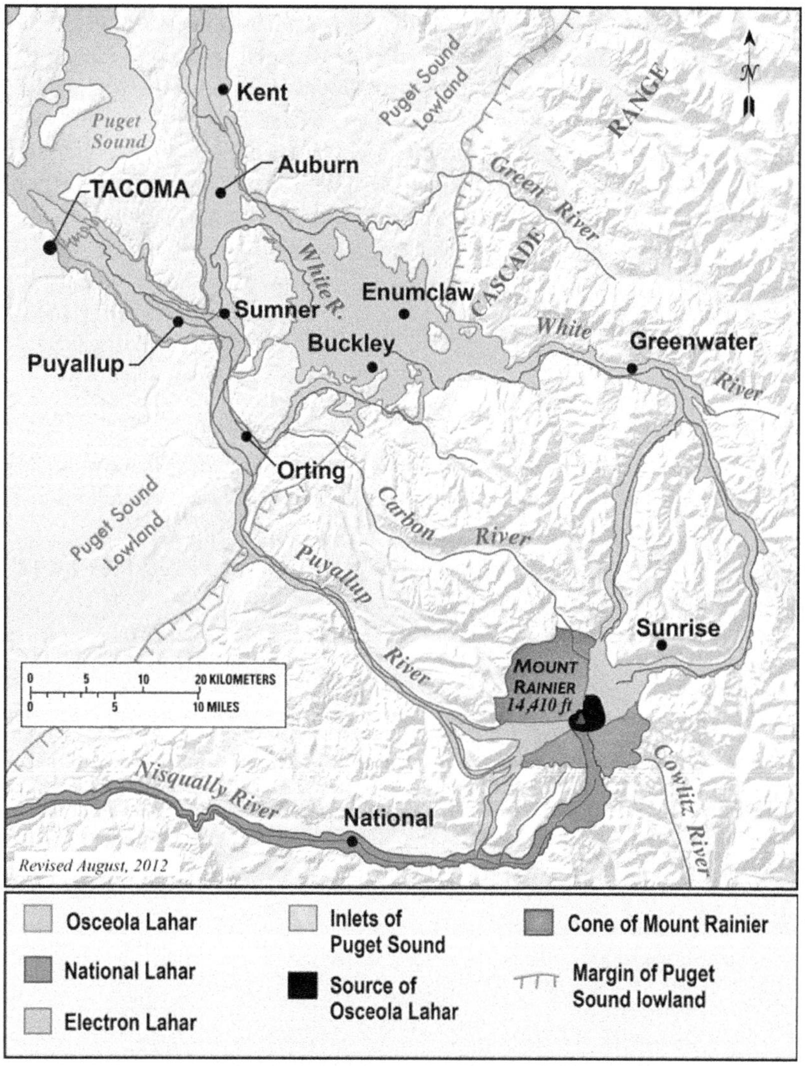

Figure 14.14 Routes of three major lahars emanating from the summit of Mount Rainier during the Holocene.

Figure 14.15 Light-brown material is hydrothermally altered rubble of a debris flow deposit. Base of Tahoma Glacier.

a solid mass of mountain near Rainier's summit had been spread out as a concrete-like slurry across the neighboring lowlands.

The Osceola mudflow was 10 times larger than any other known lahar sourced from Mount Rainier. It beheaded the summit of the mountain, leaving behind a huge amphitheater that opened to the northeast (similar to the events of the 1980 eruption of nearby Mount St. Helens). The debris deposited in the lowlands as the lahar settled out consisted of huge boulders and a massive volume of cobbles, sand, and silt, as well as battered tree trunks, all trapped in a matrix of mud. The lahar deposit buried the future sites of Buckley, Enumclaw, Sumner, Puyallup, and Auburn, which are collectively home to over 150,000 people today. The Osceola collapsed crater was partially filled over the succeeding years by lava flows and pyroclastic debris that constructed the current rounded summit of Rainier (Fig. 14.13). In time, the volcanic crown was buried beneath a thick layer of newly formed ice of the Winthrop and Emmons glaciers.

Around 1500 CE, a little over 500 years ago, another catastrophic volcanic mudflow was unleashed from the upper west flank of Rainier near Sunset Amphitheater. The **Electron mudflow** raged through the Puyallup River Valley as a clay-rich, highly fluid mass perhaps 50 m (160 ft) thick, before settling out about 100 km (60 mi) from its source an hour or two after it began (Fig. 14.14). The deposit from this lahar buried the future townsite of Orting under 6 m (20 ft) of debris, including the remains of a forest entombed by the event. The trigger for the Electron lahar is not known and may have occurred with little warning—a chilling thought for current inhabitants of the Puget Lowlands. The catalyst for the lahar may have been a small eruption, a phreatic explosion, an earthquake, or simply the sudden failure of weakened rock on a steep slope saturated with snowmelt.

The youngest lava flows on Rainier are about 2200 years old, and the most recent pyroclastic events occurred about 1100 years ago (Fig. 12.12). The Electron

mudflow occurred about 500 years ago. Reports of steam explosions or minor eruptions of pyroclastic debris happening between 1820 and 1894 are poorly documented, and physical evidence is sparse.

Hazard Potential for the Mount Rainier Region

Mount Rainier exhibits a variety of geologic indicators of its internal supply of magma and hot rock that remind us of the mountain's potential for renewed volcanic activity. Near the summit craters, fissures release intermittent bursts of steam, while **fumaroles**—vents that emit both steam and hot volcanic gases—sputter from the surface at temperatures of around 70°C (~160°F). The hot steam and gases create a unique network of interconnected ice caves in the two summit craters. These steam vents and fumaroles are surface expressions of heat flow from the body of the volcano and are part of a larger hydrothermal plumbing system that pervades the interior of the mountain. Where corrosive hydrothermal fluids infiltrate fractures within rock, they chemically interact with minerals composing the rock, converting them to clays and other alteration products (Fig. 14.15). This chemical degradation weakens the rock, making the affected area susceptible to slope failure and catastrophic collapse. The tremendous volume of ice, snow, and meltwater streams, as well as steep mountain slopes, further enhances the potential risk of far-traveling volcanic mudflows. The clay-rich deposits of the Osceola and Electron mudflows suggest that they were likely sourced in high-elevation areas of hydrothermally weakened rock.

Mapping and remote sensing have determined that the most intense **hydrothermal alteration** occurs in a narrow band that extends northeast to southwest across the summit of Rainier, likely controlled by the orientation of an internal network of fractures. The greatest concentration of crumbly, hydrothermally weakened rock resides on the upper west flank of Rainier in the area near Sunset Amphitheater. Not coincidentally, the Electron debris avalanche and mudflow were released from that part of the mountain only 500 years ago. Today, a comparable event would follow the Puyallup River (and the Carbon and Nisqually rivers to a lesser extent) toward densely populated areas within the Puget Lowlands.

Future eruptive events from Mount Rainier may include lava flows that are likely to travel no further than the park boundaries, as well as less frequent pyroclastic eruptions and pyroclastic flows. But the most probable volcanic hazard facing denizens of the region are lahars. Geologists have documented the deposits of at least 60 lahars that flowed through nearby river valleys over the past 10,000 years, and there's no reason to think that they won't happen again in the near future. The threat is high enough that the U.S. Geological Survey has teamed with local and state emergency-management agencies to establish a lahar-warning system that both detects the approach of lahars and disseminates an alarm to people in the path of the lahar. The tens of thousands of people living in the lahar pathways may have anywhere from 3 hours to as little as 40 minutes to get out of harm's way once an approaching lahar is detected, so the system needs to be accurate and trustworthy.

Obviously, protecting the public is of primary importance, but think of the infrastructure of buildings, roads, bridges, powerlines, dams, and other human-made structures that would be affected by a fluidized, clay-rich lahar racing through populated river valleys at speeds approaching 80 km/hr (50 mph). This nightmare scenario is further complicated by the lesson learned from the Electron mudflow of 1500 CE that lahars may be triggered by events that are nonvolcanic in origin and that may occur with little advance warning.

When you visit Mount Rainier National Park, remember that these volcanic hazards happen on geologic time scales that only rarely intersect with human time scales. Go for the breath-taking arrays of wildflowers in the spring and summer. Go to walk in the dense, old-growth forests of Douglas fir and western hemlock. Go to see mountain goats and marmots. Go to walk on gigantic glaciers and hike the extensive trail system. Go to climb the summit, using one of the numerous guide services if necessary. Go to view the volcanic rocks and imagine the molten fury of their origin. Just be aware that this sleeping behemoth will invariably waken at some point in the future, accompanied by an array of violent volcanic events beyond our ability to comprehend.

Key Terms

alpine glacier
arête
cirque
cleaver
continental ice sheet
fumarole
Holocene
hydrothermal alteration
Last Glacial Maximum (LGM)
Little Ice Age
moraine
outwash plain
Pleistocene Ice Ages
sublimation
valley glacier
zone of ablation
zone of accumulation

Related Sources

Harris, S. L. (2005). *Fire Mountains of the West.* Missoula MT: Mountain Press.

DriedgerC., and Scott, W. E. (2008). Mount Rainier—Living Safely with a Volcano in Your Backyard. *U.S. Geological Survey Fact Sheet* 2008-3062, 4 p.

Graham, J. (2005). *Mount Rainier National Park Geologic Resource Evaluation Report. Natural Resource Report NPS/ NRPC/GRD/NRR—2005/007.* Denver, CO: National Park Service.

Pringle, P. T. (2008). Roadside Geology of Mount Rainier National Park and Vicinity. *Washington Division of Geology and Earth Resources Information Circular* 107, 191 p. Accessed February 2021 at https://d32ogoqmya1dw8.cloudfront.net/ files/sage2yc/workshops/june2017/geologic_road_guide_ mount.pdf

Mount Rainier National Park Service: Volcanic Features and Glaciers. Accessed February 2021 at https://www.nps.gov/ mora/learn/nature/naturalfeaturesandecosystems.htm

USGS Volcano Hazards Program: Lahars and Their Effects. Accessed February 2021 at http://volcanoes.usgs.gov/ hazards/lahar/index.php

USGS Volcano Hazards Program: Mount Rainier. Accessed February 2021 at http://volcanoes.usgs.gov/volcanoes/ mount_rainier

15 Crater Lake National Park

Figure 15.1 View of Crater Lake from the south rim. The steep walls of the lake are composed of volcanic rock of ancestral Mount Mazama, which erupted catastrophically 7700 years ago and subsequently collapsed, forming the bowl-shaped lake basin. Wizard Island cinder cone rises above lake level.

Everyone who visits Crater Lake National Park in the Cascade Range of southern Oregon is astonished when they catch their first look at the lake. After a progressive climb upslope toward the lake, you reach the crest along Rim Drive with its multiple overlooks. Your first glimpse of the lake itself is startling because the water far below is such a crisp, intense shade of blue (Fig. 15.1). The picturesque cone of Wizard Island adds to the inherent natural beauty. But the placid tranquility of the lake belies the inconceivably violent origins of the basin now filled by the lake. In this chapter, you'll be forced to imagine the unimaginable—a volcanic eruption that dwarfs the 1980 event at Mount St. Helens and that was the most powerful Holocene eruption in the Cascades. Moreover, Native Americans witnessed the cataclysmic event and suffered from the widespread devastation. Modern North Americans have never experienced an eruption of such catastrophic ferocity, so an understanding of the volcanic processes that formed Crater Lake 7700 years ago provides a model for what humans might expect when these rare, large-magnitude events inevitably occur.

Crater Lake fills the 8 x 10 km (5 x 6 mi) hole in the ground left behind after the massive eruption of an ancestral volcano known as Mount Mazama (Fig. 15.2). The park is misnamed because the enormous bowl-shaped basin is not a crater, but rather a **caldera**. A **crater** is a near-circular vent in a volcano created during eruptions, whereas a "caldera" is typically a much larger depression that forms by the catastrophic vertical collapse of the volcanic crest to form a basin. It's a common misconception that the modern lake basin formed when the crest of Mount Mazama was blown skyward during the eruption. You'll learn the details in this chapter, but the lake basin actually formed through the opposite process of downward collapse *after* the main cataclysmic eruption, likely in a matter of hours. Of course, the newspaper editor who excitedly named Crater Lake in 1869 had no way of knowing its geologic origins, and so the name remains to this day.

Crater Lake is notable not only for its exceptionally clear, azure water, but also for being the deepest lake in the United States, the second deepest in North America, and the seventh deepest in the world. The lake has an average depth of around 350 m (1148 ft) and a maximum depth of 594 m (1949 ft). The surrounding steeply inclined cliffs reach over 600 m (~1970 ft) above lake level (Fig. 15.3). Note that the highest relief along the caldera walls is roughly equal to the greatest depth of the lake. Adding the two equals about 1.2 km (~4000 ft) of relief between the highest point on the rim and the deepest point in the lake. Beneath the lake floor, pyroclastic debris from the caldera-forming eruption, old lava flows, and sediment flushed into the lake by landslides are estimated to total several hundred meters in thickness. At the very bottom, beneath this pile of volcanic and sedimentary debris, lay the collapsed remnants of Mount Mazama.

The hydrology of Crater Lake is a balanced system in which the amount of water entering the lake is offset by the loss of water outward. There are no perennial streams that flow into the lake and none that allows water to exit from the saucer-shaped depression. Water is added via abundant snowmelt and rainfall directly within the lake basin, collectively averaging about 224 cm (88 in) of water per year. Minor inflow occurs from springs emanating from groundwater percolating through the caldera walls. An equivalent amount of water leaves the lake by evaporation as well as by outward seepage through a localized, permeable layer of glacial debris along the northeastern wall of the caldera. Thus, the elevation of the lake today remains stable at 1883 m (6178 ft) above sea level, with seasonal fluctuations of a few meters above and below the average. In the surrounding region, streams flow radially away from the high rim of Crater Lake.

The lack of inflowing streams to the lake is the main reason for its unusual clarity. With no sediment input from streams, the lake remains clear of suspended particles. The lack of inflowing streams also limits the amounts of nutrients entering the lake, such as phosphorus and nitrogen, that would promote the growth of algae. The combination of minimal suspended sediment and low algal growth accounts for the transparency of the water, which is arguably the most optically clear in the world. In turn, the pristine water and the great depth allow sunlight to penetrate deeply, scattering the short wavelengths that our eyes perceive as a deep shade of blue.

The walls of the caldera surrounding Crater Lake expose the interior anatomy and volcanic evolution of ancient Mount Mazama (Fig. 15.3). This rare and detailed view was abruptly revealed during the violent eruption and later collapse of the volcano 7700 years ago, which likely occurred over a few disastrous days. The story that follows is based on laborious fieldwork and laboratory analysis by geologists from universities and the U.S. Geological Survey. They spent years mapping the distribution of volcanic units, dating the various overlapping layers via radiometric methods, and synthesizing their observations and analyses to interpret the chronology of events that resulted in the growth and demise of Mount Mazama.

The geologic history of ancient Mount Mazama and its modern remnant, Crater Lake, can be roughly divided into two phases. First was a long, drawn-out stage ranging from about 420,000 years ago to 7700 years ago that was marked by the progressive growth of the volcanic mountain of Mazama. The second phase began with the climactic eruption 7700 years ago and the subsequent evolution of the deep lake that dominates the landscape today. The goal of the narrative that follows is to combine the fundamentals of volcanism introduced in Chapter 12 with the phenomenal exposures along the caldera wall to describe the growth and eventual destruction of a Cascade Volcano. Among the new concepts introduced in this chapter is the impact of enormous **caldera eruptions** on the American West, which will act as a prelude to even larger ancient eruptions addressed later in the book.

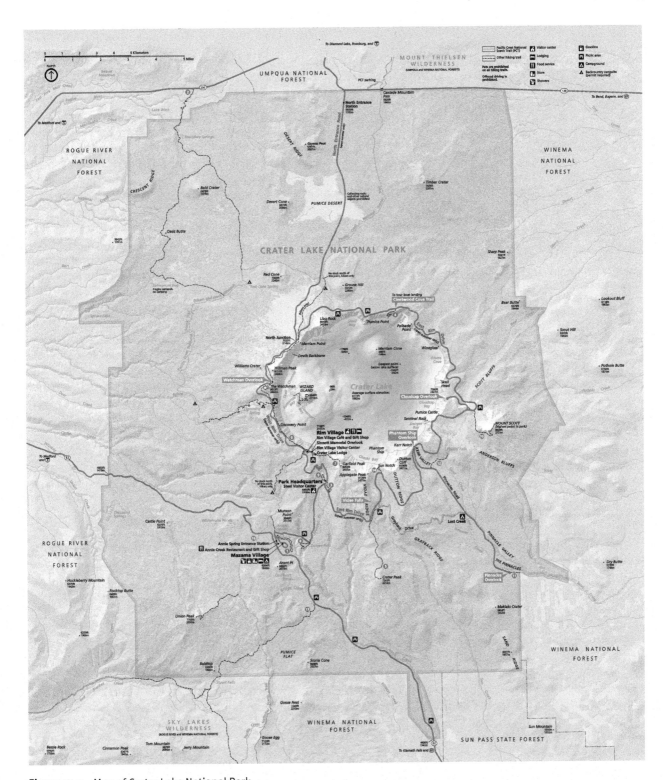

Figure 15.2 Map of Crater Lake National Park.

Growth of Mount Mazama

Like many other Cascades volcanoes, Mount Mazama grew primarily by voluminous flows of andesitic and dacitic lava, supplied by magma created along the subduction zone deep beneath the range. Later in Mazama's history, lava flows and pyroclastic eruptions became more silica-rich, reflecting the chemical evolution of the magma supply a few kilometers beneath the surface. Remember from Chapter 12 the role of silica in the eruption style of a volcano (Fig. 12.8). The higher the silica content of the magma, the higher the viscosity, the greater the trapping of gases, and thus the

Figure 15.3 Hillman Peak forms the highest point along the rim of Crater Lake. Its elevation is ›600 m (~1970 ft) above lake level at 1883 m (6178 ft). Geologists use the term *scree* to refer to the steep apron of rubbly debris covering the rock beneath.

higher the potential explosivity of eruptions. As Mazama grew episodically through time, the silica content increased from the 53–68 percent range of andesite and dacite to the 68–72 percent or so of rhyodacite. Thus, the potential for a cataclysmic pyroclastic eruption increased in concert with the long-term increase in silica content of the magma.

The story recorded in the rocks composing the walls of Crater Lake suggests that the edifice of Mount Mazama was not a single *stratovolcano*. Instead, it consisted of a series of overlapping smaller volcanoes that coalesced through time into a complex mountain estimated to have reached 3700 m (12,000 ft) in height. Recall that the *edifice* of a volcano is its physical structure, composed of lava flows, tephra, lahar deposits, and avalanche debris that accumulate through time.

Mount Mazama volcanism occurred contemporaneously with the Ice Age glaciations of the Pleistocene, introduced in Chapter 14 on Mount Rainier. During colder glacial phases, alpine glaciers carved deeply into the volcano's flanks. Lava flows descending from the summit were likely redirected around the heads of glaciers to flow along the glacier's margins, comparable to the interaction of lava and glacial ice at Mount Rainier. During warmer interglacial phases, alpine glaciers retreated upslope, permitting lava flows to funnel through glacially scoured valleys. Pyroclastic debris and lava flows accumulated during the warm phases and added mass to the edifice. The irregular pace of volcanism overlapped with the tick-tock periodicity of Ice Age glaciation, with the result being the episodic growth and decay of Mount Mazama.

The oldest remnants of Mount Mazama are about 420,000 years old (420 ka; remember that ka equals "thousands of years ago") and consist of andesite exposed in Mount Scott, east of Crater Lake (Fig. 15.4). (When considering the 420 ka ages of these rocks, remember that this is relatively young, geologically. Rocks and events recorded in parks of the Colorado Plateau (Part 1) are measured in tens to hundreds of *millions* of years, so the construction of Mount Mazama and other volcanic parks of the Cascades occurred very recently in geologic time.) Other volcanic rocks of comparable age that record the earliest phases of Mazama volcanism are the relatively small but photogenic andesitic outcrop of Phantom Ship that rises above lake level near the southeastern caldera wall (Fig 15.5). Outcrops of this volcanic unit are a tiny remnant of what used to be a much more widespread volume of rock. Postdepositional erosion and burial by younger volcanics have reduced these old Mazama rocks to small exposures.

The locus of volcanism and the growth of Mount Mazama gradually shifted westward through time. Mount Scott went extinct, but other volcanic vents opened to the west, building the edifice of Mazama laterally as well as vertically. A widespread lava flow, about 340–300 ka in age, fills and preserves an ancient U-shaped glacial valley near Sentinel Rock (Fig. 15.4) that extends broadly along the east caldera wall. By about 170–120 ka, the locus of volcanism had shifted to the northwest, as exemplified by the distinctive dark cliff of lava flows composing Llao Rock (Fig. 15.6). The thick core of the unit fills a U-shaped glacial valley about 180 m deep (~600 ft). As the flows filled the valley and built a dome on top, they spread laterally, forming the "wings" of Llao Rock. The viscous nature of the rhyodacite lava restricted the

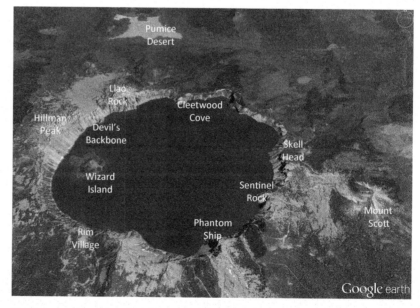

Figure 15.4 Close-up view of Crater Lake looking northward, highlighting a few key features of the landscape discussed in the text. The steep walls average ~300 m (1000 ft) in height above lake level. Note the scalloped margins of the lake created by collapse of the steep walls lining the caldera.

Llao Rock flows to only about 2 km (1.2 mi) northward of the caldera rim.

By about 70 ka, the voluminous lava flows that dominated the construction of Mount Mazama were interrupted by an explosive eruption of pyroclastic debris that spread out across the slopes of the volcano. This event is nicely preserved at Pumice Castle along the east caldera wall where a bright orange layer of *tephra* is exposed. The glassy particles of pumice were very hot as they were deposited, re-melting and fusing together to form a resistant pyroclastic rock called a **welded tuff**.

Over the succeeding tens of thousands of years, Mazama grew from eruptions of lava along its western slopes, now exposed near Hillman Peak (Fig. 15.3). Among the notable features in the area is Devil's Backbone, a *dike* of andesite that cuts vertically through older layers of lava (Fig. 15.7). Fractures in the volcanic edifice are part of the natural "plumbing system" through which magma travels on its way to the surface. These cracks widen as magma forcibly intrudes and melts adjacent rock walls. At some point, the magma solidifies in place within the widened, near-vertical fracture, forming a narrow, sheet-like dike. Upon exposure to weathering and erosion, the andesite filling the dike was more resistant than the adjacent older lava flows, so the dike is expressed topographically as an upright wall of rock projecting from the caldera rim. As you gaze at the Devil's Backbone from an overlook, think about magma churning through that fracture within the body of Mount Mazama, now hardened and exposed to view.

At this point in the approximately 400,000-year succession of eruptions at Mount Mazama, the volcano had grown to an estimated elevation of 3700 m (~12,000 ft). Eruptions after about 40 ka on Mazama are almost exclusively composed of lava and tephra of rhyodacite, indicating a shift to higher silica magma and a higher potential for an explosive eruption. This final burst of viscous, higher silica volcanism lasted till about 25 ka, when Mazama entered an extended phase of repose that spanned the Last Glacial Maximum and the beginning of the Holocene (see Chapter 14). The high-silica *lava domes* that capped vents on the volcano may have acted like corks on the magmatic plumbing system within Mazama, inhibiting the release of gases and allowing internal pressures to increase to ominous levels. It was during this 20,000-year-long dormant phase that the contents of the magma chamber physically separated, with the gas-charged rhyodacitic material rising toward the top of the reservoir, while the denser andesitic material settled to the bottom. The stage was set for a violent release of the pent-up

Figure 15.5 Phantom Ship is a small island that is part of the larger Phantom Cone andesite that forms part of the southeastern wall of the caldera. Mount Scott, another remnant of the oldest rock composing Mazama, rises in the background.

Figure 15.6 The thick, dark mass of rhyodacite composing Llao Rock is ~150 ka in age. The lava filled a U-shaped glacial valley before lapping outward, forming the "wings" of the feature. Light-colored pumice of the 7700-year-old climactic eruption crowns the peak.

Figure 15.7 Andesite dike of Devil's Backbone (50–40 ka) exposed along the northwestern wall of caldera. Magma migrated through near-vertical cracks cutting across older layers of lava to erupt as fresh lava flows on the surface. The magma in the feeder crack eventually solidified in place to form the dike, now exposed by erosion.

energy stored deep within the bowels of Mount Mazama.

Climactic Eruption of Mount Mazama

Mount Mazama woke from its long slumber with a series of relatively small precursor eruptions a century or two prior to the climactic eruption 7700 years ago. The Holocene climate was warming dramatically at this time, but the highest elevations of Mazama likely supported small alpine glaciers. The mountain was undoubtedly an important landmark for the Native Americans who hunted and foraged in the region. By our calendar, this was around 5700 BCE, and Neolithic agricultural societies were widespread throughout the world.

The caldera eruption began with an explosive release of pyroclastic debris from a single vent on the northeastern flank of Mazama (Fig. 15.8). Within hours, a thick *eruption column* of rhyodacitic pumice and ash reached several kilometers into the stratosphere. These towering, turbulent columns of pyroclastic debris characterize **Plinian eruptions** and are powered by pent-up gas pressure released from the magma chamber. The sound of the initial explosion and the roar of the eruption column must have been disorienting to Native Americans in the region. The ash cloud was blown toward the east and northeast by prevailing westerly winds, with much of the debris eventually falling back to earth as a *pyroclastic fall*, blanketing a huge area in tephra over the next few days (Fig. 15.9). (Recall that tephra consists of jagged pieces of rock and tiny, abrasive particles of glass and pumice that fall as a solid, rocky "rain" onto the surrounding landscape. As the pyroclastic debris compacts and solidifies it forms a rock known as **tuff**.) Known as the Mazama **ashfall tuff**, the tephra layer is about 20 m (66 ft) thick near its source, thinning to about 40 cm (16 in) a few hundred kilometers to the northeast and less than 5 cm (2 in) a thousand kilometers away. The pyroclastic material draping the outer walls of the caldera today consists of cobble- to sand-size pieces, whereas the particles get finer at more distal locations.

As the Plinian phase progressed, the main vent widened and the upward thrust from gas pressure dissipated (Fig. 15.10). In response, the eruption column abruptly began to collapse, with the downwelling debris crashing back on to the mountain before spreading out laterally as incandescent *pyroclastic flows* (see Chapter 13). The temperature of these flows was hot enough that large blocks and finer particles of pumice and ash melted and then fused to form thick layers of welded tuff along the northern flank

Figure 15.8 Artist's rendering of Plinian eruption of Mount Mazama during its climactic event 7700 years ago.

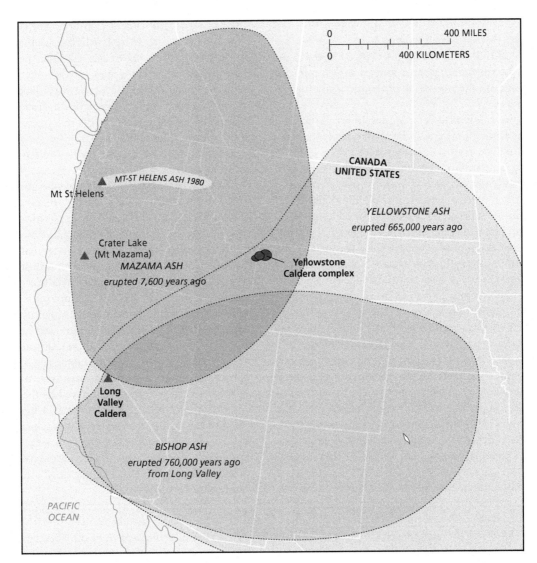

Figure 15.9 Distribution of Mazama ash. Compare the extent with the distribution of ash from the 1980 eruption of Mount St. Helens.

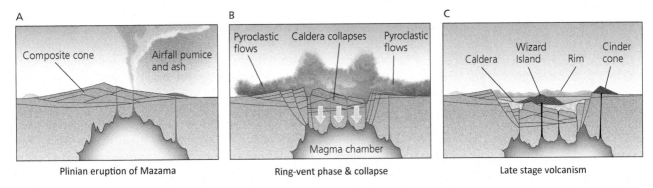

Figure 15.10 Sequence of events during the caldera eruption of Mount Mazama and the formation of Crater Lake 7700 years ago.

of Mazama. The enormous volume of Mazama ashfall tuff and overlying welded tuff were likely deposited over a few violent days, depleting the magma chamber deep within the volcano and setting the stage for the second phase of the eruption.

Formation of the caldera began as a ring of concentric fractures developed in the volcanic edifice above the rapidly emptying magma chamber, creating new conduits for the escape of magma toward the surface. During this **ring-vent phase**, the crest of the volcano began to chaotically collapse into the underlying void, acting like a plunger on the remaining magma in the chamber beneath (Fig. 15.10B). The rising gas-charged magma instantaneously chilled into pumice and ash as it was forced outward along the circular array of fractures ringing the former summit.

The ring of vents emitted a constant series of pyroclastic flows that turbulently streamed down all sides of the remnant edifice at temperatures greater than 800°C (~1500°F), incinerating any remaining form of life that may have survived the earlier blanket of Mazama ash. Pyroclastic flows may reach speeds of 140 km/hr (90 mph) and travel for great distances because they surf above a layer of superheated air trapped beneath the roiling mass of pumice and ash. One flow from Mazama rumbled through a river valley, destroying everything in its path before settling out 70 km (>40 mi) away from the volcano and leaving behind a deposit of tephra up to 100 m (330 ft) thick.

Over the remaining hours of the ring-vent phase, deeper levels of the stratified magma chamber were released as pyroclastic flows. This zonation is wonderfully exhibited at the Pinnacles, located 11 km (7 mi) southeast of the caldera rim (Fig. 15.11). The first pyroclastic flows were composed of light-colored, silica-rich rhyodacite derived from the upper layers of the magma chamber. As the pyroclastic flows erupted from the ring vents diminished, deeper levels of denser, lower-silica andesitic magma were tapped. The pyroclastic flows derived from this magma spread out on top of the earlier flows as a darker, iron-rich layer, resulting in the light-to-dark zonation seen at the Pinnacles. Gases trapped beneath the rapidly deposited flows buoyantly escaped upward as vertical plumes, creating *fumaroles* at the surface (see the Rainier chapter, Chapter 14). Minerals that crystallized from the gases

Figure 15.11 The Pinnacles, located along Wheeler Creek in the southeastern part of the park. Light-colored, rhyodacitic pyroclastic flow deposits at the bottom give way to darker, andesitic pyroclastic flow deposits above, reflecting the internal zonation of the magma chamber beneath Mount Mazama. The spires toward the top of the outcrop are composed of resistant tuff, tightly cemented by minerals crystallized from rising plumes of escaping gases soon after deposition. The outcrop is about 100 m (330 ft) high.

acted as a natural cement, solidifying the tephra as the hot gases ascended. These hardened chimneys were more resistant to the effects of later weathering and erosion than the noncemented tephra surrounding them, so they remain erect as a dense cluster of rock spires. The andesitic phase of ring-vent eruption is also responsible for the broad expanse of loose tephra of the Pumice Desert, north of the caldera (Fig. 15.4).

Imagine the landscape a few days after the eruptions ended as an eerie silence replaced the earsplitting sounds of the cataclysm. The massive volcano that had grown steadily over the preceding 400,000 years was now gone. In its place was a very large hole in the ground over a kilometer deep, surrounded by precipitous walls. A layer of still-hot pumice and ash formed an irregular veneer above the chaotically strewn blocks of rock forming the caldera floor. Plumes of steamy hot gases billowed outward from fumaroles. A gray blanket of ash and pumice turned the once-verdant countryside into a sterile moonscape. The land was devoid of life, incinerated and buried beneath countless tons of tephra. Based on cultural artifacts and bones found beneath the Mazama ash, it's likely that many Native Americans in the area were killed either by asphyxiation from the ash or starved by the environmental devastation to their food resources. People living elsewhere in the Pacific Northwest, after enduring the fallout of Mazama ash, must have wondered what produced the thunderous sounds of the previous few days. Think of the societal disruption if this magnitude of eruption were to occur in the Cascades today.

The effects of the climactic eruption of Mazama 7700 years ago likely cascaded across the planet. The unique chemical fingerprint of the Mazama ash has been detected within ice cores drilled from the Greenland ice sheet, suggesting that fine particles of the Plinian eruption column were carried aloft by eastward-moving, high-altitude winds, slowly falling back to Earth as the ash circled the globe. Temperature records derived from the ice cores suggest a decrease in global temperatures by a few degrees, likely due to the Sun-reflective properties of micron-size particles of volcanic ash high in the atmosphere. Furthermore, these tiny grains of glassy silicate refract and filter various wavelengths of sunlight passing through them at low angles, producing colorful sunsets that may last for months. It's likely that the Neolithic societies of the Middle East and China experienced lower-than-average temperatures and marveled at the lurid sunsets, without any idea of their cause half a world away.

The climactic eruption ejected 50 km³ (12 mi³) of tephra over a few furious days. The chemically distinct layer of Mazama ash erupted during the Plinian phase covers a vast area and represents a discrete point in time wherever it is found. Deposits below the ash layer are older than 7700 years and deposits above are younger. The date of 7700 years for the eruption was determined by radiocarbon dating of organic material buried just beneath the ash layer at several locales. The Mazama ash is useful for many scientific studies, from the dating of archaeological sites to the calibration of climate records, as well as for the age determination of sedimentary deposits from glaciers, lakes, and oceans. Moreover, the elemental chemistry of the Mazama ash contributes to the development of fertile soils that influence forest productivity throughout the Pacific Northwest. So the effects of the climactic eruption of Mazama impact us to this day. We'll revisit the global and regional consequences of caldera eruptions when we address the far-larger events at Yellowstone National Park in Chapter 31.

Evolution of Crater Lake over the Last Seven Millennia

The geologic evolution of the caldera floor beneath the lake was fully revealed in early 2000 when a joint team from the U.S. Geological Survey, National Park Service, and university geoscientists generated high-resolution images using sophisticated sonar mapping (Fig. 15.12). Interpretation of the bathymetric map, in concert with selective sampling of the lake floor, enabled the determination of the following sequence of volcanic and sedimentary events that occurred within the caldera over the last seven millennia.

Soon after the catastrophic eruption of 7700 years ago ended, the steep, unstable walls of the caldera collapsed as massive avalanches into the margins of the hot, steamy depression. These events left behind scalloped incisions along the caldera walls, modifying and enlarging the original smaller, circular outline of the basin. For instance, the enormous Chaski Bay landslide is estimated to have occurred about 200 years after the climactic eruption, leaving the broad embayment along the south side of the lake that we see today (Figs. 15.2, 15.12).

Subsequent volcanism and sedimentary processes further filled the depression, with most of the activity ending within a few thousand years after the climactic eruption. These volcanic flows and sedimentary deposits covered and smoothed the underlying chaotic mass of pyroclastic debris and broken blocks of volcanic rock formed during caldera collapse. As the years passed, snowmelt and rainfall within the caldera began to accumulate, initially in puddles and ponds before coalescing into a shallow lake.

Figure 15.13 Wizard Island consists of a symmetrical cinder cone surrounded by lobes of andesitic lava flows. It rises 234 m (>760 ft) above lake level and forms the crown of a much larger volcanic platform beneath. The peninsula extending toward the viewer is a lava flow that erupted from the base of the cinder cone. The crater at the top of the island is about 100 m (330 ft) across and 30 m (90 ft) deep.

Evidence from the sediments and topography of the lake-floor suggest that Crater Lake filled relatively rapidly to depths comparable to what we see today.

The earliest volcanism on the caldera floor began within a few hundred years after the climactic eruption ended as viscous, andesitic lava spilled out across the caldera floor (Fig. 15.10C). This material was likely the remnant, gas-depleted magma left behind after the main eruption emptied most of the chamber. These eruptions formed the small volcanic edifices of the central platform, Merriam Cone, and Wizard Island (Fig. 15.12).

The broad platform on which Wizard Island is perched began as a series of overlapping lava flows submerged beneath lake level. Wizard Island grew upward from the volcanic platform, gradually breaching the water surface as a cinder cone by about 7200 years ago (Fig. 15.13). A **cinder cone** is a steep-sided, symmetrical landform composed of pebble-sized volcanic particles formed as lava explodes into the air and then rapidly cools, falling back onto the cone as angular fragments called cinder. Lava flows may emerge laterally from the base of the cone because the strength of the cinder pile is not great enough to support the internal flow of molten rock. A small lava dome deep beneath water level on the Wizard Island platform is dated to ~4800 years ago, marking the last volcanic activity in the lake.

Potential for Future Eruptions?

What is the possibility of a future eruption at Crater Lake? An eruption of the magnitude of the climactic event 7700 years ago is very unlikely in the near

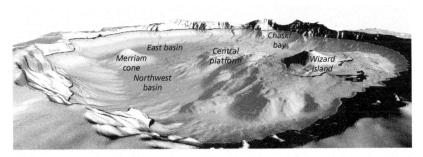

Figure 15.12 Oblique view of Crater Lake looking southeast. Color shading shows the bathymetry of the lake. Shallower depths of the lake are shown by the red and yellow shades, whereas greens and blues show deeper depths. Dimensions of the lake are 8 x 10 km (5 x 6 mi).

future. It's been loosely estimated that it takes about 20,000 years for silicic magma to replenish the chamber a kilometer or two beneath the lake, so there are several millennia yet to go before another catastrophic eruption is expected. And considering that the youngest volcanism happened 4800 years ago on the lake floor, as well as the absence of any indicators of recent magmatic activity beneath the surface, it seems unlikely that even minor volcanic events might occur in the near future. But remember that Crater Lake resides above the Cascadia subduction zone and that episodic eruptions have been occurring semicontinuously over the past 400,000 years. Thus, it seems inevitable that eruptions from the Crater Lake magma chamber will happen at some time in the distant future.

On your next visit to Crater Lake National Park, be sure to take in the splendid vistas of the sparkling blue water and stunning landscape from the overlooks along Rim Drive. But also experience the park and its wonderland of natural features by walking the numerous short trails that pass by some of the rocks that record the violent growth and destruction of Mount Mazama and the formation of the lake. Hike down the Cleetwood Trail to lake level and take a guided boat tour for a different perspective. Climb to the top of Wizard Island and peer down into the circular crater at the crest. Allow your mind to stretch back in time to imagine a massive mountain rising above you and the catastrophic eruption only a few thousand years ago that created the lake basin, now filled with the clearest of water. It's rare that geologic events of that magnitude occur so recently in the past to create such a magnificent landscape, so be sure to think in the fourth dimension of time to fully appreciate the park and its origin.

Key Terms

ashfall tuff
caldera
caldera eruption
cinder cone
crater
Plinian eruption
ring vents
scree
tuff
welded tuff

Related Sources

Bacon, C. R., Mastin, L. G., Scott, K. M., and Nathenson, M. (1997). *Volcano and Earthquake Hazards in the Crater Lake Region, Oregon. Open-file report OF-97-487.* Washington, DC: U.S. Geological Survey. Accessed February 2021 at http://vulcan.wr.usgs.gov/Volcanoes/CraterLake/Hazards/OFR97-487/framework.html

KellerLynn, K. (2013). *Crater Lake National Park: Geologic Resource Inventory Report.* Natural Resource Report NPS/NRPC/GRD/NRR—2013/719. Fort Collins, CO: National Park Service.

Klimasauskas, E., Bacon, C., and Alexander, J. (2002). *Mount Mazama and Crater Lake: Growth and Destruction of a Cascade Volcano.* U.S. Geological Survey Fact Sheet 092-02. U.S. Geological Survey. Accessed February 2021 at http://pubs.usgs.gov/fs/2002/fs092-02.

Ramsey, D. W., Dartnell, P., Bacon, C. R., Robinson, J. E., and Gardner, J. V. (2000). *Crater Lake Revealed.* U.S. Geological Survey Geologic Investigations Series I-2790L. U.S. Geological Survey.

National Park Service: Crater Lake Geology. Accessed February 2021 at https://www.nps.gov/crla/learn/nature/geologyhome.htm

USGS Digital Data Series 72 of Crater Lake bathymetric map. Accessed February 2021 at https://pubs.usgs.gov/dds/dds-72/site/persp.htm

USGS Fly-By Movie of the bathymetry of Crater Lake. Accessed February 2021 at https://pubs.usgs.gov/dds/dds-72/site/flyby.htm

USGS Volcano Hazards Program: Crater Lake. Accessed February 2021 at http://volcanoes.usgs.gov/volcanoes/crater_lake

Figure 16.1 View of Lassen Peak lava dome from the northeast. Note the steep flanks and surface dominated by loose, rocky debris.

It's not every national park where you can hike a smooth trail to the top of an active volcano, but you can at Lassen Volcanic National Park in northern California. Prepared with appropriate footwear, a few snacks, and the proper attitude, any reasonably fit person can hike the 4 km (2.5 mi) series of switchbacks to the 3187 m (10,457 ft) summit of Lassen Peak, the volcano that forms the topographic centerpiece of the park (Fig. 16.1). The only problem is that Lassen Peak just doesn't look like a volcano. Instead, it looks like a ragged mass of loose, rocky debris with a few outcrops of solid rock poking above the mantle of loose rubble. It's not till you wander around the crest that you recognize two craters that identify it as a volcanic edifice.

Lassen Peak is not a *stratovolcano* like the others that characterize the Cascade Range. Rather, it's a massive *lava dome*, a type of volcano formed from the effusive (nonexplosive) eruption of thick, pasty lava. Lassen Peak is perhaps the largest lava dome on the planet and provides an interesting contrast with the other Cascades volcanic national parks and monuments you read about in the previous chapters of Part 2. The entire area of the national park is volcanic in origin; Lassen Peak is the largest of more than 30 lava domes that have erupted in the region over the past 300,000 years.

Lassen Peak is the southernmost volcano of the Cascades chain and is located due east of the southern margin of the Cascadia subduction zone (Fig. 12.4). The relatively young Cascades volcanic rocks overlap the older granitic rocks of the Sierra Nevada to the south, with the exact boundary being difficult to determine since the region is heavily forested. To the east lies the arid Basin and Range Province, whereas to the west are the rainy Klamath Mountains and northern Sacramento Valley (Fig. 12.1). Lassen Volcanic National Park resides at the junction of these geologically and ecologically distinct provinces, creating a diversity of environmental conditions.

The western part of the national park is dominated by rugged highlands, including such notable features as Lassen Peak, Brokeoff Mountain, and Chaos Crags (Fig. 16.2). Ice Age glaciers have smoothed the rough volcanic edges of the mountains, creating a scalloped landscape of cirques, glacial valleys, arêtes, and several

beautiful glacial lakes. Streams radiate outward from the mountains, eventually draining into the Sacramento River system to the southwest. In contrast, the eastern part of the park is characterized by lower-elevation volcanic hills and plateaus, which are crossed by streams and dotted with lakes. Seven symmetrical cinder cone volcanoes form distinct topographic features in the central and northeastern parts of the park. Pine and fir forests, interspersed with grassy meadows, cover the volcanic terrain.

Access to the park is mainly along circuitous Route 89, which passes north–south through the western side of the park. A graded, unpaved road leads to a perfectly formed cinder cone and associated volcanic features in the northeast corner of the park, whereas other unpaved roads lead to Drakesbad Guest Ranch in the south and Juniper Lake in the southeast. All of these roads connect with a network of trails, including the iconic Pacific Crest Trail and sections of the historic Nobles Emigrant Trail, which wind through the park and provide access to the backcountry.

Even though Lassen Peak doesn't have the stratovolcanic stature of Mount Rainier, or the raw, recently erupted profile of Mount St. Helens, or the stately tranquility of Crater Lake, Lassen Volcanic National Park has a greater diversity of volcanic features than any of the others.

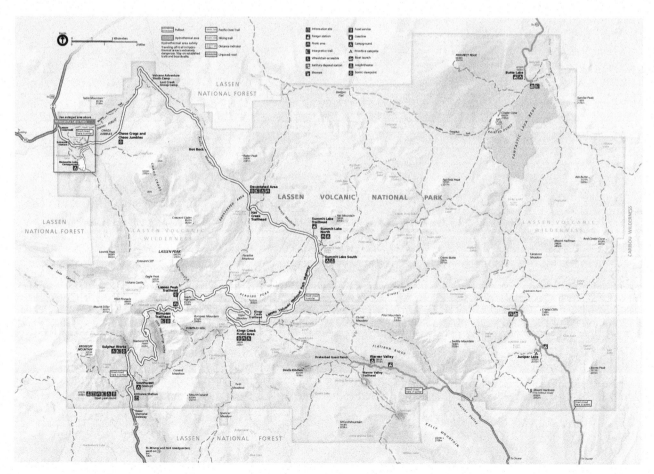

Figure 16.2   Map of Lassen Volcanic National Park. Note the contrasting topography between the western and eastern sides of the park.

For starters, the park has examples of four different types of volcanoes within its boundaries. Lassen Peak and Chaos Crags are the most notable lava dome volcanoes, and several others are concentrated in the western part of the park. No actual stratovolcanoes are found in the park, but deeply eroded remnants of an ancient *stratovolcano* occupy the southwestern corner. *Cinder cone* volcanoes dot the landscape and look as if they were built just last week. *Shield volcanoes* form the fourth type and appear as broad, forested mounds along the northern and southern boundaries.

In addition to the diversity of volcanic edifices, a distinctive feature of Lassen Volcanic National Park is the widespread evidence of youthful volcanism. Lava flows are unvegetated and look freshly cooled; scoured mountainsides are the scars of recent *pyroclastic flows*; and stream valleys show the effects of *lahars* that swept through in a matter of hours. Perhaps the main volcanic feature that distinguishes Lassen from the other Cascades parks, however, is the presence of large-scale hydrothermal activity. Hot springs, gurgling mudpots, and steaming *fumaroles* reflect the presence of an active body of magma deep beneath the surface.

This widespread evidence of recent volcanism is understandable since Lassen Peak last erupted during a series of events from 1914 to 1917, peaking with a major *Plinian eruption* in 1915. Until the eruption of Mount St. Helens in 1980, the early 20th-century events at Lassen Peak were the most recent in the contiguous 48 states. The volcanic rocks that formed during the 1914–1917 eruptions are the youngest rocks in California. Not that much older are the rocks composing the unimaginatively named Cinder Cone, which erupted in 1666, accompanied by fluid lava flows and ashfalls. And just a bit older, on geological time scales, is the jagged mass of Chaos Crags, which formed during a series of effusive, viscous lava flows only 1050 years ago. Reaching further back in time, at least 70 eruptions have occurred from various vents in the park in the past 100,000 years.

Lassen Peak is relatively quiet today, other than the burbling and smelly hydrothermal activity and a few small magma-related earthquakes. But the volcano is currently in a geologically active phase of its longer-term history. There is no reason to think that Lassen Peak will not erupt again sometime in the near future. Indeed, because of the seasonal presence of park visitors, the proximity of human-made structures, and the recent volcanism at Lassen Peak, the U.S. Geological Survey designates the entire Lassen volcanic region as posing a "high threat."

In the previous chapters in Part 2 on the Cascades, you learned several basic concepts about volcanism. These include the tectonic setting of Cascadia, the main types of volcanic rocks, the role silica plays in the viscosity of magma and thus the potential explosivity of eruptions, how stratovolcanoes grow, and how some of them erupt catastrophically to leave behind calderas. You've also learned about a variety of volcanic hazards such as pyroclastic falls, pyroclastic flows, debris avalanches, and lahars. In this chapter, you'll be introduced to a few new concepts that will add to your knowledge of volcanic systems.

Pleistocene Volcanism in the Lassen Region

The diversity of volcanic landforms and volcanic deposits that compose the park mostly formed within the past 825,000 years, creating a broad region of volcanism known as the "Lassen Volcanic Center." Like many of the other Cascades volcanoes, the geologically youthful features of the Lassen Volcanic Center have pierced through an underlying foundation of older, deeply eroded volcanic rocks that reflect earlier phases of Cascadia volcanism in the region. Hundreds of eruptions have occurred from several volcanic vents in the park over the last 825,000 years of the Pleistocene, all related to an active magma system located 8–10 km (5–6 mi) beneath the surface. Recall from Chapter 14 on Mount Rainier that the Pleistocene Epoch began about 2.6 Ma and ended around 12,000 years ago and is characterized by the growth and decay of enormous continental ice sheets and smaller alpine glaciers. In the Lassen region (as in the Cascades as a whole), Pleistocene volcanism occurred contemporaneously with alpine glaciation to create the mountainscape we see today.

Activity within the Lassen Volcanic Center occurred in three distinct phases. The first phase extended from 825 ka to ~600 ka, culminating in what was likely a massive caldera eruption that is recorded today by the widespread distribution of an *ashfall tuff* known as the Rockland Ash (Fig. 16.3). Although no remnant of the volcano that produced the ash exists today, the eruption must have been enormous since the Rockland Ash is recognized across a broad area of northern California, Nevada, and southern Idaho. The total volume of the ash deposit is calculated to be about 50 km³ (12 mi³), comparable to the volume of Mazama ash produced by the climactic eruption that created Crater Lake, and 50 times the volume of tephra produced during the 1980 eruption of Mount St. Helens.

The second phase of Pleistocene volcanism at Lassen ranged from about 600 ka to 390 ka and was centered on an ancient stratovolcano called Brokeoff Volcano (known as Mount Tehama in the older literature), located just south of modern Lassen Peak. Erosional remnants of the Brokeoff stratovolcano include Brokeoff Mountain, Mount Diller, Pilot Pinnacle, and Mount Conard (Fig. 16.4). An estimate of the dimensions of ancient Brokeoff Volcano can be determined by connecting these topographic landmarks. The volcano was about 20 km (12 mi) in diameter and rose upward to an elevation of ~3350 m (11,000 ft), higher than nearby Lassen Peak (16.5). Brokeoff Volcano likely grew above the remains of the caldera left behind after the massive eruption that produced the Rockland Ash, fed by the same magma chamber deep beneath the

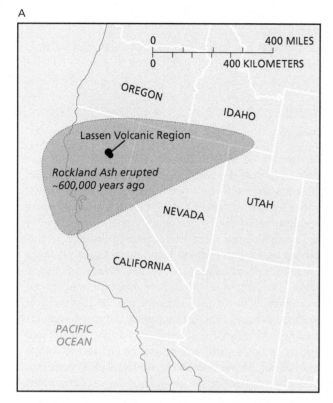

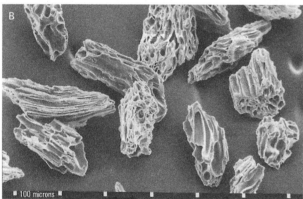

Figure 16.3 A) Map of distribution of pyroclastic fall of the Rockland Ash. B) Photomicrograph of glassy shards of volcanic ash showing elongate holes created by bubbles of gas that expanded rapidly during eruption. Each particle is magnified about 70 times its actual size.

surface. Numerous andesitic to dacitic lava flows and tephra deposits from ancient Brokeoff Volcano extend across much of the western part of the park and obscure remnants of the underlying Rockland caldera.

Volcanic activity in the Lassen Volcanic Center waned between 390 ka and 300 ka. During this phase, alpine glaciers repeatedly carved into the edifice of Brokeoff Volcano, re-sculpting the landscape and transporting the erosional debris away from the region. The effectiveness of glacial erosion was enhanced by the *hydrothermal alteration* of rock to weak clay minerals by hot, acidic fluids that flowed through the natural plumbing system beneath the edifice. The active hydrothermal areas within the boundaries of Brokeoff Volcano today, such as Sulphur

Works, Bumpass Hell, and Hot Springs Valley (Fig. 16.4), are surface expressions of the network of conduits that connect with the magma chamber at depth. When you drive the main road through the southwestern corner of the park and stop intermittently to visit the hydrothermal areas, remember that you are traversing the deeply eroded interior of Brokeoff Volcano.

The third phase of active volcanism began around 300 ka and continues to this day. The locus of volcanism shifted to the north of ancient Brokeoff Volcano and consists of a series of steep-sided lava domes composed of dacite. These three phases of volcanism at Lassen—the Rockland caldera, the Brokeoff stratovolcano, and the lava domes of today—were all produced from the same magma chamber whose chemistry evolved to become slightly more silica-rich through time.

The climax of the third phase occurred 27,000 years ago, when the Lassen Peak dacite lava dome was born as magma escaped from a vent north of its ancestral precursor, Brokeoff Volcano. Dacite, as you recall from Chapter 12, is composed of ~65 percent silica, which imparts a sticky, viscous texture to the magma (Fig. 12.8). The Lassen Peak dacitic magma was depleted of gas when it effusively oozed like hot toothpaste from the vent, with the thick, gooey lava accumulating close to its source. The oldest lava flows were pushed outward as younger lava flows erupted from the central conduit. As the older lava flows along the flanks cooled and solidified, they broke into fragmental blocks that tumbled down the steep flanks. So lava domes like Lassen Peak grow from the interior core outward, with the oldest flows forming the outer margins and the youngest flows forming the central core of the dome.

In contrast to conical stratovolcanoes that grow by the progressive upward accumulation of younger layers on top of older layers, lava domes grow into a disorganized, massive edifice as the viscous lava slowly exudes from the central crater. The processes involved in the creation of dacitic lava domes were introduced in Chapter 13 on Mount St. Helens where a lava dome of gas-depleted dacitic magma accumulated as a plug in the vent at the summit of the volcano after its eruption in 1980.

The rubbly appearance of Lassen Peak is not only the result of its irregular growth from dacitic lava flows, but was also strongly modified by alpine glaciation. The peak of the Last Glacial Maximum occurred around 20 ka, only a few millennia after Lassen Peak first arose. The angular edges of the Lassen lava dome were smoothed and rounded as glaciers abraded the bedrock, leaving behind the mantle of rocky scree that we see today (Fig. 16.1). Other evidence of alpine glaciation on Lassen Peak includes a huge cirque on its northeastern flank, as well as grooves and scratches on rock faces created as sand embedded in the base of glacial ice flowed across the underlying rock like a giant rasp.

Even though no glaciers exist in the park today, the effects of Ice Age alpine glaciation are widely expressed. The aptly named Brokeoff Mountain was not truncated along its eastern flank by a caldera eruption associated with the

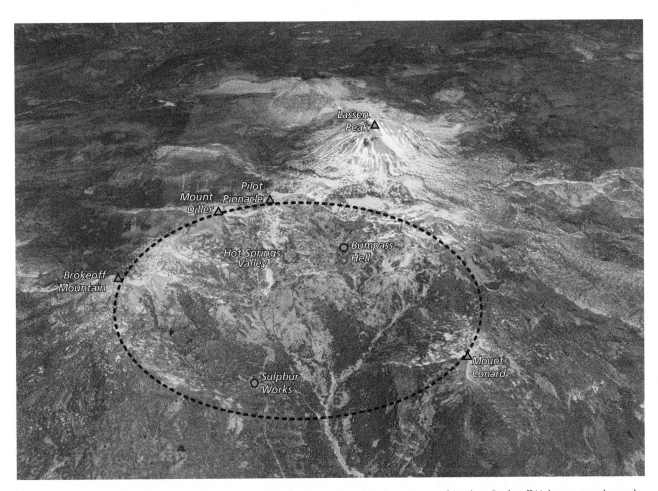

Figure 16.4 Low-angle oblique view north toward Lassen Peak. Erosional remnants of ancient Brokeoff Volcano are shown by triangles, with the dotted oval estimating the areal extent of the Brokeoff edifice.

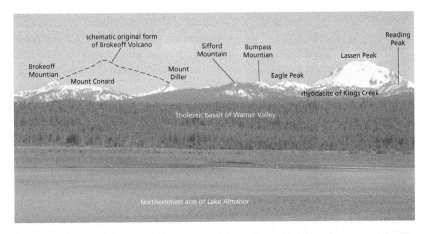

Figure 16.5 View northwest from Lake Almanor toward the reconstructed profile of ancient Brokeoff Volcano and modern Lassen Peak. After Brokeoff volcanic activity waned, hydrothermal alteration weakened the rock of the edifice. Alpine glaciers scraped away at the decomposing rock, transporting the debris downslope and leaving behind the erosional remnants of Brokeoff Mountain, Mount Diller, and Mount Conard.

whereas *morainal ridges* of glacial debris mark the terminus of glaciers that extended far downslope. Most of the lakes in the park are attributed to Ice Age glaciation as well, either formed within the scoured base of cirques or created as meltwater was impounded behind moraines.

Holocene Volcanism in the Lassen Region

After the effusive eruptions that produced Lassen Peak around 27 ka, the region experienced 26,000 years of quiescence before volcanism resumed. Lassen has been particularly active during the last millennium, with three significant eruptions attesting to the late Holocene reawakening. Around 1050 years ago (~950 CE), six overlapping lava domes of rhyodacite formed Chaos Crags, north of Lassen Peak. In 1666 CE, Cinder Cone formed in the northeastern corner of the

ancestral Brokeoff stratovolcano as you might suspect, but rather was cut by glaciers scraping away at its edges. The highlands of the southwestern part of the park are characterized by a scalloped landscape of *cirques* and *arêtes*,

park, accompanied by lava flows emanating from its base. And in 1914, Lassen Peak began a series of eruptions that lasted through 1917. The 20th-century eruptions of Lassen Peak were photographed and documented by newspapers in northern California, leading to the preservation of the region as a national park in 1916.

The formation of Chaos Crags (Fig. 16.6) began 1050 years ago with the explosive release of rhyodacitic pyroclastic debris that blanketed the nearby region. The tephra from this eruption was derived from the same magma system that supplied the material for the nearby Lassen Peak lava dome, but the magma had evolved into a more silicic fluid by the late Holocene. Gases dissolved in the viscous, high-silica magma drove these early explosive pyroclastic eruptions. The gas pressure dissipated with each eruption, however, leading to the effusive oozing of sticky lava from a series of aligned vents. As with Lassen Peak, the six overlapping domes of Chaos Crags grew from the inside out as younger flows pushed older flows away from the central vent, forming steep flanks along the margins. The precipitous walls of the actively rising domes were inherently unstable and would occasionally collapse, forming turbulent pyroclastic flows consisting of blocky debris mixed with ash that swept outward from the dome.

Long after the eruptions of Chaos Crags ended, the steep western margin of one of the domes suddenly collapsed, triggering a series of *rock avalanches* that rumbled violently to the west. The abruptness of the collapse trapped a cushion of air beneath the avalanche debris, reducing the friction and enabling the mass to rapidly cover 7 km² (~2 mi²) with an elongate lobe of angular rock fragments called the Chaos Jumbles (Fig. 16.6). The rubble settled out across Manzanita Creek, damming it and backing up the water to form Manzanita Lake. Trees that drowned and died as the lake level rose yield radiocarbon ages that reveal a date of 1672 CE for the catastrophic avalanche, about 350 years ago.

Cinder cone volcanoes are common in Lassen Volcanic National Park, but the youngest and most notable is Cinder Cone in the northeastern corner (Fig. 16.7). The volcano itself is a symmetric accumulation of countless pebble-sized particles of **scoria** (aka *cinders*), formed as lava was propelled outward from a central vent by moderate gas pressures. This style of volcanism is called a **Strombolian eruption**, named after the Italian volcanic island, Stromboli, which has been continuously active for the past two millennia (Fig. 16.8). Scoria particles are formed as droplets of incandescent lava cool and solidify as they come into contact with the atmosphere during eruption. Tiny holes called **vesicles** pock each particle where bubbles of gas were released during cooling. The innumerable fragments accumulate around the central vent, building the conical volcano upward. The cooled scoria particles tumble down the flanks, eventually settling out at a stable angle of repose around 35°. Two concentric craters at its summit record two phases of eruptive activity. Cinder Cone rises about 215 m (700 ft) above the surrounding land, comparable in size to the Wizard Island cinder cone at Crater Lake National Park. It's probable that Cinder Cone, like many cinder cones, was formed over only a few months of volcanic activity.

Scoria composing Cinder Cone are dark particles of basaltic andesite, with silica contents of 53–57 percent, significantly lower than the higher-silica dacite that forms lava domes in the western side of the park. This chemical difference may relate to the ascent of lower-silica magma from deeper sources closer to the mantle. A network of north- to northwest-oriented faults and fractures exists in the region, created by extensional stresses in the adjacent Basin and Range Province to the east. These faults and fractures provide

Figure 16.6 Low-angle oblique view to the south toward Lassen Peak. Chaos Crags consists of six overlapping lava domes. Chaos Jumbles is a huge rock avalanche deposit derived from the catastrophic failure of the steep, west-facing flank of one of the Chaos Crags domes.

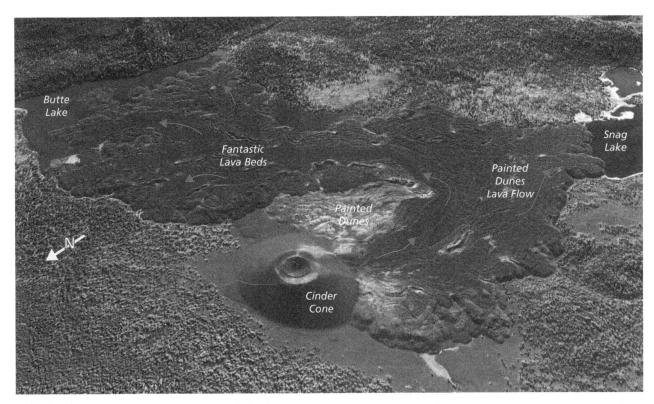

Figure 16.7 Low-angle oblique view to southeast across Cinder Cone and related lava flows. Arrows show directions of lava flows.

Figure 16.8 A) Strombolian eruption, Stromboli Volcano, Italy. B) Scoria is the solidified product of incandescent droplets of lava ejected during Strombolian eruptions.

planar conduits for the upward movement of lower-silica magma from deeper sources lurking below the magma chamber feeding the Lassen Volcanic Center.

Low-silica magma can also be released effusively to form lava flows. These low-viscosity, fluid lavas tend to flow laterally for long distances, forming broad edifices called **shield volcanoes**. Four shield volcanoes occur in Lassen Volcanic National Park and are expressed as broad, low-sloping mounds, commonly with a cinder cone at the apex. They tend to be heavily forested, and their low profile makes them difficult to see from the ground. The fluid style of volcanism that produces shield volcanoes will be explored more completely in Chapter 32 on Hawaii Volcanoes National Park.

The loose pile of scoria particles composing a cinder cone tends to be structurally weak. As a result, later lava flows may erupt from the lower flanks of the edifice rather than from the unstable summit crater. Cinder Cone in Lassen is no different, with syrupy basaltic andesite lavas emanating from a vent along the southeastern corner (Fig. 16.7). The oldest lava flows migrated southward where they blocked a creek; water impounded behind the lava dam to form Snag Lake. As these lava flows were dissipating, a burst of gas-charged ash erupted that fell onto the still-hot lava, reacting with oxygen in the atmosphere to form the multicolored hues of the Painted Dunes south and east of Cinder Cone. Very soon after, a second series of lava flows traveled to the northeast, covering part of the earlier Painted Dune lava beds and spilling into the preexisting Butte Lake. These are the Fantastic Lava Beds—blocky, angular piles of rock devoid of vegetation. Radiocarbon ages and tree-ring analysis of remnants of trees killed during the eruptions indicate that Cinder Cone and its related lava flows occurred in 1666 CE, making Cinder Cone the youngest volcano in the Cascades chain. Native Americans who hunted and foraged in the region likely witnessed the eruptions.

1914–1917 Eruptions from Lassen Peak

The ultimate Holocene eruption at Lassen Volcanic National Park began when Lassen Peak reawakened in 1914 with a series of explosions of steam mixed with small amounts of ash. (Residents of the region reported eruptions of "black smoke," misidentifying the ash-laden steam as the product of burning. There's never any "smoke" or "fire" associated with volcanic eruptions, even though these terms are commonly used descriptors. There may be burning of local vegetation that produces fire and smoke along the flanks of a volcano, but no actual combustion is associated with volcanism. The "smoke" that people see emanating from the summit of volcanoes in eruption is typically a mixture of steam and dark pyroclastic debris.) For the next several months, heat from the rising plume of magma interacted with groundwater to generate over a hundred *phreatic eruptions*, opening a crater at the summit of Lassen Peak. The volcano that had been asleep for 27,000 years was "clearing its throat" in preparation for the main event.

By mid-May 1915, a dacite lava dome had accumulated within the crater at the summit of Lassen Peak. Pent-up gas pressure generated a steam explosion that shattered the dome, triggering an avalanche of hot rock that mixed with the deep snowpack to form dense mudflows that barreled down the northeastern flank before funneling into nearby stream valleys. This surge of lahars traveled over 48 km (30 mi) downstream, uprooting trees and damaging ranches in the floodplain but causing no casualties. Overnight, viscous dacite lava surged through the reopened vent at the summit, spilling over the rim as two separate lobes.

Figure 16.9 Plinian eruption column from Lassen Peak as seen from Red Bluff, California, May 22, 1915.

A few days later, on May 22, 1915, Lassen Peak exploded in a Plinian eruption, blasting a column of pyroclastic debris 9 km (~6 mi) into the atmosphere (Fig. 16.9). Particles of pumice and ash rained down across the region as a pyroclastic fall, with the wind blowing the plume eastward as far as Elko, Nevada, 450 km (280 mi) away where it fell as a fine ash.

As the upward thrust of gas pressure began to dissipate, the eruption column partially collapsed, generating a pyroclastic flow that scoured the northeastern flank, already laid bare by the avalanche and lahars of a few days before. The scorching hot flow melted snow, rapidly transforming into a massive lahar that retraced the path of the earlier lahars, flushing through the valleys of Lost Creek and Hat Creek for tens of kilometers. The combined effects of the debris avalanche, pyroclastic flow, and surge of lahars ravaged the northeastern side of Lassen Peak, which today is aptly called the Devastated Area (Fig. 16.10). Hat Lake, along the edge of the Devastated Area, was formed as water from Hat Creek backed up behind a natural dam of mudflow debris.

Over the next few years, remnant magma beneath Lassen Peak mixed with groundwater to produce a series of relatively small phreatic eruptions, some of which were significant enough to create new vents at the summit in 1917. Activity waned, however, with the final small releases of steam occurring in 1921. The total volume of material ejected during the 1915 eruption of Lassen Peak was about 0.03 km³ (0.007 mi³), a very small amount relative to the 1980 eruption of Mount St. Helens, which emitted about 1 km³ (0.24 mi³). Lassen Peak is quiet today, but its active hydrothermal systems, fueled by the energy of a magma chamber deep beneath the surface, continually remind us of its future potential for eruption.

Modern Lassen Landscape

Of all the diverse volcanic characteristics of Lassen Volcanic National Park, perhaps the one that distinguishes it the most from other parks of the Cascades is its extensive

Figure 16.10 Devastated Area along the northeastern flank of Lassen Peak, formed during the May 1915 eruption of Lassen Peak. The large amphitheater near the summit (dashed lines) was formed by a combination of debris avalanches and pyroclastic flows. The fragmented rock and ash from these events mixed with snow to generate lahars that scoured the slope before funneling into the Lost Creek and Hat Creek drainages (arrows).

somewhat intermediate in that they have less water than hot springs but more water than fumaroles. Mudpots are filled with gray, burbling mud derived from the breakdown of volcanic rock beneath the surface by hot, acidic waters. These hydrothermal features will be addressed in more detail in Chapter 31 on Yellowstone National Park.

The energy that feeds the network of hydrothermal features is a body of magma surrounded by hot rock that resides about 8–10 km (5–6 mi) beneath much of the southern part of the park (Fig. 16.12). Water derived from the abundant snowpack as well as from rain percolates downward through fractures and pores in the underlying volcanic rock. A few kilometers beneath the surface, the cold water is heated by contact with hot rock overlying the active magma chamber at depth. The heated water expands and buoyantly rises upward, following a natural plumbing system of fractures and pipes. Along the way, sulfur and other elements are dissolved from the rock, making the water slightly corrosive. As the water rises, the pressure from the weight of overlying rock is reduced, permitting the gases in the element-rich water to expand and boil into water vapor and other gases. Fumaroles occur where steam and smelly sulfurous gases are released to the atmosphere. In other places, the water vapor and gases interact with groundwater near the surface to feed hot springs. And in yet other places, the gas-charged water is acidic enough to decompose solid rock to clay minerals; the clay-rich slurry bubbles upward to emerge as mudpots.

The flow of hot, acidic waters permeates much of the solid volcanic rock near the surface, transforming it into clays and silica minerals over time. This in-place *hydrothermal alteration* of rock imparts the light color to the land surface around hydrothermal areas and makes the

system of hydrothermal features. As the name implies, "hydrothermal" refers to features and processes that involve hot water. The most varied hydrothermal area is Bumpass Hell, reached by a pleasant hike from the main park road (Fig. 16.11), but other areas include Sulphur Works, Little Hot Springs Valley, Devils Kitchen, and Boiling Springs Lake. Each of these hydrothermal sites likely represents the exposed remains of an ancient conduit that occupied the deep interior of Brokeoff Volcano.

Bumpass Hell comprises over 75 hot springs, fumaroles, and mudpots. **Hot springs** commonly form colorful pools where the water temperatures may be greater than 90°C (~200°F). *Fumaroles* are vents in the surface from which steam and sulfurous gases are emitted. Big Boiler, a fumarole in the Bumpass Hell area, has temperatures measured as high as 161°C (322°F), among the hottest in the world. **Mudpots** are

Figure 16.11 Bumpass Hell hydrothermal area. The colorful ground surface is composed of clay- and silica-rich minerals formed by the in-place alteration of volcanic rock by hot, acidic fluids. A boardwalk provides safe access for visitors.

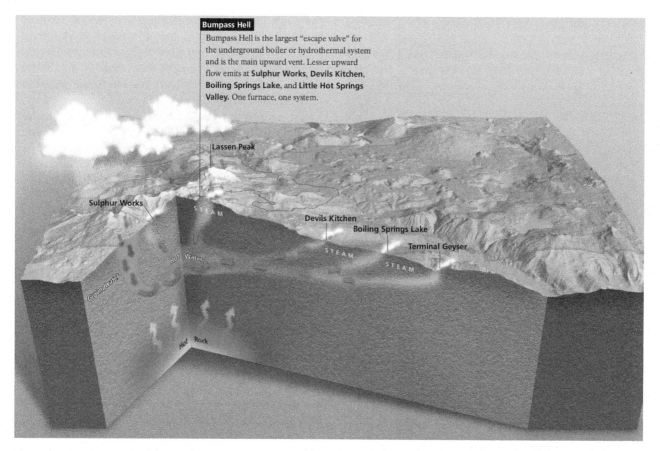

Bumpass Hell

Bumpass Hell is the largest "escape valve" for the underground boiler or hydrothermal system and is the main upward vent. Lesser upward flow emits at **Sulphur Works, Devils Kitchen, Boiling Springs Lake,** and **Little Hot Springs Valley.** One furnace, one system.

Figure 16.12   Illustration of Lassen hydrothermal system showing the interaction of snowmelt and rainwater with the heat energy of the magma chamber at depth.

rock susceptible to weathering and slope failure. The landscape near hydrothermal areas is commonly rounded and smoothed as the weak clays and silica minerals are exposed to surface weathering (Fig. 16.11). The acidic waters are also destructive to boardwalks and roads, creating a headache for park managers and a potential hazard to visitors.

Primitive single-celled microbes called **archaea** are superficially similar to heat-loving bacteria and aid in the production of sulfuric acids in the hydrothermal groundwater. Different types of archaea live within hot springs, with the species determined by the water temperature. The bold aquamarine to emerald colors of some hot springs are attributed to the presence of a specific type of archaea, as well as to the composition of dissolved elements. These primitive microbes are the focus of study for **astrobiologists**, who search for organisms on Earth living under extreme conditions thought to be comparable to those of Mars or other planetary bodies.

Other than hydrothermal areas, Lassen's modern landscape is notable for its glacial features, hundreds of lakes, and meadows seasonally strewn with wildflowers. Lassen is a bit too low in elevation to sustain glaciers, but the erosional and depositional effects of Ice Age glaciation are present throughout the park. Most lakes in the park are the

products of glacial activity, but as you've seen, they can be formed in other ways as well. Some lakes are impounded behind natural dams of rubble from rock avalanches (e.g., Manzanita Lake), whereas others form behind the debris left from lahars and pyroclastic flows (e.g., Hat Lake). Yet other lakes are confined behind lava flows (e.g., Snag Lake). The streams feeding these lakes transport sediment from nearby highlands, eventually filling in the lakes to form wetlands. As organic material from the decay of plants fills in the wet marsh, it evolves into a true meadow.

Lassen's volcanic history and landscape share many features with other parks and monuments of the Cascades. The 1914–1917 sequence of eruptive events at Lassen exhibits many similarities with the events at Mount St. Helens in 1980. The growth of a dacite lava dome in the crater at Mount St. Helens is comparable to the lava domes in Lassen. And the slow, inexorable pace of environmental succession in the decades following the eruptions at both volcanoes provides a living laboratory for ecologists. The ancient caldera at Lassen that produced the Rockland ash is very similar to the caldera eruption at Crater Lake National Park that produced the Mazama ash. The Wizard Island cinder cone within Crater Lake was created in identical fashion to Cinder Cone in Lassen, including the effusive release of lava flows from the base of each edifice. Lassen Peak also shows similarities with Mount Rainier, including the role of snowpack and glacial ice in the formation of lahars, as well as the hydrothermal alteration of rock by percolating hot fluids. All of these parks and monuments of the Cascades were profoundly affected by Pleistocene glaciation that overlapped in time with active volcanism and that sculpted the distinctive landscapes we see today.

After learning about three volcanic parks and a monument in the Cascades in Part 2, you're prepared to take a volcano road trip. Be sure to visit those volcanoes that aren't national parks, such as Mount Shasta, Newberry Crater, Mount Hood, and Mount Adams. You'll understand what you're looking at because by now you've learned a lot about volcanoes and volcanic processes. When you visit Lassen, however, be sure to take the hike to the top. It's not every day that you get to climb a volcano.

Key Terms

archaea
astrobiologist
hot spring
mudpot
scoria
shield volcano
Strombolian eruption
vesicles

Related Sources

Clynne, M. A., Christiansen, R. L., Felger, T. J., Stauffer, P. H., and Hendley, J. W., II. (1999). *Eruptions of Lassen Peak, California, 1914–1917*. Fact sheet 173-98. U.S. Geological Survey. Accessed February 2021 at http://pubs.er.usgs.gov/publication/fs17398

KellerLynn, K. (2014). *Lassen Volcanic National Park: Geologic Resources Inventory Report*. Natural Resource Report NPS/NRSS/GRD/NRR—2014/755. National Park Service.

Lassen Volcanic National Park Service: Volcanoes Accessed February 2021 at https://www.nps.gov/lavo/learn/nature/volcanoes.htm

USGS California Volcano Observatory: Lassen Accessed February 2021 at http://volcanoes.usgs.gov/volcanoes/lassen_volcanic_center/

17 Sierra Nevada Geologic Province

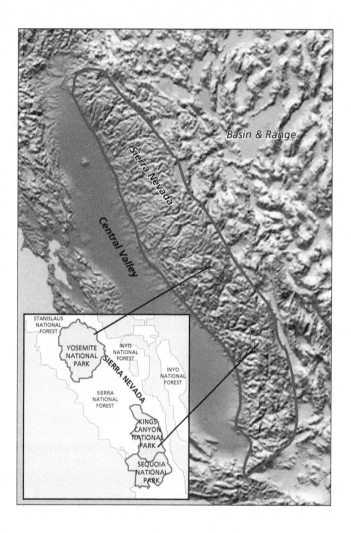

Figure 17.1 Locations of national parks of the Sierra Nevada Province addressed in Part 3.

The magnificent Sierra Nevada Geologic Province extends northwest–southeast for over 650 km (~400 mi) along the eastern flank of California (Figs. 2.1, 17.1). The mountains are highest in the southern part of the range and progressively get lower in elevation toward the northwest where they merge imperceptibly into the southernmost Cascades. The "High Sierra"—that portion comprising the highest peaks in the range—extends from north of Yosemite National Park to south of Mount Whitney in Sequoia National Park. At 4421 m (14,505 ft), Mount Whitney is both the highest peak in the Sierra and the highest point in the contiguous United States. Eleven other peaks in the High Sierra are greater than 4200 m (14,000 ft) in elevation.

The Sierra Nevada forms an asymmetric, wedge-shaped block, gentle on the west and steep on the east (Fig. 17.2). The topography rises gradually eastward from the flat expanse of the Central Valley into a broad belt of foothills in the western Sierra. The elevations culminate along the crest of the eastern margin, where the highest summits form a jagged spine. The eastern flank of the Sierra is a steep escarpment that drops off precipitously (Fig. 17.3). (Recall from Chapter 2 that an *escarpment* is a steep cliff face that extends for tens of kilometers, forming an abrupt break between highlands and adjacent valleys.)

A complex network of faults bound the range along the base of its eastern flank, separating the Sierra Nevada from the Basin and Range Geologic Province to the east (Fig. 17.4). During countless earthquakes over the last five million years or so, the rocks of the crust to the east of the fault system were dropped downward, while the rocks of the Sierran crustal block jerked upward. The eastern escarpment of the Sierra Nevada is the deeply eroded exposed surface of the fault planes. Uplift of the modern Sierra Nevada is commonly compared to the episodic opening of an enormous trapdoor, with the hinge somewhere deep beneath the Central Valley. The Sierra is the largest **tilted fault-block** mountain range in North

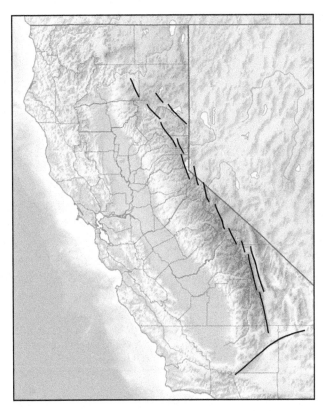

Figure 17.4 Simplified map of Eastern Sierra fault system showing the network of faults that extend along the eastern margin of the range. To the east is the Basin and Range Province.

America. The specific nature of the faulting and the mechanics of uplift will be addressed later in this chapter.

The importance of the Sierra Nevada to the well-being and prosperity of California cannot be overstated. It was the discovery of gold in the Sierran foothills that drew emigrants to the region in the mid-19th century. Imagine the dismay of the early settlers who, having just crossed the barren expanse of the Nevada desert, now faced the imposing eastern escarpment of the Sierra Nevada! The snowpack that accumulates in the Sierra during the winter slowly melts through the spring and summer, providing a steady supply of water for agricultural and urban use.

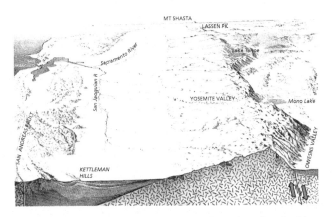

Figure 17.2 Vertically exaggerated perspective view looking northward along the Sierra Nevada and Central Valley.

Figure 17.3 Eastern escarpment of the Sierra Nevada, looking due west. A network of subparallel faults is buried beneath alluvium at the base of the escarpment. During earthquakes, the Sierran block rises upward, whereas the adjacent block to the east drops downward. The escarpment itself is the deeply eroded fault surface. The Owens Valley in the foreground has elevations of around 1200 m (4000 ft), whereas the highest peaks of the Sierra reach 4200 m (14,000 ft). That equates to a total relief of about 3000 m (~10,000 ft) on this part of the escarpment.

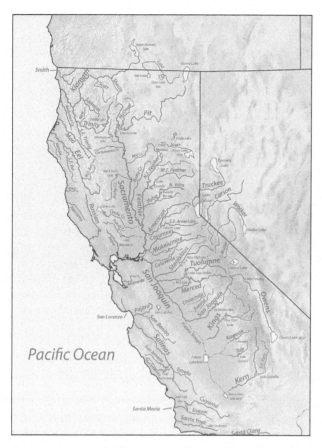

Figure 17.5 A comb-like network of rivers drain the western slopes of the Sierra Nevada, eventually flowing into the San Joaquin and Sacramento rivers.

The rivers that drain the gentle western slopes of the range feed into the San Joaquin River flowing northward and the Sacramento River flowing southward (Fig. 17.5). These two great rivers converge in the huge estuary that Californians call the "Delta." From there, much of the freshwater is redistributed via canals and aqueducts to the agricultural lands of the San Joaquin Valley as well as to urban and industrial users throughout the state. This water is truly the lifeblood of California and enables the economic vitality of the state.

Prior to the building of dams, reservoirs, canals, and tunnels necessary to capture and redistribute the snowmelt, the rivers transported eroded sediment out of the Sierra to redeposit it along the floodplains of the rivers and sloughs of the Central Valley. This sand, silt, and clay combined with organic matter derived from microbes and decaying vegetation to create some of the most fertile soils on the planet. The Sierra and the Central Valley are inextricably linked via the gravitational transfer of water and sediment downslope. Beyond the gold, the water, and the agricultural fecundity sourced in the Sierra Nevada, the range also contributes mightily to tourism. People flock to its national parks, its ski resorts, its whitewater rivers, its fish-filled lakes, and simply to revel in its inherent beauty, bringing prosperity to the region and the state. Rarely

does a natural feature enable such enormous economic well-being as the Sierra Nevada does for California.

This introductory chapter on the Sierra Nevada lays the groundwork for deeper exploration of Yosemite and Sequoia/Kings Canyon national parks. The geologic concepts addressed in this chapter include mechanisms of mountain-building, plutonism and the formation of granite, isostatic uplift, and the role of convergent margin tectonism in the origin of the American West. And we'll dig deeper into the effects of Ice Age glaciation on the evolution of landscapes.

Granitic Roots of the Sierra Nevada

Visitors to the higher elevations of the Sierra Nevada are commonly awestruck by the broad expanse of whitish-gray rock that forms the stunning landscape (Fig. 17.6). Most people don't realize that they are seeing the solidified remnants of molten rock that formed a few kilometers beneath the surface within magma chambers during the Mesozoic. Chapters 1 and 12 (on the Cascades) define rocks that form from the cooling and solidification of a molten fluid as *igneous* rocks. In the Cascades, most of the igneous rocks formed on the landscape from lava or pyroclastic debris that erupted from volcanoes. In the Sierra Nevada, however, most of the igneous rocks crystallized several kilometers beneath the surface as magma slowly released its heat to the surrounding country rock. The granitic rocks of the Sierra are thus *plutonic igneous rocks*. (Recall that lava and magma are both molten rock, with the difference being simply their location. Lava occurs on the surface, whereas magma occurs beneath the surface.)

Magma within a chamber consists of elements, dominated by oxygen, silicon, sodium, aluminum, and several others, all swirling about in a viscous, molten mixture. The intense temperatures necessary to keep rock in a molten state cause the individual atoms to vibrate at very high rates, with atoms colliding with one another and rebounding away to immediately slam into other atoms, precluding the formation of atomic bonds. But over time, heat from the magma chamber is slowly released to the surrounding solid *country rock*, and with cooling the atoms vibrate a little more slowly. As heat is gradually lost through time, perhaps over a few hundreds of thousands of years, the atoms eventually link up via chemical bonds, forming an orderly atomic arrangement of mineral crystals.

Crystallization of specific minerals occurs sequentially as the molten fluid progressively loses heat energy and the temperature decreases from perhaps 1250°C to less than ~600°C over time. The result is the interlocking mosaic of tightly welded minerals that comprise plutonic igneous rocks like *granite* (Fig. 1.20). Because the heat loss is very slow, the individual crystals have lots of time to grow larger, slowly adding elements from the remaining melt onto the outer crystal surface. Plutonic igneous rocks like granite tend to have mineral crystals that are visible to the naked eye and are much larger than the crystals

Figure 17.6 View of Yosemite high country, showing the enormous mass of light-colored granitic rock that dominates the landscape, giving rise to John Muir's name for the Sierra Nevada—the "Range of Light."

that form volcanic igneous rocks where the cooling rate is much, much faster. (Review the section on igneous rocks in Chapter 1 to fully understand the process of *plutonism* and the formation of granitic rock.)

Some of the magma in the chamber migrates upward to the surface where it erupts via volcanism. But most of the magma remains trapped within the chamber, slowly releasing its heat energy to the surrounding rock. When a magma chamber cools completely, it transforms into a large mass of igneous rock called a *pluton* (Fig. 1.22). During the Mesozoic, a few hundred individual magma chambers intruded into the overlying rock of the crust near the site of the future Sierra Nevada where they slowly cooled to become granitic plutons. Most of these plutons crystallized over a time span ranging from about 140 to 80 Ma.

Over this 60 m.y. period, some magma chambers were emplaced early and then solidified, while others came up later, intruding into and overlapping with the older plutons. Each pluton was once an individual magma chamber with a distinctive chemistry, resulting in a unique assemblage of minerals within each pluton. The rocks that compose each pluton are distinct enough in their chemical and mineral composition to warrant specific names like tonalite or granodiorite. For our purposes, however, we'll simply refer to the wide variety of plutonic igneous rocks that dominate the Sierra Nevada as "granite."

The magma chambers that would eventually become the individual plutons fed hundreds of aligned volcanoes that marked the crest of the **ancestral Sierra Nevada** during the Mesozoic. The modern Andes Mountains of South America and the Cascades Range of the Pacific Northwest provide good analogs for visualizing the ancestral Sierra. So there are two phases of the Sierra Nevada: a Mesozoic phase when the granitic plutons were formed and a late Cenozoic phase when the **modern Sierra Nevada** was

uplifted, beginning the creation of the landscape we see today. Each of these versions of the Sierra Nevada represents unique stages in the tectonic evolution of the American West, and we'll explore each of these later in this chapter.

In the modern Sierra Nevada, the hundreds of plutons overlap with one another to form a single enormous mass of granitic rock called the **Sierran batholith** (Fig. 17.7). The Sierran batholith forms the backbone of the range, with the highest peaks consisting of granitic rock. Thinking in three dimensions, visualize the huge volume of granitic rock extending downward into the crust for perhaps 15–25 km (9–16 mi). On the surface, the batholithic rocks are commonly juxtaposed against older metamorphic rocks that were intruded by the magma chambers during the Mesozoic. We'll address these metamorphic country rocks in greater detail when we visit Sequoia and Kings Canyon national parks in Chapter 19.

When you visit a national park in the Sierra Nevada and look at the expanse of whitish-gray granitic rock, remember that you're seeing the remnant, solidified interiors of

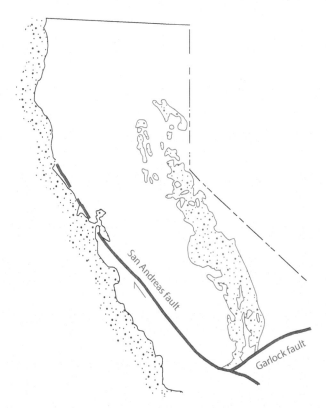

Figure 17.7 Map showing the exposed portions of the Sierran batholith in stipple pattern. The batholith is much larger in extent and underlies much of the Sierra Nevada but is covered by younger rocks. The batholith is composed of over a hundred individual plutons that mostly formed between 140 and 80 Ma.

Mesozoic magma chambers. Over the tens of millions of years since the plutons first formed, they were uplifted toward the surface by mountain-building processes. With uplift, the overlying country rocks were mostly removed by weathering and erosion, exposing the formerly deeply rooted granitic rocks. We see the rocks today with waterfalls flowing over their surface and random veneers of soil and vegetation, but try to visualize them as molten masses of magma deep beneath the surface of the ancestral Sierra Nevada.

Tectonic Origin of the Sierran Batholith

Where did the enormous volume of magma that became the Sierran batholith come from? About 200 Ma, a slab of oceanic lithosphere known as the **Farallon tectonic plate** began to subduct beneath the western margin of North America along a continental-oceanic convergent margin (Figs. 1.13, 17.8). This compressional tectonic setting would control the geologic evolution of the American West for the following 160 m.y. or so, and you'll read much more about the effects of Farallon convergence through the rest of the narrative in this book. Recall from Chapters 1 and 12 on the Cascadia convergent margin that oceanic lithosphere is consumed by subduction, ultimately to be buried in the mantle. The effects of subduction are wide ranging and include the generation of magma, the episodic rattle of large earthquakes, the formation of elongate chains of volcanic mountains, and the creation of deep oceanic trenches along the boundary between the two converging plates.

By about 140 Ma, enough water had been driven from the subducting Farallon plate to partially melt rock in the overlying mantle, with the magma rising buoyantly toward the surface (Fig. 12.7). Magma forced its way into

the compressional crush of the growing mountain belt along the continental edge of western North America, just as is happening today in Cascadia and in the Andes Mountains. A small portion of the magma leaked upward to erupt from a chain of volcanoes that marked the crest of the ancestral Sierra Nevada. But most of the magma stayed below the surface, slowly cooling within a vast network of overlapping magma chambers. From about 140 to 80 Ma, several generations of magma pushed their way into the bowels of the ancient mountain chain, coalescing into the plutons that compose the Sierran batholith. Contemporaneous Mesozoic batholiths of granitic rock exist to the north in Idaho and British Columbia, as well as to the south in Southern California and Baja Mexico, recording the crustal roots of the volcanic arc that extended along the entire margin of western North America. We'll encounter these granitic rocks in several other national parks later in the book.

A paleogeographic map illustrates the broad lateral extent of the subduction zone along western North America around 90 Ma (Fig. 17.9). Sea level was high at the time, and an elongate seaway, about 750 m (~2500 ft) deep, connected the Gulf of Mexico with the Arctic Sea. In California, the shoreline extended along what are today the foothills of the Sierra Nevada. Not much else of California existed at the time, partly because it was submerged beneath the oceans and partly because fragments of the state were accumulating far offshore, waiting to be tectonically assembled later in time. Dinosaurs migrated along the shorelines of this fragmented Mesozoic continent, and voracious marine reptiles occasionally surfaced for air in the surrounding ocean.

Isostatic Uplift of the Sierran Batholith

About 80 million years ago, the angle of descent of the Farallon plate began to decrease for reasons that are not entirely clear. Perhaps an underwater volcanic plateau of basaltic rock was consumed along the subduction zone, lowering the density of the slab and lifting it enough so that the angle of descent was flatter than before. Whatever the cause, the flatter slab beneath the ancestral Sierra Nevada didn't descend deep enough to release water and trigger the melting of rock, resulting in a shutdown in the magma supply and the cessation of volcanism. Without the intrusion of magma at depth, the chambers cooled and formed the very youngest plutons.

At this point around 80 Ma, the ancestral Sierra had reached their peak of uplift, with towering volcanoes

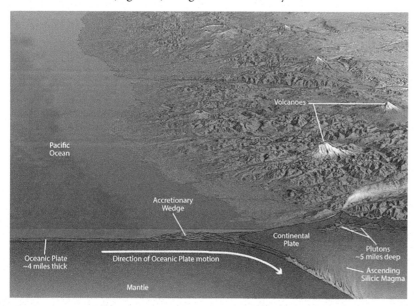

Figure 17.8 Convergent margin tectonic setting of Mesozoic western North America and the ancestral Sierra Nevada.

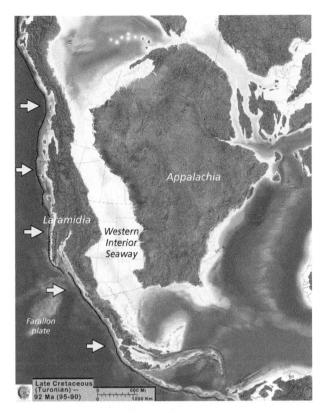

Figure 17.9 Paleogeographic map showing the extensive convergent tectonic boundary along the margin of western North America around 90 Ma. The Western Interior Seaway bisected North America at this time. Arrows show the orientation of convergence. Approximate extent of Sierran batholith shown by dashed outline.

crowning the range. The mountains were composed of deformed masses of rock surrounding granitic plutons of the Sierran batholith (Fig. 17.10A). The mass of mountainous rock acted as a load on the underlying lithosphere, causing it to sag downward. The asthenosphere beneath plastically migrated away from the overlying load of rock.

By the latest Mesozoic, the rate of erosion outpaced the slowing rate of uplift, and the mountains began to wear away. The erosional debris was swept westward into the nearby ocean, filling the ancestral deep-water precursor to today's Central Valley of California (Fig. 17.10B). The volcanic crest of the ancestral Sierra Nevada was almost completely removed by erosion. Relatively small volcanic remnants remain in the Sierra, such as at Mount Ritter east of Yosemite National Park. Other evidence of the volcanic crest is the eroded volcanic particles composing sedimentary rock found in deep wells and outcrops of late Mesozoic rock in the Central Valley.

As material eroded from the ancestral Sierran highlands and was redeposited in basins along the flanks, there was less weight bearing down on the thick crustal root (Fig. 17.10C). The underlying crustal rock began to slowly rise upward in a process known as **isostatic uplift**. With more erosion, the mass of the ancestral Sierra was diminished and the thick crustal root buoyantly rose in response, with the underlying mantle slowly rising to compensate. By the early Cenozoic, the deeply buried granitic plutons of the Sierran batholith were completely exhumed and exposed to weathering and erosion. The sedimentary debris deposited in basins along the flanks of the ancestral Sierra Nevada imposed a load on the crust, enhancing basinal subsidence, especially in the future Central Valley. By about 50 Ma, the Sierra region was a series of low rolling hills that sloped gently to the west, crossed by a drainage network of sediment-choked, braided rivers flowing from a highland plateau to the east in Nevada.

Icebergs provide a good model for the concept of **isostasy** and the isostatic uplift of mountain ranges like the ancestral Sierra. Icebergs float because the freshwater ice is about 15 percent less dense than the seawater that surrounds them, so about 15 percent of the iceberg is exposed above the waterline. As the uppermost ice melts due to the Sun's warmth, the underlying ice bobs upward in response, with the denser seawater supporting the mass from below. In an analogous fashion, the rock of the continental crust is about 15 percent less dense than the rock of the underlying mantle. As mass is removed erosionally from the higher elevations, the continental crust buoyantly rises, pushed upward by the denser mantle beneath. The net effect is that

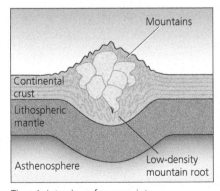

Time 1: Intrusion of magma into country rock (Mesozoic)

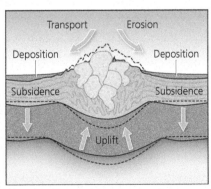

Time 2: cooling & solidification of magma chambers and erosion of overlying rock (Mesozoic)

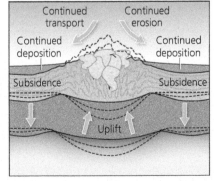

Time 3: Isostatic uplift of granitic rock as the weight of overlying rock is removed (Cenozoic)

Figure 17.10 Sequence of events illustrating isostatic uplift and exhumation of granitic plutons of the Sierra batholith.

deeply rooted granitic plutons are exhumed to the surface and become part of the landscape. Counter to your intuition, weathering and erosion not only remove rock from the highest elevations, but they also enable the slow isostatic uplift of mountains at the same time. Isostasy plays an important role in crustal uplift and mountain-building and has widespread applicability to the geology and landscape evolution of our national parks.

Uplift of the Modern Sierra Nevada

The modern Sierra Nevada tilted fault-block began to rise about 5 m.y. ago as the system of faults along the eastern flank became active. These faults are *normal faults* (which were briefly introduced in earlier chapters on Bryce Canyon and Canyonlands, Chapters 5 and 8, respectively). Normal faults occur in response to extensional stress where rocks of the rigid crust are stretched by tectonic-scale forces (Fig. 17.11). Recall that a fault is simply a planar fracture in rock along which displacement has occurred. Faults may be smoothly planar, or they may have some degree of curvature. The orientation of the fault plane may range from vertical to horizontal.

As extensional forces pull apart a body of rock, large fractures form that evolve into normal fault planes. When the accumulation of extensional stress eventually exceeds a threshold, a block of rock on one side of the fault plane will abruptly slide downward relative to the block of rock on the other side of the fault. This rapid release of pent-up tectonic stress is called an *earthquake*, and rupture along the fault plane sends out seismic energy in all directions. As these seismic waves reach the surface from their

deep-seated origin along the fault plane, they produce the groundshaking associated with earthquakes.

Repeated cumulative movement along the Eastern Sierra fault system (Fig. 17.4) over the last 5 m.y. or so has resulted in the uplift and asymmetric westward tilt of the Sierran block. Downdropping of rock along the eastern side of the fault system produced a series of north–south valleys lining the base of the Sierran escarpment. The Owens Valley, Mono Lake Valley, Carson Valley, and several others act as basins for the accumulation of sediment eroded from the adjacent highlands. The steep eastern escarpment of the Sierra Nevada is the exposed surface expression of the normal fault planes, now deeply eroded by streams, glaciers, and landslides (Fig. 17.3). Only a few mountain passes afford the opportunity to drive across the eastern escarpment. One of these roads (Route 120) follows a glacial valley from the Mono Lake Basin at 2100 m (~6900 ft) elevation to the Tioga Pass entrance to Yosemite at 3000 m (~9900 ft). Over the 13 km (8 mi) distance, you climb almost a kilometer in elevation!

Countless earthquakes have occurred over the past 5 million years along the Eastern Sierra fault system, and it is still active today. The most recent large earthquake occurred in 1872 and was located near the Owens Valley town of Lone Pine. The earthquake had an estimated magnitude near 7.8 and was accompanied by extreme groundshaking, the collapse of adobe structures, and 27 deaths. Vertical displacement along the fault ranged up to 6 m (20 ft), and the exposed fault scarp is clearly visible on the surface today. There was also a considerable amount of lateral displacement along the fault of up to 12 m (40 ft). Taken together, this part of the Sierran crustal block rose upward and shifted to the northwest by several meters over the terrifying tens of seconds of the earthquake. Extrapolating from this historical example, we see that thousands of large-magnitude earthquakes and countless smaller quakes accompanied the uplift of the Sierra Nevada over the last 5 m.y. or so. The origin of the extensional and shear forces that drive uplift will be addressed in Part 5 on national parks of the Basin and Range Province.

Measurements by the global positioning system of satellites show that the Sierra Nevada are currently rising at an average rate of 1–2 mm/yr (~0.08 in/yr). (You can think about that rate as 1–2 m per thousand years, or 1–2 km per million years.) Remember that as the mountains are episodically uplifted, weathering and erosion constantly wear away at the rising mass of rock. During the Pleistocene Ice Ages, much of the erosional debris was transported downslope by glaciers and their meltwater streams, ultimately to replenish soils of the Central Valley. Today, most of

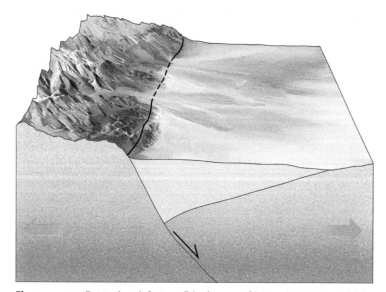

Figure 17.11 Extensional forces (block arrows) generate normal faults where the rock on one side of the fault plane slides downward relative to the rock on the other side. The fault plane may be steeply inclined near the surface but bends to a lower angle with depth. The downthrown side of the fault forms a basin that fills with eroded sediment derived from highlands on the upthrown side of the fault.

the loose particles carried by the major Sierran rivers are trapped within reservoirs behind dams.

Effects of Pleistocene Glaciation on Sierran Landscapes

The modern landscape of the Sierra Nevada reflects the influence of Ice Age glaciation perhaps more than any other single factor (Fig. 17.12). In Chapter 14 on Mount Rainier National Park, you were introduced to the impact of Pleistocene glaciation on landscape development as well as the mechanics of how modern *alpine glaciers* work. It's a good idea to go back and refresh your memory of that material. We'll build on that knowledge in this chapter and the next few, adding new glacial landforms to our portfolio as we explore the iconic landmarks of Yosemite and Sequoia/Kings Canyon national parks.

Globally over the 2.6 m.y. of the Pleistocene, there were multiple pulses of widespread glaciation that alternated with warmer interglacial phases. In the Sierra Nevada, four major advances of alpine glaciers are recognized, with the oldest being 1.5 million years old. Evidence of

previous Pleistocene glaciations was completely removed by the scouring action of the more recent events. The most extensive glaciation occurred about 800,000 years ago, with a slightly less widespread glaciation recognized and dated to 80–140 ka (ka = thousands of years ago).

The most recent glaciation peaked during the *Last Glacial Maximum* around 18–26 ka and is responsible for most of the glacial features seen in Sierran national parks. A highland **ice cap** extended for over 450 km (280 mi) in length and crowned the highest elevations of the range (Fig. 17.13). The ice reached a maximum thickness of about 700 m (2300 ft), with only the highest peaks poking up as rocky islands above the white expanse of ice. Along the east and west margins of the ice cap, long, thin alpine glaciers extended as tongues down what were previously river valleys. Alpine glaciers reached down the western slope from the Sierran crest for as much as 100 km (60 mi), but they did not extend as far down the steeper eastern side of the range. Glaciers wax and wane not only because of global and regional temperatures, but also in response to the availability of precipitation. The alpine glaciers on the west side of the Sierra extended farther downslope simply because of the higher precipitation relative to the much drier eastern slopes.

A variety of techniques are used to date ancient glacial events, including qualitative methods like the degree of weathering of granite boulders and the depth of soil development on *moraines*. Quantitative techniques include the radiocarbon dating of trees and other organic matter entrapped within glacially deposited debris as well as "surface exposure dating," which utilizes radioactive elements

Figure 17.12 Scalloped, jagged texture of the Sierra crest, formed as a result of Pleistocene glaciation. Mono Lake lies in the distance of this northward-looking view.

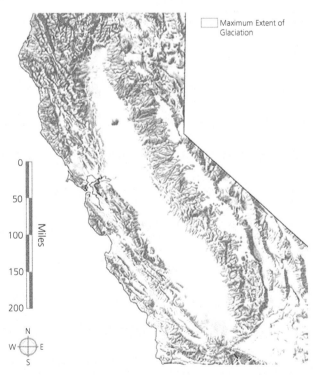

Figure 17.13 Maximum extent of Ice Age glaciation in the Sierra Nevada ~20,000 years ago.

produced as rock surfaces are bombarded with high-energy charged particles produced by cosmic rays from space. A simple feature on the landscape that is readily recognizable is the **trimline**—the near-horizontal boundary along the rock walls of glaciated valleys that marks the highest extent of glacial ice. In general, rock above the trimline is angular and jagged, whereas rock below the trimline has been smoothed and rounded by glacial erosion. The trimline isn't used to date glacial events, but it does provide an estimate of the height and thickness of alpine glaciers.

All of the Pleistocene alpine glaciers in the Sierra Nevada melted long ago as Earth entered the warm interglacial phase of the Holocene. Yet today there are 13 named glaciers as well as numerous smaller glaciers and perennial icefields, most occupying north-facing cirque basins at high elevations. All originated during the cold spell called the Little Ice Age that spanned the 500 years between 1350 to about 1850 CE. Since then, glaciers in the Sierra have generally thinned and retreated upslope as the rate of *ablation* exceeds the rate of *accumulation* in a progressively warming world.

Comparisons with Other Geologic Provinces

In Part 2 on the national parks of the Cascades, you learned that the age of the volcanic rocks and the age of the mountains themselves are the same since it was relatively recent volcanism that produced both. In this Part, we return to the format introduced in Part 1 on the rocks of the Colorado Plateau, where the rocks are old but the landscape is young. In similar fashion, the granitic core of the Sierra is Mesozoic in age, but the mountainous landscape we see today was sculpted over the last few million years by uplift and glacial modification. So, armed with an awareness of the geologic history of the Sierra Nevada as a whole, you're ready for a deeper understanding of the landscape evolution of Yosemite and Sequoia–Kings Canyon national parks.

Key Terms

batholith
Farallon tectonic plate
ice cap
isostasy
isostatic uplift
tilted fault-block
trimline

Related Sources

Alt, D., and Hyndman, D. W. (2016). *Roadside Geology of Northern and Central California.* Missoula, MT: Mountain Press Publishing Co.

Guyton, B. (1998). *Glaciers of California: Modern Glaciers, Ice Age Glaciers, the Origin of Yosemite Valley, and a Glacier Tour in the Sierra Nevada.* California Natural History Guide 59. Oakland: University of California Press.

Hill, Mary. (2006). *Geology of the Sierra Nevada.* California Natural History Guide 80, Oakland: University of California Press.

Jones, C. H. (2017). *The Mountains That Remade America: How Sierra Nevada Geology Impacts Modern Life.* Oakland: University of California Press.

18 Yosemite National Park

Figure 18.1 View from Glacier Point in Yosemite National Park, showing the enormous mass of granitic rock that dominates the landscape. Yosemite Valley lies in the foreground, and Tenaya Valley extends away from the viewer beneath the monolith of Half Dome.

Yosemite National Park is invariably on the short list of must-see natural attractions for visitors to California. With its immense whitish-gray cliffs and domes, resplendent waterfalls, towering sequoia groves, wide-open meadows, and diverse array of hiking trails, the park attracts millions of visitors every year. Most converge on Yosemite Valley because of its jaw-dropping scenery and iconic landmarks (Fig. 18.1). But an equally sublime experience awaits those who venture into the Yosemite high country where the people are fewer and the wonders of the landscape are more accessible. All of this phenomenal beauty is enabled by the geology, of course, which involved a variety of processes interacting through time, including plutonism, mountain-building, stream incision, and glaciation. We'll begin by describing the physical characteristics of the park, then move on to understand its rocks, and finish with the evolution of the Yosemite landscape.

Topography and River Systems

Yosemite National Park is located in the central Sierra Nevada (Fig. 17.1). Elevations range from ~650 m (2100 ft) near the western approaches to Yosemite Valley to almost 4000 m (13,114 ft) at the top of Mount Lyell along the Sierran crest near the eastern boundary of the park (Fig. 18.2). Mount Dana and Kuna Peak are almost as high as Lyell, both exceeding 3960 m (13,000 ft). Twenty-six other peaks of the High Sierra in Yosemite reach 3660 m (12,000 ft). The north flank of Mount Lyell supports a small glacier, the second largest in the Sierra Nevada, although it is rapidly losing mass in a warming climate. A remnant of the Little Ice Age, the Lyell glacier and most other glaciers of the Sierra Nevada are likely to disappear completely by the end of the century in a progressively warming world.

Yosemite is drained by the Tuolumne and Merced rivers, which originate in the High Sierra along the eastern boundary of the park before flowing down the westward tilt of the range (Fig. 18.3). Both rivers carved deeply into the granitic rock of the park as their gradients increased

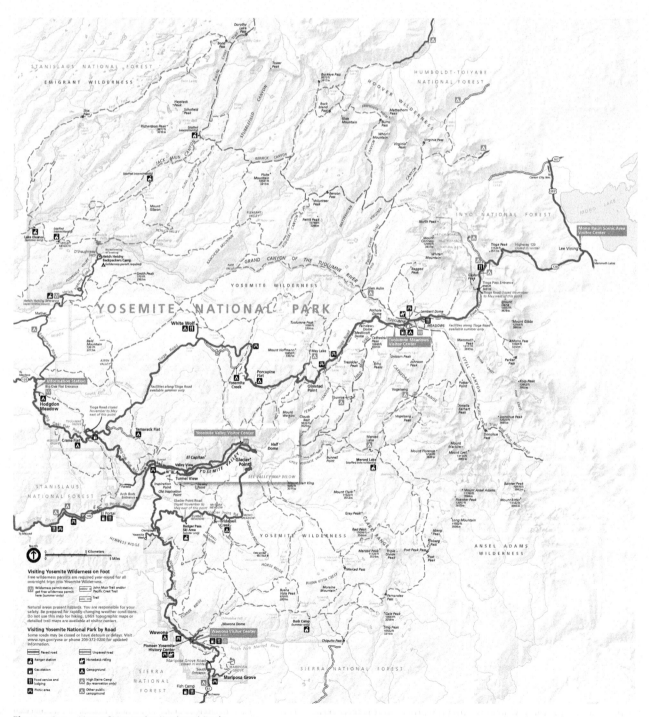

Figure 18.2   Map of Yosemite National Park.

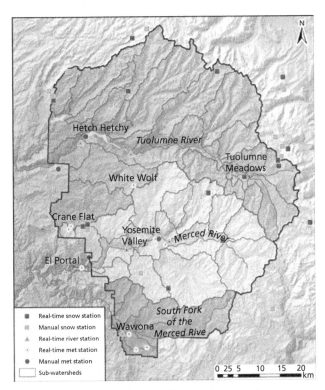

Figure 18.3 Watersheds of the Tuolumne River (green), Merced River (light blue), and South Fork of the Merced River (darker blue). Each river drains westward from its head along the Sierran crest. Small symbols show meteorological stations. Note the location of Hetch Hetchy Reservoir on the Tuolumne River and Yosemite Valley on the Merced River.

in response to late Cenozoic uplift of the range. The erosional activity of the ancestral Tuolumne and Merced river systems set the stage for the subsequent sculpting of the landscape by Pleistocene glaciers.

The Tuolumne River begins in streams emanating from Mount Lyell and Mount Dana, which meet in the Tuolumne Meadows high country. The river cuts a deep gorge, the Grand Canyon of the Tuolumne, before backing up to fill Hetch Hetchy reservoir behind O'Shaughnessy Dam, built in 1923 after a contentious legal battle (Fig. 18.4). Some of the water is released downstream where, just outside the park boundaries, the Tuolumne attains Wild and Scenic River status and is a popular rafting and kayaking destination. Much of the water in the reservoir, however, is diverted to aqueducts that supply pristine water to the San Francisco Bay Area. The Hetch Hetchy Valley rivals Yosemite Valley in its glaciated beauty and, if not flooded behind the dam, would help to alleviate overcrowding by giving visitors another spectacular destination in the park. As

it is, Hetch Hetchy is well worth a visit, even with the reservoir filling the valley.

The Merced River watershed is smaller than that of the Tuolumne, but it drains a significant area of the park on its way west from its headwaters in the High Sierra. The Merced and its tributaries cut a number of narrow gorges in the bedrock of the Yosemite high country before tumbling over Nevada Falls and Vernal Falls into Yosemite Valley. The river meanders through the meadows and woodlands of the valley before exiting the park on its way toward its confluence with the San Joaquin River in the Central Valley.

The Merced River in Yosemite Valley typically flows at about 350 cfs, with significant variation through the seasons. (Cfs, cubic feet per second, is the standard measure of *discharge*, the volume of water passing by any one area on the stream at any second. See Chapter 4.) Don't let the Merced's serene, picturesque beauty deceive you, however. After a winter of heavy snowfall followed by a warm, "Pineapple Express" storm in January 1997, the Merced raged at 24,600 cfs, inundating a huge area of Yosemite Valley, destroying campsites, roads, and bridges, as well as flooding buildings and rearranging the course of the river. Floods such as this must have been common in the geologic past. The ancestral Merced and Tuolumne rivers, together with their tributaries, transported huge amounts of sand, gravel, and boulders during floods that banged against the channel walls, incising deeply into the bedrock. The V-shaped gorges and valleys cut by these river networks formed natural pathways for alpine glaciers that waxed and waned through the Pleistocene. The raw beauty of the Yosemite landscape is typically attributed to glacial activity, but the effectiveness of sediment-charged flowing

Figure 18.4 Artistic view looking due east along Hetch Hetchy Reservoir and Yosemite Valley.

water likely played an equivalent role in sculpting the bedrock through time.

Granitic Rocks of Yosemite

Yosemite National Park is underlain almost entirely by granitic rocks of the Sierran batholith, described in Chapter 17 on the geology of the Sierra Nevada (Fig. 18.5). You understand from the previous chapter that the whitish-gray rocks that form the magnificent scenery are the solidified remnants of Mesozoic magma chambers that crystallized over time to become plutons. In the Yosemite region, several distinct plutons form a complex array of overlapping bodies created by the intrusion of one magma chamber into another during the Mesozoic. Each of the plutons is composed of a unique assemblage of minerals that distinguishes one from the other, but all are broadly granitic in composition. Although some granitic plutons are as old as 220 Ma, the majority of plutons in the Yosemite region were emplaced between 105 and 85 Ma. The granitic magmas intruded into older Paleozoic and early Mesozoic sedimentary and volcanic rocks, now thoroughly metamorphosed. The erosional remnants of these metamorphic rocks crop out along the margins of the park (Fig. 18.5).

Among the most widespread granitic rocks of Yosemite are the Half Dome Granodiorite, the El Capitan Granite, and the Cathedral Peak Granodiorite (Fig. 18.6). The minerals composing these rocks form an interlocking mosaic of

Figure 18.6 A) Hand sample of Half Dome Granodiorite. B) Outcrop photo of Cathedral Peak Granodiorite.

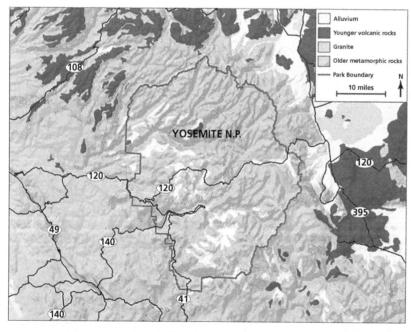

Figure 18.5 Simplified geologic map of the greater Yosemite region. A geologic map shows the distribution of various rock types across an area. The "granite," shown in pink, comprises several different overlapping plutons, each with its own unique composition and age. The tan areas marked as "alluvium" are a thin veneer of loose sand and gravel found in stream valleys and meadows. The green areas along the margins of the park are the eroded remains of older metamorphic rocks that formed the country rock intruded by the younger granitic magmas.

medium- to coarse-grained crystals that are visible to the naked eye but are best viewed with a hand lens. Lighter colored minerals such as feldspar and quartz, and darker colored minerals such as hornblende and biotite collectively impart the typical "salt-and-pepper" appearance of granitic rocks. Large symmetric crystals of pink potassium-rich feldspars floating in a salt-and-pepper matrix distinguish the remarkable Cathedral Peak Granodiorite. The larger crystals crystallized first in the magma chamber and then literally floated in the remaining mush until enough cooling occurred to solidify the surrounding matrix. These large crystals are typically more resistant to weathering than the smaller crystals composing the rest of the rock and thus project out in relief, creating ideal holds for rock-climbers. The Cathedral Peak Granodiorite is best seen in the Yosemite high country near Tuolumne Meadows.

The soaring cliffs lining Yosemite Valley are an ideal place to see the granitic rocks up close and visualize

imagery with climbing and rappelling to see the rocks up close (Fig. 18.8). The monolithic wall is composed of a variety of granitic rocks, dominated by El Capitan Granite. The dark splotch in the center of the cliff is appropriately known as the North America Wall and is composed of a plutonic rock called diorite, which is characterized by the darker minerals hornblende and biotite. Angular blocks of El Capitan Granite up to 50 m (~160 ft) across detached from the chamber walls during forcible intrusion of the dioritic magma, then floated within the liquid melt. In time, the magma cooled and crystallized into the dark diorite, engulfing the unmelted chunks of lighter colored granite (Fig. 18.9). These granitic blocks are called **xenoliths** and are useful for determining the relative chronology of intrusions. The complexity of *cross-cutting relationships* exposed on the face of El Capitan indicates that granitic plutons are anything but homogeneous masses. Rather, plutons may be composed internally of a variety of plutonic rocks that reflect a long history of changing magma chemistry and multiple episodes of intrusion.

Landscape Evolution of Yosemite: Jointing

The rocks of Yosemite are the raw materials on which the landscape is sculpted. Four main processes conspired over time to shape the Yosemite terrain and its famous landmarks: river incision, jointing, glaciation, and rockfalls. You read earlier how the Tuolumne and Merced river systems carved deeply into the granitic bedrock of Yosemite, creating V-shaped canyons and gorges coincident with late Cenozoic uplift of the Sierra Nevada and setting the stage for subsequent glacial modification. In this section, we'll address the role of *jointing* in the evolution of the Yosemite mountainscape. Jointing was introduced in Chapter 4 on Zion National Park and was expanded upon through the rest of the chapters on parks of the Colorado Plateau because it plays such a fundamental role in shaping the architecture of the landscape. Recall that joints are nothing more than fractures that penetrate through bodies of rock, commonly as multiple near-parallel planes of weakness. They are important modifiers of the landscape in that they permit water to seep deeply into a body of rock, enhancing chemical weathering and physical breakdown.

1. Half Dome	2. Vernal & Nevada Falls	3. Yosemite Falls
4. Glacier Point	5. El Capitan	6. Bridalveil Fall

Figure 18.7 Artistic view looking due east into Yosemite Valley showing a few of the iconic landmarks referred to in the text.

the relationships between plutons as well as within individual plutons (Fig. 18.7). The valley is about 12 km (7 mi) long from its entrance near Tunnel View to the confluence with Tenaya Canyon at the head of the valley. The Merced River flows along the relatively flat floor of the valley, which is up to 1.5 km (~1 mi) wide. The near-vertical cliffs bounding the valley rise upward for 900 to over 1400 m (~3000–4700 ft), providing a rare perspective on the internal complexity of plutons and the bewildering array of intrusive processes that occur in a magma chamber. Plutons in Yosemite are commonly named for the prominent landscape feature they form and the composition of the rock.

The great mass of granitic rock exposed on the kilometer-high face of El Capitan has recently been mapped in detail by intrepid geologists who merged high-resolution

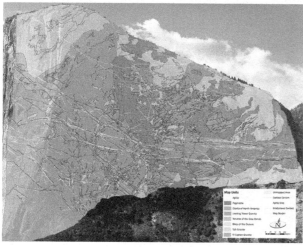

Figure 18.8 Photo of El Capitan paired with its geologic map. The monolith rises 950 m (3100 ft) above the valley floor. The darker rock in the center of the photograph is the "Diorite of North America" and is shown in green on the accompanying map.

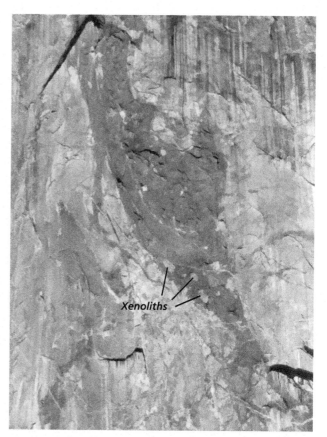

Figure 18.9 Dark mass is the Diorite of North America showing light-colored, angular xenoliths of El Capitan Granite derived from the surrounding country rock when the diorite was still in a molten state. The dark vertical streaks are waterlines, discolored over time by lichens and various oxide minerals.

In Yosemite, two main orientations of joint sets occur at right angles to one another: one set trending northeast and another trending northwest. These joint systems are widespread through Yosemite and likely formed in response to stresses imposed on the rock during uplift of the Sierran tilted faultblock. The orientation of many iconic landforms in Yosemite Valley is controlled by these two joint orientations, including the sheer northwest-facing cliff of Half Dome, the cliff faces of Vernal and Nevada Falls (Fig. 18.10A), and the parallel cliffs of Cathedral Rocks on the south side of the valley near Bridalveil Fall. These cliff faces are the remnant joint planes left behind after the adjoining rock was completely removed by a combination of rockfalls and glaciation.

Joints also control the orientation of many streams in Yosemite, including linear stretches of the Tuolumne and Merced rivers as well as the straight, northeast trend of upper Yosemite Creek near Tioga Road. Although many joint sets are near vertical, some are inclined to varying degrees, such as the west faces of Three Brothers where overlying rock has been eroded away along the tilted surface (Fig. 18.10B). The granite comprising the bulk of El Capitan is massive and relatively unjointed, but it's hard to imagine that the near-vertical cliffs of the monolith are not joint planes that have been modified through the years by countless rockfalls.

A different type of jointing, called exfoliation, forms concentric sheets in granitic rocks, giving an "onion-skin" appearance to the surface (Fig. 18.11). **Exfoliation joints** in granite are traditionally thought to be produced by the removal of confining pressure of overlying rock during exhumation of deeply rooted plutons (Fig. 18.12). As the body of rock isostatically rises to the surface in response to the removal of the overburden, the uppermost carapace of granite separates slightly along a curving fracture. With continued uplift, more rock separates along near-parallel exfoliation joints, forming a set of closely spaced stacked sheets, which are typically concordant to the topography. An alternative explanation proposed for the formation of exfoliation is that tectonic forces compress the body of rock horizontally, opening curving joint planes that parallel the convex-upward surface of the pluton.

Regardless of the mechanism that generates exfoliation jointing, the fractures between sheets permit the infiltration of water, weakening the rock and setting the stage for slabs to break away from the underlying mass. Tree roots may extend into the cracks in their search for water,

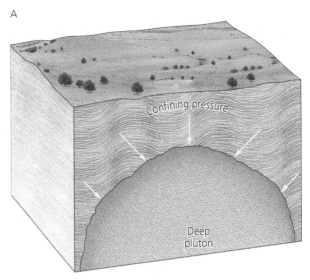

Figure 18.10 A) Vernal Falls plummets over a northwest-facing joint plane. B) West-tilted joint planes of Three Brothers. Vertical joints widened by weathering and erosion separate the three blocks.

Figure 18.12 Exfoliation jointing forms by the release of the overlying load of rock due to weathering and erosion. The granitic plutons expand slightly in response to the release of confining pressure, creating rounded fractures near-parallel to the exposed surface.

Figure 18.11 Exfoliation jointing forms "onion-skin" sheets of rock, here exposed near Sunrise Lake, Yosemite high country. Tabular blocks have detached from sheeted layers and slid downslope to pile up at the base.

physically wedging apart the rock and chemically breaking it down with their organic acids. Exfoliation joints are also instrumental in the "quarrying" action of glaciers, permitting sheets of rock to peel away as glaciers override bedrock. In Yosemite and many other places in the Sierra Nevada, the primary landform produced by exfoliation jointing is the **dome**, such as the intact southeast part of Half Dome as well as more symmetrical domes like North and Basket domes (Fig. 18.13). Royal Arches, located at the base of the cliff near the entrance to Tenaya Valley, reveals a unique view of vertically oriented exfoliation sheets developed in the cliff face. As individual sheets weakened through time, pieces broke away to fall onto the valley floor, leaving behind the arcuate

Figure 18.13 Domes formed by exfoliation jointing: North Dome (left) and Basket Dome (right).

Figure 18.15 Angular block of talus derived from a prehistoric rockfall.

edges of sheets. Glaciers eventually carried away the debris as they scoured the valley floor, creating the appearance of concentric arches. Slice an onion first horizontally and then vertically to better visualize Royal Arches.

Landscape Evolution of Yosemite: Rockfalls

Planar joints and exfoliation joints create planes of weakness in rocks that makes them susceptible to detachment from the main mass. Over time, as the final points of contact holding the slab to the rock face disintegrate, the fragment will succumb to the force of gravity and collapse downward as a rockslide or rockfall (Fig. 18.14). As the name implies, a **rockslide** occurs when a mass of rock slides down a slope before abruptly fragmenting upon impact with the valley floor. Rockslides commonly occur on domes, where slabs

of rock detach along exfoliation joints and slide along the curving surface. *Rockfalls* occur as a mass of rock free-falls or bounces down the side of a vertical cliff, exploding into pieces at the base. The debris from rockfalls accumulates as large, angular blocks of *talus* at the base of cliffs, a common feature in Yosemite Valley (Fig. 18.15). *Rock avalanches* are a third type of mass movement, driven by gravity. These may begin as a rockfall that maintains momentum upon impact, evolving rapidly into a chaotic, tumbling flow of debris that spreads outward across the ground. The talus from a prehistoric rock avalanche is recognized in El Capitan Meadow that reaches almost 670 m (2200 ft) away from the base of the El Capitan cliff.

From a landscape perspective, rockfalls and comparable mass movements act to maintain the vertical cliff faces of Yosemite Valley and many other precipitous cliffs in the park. Look closely with binoculars at any of the cliffs in the valley and you'll see light patches where a slab of rock detached from the main face as a rockfall. The ubiquitous piles of talus at the base of most cliff faces are also silent reminders of past rockfalls. The catalyst for rockfalls might be an abrupt event such as earthquake shaking. Or rocks can detach slowly but surely by the freezing and thawing of ice along a joint plane that physically wedges apart the rock. Spring runoff of snowmelt or perhaps intense rainfall may lubricate joint planes and disaggregate the rock beyond a threshold, leading to rockfall. But many rockfalls seem to occur during warm, sunny days without any obvious trigger.

An unexpected spark for rockfalls in Yosemite was recently documented to occur by the thermal expansion and contraction of rock. Sensors were installed within the exfoliation joint behind near-vertical slabs of rock to measure any movement. During the

Figure 18.14 Dust and debris created during the Happy Isles rockfall, 1996. The fall originated from Glacier Point on the upper right.

heat of the day, the massive slab of rock would expand outward by several millimeters and then contract during the cool of the evening. It's as if the rock were breathing, exhaling during the day and inhaling at night. Over time the exfoliation cracks open wider, eventually weakening to the point of failure and resulting in a rockfall for no apparent reason. Add daily and seasonal temperature fluctuations to the list of known triggers for rockfalls in Yosemite.

From a human perspective, rockfalls are a real and ongoing danger to visitors and park employees. On average, about 50 rockfalls happen each year in Yosemite, with most occurring along the cliffs of the valley. Over the years, 16 deaths and at least 85 injuries have been attributed to rockfalls. The latest fatality occurred in 2017 when a 1300-ton slab of rock unexpectedly broke loose from high on El Capitan and fell 530 m (1750 ft) to the valley floor, leaving behind a white scar on the face of the cliff. The following day, a slab of rock almost 20 times larger fell from the same part of the cliff, thankfully with no injuries. The Park Service has done an admirable job of documenting and mapping over 900 distinct rockfalls and other mass movements and has developed a master plan for reducing the risk as much as feasible to protect visitors as well as infrastructure.

Many well-documented examples of large rockfalls could be described from Yosemite Valley, but here we'll focus on just one that illustrates their dynamic and unpredictable nature. In June 1996, a slab of rock suddenly detached from the cliff beneath Glacier Point, first sliding for a few hundred meters along the slightly inclined slope before launching outward from a lip of rock. The mass, estimated to be tens of meters long and a few meters thick, plummeted 550 m (1800 ft) before impacting the talus slope near the base of the cliff near Happy Isles (Fig. 18.14). The mass of rock compressed air as it fell, generating an air blast upon impact that moved outward at speeds of over 400 km/hour (~250 mph). Imagine the force of the air blast in combination with the chaotic explosion of rock and dust! Trees were snapped in half up to 90 m (300 ft) away. One person was killed and several others were injured. It's no wonder that the National Park Service devotes lots of time and resources to mitigating the risk posed by rockfalls in Yosemite Valley.

Landscape Evolution of Yosemite: Glaciation

Recall from Chapter 14 on Mount Rainier as well as from Chapter 17 that *alpine glaciers* are frozen rivers of ice that move ponderously downslope, driven by gravity (Fig. 14.5). They migrated across Yosemite's terrain at least

four times during the Pleistocene, modifying the landscape into the magnificent vistas we enjoy today. The most recent glaciation occurred during the *Last Glacial Maximum* between 18 and 26 ka and is known as the Tioga glaciation, named for glacial features of that age at Tioga Pass near the east entrance to the park.

The Tioga glaciation put the final touches on the glacial landscape in Yosemite, not because it was particularly widespread, but rather because it erosionally modified many of the landforms of earlier glaciations. The oldest alpine glaciers followed the paths of V-shaped valleys and canyons carved by the Tuolumne and Merced river networks; later alpine glaciers traced the routes of those earlier glaciers, modifying the landscape as they advanced forward as well as during their retreat. In the following sections, we'll trace the movement of the Tioga glaciers from their origins in the Yosemite highlands westward to their distal ends in Yosemite Valley, describing the wide variety of glacial landforms along the way.

The heads of the Tioga glaciers began along the Sierran crest that lines the east side of the park. Enormous volumes of snowfall accumulated in the highlands during the Last Glacial Maximum, fed by storm fronts from the Pacific that rose up the western flank of the Sierra. Over centuries to millennia, the snows transformed into dense ice in the harsh glacial conditions, eventually becoming massive enough to flow as a plastic solid downslope under the force of gravity.

Individual alpine glaciers coalesced in the high country around Tuolumne Meadows to form a continuous ice field that reached over 600 m (2000 ft) in thickness (Fig. 18.16).

Figure 18.16 Illustration of Tioga-age alpine glaciers extending outward from the ice field centered over the Tuolumne Meadows area during the Last Glacial Maximum around 18–26 ka.

Figure 18.17 View looking southeast from Tuolumne Meadows area. Dashed line marks approximate position of the trimline, the upper limit of glacial ice. Unicorn Peak on left and Cockscomb in center rose above the Tuolumne ice field as nunataks.

Figure 18.18 Glacial striations and erratic boulders, Olmsted Point. The parallel alignment of striations along the slope of the outcrop records the direction of glacial movement. Note the narrow slots formed by weathering along the two main orientations of joint planes cutting across the exposure. Clouds Rest in the background.

The grinding movement of the ice rounded the terrain into domes and broad, smooth bedrock surfaces. Only the highest peaks rose above the expanse of white ice. These rock "islands" are called **nunataks**, and examples in the Yosemite high country include Cathedral Peak, Cockscomb, and Unicorn Peak (Fig. 18.17). The thickness of the ice near these landmarks is estimated by the position of the *trimline*. Below, the rocks are smooth and rounded, but above the trimline the rocks form spires and pinnacles, angular features untouched by the ice.

As the enormous mass of ice grinds over the underlying bedrock in granitic highlands like Tuolumne Meadows, the efficiency of the glacier to erode varies with the orientation and density of joints in the rock. The weight of the ice, as well as friction between the basal ice and rock, tends to melt the lowermost part of the glacier. This meltwater infiltrates into joints within the underlying bedrock where it may refreeze and expand, slowly wedging blocks of rock away from the bedrock in a process called **glacial plucking** (also known as **quarrying**). At some point, a slab of rock may detach from the bedrock and become incorporated into the base of the moving ice, and then transported downslope. A particularly closely spaced set of joints in the granites near Tuolumne Meadows made the area highly susceptible to glacial plucking. In contrast, the granitic rocks composing the nunataks and other elevated mountains in the same area have a lower density of joints, leaving them relatively intact and high-standing. Thus, the smoothly rounded granitic terrain of Tuolumne Meadows contrasts with the nearby granitic highlands due solely to the concentration of joint planes and their influence on glacial erosion.

Ice alone doesn't provide the frictional resistance necessary to erode granitic bedrock. But ice entrains enormous amounts of gravel, sand, and silt within the body of the glacier as the basal ice continually melts and refreezes with movement. The gritty particles within the basal ice act collectively as a rasp on the underlying bedrock, abrading it as the sediment-laden ice grinds across the surface. Erosion due to **glacial abrasion** tends to smooth the sharp

edges of rock, creating the rounded, undulating textures of glaciated landscapes (Fig. 18.17). The newly abraded particles add to the entrained grit at the base of the ice, enhancing its effectiveness and commonly scratching linear **striations** into the underlying bedrock (Fig. 18.18). The abrasive action of the gritty base of the glacier also creates a shiny surface known as **glacial polish**.

A distinctive glacial landform in the Tuolumne high country carved by a combination of abrasion and plucking are **roche moutonnee** (French for "sheep rock"), exemplified by the asymmetric Lembert Dome (Fig. 18.19). As the

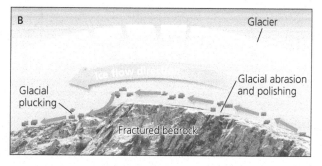

Figure 18.19 A) Lembert Dome, a roche moutonee near Tuolumne Meadows. B) Glacial abrasion dominates the gentle upflow side, whereas glacial plucking dominates the side downflow.

advancing glacier encountered the margins of the granite dome, the ice piled up against it, increasing pressure along the base of the glacier. The basal ice melted in response, lubricating the contact with the rock and allowing the grit at the base of the glacier to abrade a gentle slope into the east side of the monolith where people today hike to the top. As the glacier advanced over the crest of the dome, however, pressure along the base of the ice was alleviated, allowing much of the basal ice to refreeze. As seasonal meltwater penetrated downward into joints along the west side of the dome, slabs of rock were quarried away, resulting in the abrupt, sharp-sided west face.

Abrasion and plucking acted in concert to create some of the most ubiquitous glacial features in Yosemite (Fig. 18.20). *Cirques* are bowl-shaped amphitheaters near the heads of glaciers, formed as jointed rock was plucked away along the sides and base of the ice (Chapter 14). Rockfalls contribute to the near-vertical steepness of the headwalls of cirque basins. As glaciers melt, exposing the rock walls of the cirque to view, meltwater may accumulate in the deepest part of the cirque basin, forming a glacial lake called a **tarn**. These small alpine lakes are widespread across the Yosemite landscape, occupying the deeply quarried bottoms of many cirque basins. Other common

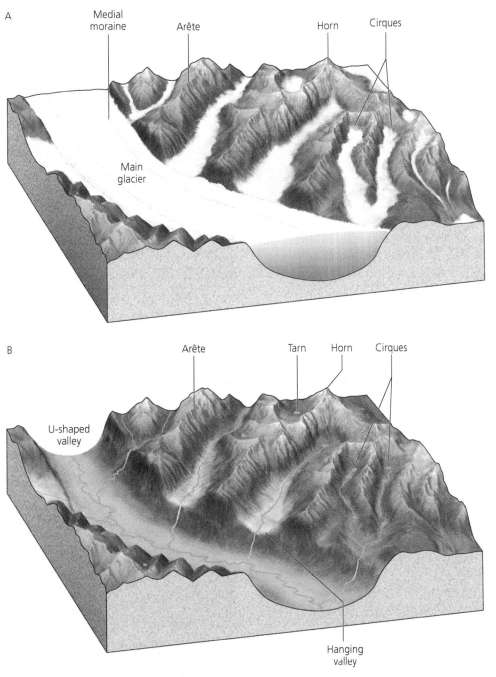

Figure 18.20 Landscape features created by the erosive effects of alpine glaciers.

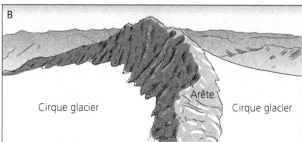

Figure 18.21 Matthes Crest in the Yosemite high country is a narrow fin of rock called an *arête*, left behind as glaciers plucked rock from each side. Exfoliation jointing formed along the sides as the rock was quarried away and as the weight of glacial ice was removed during deglaciation.

glacial features associated with the glacial plucking of rock are *arêtes*, knife-like ridges that once separated adjacent glaciers, now left behind as erosional remnants (Fig. 18.21).

As the Pleistocene glaciers migrated downslope, they abraded and plucked the bedrock walls of stream valleys, transforming them from their previous V-shape into wider, deeper **U-shaped valleys.** (Note the shape of Tenaya Valley in Fig. 18.1). The growth and decay of Ice Age glaciers in Yosemite Valley acted like a sawblade on the bedrock, incising deeper and deeper with each glacial cycle. Multiple episodes of glacial advance and retreat tended to straighten the valleys via erosion of masses of rock that protruded from the bedrock walls toward the center of the valley. Rather than curve around the rocky knob, alpine glaciers rode directly over the protrusion, abrading and plucking the rock until it was eventually removed and no longer impeded the flow of glacial ice.

Throughout Yosemite, deep depressions formed by glacial plucking within U-shaped valleys acted as natural basins for the accumulation of meltwater, creating elongate glacial lakes such as Mirror Lake in Tenaya Valley and Tenaya Lake near Tuolumne Meadows (Fig. 18-22). These are among the most picturesque and tranquil glacial features of the entire park. In time, these lakes will fill with sediment, forming the mosaic of meadows and woodlands that cover valley floors in Yosemite today.

Yosemite Valley proper is somewhat anomalous in that its valley floor is relatively flat, as opposed to the

Figure 18.22 Tenaya Lake fills a glacially deepened trough within a U-shaped glacial valley. Note the exfoliation domes at the east end of the lake.

concave-downward base of typical U-shaped valleys (Fig. 18-23). The actual U-shaped **bedrock** floor of Yosemite Valley is almost 600 m (~2000 ft) below the level, forested surface, filled above by sediments derived from glacial debris, meltwater *outwash*, and talus from rockfalls (Fig. 18-24). Deepening of the granitic bedrock within Yosemite Valley likely occurred during earlier, more intense glacial episodes than the Tioga phase of the Last Glacial Maximum. Less severe glacial phases released massive volumes of sedimentary outwash that filled in the deep bedrock valley as they melted and receded to higher elevations. As the final Tioga-age glaciers retreated, the valley was likely comprised of a mosaic of small lakes and marshy wetlands impounded behind a *morainal ridge* that formed a natural dam near the mouth of the valley. The low areas in the valley filled with the final dregs of glacial outwash, building to the level surface we see today. Eventually, the Merced River breached the natural morainal dam and flowed unimpeded on its way down the west slope of the Sierra.

Figure 18.23 Yosemite Valley from the Wawona Tunnel viewpoint. Note the flat-bottomed profile of the valley and the hanging valley above Bridalveil Fall on the right.

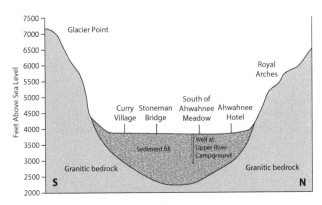

Figure 18.24 Yosemite Valley cross section interpreted from seismic imaging and drillhole data. The bedrock floor of the valley has the typical U-shaped profile due to glacial erosion. Later infilling of the valley by sediment created the flat surface of the valley floor.

Why is Yosemite National Park marked by an abundance of graceful waterfalls? The massive glaciers filling the main valleys of the Tuolumne and Merced rivers had greater erosive power than their smaller tributary glaciers, so they cut much deeper valleys. As the Pleistocene glaciers retreated during interglacial warming phases, the tributary glaciers were abandoned high on the valley walls (Fig. 18.20). As meltwater streams flowed through these U-shaped **hanging valleys**, the water plummeted over the edge into the much deeper main valley, creating the famous waterfalls of Yosemite Valley. Bridalveil Fall (Fig. 18.23) and Yosemite Falls are iconic examples of waterfalls emanating from hanging valleys in Yosemite.

The paired waterfalls of Nevada and Vernal falls in the eastern reaches of Yosemite Valley evolved in a somewhat different fashion (Fig. 18.10A). The granitic bedrock at the base of the steep valley is marked by a stepped series of eroded joint faces that were created by plucking as glaciers flowed over the rocks through the steep gorge feeding in to Yosemite Valley. Nevada Fall plummets over a north-east-oriented joint face, whereas Vernal Fall drops over a joint face oriented northwest.

During glacial phases in Yosemite Valley, the ice was several hundred meters thick, with Half Dome and a few of the highest peaks projecting >200 m (~700 ft) above the ice as nunataks (Fig. 18-25). Deep beneath the ice, glacial plucking and abrasion were sculpting the valley floor and smoothing its margins. Alpine glaciers poured through tributary valleys, including one where Yosemite Falls exists today. The scene was closer to one you'd see in modern Alaska or coastal Greenland rather than central California.

As climates warmed during interglacial phases, however, glaciers left behind a number of depositional features in their wake. **Glacial erratics** are ubiquitous huge boulders found resting on bedrock surfaces, left behind as blocky debris melted out of receding glaciers (Fig. 18.18). In Yosemite, erratic boulders are commonly composed of

Figure 18.25 Reconstruction of glacial extent in Yosemite Valley during past glacial maxima. The higher elevations rose above the ice as nunataks.

Figure 18.26 Chaotic pile of glacial till left by a retreating glacier.

an entirely different type of granite than the bedrock beneath, attesting to their transportation within or on top of a glacier.

Another common depositional feature of receding glaciers is **glacial till**, unsorted piles of sedimentary debris ranging in size from huge boulders to fine silt that washes out of the ice as it melts (Fig. 18.26). When glaciers reach their maximum extent and begin to recede, the entrained rocky debris is released from the melting ice and builds up as a ridge along the toe of the glacier. This ridge of glacial till is called a **terminal moraine** and marks the furthest advance of an alpine glacier (Fig. 14.7). In Yosemite Valley, the Tioga-age terminal moraine curves across Bridalveil Meadow near the west end of the valley.

Figure 18.27 Half Dome as seen from Glacier Point. Clouds Rest rises in the near background.

Glaciers commonly recede in pulses, depositing till as a series of **recessional moraines** marking the temporary position of the toe of the glacier as it retreats. (Remember that it's really just the position of the toe of the glacier that retreats upslope, not the entire glacier itself.) A Tioga-age recessional moraine is preserved in El Capitan Meadow. Eventually these recessional moraines were breached by the Merced River, which deposited outwash downstream beyond the morainal ridge.

Final Thoughts on the Yosemite Landscape

In this chapter, you've read how river incision, jointing, rockfalls, and glaciation combined through time to create the magnificent landscape of Yosemite. The integrated effects of these processes are nicely illustrated by Half Dome, the iconic landmark that graces the official California state quarter (Fig. 18.27). Rising over 1400 m (4700 ft) above the floor of Tenaya Valley, Half Dome looms in unglaciated splendor. Viewed from certain perspectives, Half Dome is really an elongate, monolithic ridge—much more than half a dome. The rounded summit and steep, curving walls are the result of exfoliation, modified as rockslides and rockfalls broke away through time. The sheer northwest face is a vertical joint plane, exposed by countless rockfalls. The Pleistocene glaciers that regularly filled Tenaya Valley may have undercut the northwest vertical face of Half Dome. As the glaciers receded, they left the rock hanging in space and highly susceptible to collapse. Vertical exfoliation sheets, like those forming the steep southeast sides of the monolith, occasionally detached from the northwest face to free-fall hundreds of meters to the valley floor below. The rockfall talus beneath Half Dome was likely picked up during the next glacial advance and carried downslope to be dropped as glacial till and reworked by innumerable floods of the Merced River. The end result of all that geologic activity is the instantly recognizable monolith that delights the eye of every visitor to Yosemite Valley.

Think in four dimensions when you next visit Yosemite National Park. Visualize the architecture of the granitic plutons hidden from view beneath the landscape and imagine their Mesozoic tectonic origins. Then fast forward to the late Cenozoic to experience the earthquakes that accompanied uplift along the fault system lining the eastern Sierra, quickening the rate of downcutting by the west-flowing river systems. Finally, envision the gleaming expanse of glacial ice that engulfed the terrain beneath, coming and going in tune with the rhythms of Ice Age climates. Think this way and your visit to Yosemite will be even more inspirational and rejuvenating than you thought possible.

Key Terms

bedrock
dome
exfoliation jointing
glacial abrasion
glacial erratic
glacial plucking
glacial polish
glacial till
hanging valley
nunatak
recessional moraine
roche moutonnee
rockslide
striations
tarn
terminal moraine
U-shaped valley
xenolith

Related Sources

Glazner, A. F., and Stock, G. M. (2010). *Geology underfoot in Yosemite National Park*. Missoula, MT: Mountain Press.

Graham, J. (2012). *Yosemite National Park: Geologic Resources Inventory Report*, Natural Resource Report NPS/NRSS/GRD/NRR—2012/560. Fort Collins, CO: National Park Service.

Huber, N. K. (1989). *The Geologic Story of Yosemite National Park*. Yosemite Association, Yosemite National Park, California. (Previously published in 1987 as *U.S. Geological Survey Bulletin* 1595). Accessed February 2021 at http://www.yosemite.ca.us/library/geologic_story_of_yosemite/final_evolution.html

Google Earth animation of the bedrock geology of Yosemite Valley. Accessed February 2021 at https://www.youtube.com/watch?v=5RXwZCeO78E

National Park Service. 2009. *Yosemite Panoramic Imaging Project*. Accessed February 2021 at https://www.nps.gov/yose/naturescience/panoramic.htm

Yosemite National Park geology. Accessed February 2021 at https://www.nps.gov/yose/learn/nature/geology.htm https://www.nps.gov/yose/learn/nature/granite.htm https://www.nps.gov/yose/naturescience/rockfall.htm

19 Sequoia and Kings Canyon National Parks

Figure 19.1 Great Western Divide at Kaweah Gap, Sequoia National Park. Hamilton Lake in the foreground. The path halfway up the mountainside on the right is the High Sierra Trail.

Sequoia and Kings Canyon national parks are located in the rugged southern Sierra Nevada (Fig. 17.1). The jagged topography and extreme relief in the eastern portion of the parks are among the most remote of anywhere in the contiguous United States but provide a mecca for solitude-seeking backpackers. The western approaches to the park are easily accessible, however, and include a broad range of stunning landscapes (Fig. 19.1). Elevations in the parks range from about 420 m (1370 ft) near the western entrance to Sequoia National Park to a high of 4421 m (14,505 ft) at Mount Whitney, the highest peak in the Sierra as well as in the entire lower 48. The eastern boundary of the two parks skirts the very crest of the eastern escarpment of the Sierra Nevada, about 3 km (~2 mi) above the adjacent Owens Valley (Fig. 17.3).

Sequoia National Park was established in 1890, making it the second oldest park in the National Park System. The adjoining Kings Canyon National Park to the north was established in 1940, and the parks have been jointly administered since 1943 (Fig. 19.2). The initial motivation for creating the parks was the preservation

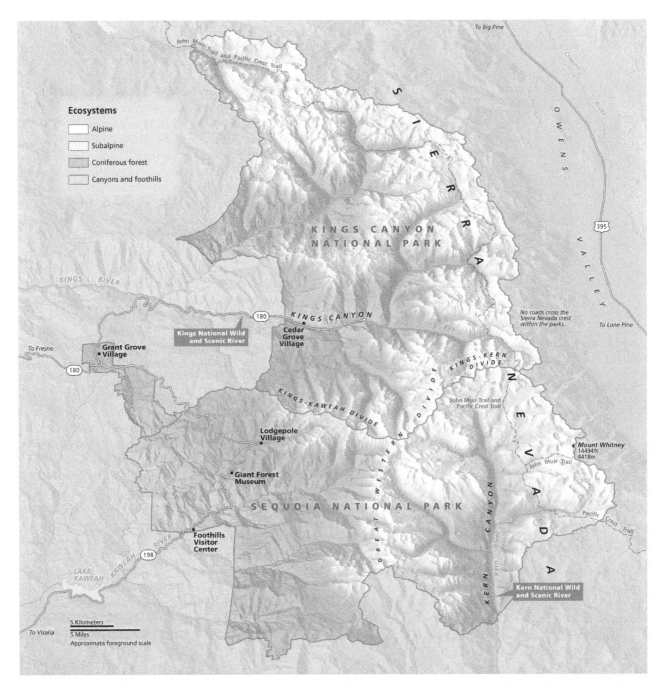

Figure 19.2 Map of Sequoia–Kings Canyon national parks highlighting the major river systems and the major ecosystems of the two parks.

of the magnificent groves of giant sequoia trees that thrive at intermediate elevations in the western portions of both parks (Fig. 19.3).

By volume, giant sequoias are the largest trees on the planet. The General Sherman tree in the Great Forest of Sequoia National Park earns the honor of Earth's largest living tree and is estimated to be over 2200 years old. Another distinctive feature of Sequoia–Kings Canyon is the 200 or so caves that lie hidden beneath the landscape. The caves evolved over time in the metamorphic marbles that exist as elongate pods within the dominantly granitic bedrock. We'll address the origin of the metamorphic rocks

and the formation of caves and their ornate decorations later in this chapter.

The northern boundary of Kings Canyon National Park is only ~60 km (~40 mi) southeast of the southern boundary of Yosemite National Park. As you might expect, there are many similarities in the geology and terrain of Sequoia, Kings Canyon, and Yosemite. High elevations, deep canyons, granitic rocks, and glaciated, alpine terrain characterize all three parks. But there are distinct differences that make Sequoia and Kings Canyon unique relative to their more popular cousin to the northwest. For the following discussion, we'll treat the two adjoining parks as a

Figure 19.3 The General Sherman giant sequoia is Earth's largest living tree, with an estimated age of >2200 years.

single geologic entity. To make the most of this chapter, it's important to review Chapters 17 and 18.

Topography and River Systems

The highest peaks in Sequoia–Kings Canyon are aligned north–south along the eastern crest of the Sierra Nevada (Fig. 19.2). Twelve peaks rise above 4267 m (14,000 ft), and several others are higher than 3962 m (13,000 ft). Other prominent ranges west of the Sierran crest include the Great Western Divide (Fig. 19.4), the Monarch Divide, and the Cirque Crest. All of these ranges form barren, serrated highlands with arctic-alpine habitats. No roads penetrate into the intimidating terrain of the Sequoia–Kings Canyon high country. The three main roads through the parks (the Kings Canyon Scenic Byway, the Generals Highway, and the Mineral King road) reach elevations of about 2400 m (~7800 ft), so most visitors only catch distant views of the mountains to the east. Backpackers and people using pack animals access the extensive and well-maintained trail system from several trailheads in the western approaches as well as through a few

high passes along the Sierran crest to the east. The famous John Muir Trail extends northward from Mount Whitney through the High Sierra backcountry all the way to Yosemite.

The southernmost glaciers in North America are found tucked into cirque basins in the northern part of Kings Canyon National Park. Small remnants of the Little Ice Age, they are located in the vicinity of Mount Goddard and Mount Darwin and along the Palisades Crest. The Palisade Glacier lies just outside the park boundaries on the east side of the Sierran crest but is the largest glacier in the Sierra Nevada, with an area of less than 1 km² (0.4 mi²).

The mountainous highlands and upland plateaus of Sequoia–Kings Canyon are the headwaters for three major rivers whose names serendipitously begin with K: the Kaweah, the Kings, and the Kern (Fig. 19.2). The Kaweah River originates in the mountains of the Great Western Divide and incises deeply through the western part of Sequoia National Park. Several canyons of the Kaweah River network have relief of over 1200 m (~4000 ft).

The water for the Kings River and its tributaries inside Kings Canyon National Park are derived from snowmelt on the Sierran crest and Great Western Divide. Inside the park, the canyons of the Kings River system are glacially widened. Just outside the park boundaries to the west, however, the main canyon of the Kings River (i.e., Kings Canyon) becomes one of the deepest in North America. Pleistocene glaciers never reached this far westward, so Kings Canyon is V-shaped and deeply incised. Kings Canyon has relief of about 2400 m (~8000 ft), measured from the highest peak on the north side of the canyon to the lowest elevation at the confluence of the South and Middle forks of the Kings River. This spectacular gorge is best viewed from Junction View or Yucca Point along the Kings Canyon Scenic Byway outside the park to the west.

The Kaweah and Kings rivers drain westward across the foothills toward the Central Valley. But the Kern River flows due south through Sequoia National Park deep within a glacially modified gorge, located between

Figure 19.4 Great Western Divide viewed from the south. Note the broad expanse of bare rock exposed above treeline, the upper limit of tree growth.

the Great Western Divide and the Sierran crest. The discordant north-to-south path of the river is controlled by the near-vertical, inactive Kern Canyon fault that focuses runoff along the weakened rock lining the fault plane. The linear canyon is over 600 m (~2000 ft) deep in places. The Kern eventually bends to the southwest outside the southern boundary of Sequoia National Park, flowing down the slope of the southern Sierra Nevada.

Prior to the human reorganization of California's water system in the 19th and 20th centuries, the Kern, Kaweah, and the Kings flowed into closed lake basins in the southern San Joaquin Valley. Today the three rivers are dammed and diverted for agricultural use and for recharging of groundwater basins; the former lakebeds are now valuable irrigated farmland. Upslope in the Sierra, parts of the Kings and Kern rivers have earned Wild and Scenic River status.

Origin and Age of Rocks in Sequoia–Kings Canyon

Similar to Yosemite National Park, granitic rocks of the Sierran batholith dominate the bedrock in Sequoia–Kings Canyon (Fig. 19.5). A few of the plutons beneath the park

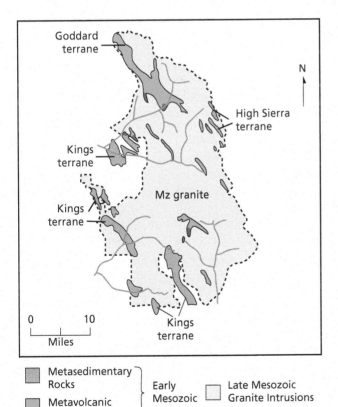

Figure 19.5 Simplified geologic map of Sequoia–Kings Canyon national parks. The landscape is mostly composed of granitic rocks of the Sierran batholith. Several roof pendants of metamorphic rocks trend northwesterly, enclosed by granitic plutons. Related groups of these pods of metamorphic rock are called terranes.

are among the largest in the entire Sierra, covering a few hundred square kilometers. The Paradise and Whitney granodiorite plutons extend over large areas of the high country and are very similar to granitic rocks in Yosemite. The Paradise Granodiorite exhibits huge crystals of pink, potassium-rich feldspar that are highly resistant to weathering, remarkably similar to the Cathedral Peak Granodiorite in Yosemite (Fig. 18.6B). The Mesozoic age and origin of these granitic rocks were thoroughly explained in Chapters 17 and 18.

The unique aspect of the bedrock geology of Sequoia–Kings Canyon, and the one that distinguishes these parks from Yosemite, is the presence of elongate pods of metamorphic rock that extend northwest–southeast across the terrain (Fig. 19.5). These metamorphic rocks are remnants of the country rock that was already present when the granitic magmas rose up beneath during the mid- to late Mesozoic. They were intruded by the magmas and formed the roofs and walls of magma chambers. These masses of metamorphic rock are called **roof pendants** because some of this rock may have hung downward, pendant-like, from the roofs of magma chambers, whereas other bodies of metamorphic rock may have formed wall-like barriers between magma bodies. Rocks of the roof pendants compose about 20 percent of the bedrock within Sequoia–Kings Canyon national parks. Exposures of these rocks are the remnants left behind after they were exhumed during the late Cenozoic uplift of the Sierra Nevada, followed by weathering and erosion that removed what used to be a much larger extent of metamorphic country rock.

Metamorphic rocks are created from previously existing rocks that have been subject to extreme temperatures and pressures (Fig. 1.30). The elements composing the rock are slowly rearranged in the solid state (i.e., without melting) until they achieve a more stable atomic configuration that is compatible with the ambient pressure and temperature conditions. Metamorphism may occur deep within the cores of growing mountain ranges, under the high pressures associated with subduction zones, or when rocks are subjected to the intense heat of magma.

Each of these three settings contributed to the metamorphism of preexisting rocks of the ancestral Sierra Nevada because of the tectonic location along a Mesozoic convergent plate boundary. Recall from Chapter 17 that the western margin of North America was an active subduction zone beginning about 200 Ma as the oceanic Farallon plate descended eastward beneath the continental margin (Fig. 17.8). Intense pressures and temperatures associated with subduction, compressional mountain-building, and magma emplacement all affected the preexisting country rocks to some degree (Fig. 19.6). The roof pendants of Sequoia–Kings Canyon preserve this long tectonic history, although it is very difficult to distinguish the different phases of metamorphism. Especially vexing is that many of the fossils that once existed within the rocks were mostly obliterated during metamorphism, making age determinations difficult. What we do

Figure 19.6 Deformed metamorphic rocks of the Sequoia–Kings Canyon region. Roadcut located along Route 180 east of Boyden Cave. Height of outcrop is ~6 m (~18 ft).

Figure 19.7 Roof pendant (reddish brown) in Mineral King Terrane, juxtaposed against gray granitic rock, Franklin Lake, Sequoia National Park. Dashed line marks the contact between the two masses of rock.

know from radiometric age dating is that metamorphism occurred during the Mesozoic, affecting rocks that were Paleozoic and earlier Mesozoic in age.

Subtle hints within the metamorphic rocks of the roof pendants suggest that they were originally sedimentary and volcanic in origin. *Marbles* within the pendants were once limestones, and *quartzites* were once pure sandstones (Chapter 1). *Slates*, *phyllites*, and *schists* were likely once sedimentary shales. Other roof pendants show evidence that the precursor rock was composed of pyroclastic debris derived from explosive volcanism. Yet others suggest that the original rock prior to metamorphism was volcanic basalt or andesite. The scarcity of evidence for relative ages based on fossils is compounded because the heat and pressure associated with metamorphism tend to reset the radiometric clocks within igneous rocks, precluding the measurement of numerical ages for the original rock. Enough fragmentary evidence exists, however, so that the ages of the preexisting rocks composing the roof pendants can be narrowed down (described in the next section).

Accreted Terranes in Sequoia–Kings Canyon

Roof pendants in Sequoia–Kings Canyon national parks commonly stand out as darker outcrops surrounded by whitish-gray granitic rocks (Fig. 19.7). Dozens of roof pendants are recognized in the parks, but they can be grouped into three main belts oriented northwest–southeast based on distinct similarities in composition and origin. These belts of genetically related roof pendants define larger entities called **terranes** that share a common tectonic history. The tectonic term *terrane* (not to be confused with the landscape term *terrain*) can be thought of as a discrete

body of rock that originally formed somewhere else, but that was tectonically transported and eventually accreted to a continental margin. **Accretion** is the physical attachment of a large body of rock to a continental margin by lateral tectonic transport. Continents commonly grow outward by the sequential attachment of **accreted terranes** to their outer edges. Accretionary tectonics is a fundamental process in continental growth and will be addressed in greater detail in several later chapters.

Three distinct terranes are recognized within the boundaries of Sequoia–Kings Canyon national parks and enough evidence exists to broadly estimate their ages (Fig. 19.5). The easternmost High Sierra Terrane is considered to have formed fairly close to its current location and likely represents deposition along the Paleozoic continental shelf that existed prior to the Mesozoic convergent margin. The Goddard Terrane is Mesozoic in age and appears to be the remnant of a chain of volcanoes. The coincidence in age with the Sierran batholith suggests that the Goddard rocks may be the metamorphosed volcanic crown of the ancestral Sierra Nevada, fortuitously preserved as erosional remnants. The westernmost Kings Terrane in the parks was originally sedimentary rock deposited in a shallow marine basin during the late Paleozoic to early Mesozoic. Fragments of these rocks were tectonically transported potentially thousands of kilometers before accreting onto the continental margin.

The roof pendants are all that remains of what was likely a much larger mass of accreted terranes extending across a broad area of the Sierra Nevada. They were part of the Mesozoic North American continent prior to and coincident with the intrusion of Sierran magma during the mid- to late Mesozoic (Fig. 17.9). Huge volumes of magma rose up into the country rock, melting much of it and assimilating the newly formed liquid into the original magma, changing its composition. Country rock that remained solid along the margins of the magma chambers was metamorphosed upon contact with magmatic heat. After magmatism ended in the ancestral Sierra Nevada around 80 Ma, erosion, *isostatic uplift*, and late Cenozoic uplift of the modern Sierra Nevada tilted fault-block stripped away most of the metamorphosed rock of the terranes (Fig. 17.10), exposing the granitic rocks beneath and leaving behind random pods of roof pendants. Who would have thought that this motley assemblage of metamorphic rocks would tell such a strange and complicated geologic history?

Caves and Cave Deposits

The Kings Terrane in the parks was originally composed of sedimentary shale, sandstone, and limestone. Subsequent metamorphism changed these rocks into their current composition of slate, schist, quartzite, and marble. A thick slab of marble, the Boyden Cave roof pendant, juts skyward above the South Fork of the Kings River and can be viewed from Horseshoe Bend along the Scenic Byway (Fig. 19.8). The original sedimentary beds, ~610 m thick (2000 ft), were flipped vertical during mountain-building associated with the ancestral Sierra Nevada. Marble (and its limestone precursor) is composed primarily of the mineral calcite, which dissolves in mildly acidic water. Groundwater percolating through the marble of the roof pendant dissolved some of the calcite to form Boyden Cave, just outside the western boundary of Kings Canyon National Park. Over 275 caves have been discovered in marble of the Kings Terrane in the two parks. Lilburn Cave is the longest in California, comprising 27 km (17 mi) of passageways and colorful mineral formations. Crystal Cave is the only cave in the parks open to the public and is notable for its multiple levels and ornate cave deposits (Fig. 19.9).

Figure 19.9 Crystal Cave is formed in marble of the Sequoia roof pendant of the Kings Terrane. Stalactites grow downward like icicles, whereas stalagmites grow upward from dripwater.

Caves and their decorative deposits form by a two-step process that involves the position of the **water table**, a boundary beneath the ground that roughly parallels surface topography (Fig. 19.10). Above the water table, pores and cavities in the rock are filled with air; below the water table, pores and cavities in the rock are filled with water. The first step involves the formation of openings and passageways within the fractured marble. Rainwater is slightly acidic because of interaction with carbon dioxide in the air, as well as from organic-rich soil that the water passes through as it percolates into the ground. In the marble roof pendants of the Kings Terrane, surface water penetrates downward through joint systems and along tilted bedding planes, passing through the air-filled passages until it accumulates in the water-filled cavities below the water table. The calcite of the marble is dissolved upon prolonged contact with the slightly acidic water, resulting in a network of passageways forming along joints and bedding planes within both the air-filled and water-filled zones. As long as the acidic waters are chemically balanced toward dissolving calcite, openings and passageways will develop, regardless of whether the water is above or below the water table.

Water tables rise or fall in tune with the seasons, as well as in response to long-term changes in climate or surface movements such as subsidence or uplift. If a shift to a drier climate or tectonic uplift causes the water table to fall, overlying cave openings are left empty, filled with air. This initiates the second step, the formation of cave deposits called **speleothems**. Slightly acidic rainwater and snowmelt dissolves calcite as it percolates downward toward the open cave system, becoming saturated with calcite in solution. As the water comes in contact with the air of the cave, say from a fracture opening in the cave ceiling, it releases carbon dioxide. This causes

Figure 19.8 The Boyden Cave roof pendant (bluish gray) is composed of marble, produced by the metamorphism of limestone.

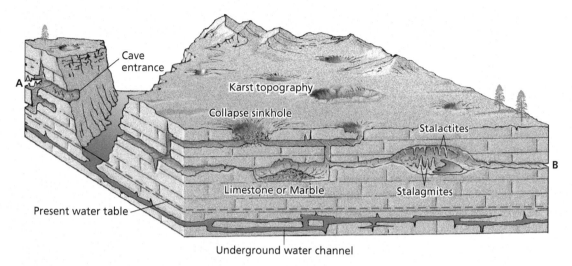

Figure 19.10   Caves and their ornamentation form in two phases, related to the position of the water table. When the water table was near levels A and B, dissolution dominated and cave networks evolved. As the water table fell, precipitation of speleothems within the open caves became dominant, although dissolution still occurred if the infiltrating groundwaters were appropriately acidic. Below the present water table, dissolution and the opening of passageways are the dominant processes.

a chemical reaction that results in the **precipitation** of calcite near the ceiling fracture of the cave. These speleothems include common icicle-like forms such as **stalactites** that grow downward from the ceilings of caves (Fig. 19.9). If water drips from the tips of stalactites to the floor of the cave, the droplet of water will release its carbon dioxide upon splashing, causing calcite to precipitate at the point just below the stalactite. Continued precipitation from dripping, calcite-saturated water causes the growth of a **stalagmite**, an upward-pointing speleothem. In time, as the stalactite grows downward and the stalagmite grows upward, they may merge to form a **cave column**. Groundwater charged with dissolved calcite may flow over the surface of a wall of marble, release its carbon dioxide, and then precipitate a wavy sheet of calcite called **flowstone**.

Most caves and cave deposits form in the general manner described above, with many variations on the theme. Crystal Cave in Sequoia National Park formed in a slightly different manner, however, that was not dependent only on the position of the water table. The ridge of marble and schist in which Crystal Cave is located separates Yucca Creek on one side from Cascade Creek on the other. Water from Yucca Creek carved passageways along a contact between the soluble marble and insoluble schist as it infiltrated underground on its way downslope toward Cascade Creek. Through time, the creeks incised deeper into the terrain on either side of the ridge, lowering the level of cave formation and leaving the original upper caves filled with air. Snowmelt and rainwater penetrated deeper through the ridge of marble, dissolving the rock and thus enriching the groundwater in the elements composing calcite. Elaborate speleothems were precipitated from these calcite-saturated groundwaters within the open voids of the Crystal Cave system.

Over time, weathering and erosion may lower the land surface to the level of the cave systems riddling the marble. Continued **dissolution** and cave collapse create a chaotic, irregular topography called a **karst landscape** (Fig. 19.10). The Mineral King area of Sequoia National Park is considered to be one of the premier examples of alpine karst landscape in the National Park System. Multiple caves, sinkholes, springs, and "disappearing streams" that flow into the subterranean network of caves are common in this area of the park. The formation of caves, cave deposits, and karst will be expanded upon later in the book in Chapter 39 on Mammoth Cave National Park.

Effects of Jointing and Glaciation on the Sequoia–Kings Canyon Landscape

As in Yosemite, jointing controls the architecture of many of the landforms in Sequoia–Kings Canyon national parks. There are two predominant orientations of joints in the parks, one trending east–west, the other north–south. At the eastern boundary of Sequoia National Park, the east face of Mount Whitney and its adjoining angular peaks nicely express the intersection of these orthogonal joint sets (Fig. 19.11). Many other cliff faces and mountainsides in the parks illustrate the importance of vertical jointing on landscape evolution.

Curving exfoliation joints split the granitic bedrock as well, roughly paralleling the topography (Fig. 18.12). Layered slabs of rock break away from the surfaces of plutons as rockslides and rockfalls, creating spectacular *domes* such as Moro Rock (Fig. 19.12) and Tehepite Dome (19.13). The view from the top of Moro Rock is outstanding and can be reached via a 400-step stairway. Moro Rock is located near the margin of the Sierran batholith where it abuts the Sequoia roof pendant to the southwest. The rocks of the roof

Figure 19.11 Summit of Mount Whitney (W) from the east, with Keeler Needle (K), Day Needle (D), and Third Needle (T) to the south. The angularity of the needles and the steepness of the buttresses are related to the intersection of two joint sets.

Figure 19.13 Tehipite Dome (T) rises almost 1100 m (3600 ft) above the U-shaped glacially modified valley of the Middle Fork of the Kings River, Kings Canyon National Park. View up-river from the southwest.

Figure 19.12 Moro Rock, a dome formed by exfoliation of granitic rock of the Sierran batholith. View from the southwest.

pendant—mostly slates and schists in this area—are less resistant to erosion than the adjacent granitic rocks and are rapidly being worn down by the Kaweah River system. This differential erosion creates a sharp dropoff in the topography of the region from the highlands where Moro Rock stands to the Kaweah River Valley to the southwest. Over a horizontal distance of about 4 km (2.5 mi), the topography drops over 1000 m (~3300 ft). The Generals Highway circuitously switchbacks down this precipitous grade as it descends from the granitic rocks into the metamorphic rocks of the pendant. Think about the effects of *differential erosion* on the topographic relief as you gaze outward from the top of Moro Rock.

The impact of Pleistocene glaciation is widespread throughout the high country of Sequoia–Kings Canyon.

Many features are comparable in grandeur to those at Yosemite but are not as easy to get to by car due to the limited road access and rugged topography. *Cirques, arêtes, U-shaped valleys, hanging valleys* with waterfalls, *tarn lakes, glacial polish, striations,* and *glacial erratics* are all common features in Sequoia–Kings Canyon national parks (Fig. 19.14). As in Yosemite and elsewhere in the Sierra Nevada, rivers that cut V-shaped canyons through the highlands set the geomorphic stage. During Pleistocene glacial maxima, alpine glaciers naturally followed the river canyons, *abrading* and *plucking* rock as the ice slowly crept downslope. The meandering V-shaped river valleys were straightened and re-sculpted into U-shaped glacial valleys (Fig. 19.13). The maximum downslope extent of alpine glaciation can be determined by the locations of the transition from straight U-shaped valleys upstream to sinuous V-shaped valleys downstream. The Middle and South Forks of the Kings River as well as the Kern River exhibit this change in morphology, defining the farthest western reach of glaciers from their high-country source.

One of the other primary ways to determine the extent of glaciation is to map the locations of *terminal moraines,* the curving ridges of glacial till dropped from the toes of melting glaciers (Fig. 14.7). Moraines from the Tioga glaciation (18–26 ka, the Last Glacial Maximum) and the preceding Tahoe glaciation (~125–150 ka) are both recognized in the parks. Evidence for the earlier glaciations identified elsewhere in the Sierra Nevada was erased by the two latest advances. As in Yosemite, alpine glaciers grew outward from the Sierran crest, coalescing into an ice field that covered much of the Sequoia–Kings Canyon high country (Fig. 19.15). Tongues of ice flowed down the main channels and tributaries of the Kings, Kaweah, and Kern river systems, reaching elevations as low as 1200 m (4000 ft) along the western boundaries of the parks.

Alpine glaciers were much smaller along the drier, eastern side of the Sierran crest, extending downslope along the escarpment to elevations of about 1800 m (6000 ft). Pleistocene glaciers were not as thick or widespread in

Figure 19.14   Glaciated topography south of Mount Whitney (W).

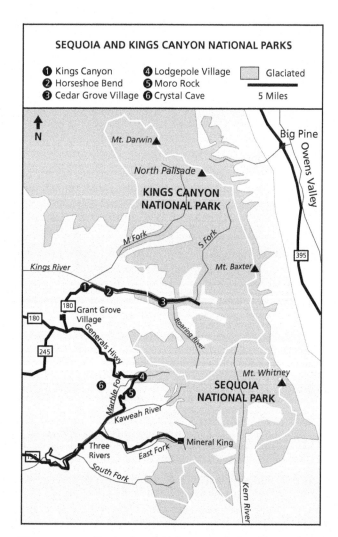

Figure 19.15   Maximum glacial extent during Tioga phase (~20 ka) in Sequoia–Kings Canyon.

Sequoia–Kings Canyon as they were in Yosemite because of the more southerly latitude and slightly warmer average temperatures. During the peak glacial advances, over half of the park area was covered in glacial ice, with large *nunataks* of rock sticking up above the white expanse. The Tioga-age glaciers reached just beyond the southern boundary of Sequoia National Park, marking the southernmost extent of the Sierra Nevada glaciation during the Last Glacial Maximum.

Final Thoughts on Sequoia–Kings Canyon

Unfortunately, the spectacular beauty of Sequoia–Kings Canyon national parks is commonly obscured by air pollution that originates in the Central Valley and San Francisco Bay Area. Warm winds rise up the flanks of the Sierra, transporting smog and ozone to the parks where it settles as a yellowish-gray haze within valleys and canyons. Currently, Sequoia–Kings Canyon has the unwanted distinction of having the worst air quality in the National Park System. Moreover, ozone stresses Jeffrey and ponderosa pines in the parks, causing yellowing of their needles and enhancing the fire hazard. The air quality is usually much better at the higher elevations of the parks, making vistas clearer for those lucky people hiking the high country. But the majority of visitors to the parks follow the main roads along the western boundaries where the smog is most apparent. The National Park Service monitors air quality in the parks, participates in policy discussions with regional, state, and federal agencies, and educates the public on the nature of the issue, but the problem persists.

Don't let this dissuade you from visiting Sequoia–Kings Canyon national parks. Go during the cooler off-season when the air quality is better and the tourists are fewer. Get out of your car at a trailhead and venture into the backcountry, making sure that you're well prepared with appropriate maps, gear, and fitness. You'll be rewarded by the breathtaking views and rejuvenation of spirit that accompany any sojourn into the wilderness.

Key Terms

accreted terrane
accretion
cave column
dissolution
flowstone
karst landscape
precipitation
roof pendant

speleothem
stalactite
stalagmite
terrane
water table

Related Sources

Konigsmark, T. (2007). *Geologic Trips, Sierra Nevada*. Gualala, CA: GeoPress, 320 p.

Moore, J. G. (2000). *Exploring the Highest Sierra*. Stanford, CA: Stanford University Press, 427 p. (This book provides a wonderfully detailed account of the early exploration of the southern Sierra Nevada near Sequoia–Kings Canyon and is a must-read for anyone wanting deeper insight into the history and geology of the region.)

Sequoia–Kings Canyon cave/karst systems Accessed February 2021 at http://www.nps.gov/seki/learn/nature/cave.htm

Sequoia–Kings Canyon National Parks geology overview. Accessed February 2021 at http://www.nps.gov/seki/learn/nature/geology_overview.htm

PART 4
National Parks of the Rocky Mountains

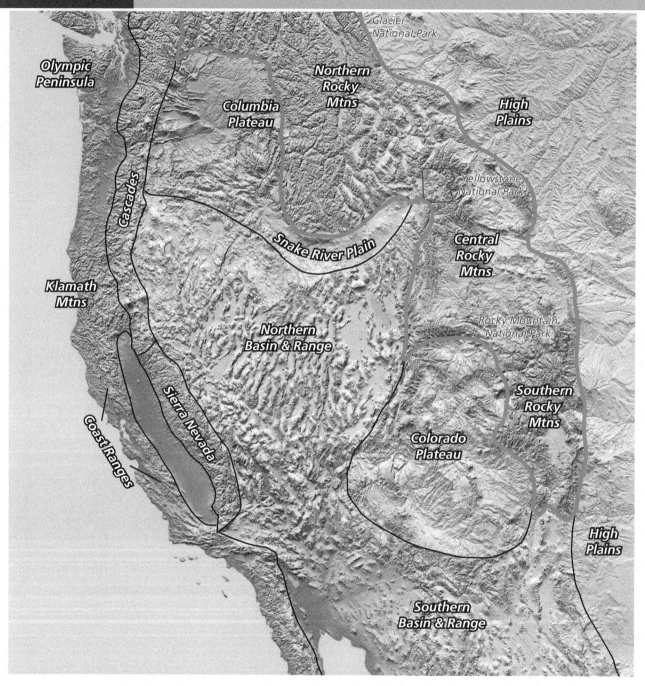

Figure 20.1 Physiographic provinces of the American West, with the Rocky Mountain Province highlighted. Locations of Rocky Mountain and Glacier national parks addressed in Part 4 outlined, plus Yellowstone National Park, addressed in Part 7.

The Rocky Mountain Geologic Province forms the eastern flank of the broad, mountainous topography of the American West (Fig. 20.1). The province is composed of a discontinuous chain of mountain ranges that extend northwest from New Mexico through Colorado, Wyoming, Idaho, and Montana, and then continues deep into Canada. In the United States, the Rockies are further divided into southern, central, and northern sections. Each of these subprovinces is internally composed of individual mountain ranges and intervening basins, with each subprovince defined by topographic boundaries and distinct geologic features. Rocky Mountain National Park is located in Colorado in the Southern Rockies, whereas Glacier National Park is located in Montana in the Northern Rockies.

A physiographic map of North America (Fig. 20.2) shows that the middle and eastern parts of the continent are relatively flat, with just a few elongate wrinkles representing the Appalachian Mountains. Western North America, however, is a rumpled mass of mountains that stretches north–south from Alaska to Central America and east–west from the Pacific Ocean to central Colorado. This chaotic chain of highlands is called the **North American Cordillera**, and many of the national parks discussed in this book are located among the mountains, deserts, plateaus, and intermontane basins of this widespread region. (The term *cordillera* was originally applied to the mountain chain by Spanish explorers and is derived from their word for "cord," eliciting an image of a rope, knotted in places.) The geologic provinces composing the U.S. portion of the Cordillera provide the broader framework for understanding the origins of national parks of the American West.

The Rocky Mountains and the entire North American Cordillera have been mapped in great detail, much of it by geologists with the U.S. Geological Survey. But even though the maps are accurate representations of what can be seen at the surface, geoscientists still aren't quite certain that they fully understand the origin of the Rockies and the

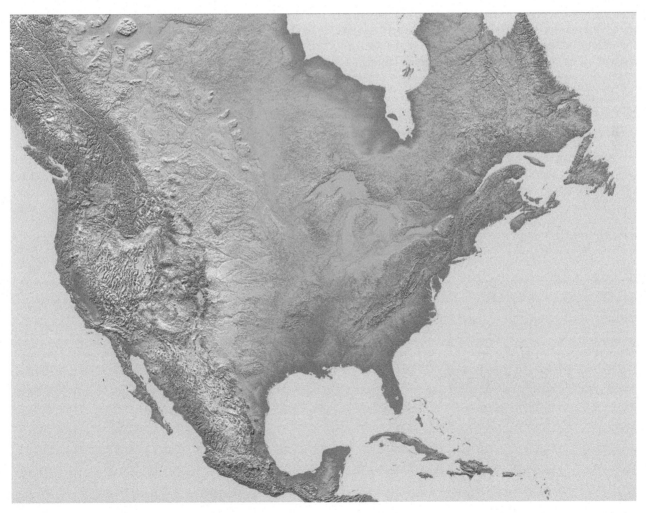

Figure 20.2 Physiographic map of North America contrasting the North American Cordillera of the American West with the Appalachian Mountains of eastern North America.

Cordillera as a whole. What we are sure of is that uplift of the Rockies was achieved primarily by *compression*, where the tectonically driven stresses are oriented toward one another. Compressive forces were addressed in Chapter 1 in the context of convergent plate boundaries. In the Rockies, those compressive forces were driven by the subduction of the Farallon plate along the western margin of North America, introduced in previous chapters on parks of the Sierra Nevada (Fig. 17.8). Convergent margin tectonism dominated the American West from about 200 Ma to about 40 Ma.

The Rockies are a bit of an enigma because they are located so far inland from the convergent plate boundary. Continental-oceanic subduction systems usually affect only the outer margin of the continent, raising a coastal welt of mountains cored by massive granitic plutons and crowned by volcanoes (Fig. 1.13). The Cascades and the ancestral Sierra Nevada exemplify this type of convergent system, as does the Andes Mountains of South America. But why did the uplift of the Rockies happen 1600 km (~1000 mi) inland, far from the subduction zone? The current prevailing model involves a decrease in the angle of descent of the Farallon plate beginning around 80 Ma, effectively shutting off magmatism in the ancestral Sierra Nevada and generating a wave of uplift and deformation that swept across the western United States. This model for mountain-building through the interior of the American West is called **flat-slab subduction** and seems to explain the origin of the Rockies reasonably well, with a few notable questions remaining.

Flat-Slab Subduction and the Laramide Orogeny

One important piece of evidence for a flattening of the Farallon plate is that the youngest granitic plutons in the Sierra Nevada are about 80 m.y. old (Chapter 17). This

suggests that the supply of magma shut down at that point, and the Sierra entered a phase dominated by erosion and isostatic uplift. The reason for the halt in magma production is related to an inferred decrease in the angle of descent of the Farallon plate. The flattened oceanic plate was simply not deep enough to release the water necessary to trigger the melting of rock, so the volcanism of the ancestral Sierra Nevada ended due to the halt in magma generation.

Inland, mountains were raised sequentially eastward, culminating in the uplift of the Rocky Mountains (Fig. 20.3). This pulse of mountain-building ranged from ~80 Ma to ~40 Ma and is called the *Laramide orogeny*. (The Laramide orogeny may have begun as late as ~70 Ma and ended ~30 Ma; age ranges for the event vary in the literature. Most workers generally agree on the 80 to 40 Ma interval.) During the Laramide, the North American plate was subject to compressional stress as the Farallon oceanic plate, subducting at a low angle, scraped along the base of the overriding continental plate. The Laramide orogeny played a fundamental role in the uplift of mountains in Glacier and Rocky Mountain national parks.

An elementary way of visualizing the mechanics of Laramide mountain-building is to think of a rigid piece of plywood being slid horizontally beneath a throw rug. As the sheet of wood is pushed laterally, parts of the overlying rug form compressional rumples and ridges due to friction between the two materials. The zone of rumpling in the rug will appear to migrate as the wood is pushed further along.

Why did the Farallon plate flatten? Something had to lower the buoyancy of the Farallon so that it "floated" upward during subduction, shallowing the angle of descent into the mantle. The principle of *actualism* helps us understand how this might occur (Chapter 1). The modern ocean floor is marked by widespread submarine

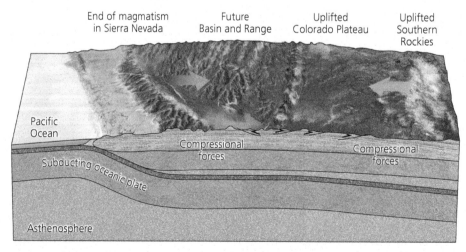

Figure 20.3 Proposed model of flat-slab subduction of the Farallon plate beneath the North American plate during the Laramide orogeny. Arrows show the orientation of compressional stress within the upper plate. Compare this image with the steeper subduction angle shown in Figure 17.8.

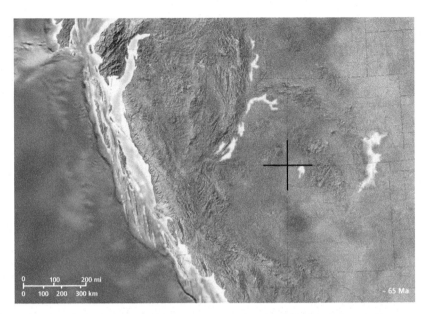

Figure 20.4 Paleogeographic map of the American West during the Laramide orogeny ~65 Ma. Rivers drained highlands in Colorado, Nevada, and central Arizona, depositing their sedimentary load in rivers, floodplains, and lakes of an intermontane basin occupying what is today the Colorado Plateau. The American West was narrower than today because Basin and Range extension had yet to happen over Nevada and Utah. Dark line just off west coast is the Farallon subduction plate boundary. Cross marks Four Corners region.

plateaus formed by voluminous outbreaks of basaltic lava along the seafloor. These volcanic plateaus thicken the oceanic crust and lower the overall density of the oceanic lithosphere (because basalt has a lower density than the deeper crust and lithospheric mantle). As oceanic lithosphere capped by volcanic plateaus are consumed within subduction zones, the lower density lithosphere may buoyantly rise upward, flattening the angle of subduction. Seismic imaging of a portion of the modern Andean subduction zone suggests that this very process is occurring today, with mountain-building happening far inland from the plate boundary. Perhaps the flat-slab subduction of the Farallon plate beginning ~80 Ma was related to the consumption of an oceanic plateau, now submerged deep within the mantle beneath North America.

The Laramide orogeny was previously addressed in several chapters in Part 1 on the Colorado Plateau. Late Mesozoic–early Cenozoic Laramide compressional tectonism gently lifted the Colorado Plateau a few kilometers in elevation, forming a high intermontane basin between the even higher elevation mountains surrounding it (Fig. 20.4). The Precambrian crystalline basement was raised along high-angle *reverse faults*, with the overlying Paleozoic and Mesozoic sedimentary rocks flexing into *monoclinal folds* above the buried faults. Early Cenozoic rivers draining the adjacent highlands surrounding the Laramide Colorado Plateau meandered across the high-elevation landscape, depositing a blanket of sand and gravel. Large lakes formed on the high plateau, eventually

filling with erosional debris such as the early Cenozoic Claron Formation, lakebed deposits now sculpted into the hoodoos of Bryce Canyon National Park. The mechanics of uplift of the Colorado Plateau will be further developed in Chapter 33.

As opposed to the mild deformation that affected the Colorado Plateau, the adjacent Rocky Mountain Province was subjected to intense folding and faulting during Laramide uplift. These deformational features are beautifully exhibited in Rocky Mountain and Glacier national parks and form the underlying controls on their distinctive landscapes. These parks share many similarities, but with enough variability to make each of them unique.

Both parks are located along the Front Range of the Rockies, the huge rampart of mountains that rise upward above the High Plains to the east. Precambrian rocks dominate both parks, although the exact age and origin of the rocks in each park are significantly different. The highlands in each park were uplifted during the late Mesozoic–early Cenozoic Laramide event, but the style of mountain-building is expressed differently. The landscapes of each park were heavily influenced by Ice Age glaciation, but Glacier is located almost 8° of latitude to the north of Rocky Mountain National Park and therefore was more deeply incised by alpine glaciers, magnifying the steepness of the topography. Moreover, the gigantic continental ice sheets that extended into the region from the north impacted the Glacier mountainscape, leaving behind its own unique set of glacial features. As we explore these two parks, we'll deepen our understanding of earlier concepts and add to our ongoing narrative of the origin of the American West.

Key Terms

Cordillera
flat-slab subduction

Related Sources

Abbott, Lon, and Cook, Terri. (2012). *Geology Underfoot along Colorado's Front Range*. Missoula, MT: Mountain Press.
Meldahl, Keith Heyer. (2013). *Rough-Hewn Land: A Geologic Journey from California to the Rocky Mountains*. Berkeley: University of California Press.
Animation of Pleistocene Ice Ages over last 180 ka. Accessed February 2021 at http://www.nps.gov/features/romo/feat0001/BasicsIceAges.html

Rocky Mountain National Park

Figure 21.1 Alpine tundra high country, Rocky Mountain National Park.

Colorado is notable for its stunning mountainous terrain, its charismatic wildlife, and its massive amounts of snowfall. Many places in the state exhibit these features, but perhaps the easiest place to access all of them is Rocky Mountain National Park (Fig. 21.1). This is a high-elevation park, with one-third of the land area rising above treeline, the upper limit of tree growth that begins around 3500 m (~11,500 ft) in the park. Terrain above the treeline exhibits a windswept alpine tundra ecosystem where only hardy, ground-hugging plants and lichens survive.

But don't think of this ecosystem as accessible to only robust backpackers; you can drive through it on Trail Ridge Road, the highest paved road in the National Park System. The road begins in the east near Estes Park at about 2300 m (~7500 ft) and reaches 3713 m (12,183 ft) at its highest point before descending back to lower elevations at the western entrance (Fig. 21.2). Eighteen km (11 mi) of the total length of the road in the park traverses the alpine tundra zone, allowing access to environments more typical of the Arctic than of the lower 48. As you might expect, the higher elevation parts of the road are closed from mid-October to the end of May, buried under snowfall and snowdrifts.

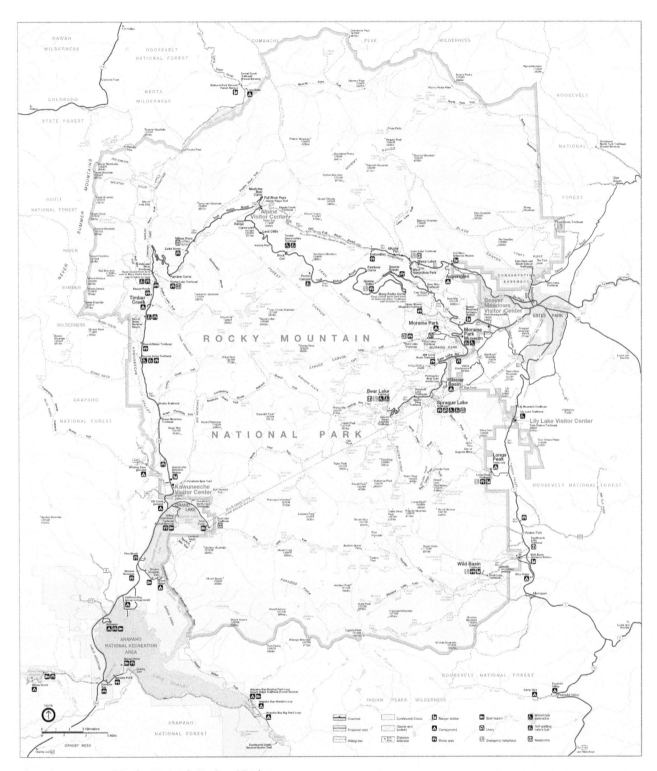

Figure 21.2   Map of Rocky Mountain National Park.

Rocky Mountain National Park is located northwest of Boulder in the Front Range of the Rockies, the easternmost of the tens of ranges that compose the Colorado Rockies (Fig. 21.3). Forty-two peaks rise above 3657 m (12,000 ft) in the Front Range, with Longs Peak, the highest in the park, rising to 4346 m (14,259 ft). The steep rampart of the Front Range was intimidating to the pioneers and their ox-drawn wagons, and so they wisely detoured north or south to less rugged terrain. The Front Range itself is divided into several smaller ranges in the park such as the Mummy Range and the wonderfully named Never Summer Mountains.

The **continental divide** bisects Rocky Mountain National Park, separating drainage to the Atlantic Ocean

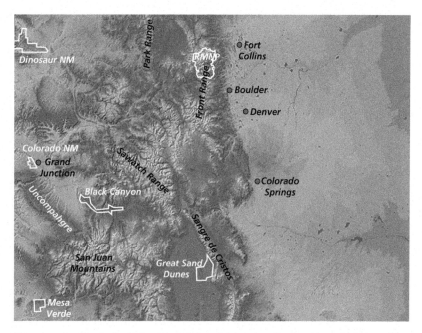

Figure 21.3 Shaded relief map of Colorado showing the location of Rocky Mountain National Park (RMNP) within the Front Range of the Colorado Rockies. A few other national parks and monuments are also shown.

and across the continental divide to the Cache la Poudre River, which flows to the eastern plains.

A second major conveyance system moves water *under* the mountains of the park through a 21 km (13 mi) long tunnel. In the 1930s, eastern Colorado was parched by the drought of the Dust Bowl. The U.S. Bureau of Reclamation built the Adams Tunnel beneath the mountains of the continental divide, diverting Colorado River water to irrigate the eastern plains. Today, with the rampant increase in population along the eastern slope, 65 percent of the water from this diversion goes to thirsty cities and industries, whereas 35 percent goes to irrigate farmland. Rocky Mountain National Park has a greater water diversion infrastructure than all other national parks combined.

from drainage to the Pacific. Rivers and streams on the east side of the divide flow into the South Platte River, which eventually flows into the Missouri–Mississippi watershed. Rainfall and snowmelt on the west side of the divide create the headwaters of the Colorado River system that flows west across the Colorado Plateau and the desert southwest. The west side of the divide receives most of the precipitation due to prevailing westerly winds, whereas the east side tends to be much drier. Unfortunately, 80 percent of Colorado's population lives in the metropolitan corridor that lines the edge of the High Plains flanking the Rockies to the east. Because of this disparity between where the precipitation falls and where the population lives, water in Colorado doesn't necessarily follow the inexorable force of gravity. Rather, it sometimes flows uphill through human-made constructs toward the people and businesses who need it.

Twelve different conveyance systems divert water from the Colorado River across the continental divide to homes, industries, and farms in eastern Colorado. In the Front Range near Rocky Mountain National Park, two major water diversion networks redirect flows eastward (Fig. 21.4). Ninety percent of the Colorado River's headwaters come from snowmelt in the Never Summer Mountains that form the northwestern boundary of the park. A third of that water is intercepted by the Grand Ditch, a human-made waterway that redirects the flow northward through the park

Origin, Age, and Uplift of Rocks in Rocky Mountain National Park

Recall from earlier chapters that the principle of *cross-cutting relations* dictates that rocks have to already exist before they can be uplifted as a mountain range or high plateau. The geologic story at Rocky Mountain National Park begins almost 1.8 billion years ago with the origin of the Precambrian rocks that dominate the park. These ancient rocks were raised as mountains much later, primarily during the Laramide orogeny beginning about 80 million

Figure 21.4 Oblique view looking due west across Rocky Mountain National Park, outlined in green. The blue line is Trail Ridge Road; red dashed lines are water diversions addressed in the text. The lighter shades on the image represent alpine terrain above treeline.

years ago. That's the essence of the story. But the full narrative involves two pre-Laramide episodes of uplift, with the remains of those ancient mountains preserved only by their eroded sedimentary debris. Even the Laramide-age Front Range was eventually buried in its own erosional remains, only to reemerge late in the Cenozoic as the mountains we see today.

In the Front Range of the Colorado Rockies, multiple phases of mountain-building occurred through time, only to wear away and set the stage for the next round of uplift. Rocks were transformed during each of these episodes of uplift and erosion, tracing a path through the rock cycle (Fig. 1.32). Humans perceive mountains as permanent and timeless, but in Rocky Mountain National Park and the Colorado Rockies as a whole, mountain ranges grow and decay with geologic regularity. The simplest way to recount this complex geologic history of the park is to tell it chronologically.

PRECAMBRIAN ORIGINS OF COLORADO AND GROWTH OF NORTH AMERICA

The oldest rocks composing the terrain of Rocky Mountain National Park are metamorphic *schists* and *gneisses* as old as 1.78 Ga (Ga = billions of years). They make up a large part of the crust of Colorado and record the original assembly of the state as a physical entity. Today these rocks are exposed far and wide throughout the Colorado Rockies, but they originated deep within the roots of actively growing mountain ranges during the Precambrian.

The original Precambrian precursor rocks were progressively subjected to higher and higher pressures and temperatures, perhaps at depths of 15–25 km (9–16 mi), transforming into the banded rocks exposed today in the park (Fig. 21.5). Under the extreme conditions of metamorphism (Fig. 1.30), the original suite of minerals changed into a new suite of minerals and reorganized into a parallel alignment, forming a banded texture in the rock called **foliation**. With increasing pressures and temperatures, the rocks of Rocky Mountain National Park evolved into schists, with their characteristic foliation of platy biotite crystals, and gneisses, marked by foliated bands of alternating lighter and darker minerals. At a certain threshold of pressure and temperature, the rocks began to deform plastically, with the foliation smeared out along tight folds within the rock. (A more thorough explanation of metamorphism is provided in Chapter 1.) These ancient metamorphic rocks are readily seen along the western sections of Trail Ridge Road and near many of the trails in the backcountry that traverse these fragments of the original crust of Colorado. What might have been the original source and composition of the precursor rocks prior to metamorphism?

The tectonic origins of these old metamorphic rocks are related to the physical growth of the North American continent. Prior to ~1.8 Ga, North America consisted of a large part of Canada, the upper American Midwest, and parts of Montana and Wyoming. The southwestern edge

Figure 21.5 Precambrian gneiss in Rocky Mountain National Park. Note the plastic flow of foliated banding created during metamorphism.

of North America at that time ran along what is today the Wyoming–Colorado state line (Fig. 21.6A). Not much of Colorado existed prior to 1.8 Ga, but volcanic island arcs extended across the region, bordering a subduction zone. These islands shed sedimentary debris into the adjacent oceanic basins to create the precursor materials that would eventually be metamorphosed into the crust of Colorado.

The volcanic islands migrated tectonically toward the southern margin of North America, burying, squeezing, and deforming the intervening sedimentary rocks of the rapidly closing ocean basin. The chaotic mass of oceanic sedimentary rock and volcanic islands slowly "docked" onto the edge of the continent as a series of *accreted terranes*, raising a coastal welt of mountains. (Accretionary tectonism was introduced in Chapter 19.) It was deep within this ancient mountain range along the convergent plate boundary that the original metamorphic rocks of Colorado's crust were formed.

These 1.8- to 1.7-billion-year-old rocks extend in a broad swath from northwest Mexico through northern Colorado, then northeasterly beneath Nebraska to the Great Lakes region. They represent the accretion of new continental crust during the middle Precambrian and the outward growth of the North American continent by about 500 km (~300 mi). Other volcanic island chains and continental fragments would subsequently accrete to the 1.8 Ga crust like an ultra-slow-motion trainwreck, adding even more rock to the outer margins of North America. All continents were assembled in this fashion, tectonically suturing together bits and pieces of continental and oceanic rock along convergent margin boundaries. A map of crustal ages of the North American continent looks like

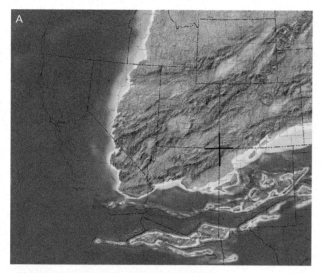

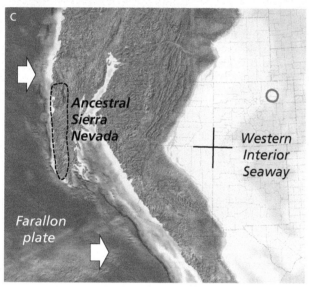

Figure 21.6 A) Southwestern United States during the Precambrian at 1700 Ma. Island chains and small landmasses along the southern coastline are poised to become accreted terranes in Colorado and elsewhere. B) Ancestral Rocky Mountains of Uncompahgria and Frontrangia ~308 Ma. C) American West ~90 Ma during inundation by the Western Interior Seaway. On all three maps, crossed lines mark Four Corners region; red circle marks location of Rocky Mountain National Park.

a patchwork quilt of irregular odds and ends, arranged as belts surrounding a crustal core in northern Canada as old as 4.0 Ga. The assembly of North America will be more fully addressed in Chapter 35 on the origin of the Appalachians.

In the American Midwest, perhaps a kilometer or so beneath the farmland of Iowa and Nebraska lies this 1.8 Ga crust, overlain by younger sedimentary rocks and a thin veneer of soil. You touch this very same rock in the high country of Rocky Mountain National Park and throughout the Colorado Rockies. Incidentally, if this Precambrian story sounds familiar to you, it's because you read the same tale in Chapter 3 on the crystalline rocks of the basement of the Grand Canyon. The schists and gneisses of the deepest gorges of Grand Canyon are

the same age as those in Rocky Mountain National Park and originated as part of the same accretionary tectonic event (Fig. 21.6A). Furthermore, the Precambrian rocks of Colorado National Monument and Black Canyon of the Gunnison National Park are the exact same rocks and reflect the same tectonic origin. It's advisable to review the relevant parts of Chapters 3, 10 and 11 to connect their related geologic histories and thus gain a deeper understanding of the Precambrian evolution of the American West.

The story of the Precambrian rocks of Rocky Mountain National Park becomes even more complicated between 1.6 and 1.4 Ga when a series of granitic magma chambers intruded into the metamorphic country rock. This occurred a few hundred million years or so after accretion as magma was generated deep along subduction zones, buoyantly rising into the overlying accreted terranes. The granites commonly occur along **suture zones**, the tectonic boundaries between accreted terranes, suggesting that the magma used the fractured and faulted rock along the suture zone as a conduit to rise upward. The Silver Plume Granite (Fig. 21.7) is widespread across the park and is particularly well exposed along the eastern part of Trail Ridge Road. These Precambrian igneous and metamorphic rocks constitute more than 90 percent of the surface of Rocky Mountain National Park. Below treeline, a thin veneer of soil and vegetation may cover these rocks, but in the alpine tundra of the highest elevations, much of this crystalline rock is fully exposed.

A long phase of erosion lasting several hundred million years dominated the rest of Precambrian time in the region around Rocky Mountain National Park, reducing the mountains to lowland plains. Rivers transported the erosional debris across the barren, unvegetated landscape of late Precambrian North America, ultimately

Figure 21.7 A) Glacial headwall of Longs Peak, which consists primarily of 1.4 Ga Silver Plume Granite. B) Close-up view of Silver Plume Granite showing the interlocking texture of feldspar, quartz, biotite, and muscovite composing the rock.

depositing it in the surrounding oceans. Many geologic events undoubtedly occurred during this prolonged phase, but no evidence for these events exists in neither the Colorado Rockies nor elsewhere in the American Southwest. Much of this long interval is lost in time along the widespread *Great Unconformity* that caps Precambrian crystalline basement in parks of the Colorado Plateau (e.g., Fig. 3.16).

By the beginning of the Paleozoic ~500 Ma, the exposed erosional surface was transgressed by coastal depositional environments driven by rising sea level. Sedimentary rocks laid down in shallow seas accumulated along the broad, tectonically quiet margin of western North America. Many of the sedimentary rock layers exposed in the upper Grand Canyon extended all the way across Colorado and far to the east, deposited as sea level rose and fell through time. None of these Paleozoic rocks are preserved in Rocky Mountain National Park today because they were eroded away during subsequent uplift of the Front Range. But they exist in other parts of Colorado and throughout the American West, recording the widespread extent of early to middle Paleozoic seas.

In summary, the Precambrian crystalline rocks you see today forming the highest peaks in Rocky Mountain National Park were originally formed several kilometers beneath the surface in the roots of an ancient mountain range. They were uplifted to the surface and planed off by erosion by the end of Precambrian time, forming coastal lowlands. As sea level rose and fell through the early to mid-Paleozoic, the crystalline basement was buried beneath a kilometer or two of sedimentary rock. How they were brought up to their current rarified elevations is the focus of the rest of this chapter.

LATE PALEOZOIC ANCESTRAL ROCKIES

From ~315 Ma to 300 Ma, a widespread pulse of mountain-building occurred worldwide as continents converged to build the supercontinent of Pangea. The Appalachians were lifted to heights rivaling the modern Swiss Alps by *continent–continent collision* with northwest Africa, while smaller mountain belts circumscribed the growing North American landmass. Volcanic island arcs were accreting along the western margin of late Paleozoic North America, building the continent outward.

In Colorado, two elongate mountain ranges rose, forming highlands that geologists informally call Uncompahgria and Frontrangia (Fig. 21.6B). These highlands were located very close to their modern counterparts and are thus referred to as the Ancestral Rockies. (The Ancestral Rocky Mountain orogeny was described in Chapter 10.) These mountains likely rose along old faults originally related to Precambrian uplift of the region; these faults were reactivated in response to late Paleozoic compressional tectonic stress. The Ancestral Rockies were beveled back to sea level through the rest of the Paleozoic and into Mesozoic time. So how can we tell they even existed?

The evidence resides in the erosional debris that accumulated along the flanks of the ancient highlands, outside the boundaries of Rocky Mountain National Park and the Front Ranges. The reddish sandstones and *conglomerates* of the 300 m.y. old **Fountain Formation** are exposed today in elongate bands paralleling the margins of their ancestral source in the highlands. Ephemeral streams deposited the course sediments as a thick wedge of debris that bordered the Ancestral Rockies to the east and west. Many pebbles and cobbles in conglomerates of the Fountain Formation are composed of gneiss and granite derived from the Precambrian crystalline rocks of the adjacent highlands.

Away from the ancient mountainfront, coarse clastic sediments of the Fountain merge with beds of limestone containing marine fossils, evidence for shallow seas surrounding the mountains and their wedge of sedimentary debris. Sedimentary rocks overlying the Fountain Formation are finer grained sandstones, suggesting that the Ancestral Rockies were reduced to lowlands by the end of the Paleozoic. No actual evidence for this uplift event exists in Rocky Mountain National Park, but geologists deduce the existence of the mountains from the erosional by-products of weathering and erosion left behind as the Fountain Formation. We'll return to the Fountain redbeds as the third episode of mountain-building is addressed in the following section.

This story of the Ancestral Rocky Mountain orogeny connects with earlier chapters on parks of the Colorado Plateau. The Precambrian rocks exposed in Colorado National Monument and Black Canyon of the Gunnison National Park were uplifted in the late Paleozoic as part of the Uncompahgre highlands of the Ancestral Rockies (Fig. 21.6B). The rise of the Ancestral Rockies was the original cause for the joint-controlled landscapes we see today in many parts of the Colorado Plateau.

It's important to understand that episodes of regional uplift always have a reciprocal depositional response elsewhere in the general region. For instance, many widespread layers of late Paleozoic clastic sedimentary rock that form landscapes throughout the American West were deposited in response to the uplift and erosion of the Ancestral Rockies. The shales, sandstones, and conglomerates of the Cutler Group in Canyonlands (Fig. 8.6), the Weber Sandstone in Dinosaur National Monument (Fig. 9.5), and the river-deposited clastics of the Supai Group in the upper Grand Canyon (Fig. 3.13) were all derived from the erosional breakdown of the Ancestral Rockies. They were deposited contemporaneously with the Fountain Formation in Colorado in a variety of continental depositional environments. The point to remember is that the rock of mountainous terrain is always slowly recycled into sedimentary layers of erosional debris in nearby lowlands.

LATE MESOZOIC–EARLY CENOZOIC LARAMIDE OROGENY

Around 200 Ma, the region of the Colorado Rockies was reduced to lowland plains. Far to the west, the ancestral Sierra Nevada was beginning to rise as the Farallon plate began its slow descent beneath western North America. By late Mesozoic time, inland from the Sierra, a north–south belt of highlands extended from Mexico to Canada, rising in response to compressional tectonism (Fig. 21.6C). These mountains created a load on the underlying lithosphere that depressed the adjacent crust to the east, forming the **Western Interior Seaway** that connected waters of the Gulf of Mexico with the Arctic Sea (Fig. 17.9). Basins formed by lithospheric downwarping due to the weight of nearby mountain belts are called **foreland basins**. Because of their proximity, they capture the sediment eroded from the adjacent mountains.

Dinosaurs migrated along the coastal plain bordering the seaway, leaving behind trackways in the soggy mud, now turned to rock. Other tracks were left behind by small mammals that scurried about, trying to avoid becoming a snack. Giant marine reptiles swam the waters, hunting for fish, while huge winged reptiles called pterosaurs roamed the skies above. The region that would become Rocky Mountain National Park was submerged deep beneath this seaway, with layers of mud and sand accumulating as the shoreline migrated laterally with changes in sea level.

This scenario changed dramatically toward the end of the Mesozoic with the advent of the Laramide orogeny. *Flat-slab subduction* seems to have begun beneath the ancestral Sierra Nevada sometime around 80 Ma, but the wave of mountain-building didn't reach inland to Colorado till ~67 Ma (Fig. 20.4). Compressional stresses reactivated the zones of weakness along old faults originally created during the Precambrian, raising the Front Ranges and the rest of the Colorado Rockies for a third time. Most of the compression was directed eastward from the convergent boundary on the west, with the solid crust of the continental interior acting as an anchor to the east of the Front Range.

The crystalline rocks of the Precambrian basement rose upward along *high-angle reverse faults*, a style of faulting introduced in Chapter 10 on Colorado National Monument (Fig. 10.7). To reiterate the mechanics of reverse faulting, the block of rock resting above the inclined fault plane is squeezed upward and above the block of rock below the fault plane. Movement along the fault occurs abruptly during earthquakes when the fault plane ruptures due to accumulated compressional stress. Mountains can grow slowly due to isostatic uplift in response to erosion, but they also grow spasmodically during countless earthquakes.

During Laramide uplift, layers of Paleozoic and Mesozoic sedimentary rock overlying the crystalline basement were bowed upward as a *structural arch*, exposing them to the unavoidable effects of weathering and erosion (Fig. 21.8). Several hundred meters of these sedimentary

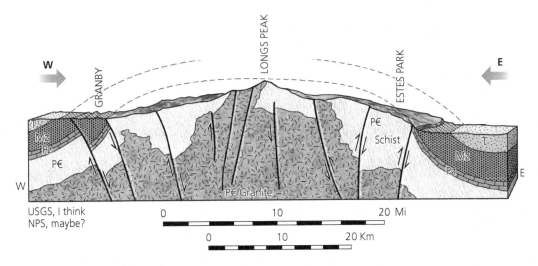

Figure 21.8 East–west profile across Rocky Mountain National Park showing the high angle of reverse faults that raised the Precambrian crystalline rocks to their current elevations. The overlying structural arch of Paleozoic and Mesozoic sedimentary rocks shown by the dashed lines were completely eroded from the park, with the remnants surviving as uptilted layers along the flanks of the Front Range.

rocks are preserved along the margins of the Front Range, providing an estimate of the thickness of rock stripped away from the crest of the structural arch above the Precambrian core. Laramide mountains like the Front Range and the rest of the Colorado Rockies are called **basement-cored uplifts** because of the high elevations attained by Precambrian crystalline rock as they were shoved upward along reverse faults. Recall that these metamorphic rocks originally formed 15–25 km (9–16 mi) beneath the surface but are now found at elevations exceeding 4 km (2.5 mi).

A network of high-angle reverse faults bounds the Front Range on its east and west sides, but they are mostly buried beneath alluvial debris shed from the highlands. To the east of Rocky Mountain National Park, these reverse faults descend westward beneath the mountainfront, bowing outward to the east (Fig. 21.9B). The 300 m.y. old redbeds of the Fountain Formation, composed of erosional debris shed from the Ancestral Rockies, are tilted upward against the flanks of the basement-cored uplift like surfboards stacked against the side of a van. These boldly colored rocks extend laterally along the mountainfront, forming the "flatirons" west of Boulder, the scenic backdrop of Red Rocks Amphitheater west of Denver, and the Garden of the Gods near Colorado Springs. The mid-Mesozoic Dakota Formation is composed of resistant sandstone and forms a prominent, east-tilting *hogback ridge* parallel to the mountainfront. These sandstones stand out in relief due to their resistance to erosion relative to the adjacent layers of less resistant shales and siltstones.

Coincident with late Mesozoic–early Cenozoic uplift, the Laramide Rockies shed erosional debris into basins lining the east and west sides of the Front Range. Particles in these sedimentary rocks include fragments of gneiss and granite, reflecting their Precambrian source rocks in the Laramide highlands. Dating of these rocks with fossils and the occasional volcanic ash helps to constrain the age of Laramide uplift of the Front Range from 67 Ma to around 37 Ma. By the end of active uplift in the mid-Cenozoic, the Front Range had buried itself in its own erosional debris. At the time, the surface of the High Plains to the east was a gently sloping ramp of sediment that rose upward to the low crest of the buried Laramide Front Range. The modern Front Range that we visit today in Rocky Mountain National Park underwent a final, fourth phase of uplift during the latest Cenozoic, the subject of the next section in this chapter.

LATE CENOZOIC EXHUMATION

By the late Cenozoic, the Colorado Front Range was dominated by the forces of erosion, with sedimentary debris deposited as an eastward-thinning wedge across the adjacent High Plains. The Front Range and Rocky Mountain National Park would have been unrecognizable, being merely the continuation of the sediment-filled ramp rising upward from the east. Paleobotanical evidence from fossil leaves suggests that mean elevations of the Front Ranges at the crest of the ramp were around 1800 m (~6000 ft) at this time.

The underlying trigger is not well understood, but the Colorado Rockies began a rejuvenated phase of uplift around 5 Ma. The gradients of rivers and streams flowing eastward increased with the rise of the Front Range, stripping away the veneer of loose debris above the crystalline basement and the adjacent High Plains. The ancestral South Platte River and other stream networks cut through the softer sediment blanketing the crystalline rock beneath, transporting it eastward away from the mountainfront. The solid rock of the exhumed Precambrian basement was more resistant to erosion, however, and was

A

B

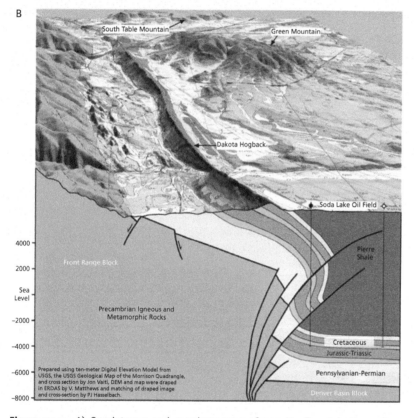

Figure 21.9 A) Sandstones and conglomerates of 300 Ma Fountain Formation tilted at high angles along the flank of the Front Range, Roxborough State Park. B) Cross section of eastern flank of Front Range showing high-angle reverse faults buried beneath the surface and the upturned Mesozoic rocks tilting downward away from the basement-cored uplift. Differential erosion creates the elongate ridges of resistant sandstones and conglomerates.

could possibly have caused the regional uplift?

The working hypotheses for late Cenozoic uplift and exhumation of the Colorado Rockies fall into two camps. One explanation suggests that regional uplift of the Front Range and the entire Colorado Rockies may be related to a mass of hot, buoyant asthenosphere that reaches relatively shallow depths below the crust. Geophysical evidence indicates that the hot rock of the mantle is anomalously shallow beneath Colorado. In this interpretation, the high elevations of the Colorado Rockies were reached as the crust "floated" on the welt of hot asthenosphere beneath. It may be that the plug of hot asthenosphere plastically rose up under the region during the late Cenozoic, triggering renewed uplift. But no one really knows the timing of the rise of the underlying asthenosphere, only that it exists today and supports the rarified elevations of Colorado.

The alternative explanation for rejuvenated uplift of the Front Ranges relates to a global shift in climate since 5 Ma. The planet has been progressively cooling over the last 30 m.y. but seems to have accelerated over the past few million years, culminating in the Pleistocene Ice Ages. The wetter and cooler climates of the American West since 5 Ma may have enhanced river incision and downcutting, resulting in the regional isostatic uplift and exhumation of previously buried crystalline rock. This idea is supported by the recognition that many mountain ranges worldwide seem to have experienced a renewed pulse of uplift since 5 Ma.

It may be that both mechanisms acted in concert to generate the exhumation of the Laramide Front Ranges and the high elevations of Rocky Mountain National Park. Whatever the answer, it's important to understand and accept the inherent uncertainty in the sciences. Unresolved questions like the trigger for rejuvenated uplift of mountains worldwide over the last few million years promote the questioning of assumptions and catalyze new ideas. Geoscientists thrive on these sorts of problems, and our comprehension of how the planet operates is slowly but surely deepened as more evidence is revealed. Remember that science is a dynamic, ongoing process, not a static set of facts.

thus left elevated above the plains to the east. Elevations today are well over 3600 m (~12,000 ft), double that of only 5 m.y. ago.

The only problem with this scenario is that rejuvenated uplift and exhumation of the crystalline rocks of the Front Ranges occurred in the absence of tectonic compression. The convergent margin that bordered the western edge of the continent from the Mesozoic through mid-Cenozoic was long gone by 5 Ma, and there is no evidence of renewed activity along the Laramide reverse faults. What

Glaciation and Landscape Evolution in Rocky Mountain National Park

Rocky Mountain National Park is notable for its variety of glacial landforms such as U-shaped valleys, hanging valleys, cirques, arêtes, roche moutonnee, glacial lakes, and moraines. Several previous chapters addressed the mechanics of alpine glaciers, the Pleistocene Ice Ages, the Last Glacial Maximum, and many glacial landforms that developed as glaciers grew and receded over the last 2.6 m.y. If necessary, review Chapters 14, 17, and 18 on Mount Rainier National Park, the Sierra Nevada, and Yosemite National Park, respectively, to refresh your understanding of the effects of glacial modification of landscapes. The back and forth motion of glaciers, advancing during cold glacial maxima and retreating during warmer interglacial phases, acted like a buzzsaw on the terrain. Glacial ice and its load of sedimentary debris did the work during cold phases, whereas meltwater streams charged with glacial outwash modified landscapes during warm phases.

Evidence of three Pleistocene glaciations is recognized in the Front Ranges around Rocky Mountain National Park. The oldest, called pre-Bull Lake, must have been the largest of the three since proof for its peak extent lies at lower elevations outside the park boundaries. The penultimate glaciation, the Bull Lake, likely ended around 127 ka and was followed by a warm interglacial phase. The Last Glacial Maximum peaked in the area around 23 to 21 ka. This final glaciation is called the **Pinedale** and is responsible for obscuring the evidence for much of the earlier glacial events as it re-sculpted the terrain into the landscape we see today. (The Pinedale is equivalent to the Tioga glaciation described in Chapters 17–19 on parks of the Sierra Nevada.)

During the Last Glacial Maximum, alpine glaciers flowed downslope from their heads along the continental divide (Fig. 21.10). At their peak extent, the toes of glaciers reached elevations around 2300 m (~7600 ft). The upper limit of glaciation is approximated by the upper limit of tree growth at around 3500 m (~11,500 ft). In Rocky Mountain National Park, the treeline also marks the *trimline*, the upper limit of glacial ice. Why were the highest, coldest elevations in the park not covered in glacial ice? Because the constant winds that sweep the alpine tundra above treeline precluded the accumulation of significant thicknesses of snow necessary to compact into glacial ice. The gently rolling terrain above treeline appears much as it did during glacial phases when it formed broad *nunatak*s above the level of glacial ice (Fig. 21.11).

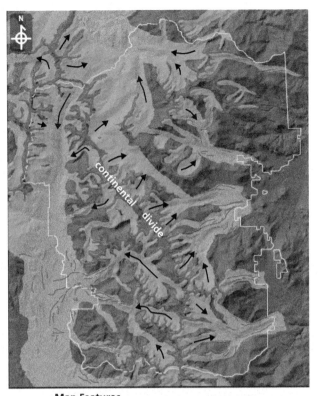

Map Features

⌒ Cirques		Glacially
● Glacial valleys		related
● Glacial sediment		landforms.
∿ Moraines		

Figure 21.10 Extent of alpine glaciers in Rocky Mountain National Park during the Last Glacial Maximum. Many of the glaciers began along the continental divide as evidenced by the concentration of cirque headwalls (dark blue). Glacially carved valleys are shown in light blue, and glacially deposited outwash is shown in yellow. Moraines are marked by the red lines. Black arrows show a few of the glacial flow directions.

Figure 21.11 The alpine tundra above the treeline in Rocky Mountain National Park roughly marks the upper limit of alpine glaciers. Below the treeline, glaciers accumulated enough mass to carve glacial valleys and modify the landscape.

Figure 21.12 The broad valley of Moraine Park. The forested lateral moraine was left behind as the enormous glacier receded back toward the highlands in the distance. The flat floor of the valley formed in response to the infilling of meltwater ponds and wetlands by glacial outwash.

As you traverse the tundra ecosystem along the highest elevations of Trail Ridge Road, visualize the white expanse of ice below the stark alpine terrain. While you're there, think about what the Pleistocene environment may have looked like. Rivers flowed milky white with entrained silt derived from the melting toes of alpine glaciers. Glacial lakes glowed a turquoise green from the refraction of sunlight off the countless particles of glacial silt suspended in their waters, just as they do today in higher elevations. Along the margins of the glaciers and in the surrounding lowlands, saber-tooth cats and wolves stalked herds of mammoth and bison. Around 12,000 years ago, as the ice retreated into the highlands after the Last Glacial Maximum, the earliest Americans hunted and gathered in the region. Very few places transport you back in time to a different world as does Rocky Mountain National Park.

A wide variety of textbook glacial features are visible from multiple vantage points in the park, especially along Trail Ridge Road. This section describes a few familiar landforms of the glaciated terrain along an east-west traverse. Moraine Park, just inside the east entrance, is a broad, flat valley bounded by tree-covered **lateral moraines**, linear ridges of glacial till deposited along the upper flanks of alpine glaciers (Fig. 21.12). Several glaciers converged from the highlands to create a single large trunk glacier that spread out across the valley,

widening it through time. As the Pinedale-age glacier retreated at the very end of the Pleistocene, it left behind a *terminal moraine* that acted as a dam, blocking the outflow of meltwater. The array of ponds and wetlands that occupied the valley filled with glacial outwash and organic debris, transforming it into the wildflower-filled meadow we see today. Looming off in the distance south of Moraine Park, the northeastern face of Longs Peak is a 600 m (~2000 ft) high sheer headwall of a glacial *cirque* known as The Diamond (Fig. 21.7A). Chasm Lake is a *tarn* that fills the deepest bowl of the cirque far below the peak.

Near the highest elevations along Trail Ridge Road, the Forest Canyon overlook provides a view of merging *U-shaped valleys* beneath the row of peaks anchored by Terra Tomah Mountain (Fig. 21.13). Alpine glaciers flowed 21 km (13 mi) down Forest Canyon and reached a thickness over 300 m (~1000 ft). Terra Tomah Mountain itself exhibits a classic cirque basin, bounded by *arêtes*, that opens into the adjacent glacial valley (Fig. 21.14). Broad saddles in the knife-like arête ridges are known as **cols**, formed as adjoining glaciers eroded the narrow wall of rock separating them. Nearby Hayden Spire is a **horn**, a pyramid-shaped peak left behind

Figure 21.13 Paired images of Forest Canyon from the Alpine Ridge Trail showing modern landscape features and a reconstruction of glacial flow during the Last Glacial Maximum. Mount Ida marks the continental divide in the area.

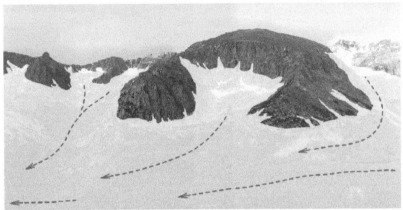

Figure 21.14 Paired images of Terra Tomah Mountain from the Forest Canyon Overlook showing glacial landforms and a reconstruction of glacial flow during the Last Glacial Maximum. Note how treeline approximates the upper limit of glaciation and the rounded, unglaciated terrain near the crest of Terra Tomah.

by *glacial plucking* where the heads of three glaciers converged.

At the west end of the park, the headwaters of the Colorado River originate in the Never Summer Mountains. The sinuous channel of the Colorado River flows southward within the broad, U-shaped Kawuneeche Valley. Visualize the largest alpine glacier in the park occupying this valley during the Last Glacial Maximum, reaching 52 km (20 mi) downvalley (Fig. 21.10). Just outside the western park boundary, Grand Lake is a natural lake impounded behind a series of moraines. Nearby reservoirs redirect Colorado River water into Grand Lake where it's pumped into the Adams Tunnel for its underground journey to Estes Park and points east.

Modern Glaciers and Climate Change in Rocky Mountain National Park

Eight small glaciers and several permanent snowfields remain today in Rocky Mountain National Park, where they occupy cirque basins on shadowed, north-facing slopes. A chain of small glaciers and snowfields tracks the east side of the continental divide and a few small glaciers are found in the Mummy Range in the north end of the park. Similar to glaciers elsewhere in high elevations of the American West, they are remnants of the Little Ice Age that spanned about five centuries from 1350 to 1850. The cirque glaciers of Rocky Mountain National Park grew slightly during a few decades of cooler summers in the 1970s and 1980s but have been in retreat since the 1990s as global temperatures creep upward (Fig. 21.15).

The National Park Service is working to understand the environmental and ecological consequences of climate change in Rocky Mountain National Park. Some of the implications of a warming climate in the park include more precipitation in the form of rain rather than snow, changes in the populations of mammals related to slowly shifting habitats, increases in the threat of wildfire, and a reduction in alpine ecosystems as the treeline creeps upward into higher elevations. At the broader scale, water agencies are closely monitoring the changing snowpack throughout the entire Rocky Mountain Province. Spring runoff has been happening earlier in the last few decades, reducing the frozen reservoir of water in the winter snowpack along the length of the Rockies.

Meltwater from Rocky Mountain snowpack provides a substantial portion of the water supply for over 70 million people in the western United States, much of it derived from the Colorado River. The warming climate is creating

Figure 21.15 A) Andrews Glacier, RMNP, 1989. B) Andrews Glacier, RMNP, 2013.

challenges for watershed management and the redistribution of much-needed water throughout the American West.

Key Terms

basement-cored uplift
col
continental divide
foliation
foreland basin
horn
lateral moraine
suture zone

Related Sources

Abbott, Lon, and Cook, Terri. (2012). *Geology Underfoot along Colorado's Front Range*. Missoula, MT: Mountain Press.

KellerLynn, K. (2004). *Rocky Mountain National Park Geologic Resource Evaluation Report.* Natural Resource Report NPS/NRPC/GRD/NRR -2004/004. Denver, CO: National Park Service.

Raup, O. (1996). *Geology along Trail Ridge Road*. Helena, MT: Rocky Mountain Nature Association.

Climate Change in Rocky Mountain National Park. Accessed February 2021 at http://www.nps.gov/romo/learn/management/upload/climate_change_rocky_mountain2.pdf

Gateway to Glaciers and Glacier Change of Rocky Mountain National Park. Accessed February 2021 at http://www.nps.gov/features/romo/feat0001/index.html

National Park Service—Geologic Activity, Rocky Mountain National Park. Accessed February 2021 at http://www.nps.gov/romo/learn/nature/geologicactivity.htm

Figure 22.1 Aerial view of Glacier National Park showing glacially sculpted terrain of jagged mountains, lush U-shaped valleys, and glacial lakes.

Don't visit Glacier National Park in Montana expecting to see huge glaciers cascading across the mountainscape. You'd be about 20,000 years too late. Yes, about 25 active glaciers exist today, but they are modest in size and restricted to north- and northeast-facing cirques where they are shadowed by high mountain peaks. The park gets its name from its spectacular ice-carved scenery, a remnant of the Pleistocene Ice Ages when glaciers covered the terrain, growing and retreating in response to climate cycles over the past 2.6 m.y. (Fig. 22.1).

The dramatic topography in Glacier National Park forms the backdrop for a variety of ecosystems ranging from lowland prairie to highland arctic tundra. In turn, the ecosystems support an abundance of wildlife, including grizzly bear, wolves, lynx, mountain lion, moose, elk, bighorn sheep, and ubiquitous mountain goats. The biological diversity is so exuberant that the park was named a Biosphere Reserve to sustain its status as a sanctuary for wildlife. Only a few places in North America express such a vast panorama of geological and biological richness.

Glacier National Park is located in the Northern Rocky Mountains of northwestern Montana (Fig. 20.1). Waterton Lakes National Park in Canada abuts Glacier along the 49th Parallel, and the two parks jointly form Waterton-Glacier International Peace Park (Fig. 22.2). The parks are administered independently, but wilderness respects no political boundaries, so the two parks work together to preserve their shared natural and cultural resources.

The main access to Glacier National Park is via Going-to-the-Sun Road, a marvel of engineering that circuitously winds its way east–west across the park. The majority of visitors see the park from the many viewpoints along the road, but over 1100 km (700 mi) of trails permit exploration deeper into the wilderness.

Glacier National Park is bisected by the *continental divide*, which neatly separates streams flowing eastward

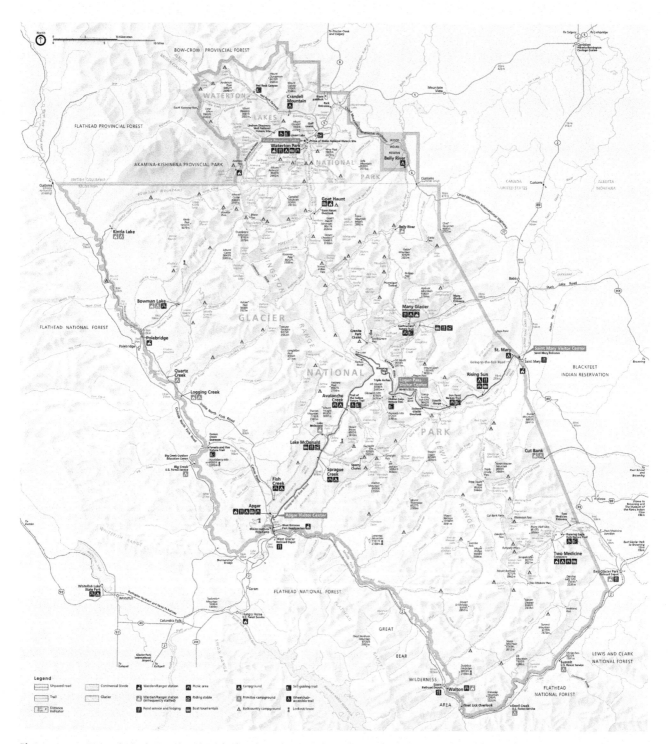

Figure 22.2 Map of Glacier National Park showing Going-to-the-Sun Road and main physiographic features. Continental divide shown by yellow line. Note how streams southwest of the divide flow in the opposite direction of streams northeast of the divide.

from those flowing westward (Fig. 22.2). Pleistocene glaciers emanated outward from the narrow crest of the divide, so the streams flow through broad, U-shaped glacial valleys, commonly draining into elongate finger lakes. There are about a dozen large lakes and over 650 smaller lakes in the park, almost all of them formed in response to glacial activity. Many of the lakes fill deep gouges created by *glacial plucking*, while others are dammed behind *moraines* left during glacial retreat.

The continental divide follows the crest of the Lewis Range through most of the park, except in the north where it shifts westward to the Livingstone Range (Fig. 22.3). Runoff from rainfall or snowmelt in the park may end up in one of three different ocean basins because the topographic divides between three major watersheds converge in the park. Streams along the west side of the continental divide form the headwaters of the Flathead River system, which flows to the Columbia River and out to the Pacific. Streams along the southeast side of the divide eventually flow to the Missouri River system and then on to the Atlantic. Runoff along the northeast side of the divide drains toward the Saskatchewan River system that flows northeast across Canada into Hudson Bay. The **triple divide** where the three watersheds meet is appropriately called **Triple Divide Peak**, located in the southeastern part of the park. The continental divide in Glacier National Park also acts as a barrier to moisture-laden weather systems that move in from the west. Most of the several meters of annual snow falls to the west of the divide, whereas the terrain east of the divide is windswept, drier, and colder.

Even though Glacier National Park follows the spine of the Northern Rockies, the elevations are not as high as in the Southern Rockies of Colorado. In contrast to the numerous peaks over 4200 m (~14,000 ft) in the Colorado Rockies, only six peaks reach above 3050 m (10,000 ft) in Glacier National Park, with a high of almost 3200 m (10,466 ft) on Mount Cleveland. The terrain in Glacier National Park may lack high elevations, but it more than compensates in the sheer, jagged relief of the landscape. Since Glacier is about 8° of latitude farther north than the Colorado Rockies, the Pleistocene alpine glaciers were thicker and incised more deeply into the underlying rock. They left behind a scalloped landscape of cirques, with near-vertical headwalls thousands of meters high, narrow arêtes, and spiky *horns* separated by deep, verdant U-shaped valleys. The higher latitude, harsh winters, and steep relief of Glacier National Park constrain treeline to around 2100 m (7000 ft). Below treeline, evergreen forests dominate the park; above treeline, an arctic-alpine ecosystem of ground-hugging plants forms a mosaic with slopes of talus and perennial snowfields.

The mountainous landscape of Glacier National Park is the result of a distinct series of geologic events that form a familiar pattern, similar to earlier chapters. The first step involves the creation of the rocks that form the raw material for all subsequent events to act on. In Glacier National Park, late Precambrian sedimentary rocks dominate the terrain. Our description of the origin of these rocks will include evidence of some of Earth's earliest forms of life, the strange conditions of the Precambrian world, and a discussion of the breakup of an ancient supercontinent.

The second step entails the uplift of the rocks to create highlands. Mountain-building occurred in the Glacier region during the Laramide orogeny, but with a distinctly different style than in the Colorado Rockies. This section of the chapter will expand on previous discussions of how

Figure 22.3 Low-angle oblique Google Earth image of Glacier National Park looking due west. Continental divide shown by blue dashed line, park boundaries by green.

mountains form and will introduce the concepts of folding, thrust faulting, and the fold-and-thrust belt of compressional mountain chains.

The third and last step is the creation of the modern landscape by the erosional serendipity of Ice Age glaciers, the singular event for which Glacier National Park derives its name. We'll expand on the numerous glacial landforms created by alpine glaciers described in earlier chapters and introduce the dynamics of continental ice sheets that reached their southernmost extent in the region during the Last Glacial Maximum. Geologic events and concepts addressed in this chapter will add to our overall theme of the evolution of the mountainous landscapes of the American Cordillera.

Origin and Age of Rocks in Glacier National Park

The layered texture of mountainside exposures in Glacier National Park indicates that most of the park is composed of sedimentary rocks (Fig. 22.4). This colorful stack of sandstones, siltstones, shales, and limestones is part of the **Belt Supergroup**, a thick Precambrian succession recognized across a broad swath of Montana, Idaho, Washington, and north into Canada. ("Supergroup" is simply a term of convenience for an assemblage of related formations.) In Glacier National Park, about 2900 m (~9500 ft) of Belt rocks are exposed along the mountainsides, although the Belt reaches a thickness of ~5500 m (18,000 ft) to the west.

These rocks are the preserved record of tens of millions of years of changing depositional environments ranging from coastal streams and floodplains to broad tidal flats, deltas, and deeper water settings below the reach of waves and tides. The Belt sediments were eventually buried, compacted, and cemented to solidify as sedimentary rocks. With deeper burial through time, the Belt rocks were subjected to higher pressures and temperatures, but the degree of metamorphism was minimal. Rocks of the Belt Supergroup are more accurately described as **metasedimentary**, but they are remarkably well preserved and retain many of their original sedimentary features.

Many depositional structures exhibited in Belt rocks provide evidence for shallow water environments. Interpretations of these sedimentary features are supported by fieldwork at modern depositional settings that provide actualistic analogs for ancient environments. **Mudcracks** are common within siltstones and clay-rich shales and indicate the episodic exposure and desiccation of the sediment surface, likely on muddy tidal flats (Fig. 22.5). The muddy sediment contracts as water evaporates during prolonged exposure, with a polygonal network of cracks developing.

Figure 22.4 Sedimentary layering of the Belt Supergroup, Bearhat Mountain

Figure 22.5 A) Mudcracks in muddy siltstone of Belt Supergroup. B) Modern mudcracks formed by desiccation on a tidal flat. Polygons in both images are several centimeters across.

Later flooding by high tides transports silt and sand that fills the cracks, preserving the entire layer. After uplift and erosion, the mudcracked surface of the ancient tidal flat is re-exposed to view.

Wavy surfaces are evident along bedding planes of many sandstones in the Belt (Fig. 22.6). These **ripplemarks** preserve the movement of grains by waves and tides, perhaps on beaches, sandflats, or river floodplains. Some quartzites (slightly metamorphosed sandstones) exhibit *cross-bedding*, indicative of deposition within fast-moving streams or wave-washed beaches. Even the color of clastic rocks reveals information about the relative water depths where they accumulated, primarily due to the chemical state of iron within the sediments during deposition. Reddish or maroon colors indicate the presence of small amounts of **hematite**, a mineral that forms when iron comes in contact with oxygen in the atmosphere or in shallow water. Clastic rocks deposited in streams and floodplains, sandflats, and other oxygenated coastal environments are commonly a ruddy red shade. In less oxygenated, deeper water, iron takes on a "reduced" form. After burial and mild metamorphism, the reduced iron contributes to the formation of a greenish mineral called **chlorite**, imparting an olive hue to the rocks. Sedimentary rocks with a greenish cast thus indicate deposition in deeper, poorly oxygenated water.

Figure 22.6 A) Ripplemarks in sandstone of Belt Supergroup. B) Modern ripplemarks formed by tidal waters moving across a sandy surface.

Limestones typically form from the accumulated remains of calcite-shelled marine organisms (Chapter 1). But in the Precambrian, marine invertebrates like corals, clams, oysters, and other calcite-secreting animals had not yet evolved. Some limestones likely formed inorganically from the Precambrian seas as mud-sized particles of calcite ($CaCO_3$) precipitated directly from seawater saturated with the necessary elements. Other limestones accumulated from the activity of photosynthesizing algae and related microbes called **cyanobacteria** that proliferated in the shallow waters of Precambrian oceans. Algae and cyanobacteria commonly formed slimy, carpet-like mats in very shallow coastal waters to maximize exposure to sunlight.

How did living organisms like algae and cyanobacteria create layers of limestone during the Precambrian? During the process of photosynthesis, these organisms would consume carbon dioxide from seawater, releasing oxygen as a waste product. The removal of carbon dioxide would trigger the chemical precipitation of tiny particles of calcite from seawater that would settle out along the tops of cyanobacterial mats, covering them and inhibiting photosynthesis. The resilient microbes would send filaments upward through the coating of calcite particles to recolonize the surface with a sticky mat of slime, restarting the process. Over time, the microbial colonies would build up mm-scale laminations of trapped and bound calcite particles alternating with thin mats of dead organic matter. These laminated structures, formed by the photosynthetic activity of microbes, are called *stromatolites*, introduced in Chapter 3 (Fig. 22.7).

Stromatolites are common in limestones of the Belt Supergroup and may exhibit a variety of forms, from planar laminations, to upright columns, pointed cones, and rounded, domal "cabbage-heads." Stromatolites are another indicator of shallow water depositional environments because algae and cyanobacteria live in coastal waters where the penetration of sunlight is greatest and the conditions for photosynthesis are optimal. As marine invertebrates evolved and diversified in the early Paleozoic, animals like snails would graze the cyanobacterial mats, forcing the microbes into harsher environments such as hypersaline tidal flats where the grazers couldn't survive. Today, active stromatolites grow in shallow tropical settings marked by high salinities (such as Shark Bay in Western Australia, Fig. 22.7C) or in high-velocity tidal channels (such as Lee Stocking Island in the Bahamas), environments too harsh for grazing invertebrates.

AGE OF THE BELT SUPERGROUP AND THE PROTEROZOIC WORLD

Reputable sources variously ascribe the Precambrian age range of the Belt Supergroup to 1600–800 Ma, 1470–1400 Ma, 1325–900 Ma, and 1450–1250 Ma. One reason for the discrepancy in ages for the Belt Supergroup is that the sediments accumulated prior to the advent of hard-shelled animals, those organisms that leave behind a fossilized external skeleton useful for relative age dating. Another reason is that relatively few igneous units cross-cutting

Figure 22.7 A) Bedding plane view of stromatolite "cabbage-heads," (exposed by glacial erosion near Grinnell glacier). B) Millimeter-scale laminations within stromatolite, Glacier National Park. C) Modern stromatolites, Shark Bay, Western Australia.

the layering could provide a numerical age via radiometric dating methods. Regardless of the exact age range, for our purposes the Belt Supergroup was deposited in the late Precambrian accounting for hundreds of millions of years of sediment accumulation in the region.

Just because geologists pack about 4 billion years into the Precambrian doesn't mean that we can't discern any major events through that vast expanse of time. Indeed, abundant

evidence exists for the first microbial origins of life, the evolution of photosynthesis as a metabolic process, multiple episodes of widespread glaciation, the assembly and breakup of supercontinents, and many other seminal events in Earth history. In particular, the emergence of photosynthesis by single-celled microbes led inexorably to a progressively more oxygenated atmosphere, a critical event dated to ~2.5 Ga. Precambrian rocks worldwide exhibit evidence for a mildly oxygenated atmosphere beginning around this time that enables subdivision of the Precambrian.

Earth history from 4.0 to 2.5 Ga is known as the **Archean**, the age of many crystalline rocks of the continental crust around the world. (We'll encounter Archean rocks in Grand Teton National Park.) The expanse of geologic time from 2.5 Ga to ~540 Ma is known as the **Proterozoic**, which includes the range of ages for the Belt Supergroup. So instead of *late Precambrian*, we'll begin to use the more precise term *Proterozoic* to describe rocks and events of this age (Fig. 1.37).

The Proterozoic Earth must have been a very strange place and would appear utterly alien if a human were to somehow be transported back in time to that world. The continents were lifeless expanses of reddish-gray rock, covered randomly with a thin coating of loose sediment derived from the weathering of underlying rock. Soil, defined in today's world by its rich organic content, would have been rare due to the absence of plants. Rivers meandered wildly with each rainstorm due to the lack of bank-stabilizing vegetation. The only sounds would have been from the wind, the gurgling of water within streams and rivers, the plik-plik of rain on rock, and waves lapping against the shoreline.

The Proterozoic oceans would have been blue expanses of lifeless water, tinged green along the coast by slimy mats of photosynthesizing algae and cyanobacteria. Stromatolites, the only tangible evidence of Proterozoic life prior to ~700 Ma, would have flourished along the shallow margins of the sea and within lakes on land. Oxygen, a by-product of photosynthesis, would have been slowly accumulating in the atmosphere and in the shallow ocean. It combined readily with iron in rocks, creating rusty hematite and coloring the landscape in shades of red. The slow increase in atmospheric oxygen would eventually result in the formation of an ozone layer (O_3) that would block harmful ultraviolet radiation from the Sun, enabling eventual colonization of the land as plants and animals evolved. The region around what is today Glacier National Park would have been a profoundly foreign place when the rocks of the Belt Supergroup were accumulating.

SIGNIFICANCE OF THE BELT SUPERGROUP

The Belt Supergroup of rocks is characterized by abrupt changes in thickness over relatively short distances. This is interpreted to reflect deposition occurring at the same time as tectonism, with different sediment thicknesses recording accumulation on opposite sides of an active fault. This pattern suggests that the overall basinal setting

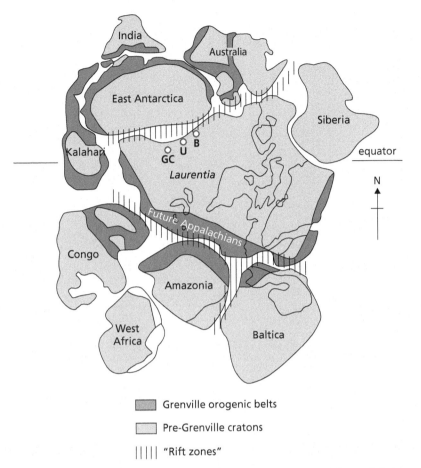

Figure 22.8 Reconstruction of the continental configuration of the Rodinian supercontinent ~1 Ga. The ancestral precursor to North America is called Laurentia. In this reconstruction, Laurentia is rotated ~90° clockwise, and the equator extends through the middle. The "rift zones" show the location of the breakup of Rodinia later in the Precambrian. Small circles show approximate locations of Belt, Uinta, and Grand Canyon rift basin rocks.

90° from its current orientation, with some models suggesting that East Antarctica and eastern Australia lay to the north, sutured to what is today the west coast of North America.

Breakup of Rodinia was well underway by 800 Ma, with the thick packages of Proterozoic rocks of the Belt, Uinta, and Grand Canyon marking the locations of continental rift basins. A comparable rifting event occurred along the future east coast of North America where fragments of Africa, South America, and Scandinavia broke away. The continental fragments disassembled through the end of Proterozoic time, with the intervening continental rifts growing into full-blown ocean basins. This story will be developed more completely in Chapter 33 on the evolution of the American Cordillera and in Chapter 35 on the origin of the Appalachians.

Laramide Uplift

In earlier chapters, you learned several ways that mountains of the North American Cordillera are created. The modern Cascades Range formed by in-place growth of mountains by active volcanism related to a subduction zone. The modern Sierra Nevada is a *tilted fault-block* mountain range, uplifted along normal faults generated by extensional forces along its eastern boundary. The ancestral Sierra Nevada of the late Mesozoic was raised by compression and magmatism as the Farallon plate was consumed along the North American western margin. And the Southern Rocky Mountains of Colorado were pushed upward along *high-angle reverse faults* as *basement-cored uplifts* during Laramide compressional tectonism.

In the Northern Rockies, however, the style of Laramide faulting and uplift was very different, even though the rocks were subjected to the same compressional tectonic forces. Uplift of highlands in Glacier National Park occurred along **thrust faults**, a type of *low-angle* reverse fault that places older rocks above younger rocks (Fig. 22.9).

Rigid, horizontally layered sedimentary rocks like those of the Belt Supergroup are particularly prone to breaking and sliding along weak bedding planes when subjected to compressional forces. Tectonic squeezing may initially cause the stack of layered rock to buckle into a convex-upward *fold* called an **anticline**. But with continued stress, the rock may break along one or more of the folds, with the overlying slab of rock sliding forward along a bedding plane up and over the adjacent block of rock. The end result is that older rocks are displaced over younger rocks,

in which the Belt sediments were deposited was likely a fault-bounded *rift basin* where tectonic stresses are extensional. The most active deposition occurred within the downdropped block between *normal faults*. Continental rift basins are a type of divergent plate boundary that was introduced in Chapter 1 on the birth of ocean basins. A modern example of a continental rift system is East Africa where the landmass is actively tearing apart (Fig. 1.12).

Similar rift settings were addressed in Chapter 3 on the Grand Canyon Supergroup and Chapter 9 on the Uinta Mountain Group in Dinosaur National Monument. These two successions of Proterozoic metasedimentary rock are roughly contemporaneous with the Belt Supergroup and are collectively interpreted to record the rifting and breakup of a supercontinent called **Rodinia** (Fig. 22.8). Ancient North America (given the name Laurentia by geologists) formed the core of the supercontinent, which had assembled by *accretion* of large continental fragments about 1 Ga. (The tectonic assembly of North America described in the previous chapter on crystalline basement rocks of Rocky Mountain National Park is part of the accretionary history of Rodinia.) Laurentia was rotated clockwise about

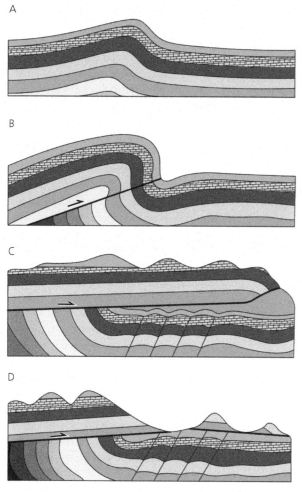

Figure 22.9 Evolution of the Lewis Thrust in Glacier National Park. Cross section is about 65 km (40 mi) across. A) Tectonic compression causes rigid sedimentary rocks to flex into folds. B) The folded rock eventually breaks, with a slab of rock sliding along a thrust fault. C) The thrust plane maintains a low angle, juxtaposing older rock above younger. D) Erosion acts to expose the thrust contact and modify the landscape.

Figure 22.10 The Lewis Thrust (black line) places the Proterozoic Altyn Formation above late Mesozoic (Cretaceous) shales, Curly Bear Mountain, Glacier National Park. The near-horizontal thrust plane is mostly buried beneath talus but is mapped along the boundary between the resistant Altyn and slope-forming shale beneath. This arrangement of older-over-younger is in conflict with the principle of superposition and thus helps to identify the location of the fault.

an arrangement at odds with the principle of *superposition*. This anomalous arrangement of older over younger rocks helps with the identification of the thrust fault in the field. **Offset** (aka *displacement*) along thrust faults occurs episodically during earthquakes, with movement ranging from a few centimeters to several meters per event. After countless earthquakes, total offset may reach several tens of kilometers if the tectonic stress is long-lived and the geologic conditions remain conducive to thrusting.

In Glacier National Park and the Northern Rockies as a whole, rocks of the Belt Supergroup were displaced 80 km (50 mi) to the east along the **Lewis Thrust** during the Laramide (Fig. 22.10). The easternmost front of the Lewis Thrust extends north–south for 450 km (280 mi) along its total length from south of Glacier National Park to its northern tip near Banff National Park in Canada. The glide plane along which thrusting occurs is the base of the Altyn Formation of the lower Belt Supergroup, a rigid layer composed of limestone with interbedded sandstone. The Altyn and all overlying rocks of the Belt (plus Paleozoic and Mesozoic

rocks now removed by erosion) were displaced over and above clay-rich shales of late Mesozoic (Cretaceous) age. Imagine the enormous mass of the slab of rock, pushed upward and eastward for tens of kilometers during countless earthquakes over the duration of the Laramide orogeny. Recall from Chapter 20 that Laramide tectonism relates to the flattening of the Farallon plate beneath western North America during the late Mesozoic to mid-Cenozoic.

The Mesozoic shales beneath the thrust were deposited about 90 m.y. ago within the Western Interior Seaway, which extended north–south through western North America (Fig. 21.6C). By about 70 Ma, Laramide compression reached inland to the Montana region, squeezing the rocks into folds before thrusting the Belt Supergroup and all younger rocks eastward over and above the late Mesozoic shales. So the Glacier region was submerged beneath the Western Interior Seaway about 80--90 Ma but had transformed into mountainous highlands by about 60–70 Ma, a radical transformation over a relatively short period of geologic time.

In Glacier National Park, the rugged landscape we see today is a product of Ice Age glaciation combined with differential erosion, both superimposed on Laramide folded and thrusted rocks. Limestones and quartzites of the Belt tend to form steep cliffs, whereas siltstones and shales weather more readily, forming slopes between the cliff-formers. The eastern front of the Lewis Thrust, marked by the high elevations of resistant Belt rock above the easily weathered Mesozoic shales, accounts for the abrupt rampart of the Northern Rockies rising from the prairie to the east (Fig. 22.3).

A simplified east–west cross section spanning Glacier National Park illustrates the gigantic sheet of rock that slid eastward along the thrust surface one earthquake at a time in response to Laramide compression (Fig. 22.11). The Lewis thrust plane juxtaposes rock about 1600 m.y. old over rock only 90 m.y. old. Also shown in the cross section is a

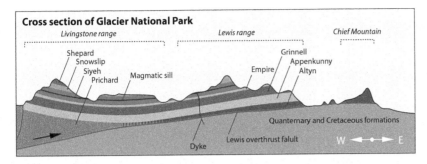

Cross section of Glacier National Park

Livingstone range   *Lewis range*   *Chief Mountain*

Shepard
Snowslip
Siyeh   Grinnell
Prichard   Magmatic sill   Empire   Appenkunny
Altyn

Quanternary and Cretaceous formations

Dyke   Lewis overthrust falult   W ← • → E

Figure 22.11 East–west cross section across Glacier National Park. The farthest extent of thrusting is at the east edge of Chief Mountain, an isolated remnant of Belt rocks left detached from the main mass by erosion. Note the broad synclinal shape to the Belt rocks above the thrust.

broad concave-upward fold known as a **syncline**. The Akamina syncline extends northward through Glacier National Park and likely formed contemporaneously with thrusting. The trend of the syncline loosely follows the continental divide and can be seen by anyone traversing Going-to-the-Sun Road. Along the west side of the park, rocks of the Belt Supergroup tilt gently toward the east, whereas along the east side of the park the rocks tilt gently westward. The axis of the syncline where the tilted rocks converge occurs just to the west of Logan Pass, midway along Going-to-the-Sun Road. Mountains composed of stacked sheets of thrusted and folded sedimentary layers like those of the Northern Rockies at Glacier National Park are called **fold-and-thrust belts**. This style of mountain architecture is common though the inland parts of the North American Cordillera and many mountain belts worldwide. We'll encounter fold-and-thrust belts once again in Part 9 on national parks of the Appalachians.

The cross section (Fig. 22.11) simplifies the bigger picture of thrust-faulted mountains, however, because it only shows the configuration of units visible along the landscape. What is not shown are the kilometers of younger rock that have been eroded above the remaining slab of Belt rocks. Moreover, the architecture of the rocks deep beneath the surface is not shown. Detailed mapping throughout the Northern Rockies, combined with seismic imaging in the region related to oil exploration, reveals a thick stack of thrust sheets superimposed one above the other buried beneath the surface. The Lewis Thrust is just one of several thrusts that juxtapose slabs of rock, thickening the crust and creating mountainous highlands. Many mountain ranges worldwide exhibit the vertical stacking of thrust sheets by thrust faulting, including the Himalayas, the Alps, and the Appalachians.

Glaciation and Landscape Evolution in Glacier National Park

As with all of the other high-elevation national parks, Ice Age glaciation created the modern landscape we enjoy today at Glacier National Park, sculpting the mountains in the same manner as at national parks of the Cascades, the Sierra Nevada, and throughout the Rocky Mountains. All of the main glacial features are evident at Glacier, including broad U-shaped valleys, cirques, arêtes, cols, horns, hanging valleys, moraines, and hundreds of glacial lakes. Many of these are clearly expressed on the aerial image at the beginning of this chapter (Fig. 22.1).

Ice Age glaciers came and went multiple times in the Glacier region, but the most recent glaciation erosionally modified the effects of previous glacial events and is thus the most evident in the park. This is the Pinedale glaciation, which reached its peak ~20 ka during the Last Glacial Maximum, the same event as in Rocky Mountain National Park. Virtually the entire park was covered in ice, with only the highest peaks and ridges rising above as nunataks. These islands of rock are visible today above the treeline as the highest glacial horns and the elongate crests of arêtes.

The landscape in the park was carved primarily by alpine glaciers, but the southern terminus of two huge *continental ice sheets* reached the surrounding region near the park ~20 ka (Fig. 22.12A). Today, continental ice sheets are restricted to Greenland and Antarctica, but during glacial phases of the Pleistocene, continental ice sheets extended southward across Canada into the northern latitudes of the United States. The Laurentide ice sheet spread out from its core near Hudson Bay, whereas the Cordilleran Ice Sheet ranged from Alaska to Montana, growing outward from the mountainous spine running along its north–south axis. The two ice sheets were at times separated by an ice-free corridor that ran along the east side of the Canadian Rockies. The ice sheets attained thicknesses of 3–4 km (2-2.5 mi), thinning to a few hundred meters near their southern terminus. Imagine the imposing wall of ice that loomed above the mammoths, bison, and saber-tooth cats that roamed the region at the time. The vast ice sheets waxed and waned multiple times with global changes in climate through the 2.6 m.y. of the Pleistocene.

Near Glacier National Park, alpine glaciers spread laterally outward from their heads along the continental divide. To the west, they merged with an elongate tongue of the Cordilleran ice sheet (Fig. 22.12B). To the east, alpine glaciers spread into the southern mouth of the ice-free corridor, several tens of kilometers distant from a lobe of the Laurentide ice sheet. Pinedale-age glaciation ended in the region by 10 ka, with alpine glaciers retreating to their cirques along the continental divide. These remnant glaciers were entirely gone by 6 ka as Earth entered a warm and dry phase of Holocene climate. These dates of deglaciation are well constrained by careful mapping combined with tree-ring dating, radiocarbon dates of organic matter in moraines and glacial lakes, and the fortuitous presence of datable volcanic ash layers from eruptions in the Cascades to the west. The 25 remaining active glaciers in the park today likely formed during Little Ice Age global cooling between about 1350 and 1850 CE (Fig. 22.13).

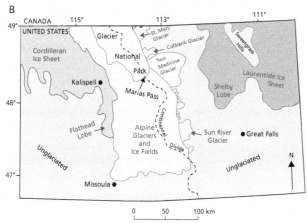

Figure 22.12 A) Maximum extent of continental ice sheets during Last Glacial Maximum ~20 ka shown in blue. Older ice sheets (light purple) reached farther north and south. B) Extent of alpine glaciers and the southern terminus of continental ice sheets in the area around Glacier National Park.

former locations of tributary glaciers that merged with the main glacier whose deeply carved valley is now filled by Saint Mary Lake. The main glacier, being more massive, eroded deeply into the underlying rock. The smaller tributary glaciers scoured less deeply, meeting the main glacier along its upper surface. As the ice melted, the now empty tributary valleys were left hanging high above the floor of the main valley.

Going-to-the-Sun Road follows the glacially sculpted U-shaped valleys of McDonald Creek in the west and St. Mary Valley in the east. The two valleys meet at Logan

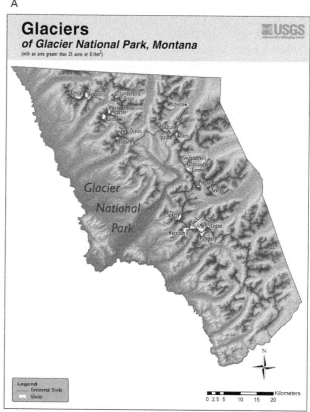

Figure 22.13 A) Map of active glaciers in Glacier National Park. Note how most are constrained to north-facing cirques along the continental divide. B) Sperry Glacier, the second largest in the park after Grinnell Glacier.

Visitors to the park can see many glacial landforms along Going-to-the-Sun Road, beginning with the enormous glacial lakes that anchor each end (Fig. 22.2). On the west, Lake McDonald (Fig. 22.14) is bound on its northwestern and southeastern sides by *lateral moraines* whose linear crests reach about 470 m (1550 ft) above lake level. The lake itself is about 140 m (470 ft) deep, so the combined elevations of over 610 m (~2000 ft) provide an estimate of the total ice thickness that filled the valley during the Last Glacial Maximum. Glacial plucking along the base of the glacier gouged the trough, now filled by the lake, while deposition of glacial debris along the margins formed the lateral moraines.

At the east end of Going-to-the-Sun Road, Saint Mary Lake exhibits classic *hanging valleys* emanating from its southern margin (Fig. 18.20). These valleys mark the

Figure 22.14 Lake McDonald is about 140 m (470 ft) deep and is bounded along its flanks by tree-covered moraines.

Figure 22.15 The continental divide runs along the spine of the Garden Wall, an arête that shields the Grinnell and Salamander glaciers on its northeast flank.

like sandpaper on the underlying rock. The glacial silt is discharged with meltwater and is fine enough to remain suspended in lakes where it reflects and refracts light to produce the characteristic bluish sparkle.

Many glacial valleys in Glacier National Park feature chains of lakes connected by narrow channels that funnel snowmelt from one lake to the next (Fig. 22.16). These **paternoster lakes** may form within the lowest, glacially plucked depressions of the glacial valley, or they may back up behind moraines that block downstream flow. These are just a few of the obvious glacial landforms visible from Going-to-the-Sun Road and the road to the Many Glacier area, but countless others await the intrepid explorer in less visited corners of the park.

Modern Glaciers and Climate Change in Glacier National Park

It's ironic that a national park named for its spectacular glacial landscape will likely have no remaining glaciers by the year 2030, but extrapolation of rates of glacial retreat measured over the past few decades by the U.S. Geological Survey and National Park Service suggest that this is likely to occur. In 1910, when Glacier National Park was first established, over 150 glaciers were identified in the park. Today, there are 37 named glaciers, but only 25 are considered to be actively moving, with the remainder reduced to mere snowfields.

Grinnell Glacier, the largest in the park, has retreated significantly over the past several decades, accompanied by the growth of a large meltwater lake (Fig. 22.17). Of all the national parks in the coterminous United States, Glacier is ideally suited to maintain alpine glaciers due to its

Pass, a *col* formed where the heads of the two glaciers merged along the continental divide. From the Logan Pass Visitor Center, you can see the pyramid-shaped *horns* of Reynolds Mountain and Clements Mountain as well as the spine-like arête of the Garden Wall (Fig. 22-15). On the western approach to Logan Pass, Bird Woman Falls plummets from a textbook hanging valley.

The abundant snowmelt supplies many lakes in the park with water that is typically too cold for the growth of plankton, preserving a remarkable clarity. Other lakes are charged with **glacial flour**—fine particles of silt that imparts a milky turquoise color to the waters. The powdery glacial flour is derived from the abrasive action of sand- and gravel-size sediment entrained along the base of glaciers, which act

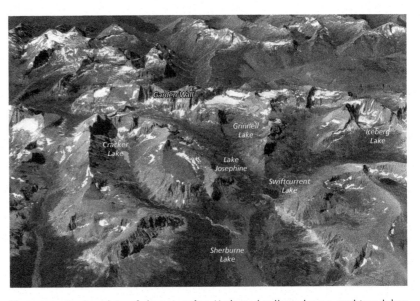

Figure 22.16 A variety of cirques, arête, U-shaped valleys, horns, and tarn lakes are evident in this oblique view westward in the Many Glacier area of the park. Chains of glacial lakes connected by narrow outlets (such as Grinnell to Sherburne) are called paternoster lakes.

1938   1981   1998   2009

Figure 22.17 Oblique views of Grinnell Glacier showing the recession of the glacier's area over the past several decades. The Salamander Glacier on the upper ledge is shielded from the sun by a section of the Garden Wall and remains relatively stable, although it is thinning.

Figure 22.18 Hiking the Highline Trail with Going-to-the-Sun Road in the middle distance and the horn of Reynolds Mountain in the background.

Key Terms

anticline
Archean
chlorite
cyanobacteria
fold-and-thrust belt
glacial flour
hematite
metasedimentary
mudcracks
offset
paternoster lakes
Proterozoic
ripplemarks
Rodinia
syncline
thrust fault
triple divide

Related Sources

Alt, D. D., and Hyndman, D. W. (1973). *Rocks, Ice and Water—The Geology of Waterton-Glacier Park*. Missoula, MT: Mountain Press, 104 p.

Raup, O. B., Earhart, R. L., Whipple, J. W., and Carrara, P. E. (1983). *Geology along Going-to-the-Sun Road, Glacier National Park, Montana, West Glacier, Montana*. West Glacier, MT: Glacier Natural History Association.

Thornberry-Ehrlich, T. (2004). *Glacier National Park Geologic Resource Evaluation Report*. Natural Resource Report NPS/NRPC/GRD/NRR—2004/001. Denver, CO: National Park Service.

National Park Service—Geology, Glacier National Park Accessed February 2021 at http://www.nps.gov/glac/learn/education/geology.htm

National Park Service—Geologic Formations, Glacier National Park Accessed February 2021 at http://www.nps.gov/glac/learn/nature/geologicformations.htm

high latitude that permits cool summers and its high elevations that arrest moist weather systems moving in from the Pacific. But the glaciers are fighting a losing battle, as are most glaciers worldwide. As the climate warms, the rate of ablation at the toe of glaciers exceeds the rate of accumulation at the head of glaciers. The change in the mass balance results in the progressive shrinkage of the volume of glaciers worldwide.

In Glacier National Park, the average rate of retreat ranges from 6 to 17 m (20 to 56 ft) per year. A warming planet is a reality that we have to deal with, first and foremost by reducing the amount of human-related greenhouse gases in the atmosphere. In the meantime, as humanity grapples with a rapidly changing climate, put on your boots and visit Glacier National Park to revel in its awe-inspiring vistas and to contemplate the lost world of the Pleistocene (Fig. 22.18).

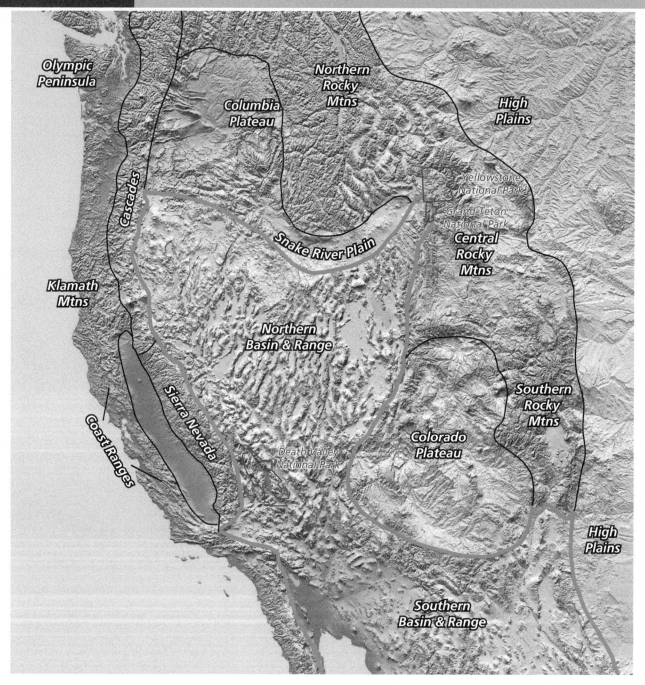

Figure 23.1 Physiographic provinces of the American West, with Basin and Range Province highlighted. Locations of two national parks addressed in Part 5 outlined, plus Yellowstone National Park, addressed in Part 7.

The largest geologic province of the North American Cordillera in the coterminous United States is the Basin and Range, an expanse of mountains and valleys that extends 700 km (430 mi) from the eastern escarpment of the Sierra Nevada to the Wasatch Mountains of central Utah (Fig. 23.1). The province reaches as far north as southern Oregon and southern Idaho and extends far south into Mexico. It encompasses the entire state of Nevada, western Utah, southeastern California, southern Arizona, southern New Mexico, and West Texas.

The physiography of the region is marked by hundreds of elongate mountain ranges and intervening basins oriented more or less north–south (Fig. 23.2). The alignment of basins and ranges throughout the province has been variously compared to "an army of caterpillars marching toward Mexico" or "a fleet of battleships in a desert sea." These metaphors aptly describe the striking uniformity of the terrain evident in high-altitude images of the province. As you'll find out, the north–south orientation relates to the east–west direction of *tectonic extension* that has dominated the region since the mid-Cenozoic. The two parks that we'll visit in Part 5 are Death Valley and Grand Teton, located at opposite corners of the province. But other fascinating national parks in the Basin and Range include Great Basin, Guadalupe Mountains, Carlsbad Caverns, and Big Bend.

Many people develop a perception of the Basin and Range Province during the interminable drive along Route 80 that connects Reno, Nevada, to Salt Lake City or along Route 50 across Nevada, justifiably labeled "The Loneliest Road in America." The traveler drives for kilometers across windswept, sage-covered valleys, then gradually climbs the slopes of ranges to a pass before descending, starting the rhythmic pattern over again (Fig. 23.3). Seen from a car, the terrain looks forbidding and empty, leading

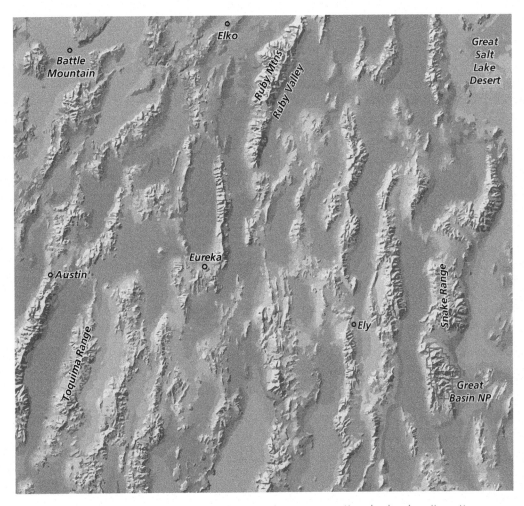

Figure 23.2   Relief map of Basin and Range centered over eastern Nevada showing alternating ranges (whitish-gray) and basins (browns and greens).

Figure 23.3 View west across sagebrush-covered Ruby Valley to Ruby Mountains, Nevada - a typical vista within the Basin & Range province.

many to never consider leaving the safety of the highway and the isolated but convenient towns along the way.

But what the highway traveler misses is a place of otherworldly beauty marked by serene silence and stark vistas. The higher elevations of ranges are surprisingly lush, studded with pinyon–juniper forests at intermediate elevations that give way to pine forests above. Some of the highest ranges, like the White Mountains of eastern California, unexpectedly exhibit alpine-tundra ecosystems along the crest of the range.

The arid *intermontane valleys* are less vegetated but equally varied in their ecology and diversity of environments. Normally dry washes may rage with flash floods charging out from adjacent mountain canyons, providing just enough moisture for riparian vegetation to live another few years. Springs pop out of the ground in unexpected places, nourishing an isolated wetland ecosystem on an otherwise arid plain. Many of the springs are warm, perfect for a luxurious soak after a hard day exploring, while other springs are far too hot and better left alone. The adventurous way for the prepared traveler to see the Basin and Range is to drive the paved backroads and washboard gravel roads that cross the basins and reach deeply into remote parts of the ranges. That's when you'll develop a deeper appreciation of the inherent natural beauty of the region.

The Basin and Range is rugged high country with an average elevation of ~1400 m (4600 ft). Elevations reach a low of -86 m (-282 ft) in Death Valley to a high of 4344 m (14,252 ft) at White Mountain Peak, the highest point fully in the province. Basins typically reach elevations well above 1200 m (~4000 ft), with ranges commonly attaining heights above 3000 m (~10,000 ft). The Basin and Range spans over 15° of latitude, so it's an oversimplification to summarize the hydrology and climate of the entire province as if it was a single entity. We'll briefly address the southern Basin and Range, then focus on the northern part because both Death Valley and Grand Teton national parks are located there.

Low-elevation deserts such as the Mojave, Chihuahua, and Sonora characterize the southern Basin and Range, which extends from southeastern California to West Texas. Two major rivers drain the ranges and basins in the southern part of the province. The Colorado River forms the California–Arizona border after it drops off the Colorado Plateau on its way to the Gulf of California (Fig. 2.4). The Rio Grande flows south through central New Mexico, then forms the Texas–Mexico border in the southeastern part of the province.

The northern Basin and Range centered over Nevada is also arid, but it's higher in elevation and cooler than its southern counterpart. Variations in climate in the northern Basin and Range tend to differ with elevation and latitude, but in general the aridity is governed by the **rain shadow** created by the Sierra Nevada to the west that blocks Pacific weather systems from reaching inland. Prevailing winds from the ocean transport moisture that rises along the western slopes of the Sierra where it cools and condenses to fall as rain and snow. The remaining air mass is sapped of water as it descends the steep eastern side of the Sierra, creating a shadow of aridity in the Basin and Range. Much of the precipitation in the northern Basin and Range falls as snow in the higher elevations, which melts through the spring and summer to provide water for streams and rivers.

The northern Basin and Range loosely overlaps in area with the **Great Basin**, a vast hydrologic system marked by **internal drainage** where rivers remain trapped without an outlet to the ocean (Fig. 23.4). Countless streams direct snowmelt and rainfall from the ranges into adjacent basins where the streams merge to form small rivers. The discharge from these rivers eventually drains into closed-basin lakes, or it simply seeps into the ground in the lowest parts of basins called **sinks**. Closed-basin lakes in the Great Basin include the Great Salt Lake in Utah, Pyramid Lake in Nevada, and Mono Lake in California.

The boundaries of the Great Basin are not coincident with the northern Basin and Range. For example, Lake Tahoe is geographically part of the Sierra Nevada Province, but it drains via the Truckee River to Pyramid Lake in Nevada and is thus hydrologically part of the Great Basin watershed. The largest watershed within the Great Basin is that of the Humboldt River, which flows east to west for 530 km (330 mi) across Nevada. Along the way, the Humboldt becomes progressively smaller and more saline as it loses water to evaporation during its passage across the arid landscape. The Humboldt ends by simply disappearing via seepage and evaporation in the Humboldt Sink (Fig. 23.5). Many sinks and valleys in the province have salt-encrusted dry lakebeds called **playas** spanning the lowest areas, evaporative remnants of what were once salty lakes that existed during wetter climates. The internally drained Great Basin watershed, physically and meteorologically separated from the Pacific by the Sierra Nevada, is arguably the most distinguishing characteristic of the northern Basin and Range geologic province (Fig. 23.6)

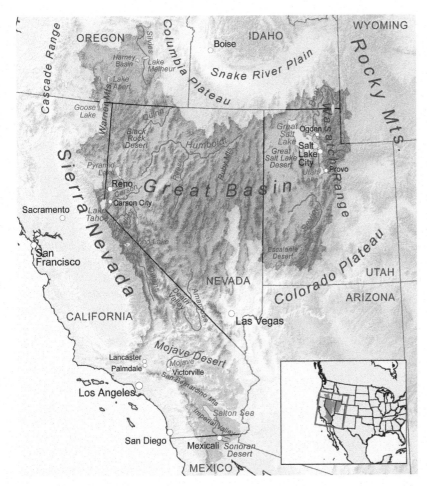

Figure 23.4 The Great Basin is characterized by internal drainage where streams and rivers have no outlet to the open ocean, but rather flow into closed-basin lakes and dry sinks. The boundaries of the Great Basin do not exactly coincide with the northern Basin and Range Geologic Province.

Figure 23.5 View northward across the dry playa of the Humboldt Sink, the final destination for the Humboldt River. During wet years, a shallow pan of water forms in the lowest part of the basin. The sink is a remnant of Pleistocene Lake Lahontan.

Origin and Age of Rocks Composing the Basin and Range

Rocks in the northern Basin and Range are dominantly sedimentary and igneous, with lesser amounts of metamorphic rocks. They are arranged, as you might expect,

as a Precambrian *crystalline basement* overlain by a veneer of Paleozoic and Mesozoic sedimentary rock and Cenozoic volcanic rock. The Precambrian rocks are *Proterozoic* in age and reflect the accretionary assembly of the North American continent described in earlier chapters. Some of the Paleozoic rocks are part of *accreted terranes*, originally formed far out in the ancestral Pacific Ocean, then tectonically rafted in to dock against the western margin of the continent. Most of the Paleozoic and Mesozoic sedimentary rocks, however, were deposited along the broad continental margin during the numerous transgressions and regressions that occurred in response to changes in sea level through time. The Paleozoic rocks are related to those of the upper Grand Canyon, whereas the Mesozoic rocks were deposited contemporaneously with those that dominate parks on the Colorado Plateau. Many of these sedimentary layers originally exhibited broad *lateral continuity*, but today they are geographically fragmented by mountain-building and erosion into isolated exposures along the flanks of ranges.

The Cenozoic volcanic rocks deserve special consideration because they reflect one of the most voluminous and violent phases of eruptions in Earth history. Between ~43 and 18 Ma, the American West was subject to a prolonged phase of explosive *caldera eruptions* known as the **ignimbrite flare-up**. Ignimbrites are a type of high-silica *volcanic tuff* formed from powerful pyroclastic flows that travel far distances as a turbulent, ground-hugging cloud that eventually settles out as a blanket of hot ash and volcanic debris. The remnant volcanic heat within the deposit causes it to rapidly weld together, forming the solid rock of the ignimbrite. (See Chapter 12 on Cascades volcanism for a review of volcanic processes and products and Chapter 15 for a review of caldera eruptions.)

Seventeen of the forty largest eruptions identified in the entire world occurred during the flare-up in the American West. Some of the most violent volcanism occurred in Colorado where multiple calderas spewed enormous volumes of pyroclastic debris across the mid-Cenozoic landscape, burying the eroded remnants of the Laramide Rocky Mountains. Other major caldera complexes and related ignimbrite layers are recognized in the Basin and Range of Nevada and Utah.

Recent eruptions in the Cascades, like those from Mount Lassen and Mount St. Helens, are puny in

Figure 23.6 Major North American watersheds with main directions of water flow shown by arrows.

thick volcanic layers manifest a series of widespread timelines, and many of the deposits have been radiometrically dated, constraining the age of Basin and Range faulting. The tectonic controls on the ignimbrite flare-up will be addressed below.

Extensional Mountain-Building in the Basin and Range

The geologic provinces and the national parks that we've visited so far in the book have been tectonically related to *compressional stress*. Uplift in the Cascades is related to volcanism along the compressional Cascadia subduction zone. Uplift of the ancestral Sierra Nevada, the Rocky Mountains, and the Colorado Plateau occurred in response to compression induced by subduction of the Farallon plate. In contrast, here in the Basin and Range, the mountains are *tilted fault-blocks* that formed in response to *extensional stress* generated as the tectonic regime of the American West evolved through time. Lithospheric stretching in the Basin and Range began around 16 Ma and continues to this day. Uplift of tilted fault-block mountains was originally addressed in Chapter 17 on the modern Sierra Nevada, but we'll expand on the controls and mechanisms underlying extensional uplift in this chapter.

The Basin and Range is subject to extensional tectonic forces where the lithosphere is stretched in an east–west fashion. The rigid upper crust breaks along *normal faults*, with a block of rock on one side rising upward, while the block of rock on the other side drops downward, typically coincident with earthquakes (Figs. 5.12, 8.12, 17.11). Through time, as adjacent masses of rock move relative to one another along normal faults, a characteristic pattern of elongate ranges and intervening basins evolves perpendicular to the orientation of extensional stress (Fig. 23.7). The landscape formed by this process is called *horst-and-graben* topography, introduced in Chapter 8 on Canyonlands National Park. The upraised blocks are called *horsts* and form elongate mountain ranges, modified by erosion. The downdropped fault blocks result in linear rift valleys called *graben* that receive the erosional debris derived from the adjacent uplifted horst blocks. This is the (over) simplified explanation for the hundreds of basins and ranges that characterize the province. The actual geometry and mechanics of the normal faulting that controls

comparison, thousands of times smaller than those of the flare-up. Even the eruption of Krakatua in Indonesia in 1883, which is among the most catastrophic eruptions in modern history, was a mere burp compared to the magnitude and ferocity of those that occurred during the flare-up. Erupted in deafening blasts, superheated clouds of gas-charged pyroclastic debris flowed at over 100 km/hr (60 mph) along the ground surface, incinerating every plant and animal in their path. Pyroclastic flows and other types of volcanic eruptions occurred countless times over the several million years of the flare-up event.

When the prolonged volcanism of the flare-up finally came to an end, it left a thick blanket of volcanic lava and tuff across the surface of the future northern Basin and Range and nearby regions. The mid-Cenozoic landscape of Nevada during the flare-up was a highland, based on evidence from plant fossils. The terrain was likely a gently rolling upland plateau due to the smoothing cover of volcanic debris. The widespread volcanic deposits were fragmented later in the Cenozoic during Basin and Range extensional tectonism but have been pieced together by careful mapping and chemical fingerprinting. These

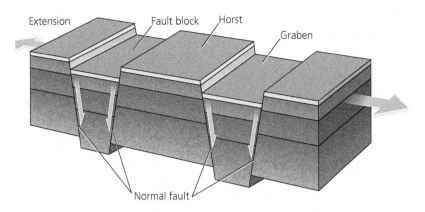

Figure 23.7 Horst-and-graben topography formed by normal faulting. The masses of rock between faults are called "fault blocks." Broad arrows depict extensional tectonic stress.

the landscape of the Basin and Range is more complex and requires a brief discussion of the way rocks respond to tectonic forces.

Types of tectonic stresses include compression, extension, and shear, each of which was introduced in Chapter 1. The physical response of a body of rock to stress is called **deformation**. A few of the styles of deformation you've read about so far in our tour of national parks include *joints, faults, folds,* and *metamorphic foliation.* When a body of rock is exposed to tectonic stresses, it will respond by bending, breaking, or flowing. If a body of rock breaks, say along a joint or a fault, then that rock exhibits *brittle* behavior. Take a *cold* stick of chewing gum and flex it until it snaps in half—that's an example of a brittle material responding to stress. If a body of rock under stress bends into a fold or flows to create a foliated texture, then that rock exhibits **ductile** behavior. Take a *warm* stick of chewing gum and flex it until it bends without breaking— that's ductile behavior. Squeezing and rolling the warm stick of gum between your hands until it flows into a rounded blob is also an example of ductile behavior. (Ductile deformation is also referred to as *plastic* deformation, addressed in earlier chapters in the context of mantle flow, salt migration, and metamorphism.)

Whether a body of rock behaves in a brittle or ductile fashion depends on a variety of factors, including temperature, pressure, rate of deformation, and composition and strength of the rock under stress. Rock buried deep in the crust is warm and subject to the considerable pressure of overlying rock. It typically responds to stress by ductile bending or plastic flow. Rock composing the upper crust is cooler and under less overburden pressure. It typically responds to tectonic stress by brittle breakage along joints and faults. The boundary within the crust that separates brittle rock above from ductile rock below is called the **brittle-ductile**

transition and is located about 13–18 km (8–11 mi) beneath the surface. (Remember that continental crust is on average about 35–40 km (22–25 mi) thick.)

What does all this have to do with faulting and uplift of mountains in the Basin and Range? You may be asking yourself what happens to normal faults at depth. Do they just end? When a block is uplifted, what fills in the gap beneath? What does a downdropped fault block fall into? Figure 23.8 provides a more realistic image of the formation of tilted fault-block mountains and their intervening basins. Above the brittle-ductile transition zone, rocks act in a brittle manner, such as breaking along faults in response to extensional stress. Below the transition zone, the hotter rock behaves in a ductile fashion, bending or flowing in response to extensional stress. In the Basin and Range, normal faults typically are steeply angled near the surface, but they curve and flatten at depth toward the brittle-ductile transition. Individual fault-blocks tend to rotate along the curved fault plane, with multiple faults merging into a single master fault along the transition zone 13–18 km (8– 11 mi) beneath the surface. Normal faults with planes that curve and flatten with depth are called **listric normal faults**. Geologists know this geometry exists deep beneath the ground due to seismic reflection profiling, a technique that uses the echoes of sound waves to create a ghostly image of the arrangement of rock beneath the surface.

Rotation of fault-blocks along listric normal faults occurs episodically as faults "rupture" in response to extensional stress, triggering earthquakes. During each quake, the fault block on the downdropped side will rotate as it slides, incrementally creating a variety of features. (1) The "uprotated" part of the fault-block tends to stand high and form ranges, whereas the "downrotated" part of the fault block tends to form **half-graben basins**

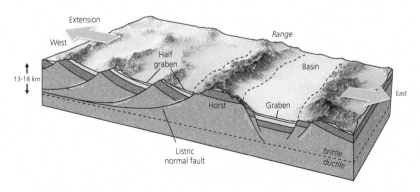

Figure 23.8 Fault-blocks rotate along listric normal faults, creating tilted fault-block mountain ranges along the "uprotated" corner and basins above the "downrotated" corner. Dashed lines show the trace of the fault along the boundary between ranges and adjacent basins. The depth where the fault planes flatten is the brittle-ductile transition. Broad arrows show extensional tectonic stress directions.

Figure 23.9 Sedimentary layers dip away from the viewer in this tilted fault-block mountain range, Grapevine Mountains, Death Valley National Park. Episodic movement along a normal fault hidden beneath the alluvial fans at the base of the range exhumes the rocks to the surface, while simultaneously dropping the adjacent block downward to form a sediment-filled basin. A gravel plain marks the surface of the basin in this image.

that fill with sediment. So a single tilted fault-block may form both a range and an adjacent basin. (2) Sedimentary or volcanic strata within individual fault-blocks may have been horizontally layered prior to faulting but will progressively rotate with each earthquake along the fault so that the bedding tilts at an angle (Fig. 23.9). In this manner, rock that was once deeply buried becomes exhumed within a mountain range as it rotates upward through time.

(3) Ranges may develop an asymmetric shape, with a steep escarpment forming along the exposed fault plane and a gentler slope forming the opposite side. (The Sierra Nevada has this geometry due to rotation along the fault system bounding the eastern flank.) (4) The sediment filling the half-graben basin may be wedge-shaped, with the thickest accumulation near the fault plane due to the episodic creation of space as the basin drops a few meters or so during earthquakes (Fig. 17.11). Sediment eroded from the adjacent range readily fills the elongate trough at its base. All of these four features are prominently displayed in Death Valley and Grand Teton national parks, as well as throughout the Basin and Range.

Recall that "uplift" is simply the raising of sections of Earth's crust to higher elevations. The higher the elevation, the more prone the rock is to attack by the inevitable effects of erosion. If the rate of uplift exceeds the rate of erosion, the mountains remain high and provide a source of sedimentary debris that collects in adjacent basins (Fig. 23.10). If the rate of uplift slows so that erosion becomes dominant, the elevations of ranges may decrease as they bury themselves in their own rubble. The Mojave subprovince of the Basin and Range in Southern California is characterized by erosion outpacing uplift, resulting

in relatively low-elevation ranges rising above broad, sediment-filled basins.

Countless earthquakes accompany the active uplift of mountains in the Basin and Range. The relationship between normal faulting and *seismicity* was addressed in Chapter 17 on uplift of the modern Sierra Nevada and will only briefly be reviewed here. The massive blocks of brittle rock on either side of normal faults in the Basin and Range remain locked for long periods of time due to friction along the fault plane. Extensional stress builds over hundreds to even thousands of years until the frictional strength holding the two blocks together is exceeded. Once this threshold is breached, the fault will rupture, releasing pent-up energy in the form of seismic waves. As the pulses of seismic energy reach the surface, the earthquake announces its arrival through **groundshaking**. Rupture along the fault plane results in the abrupt downdropping of the basin and the relative uplift of the range, perhaps creating a **fault scarp**, a sharp step in the topography that may extend a few tens of kilometers along the base of the range where the fault plane breaches the surface. Fault scarps are common surface features in both Death Valley and Grand Teton national parks due to their active seismicity.

Depending on the magnitude of the earthquake, movement along the fault may be as little as a few centimeters or as large as several meters. So in order to raise mountains thousands of meters above the adjacent basin, tens of thousands of significant earthquakes must have occurred along each range-bounding fault in the province since the mid-Cenozoic. Basins and ranges in the province today are actively growing and earthquakes are common, an expression of geologic dynamism not commonly witnessed on human time scales.

Tectonic Model for Extension in the Basin and Range

Geophysical measurements reveal that the modern Basin and Range is characterized by relatively thin crust (~30km/19 mi, Fig. 1.3) and a high rate of **heat flow** emanating from the surface. Earth emits heat everywhere, but the amount varies with geologic conditions. Where the crust is thinnest in central Nevada, heat escapes from the surface at twice the global average. Both the thin crust and high heat flow reflect stretching of the lithosphere and the rise of hot, plastic asthenospheric rock beneath the province over the last 16 m.y. or so (Fig. 23.11). The hot rock of the asthenosphere is buoyant and pushes the overlying lithosphere

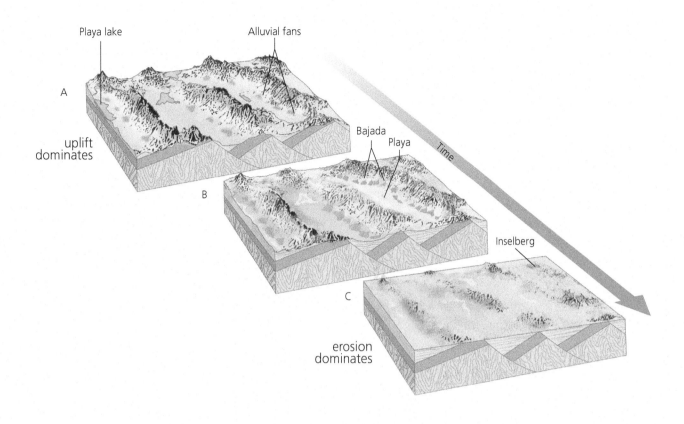

Figure 23.10 The elevation of mountains in the Basin and Range is determined partly by the rate of uplift relative to the rate of erosion.

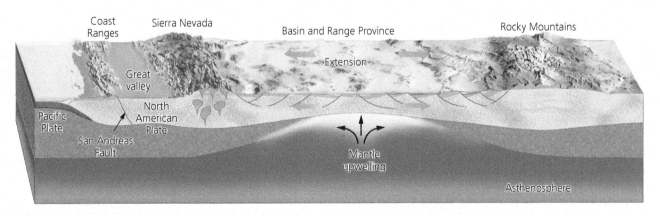

Figure 23.11 Extension in the Basin and Range over the past 16 m.y. is related to the rise of hot, buoyant asthenosphere that flexes and stretches the overlying lithosphere. The thin crust, high heat flow, and high elevations of the province are related to the bulge of plastically flowing asthenosphere beneath.

upward to create the high average elevation of the northern Basin and Range. The upward flexure creates extensional stresses in the brittle rock of the upper crust, generating normal faults, earthquakes, and the formation of basins and ranges. Estimates of total extension across the northern Basin and Range vary, but the consensus seems to converge

on 100 percent; the distance from Reno to Salt Lake City has doubled in the 16 m.y. of stretching in the region.

Active extension can be measured directly today by the **global positioning system** of satellites (GPS) and averages a little over 1 cm/yr. (One cm/yr translates to 10 m/k.y. and 10 km/m.y., so given enough geologic time, two locations can separate by a significant distance.) In addition to real-time GPS measurements, evidence for active extension takes the form of a constant beat of small to moderate earthquakes, steep fault scarps along the base of mountains, thousands of hot springs, and youthful volcanism occurring over the past few hundred years. Add the quantifiable characteristics of a thin crust and high heat flow and it's evident that the Basin and Range is an active *continental rift system*, somewhat comparable to the East African Rift described in Chapter 1. It remains an open question as to whether the American West will split apart in the next few tens of millions of years because continental rifting doesn't always continue to the point of full separation.

TRANSITION FROM COMPRESSION TO EXTENSION

Several tectonic models for the origin of extensional uplift in the Basin and Range have been proposed to explain the physical characteristics described above. To provide context, we'll begin our geologic narrative of the evolution of the Basin and Range in the mid-Mesozoic around 100 Ma (Fig. 23.12A). At that time, volcanoes lined the crest of

the Sierra Nevada, fed by multiple magma chambers supplied with molten rock from the subducting Farallon plate far below. Through the rest of the Mesozoic and into the Cenozoic, a wave of compressional tectonism swept eastward from the ancestral Sierra, raising a high-elevation *fold-and-thrust belt* throughout Nevada and Utah (see the paleogeographic map, Fig. 17.9).

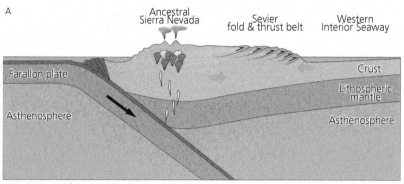

120–80 Ma peak Sierran magmatism

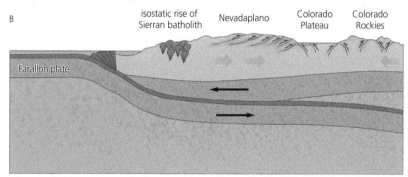

80–40 Ma Laramide orogeny

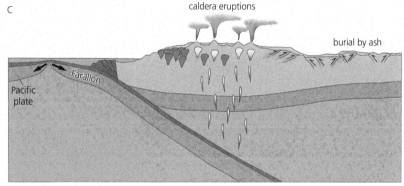

40–20 Ma Ignimbrite flare-up

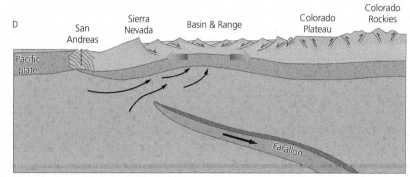

20–0 Ma convergent to transform tectonism

Figure 23.12 A) During the peak of Sierran magmatism, a wave of compressional tectonism swept eastward, forming a fold-and-thrust belt of highlands in what is today Nevada and Utah. The weight of the thickened crust depressed the adjacent crust, bowing it downward to create the Western Interior Seaway. B) With the flattening of the Farallon plate beneath the western North American plate, uplift progressed eastward to the Colorado Plateau and Rocky Mountain provinces. Granite of the Sierran batholith rose isostatically to form the western shoulder of a highland plateau called the Nevadaplano. C) Between ~40 and 20 Ma, the ignimbrite flare-up spread far and wide across the American West. A gradual steepening of the Farallon plate likely triggered the renewed magmatism. To the west, the mid-ocean ridge between the Farallon and Pacific plates migrated toward the subduction zone. D) By about 16 Ma, the Farallon plate had separated from the Pacific plate and continued its downward journey beneath North America. Hot rock of the asthenosphere rose into the gap and flexed the Nevadaplano upward, eventually to form the Basin and Range.

Compressional uplift reached a crescendo between 80 and 40 Ma with the Laramide orogeny and uplift of the Rocky Mountain and Colorado Plateau provinces (Fig. 23.12B). The increase in frictional contact between the Farallon and overriding North American plate generated the compressional stresses of the Laramide orogeny. Recall that ~80 Ma, magmatism in the ancestral Sierra Nevada ended, likely due to a flattening of the angle of subduction of the Farallon plate. The Sierran granitic batholith rose isostatically as the weight of overlying rock was weathered away. The topography of the ancestral Sierra Nevada was likely rolling hills that rose eastward into the fold-and-thrust belt that extended across Nevada and Utah. This upland plateau has been called the **Nevadaplano**, based on similarities with the high-elevation Altiplano of today's Andes.

The widespread Laramide phase of mountain-building ended about 40 Ma, roughly coincident with the outbreak of explosive volcanism of the ignimbrite flare-up (Fig. 23.12C). The catalyst for the flare-up was likely related to an increase in the angle of subduction of the Farallon plate beneath the North American continent, bringing an end to Laramide compression and reigniting the generation of magma. The Farallon plate slowly steepened, as if on a hinge. When parts of the slab reached depths of about 150 km (~90 mi), water would be released that triggered melting of the overlying asthenosphere. Magma generated in this fashion rose upward in a broad swath across the future Basin and Range and parts of the Colorado Rockies. The resulting caldera eruptions blanketed the region in ash and coarser pyroclastic debris, smoothing the topography.

During the mid-Cenozoic, the ignimbrite flare-up was in full volcanic fury from Nevada to Colorado, blanketing the Nevadaplano and the remnants of the Laramide Rockies in pyroclastic debris. Just off the west coast, the East Pacific mid-ocean ridge separating the Farallon plate from the Pacific plate was approaching the subduction zone (Fig. 23.13). The American West was a few hundred kilometers narrower at the time because Basin and Range extension had not yet begun. But the tectonic stage was set for the transition from compressional tectonics to an extensional regime.

Around 28 Ma the mid-ocean ridge came into contact with the western North American plate near what is today the Southern California coast. The hot and buoyant ridge resisted subduction, however, with the

Farallon plate detaching and continuing its descent into the mantle. The ignimbrite flare-up continued to rage on the highlands of Nevada, Utah, and Colorado for several million years as the descending Farallon plate generated magma deep below. With the demise of the Farallon plate along the western continental margin, the convergent tectonic setting that had existed since about 200 Ma came to a close. The Pacific plate now abutted the western North American plate but could not subduct due to the buoyancy of the hot rock along the now defunct mid-ocean ridge. *Transform faults* (Chapter 1) bisecting the jammed mid-ocean ridge began to dominate the plate boundary.

In time, the Pacific plate started to move northwestward along the newly evolved **San Andreas transform fault**, converting what was once a compressional stress regime along the western margin of the United States to one dominated by shear (Fig. 23.12D). Inland, the alleviation of compressional stress allowed for extensional forces to take over. By ~16 Ma, the elevated Nevadaplano began to rift apart, collapsing through the rest of the Cenozoic to become the northern Basin and Range of today. This mid-Cenozoic transition from the long-lived Farallon convergent boundary to the modern San Andreas transform boundary will be addressed in greater detail in Part 6 on parks of the active Pacific Margin of North America.

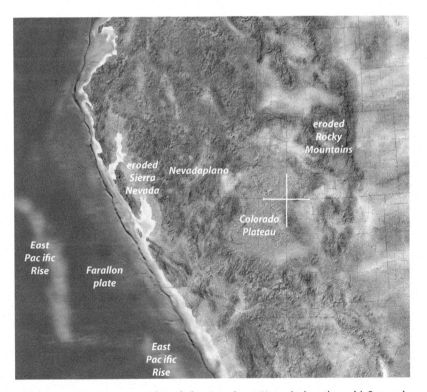

Figure 23.13 Paleogeography of the American West during the mid-Cenozoic ~35–30 Ma. The deeply eroded ancestral Sierra Nevada merged into the Nevadaplano uplands to the east, cenered over Nevada. The Rockies and the Colorado Plateau were undergoing burial by sedimentary debris and ash from the ignimbrite flare-up. The East Pacific Rise spreading center (light blue band) was closing in on the subduction zone. Dark line just off west coast is the Farallon subduction plate boundary.

Extensional stretching in the Basin and Range over the past 16 m.y. triggered ductile flow in the lower crust and brittle faulting in the upper crust, resulting in the landscape of alternating basins and ranges that we see today. Crustal thinning is likely a response to the combination of ductile flow and listric normal faulting that broadened the region. The high elevations and high heat flow of the province are attributed to the upwelling of buoyant asthenosphere through the gap created by detachment of the Farallon plate. (Sophisticated seismic imagery suggests that the tail end of the Farallon plate currently resides several hundred kilometers beneath the American Midwest.) Crustal extension in the Basin and Range continues today, as revealed by the rattle of countless small earthquakes, punctuated by the occasional larger jolt.

In this book, the story of the evolution of the North American Cordillera has been described chapter by chapter as specific events relate to individual provinces, which may make the story seem a bit disjointed and difficult to follow. The entire narrative will be synthesized in Chapter 33, linking together the various events previously addressed. We'll use the individual stories told by each national park and their representative geologic provinces to develop a fully four-dimensional history of the American West, integrating the landscapes we see today with key events in time.

lakes connected multiple intermontane basins in the Basin and Range. In northwestern Nevada, glacial Lake Lahontan covered an area larger than modern Lake Ontario and reached a depth of almost 150 m (~500 ft). The Humboldt Sink, mentioned earlier as the dry terminus of the Humboldt River, is a parched remnant of Lake Lahontan.

In the east, glacial Lake Bonneville covered much of western Utah. Ancient Bonneville shorelines are recognized 300 m (~1000 ft) above Salt Lake City, attesting to the copious supply of water from the surrounding mountains. Ranges projected upward as islands within the deep lake, providing isolated ecospace for resident flora and fauna. The lake drained catastrophically around 14,500 years ago when a natural dam along the northern edge was breached, triggering a massive deluge that flushed into the Snake River drainage basin. The modern Great Salt Lake is a saline puddle a few meters deep, left behind as the climate dried and Lake Bonneville shrank during the current Holocene interglacial phase.

Many playa lakebeds and salt flats veneering the lowest floors of basins in the Basin and Range are the end products of the desiccation of Pleistocene pluvial lakes. As streams drained highlands during the wetter phases of the Ice Ages, they dissolved salty elements from the rocks, transporting them in solution to the pluvial lakes.

Basin and Range during the Pleistocene

The effects of the Pleistocene Ice Ages are expressed in national parks of the Cascades, the Sierra Nevada, and the Rocky Mountains as magnificent glacially sculpted mountains. The influence of the Ice Ages in the Basin and Range is less obvious, but is just as important in creating certain features of the arid landscape we see today. The main effect was that the climate of the Basin and Range was wetter and cooler during phases of the Pleistocene, leading to the expansion of **pluvial lakes** throughout the region (Fig. 23.14). Lakes that form during glacial times within closed, landlocked basins are given the qualifier "pluvial." They fill with stream runoff during wetter phases and shrink or evaporate completely during drier phases. Many basins in the Basin and Range were filled with freshwater lakes during the wettest phases; large Pleistocene mammals roamed the lush shorelines surrounding the water.

At the peak of precipitation during the late Pleistocene, two enormous

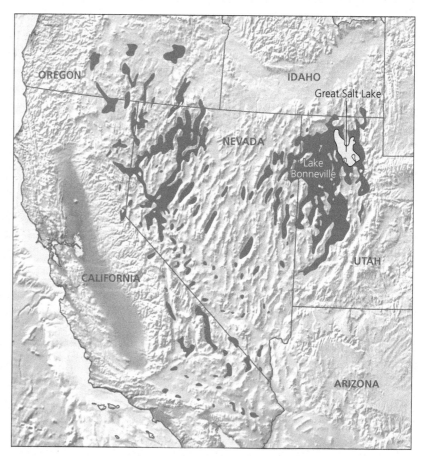

Figure 23.14 Maximum extent of pluvial lakes in the Great Basin during the late Pleistocene.

The water in the lakes maintained its chemical balance as long as the lake water overflowed into an outlet stream. When the climate dried, however, the level of the lakes fell due to enhanced evaporation and a reduced supply of runoff. If the lakes became too low to overflow into an adjoining basin, salts became concentrated by evaporation, causing the remaining water to increase in salinity. With further desiccation and evaporation, the salts precipitated and accumulated along the lake floors. After all the water evaporated through time, a playa lakebed veneered by a layer of salty mud is all that remained. The next chapter will describe the growth and demise of the pluvial lake that filled Death Valley, leaving behind the salt flats that fascinate visitors to the park.

Key Terms

brittle-ductile transition
deformation
ductile
fault scarp
Global Positioning System
groundshaking
half-graben basin
heat flow
ignimbrite
internal drainage
listric normal fault
playa
pluvial lake
rain shadow
sink

Related Sources

Baldridge, W. S. (2004). *Geology of the American Southwest.* Cambridge: Cambridge University Press.

Fiero, Bill. (1986). *Geology of the Great Basin.* Reno: University of Nevada Press.

McPhee, John. (1981). *Basin and Range.* New York: Farrar, Straus, Giroux.

Meldahl, Keith Heyer. (2013). *Rough-Hewn Land: A Geologic Journey from California to the Rocky Mountains.* Berkeley: University of California Press.

Animations of Basin and Range faulting: IRIS. Accessed February 2021 at http://www.iris.edu/hq/files/programs/education_and_outreach/aotm/15/BasinRange_Background.pdf

National Park Service animation of Basin and Range Formation. Accessed February 2021 at https://www.youtube.com/watch?v=Owjt_WJKrBo

24 Death Valley National Park

Figure 24.1 The Panamint Mountains reach over 3350 meters (11,000 ft) in elevation above the dry salt flat of the Badwater Basin in the foreground.

Only a few national parks rival the raw landscape of Death Valley. The extreme aridity inhibits the growth of most plants, fully exposing the rocks of the mountains and the salt flats of the intermontane valleys (Fig. 24.1). The heat can be intense during the summer months, regularly reaching 120°F (49°C) and once reaching 134°F (57°C). Rainfall is scarce, averaging less than 5 cm (2 in) per year. The *rain shadow* effect is powerful in Death Valley, with Pacific moisture having to get past not only the Sierra Nevada, but three other mountain ranges as well before reaching the valley. When rain does arrive, it tends to be short-lived and torrential, occasionally triggering flash floods.

But don't think of Death Valley as only a stark, desolate landscape. Rather, Death Valley is full of surprises. Spring-fed wetlands are scattered about the park, and winter brings snow to the higher elevations. In combination with the occasional rainstorm, enough water exists to nourish a profusion of wildflowers in the spring when the conditions are just right for a visit. Furthermore, the clear, dry air makes for dark, star-filled night skies. Death Valley was named a Gold Tier International Dark Sky Park in 2013, so if you've never actually seen the milky arc of our own galaxy, this is the place to visit.

Located in the Basin and Range Province of southeastern California, the topography of Death Valley is as extreme as its climate (Fig. 24.2). Several rugged mountain ranges trend slightly northwest in the park, including the Black Mountains, Panamints, Funerals, Grapevines, Cottonwoods, and Last Chance Range. Deep basins separate the ranges, including Death Valley proper, Panamint Valley, Saline Valley, and Eureka Valley. The main Death Valley is over 70 km (~45 mi) long and is notable for the white salt flats of the Badwater Basin and sub-sea-level elevations. A northern extension of Death Valley is offset to the northwest, reaches slightly higher elevations, and extends over 50 km (~30 mi) in length. The area of Death Valley National Park is greater than any other park outside of Alaska.

The Badwater salt flat contains the lowest point in North America at −86 m (−282 ft). This topographic nadir is less than 140 km (90 mi) from the highest elevation in the contiguous 48 at Mount Whitney in Sequoia National Park. Closer by in the park, Telescope Peak in the Panamint Range reaches 3368 m (11,049 ft), looming over the adjacent Badwater Basin. The difference in elevation marks the highest relief in the coterminous United States. (The Grand Canyon is "only" 1857 m/6093 ft deep, although it occurs over a shorter distance.) The extreme relief is the underlying reason for the intense heat in Death Valley. The desert surface heats up during the sunny days, but the warm air mass becomes trapped by the surrounding high mountains so that the hot air can't escape at night. Any cooler air descending from the adjacent mountains becomes compressed and heated in the low elevations of the valleys, adding to the hot winds that swirl across the basin floor.

Death Valley is part of the Great Basin, so any water that enters the region only leaves via infiltration into the ground or by evaporation. *Ephemeral streams* drain the mountains, funneling snowmelt and rainwater down to the intermontane basins. The Amargosa River originates in the Amargosa Valley to the east of the park in Nevada, fed by groundwater emanating from springs. It flows southeast, then makes a full U-turn to the northwest before dying in the salt flats of the Badwater Basin (Fig. 24.2). On rare occasions, higher than normal rainfall maintains flow in the Amargosa so that a broad, shallow lake fills the lowest elevations of the basin. The ponded water evaporates quickly, reverting back to the flat, salt-encrusted surface in a matter of weeks.

Virtually all of the characteristics of the Basin and Range Geologic Province are expressed within the expansive desert landscape of Death Valley: tilted fault-block mountains, sediment-filled basins veneered by *playas*, active faults commonly marked by fault scarps, internal drainage, and remnant shorelines of Ice Age pluvial lakes. The rocks and sediments composing the landscape span an astonishing range of time and chronicle almost all of the major events that affected the North American Cordillera. After a discussion of the origin and significance of the diversity of rocks composing the park, we'll address their deformation and uplift into the topography we see today. Then we'll take a tour of the variety of modern desert landforms for which the park is famous.

Origin and Significance of Rocks in Death Valley

Mountain ranges in Death Valley exhibit rocks of almost all ages ranging from Proterozoic metamorphic and igneous rocks over a billion years old to a thick succession of Paleozoic sedimentary rocks to youthful late Cenozoic volcanic rocks. The older Proterozoic and Paleozoic rocks were once buried deep beneath the surface, but they were exhumed by rotation and uplift along *listric normal faults* (Chapter 23). Today, they are exposed along the steep escarpments of many ranges in Death Valley, allowing us to peer into the deep history of the region (Fig. 23.9). You'll soon find out that the rocks composing the mountains of Death Valley are very old, but the mountainous landscape itself is very young and actively growing.

The valleys in the park are filled with the detritus of the ranges, washed into the basins continuously as they opened and deepened through time. Even today, wind and flash floods transport sediment onto the basin surface from adjacent highlands. The wide range of ages and rock types makes for a very complex geologic history of the region, which is great for geologists but a lot for most others to absorb. Most of the rocks of Death Valley, however, are related to geologic events previously discussed in the context of other parks, which makes their description and significance easier to describe and understand.

PROTEROZOIC ROCKS OF DEATH VALLEY
Metamorphic gneisses older than 1.7 Ga (Ga = billion years) are exposed along the steep western face of the Black Mountains facing the flat expanse of the Badwater Basin (Fig. 24.3). The 1.7 Ga age dates the time during the Proterozoic when the original rocks were metamorphosed. We can't see back through the metamorphism to date the parent rocks because the high temperatures and pressures reset the radiometric clock inside minerals within the rocks. The gneisses were

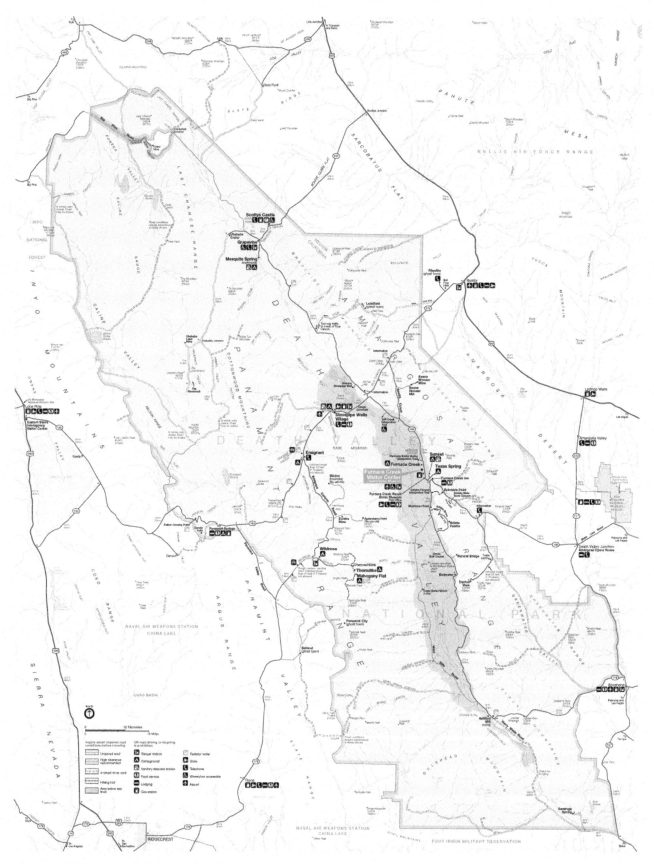

Figure 24.2   Map of Death Valley National Park.

Figure 24.3 Proterozoic gneiss of crystalline basement showing plastic flow, Death Valley National Park.

intruded by granitic magma around 1.4 Ga, and collectively these rocks form the crystalline basement within the region.

These crystalline rocks are the same age and origin as those in the lower Grand Canyon, Colorado National Monument, Black Canyon of the Gunnison National Park, and Rocky Mountain National Park. In all of these places, the age, degree of metamorphism, and subsequent granitic intrusion indicate that these events occurred contemporaneously during a Proterozoic mountain-building episode. This orogenic event is likely related to the accretionary assembly of crustal blocks throughout the entire southwestern United States, a phase of Earth history described in detail in Chapter 21 (Fig. 21.6A). The Proterozoic gneisses exposed in Death Valley National Park are thus the exhumed remains of the continental growth of ancestral North America. Recall from Chapter 22 on Glacier National Park that continental accretion culminated around 1 Ga with the assembly of the Rodinian supercontinent (Fig. 22.8).

Continental rifting and breakup of Rodinia are recorded in *metasedimentary* rocks in Death Valley deposited between 1200 and 800 Ma. Known as the **Pahrump Group**, these rocks accumulated in elongate *rift basins* along the continental margin near southeastern California. Limestones and **dolomites** (a magnesium-rich form of limestone) suggest deposition in warm, tropical seas, while abundant *stromatolites* indicate that portions of the Proterozoic rift basin were shallow enough for photosynthesizing microbes to flourish. Other metasedimentary rocks of the Pahrump Group include conglomerates, made of a mixture of rounded sand- to cobble-sized fragments, and mudstones, made of clays and silts

(Chapter 1). The sediment composing these rocks was derived from the erosion of the uplifted crystalline basement rocks of the Rodinian supercontinent during continental breakup.

The deposition of the Pahrump Group was contemporaneous with the Grand Canyon Supergroup, the Uinta Mountain Group (Dinosaur National Park), and the Belt Supergroup (Glacier National Park). All of these rocks, well exposed in national parks of the American West, accumulated within continental rift basins and record the tectonic separation of ancestral North America from the Rodinian supercontinent.

PALEOZOIC ROCKS OF DEATH VALLEY

Most of the ranges in Death Valley National Park are composed of sedimentary rocks of Paleozoic age (Fig. 24.4). These limestones, dolomites, sandstones, and shales reach about 6 km (~4 mi) in total thickness and form the greatest volume of rock in the Death Valley region. Individual sedimentary formations can be correlated across a wide area of the American West, attesting to their original lateral continuity.

The earlier Proterozoic rifting event continued to the point where, during the earliest Paleozoic, the western margin of the North American continent fully split from the continental fragments on the other side of the rift, which may have been what is today East Antarctica (Fig. 22.8). As seafloor spreading occurred along a mid-oceanic ridge within an ever-widening ocean basin between the diverging continents, a broad, shallow **continental shelf** developed along the margin of western North America (Fig. 1.11C). A continental shelf is the seaward continuation of the coastal plain of continents and may extend outward for tens to hundreds of kilometers. They slope gently seaward until water depths of ~140 m

Figure 24.4 Pyramid Peak, Death Valley National Park, is composed of Paleozoic sedimentary rocks that record transgressions and regressions of the ancient shoreline.

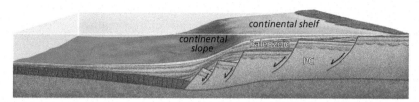

Figure 24.5 Cross section of the Paleozoic continental shelf and slope of western North America. Sedimentary rocks accumulated during transgressions and regressions of the sea above the rifted Precambrian basement. Continental shelves far removed from a tectonic plate boundary are called passive margins. Much later, during the late Cenozoic, these rocks were uplifted within tilted fault-block mountains seen today in Death Valley National Park.

(~460 ft) are reached where the seafloor abruptly increases onto the **continental slope** (Fig. 24.5). Continental shelves are shallow-water marine basins where sediment derived from the erosion of continents accumulates, commonly to great thicknesses.

Continental shelves that are far removed from tectonic plate boundaries are called **passive margins**. The modern Atlantic continental shelf is a passive margin and is characterized by minimal tectonism because the nearest plate boundary (the Mid-Atlantic Ridge) is thousands of kilometers away (Fig. 1.9). Continental shelves subside slowly under the weight of overlying sediment, with the underlying asthenosphere migrating laterally to accommodate the load (see Chapter 1 on sedimentary basins).

The broad Paleozoic passive margin of western North America was repeatedly submerged and exposed as sea level rose and fell through time. With each transgression and regression, sediment was deposited along the top of the continental shelf, driving *subsidence* and creating even more space for sediment to accumulate. These rocks form a seaward-thickening wedge, with the thinnest accumulations occurring near the Grand Canyon and further eastward toward Colorado and the Midwest. Near Death Valley, the Paleozoic passive margin sequence of rocks is several kilometers thick, as measured from mountainside exposures. The long-term accumulation of sediment along continental shelves is a fundamental way that continents grow laterally outward. The Paleozoic passive margin will be more fully incorporated into the larger narrative of the evolution of the American West in Chapter 33.

MESOZOIC ROCKS OF DEATH VALLEY

Chapter 23 on the evolution of the Basin and Range Province addressed the tectonic setting of the American West during the Mesozoic. Around 200 Ma, Farallon subduction began to dominate the western margin of North America. As the volcanoes of the ancestral Sierra Nevada pumped ash into the atmosphere from their deep supply of magma, compressional stress reached inland across the future northern Basin and Range, raising the Nevadaplano highlands that extended across Nevada and Utah (Fig. 23.12).

In the Death Valley area, granitic plutons intruded the crust, supplied from the same deep reservoirs of molten rock that fed the *Sierran batholith*. These Mesozoic granites were exhumed during later extensional uplift of the Panamint and Cottonwood mountains in the park. Tight folds and thrust faults exposed in some of the ranges near Death Valley attest to tectonic compression related to Farallon subduction. Many Paleozoic rocks were squeezed to such an extent that the folds were smeared out into complex shapes, an expression of plastic deformation under high pressures and temperatures deep beneath the surface. The folds and thrusts in Death Valley provide evidence that the Mesozoic was a time of deformation and erosion in the region, rather than a time of deposition.

Cenozoic Rocks and Extensional Uplift in Death Valley

Recall from the previous chapter that during the mid-Cenozoic the American West experienced the *ignimbrite flare-up*, a widespread phase of explosive volcanism that smothered the area with a thick blanket of welded ash (ignimbrite). From 27 to ~20 Ma, volcanic eruptions to the northeast of Death Valley spread hot clouds of rhyolitic ash across the region, filling depressions in the terrain and creating a subdued volcanic landscape. These datable layers of tuff reach a thickness of 370 m (~1200 ft) in the Death Valley region and provide an important constraint on the age of extensional uplift in the Basin and Range because the tuffs are cross-cut by later normal faulting.

Late Cenozoic rocks of Death Valley tell the story of extensional uplift in the region because they were deposited in rapidly downdropping basins concurrent with the rise of tilted fault-block mountains (Fig. 24.6). Basin and Range extension began in the Death Valley region around 16 Ma, as recorded by the sedimentary and volcanic rocks of the Furnace Creek and Funeral formations, as well as other late Cenozoic rock units.

Rotation of tilted fault-blocks along listric normal faults over the last 16 m.y. exhumed old, deeply buried rocks like those of the crystalline basement, the Pahrump Group, and the Paleozoic sedimentary succession upward toward the surface. As the uprotated edge of each fault-block was lifted higher during each earthquake, their erosional debris washed downslope to accumulate in alluvial fans and lakes, filling the adjacent downdropped basins. Magma occasionally rose through the thinned crust along normal faults, erupting as gigantic clouds of pyroclastic particles that rained down on the landscape. This sedimentary debris and volcanic ash compose the late Cenozoic rock filling basins in the Death Valley area. Some of these semiconsolidated rocks were lifted upward to the surface by later faulting, exposing them to view (Fig. 24.7).

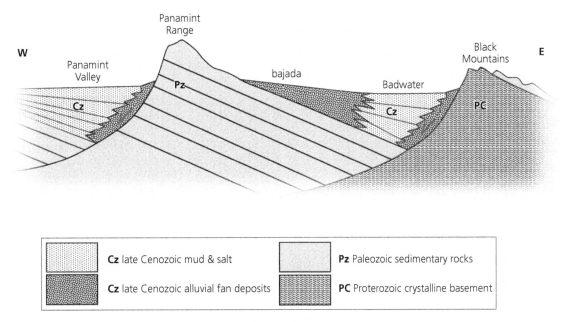

Figure 24.6 Simplified cross section from the Panamint Valley to the Black Mountains, showing tilted fault-block ranges and half-graben basins.

Figure 24.7 View from Zabriskie Point across the badlands formed in the late Cenozoic Furnace Creek Formation. These "soft" sedimentary layers have been rotated during episodic fault movement to a steep inclination downward to the right. The Panamint Range rises in the background.

The climate and ecology of Death Valley during the late Cenozoic was very different from what it is today, as revealed by footprints of camels, horses, and mammoths found along bedding planes.

STRIKE-SLIP FAULTING AND THE BADWATER BASIN

Ranges grew higher and basins dropped lower over the past 16 m.y. as extension stretched the crust, triggering rupture along normal faults (Fig. 24.8). **Strike-slip faults** are an equally important style of faulting in the park and are responsible for the formation of the Badwater Basin in the southern part of Death Valley. Faults with strike-slip motion tend to be vertical or near-vertical planes with the blocks of rock on either side moving laterally in opposite directions to one another (Fig. 24.9A). They form in response to *shear stress*, where the orientation of forces is parallel to the fault plane (Chapter 1). Similar to the other types of faults, strike-slip faults typically move abruptly when the accumulated shear stress exceeds the frictional strength of the rock along the fault plane, triggering earthquakes.

The orientation of offset along strike-slip faults further defines the mechanics of movement. If the block of rock on the side of the fault opposite the observer moves to the right, the type of slip is called **right-lateral**. If the block of rock moves to the left, the fault exhibits **left-lateral** motion. The point of view of the observer doesn't matter; the sense of slip is the same regardless of which side of the fault the observer views from.

The two main strike-slip faults in the park are the Northern Death Valley Fault Zone (NDVFZ) and the Southern Death Valley Fault Zone (SDVFZ), both of which exhibit right-lateral motion (Fig. 24-8). Both faults show offset of landscape features totaling several tens

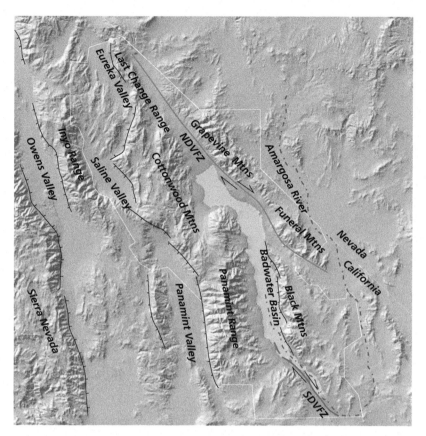

Figure 24.8 Major faults of the Death Valley region. Strike-slip faults shown in red. Normal faults shown in black, with tick marks on downthrown side. Many other faults are not shown for simplicity. Park boundaries shown in white. NDVFZ = northern Death Valley fault zone; SDVFZ = southern Death Valley fault zone. Gray shading shows the area of the park below sea level. Amargosa River shown as blue dashed line.

of kilometers of total displacement; this suggests that they've been active for several million years. Evidence for right-lateral motion along the SDVFZ is exhibited where a young *cinder cone* has been split and offset about a hundred meters by the fault (Fig. 24.9B). About 300,000 years ago, magma rose upward along the fault plane, erupting particles of cinder to build the cone in a geologically short time. Since then, the fault has continued its right-lateral motion, ripping the cinder cone apart one earthquake at a time. Dividing the amount of offset by the age of the cinder cone suggests that the long-term **slip rate** averages about a meter every 3000 years or so. Slip on the fault occurs during short, violent earthquakes, with the fault remaining locked for the rest of the time.

Right-lateral motion along the NDVFZ and SDVFZ occurs subparallel to one another. The region in between the faults is therefore subject to extensional stress as the block of rock is stretched in opposite directions (compare Figs. 24.10 and 24.8). About 4 Ma, normal fault motion was initiated along the Black Mountain fault extending along the base of the mountains where it abuts the Badwater Basin of Death Valley. The network of normal faults along the west side of the Black Mountains connects the two strike-slip faults and accommodates the extensional stress

in between. As the two strike-slip faults moved through time, the Badwater Basin deepened as the Black Mountain fault episodically slipped. These types of basins, opened along normal faults born in response to subparallel strike-slip faulting, are called **pull-apart basins**.

The Badwater Basin and the adjacent Black Mountains are younger than most of the other basins and ranges in Death Valley and show abundant evidence of youthful activity. The sub-sea-level elevations on the Badwater salt flat attest to active downdropping along the normal faults lining the Black Mountains. Geophysical measurements indicate that the Badwater Basin is an asymmetric *half-graben*, with 2700 m (~9000 ft) of sedimentary fill along the Black Mountain fault, thinning to a feather edge westward toward the Panamint Range (Fig. 24.6). The thick accumulation of sedimentary debris beneath the steep front of the Black Mountains reflects active movement along the normal fault. Sharp *fault scarps* almost 6 m (20 ft) high cut the alluvial fan near Badwater and are estimated to be about 2000 years old (Fig. 24.11). It's just a matter of time before the next large earthquake rattles the Black Mountains and the adjacent southern Death Valley.

RECENT VOLCANISM IN DEATH VALLEY

One of the main pieces of evidence for active extension in the Basin and Range is the occurrence of recent volcanism. Several basaltic eruptions have occurred in the park during the *Holocene*, the last 12,000 years or so of Earth history. The most recent eruption formed Ubehebe Crater, part of a volcanic field in northern Death Valley that may be only 300 years old (Fig. 24.12). The crater lies along the trace of a normal fault, suggesting that the basaltic magma used the weakened fault plane as a conduit to reach the surface. When the molten fluid encountered groundwater at shallow depths, the water flashed to steam, triggering an explosive release of black, sand- to pebble-sized cinders that buried the surrounding landscape. These types of eruptions that involve the volatile mixing of magma and water are called *phreatic eruptions* (introduced in Chapter 13).

The blast opened the 200 m (700 ft) deep main crater of Ubehebe, revealing layers of conglomerate, stained orange due to interaction with hot, acidic fluids prior to the main explosion. Native Americans likely witnessed the eruption and were probably stunned by the ear-splitting sound and fiery violence of the event. Other, smaller eruptions

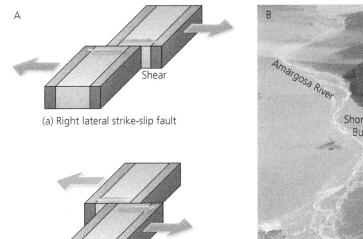

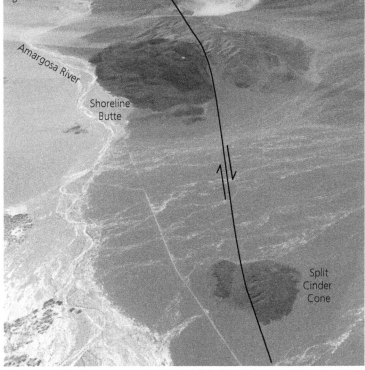

Figure 24.9 A) Block diagrams showing lateral offset along strike-slip faults. Broad arrows show the shear forces acting on the fault plane. B) Right-lateral Southern Death Valley fault cutting Split Cinder Cone and the west side of Shoreline Butte. Google Earth view looking south. White line is the West Side Road.

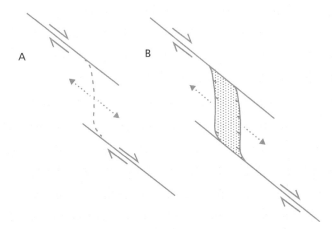

Figure 24.10 Mechanics of pull-apart basins. A) Two subparallel strike-slip faults create extension in the area in between as episodic slip occurs. B) In time, normal faults develop to accommodate the extension, creating the downdropped depression of the pull-apart basin (stippled parallelogram).

created the satellite craters that surround the main crater. Volcanism in Death Valley tends to be basaltic because of the proximity of the low-silica rock of the asthenospheric mantle just below the thinned lithosphere. As the asthenosphere slowly flows upward beneath the Basin and Range, the weight of overlying rock lessens, reducing the pressure on the rising hot rock. The lower pressure permits the rock

to partialy melt, creating low-silica basaltic magma. Faults in the overlying crust provide multiple pathways for the low-viscosity fluid to escape to the surface.

Pleistocene Lake Manly

During the glacial-interglacial phases of the Pleistocene, the climate of Death Valley and the entire Basin and Range alternated between wetter and drier conditions. During phases of high precipitation, large *pluvial lakes* formed in the intermontane valleys, with many overflowing and connecting hydrologically with other lakes (Fig. 23.14). Lakes Bonneville and Lahontan were the two largest lake networks, but Death Valley was also part of a larger system of connected pluvial lakes in eastern California (Fig. 24.13). Much of the water was derived from melting alpine glaciers in the Sierra Nevada that drained into lakes in the Owens Valley at the base of the eastern escarpment. Those lakes spilled over low passes into other lakes, including one filling Panamint Valley, before finally draining into the lowest elevations of Death Valley. Water was also sourced from the spring-fed Amargosa River, which first drained into Lake Tecopa as it wound its way toward Death Valley. The temperate climate and large supply of fresh water supported a fertile ecosystem in the surrounding landscape. When the regional climate turned drier, the pluvial lakes of eastern California evaporated, leaving behind broad layers of mud and salt on the floors of the intermontane basins.

Figure 24.11 Two fault scarps cut the alluvial fan near Badwater (arrows). Badwater Spring (out of view in the foreground) is fed by groundwater migrating upward along the fault planes. The alluvial fan is a youthful feature, so the cross-cutting faults must be even younger.

Figure 24.12 Ubehebe Crater (upper left) forms the largest of 13 craters formed as basaltic magma interacted explosively with groundwater.

left behind as narrow terraces along mountainsides. Lake Manly wave-cut benches are common horizontal features on mountainsides facing Death Valley proper, and they are particularly visible on Shoreline Butte (Fig. 24.9B). This mound of volcanic rock was a small island in southern Lake Manly and was battered by waves driven by winds sweeping southward across the lake.

The difference in elevation between the highest wave-cut bench on Shoreline Butte and the lowest point in the Badwater Basin indicates that the maximum depth of the lake was about 180 m (600 ft). Considering the stark landscape of Death Valley today, it's hard to visualize what it must have looked like only 10,000 years ago. Marshes and woodlands surrounded an elongate trough of blue water that stretched north–south for over 100 km (60 mi). Large grazing mammals and the predators that tracked them roamed the fertile terrain. Rapidly flowing streams carved narrow canyons into the steep mountainfronts, transporting muddy sediment and dissolved salts to the lake below. The blanket of snow on the Panamints and other nearby ranges lasted longer through the year and reached farther downslope. Contrast that image with the sere landscape of today's Death Valley and you get a feel for the volatility of climate change through the Pleistocene.

With Holocene warming, Lake Manly shrank as its water was sucked into the arid atmosphere, leaving a flat, salt-encrusted playa. Slightly wetter conditions about 2000 to 4000 years ago rejuvenated streams that filled the lakebed with about 10 m (30 ft)

Drill cores through the sediment of the Badwater Basin reveal that a large lake, called Lake Manly, filled Death Valley two times in the recent past that coincide with glacial phases. The first lake existed from 186 to 120 ka before drying up. The most recent Lake Manly filled the valley between 35 and 10 ka, overlapping in time with the *Last Glacial Maximum*. The level of Lake Manly fluctuated as temperature and precipitation varied through time. During phases of stable lake levels, waves bashing into the rock of the surrounding ranges would erode horizontal notches called **wave-cut benches**. As the lake level fell during drier phases, the wave-cut benches were

of water. Salt along the floor of the lake was dissolved, making the shallow puddle of water somewhat briny. The ephemeral lake evaporated as rainfall decreased over the last 2000 years, re-precipitating a meter-thick layer of salt on the playa surface. The salt is composed of various chemical compounds, but *halite*—NaCl, common table salt—is most common. The sodium and chloride ions were derived from the chemical weathering of rocks in the surrounding ranges. When you wander around the Badwater salt flat or the Devil's Golf Course, you are walking on the desiccated remnants of the Holocene version of Lake Manly.

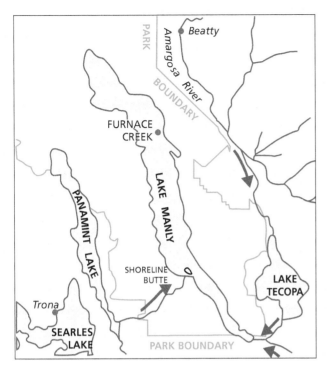

Figure 24.13 Lake Manly filled Death Valley to a maximum depth of about 180 m (600 ft). During its peak, it was hydrologically connected to other closed-basin lakes in the region.

Desert Landforms in Death Valley

Death Valley National Park is perhaps best known for its iconic desert landscape. Deserts are defined, however, by the scarcity of precipitation, not the stereotypical image of a vast sea of sand dunes stretching to the horizon. Sand dunes cover less than 1 percent of the land surface of Death Valley, with most of the surface composed of bare rock, gravel plains, or salty playa lakebeds. A common

Figure 24.15 Wind-rippled surface of Mesquite Flat sand dunes near Stovepipe Wells, Death Valley. The Grapevine Mountains rise in the distance.

feature that is widespread across the valley floor is **desert pavement**, an armored surface of angular pebbles and cobbles (Fig. 24.14). Pavements were once thought to be the heavier gravels left behind after powerful winds blew away surrounding grains of clay, silt, and sand. It turns out, however, that the surface layer of platy chips of rock is underlain by a few centimeters of silt. It's now thought that the silt filters downward, while the coarser gravel moves upward due to vibrations caused by the patter of rainfall, the wetting and drying of the ground, or the freezing and thawing of the surface.

The relatively rare fields of sand dunes in Death Valley tend to be isolated accumulations of sand trapped along the margins of valley floors where they abut steep mountainfronts (Fig. 24.15). The mountains serve a dual purpose by both supplying the sand for the dunes as well as protecting the dunes from being blown away by prevailing winds. Dune fields in Death Valley tend to be relatively stationary, changing form with the seasons but remaining in place. Sand for the dune fields is originally derived from the erosional breakdown of rock, which is then transported by flash floods to alluvial fans lining the base of the mountains. From there, winds transport sand out onto the margins of the valley floor where it is sculpted into sand dunes.

The most accessible dune fields in Death Valley are the Mesquite Flat dunes near Stovepipe Wells. Other dune fields include the impressive Eureka Dunes in the far north corner of the park where the steep front of the Last Chance Range prevents them from migrating. The Eureka Dunes are the highest dunes in California (~210 m/700 ft), and visitors who clamber about on the dunes will be

Figure 24.14 Desert pavement—a mosaic of fitted angular stones that forms the surface of many gravel plains in Death Valley.

Figure 24.16   The Badwater alluvial fan forms at the mouth of a narrow canyon that drains the Black Mountains. Fault scarps (also shown in Fig. 24.11) cut across the highest slopes of the fan but have been dissected by flowing water near the mouth of the canyon. The outer fringe of the alluvial fan merges with the white salt flats of the playa. Badwater Road winds around the outer margins of the fan.

load, so the largest boulders and cobbles are dumped near the head of the fan with the finest particles carried in a network of small channels down to the distal edges of the fan. Particularly powerful flash floods may carry mud all the way out onto the adjoining playa surface. Over time, channels on the fan become clogged with sedimentary debris, forcing later floods to flow along easier paths. The fan builds a symmetrical cone over time as thousands of debris flows deposit loose sediment back and forth across the fan surface.

Alluvial fans lining the western base of the Black Mountains are isolated, discrete features, but alluvial fans lining the eastern flank of the Panamint Range to the west coalesce to form a single huge apron of sedimentary debris called a **bajada** (Fig. 24.17). The difference can partly be attributed to the greater height of the Panamints and thus the larger sediment load that they supply to the adjacent basin. But the main reason for the difference in morphology and size of alluvial fans is the rapid rate of slip along the Black Mountain normal fault. Recall that the Badwater Basin is an asymmetric half-graben, with the thickest accumulation of sediment occurring along the base of the Black Mountains (Fig. 24.6). The small alluvial fans at the base of the Black Mountains are continually

enchanted by the mysterious "singing" sound produced as the bone-dry grains rub against one another as they avalanche downslope. Other dune fields in the park are located in northern Panamint Valley, Saline Valley, and near Saratoga Springs. These fields of sand dunes are small-scale modern analogs for ancient cross-bedded sandstones such as those that compose the Coconino and Navajo formations (both found throughout the Colorado Plateau). Recall how dunes form and migrate as sand grains skitter up the gentle windward side of the dune to the crest, then cascade down the steep lee face of the dune to produce the cross-beds (Fig. 4.5).

Other common features of the Death Valley landscape are **alluvial fans**, cone-shaped aprons of sedimentary debris that accumulate at the mouths of canyons near the base of mountainfronts (Fig. 24.16). Occasional intense rains rapidly run off the steep, rocky slopes of mountains, collecting in narrow canyons before flowing through the canyon as a *debris flow*. Recall from Chapter 3 that debris flows are powerful slurries of water choked with rock fragments ranging from tiny clay particles to enormous boulders that flush through channels as flash floods. When the debris flow reaches the mouth of the canyon, the flow velocity suddenly decreases as it encounters the upper surface of the fan. The slower flow becomes incapable of transporting the sediment

Figure 24.17   Low-angle oblique Google Earth image looking northward along the axis of the Badwater Basin of Death Valley. Note the difference in the morphology of discrete alluvial fans lining the base of the Black Mountains versus the broad bajada along the base of the Panamint Range.

Figure 24.18 Badlands topography of rills and gullies carved into easily eroded mudstone. Zabriskie Point, Death Valley.

being downdropped and buried before they can build outward onto the playa surface. In contrast, the bajada along the eastern flank of the Panamints can expand across the flat basin floor because of minimal downdropping on the west side of the valley.

Common landscape features in desert settings like Death Valley are denuded hills and ridges of muddy, semi-consolidated sedimentary rock and weathered volcanic ash called **badlands** (Fig. 24.18). Many rocks in Death Valley are weakly cemented *mudstones*, composed of clays and silts deposited in ancient lakebeds. Other clays are derived from the chemical weathering of layers of volcanic ash. The Furnace Creek Formation, discussed earlier, is a good example of lake-deposited mudstones mixed with altered volcanic ash (Fig. 24.7). Clay minerals are shaped like miniscule plates that, when compressed beneath overlying sediment, align to form an impermeable barrier to the flow of water. The aridity of Death Valley, in concert with the impermeable mudstones, precludes the development of soils that might sustain a thin veneer of vegetation. When sporadic intense rainstorms hit these soft rocks, sheets of water quickly drain off the surface, carrying particles of clay downslope while carving a network of rills and gullies into the rounded mass of mudstone. Badlands are constantly reshaped by storms and flash floods that erode and transport the muddy sediment further downslope to be deposited on the basin floor. Badlands created by this combination of processes are among the most photogenic features in the park.

The vast **salt flats** of Death Valley may look timeless, but they are ever-changing (Fig. 24.19). The occasional wet winter will recharge the springs feeding the Amargosa River, enhancing the flow volume so that the river actually reaches its terminus in the Badwater Basin. Supplemented with runoff from the adjacent mountains, the playa floor may flood to a depth of a half-meter or less, creating an

ephemeral lake. During the short life of the lake, the playa surface of salt dissolves in the freshwater, turning it briny. While the temporary puddle of floodwater bakes in the desert sun, a thin layer of mud may settle onto the lakebed. As the lake dries up, halite crystals precipitate from the remnant brine, forming a new crust of salt above the muddy layer below. This scenario has happened countless times, filling the deepest parts of the basins in Death Valley with a complex sequence of mud layers mixed with beds of salt. The combination of an evaporative climate, internal drainage, and an abundant supply of salty elements derived from the rocks of the surrounding ranges makes Death Valley National Park ideal for the formation of salt flats and mud-covered playas.

The desiccating aridity of Death Valley causes the layers of mud on playa surfaces to contract, creating polygonal patterns bound by *mudcracks*. Polygons also form on the salt flats and likely mimic the pattern in underlying layers of mud (Fig. 24.19). Rainfall and temporary flooding may trigger the dissolution of salt, with some of the saline water filling the underlying mudcracks. As the halite re-precipitates during drying, crystals grow preferentially along the margins of polygons. Where salt crystals meet, they bow upward along the contact between adjacent polygons, forming double-sided walls. Rainwater from the next storm collects within the miniature polygonal ponds, dissolves some of the salt, and re-precipitates it along the margins, propagating the growth of raised polygon edges. The next major flooding event may dissolve the entire surface, smoothing it and starting the entire process over again.

The Devil's Golf Course is slightly higher in elevation than the rest of the Death Valley salt flat, so it seldom

Figure 24.19 Salt polygons, central Death Valley.

Figure 24.20 Low-angle perspective looking north across Racetrack Playa. The lakebed is about 4 km long and 1.5 km wide (2.5 x 1 mi). Ubehebe Peak rises to the west, while the Cottonwood Mountains form the eastern boundary. The tiny outcrop of plutonic rock called the Grandstand peaks above the playa surface near the northern edge.

Figure 24.21 Angular "sliding stone" of Racetrack Playa. Note the winding trail left behind as the boulder slid across the mud-cracked playa surface.

floods. Without the resurfacing effects of dissolution and re-precipitation, the salt crystals slowly grow upward into bizarre mounds and pinnacles. Vertical growth is driven by evaporation that draws salty water upward from underlying layers of mud. The salt mounds and pinnacles are constantly modified by wind and the occasional storm.

Perhaps the most surreal feature of the Death Valley landscape is found in Racetrack Playa, a remote lakebed tucked in the western shadow of the Cottonwood Mountains (Fig. 24.20). The flat playa surface is covered with a layer of mud, deposited within ephemeral lakes that form during infrequent floods. As the playa bakes in the sun, the mud contracts into a mosaic of small polygons outlined by mudcracks. The strangest aspect of the playa surface, however, is the presence of hundreds of cobbles and small boulders of limestone and dolomite that rest at the end of long furrows gouged into the playa surface (Fig. 24.21). These are the famous sliding stones of Racetrack Playa, sometimes called *sailing stones*, *slithering stones*, or even *playa scrapers*. The inescapable conclusion is that the rocks somehow slid across the surface, perhaps pushed along by gusts of wind when the muddy playa was slick with water or a thin veneer of icy slush. Or at least that was the initial assumption.

The rocks were originally derived from a ridge of limestone and dolomite that rims the southern edge of the playa, having tumbled down onto the mudcracked surface long ago. Geologists trying to understand how the rocks moved across the playa surface whimsically gave several of them women's names in order to track them more easily through time, so the movements of "Karen" and "Hortense" have been followed for several decades. The problem was that, despite Global Positioning System (GPS) monitoring and frequent visits by researchers, no one had ever witnessed the stones sliding across the surface, likely due to the wet, cold, and windy conditions believed necessary to trigger movement.

Recently, geologists filmed GPS-instrumented rocks moving across the surface during the winter of 2013–2014. A shallow pond about 10 cm deep had formed in the southern end of the playa from rain and snowmelt. A layer of ice, less than a cm thick, froze along the top of the pond. Sunlight early in the day fractured the ice into separate sheets that glided across the pond as a relatively gentle wind blew. When some thin panels of ice encountered a boulder, they broke into shingles that built up against the rock. The tilted and stacked panels of ice captured the wind like sails and pushed the rock along in front. In all, about 60 rocks were recorded slowly sliding across the playa surface in response to what's called "ice shove," with some rocks moving as much as 224 m (735 ft).

The enigma may be solved. But scientists are unapologetically skeptical, particularly geologists, and the recent explanation is far from decided in the minds of many. Maybe the sliding stones are better left unresolved as an eccentricity of nature. Regardless, it's worth the drive along the sandy road from Ubehebe Crater to see the stones for yourself. Be sure to stop at Teakettle Junction and read the inscriptions on the kettles left behind by visitors from everywhere in the world who converge on this one particular place. It's just one more weird and wonderful oddity of Death Valley National Park, a place full of surprises for those curious enough to explore the vast desert landscape.

Key Terms

alluvial fan
badlands
bajada
continental shelf
continental slope
desert pavement
dolomite
left-lateral slip
passive continental margin
pull-apart basin
right-lateral slip
salt flat
sliding stone
slip rate
strike-slip fault
wave-cut bench

Related Sources

Collier, M. (1990). *An Introduction to the Geology of Death Valley*. Death Valley Natural History Association. Death Valley, CA.

Miller, M. B., and Wright, L. A. (2004). *Geology of Death Valley National Park*, 2nd ed. Dubuque, Iowa: Kendall/Hunt Publishing Co.

Sharp, R. P., and Glazner, A. F. (1997). *Geology Underfoot in Death Valley and Owens Valley*. Missoula, MT: Mountain Press.

National Park Service—geology, Death Valley National Park. Accessed February 2021 at https://www.nps.gov/deva/learn/nature/geology.htm

National Park Service—geologic formations, Death Valley National Park. Accessed February 2021 at http://www.nps.gov/deva/learn/nature/geologicformations.htm

Geology of Death Valley through photos. Accessed February 2021 at http://www.marlimillerphoto.com/dvpics.html

Grand Teton National Park

Figure 25.1 View from the flat valley floor of Jackson Hole westward toward the Teton Range. Highest peak left of center is Grand Teton. The National Park Service maintains a bison herd in the southeast part of the park.

Nowhere in America does the juxtaposition of serrated mountains, expansive valleys crossed by sinuous rivers, and crystalline lakes occur in such a compact area as Grand Teton National Park. The vista of the Teton mountainfront rising abruptly from the valley floor of Jackson Hole is one of the most spectacular views in North America (Fig. 25.1). You've seen it, even if you've never been there, because many a Hollywood western uses the Tetons as a dramatic backdrop. Add an abundance of large charismatic fauna like elk, moose, mule deer, pronghorn, bison, and grizzly bear and the popularity of the park becomes evident.

Grand Teton National Park is located in the northwestern corner of Wyoming, just south of Yellowstone National Park (Fig. 25.2). Geographically, the park resides within the Middle Rocky Mountain Province, but geologically it's a Basin and Range *tilted fault-block* mountain range (Fig. 23.8). A series of tilted fault-block mountains continues westward into Idaho, with their northern extensions

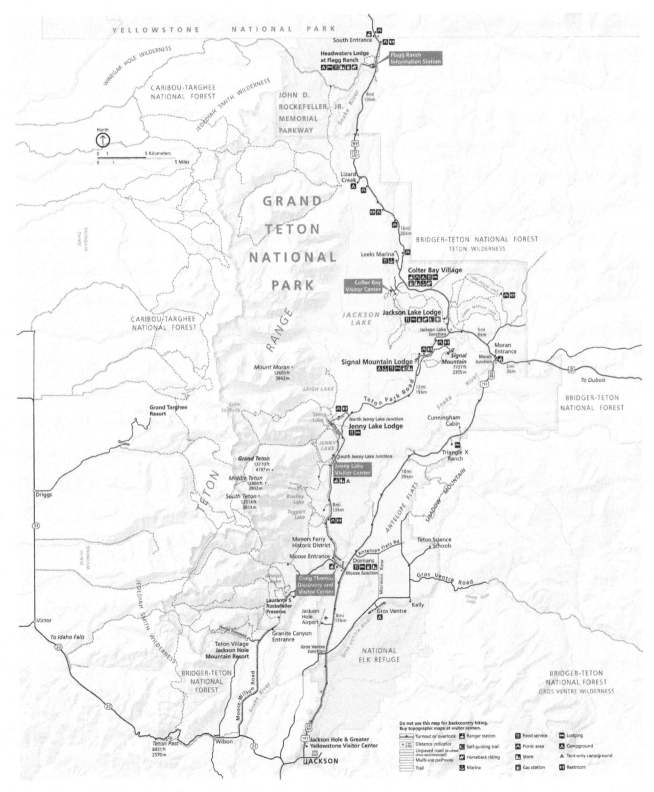

Figure 25.2 Map of Grand Teton National Park in northwest Wyoming.

terminated at the Snake River Plain. We might think of the Tetons as the northeastern corner of the Basin and Range where it overlaps with the Rocky Mountains (Fig. 23.1). Originally named by forlorn French-Canadian fur-trappers in the early 19th century, the range and the lakes near the base became a national park in 1929. After a contentious tussle with private landholders, the park boundaries were extended to include much of the adjacent Jackson Valley in 1950.

The Teton Range tilted fault-block has an asymmetric geometry, with a very steep eastern escarpment and a broadly sloping western flank that extends toward the Idaho border (Fig. 25.3). The range is about 70 km (~40 mi) long and only about 16 km (~10 mi) wide, covering a relatively small area. But the jagged, sharp-edged relief of the range more than compensates for its compact size. The highest peak, Grand Teton, is almost 4200 m (13,770 ft) in elevation, with six other peaks reaching above 3660 m (~12,000 ft). The surface elevation of Jackson Lake, at the base of the range, is 2064 m (6770 ft). So the maximum relief of the Tetons relative to the adjacent valley is over 2100 m (~7000 ft), occurring over the short distance of about 6 km (3.7 mi) – a slope of 35 percent! For perspective, Death Valley National Park has a higher maximum relief, but the slope between the highest and lowest points is only ~13 percent. The highest slope along the eastern escarpment of the Sierra Nevada is about 26 percent, whereas the slope in the Grand Canyon from the South Rim to the Colorado River is about 27 percent. So the steepness of the east flank of the Tetons is among the most extreme in the contiguous 48 states.

The ruggedness of the Teton Range is in part due to weathering and erosion from fast-flowing streams, *freeze–thaw cycling* along joint planes, and rockfalls and other mass movements. But three other factors further influence the jagged relief of the range. First, the rapid rate of uplift currently outpaces the rate of erosion (Fig. 23.10). Second, the crystalline rocks that compose much of the range are highly resistant to erosion. Finally, the greatest contribution to creating the serrated profile of the Tetons was made by Ice Age alpine glaciers that scoured deep, U-shaped valleys and cirques into the range. The sharp horns of mountain peaks, the narrow arêtes lining hanging valleys, and the numerous sparkling lakes are among the many landscape features left behind after the Last Glacial Maximum about 20,000 years ago. Today, nine named glaciers occupy north- and east-facing cirque basins in the shadows of the highest peaks (Fig. 25.4). These small glaciers are vestiges of the Little Ice Age (~1350–1850 CE) and are all currently receding despite the 11 m (450 in) of annual snowfall.

The broad, sagebrush-covered valley to the east of the range is called Jackson Hole (Fig. 25.3). Any high-elevation valley surrounded by mountains was termed a "hole" by the mountain men who trapped beaver throughout the Rocky Mountains. The relatively flat surface of Jackson Hole tilts gently westward toward the base of the mountains, with no intervening foothills. A necklace of stunning glacial lakes resides in the low wedge between the tilting valley floor and the range, with Jackson Lake being the largest. Jackson Lake is a natural body of water that fills an elongate glacial scour left behind by a lobe of ice that flowed southward from the Yellowstone high country. The lake reaches a maximum depth of 133 m (437 ft) within the western trough near the base of the

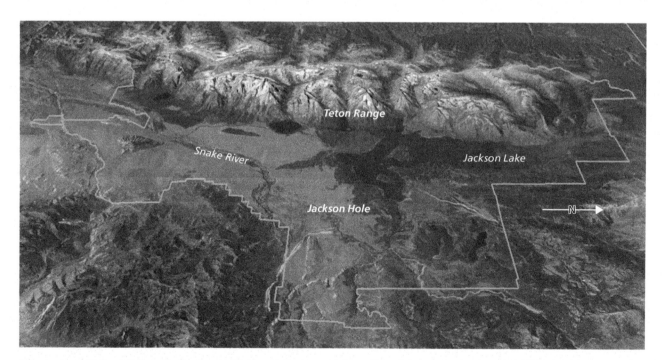

Figure 25.3 Oblique Google Earth view looking westward across the valley of Jackson Hole toward the Teton Range. Green line marks the park boundary.

Figure 25.4 Middle Teton Glacier, one of nine named glaciers in the park. The ice extent is about 40 acres and shrinking.

Tetons. Today, the lake is dammed, partly for flood control, partly to supply summer irrigation water for farmland in Idaho.

Jackson Lake is fed by the Snake River that originates north of the park (Fig. 25.5). The river exits the lake through a dam along the east side before curling southward through Jackson Hole. The low gradient of the valley causes the Snake to meander in some reaches and split into multiple **anastomosing streams** in others. Toward the southern part of Jackson Hole, the river and its floodplain are constrained by levees, built for flood control. The Snake River swings westerly into Idaho where it meanders across the broad Snake River Plain. The river eventually turns northward, forming the boundary with Oregon, before it merges with the Columbia River in southwestern Washington. So snowmelt and rainfall that originate in Grand Teton National Park eventually ends up in the Pacific Ocean.

The Teton region contains one of the most complete records of geologic time in the national park system, with evidence for *Archean* continental assembly, transgressions and regressions of the Paleozoic and Mesozoic seas, and Mesozoic and Cenozoic pulses of mountain-building. Cenozoic time is further represented by thick accumulations of sedimentary rock deposited in continental settings, as well as by voluminous outpourings of volcanic material from the Yellowstone country to the north. Basin and Range extensional tectonics are beautifully expressed in the Grand Tetons, as are the Pleistocene Ice Ages. Very few places in the American West exhibit such a high-fidelity record of events through time as does Grand Teton National Park. The following sections will address the origin and age of the rocks composing the park, as well as the mechanics of uplift of the Tetons. We will conclude with the influence of the Pleistocene glaciations on the stunning landscapes within the park.

Uplift and Exposure of Rocks in Grand Teton National Park

The greatest mass of the Teton Range is composed of Precambrian crystalline rocks. A kilometer and a half of Paleozoic sedimentary rocks overlie these igneous and metamorphic rocks of the basement and form a broad, westward-tilted stack of beds (Fig. 25.6). This architecture was created as the Teton Range was uplifted along the **Teton fault**, a *normal fault* that extends along the eastern base of the range. The Tetons are a very young mountain range, with uplift along the Teton fault originating only about 9–10 Ma and continuing to today. During each of the tens of thousands of major earthquakes along the Teton fault, the range would rise perhaps a meter or two while the adjacent block of Jackson Hole would drop downward an equal amount, accompanied by violent groundshaking.

You might visualize uplift of the Tetons as a giant trapdoor opening, similar to the rise of the modern Sierra Nevada and other tilted fault-block mountains (Fig. 17.11). The end result of 9–10 m.y. of uplift along the Teton fault is the extreme vertical relief of the eastern escarpment of

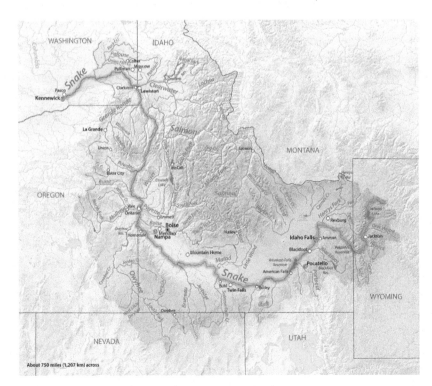

Figure 25.5 Watershed of the Snake River.

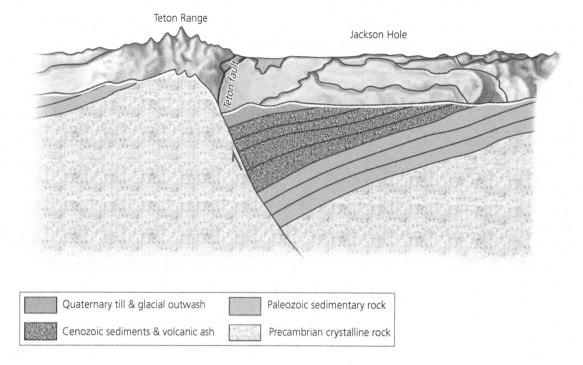

Teton Range

Jackson Hole

Quaternary till & glacial outwash	Paleozoic sedimentary rock
Cenozoic sediments & volcanic ash	Precambrian crystalline rock

Figure 25.6 Simplified cross section across the Teton Range and Jackson Hole. The old Precambrian crystalline rocks and their blanket of Paleozoic sedimentary rocks were uplifted and exhumed within the Teton Range during tens of thousands of huge earthquakes along the Teton fault. The half-graben beneath Jackson Hole is filled with late Cenozoic clastic sediments and layers of volcanic ash from the nearby Yellowstone system.

the Tetons juxtaposed against the relatively flat floor of Jackson Hole. Enough uplift has occurred that the ancient crystalline rocks of the basement were exhumed along the steep east face of the range. As the Precambrian complex was lifted upward along the Teton fault, the overlying layers of Paleozoic sedimentary rock were tilted westward so that today they slope toward Idaho like a massive stack of shingles.

Deep beneath the surface of Jackson Hole lie the rocks of the Precambrian basement and its blanket of younger sedimentary rocks, dropped and rotated downward along the Teton fault (Fig. 25.6). An asymmetric *half-graben* of late Cenozoic clastic sediments derived from the erosion of the adjacent Teton block unconformably overlies the older rocks. The Cenozoic sediments are thickest near the Teton fault where offset during earthquakes created the greatest space for sediments to accumulate. The wedge tapers eastward to a thin margin. The uppermost unconsolidated sediments in Jackson Hole were deposited during the last 2.6 m.y. of the Quaternary. It's worthwhile to review the mechanics of uplift of tilted fault-block mountains along normal faults in Chapter 23 on the Basin and Range. We'll dig deeper into the specifics of uplift along the Teton fault later in the narrative, but for now you should understand the setting in which the rocks of Grand Teton National Park are exposed.

Origin and Age of Rocks in Grand Teton

PRECAMBRIAN ROCKS OF THE GRAND TETONS

Some of the oldest rocks in the National Park System are exposed in the highest peaks of the Teton Range (Fig. 25.7). These are gneisses and schists dated to 2.7 Ga, which is

Figure 25.7 Grand Teton, the highest peak in the range at 4198 m (13,770 ft), is composed of Archean granite and gneiss. These ancient rocks are the remnants of a long-vanished mountain range that existed in the region ~2.7 b.y. ago, now exhumed as part of the modern Teton Range.

the age when the original rocks were metamorphosed. The precursor rock, prior to its deformation and high-grade metamorphism into the strongly foliated rock seen today, may be significantly older. These rocks are Archean in age, a long interval of the Precambrian that spans 4.0–2.5 Ga. (The Archean was introduced in Chapter 22 on Glacier National Park.) The metamorphic rocks were intruded by granitic magma around 2.4 Ga. These metamorphic and magmatic events occurred deep in the cores of ancient mountains, now long vanished.

Earlier chapters on *Proterozoic*-age crystalline basement in Grand Canyon National Park, Colorado National Monument, Black Canyon of the Gunnison National Park, and Rocky Mountain and Death Valley national parks addressed how metamorphic and intrusive events dated around 1.8–1.4 Ga reflect the accretionary assembly of the North American continent. The Archean crystalline rocks in the Grand Tetons record an even earlier episode of continental growth. The highest peaks in the Tetons are composed of these ancient Precambrian rocks. To view them at these rarified elevations in the Tetons attests to the geologic dynamism of our planet. As you hike across these rocks on a trail in the Tetons, it is remarkable to think about how they were formed deep in the crust billions of years ago, but have only recently been raised to their current position high in the mountains.

Late in the Proterozoic, sometime between 1.3 Ga and 770 Ma, magma was injected along fractures within the gneiss, cooling and solidifying to form dark-colored *dikes*. The most obvious dike cuts vertically through Mount Moran in the northern Tetons (Fig. 25.8). As described in Chapter 11, a dike is a tabular body of igneous rock that formed by intrusion into older rock. In the Teton area, the dikes likely represent the solidified "plumbing system" of the magma supply within the body of an ancient volcano that existed during the late Proterozoic. The dark rock composing the dike is chemically similar to *basalt* and likely solidified about a kilometer or so beneath the surface. The dike was raised to its current position during the late Cenozoic uplift of the Tetons where it was exposed by weathering and erosion.

PALEOZOIC AND MESOZOIC ROCKS OF THE GRAND TETONS

Paleozoic and younger sedimentary rocks form a broad, sloping surface of tabular sandstone, shale, and limestone strata tilted westward along the west side of the Tetons (Fig. 25.6). As in all other parts of the American West, the lowermost Paleozoic rocks overlie the Precambrian basement rocks along an erosional surface known as the Great Unconformity. By ~500 Ma, the crystalline rocks were beveled down to a relatively flat surface, which was transgressed by sandy beaches of an encroaching sea. In the Grand Teton region, this beach deposit is known as the Flathead Sandstone. The unconformity and a thin remnant of the Flathead are exposed on the crest of Mount Moran (Fig. 25.8). The *Great Unconformity* separating the 2.7 Ga gneiss from the 500 Ma sandstone spans 2.2 billion years of time, much longer than the Great Unconformity exposed in the Grand Canyon, Colorado National Monument, and Black Canyon of the Gunnison National Park.

Younger Paleozoic and Mesozoic sedimentary rocks were laid down as the shoreline transgressed and regressed in concert with rising and falling sea level. These rocks are the northern equivalents of the Paleozoic and Mesozoic rocks exposed in national parks described in earlier chapters. For example, the Flathead Sandstone represents the same early Paleozoic beach sandstone as the Tapeats in the Grand Canyon and the Lodore in Dinosaur National Monument. The cliff-forming Madison Limestone in Grand Teton manifests deposition in the same shallow tropical sea as the Redwall Limestone at Grand Canyon. The *lateral continuity* of these rock units across the American West attests to the extensive nature of their original depositional environments, which migrated back and forth across the Paleozoic *passive continental margin* in response to fluctuations in sea level.

The highest sea levels were reached during the late Mesozoic when the Western Interior Seaway bisected the North American continent (Fig. 17.9). The Teton area was located along the western shoreline

Figure 25.8 Mount Moran (3842 m/12,605 ft) looms above Jackson Hole. A late Proterozoic dike cuts vertically across the mountainfront, but its lower part is hidden behind Falling Ice glacier. Skillet Glacier descends to the right of the peak. A thin layer of Flathead Sandstone forms a low, rounded knob along the very crest. The Great Unconformity separates the Flathead from the crystalline rocks beneath.

of the inland sea, and dinosaurs must have lived in the area, based on fossils found in late Mesozoic rocks. By the latest Mesozoic and early Cenozoic, compressional tectonism of the Laramide orogeny uplifted nearby mountains such as the Wind River Range and the Gros Ventre Range but only mildly affected the rocks of the Teton region.

CENOZOIC ROCKS OF THE GRAND TETONS

By the mid-Cenozoic, the tectonic regime of the American West had changed from compressional to extensional. (See Chapter 23 on Basin and Range tectonics.) Basin and Range stretching began about 16 Ma, but active uplift of the Tetons along the Teton fault didn't begin till about 9–10 Ma.

How do we know the timing of initial motion along the Teton fault and thus the age of the Teton Range? Just prior to the beginnings of movement along the Teton fault, clastic sediments and volcanic ashes of the Colter Formation were deposited as horizontal layers in lakes in the Teton region. Sediment accumulation of the Colter Formation eventually ended about 13 Ma. By ~10 Ma, the Teton fault became active and the ancestral Jackson Hole basin began to subside. With each earthquake, the distinctive layers of the Colter Formation were tilted down to the west toward the fault. Limestone, mudstone, conglomerate, and volcanic ash accumulated in the actively subsiding basin over several million years. This wedge-shaped succession of sedimentary rocks is known as the Teewinot Formation and forms horizontal beds above the inclined layers of the Colter lake deposits (Fig. 25.9). This type of discordant contact is called an *angular unconformity* (Chapter 3) and helps to date the beginnings of uplift of the Tetons along the Teton fault. Because the Colter lakebeds are tilted while the Teewinot lakebeds are horizontal, the Teton fault must have initiated movement during the time in between.

As the Jackson Hole block episodically collapsed along the Teton fault through the rest of the Cenozoic, erosion of the rising Teton Range supplied clastic sediments that were deposited in lakes, alluvial fans, and river systems. Broad layers of volcanic ash added to the sedimentary wedge beneath Jackson Hole, derived from explosive eruptions of the nearby Yellowstone volcanic system. Today, over 4 km (~14,000 ft) of Late Cenozoic sediment and sedimentary rock reside beneath the surface of Jackson Hole.

Offset and Seismicity along the Teton Fault

Cumulative vertical offset along the Teton fault is about 9 km (~30,000 ft), created episodically during tens of thousands of large-magnitude earthquakes over the past 9–10 m.y. This total displacement is determined by the

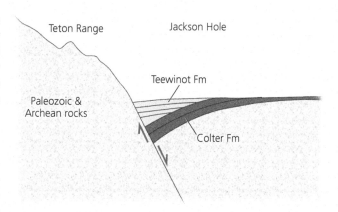

Figure 25.9 Schematic cross section showing the beginnings of movement along the Teton fault about 9–10 Ma. Episodic downdropping along the Teton fault progressively tilted the older Colter Formation lakebeds, while a wedge-shaped basin opened along the mountainfront. The sediments of the Teewinot Formation filled the newly formed basin concurrent with motion along the fault plane. The juxtaposition of the horizontally bedded Teewinot against the tilted Colter beds creates an *angular unconformity* that helps to date the beginning of motion along the Teton fault.

distance between the unconformable base of the Flathead Sandstone perched 1800 m (6000 ft) high on Mount Moran and this same surface buried ~7200 m (24,000 ft) beneath Jackson Hole, recognized from deep drilling. Since the Flathead Sandstone was once a continuous layer prior to faulting, the 9 km offset marks the total amount of uplift of the Tetons plus the total amount of downdropping of the Jackson Hole Basin (Fig. 25.6). Thus, the Teton fault, extending along the base of the range, is the primary feature that generates topography in Grand Teton National Park (Fig. 25.10). The uplifted mountains and adjoining basin are elegantly modified by glaciation, the other great sculptor of landscapes in the park.

The trace of the Teton fault along the base of the range extends north–south, an orientation that reflects east–west Basin and Range extension. Another possible source of crustal stretching in the Teton region is the Yellowstone volcanic plateau just to the north. Beneath Yellowstone, a huge plume of hot rock rises through the mantle until it becomes trapped beneath the base of the lithosphere. From there, the plume supplies magma that buoyantly flows upward, likely through fractures. The intrusion of magma into country rock and the thermal expansion of the rock due to the high heat flow force the overlying crust to rise, elevating the Yellowstone Plateau. As the crust flexes upward, it extends, weakening the rock and creating normal faults in response. The Yellowstone area is crossed by numerous active faults aligned near-parallel to the Teton fault and is the most seismically active region in the coterminous United States, other than California. It may be that crustal stretching in the Yellowstone region, perhaps acting in concert with Basin and Range

Figure 25.10 Oblique view of the Tetons, with Jackson Hole and Jackson Lake in the foreground. The Teton fault extends for 70 km (~40 mi) along the base of the range (dark line with hachures on the downthrown side of the fault).

fault plane during earthquakes creates a *fault scarp* along the base of the range, which is expressed as a sharp change in slope along the base of the mountainfront (Fig. 25.11). The scarp records the cumulative offset of several earthquakes. Visualize this small segment of the fault plane projected downward beneath Jackson Hole to a depth of perhaps 10–12 km (6-7 mi) where it curves and dissipates along the *brittle-ductile transition*.

The Teton fault scarp cuts glacial sediments left behind after the Pleistocene glaciers receded from the area around 14 ka (ka = thousands of years). Paleoseismologic studies reveal that the last major magnitude 7 earthquake along the Teton fault occurred between 7100 and 4800 years ago. (The broad range of dates is due to the limitations of the available material being dated.) This last major quake happened after an earlier quake around 7900 years ago (~6000 BCE). With lots of fieldwork, a few assumptions, and a little math, it was determined that there were nine large earthquakes along the Teton fault between 14,000 and 7900 years ago.

extension, created the Teton fault and likely contributes to its activity today.

Paleoseismologists attempt to identify past earthquakes by digging trenches perpendicular to the linear trace of faults. They estimate the timing of ancient earthquakes and the *slip rate* of the fault by radiocarbon dating pieces of wood or other organic material offset across the fault plane. Studies along the Teton fault suggest that past earthquakes of magnitude 7.0 or so would offset the range from the adjacent basin by 1–2 m (~3-6 ft) or more. Repeated rupture along the

So what can we take away from this geologic detective work along the Teton fault? It appears that earthquakes may happen in clusters, separated by long periods of seismic quiescence. Between 14 and 7.9 ka, earthquakes happened on average about every 680 years. But there hasn't been a major magnitude 7 quake in the last 7100–4800 years. Does this mean that the Teton fault is locked, slowly accumulating tectonic stress until some threshold is reached and the fault ruptures again? This is likely: The Teton fault has been ripping large-magnitude earthquakes for the past 9–10 million years, so there's no reason to think that it won't continue rupturing for the next few million years.

Authorities in Grand Teton National Park are certainly aware of the potential hazard and the inherent unpredictability of earthquakes along the Teton fault. Seismic monitoring of the Yellowstone–Teton region detects small- to moderate-size earthquakes just about every day, but few actually occur on the Teton fault. In fact, no significant quake has occurred on the Teton fault since white settlers first came to the region. But past earthquakes in the greater area provide scenarios for the immense

Figure 25.11 A sharp break in the slope marks the topographic expression of the Teton fault scarp above String Lake. (White arrows mark the steep surface of the scarp.) The scarp is 38 m (124 ft) high and cuts glacial debris from the Last Glacial Maximum.

Figure 25.12 Scar of the Gros Ventre landslide of 1925, just east of Grand Teton National Park. An earthquake with an estimated magnitude of about 4.0 destabilized the water-saturated slope.

power and potential dangers that will accompany the next large quake along the Teton fault.

One example occurred in 1925 after heavy rains and spring snowmelt had left the area saturated. An earthquake estimated to be about magnitude 4 shook the waterlogged ground in the Jackson Hole area. A huge mass of rock, soil, and vegetation separated from the side of Sheep Mountain just to the east of today's park and plunged as a chaotic mass down into the valley of the Gros Ventre River. No one was killed in the sparsely populated area. The **landslide** dammed the river, forming a large lake upstream. Two years later, however, the natural dam was breached, triggering a devastating flood of debris that rushed through the Gros Ventre River Valley, inundating ranches and the town of Kelly, Wyoming. Six were killed in the deluge. The scar of the Gros Ventre Slide is still visible just east of the park on the slopes of Sheep Mountain (Fig. 25.12).

Another earthquake rocked the region near the Grand Tetons in August 1959 when the largest historic earthquake in the Rockies occurred about 100 km (60 mi) to the north of the Tetons along the west side of the Yellowstone Plateau. The Hebgen Lake earthquake was a magnitude 7.5 event, with the fault rupturing up to 6.5 m (~22 ft) during the terrifying tens of seconds of violent groundshaking. The quake triggered an enormous landslide that swept downslope into the Madison River Canyon where it kicked up walls of water 10 m (33 ft) high. The event began just before midnight, so campers in the canyon were completely unprepared for the catastrophe. Twenty-eight people were either buried by the landslide or drowned in the resulting flash flood.

Consider the consequences if a major earthquake happened on the Teton fault during the summer tourist season when tens of thousands of visitors are driving the roads, hiking the trails, and fishing the rivers. It's no wonder that emergency management officials have prepared a detailed hazard response plan in anticipation of the next major quake on the Teton fault. Potentially, it won't be necessary for the next few hundred or even the next few thousand years. Unfortunately, it might be necessary next week.

We just don't know because the prediction of earthquakes may be an unattainable quest.

Glaciers and the Teton Landscape

Millions of years of tectonic activity on the Teton fault set the stage for the Ice Age glaciers to work their erosional magic, transforming a mere tilted fault-block mountain range into the craggy peaks and smoothly rounded valleys we enjoy today. Alpine glaciers modified the Teton Range, creating all the glacial features you read about in earlier chapters such as cirques, horns, arêtes, tarns, striations, erratics, moraines, and hanging valleys. In contrast, an *ice cap* centered over the Yellowstone Plateau transformed the lower elevation valley of Jackson Hole (Fig. 25.13). It would be useful to review the relevant sections of the chapters on the national parks of the Cascades, Sierra Nevada, and Rocky Mountains (Chapters 12, 17, and 21, respectively) to refresh your memory of glacial processes and landforms. Here we'll focus on the effects of glaciation that make Grand Teton National Park such a unique and special place.

YELLOWSTONE PLATEAU ICE CAP
The global difference in average temperature between glacial and interglacial phases of the Pleistocene was about 6–10°C (11–18°F). In high-elevation terrain such as the Yellowstone–Teton region, many glacial-interglacial cycles of the Pleistocene affected the landscape, but the erosional activity of the three most recent glacial advances over the past 250,000 years obscured the earlier events. During these recent glaciations, an ice cap developed on the Yellowstone Plateau that reached thicknesses of over 1 km (3500 ft). Ice caps are smaller than *continental ice sheets* and commonly nucleate along the crests of mountain ranges or plateaus. Ice caps are also different from alpine glaciers in that they are not constrained within high-elevation valleys.

Lobes of ice grew laterally from the Yellowstone ice cap into the surrounding region, including southward across Jackson Hole. The Bull Lake glaciation (~160–130 ka) was extensive, reaching a thickness of ~450 m (1500 ft) in the southern part of the valley. The forested ridge of Timbered Island that rises above the sagebrush plains of Jackson Hole is a remnant *lateral moraine* left behind by the Bull Lake event. Lateral moraines are elongate piles of unsorted *glacial till* that accumulate along the flanks of glaciers and then are left behind as the lobe of ice melts and recedes. Most other vestiges of the Bull Lake event were degraded, however, by the later Pinedale glaciation (~70–14 ka), the same Last Glacial Maximum event recognized in Rocky Mountain and Glacier national parks.

A tongue of ice from the Yellowstone ice cap flowed into northern Jackson Hole during the Pinedale glaciation, reaching a thickness of 600 m (2000 ft). The ice scoured deeply into the soft glacial debris left by previous glaciations, carving a trough along the east base of the Tetons

Figure 25.13 An ice cap over 1 km (3500 ft) thick was centered over the Yellowstone Plateau during the most recent glacial phases of the Pleistocene. Glacial lobes extended outward from the crest, migrating south across Jackson Hole. The contours are elevations on the top of the ice cap. The orange line marks the extent of an earlier glaciation, whereas the blue line marks the edge of the ice cap during the Last Glacial Maximum.

where the elevations were lowest along the Teton fault. As the glacial lobe receded northward during the most recent interglacial phase, it left behind a *terminal moraine* of debris that impounded glacial meltwater, creating Jackson Lake (Fig. 25.10).

Fast-flowing streams draining the retreating glacial lobe ~14 ka were choked with sedimentary debris left behind as the ice melted. Much of the sediment was deposited in the youthful Jackson Lake, but many streams of the ancestral Snake River system bypassed the lake along its eastern edge and then flowed southward across Jackson Hole. **Braided streams**, overwhelmed by huge loads of sediment, frequently overtopped their banks and spread out laterally across their floodplains, depositing coarse sand and gravel as a broad *outwash plain*. Glacial debris filled the basin of Jackson Hole to its highest postglacial

level (Fig. 25.14A). The sand and gravel of the outwash plain don't retain water very well, so today the poor soils of Jackson Hole only support sagebrush and hardy grasses.

At some point after the maximum postglacial accumulation of sediment within Jackson Hole, episodic floods of meltwater turned the Snake River into a raging torrent that incised deeply into the soft sediments of the outwash plain, leaving behind flat **stream terraces** that mark the level of former floodplains (Fig. 25.14B). Downcutting by the Snake River may have been triggered by a subtle change in climate that increased the flow velocity, or perhaps by uplift near its source on the Yellowstone Plateau, which would have increased the gradient of the river. Flow velocity and thus the ability to cut downward may also have been enhanced by a reduction in the glacial sediment load in the river. In the northern part of Jackson Hole today, the

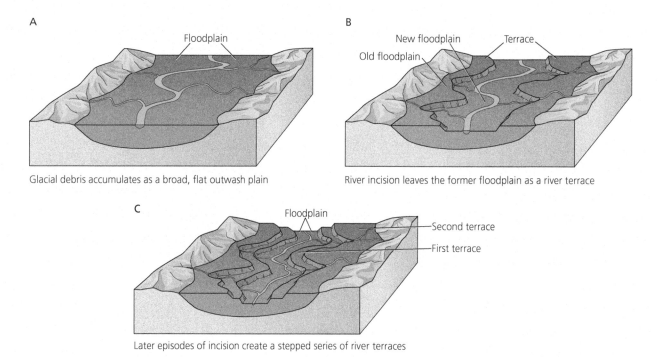

A

Floodplain

Glacial debris accumulates as a broad, flat outwash plain

B

New floodplain   Terrace

Old floodplain

River incision leaves the former floodplain as a river terrace

C

Floodplain

Second terrace

First terrace

Later episodes of incision create a stepped series of river terraces

Figure 25.14   Stream terraces like those along the Snake River form by a sequence of sediment accumulation followed by phases of stream incision. A) The process begins with the deposition of a broad outwash plain that fills the valley. The Snake River and its sediment-choked tributaries flowed along the upper surface. B) An increase in flow velocity caused the Snake River to incise downward into the underlying outwash plain, leaving the former floodplain stranded above as a stream terrace. C) Repeated episodes of incision left behind a stepped series of stream terraces above the modern Snake River.

Figure 25.15   Stream terraces (marked by arrows) left behind as the Snake River incised episodically into the outwash plain of Jackson Hole. The terraces were once continuous surfaces that acted as the Snake River floodplain prior to episodes of incision.

Snake River flows about 40 m (~140 ft) beneath the highest terrace (Fig. 25.15). Ten individual terrace levels are identified, each developed since ~14 ka as the erosive power of the Snake River varied in response to short-term changes in regional climate, uplift near its headwaters to the north, or a smaller volume of sediment.

TETON ALPINE GLACIERS AND LAKES

During the Last Glacial Maximum, the peaks of the Tetons were exposed as nunataks above the tongues of alpine glaciers that flowed east and west from the crest (Fig. 25.13). In Jackson Hole, the alpine glaciers converged with southward-flowing lobes of glacial ice from the Yellowstone ice cap. Envision the vast expanse of ice that enveloped the region 20,000 years ago, broken only by the ragged peaks of the Tetons poking up like rocky islands above a white sea. As the alpine glaciers receded in response to Holocene interglacial warming, they revealed sculpted U-shaped valleys that terminate upslope in amphitheater-like cirque basins (Fig. 25.16). The retreat of tributary glaciers left behind hanging valleys high above the main glacial valleys. Serrated arêtes mark the boundaries between adjacent U-shaped valleys, while tooth-like horns (like Grand Teton peak) compose the highest elevations.

As the alpine glaciers receded over the last 14,000 years or so, they left behind enormous amounts of rocky debris at the mouths of glacial valleys along the eastern edge of the Tetons. These lateral and terminal moraines impounded meltwater in the depression between the moraine and the mountainfront, forming pristine **moraine lakes** like Jenny Lake at the mouth of Cascade Canyon (Fig. 25.17). A necklace of azure lakes—Jackson, Leigh, Jenny, Bradley, Taggart, and Phelps—developed behind morainal dams along the base of the Tetons (Fig. 25.18). The debris composing

Figure 25.16 Computer-generated perspective of glacially carved eastern front of the Teton Range. Jenny Lake is the rounded body of water near the center of the image, and Leigh Lake extends back into the range to the right of center.

Figure 25.17 Reconstruction of the alpine glacier that once flowed through Cascade Canyon before splaying out at the base of the range. As the ice melted, it left behind a terminal moraine that impounded meltwater, forming Jenny Lake.

the morainal ridges ranges in size from boulders to clay, a mixture that retains water and supports the growth of pine forests. The green, wooded moraines can easily be identified by their contrast with the brownish, sagebrush-covered plain of Jackson Hole (Fig. 25.16). Over the long term,

the moraine lakes lining the base of the Tetons may eventually convert to wetlands and meadows as they fill with sediment from inflowing streams.

The string of lakes was originally localized by the low elevations created by movement on the Teton fault running along the base of the range. The surface of Jackson Hole tilts imperceptibly westward at less than a 1° slope toward the base of the Tetons. Episodic downdropping of Jackson Hole adjacent to the Teton fault maintains a subtle trough near the base of the range, ideal for the ponding of water within glacial lakes. Thus, the Teton fault acted in concert with Ice Age glaciation to create one of the most aesthetically pleasing landscapes in the national park system.

Grand Teton National Park is one of those iconic places that we associate with the American West. Visit the park to see very old rocks at the highest peaks of very young mountains. Gawk at the jagged terrain, the towering summits, the lush valleys, the cobalt-colored lakes, and the abundant wildlife. (My kids will never forget the moose that wandered into Cascade Creek from the opposite bank as they were wading in the snowmelt waters.) Hike the trails and catch-and-release fish in the swift-running streams. Grand Teton National Park should be on your shortlist of national parks that you absolutely have to visit.

Key Terms

anastomosing stream
braided stream
landslide
moraine lake
paleoseismology
stream terrace

Related Sources

KellerLynn, K. (2010). *Grand Teton National Park and John D. Rockefeller, Jr. Memorial Parkway.* Geologic Resources Inventory Report. NPS/NRPC/GRD/NRR—2010/230. Fort Collins, CO: National Park Service.

Love, J., Reed, J. C., Jr., and Pierce, K. L. (1971). *Creation of the Teton Landscape.* Moose, WY: Grand Teton Natural History Association. Accessed February 2021 at http://www.nps.gov/parkhistory/online_books/grte/grte_geology/contents.htm

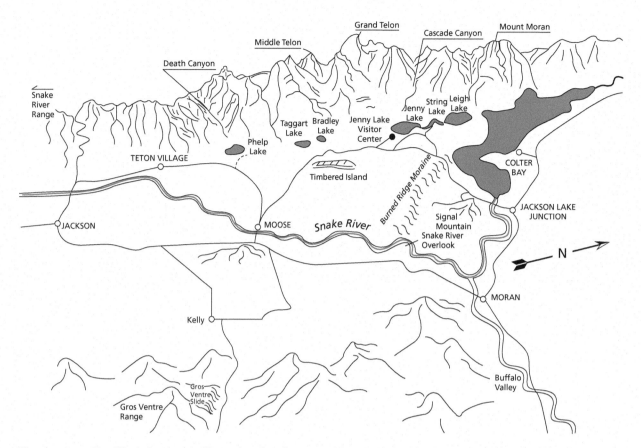

Figure 25.18 Simplified sketch map illustrating the alignment of moraine lakes that occupy the low-elevation trough along the Teton fault between the range and Jackson Hole.

McPhee, John. (1986). *Rising from the Plains*. New York: Farrar, Straus, Giroux.

Smith, R. S., and Siegel, L. J. (2000). *Windows into the Earth*. New York: Oxford University Press.

National Park Service—Geologic Activity. Accessed February 2021 at https://www.nps.gov/grte/learn/nature/geology.htm

Discover Grand Teton. Accessed February 2021 at http://www.discovergrandteton.org/teton-geology

To see animations and videos of the geology of the Tetons, go to: https://www.nps.gov/grte/learn/photosmultimedia/geo_podcasts.htm (accessed February 2021)

PART 6
National Parks of the Active Pacific Margin

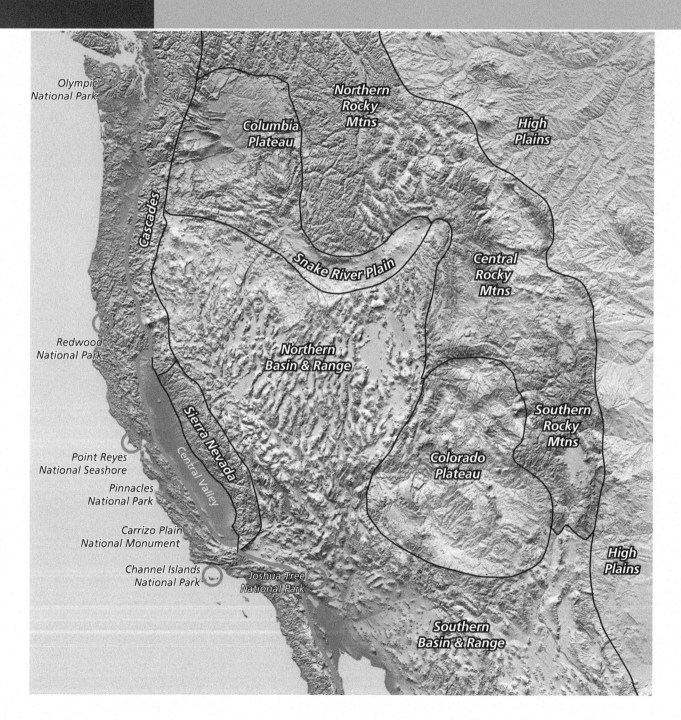

Figure 26.1 Physiographic provinces of the American West, with Pacific Margin Province highlighted. Locations of seven national park units addressed In Part 6 marked by orange circles.

The striking landscape of the Pacific Margin Province of California, Oregon, and Washington encompasses hundreds of individual ranges and intervening valleys, all bordered by a rugged coastline (Fig. 26.1). The province spans the climatic spectrum from the arid interior of Southern California to the rain-splashed Pacific

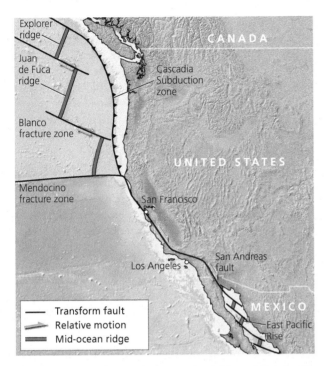

Figure 26.2 Modern active tectonic setting of the Pacific Margin of North America illustrating the right-lateral motion along the San Andreas fault in California that connects a transform-dominated divergent boundary in the Gulf of California with the Cascadia convergent margin in the north.

Northwest. The aesthetic appeal of the Pacific Margin is the result of over 200 million years of active tectonism along the continental edge. In the past, the dominant tectonic control was the subduction of the Farallon plate, whose effects on the evolution of the American West from the Sierra Nevada to the Rocky Mountains were described in earlier chapters. The active Pacific Margin today, however, is controlled by the San Andreas fault in most of California and the Cascadia subduction zone in the Pacific Northwest (Fig. 26.2).

An **active continental margin** is one where the geographic edge of a continent is roughly coincident with a tectonic plate boundary, in contrast to a *passive continental margin* where the continental edge is far removed from a plate boundary. The east coast of North America is a passive continental margin, thousands of kilometers west of the divergent plate boundary along

the Mid-Atlantic Ridge (Fig. 1.6). (Ancient passive margins were introduced in Chapter 24 on Death Valley National Park.) Active continental margins are characterized by volcanism, seismicity, and uplift, whereas passive continental margins are dominated by sediment deposition and slow, gradual subsidence due to the tectonically quiet conditions. Coastal landscapes reflect the contrast in tectonic setting. Lowland coastal plains and broad beaches typify the Atlantic shoreline, reflecting the tectonic quiescence. In comparison, the tectonically active Pacific Coast from Southern California to Washington is an elevated, ruggedly incised series of seacliffs, coves, and high terraces (Fig. 26.3).

Three of the national parks, one national seashore, and one national monument described in Part 6 are intimately related to the San Andreas fault in California (Fig. 26.4). In the south, Joshua Tree National Park is geographically part of the Mojave Desert subprovince of the Basin and Range, but the uplift of mountains within the park is primarily related to motion along the nearby San Andreas system of faults. Off the coast of Southern California, Channel Islands National Park owes its origin to uplift along a bend in the San Andreas fault. In central California, Carrizo Plain National Monument exhibits textbook landscape features created by episodic movement along the San Andreas, and Pinnacles National Park provides a critical "pinning point" for determining the total offset along the fault. A bit further to the north, Point Reyes National Seashore provides a dramatic setting for understanding shoreline processes and coastal landscapes related to the San Andreas. The two other national parks addressed in Part 6 are related to the Cascadia convergent margin. Olympic National Park in Washington offers insight into the exhumed interior of the Cascadia subduction zone, whereas Redwood National and State Parks along the northern California coast preserves the complex anatomy of a Mesozoic subduction zone.

Figure 26.3 The steep, jagged coastline of California south of Big Sur—a common landscape of the tectonically active Pacific Margin.

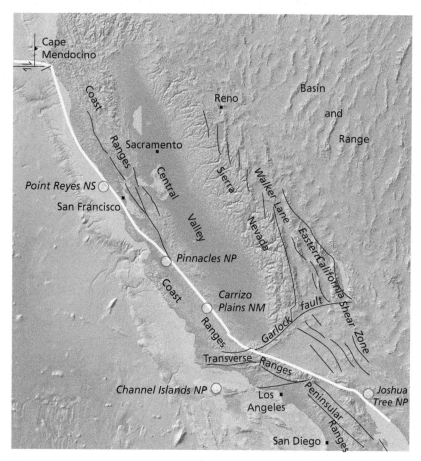

Figure 26.4 Location of select national parks, monuments, and seashores in California in relation to the San Andreas fault zone. Note the continental shelf and slope seaward of the shoreline.

In this introductory chapter to Part 6, we'll focus on the regional geologic setting of national park units along the San Andreas transform margin in California. Some of the material in this section may be a bit difficult to comprehend, mostly because the geology of California is inherently complex. It may be useful to refer back to this chapter for the bigger picture as you read the next three chapters. To avoid overwhelming you, we'll wait until Chapter 30 on Olympic and Redwood national parks to review the tectonic setting of the Cascadia convergent margin.

San Andreas Transform Fault

Many of the earlier chapters of this book focused on the long-lived convergent plate boundary where the Farallon plate descended beneath western North America along a subduction zone, giving rise to the mountainous landscapes of the American Cordillera. With this chapter we introduce *transform plate boundaries*, extensive *strike-slip fault systems* that form in response to shear stress where the tectonic forces are oriented lateral to one another (Figs. 1.8, 24.9). Along the 1300 km (800 mi) of the San Andreas transform fault, the Pacific plate is moving to

the northwest relative to the adjacent North American plate. (It would be worthwhile to review the relevant sections on transform plate boundaries in Chapter 1 to facilitate understanding of the material in this chapter.)

It's a common misconception that the transform plate boundary in California occurs directly on the San Andreas strike-slip fault. In reality, the San Andreas is only the dominant fault in a series of roughly parallel faults that span a few hundred kilometers in width. The plate boundary is better thought of as the San Andreas fault *zone*, consisting of the San Andreas itself and thousands of subsidiary faults. Each of the faults in the zone exhibits the same *right-lateral* sense of motion, with the east side of each fault moving southeast relative to the west side of each fault. (Recall from Chapter 24 on Death Valley that a right-lateral strike-slip fault is defined by the block of rock on the opposite side of the fault from the observer moving to the right.)

In Southern California, seismically active strike-slip faults like the San Jacinto, Elsinore, and Newport-Inglewood faults soak up the tectonic stresses between plates. In the San Francisco Bay Area, the San Andreas is accompanied by strike-slip faults like the Hayward, Calaveras, and Rodgers Creek faults. All of the faults in the San Andreas system collectively accommodate the massive amount of shear stress created by the northwesterly moving Pacific plate and adjacent North American plate, somewhat like a vertically oriented deck of cards smeared laterally by your hands.

The Global Positioning System of satellites permits the measurement of *absolute* rates of plate motion. The Pacific plate is moving to the northwest at an absolute rate of ~10 cm/yr (~4 in/yr), whereas the adjacent North American plate is moving in a somewhat perpendicular direction to the west–southwest at an absolute rate of ~2 cm/yr (~1 in/yr). The net product of these absolute rates and directions produces a *relative* rate of motion; the Pacific and North American plates grind past one another along the San Andreas fault zone at a relative rate of about 5 cm/yr (2 in/yr).

How do the San Andreas and its companion faults express themselves on the landscape? In some places, the San Andreas fault is easily identified as a linear gash cutting across the surface, bound by narrow ridges on either side (Fig. 26.5). The arid **Carrizo Plain National Monument** of central California is a remote but ideal place to see the influence of a major strike-slip fault on the landscape. The Temblor Range borders the Carrizo Plain to the

Figure 26.5 Linear trace of San Andreas fault looking northwest, Carrizo Plain National Monument, central California. Soda Lake is visible in upper left. Arrows show right-lateral sense of motion.

northeast, while the Caliente Range encloses the plain on the southwest (Fig. 26.6). Windswept grasslands cover the plain, with the salt-encrusted bed of Soda Lake marking the lowest part of the closed basin. After wet winters, the normally auburn grasslands erupt in a "superbloom" of wildflowers.

Where the San Andreas fault cuts across the base of the Temblor Range, dry stream channels that drain episodic runoff from the range suddenly make a sharp 90° bend where they intersect the fault (Fig. 26.7). The usually dry channels are offset over a 100 m (330 ft) before they resume their downslope path across the edge of the plain.

Offset channels are reliable indicators of the presence of a fault and can be used to measure the long-term average rate of motion along faults like the San Andreas. It's important to understand, however, that massive blocks of rock on either side of a major fault like the San Andreas rarely slide smoothly past one another in response to shear stress. Instead, they jerk violently a few meters to several meters at a time during individual earthquakes. After the abrupt stress release during the quake, the fault resumes its former state of repose, slowly accumulating tectonic stress for centuries to millennia until it ruptures once again in a furious spasm of motion.

Strike-slip faults like the San Andreas are not strictly smooth, linear planes of weakness. Slight bends in the fault plane or changes in rock type along the fault plane beneath the surface may create localized areas of compression or extension. The narrow, elongate hills that parallel the linear valley of the San Andreas in the Carrizo Plains (Fig. 26.5) are called **pressure ridges** that form due to localized areas of compression and uplift along the fault. The combination of compression and shear stresses is called **transpression**. In other places along the fault, slight bends may create small amounts of extension that result in the downdropping of small basins called **sag ponds** that may fill with water (Fig. 26.8). The combination of extension and shear stresses is called **transtension**. The hybrid stresses of transpression and transtension are fundamentally important for explaining many California landscapes. Several other surface features that reveal the trace of the San Andreas will be identified as we visit other national parks along the fault.

Seismicity along the San Andreas Fault Zone

The San Andreas and its network of accompanying faults are notable for their active seismicity, ranging from relatively common small quakes to rare catastrophic quakes. The *stick-slip* mechanics of earthquake faults was addressed in several earlier chapters in the context of individual provinces and parks. The following brief discussion adds a few more concepts and builds on those earlier descriptions.

The actual location along the fault plane at depth where rupture begins is called the **focus**, with the **epicenter** located on the surface directly above the focus. Seismologists plot the focal depths for individual earthquakes along the fault plane using seismograph records. The distribution of earthquake foci along the fault plane permits geoscientists to visualize the

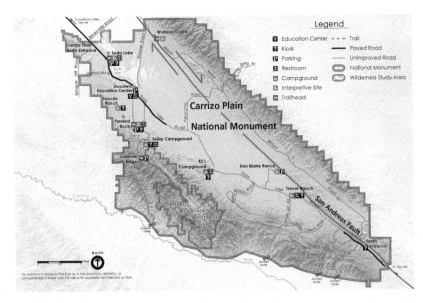

Figure 26.6 Carrizo Plain National Monument, showing the northwesterly trend of the San Andreas fault across the landscape.

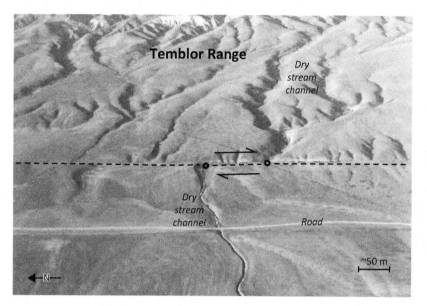

Figure 26.7 Right-lateral offset of stream channels along the San Andreas fault, Carrizo Plain National Monument. Fault trace marked by dashed line. Small circles represent the tips of the channel that were once aligned but are now offset ~100 m along the fault by the cumulative effects of several earthquakes. Arrows show right-lateral sense of motion.

orientation and dimensions of a fault deep beneath the surface. From these data, we know that the San Andreas fault is a near-vertical plane that extends at least 16 km (10 mi) into the crust. This depth is close to the *brittle-ductile transition* (introduced in Chapter 23 on the Basin and Range). Earthquakes happen where crustal rocks behave as a brittle material in response to tectonic stress. Below about 16 km or so, the rock behaves like a plastic, accommodating tectonic stress by gradual ductile flow. Recall that continental crust is about 35 km (~22 mi) thick and lithospheric plates are 100–150 km (~60–90 mi) thick. Relative to these thicknesses, the San Andreas fault zone and the accompanying seismicity are comparatively shallow features.

Over the past few centuries, three major earthquakes have occurred along the San Andreas fault (Fig. 26.9). In 1906, an earthquake with an estimated magnitude of 7.8 rocked northern California, with the epicenter located just seaward of the Golden Gate. The northern segment of the San Andreas ruptured along a 450 km (~300 mi) length, with maximum surface offset of about 6 m (20 ft). The devastating effects on San Francisco are well documented, and the quake and accompanying firestorm resulted in one of the deadliest natural disasters in U.S. history.

In 1857, a relatively uninhabited portion of the San Andreas in Southern California ruptured along 350 km (220 mi) of its length, accompanied by a maximum of 9 m (30 ft) of right-lateral displacement (recorded along

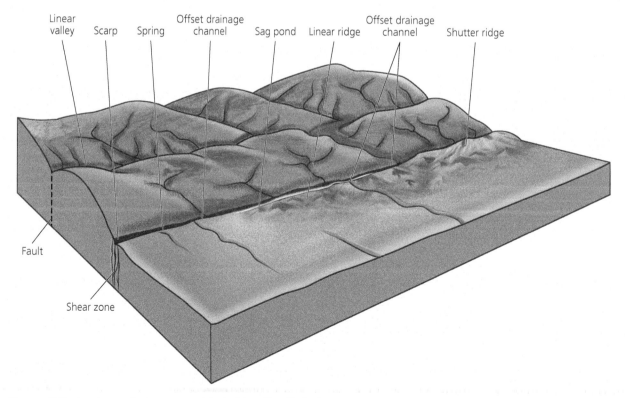

Figure 26.8 Landscape features associated with strike-slip faults

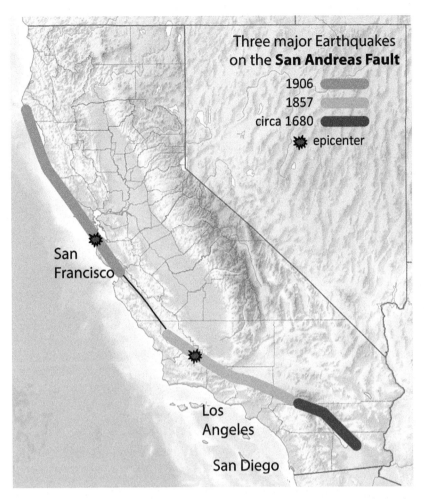

Figure 26.9 Length of rupture for the three most recent large-magnitude earthquakes along the San Andreas fault.

the offset channels at Carrizo Plains National Monument). The magnitude is estimated to have been around 7.9, and groundshaking was felt from northern California to San Diego and east to Las Vegas. Known as the Fort Tejon earthquake, this was the last "Big One" in Southern California.

The third major quake on the San Andreas is estimated from paleoseismic studies to have occurred around 1680 and originated in the Coachella and Imperial valleys in southeastern California, just west of Joshua Tree National Park. Its magnitude may have been around 7.8. This segment of the San Andreas, located just east of the Los Angeles–San Diego corridor, is considered to be the most likely location of the next major earthquake along the fault. Californians alive today have never experienced an earthquake with magnitudes like any of these three.

These earthquakes are just the latest large-magnitude events in a continuum of tens of thousands of violent quakes that have been ripping along the San Andreas for the past several million years. Each of the main segments along the San Andreas is currently locked and slowly accumulating strain energy as the Pacific plate continues its inexorable motion to the northwest. Eventually,

another segment of the San Andreas will abruptly transform its pent-up potential energy into the kinetic energy of an earthquake, releasing seismic waves toward the surface and its thin veneer of civilization.

Origin of the San Andreas Transform Boundary

The origin and evolution of the San Andreas transform plate boundary was first addressed in Chapter 23 within a broader discussion of the transition from compressional tectonism to extensional tectonism in the Basin and Range Province. The bottom-line message of that earlier account was that extension began in the American West in response to the alleviation of compression as the San Andreas transform fault evolved through the mid- to late Cenozoic.

The following description of the generally accepted model for the origin of the San Andreas adds context and detail to the earlier introduction.

By 40 Ma, the Laramide orogeny was coming to a close as the Farallon plate was consumed beneath western North America (Fig. 26.10A). Farther out to sea, volcanism along the East Pacific mid-ocean ridge was adding new oceanic lithosphere along the divergent plate boundary separating the Farallon and Pacific plates. (Be sure to review the processes associated with seafloor spreading in Chapter 1, Fig. 1.10.) The East Pacific Rise and its associated transform fault system were slowly dragged toward the subduction zone as the rate of convergence exceeded the rate of seafloor spreading along the ridge.

Around 28–30 Ma, the East Pacific Rise made contact with the subduction zone at an oblique angle somewhere near what is today the continental margin of Southern California (Fig. 26.10B). The hot and buoyant East Pacific Rise resisted descending beneath the adjacent continental plate, but the dense oceanic rock of the Farallon separated from the Pacific plate, continuing to subduct into the mantle. The preferred motion of the Pacific plate was toward the northwest, governed by the transform faults that offset the mid-ocean ridge. By 20 Ma, the evolving plate boundary between the Pacific and North American plates was a series of strike-slip transform faults that were the precursors to the San Andreas (Fig. 26.10C). The remnants of the Farallon plate were now two smaller, separate plates, each continuing to subduct beneath the North American plate. The relative motion between the Pacific

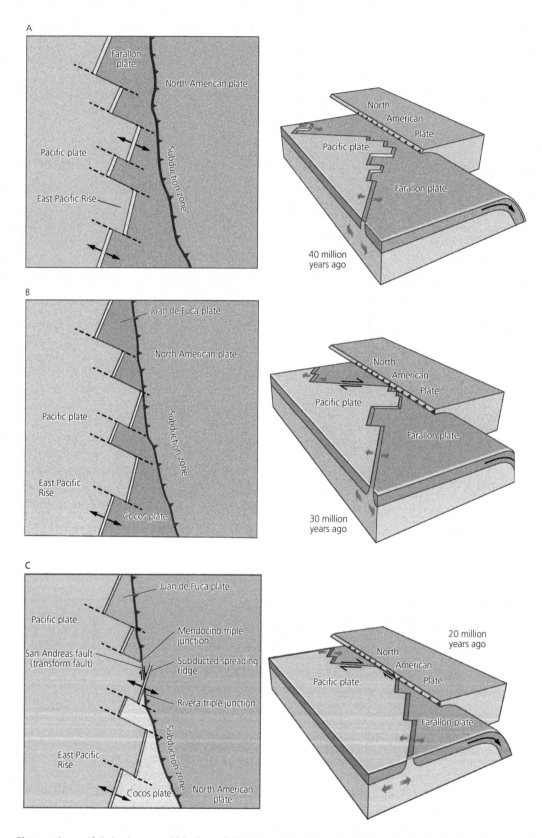

Figure 26.10   A) Paired map and block model for the origin of the San Andreas fault. Tectonic setting ~40 Ma. B) Tectonic setting ~30 Ma. C) Tectonic setting ~20 Ma.

and North American plates along the newly evolving San Andreas fault progressively changed from one of compression to one dominated by shear, accompanied by an easing of tectonic stress toward the continental interior.

As the subducting Farallon plate continued its descent into the mantle inland from the young San Andreas fault, a gap likely opened between it and the Pacific plate, which now moved northwest along transform faults (Fig. 23.12D). Plastic rock of the asthenosphere flowed into the void, eventually rising beneath the interior of the American West and buoyantly lifting the overlying continental rock. This upward flexure of the lithosphere likely began around 16 Ma, coincident with the beginning of extensional tectonism and the formation of the Basin and Range. In places, the lower pressures of the upper asthenosphere permitted the rock to melt, creating magma that leaked upward through faults to the surface.

The Pacific plate was fully juxtaposed against the North American plate by 16 Ma, with the San Andreas now becoming the dominant fault along the transform boundary. The north and south ends of the San Andreas were **triple junctions**, places where three tectonic plates intersect (Fig. 26.10C). In the north, the Mendocino triple junction marked the location where the North American plate, the Pacific plate, and the remnant of the Farallon plate met. Geologists call this northern vestige of the Farallon plate the Juan de Fuca, Gorda, and Explorer microplates, which today are consumed within the Cascadia subduction zone. To the south, the remnants of the Farallon plate are called the Rivera and Cocos plates, now descending beneath Mexico and Central America. The San Andreas fault grew both northward and southward through time as the triple junctions migrated in response to the consumption of the final fragments of the Farallon plate. You might imagine the San Andreas growing somewhat like a gigantic zipper slowly closing at both ends.

By 5–6 Ma, the sliver of continental rock composing the Baja California peninsula and much of Southern and central California was captured by the Pacific plate along the west side of the San Andreas fault (Fig. 26.11). The opening of the narrow continental rift of the Gulf of California around this time may be related to seafloor volcanism associated with the partial subduction of the East Pacific

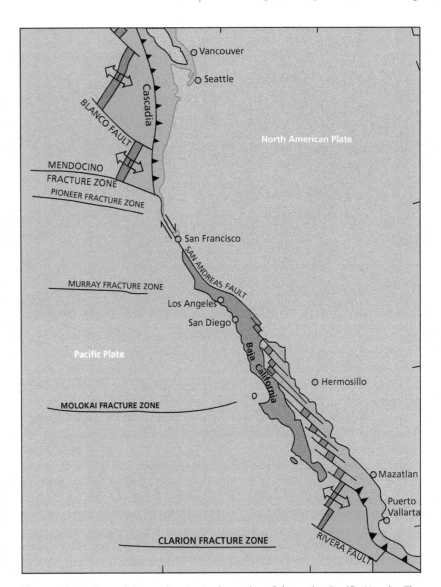

Figure 26.11 Map of the modern tectonic setting of the active Pacific Margin. The sliver of land labeled "Baja California" is continental rock attached to the edge of the Pacific plate, sliding to the northwest along the San Andreas fault.

Rise mid-ocean ridge. Regardless of the exact mechanism, the narrow slice of continental rock to the west of the San Andreas today moves to the northwest as part of the mostly oceanic Pacific plate. The Gulf of California is widening by motion along a series of parallel transform faults that are slowly ripping Baja California obliquely away from mainland Mexico. (In Chapter 34, you'll see how the opening of the Gulf of California around 5–6 Ma influenced the incision and evolution of the Grand Canyon.)

Origin of the Modern California Landscape

Over the last several million years, the Baja-to-Point Reyes sliver of continental rock has migrated over 315 km (195 mi) to the northwest along the San Andreas transform fault, dragged along as the ragged edge of the much larger oceanic part of the Pacific plate. Most of this movement

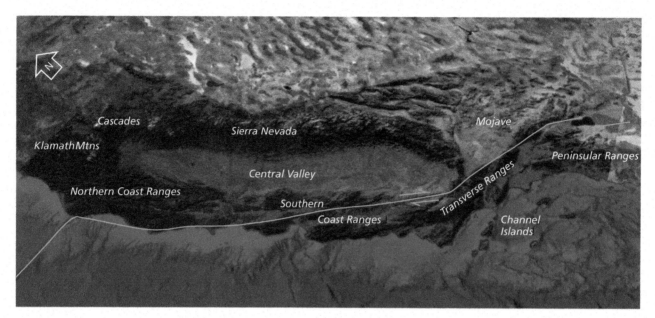

Figure 26.12   Google Earth image of California showing the major physiographic features. The thin line extending through the state is the trace of the San Andreas fault.

occurred during the tens of thousands of large-magnitude earthquakes that jolted the region through this period. It was during the last 5 m.y. of motion that many of the topographic features we see today in the national parks of California were formed.

Some areas along the network of subparallel faults of the San Andreas system experienced transpression, triggering the uplift of highlands. The **Transverse Ranges** of Southern California, including the enormous San Gabriel and San Bernardino Mountains, grew rapidly in response to transpressive stress along the San Andreas and neighboring faults (Fig. 26.12). The eastern Transverse Ranges include the mountains that compose Joshua Tree National Park, and the western edge of the range extends offshore to include four of the Channel Islands. West of the Central Valley, the northwest-oriented **Coast Ranges** were elevated by transpression along the network of faults aligned subparallel to the coastal edge of California.

In other areas along the San Andreas fault zone, the rocks were subjected to transtensional stresses, resulting in the downwarping of pull-apart basins, forming elongate valleys like Napa and Salinas, as well as San Francisco Bay. Given vast amounts of geologic time, earthquakes do the majority of the work involved in creating the modern landscape of California. Weathering and erosion act to modify highlands and lowlands, but the ultimate origin of the topography is due to the unceasing seismicity associated with the San Andreas fault zone.

Recall from Chapter 17 that the modern Sierra Nevada began their uplift about 5 m.y. ago. Extensional tectonism in the central Basin and Range of Nevada migrated westward through time to the eastern edges of the *Sierran batholith*. The great normal fault system that lines the east flank of the Sierra Nevada block began to raise the range

one earthquake at a time beginning around 5 Ma, culminating in the magnificent Sierran landscape that we see today. Smaller normal faults just east of the Sierra raised mountains of the Basin and Range in California and western Nevada, including those bounding Death Valley. Weathering and erosion smoothed the newly formed California highlands, with the transported sedimentary debris filling the intervening basins. Where rainfall was sufficient, the loose sediment combined with organic matter to create the fertile soils of the Central Valley and many other intermontane valleys throughout California. A thin veneer of grasslands and oak woodlands cover much of the terrain today, imparting the gold and green tones that color the landscape.

Rocks and Terranes of the Pacific Margin in California

Some of the rocks that compose the current landscape of national parks of the active Pacific Margin of California accumulated in place where they are found today. Most of the rocks, however, originated far away on the seafloor of the ancient Pacific Ocean and then emigrated to their current location via tectonic transport. Both the homegrown rocks and the recent arrivals came together within the *deep ocean trench* that marked the convergent plate boundary between the Farallon and North American plates. Much later, these rocks were lifted up from the seafloor to their current continental position. Here, we'll briefly address the origin of some of these rocks, specifically those that characterize the Coast Ranges that encompass Pinnacles National Park, Point Reyes National Seashore, and the recently designated **Berryessa Snow Mountain National Monument**. To gain an appreciation of what the region that would become California looked like in the past, revisit the paleogeographic maps of Figures 3.17 (~1.7 Ga), 4.8 (185 Ma), 17.9 (90 Ma), 20.4 (65 Ma), and 23.13 (35 Ma). You'll see that California didn't really exist for most of Earth history, but rather accumulated as fragmentary masses that were tectonically constructed one piece at a time.

FRANCISCAN ASSEMBLAGE

The greatest volume of rock that makes up the Coast Ranges is called the **Franciscan assemblage**, which consists of a diverse array of rock types. The Franciscan is Mesozoic through early Cenozoic in age and accumulated within the deep ocean trench where the Farallon plate descended beneath western North America. The dominant rock type of the Franciscan consists of thick sequences of interbedded shale, siltstone, and sandstone (Fig. 26.13). Energetic, seabed-hugging flows called turbidity currents transported sand, silt, and clay from the coastline to the trench where the sediments settled out as thin beds on the seafloor. These rhythmically layered sandstone, siltstone, and shale strata are known as **turbidites**, named for the turbidity current from which they were deposited. Turbidites are the only rock type of the Franciscan assemblage that accumulated close to their current location.

A second common sedimentary rock of the Franciscan is **chert**, formed primarily from the countless siliceous skeletons of radiolarians (Fig. 26.14). These zooplankton proliferated in the equatorial waters of the Mesozoic oceans, with their tiny dead bodies drifting to the seafloor to build up as a silica-rich "ooze," eventually solidifying into chert with burial and time. These chert beds were originally deposited as horizontal layers on the abyssal seafloor thousands of kilometers away, and then they rode passively atop the Farallon plate until they began their descent into the subduction zone. The edge of the North American plate acted like a gigantic bulldozer blade, scraping off the veneer of layered chert beds of the seafloor while the underlying harder rocks of the oceanic lithosphere continued downward into the mantle. The chert layers typically deformed into tight folds or were tilted into high angles as they jammed up within the deep sea trench along the subduction zone.

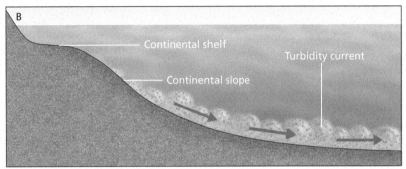

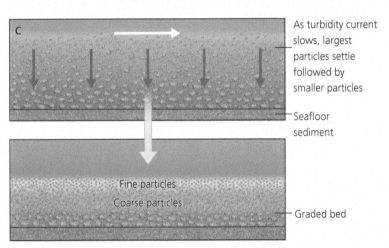

Figure 26.13 A) Alternating thin beds of shale, siltstone, and sandstone known as turbidites—Salt Point State Beach, California. B) Turbidites form from turbidity currents that sweep downslope before spreading out across the deep sea floor. C) The coarser sand particles fall first, while the finer clays settle out on top. The result is a sandstone-shale couplet representing a single turbidity current.

Other parts of the seafloor atop the Farallon plate were scraped off as well, including "pillows" of volcanic basalt (Fig. 26.15). When lavas extrude from fractures along mid-ocean ridges or from underwater volcanoes, they come in contact with the cold waters of the deep sea. This causes the outer surface of the lava flow to solidify into a glassy shell that insulates the lava within, allowing it to flow further until it chills completely in the shape of an elongate pillow. These rounded, overlapping blobs of volcanic rock are called **pillow basalts.** Because they form from *seafloor spreading,* they compose the entire ocean floor beneath a thin blanket of marine sediment. Like the bedded chert,

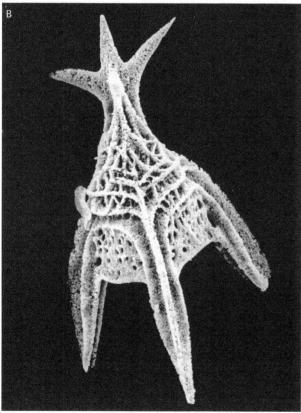

Figure 26.14 A) Outcrop of thinly bedded chert of the Franciscan assemblage tilted at a high angle from its original horizontal orientation, Marin Headlands. B) Siliceous skeleton of radiolarian of Jurassic age (~0.1 mm in length).

rock of California, is a greenish metamorphic rock derived from the chemical alteration of *peridotite*, an igneous rock that composes a large volume of the mantle (Chapter 1). The minerals composing peridotite convert to the different minerals within serpentinite when they come in contact with intensely hot fluids deep beneath the seafloor. Most of the serpentinized mantle is consumed within the subduction zone, but large masses may be sheared off and integrated into the chaotic mass of rock accumulating within the deep ocean trench.

During subduction, all of the rocks of the Franciscan assemblage piled up in a tectonic setting called the **accretionary wedge**, located along the front of the overriding continental plate where it meets the descending oceanic plate (Fig. 26.16). Through time, the detached slivers of chert, pillow basalt, serpentinite, and other remnants of the Farallon plate became intimately mixed with the land-derived turbidite beds. As subduction continued, the disparate collection of rocks was rotated and heavily deformed within the wedge, collectively forming the chaotic mass called the Franciscan assemblage. All of this deformation was happening coincident with the magmatism of the Sierra granitic batholith deep beneath the Mesozoic Sierra Nevada, as well as during the later phases of Laramide subduction that extended into the Cenozoic. The Franciscan rocks of the accretionary wedge were exhumed only in the last 5 m.y. to form the modern Coast Ranges that rose from the sea by transpression along the San Andreas network of faults. (We'll explore accretionary wedges once again when we visit Olympic and Redwood national parks in Chapter 30.)

To make things even more complex, geologically continuous strike-slip fault motion smeared out the rocks of the Franciscan along a northwest–southeast trend both east and west of the San Andreas. Fault-bounded slivers of disparate Franciscan rocks are called *accreted terranes*, each characterized by distinct rock types and fossils. (The concepts of *accretion* and *terranes* were introduced in Chapter 19 on Sequoia–Kings Canyon national parks and expanded upon in Chapter 21 on Rocky Mountain National Park.) Because each of these bodies of

slivers of pillow basalt were detached from the underlying lithosphere during their entry into the subduction zone and were incorporated into the overlying wedge of rock located along the contact between the Farallon and North American plates within a deep ocean trench.

A fourth rock type of the Franciscan comes from even deeper in the oceanic lithosphere. **Serpentinite**, the state

Figure 26.15   Pillow basalts of Franciscan assemblage, Nicasio Reservoir, Marin County, Golden Gate National Recreation Area.

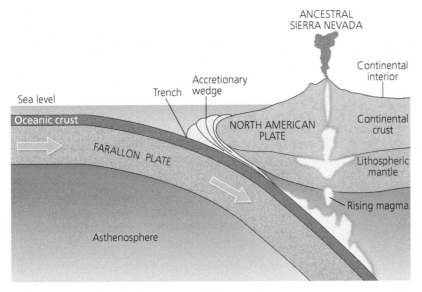

Figure 26.16   Location of the accretionary wedge along the front of the Mesozoic continental margin, directly above the subduction contact with the Farallon plate.

rock within the Franciscan experienced its own unique origin far from its current position in the Coast Ranges, accreted terranes are sometimes called "exotic" terranes, or "suspect" terranes.

SALINIAN TERRANE

Another type of accreted terrane, composed partially of Mesozoic granitic rock, records a different aspect of the assembly of the active Pacific Margin of California. The granitic rock is identical in age and composition to that of the Sierran batholith, but the question arises as to how it ended up along the west side of the San Andreas fault, inserted between slivers of Franciscan terrane (Fig. 26.17).

These belts of granitic rock are part of the **Salinian Terrane** that were sliced off the southern part of the Sierran batholith and then slid into place as part of the Pacific plate by strike-slip faulting.

Salinian granitic rocks are exposed throughout the Coast Ranges as far north as the San Francisco peninsula, Point Reyes National Seashore, Bodega Head, and ~50 km (30 mi) offshore in the Farallon Islands. This suggests that Sierran granitic rocks were transported hundreds of kilometers to the northwest from their original location by episodic motion along the San Andreas fault system. These wayward Sierran rocks have experienced a strange journey that began as magma deep along the subducting Farallon plate before rising to form the bowels of the volcano-capped ancestral Sierra Nevada. After isostatic exhumation and fault-related uplift of the Sierran batholith, these granitic bodies were detached along the San Andreas fault and were transported one quake at a time along the edge of the Pacific plate headed northwest. What a long, strange trip indeed.

The Franciscan assemblage and Salinian Terrane were introduced here not only to provide context for the rocks of Pinnacles National Park and Point Reyes National Seashore, but also to illustrate a mechanism for the outward growth of continents through time. These accreted terranes, now exposed within the California Coast Ranges, were originally assembled by ancient subduction processes and were later rearranged by strike-slip faulting along the transform plate boundary. They collectively define the westernmost rocks of the North American Cordillera in California.

When you look out across the serene vistas of the California Coast Ranges, try to think about what lies beneath the thin veneer of soil, grasses, and oak woodland. Rocks formed far away on the ancestral Pacific seafloor underlie those rolling hills, narrow ridgelines, and lush valleys. Then consider how they were amalgamated over geologic time within the crushing conditions of the accretionary wedge. All this was occurring just seaward of the enormous volcanic mountain chain of the ancestral Sierra Nevada. Later, huge blocks of granitic rock were translated to the northwest by episodic movement along the San Andreas network of faults where they were juxtaposed against the rocks of

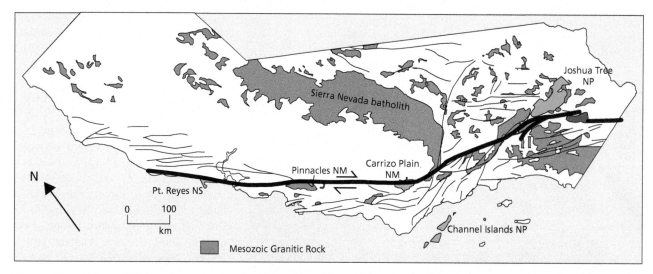

Figure 26.17   Slivers of Mesozoic granite detached from the southern Sierran batholith were transported by motion along the San Andreas fault system to positions within the Coast Ranges. These belts of displaced granite are part of the Salinian Terrane.

the Franciscan assemblage. Understanding this level of complexity will add a significant boost to your aesthetic appreciation of the coastal landscapes of California and its westernmost national parks.

Key Terms

accretionary wedge
active continental margin
chert
earthquake epicenter
earthquake focus
offset channel
pillow basalt
pressure ridge
sag pond
serpentinite
transpression
transtension
triple junction
turbidites

Related Sources

Collier, M. (1999). *A Land in Motion: California's San Andreas Fault.* Berkeley: University of California Press.

Lynch, D. K. (2015). *The Field Guide to the San Andreas Fault.* El Cajon, CA: Sunbelt Publications.

McPhee, John. (1993). *Assembling California.* New York: Farrar, Straus, Giroux.

Meldahl, Keith Heyer. (2013). *Rough-Hewn Land: A Geologic Journey from California to the Rocky Mountains.* Berkeley: University of California Press.

To see animations of projected groundshaking from earthquakes along the San Andreas Fault, go to: https://www.youtube.com/watch?v=L0vHyHLMNx8&list=PLfSGTU-Jx7YsZ-wAQH9XXtQZKFyQ6LBZDx (accessed February 2021)

To see animations of the tectonic evolution of the San Andreas transform fault by Tanya Atwater, a professor at UC Santa Barbara, go to: https://www.youtube.com/watch?v=bjU2ue-b1Rvg (accessed February 2021)

San Andreas Fault homepage: http://www.sanandreasfault.org (accessed February 2021)

27 Joshua Tree National Park

Figure 27.1 Granite monolith and Joshua trees, Old Woman Rock, Joshua Tree National Park.

The most obvious features of Joshua Tree National Park are rocky highlands, isolated monoliths of rock, and loose piles of boulders, all surrounded by broad alluvial plains dotted with Joshua trees and other desert plants (Fig. 27.1). In many ways, the terrain of Joshua Tree National Park is a hybrid of several other national parks. The desert scenery evokes many of the landforms seen in Death Valley, whereas the rounded exposures of granite are vaguely reminiscent of the rocks at Yosemite or Sequoia–Kings Canyon. The multiple sets of joints cross-cutting the massive outcrops bring to mind the highly fractured rocks of Zion or Canyonlands. And the strike-slip faults that bound the park and cut across its terrain are similar to those at Pinnacles or Point Reyes. But the surreal array of features at Joshua Tree National Park makes for an entirely unique landscape, one deserving of an extended visit to explore the desert terrain and experience the star-filled night sky.

Joshua Tree National Park is located in the **Mojave Geologic Subprovince**, easily identifiable on satellite photos by the angularity of its western corner (Fig. 27.2). The northern boundary of the Mojave block is defined by the **Garlock fault**, a left-lateral, strike-slip fault that slices across the southern margin of the Sierra Nevada. The southwestern boundary of the Mojave is the northeastern flank of the Transverse Ranges, themselves sharply bound by the San Andreas and related faults. The two faults meet at a ~50° angle, forming the wedge-shaped western tip of the Mojave. The isolated mountains and broad alluvial plains

of the Mojave subprovince are commonly included within the larger Basin and Range Province, but at this point in the geologic history of the Mojave, extensional faulting has waned and weathering and erosion dominate over uplift. The mountains in Joshua Tree National Park and the Mojave are slowly burying themselves in their own erosional debris (Fig. 23.10).

Seasonally high temperatures, sparse rainfall, dry winds, and a hardy flora and fauna characterize the Mojave Desert. In Joshua Tree National Park, the Mojave Desert ecosystem dominates the higher elevations of the western half of the park (Fig. 27.3). In the lower elevations to the east resides the hotter and drier Colorado Desert ecosystem. The namesake plant of the park, the Joshua tree, looks like it belongs in a book by Dr. Seuss. Scattered clusters of Joshua trees flourish in the cooler high desert of the Mojave, whereas the low desert of the Colorado is marked by spindly creosote, ocotillo, and cholla cactus. The plants support a surprisingly robust fauna, including bighorn sheep, bobcats, coyotes, and golden eagles.

North of Joshua Tree, the sun-baked **Mojave Trails National Monument** protects a desert wilderness of golden sand dunes, cinder cones, ancient lava flows,

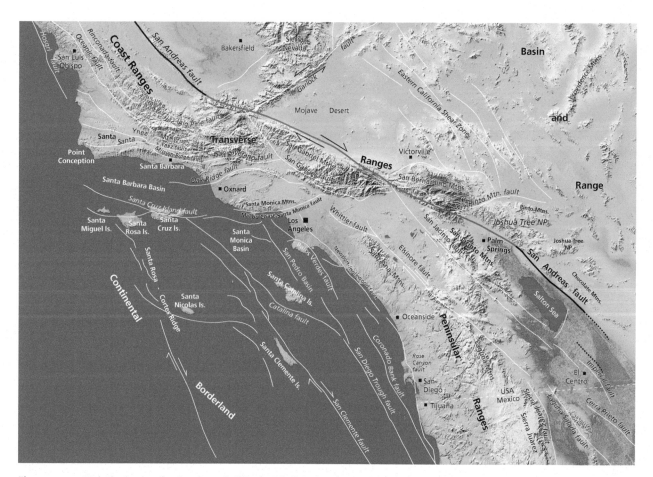

Figure 27.2 Main faults cutting Southern California relative to Joshua Tree and Channel Islands national parks. Note the wedge-shaped block of the Mojave geologic sub province, marked by the intersection of the San Andreas and Garlock faults. The "Big Bend" in the San Andreas fault is shown as a red line

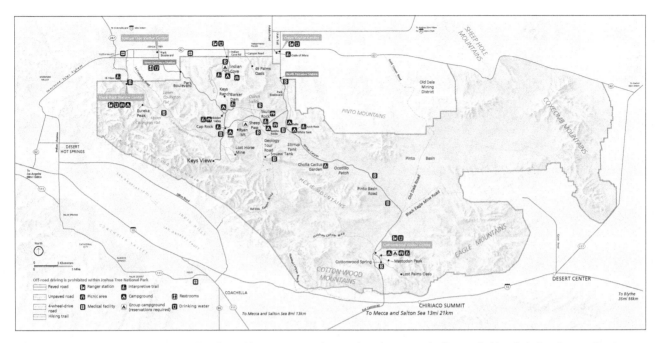

Figure 27.3 Joshua Tree National Park is bound by Route 62 in the north and Route 10 in the south. The Little San Bernardino Mountains form the western edge and the Coxcomb Mountain form the eastern edge.

natural caverns, and sharp-edged mountains (Fig. 27.4). **Sand to Snow National Monument** abuts the western end of Joshua Tree National Park and encompasses a diverse array of landscapes and ecosystems. Along with the Mojave National Preserve to the north, these interconnected National Park System units create the world's second-largest desert preserve.

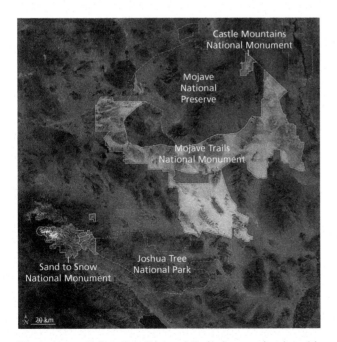

Figure 27.4 Joshua Tree National Park shares a border with the recently formed Mojave Trails National Monument. Sand to Snow National Monument adjoins Joshua Tree along the western edge.

Joshua Tree National Park is located at the eastern end of the **Transverse Ranges**, which extend generally east–west for several hundred kilometers across Southern California (Fig. 26.12). In the park, the Little San Bernardino, Hexie, and Coxcomb Mountains rise above broad, *alluvium*-filled basins, with the highest peak reaching just under 1800 m (5800 ft). The extensive *rain shadow* caused by the Transverse Ranges robs the Joshua Tree area of moisture from the Pacific, enhancing the arid conditions. The low-relief mountains in the park owe their existence to motion along the San Andreas system of faults, a topic explored more deeply later in this chapter on the uplift of the Transverse Ranges.

The vista from Keys View, a popular overlook in the park, provides the ideal spot to gain a regional perspective of the geologic setting surrounding Joshua Tree National Park (Fig. 27.5). Looking southwest across the corrugated hills of the Little San Bernardino Mountains in the foreground, we see that the Coachella Valley lies far below. The mountains looming in the distance are the Santa Rosa Mountains and San Jacinto Peak, the northern end of the larger Peninsular Ranges that continue southward along the Baja Peninsula. Off to the north, the snowcapped peak of Mount San Gorgonio (~3500 m/11,500 ft) marks the highest point in the Transverse Ranges.

Several tens of kilometers to the south and a full kilometer in elevation below Keys View lies the shimmering Salton Sea with a surface elevation of 71 m (234 ft) below sea level (Fig. 27.6). This shallow, saline lake fills the lowest part of the **Salton Trough**, an onshore continuation of the Gulf of California rift basin. The Salton Trough is a *pull-apart basin*, created by *transtension* between parallel strike-slip faults of the southern continuation of the San

Figure 27.5 Panorama of the Coachella Valley, the Santa Rosa Mountains, San Jacinto Peak, and the linear ridge that marks the trace of the San Andreas fault (dashed line) as seen from Keys View in Joshua Tree National Park. The rugged hills in the foreground are the edge of the Little San Bernardino Mountains.

Figure 27.6 Main physiographic features and strike-slip faults of the Joshua Tree region. All of the faults are right-lateral, except for the Pinto Mountain fault. Blue dot marks Keys View and green line outlines the national park. Red symbols mark epicenters of recent earthquakes near Joshua Tree National Park.

Andreas transform fault zone (see Chapter 24 on Death Valley National Park). The only reason that the sub-sea-level Salton Trough is not filled with seawater from the Gulf of California is the natural barrier created by the buildup of sediment from the Colorado River Delta.

The sharp ridgeline running up through the Coachella Valley in the middle distance from Keys View is a linear ridge that marks the trace of the San Andreas fault. Disregard the fact that the actual San Andreas transform plate boundary is over 400 km (250 mi) wide in this part of Southern California, consisting of an array of subparallel strike-slip faults (Fig. 27.2), and assume for simplicity that the San Andreas itself is the plate boundary. This means that Joshua Tree National Park is located on the very

western margin of the North American plate and that everything on the other side of the San Andreas is on the Pacific plate. Think about the entire Peninsular Ranges, much of the Transverse Ranges, the Los Angeles Basin, San Diego and Orange counties, the Baja Peninsula of Mexico, and the millions of people living on that sliver of land moving inexorably to the northwest as part of the Pacific plate.

Origin of Rocks in Joshua Tree National Park

In the following description of the rocks at Joshua Tree National Park, you'll recognize similarities in composition, age, and origin to rocks of other national parks discussed earlier in the book. We'll integrate the geologic history of the rocks composing Joshua Tree with those from other parks to see the larger picture. Much of the following should be familiar to you at this point, but review Chapter 1 as necessary to refresh your understanding of the origin of common metamorphic and igneous rocks.

The oldest rock in Joshua Tree National Park is the 1.7 b.y old Pinto Gneiss (Fig. 27.7A). It exhibits a distinct *foliation*, a parallel alignment of minerals that forms in response to the high pressures and temperatures of metamorphism. The darker bands of the gneiss are composed of minerals like biotite and hornblende, whereas the lighter bands are composed of quartz and feldspar. Metamorphic rocks of this same age were previously encountered in the basement of the lower Grand Canyon, along the steep flanks of mountains in Death Valley, and on the highest peaks of

Figure 27.7 A) Loose block of Precambrian banded gneiss, Joshua Tree National Park. Metamorphic rocks commonly exhibit a wavy striped appearance that contrasts with the salt-and-pepper homogeneity of plutonic igneous rocks B) Outcrop of Mesozoic granitic rock, cut by a near-vertical vein of quartz that stands out in relief due to its resistance to weathering.

Rocky Mountain National Park. Collectively, these rocks form a broad swath that extends from northwest Mexico, through the Mojave and Grand Canyon region, and across Colorado, all the way to the upper Midwest. They represent mountain-building events associated with the accretionary growth of the North American continent during the Proterozoic (Fig. 3.17). Seen in isolated outcrops in Joshua Tree National Park, these rocks are just another

metamorphic rock. But viewed in their larger context, these ancient rocks are the exhumed remains of the early assembly of North America.

The Joshua Tree region was likely submerged beneath the seas during the latest Precambrian and Paleozoic marine transgressions and regressions. Sedimentary rocks of these ages indicate that this happened elsewhere in the Mojave, and it's probable that they existed above the Pinto Gneiss at Joshua Tree as well but were subsequently worn away.

The ancient gneiss was intruded during the mid- to late Mesozoic by granitic magma that cooled to produce a series of overlapping plutons. A variety of plutonic igneous rock types record these magma bodies, each preserving a distinct magma chemistry and mineralogy. The plutons are commonly cut by planar *dikes*, created when magma was forcibly injected into fractures in preexisting plutonic rock, which then cooled and solidified in place. Thin **quartz veins** cut across other plutons and were formed as silica-rich fluids were squirted out of magma chambers into tight fissures where the fluids cooled and quartz crystallized (Fig. 27.7B). Upon uplift and exposure, the minerals composing the veins and dikes are commonly more resistant to weathering than the surrounding rock and may stand out in relief as discordant bands within the homogeneous granite.

Although some of the granitic rocks in Joshua Tree National Park are as old as 245 Ma, the majority ranges in age from about 140 to 75 Ma. This age range should sound familiar because this is the age of the granitic rocks of the Sierran batholith to the north. In fact, these are the same rocks, formed within the same Mesozoic tectonic setting as the Farallon plate slid beneath the North American plate (Fig. 17.8). Equivalent granitic rocks extend north to south from the Sierra Nevada, through the mountains of the Mojave, and across much of the Transverse Ranges. These granitic plutons continue southward as the **Peninsular Range batholith**, which composes many of the mountain ranges between San Diego and the Salton Trough. The Peninsular Range batholith extends 800 km (~500 mi) further south along the mountainous spine of the Baja Peninsula of Mexico. This enormous volume of granitic rock represents the southern extension of the magmatic roots of elongate, volcano-capped mountains along the west coast of North America during the Mesozoic (Fig. 17.9).

After magmatism ended in the region around 75 Ma, erosion stripped off the volcanic crest, causing the solidified plutons and their metamorphic country rocks to rise isostatically in response (Fig. 17.10). Through the Cenozoic, as the nearby plate boundary evolved from one dominated by compression to one marked by shear and extension, the Precambrian metamorphic and Mesozoic igneous rocks of Joshua Tree were exhumed to the surface. The granitic rocks in particular form much of the distinctive landscape at Joshua Tree and are well expressed in popular destinations such as the Wonderland

Figure 27.8 Granitic rock forms isolated monoliths called in-selbergs in Joshua Tree National Park. The rounded shape is due to weathering and erosion along exfoliation joints.

of Rocks, Jumbo Rocks, White Tank, Indian Cove, Lost Horse Valley, and Queen Valley (Fig. 27.8). Weathering and erosion of the light-colored granitic rocks, especially along joint surfaces, create the unique landforms of Joshua Tree, a topic addressed later in this chapter.

Uplift of Mountains in Joshua Tree National Park

The relatively small mountains in Joshua Tree were up-lifted as part of the greater Transverse Ranges that extend westward to Point Conception (Figs. 27.2). The Transverse Ranges include such giants as the San Bernardino and San Gabriel Mountains, as well as smaller mountains such as the Santa Monica and Santa Ynez. The four northern Channel Islands are also considered part of the Transverse Ranges. And it's no accident that the south-facing coast of California between Point Conception and Los Angeles broadly parallels the Transverse Ranges.

The Transverse Ranges are notable for their east–west orientation: "transverse" to the predominant north–south trend of mountains and valleys in California. The alignment of the

Transverse Ranges is due to a large bend in the San Andreas fault system called the **Big Bend**, a 290 km (180 mi) long curve in the trend of the fault (Fig. 27.9). To the north and south of the Big Bend, the San Andreas trends slightly west of north, and the Pacific and North American plates grind laterally past one another in response to the dominant shear stress. But along the Big Bend itself, the fault has a sharper northwest–southeast orientation, creating a huge kink in the trace of the fault. The continental sliver of the Pacific plate moving northwest along its subparallel system of strike-slip faults slowly rams into the continental rock of the North American plate along the Big Bend, creating a tremendous buildup of transpressive stress by the combination of shear and compression. Visualize this relationship in three dimensions, with the bend in the San Andreas as a near-vertical planar surface, rather than as just a line across a map. That gigantic surface, 290 km (180 mi) long and about 16 km (10 mi) deep, is the focus of the enormous amount of transpressive stress that accumulates between the two tectonic plates.

Transpression along the Big Bend drives the uplift of the Transverse Ranges, which rank as one of the fastest growing mountain ranges in the world. Over the past few million years, the mountains have risen at rates of up to a centimeter per year, excruciatingly slow in human terms but speedy on geologic time scales. This rapid rate of uplift contributes to the extremely steep slopes of the ranges as well as to their decay by landslides and debris flows, hazards all too common to the populace of Southern California. Uplift of the mountains of the Transverse Ranges occurs primarily along *thrust faults*. Thrust faults are a type of low-angle *reverse fault* that juxtaposes a slab of older rock above younger rock, stacking them as sheets to create elevated topography (Fig. 22.9).

The uplift that accompanies thrust faulting occurs episodically during earthquakes. Recent evidence includes the 1971 San Fernando earthquake when nearby mountains rose 2 m (6 ft) in a few seconds of shaking. The 1994 Northridge earthquake produced another 70 cm (28 in) of uplift. Although these events seem like a slow and inefficient way to raise mountains, remember that the current relief of the Transverse Ranges occurred during tens of thousands of similar earthquakes over the past few million years. Earthquakes are commonly called the "growing pains of mountains"; without them the spectacular landscape of Southern California would not exist.

Armed with the broader story of the uplift of the Transverse Ranges, we can better understand the physical appearance of Joshua Tree National Park. Mountains in the park are lower in elevation than those in the rest of the Transverse Ranges because they lie near the eastern end of the Big Bend where the southern San Andreas takes a more north-south orientation. As a result, shear stresses become dominant over compressive stresses in the Salton Trough, reducing the amount of uplift. Thrust faults are evident in Joshua Tree and contribute to the rise of mountains, but strike-slip motion is dominant among the multiple active faults that cut across the park, governed by proximity to the

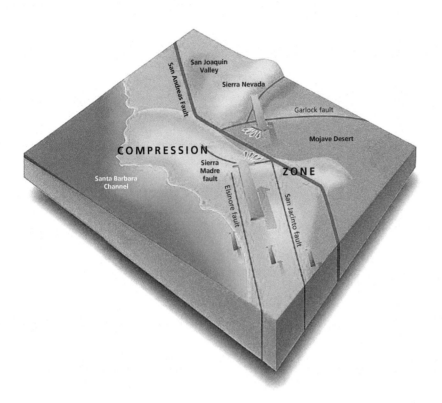

Figure 27.9 Orientation of tectonic stress along the Big Bend of the San Andreas fault. The "compression zone" in the figure corresponds to the Transverse Ranges.

along a set of right-lateral strike-slip faults cutting the Mojave Desert called the **Eastern California Shear Zone**. (The right-lateral strike-slip faults of Death Valley described in Chapter 24 are part of this major shear zone.) This zone of weakness in the crust is considered to be part of the broader transform plate boundary and accommodates some of the tectonic stress between the Pacific and North American plates (Fig. 27.2). There is no doubt that earthquakes of comparable magnitude will continue to pop off along the Eastern California Shear Zone. It might even be exciting to experience one during a visit to Joshua Tree—as long as you are in a safe place away from tumbling boulders!

But by far the larger seismic threat to the region around Joshua Tree National Park resides on the southern segment of the nearby San Andreas fault. As noted in Chapter 26, the last major earthquake to strike this part of the San Andreas was around 1680, give or take a few decades. There are no historic accounts of this quake, but it was recognized by radiocarbon dating of offset sedimentary layers within trenches dug perpendicular to the fault. Comparable *paleoseismology* studies suggest that the southern section of the San Andreas that slices through the Coachella Valley to the Salton Sea ruptures about every 200 years or so on average. Active faults don't rupture with a precise periodicity, and the 200-year recurrence estimate is nothing more than the mean duration between large quakes. But the fact that it's been over three centuries since the last major quake on the southern San Andreas is cause for concern to residents of Southern California. Unlucky visitors to Joshua Tree who happen to be there when a major quake strikes on the San Andreas will have to cope with a minute or more of sheer terror until the shaking stops.

San Andreas fault just a few tens of kilometers to the west (Fig. 27.6). Localized zones of transpression and transtension along the strike-slip faults have resulted in the uplift of mountains and the downwarping of basins in the park. The rise of the Precambrian and Mesozoic rocks to the surface exposes the roots of ancient mountain ranges that once towered above the region long ago. Ongoing erosion through the late Cenozoic has modified the fault-generated topography in Joshua Tree into the landscape we enjoy today.

The locations of some of the faults in Joshua Tree are easily recognized by the presence of desert fan palm oases aligned along springs that emanate from the ground along the fault plane (Fig. 27.10). Rock lining the fault is progressively broken and fractured with each earthquake rupture, creating a vertical zone of angular debris ranging in size from cobbles to sand to clay. As groundwater migrates through pores in the zone of shattered rock lining the fault, it may encounter an impermeable barrier of clay or solid rock that forces the water to the surface. These springs not only enable the luxuriant growth of fan palms, but also support a variety of other plants and animals in the otherwise harsh environment. This is a wonderful example of the geology influencing the hydrology, which in turn influences the ecology of the region.

With all the faults in the area, what is the earthquake risk in Joshua Tree National Park? Between 1992 and 1999, a series of moderate to large earthquakes rattled the region (Fig. 27.6). A magnitude 6.1 earthquake occurred inside the national park about 2 months before the magnitude 7.3 Landers earthquake to the north. The quakes occurred

Figure 27.10 Linear arrays of desert fan palms flourish along springs and seeps emanating from the Pinto Mountain fault, Oasis of Mara.

Desert Landforms in Joshua Tree National Park

It's tempting to view the desert landscape in Joshua Tree as a product of the current arid climate and erosional processes related to the intense heat, harsh wind, and infrequent torrential storms. But the majority of the desert landforms at Joshua Tree are relict features left behind by weathering and erosion during older phases of cooler, wetter climates. The **aridification** of the Mojave region ultimately relates to the creation of rain shadows by the late Cenozoic uplift of the Transverse Ranges, the Coast Ranges, and the southern Sierra Nevada. Prior to these mountain-building events, the region was marked by a very different climate and ecology. Geologic evidence for long-term environmental change comes primarily in the form of fossils of both plants and animals, which are found in relative abundance in Cenozoic sedimentary deposits throughout the Mojave. Fossil leaves, especially, reveal details about precipitation levels and elevation because plants can be particular about their environmental requirements. More specific evidence for ancient climatic conditions is revealed by the laborious process of searching sediments for tiny pollen grains, which are particularly diagnostic of prevailing climatic and ecologic conditions. As we explore the diverse array of landforms in Joshua Tree keep in mind the longer term perspective of climate change and the role it plays in the formation of landscapes.

Typical terrain in Joshua Tree National Park consists of dark, rocky highlands interspersed with lighter colored outcrops of granitic rock, both surrounded by broad, sediment-filled plains (Fig. 27.11). The topography partly relates to *differential erosion* between the two primary rock types composing the park. Less resistant gneiss tends to weather to dark, rounded hills incised by dry washes. The more resistant granitic rocks tend to form lighter-colored, isolated monoliths and boulder piles called **inselbergs**. *Ephemeral streams* flow through the normally dry washes only during seasonal storms, transporting weathered

sediment away from highlands out onto the broad valleys. Most of the valleys have *internal drainage*, with no connection to a larger drainage system. This means that any loose sediment carried from highlands by rain runoff or flash floods becomes trapped within adjoining closed basins. Consequently, the alluvial sediment builds up to a broad plain, filling the valleys and burying the adjacent rocky highlands and inselbergs in their own debris.

The dry winds that sweep across Joshua Tree move sand and silt particles that abrade the surfaces of rock exposures. But water is also an effective agent of weathering, erosion, and deposition in deserts. Moving water produced during intense storms can transport prodigious amounts of sediment in a short time. *Alluvial fans* and *bajadas* in Joshua Tree are formed by infrequent flash floods occurring over long spans of time in identical fashion to those at Death Valley (Fig. 24.17).

Very slowly moving water, within the soil as well as within the body of rock, also contributes significantly to the formation of desert landforms at Joshua Tree. Over the long term, water in contact with rock has the ability to decompose solid rock into smaller grains, especially along joint surfaces. *Joints* are fractures that penetrate through bodies of rock, commonly as multiple parallel planes of weakness. Jointing was addressed in Part 3 on parks of the Sierra Nevada where they cut through granitic rocks to shape the scenery. The intersecting sets of joints in the granitic rocks at Joshua Tree play an equally fundamental role in creating the panorama we enjoy today. Rock-climbers in particular take advantage of the narrow slots created by weathering along joint planes to maneuver their way upward.

Jointing in the granites at Joshua Tree exhibit three main orientations. Many inclined joint sets intersect at angles of 60° and 120°, a common orientation caused by vertical compressional stress (Fig. 27.12). Joints also cut vertically through the granites and likely reflect an older phase of extensional stress imposed on the rocks. And similar to the granitic rocks of Yosemite, *exfoliation joints* were produced as the confining pressure of overlying rock was removed during exhumation to the surface (Fig. 27.8). Exfoliation domes in Joshua Tree are considerably smaller than their counterparts in Yosemite. These three generations of jointing act to partition the granite into rectangular, cubic, or sharply angular blocks of rock enclosed within the larger mass.

Joints permit the infiltration of water into the body of the rock, initially only by a millimeter or two. But over time, the decomposition of minerals composing the rock widens the joint plane, permitting water to penetrate deeper. Different minerals respond in different ways when exposed to slightly acidic rainwater. Quartz, a highly resistant mineral, typically remains a

Figure 27.11   View westward across isolated granitic inselbergs and a broad sandy plain to rounded hills in the distance, Lost Horse Valley, Joshua Tree.

Figure 27.12 Intersecting joint sets in granitic rock creating angular outcrop. Rounded boulders were created by spheroidal weathering, Jumbo Rocks.

discrete particle. Most feldspars and some mica minerals, on the other hand, will decay entirely to weak clay minerals. As the rock disintegrates, the loosened quartz grains accumulate as a sandy residue called **grus**, which is removed by rain and snowmelt to accumulate as alluvial fill within nearby basins. The loose sand on many trails in the park is composed of grus derived from the breakdown of granitic rock.

The chemical disintegration of granite preferentially occurs along the intersecting edges of joint planes where there is more surface area for water to attack. Slowly but surely, the angular blocks of granite become slightly rounded by weathering along their jointed edges. As chemical breakdown of the rock along intersecting joint planes continues through time, the formerly angular block may become a well-rounded boulder perched precariously on the remaining body of rock (Fig. 27.12). This process is called **spheroidal weathering** and is responsible for the wide variety of rounded granite boulders that mantle many of the slopes in Joshua Tree. Once isolated, a rounded boulder may disintegrate further by thermal expansion and contraction related to the wide daily temperature swings common in deserts like the Mojave (Fig. 27.13). Repeated heating and cooling of granitic boulders results in the formation of cracks parallel to the surface. The rounded rock may break apart along these cracks, creating a pile of angular stony debris around the base of boulders.

Many of the weathering and erosion processes described here were accelerated during the cooler, wetter climates of the past, including those associated with the Pleistocene glacial-interglacial phases. In particular, the development of *soils* was much more effective during the wetter phases, relative to the very slow rate of soil development within the current arid regime. Soils are composed of particles of sand, silt, and clay mixed with organic matter derived mostly

from the decay of vegetation and the activity of microbes. Many **residual soils** develop directly above rock by the in-place breakdown of the rock (Fig. 27.14). The lowest part of the soil layer marks the transition from highly weathered rock to intact bedrock beneath.

In the Joshua Tree region and throughout the Mojave, soils that formed during earlier, wetter phases were later removed by erosion as the rate of soil formation slowed with the progressive aridification of the regional climate over the past few million years (Fig. 27.15). Loss of the organic-rich upper soil horizons left behind the remaining bedrock and its surrounding debris as inselbergs, rocky islands in a sea of alluvial sediment.

Evidence for the existence of these ancient soils is revealed in weathering features along the margins of inselbergs (Fig. 27.16). The upper surface of ancient soils is commonly marked by a notch cut into the rock by long-term contact with organic acids and water in the soil. Pits, hollows, and "honeycomb" weathering on the surface of many inselbergs indicate chemical weathering by prolonged interaction with organic-rich and water-laden soils of the past. These

Expansion cracks form parallel to the boulder surface due to repeated expansion and contraction

Rock fragments/clasts

Figure 27.13 Thermal expansion and contraction of granitic boulders cause it to disintegrate into smaller fragments.

Joints widen as root grows

Soil

Decay of organic material releases acids that promote chemical weathering.

Figure 27.14 Residual soils may form directly above the weathered surface of bedrock. Plants use the nutrients in the soil for their growth and also aid in the physical and chemical breakdown of the underlying rock.

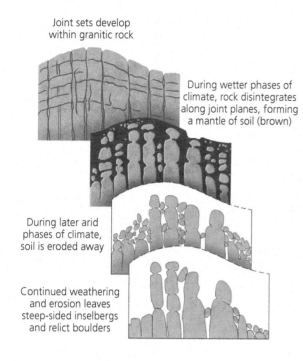

Joint sets develop within granitic rock

During wetter phases of climate, rock disintegrates along joint planes, forming a mantle of soil (brown)

During later arid phases of climate, soil is eroded away

Continued weathering and erosion leaves steep-sided inselbergs and relict boulders

Figure 27.15 Inselbergs may emerge as the overlying soil is removed during the late Cenozoic transition to a more arid climate.

strange weathering features were exposed as the soils were removed during the transition to drier conditions through the late Cenozoic.

The diverse landscapes of Joshua Tree are more than just a bizarre assortment of desert landforms, but rather the culmination of a long series of geologic events. Use your deeper understanding of the geologic origins of the park to enrich your experience as you explore this desert wonderland.

Channel Islands National Park

Why would a short summary of Channel Islands National Park be included within a longer chapter on Joshua Tree National Park? The two parks are ~400 km (250 mi) away from each other and exhibit completely different geology, landscapes, and ecosystems. The connection is that they occupy opposite ends of the Transverse Ranges and thus were formed by the same transpressive tectonic stress associated with the Big Bend of the San Andreas fault. The four northern islands of the Channel Islands are the emergent crests of an elongate underwater mountain range that forms the western extension of the Santa Monica Mountains on the mainland (Figs. 27.17).

Located ~40 km (25 mi) south of Santa Barbara, the four northern Channel Islands form an east–west chain that rises dramatically from the sea. From east to west, the four main islands in the park are Anacapa, Santa Cruz, Santa Rosa, and San Miguel (Fig. 27.18). Far to the south, the tiny island of Santa Barbara is also included in the park boundaries. (The three other main Channel Islands— Santa Catalina, San Clemente, and San Nicolas—are not part of the park.) But Channel Islands National Park is more than just the land area of the five islands; the park boundaries extend outward for almost 2 km (1.2 mi) into the ocean surrounding each island. About half of the area of the national park is above sea level, while the rest is submerged beneath the waves. The national park is included in a larger entity called the Channel Islands National Marine Sanctuary, which preserves a stunning array of marine habitats and an exuberant display of biological diversity.

Figure 27.16 The eye sockets and nostril of Skull Rock were formed by chemical weathering of granitic rock while buried by soils during wetter climate conditions. Soil formation ended during the late Cenozoic aridification of the regional climate, with the soils slowly removed by erosion, exposing the previously buried rock. The groove of the "eyebrow" may represent the former upper surface of the soil.

Origin and Uplift of Rocks Composing Channel Islands National Park

The bedrock of Channel Islands National Park composes a varied suite of Mesozoic and early Cenozoic metamorphic, igneous, and sedimentary rocks that accumulated along the convergent margin associated with Farallon subduction. The islands are dominated, however, by mid- to late Cenozoic sedimentary rocks and *pillow lavas* that accumulated on the seafloor during the transition from convergent to transform tectonism. (Recall from Chapter 26 that lava erupted under water will spontaneously form rounded mounds that stack like a pile of overlapping pillows.) Known as the Conejo Volcanics, this stack of solidified lava flows may reach thicknesses of

a few kilometers. These volcanic rocks dominate Anacapa and Santa Barbara islands.

Most of the sedimentary rocks of the Channel Islands reflect deposition in marine environments, indicating that today's islands were under water for most of the mid- to late Cenozoic. Some of the shale and sandstone strata were derived from the mainland, washed out to the deep ocean through submarine canyons by bottom-hugging *turbidity currents*. Other fine-grained sedimentary rocks record the slow suspension settling of clay and the miniscule shells of planktonic microorganisms to the seafloor. Marine deposition ended by 5 Ma as the islands were raised above sea level as part of the transpressive uplift

Figure 27.17 Oblique view westward across Santa Cruz Island, Channel Islands National Park. Santa Rosa Island lies off in the distance.

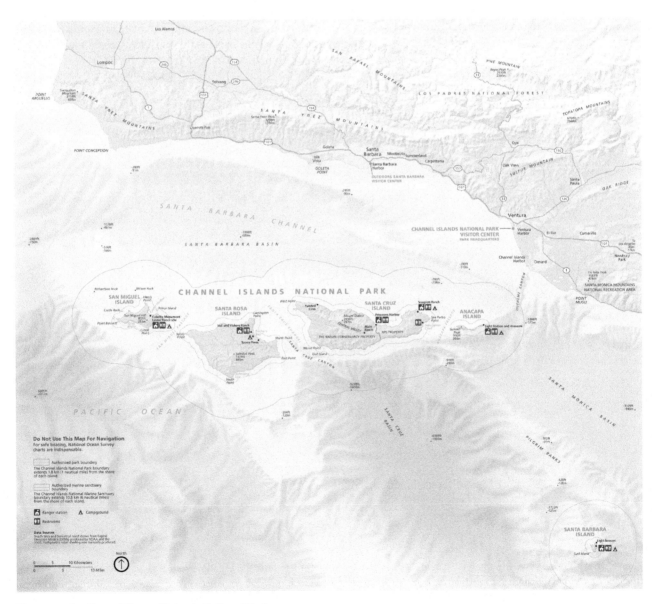

Figure 27.18 Map of Channel Islands National Park.

of the western Transverse Ranges along the Big Bend of the San Andreas fault. The marine-deposited sedimentary and volcanic rocks are deformed into broad *folds* and are cut by a variety of faults, providing evidence for the uplift of the Channel Islands over the last few million years. The faults exhibit both lateral and vertical motion, reflecting the combination of shear and compressional tectonic stress that characterizes the Transverse Ranges. Many of these faults are seismically active today, attesting to the ongoing uplift of the islands.

Continental Borderland

Recall the story from Chapter 26 on the origin of the San Andreas fault and the transition from a convergent plate boundary to the modern transform boundary in California. You learned that the shearing motion of the Pacific plate moving to the northwest ripped slivers of the continental edge away from the North American plate, including the Baja California peninsula and much of southern and central California. Add to this list the continental fragment known as the **Continental Borderland** that forms the crust beneath the Channel Islands (Fig. 27.19). The seafloor bathymetry of the Borderlands reveals a series

of fault-bounded shallow banks and ridges surrounding deep basins. In places, faulting has raised blocks of rock above the sea surface as the eight Channel Islands, providing a glimpse into the continental crust on the seafloor deep below.

How did the continental fragment of the Borderland become part of the Pacific plate, and why is it submerged beneath sea level? As the massive block of continental rock making up the Transverse Ranges was sheared and rotated along the ragged boundary between the Pacific and North American plates over the past 15–20 m.y., the adjacent area of the Continental Borderland was subject to transtensional tectonic stress. As the rock was stretched and sheared, blocks of continental crust were raised upward along faults, whereas adjacent blocks were dropped downward. Most of this movement occurred sight unseen deep beneath the sea, with only the Channel Islands protruding above the waves.

The seafloor topography of the Borderland is very similar to the modern Basin and Range landscape (Chapter 23). The reason is that both regions were affected by tectonic stretching through the mid- to late Cenozoic as the San Andreas transform plate boundary grew and evolved. The thinned continental crust of the Basin and Range is

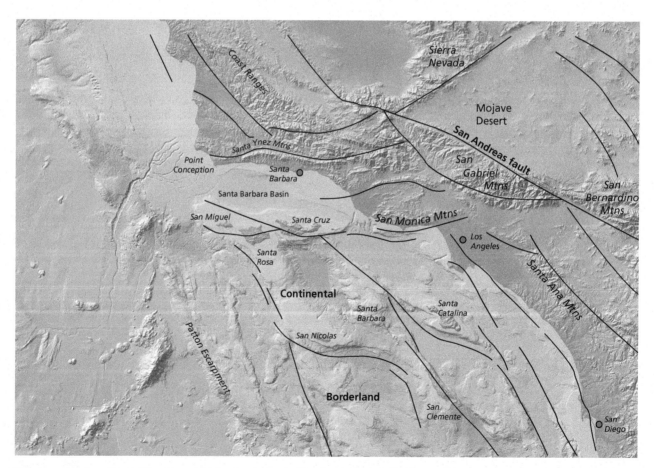

Figure 27.19 Location of the Channel Islands within the Continental Borderland, a submerged fragment of continental crust attached to the Pacific plate (shown in shades of tan and pink). The blue color of the deep ocean seaward of the Patton Escarpment is underlain by higher density oceanic rock. Select major active faults are shown as black lines.

elevated far above sea level because of a welt of hot mantle that pushes the region upward from below. The thinned continental crust of the Borderlands, however, has no underlying heat source, so it sags downward into the underlying mantle. It's strange to think that the Borderlands and the Channel Islands exhibit a tectonic history and landscape comparable to Death Valley National Park, but the similarities are real and related to the evolution of the San Andreas transform plate boundary.

Effects of the Ice Ages

You would think that the glaciations of the Pleistocene Ice Ages wouldn't affect the low-elevation, midlatitude setting of Channel Islands National Park, but the ecology and coastal landscapes were directly influenced by the rapid fluctuations in global climate of the past 2.6 m.y. As Earth becomes cooler during glacial phases, water is transferred from the world ocean via the hydrologic cycle to form the ice composing *continental ice sheets*. The result is that during glacial stages, global sea level falls, often dramatically. As continental ice sheets melt, the water is transferred back to the oceans via river systems. Thus, global sea level rises during warmer, interglacial stages. Because Earth experienced more than 20 glacial-interglacial cycles over the past 2.6 m.y., coastal regions have been repeatedly exposed and then drowned as sea level fluctuated in concert with the changes in climate.

During the Last Glacial Maximum only 20,000 years ago, global sea level fell by ~120 m (~400 ft) as water was transformed into the ice of vast continental glaciers. As the shoreline regressed seaward, the shallow shelf that surrounds the four northern Channel Islands was exposed as new land, creating a single huge island about 125 km (~80 mi) long that has been named Santarosae (Fig. 27-20). The lower sea level permitted huge Columbian mammoths to swim the narrow strait from the mainland to Santarosae where they apparently prospered. As the climate warmed and the shoreline migrated inland in response to rising sea level, Santarosae was fragmented into the four northern Channel Islands, isolating the mammoths and decreasing their food resources. Over several generations,

the mammoths adapted to the shrinking land area by decreasing their body size and thus their energy requirements. Fossils of these "pygmy" mammoths, less than 2 m (6 ft) high at the shoulder, have been found on Santa Cruz, Santa Rosa, and San Miguel islands. They went extinct about 13,000 years ago, possibly due to the arrival of humans. Archaeological proof for human activity on the islands dates to ~12,000 years, among the oldest evidence of people in North America. The indigenous Chumash people prospered on three of the northern Channel Islands for thousands of years before leaving in the early 19th century for the mainland.

Ecology of Channel Islands National Park

The Channel Islands are known as the "Galapagos of North America" because of the incredible diversity of life that abounds on the islands and the surrounding waters of the marine sanctuary. The islands are located in a unique oceanographic setting where the south-flowing cold waters of the California Current mingle with the north-flowing warmer waters of the Southern California Countercurrent. This interaction of currents, combined with the right wind conditions, promotes the upwelling of cool, nutrient-rich water from the deep sea toward the surface. Phytoplankton and zooplankton that form the base of the marine food web thrive in the upwelling bath of nutrients, providing food for higher trophic levels. The result is a profusion of marine wildlife highlighted by charismatic marine mammals such as seals, sea lions, dolphins, and whales that live off the bounty of fish and marine invertebrates in the shallow seas around the islands.

The terrestrial ecosystems of the islands are in a recovery phase because of the dramatic erosion that occurred in response to overgrazing by sheep, cattle, and other domesticated animals introduced to the islands in the late 19th and early 20th centuries. The combined effects of overgrazing, occasional drought, and harsh winds denuded the vegetation and soils from the land. In the decades since the domesticated animals were removed, especially after the islands were protected under the stewardship of the National Park Service and Nature Conservancy, the rate of erosion has decreased and landscapes have stabilized. The resurgence of native plants and animals has mostly restored the Channel Islands to their natural, windswept beauty.

Most visitors to Channel Islands National Park only stay for the day because the infrastructure of lodging and restaurants intentionally doesn't exist in order to preserve the remote wilderness. Camping is permitted, however, and is the best way to fully experience the profound isolation afforded by the island setting. Go to explore the trails, kayak the coves and sea caves, dive in the sanctuary waters, and revel in the biological diversity.

Figure 27.20 Outline of Santarosae Island during the Last Glacial Maximum ~20 ka. The eastern tip of the island reached to within ~10 km (6 mi) of the mainland.

Key Terms

aridification
grus
inselberg
pygmy mammoth
quartz veins
residual soil
spheroidal weathering

Related Sources on the Geology of Joshua Tree National Park

Decker, B., and Decker, R. (2006). *Road Guide to Joshua Tree National Park.* Mariposa, CA: Double Decker Press.

Sharp, R. P. (1993). *Geology Underfoot in Southern California.* Missoula, MT: Mountain Press.

Trent, D. D.. and Hazlett, R. W. (2002). *Joshua Tree National Park Geology.* Twentynine Palms, CA: Joshua Tree National Park Association.

National Park Service—geologic formations, Joshua Tree National Park. Accessed February 2021 at https://www.nps.gov/jotr/learn/nature/geologicformations.htm

USGS Geology of National Parks, 3D and Photographic Tours, Joshua Tree. Accessed February 2021 at http://3dparks.wr.usgs.gov/jotr

Related Sources on the Geology of Channel Islands National Park

Meldahl, Keith Heyer. (2015). *Surf, Sand, and Stone: How Waves, Earthquakes, and Other Forces Shape the Southern California Coast.* Berkeley: University of California Press.

Sylvester, A. G., and Gans, E. O. (2016). *Roadside Geology of Southern California.* Missoula, MT: Mountain Press.

National Park Service—geologic formations and animations of the geologic history of Channel Islands National Park https://www.nps.gov/chis/learn/nature/geologicformations.htm (accessed February 2021)

28 Pinnacles National Park

Monterey Bay

San Juan Bautista

Hollister

Salinas

Gabilan Range

Diablo Range

Monterey

Salinas Valley

Pinnacles National Park

Santa Lucia Range

Soledad

San Andreas Fault

Figure 28.1 Location of Pinnacles National Park within the southern Gabilan Range of central California.

The steep, craggy terrain of Pinnacles National Park rises abruptly above the rolling hills of the southern Gabilan Range in the Coast Ranges of central California (Fig. 28.1). A few kilometers northeast of the park boundary, the San Andreas fault cuts through the San Benito Valley. To the west, farmland of the broad Salinas Valley separates the Gabilan and Santa Lucia Ranges. The northwest-trending texture of the elongate mountains and valleys composing the central Coast Ranges reflects millions of years of motion along the subparallel network of faults associated with the San Andreas transform plate boundary (Chapter 26).

The rocky landscape of the park (Fig. 28.2) contrasts sharply with the farmlands, grassy valleys, and chaparral-covered rolling highlands that surround it, giving it the appearance of being geographically misplaced. It turns out that the surreal volcanic terrain of Pinnacles is indeed foreign, having originated far to the south before migrating to its current location by tectonic movement along the San Andreas fault. In fact, all the land west of the San Andreas has been displaced hundreds of kilometers northwest as it rides along as part of the Pacific plate. The geologic significance of the Pinnacles will be the focus of much of the body of this short chapter.

Pinnacles became a national park in 2013, deservedly upgraded by congressional legislation from its former status as a national monument (Fig. 28.3). Even though rocky spires and precipitous cliffs are the most distinctive features of the park, much of the landscape is composed of rolling hills, some reaching 1000 m (~3300 ft) in elevation. In addition to the rugged scenery, the park is notable as an island of biodiversity, including a healthy population of California condors. The chaparral ecosystem that dominates the park comprises a dense, low-growing community of drought-resistant shrubs and brush ideally adapted to the hot, dry summers, seasonal rainfall, and occasional wildfires. A winding, narrow road provides access to the park from the west and another road enters from the east, but no roads pass entirely through the park. Pinnacles is best experienced via the network of trails that meander through the park from trailheads along the main roads.

Creeping Segment of the San Andreas Fault

The San Andreas fault maintains a linear trace as it cuts through the Coast Ranges from Carrizo Plains National Monument to just north of Pinnacles National Park (Fig. 26.4). Many of the surface features associated with right-lateral offset along the fault expressed in the Carrizo Plains are also present along the San Andreas near Pinnacles, including narrow linear valleys developed within the weakened rock along the fault plane, *sag ponds, offset stream channels,* and linear ridges (Chapter 26). But there is a significant difference between the behavior of the San Andreas in the Carrizo Plains and the region near Pinnacles. In the Carrizo Plains, as has been noted, dry streambeds were abruptly displaced by several meters during large earthquakes such as the 1857 Fort Tejon event. In contrast, the 175 km (~110 mi) length of the San Andreas from the small farming community of Parkfield northwest to San Juan Bautista is marked by **fault creep**, meaning that the massive blocks of rock on either side of the fault plane slide at a relatively constant, agonizingly slow rate. This creeping segment of the San Andreas near Pinnacles hasn't experienced a large-magnitude earthquake in historic times, as opposed to the three major events that ruptured other segments of the fault in ~1680, 1857, and 1906 (Fig. 26.9).

The *slip rate* of fault creep along this segment of the San Andreas has been precisely measured by the Global Positioning System of satellites for the past 30+ years to average ~2.5 cm/yr (~1 in/yr). Most movement occurs during countless micro-earthquakes that are detectable by seismometers but are generally not felt by people. Occasional small- to moderate-size earthquakes do occur, however, and the possibility of significant groundshaking exists in the region around Pinnacles National Park. Apparently, friction along the San Andreas fault plane in the creeping segment is reduced by the presence of fibrous minerals that grow parallel to the surface of the fault, permitting the massive blocks of rock on either side to slide smoothly past one another.

The near-constant creeping motion impedes the stick-slip behavior that leads to large-magnitude earthquakes. The segments of the San Andreas north and south of the creeping section are both considered to be "locked," with only small- to moderate-sized earthquakes alleviating the accumulating tectonic stress. Those locked segments of the San Andreas will invariably release their pent-up potential energy as major earthquakes over the next few decades to centuries. The area around Pinnacles National Park will certainly feel the groundshaking from future

Figure 28.2 Massive spires and monoliths of rock, Pinnacles National Park. Note people in lower right and climber (circled).

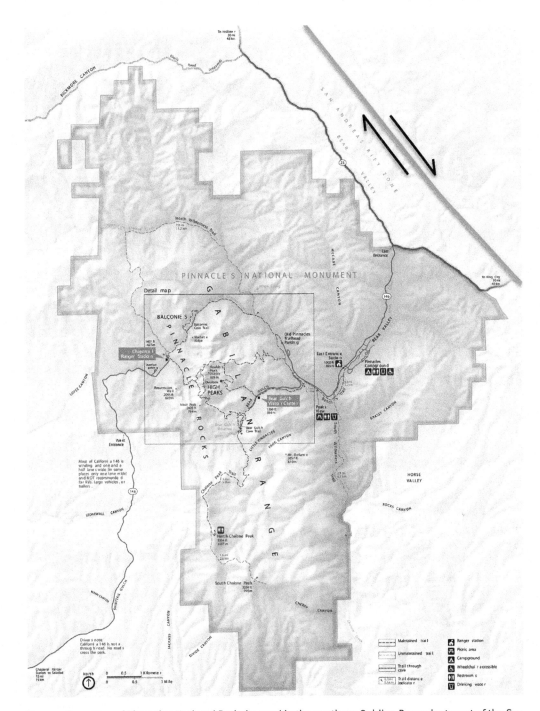

Figure 28.3 Map of Pinnacles National Park, located in the southern Gabilan Range just west of the San Andreas fault.

earthquakes to the north and south, but the probability of a large-magnitude earthquake along the creeping segment is relatively low.

Origin and Tectonic Significance of Rocks at Pinnacles

Much of the Gabilan Range is composed of coarse-grained granitic rock that erodes to gently rolling highlands. The age of the granite varies from about 100 to 78 Ma, a range

that should ring familiar to you as the primary age of granite composing the *Sierran batholith*. Granitic rocks in the Transverse and Peninsular Ranges of Southern California are also mid- to late Mesozoic in age. All of these granitic rocks originated within the *continental volcanic arc* created by subduction of the Farallon plate, discussed in several earlier chapters. The granites of the Gabilan Range are part of the *Salinian Terrane*, interpreted to have been detached from southern sections of the Sierran batholith and then displaced northward by San Andreas strike-slip faulting over millions of years (Chapter 26).

Figure 28.4 The massive spires and cliffs of volcanic rock composing Pinnacles National Park are distinct from the chaparral-covered hills of Salinian granite composing the main mass of the Gabilan Range in the background. Balconies Cliffs from High Peaks Trail.

The jagged exposures of barren, rust-colored rock composing Pinnacles National Park stand out in stark contrast to the chaparral-covered Salinian granites (Fig. 28.4). The rocks at Pinnacles are volcanic in origin, with a composition that is mainly rhyolitic, indicative of highly explosive eruptions. (The basics of volcanism, volcanic rocks, and the role of silica in determining the explosivity of an eruption were comprehensively addressed in Chapter 1 as well as in the five chapters of Part 2 on the Cascades geologic province.) Age dating of minerals within these volcanic rocks at Pinnacles indicates that they erupted around 23 Ma, with the magma rising through fissures in the underlying Mesozoic granitic rock. The erosional remnants of the mid-Cenozoic volcano rest unconformably on top of Salinian granites within Pinnacles, with over 50 m.y. of time "missing" along the boundary.

Some of the rhyolitic rock at Pinnacles formed as viscous lava flows, whereas others were violently erupted as pyroclastic debris. The most widespread and distinctive rock at Pinnacles is **volcanic breccia**, which is composed of angular, cobble-sized fragments of volcanic rock embedded within a finer-grained sedimentary matrix (Fig. 28.5). Volcanic breccias typically form as fragments of volcanic rock tumble downslope as a chaotic mass along the margins of volcanic vents before hardening into rock. Some of

the volcanic breccias exposed in Pinnacles suggest rapid deposition under water along the flanks of the volcano. Other volcanic breccias may form when angular pieces of rock ejected from the throat of the volcano become embedded within finer pyroclastic debris during explosive eruptions. These breccias reflect the violent and abrupt deposition of semi-consolidated material that occurs along the steep slopes of volcanoes. The breccias at Pinnacles form thick, massive layers interbedded with lava flows and pyroclastic deposits, each representing different eruptive events (Fig. 28.4).

Rhyolitic rocks of identical age, chemistry, and appearance are recognized in the Neenach volcanic field near Lancaster, California, 315 km (195 mi) to the south (Fig. 28.6). Whereas the Pinnacles volcanics reside on the Pacific side of the San Andreas fault, the Neenach volcanics are located on the North American side. The logical conclusion is that the original volcanic field arose 23 m.y. ago directly along the youthful San Andreas fault and later was ripped in half by right-lateral strike-slip motion. The Pinnacles side of the volcanic field slid one earthquake at a time 315 km to the northwest along the Pacific side of the San Andreas, whereas the Neenach side of the volcano remained attached to the North American plate.

This offset of the Pinnacles volcanics provides a precise **pinning point** for determining the total offset of rock along the San Andreas. The Salinian granite on which the Pinnacles volcanics rest must also have been displaced the same distance along the fault. Other geologic pinning

Figure 28.5 Boulder of volcanic breccia, Pinnacles National Park. Note the angularity of the cobble-size fragments embedded within the finer-grained matrix.

points along the length of the San Andreas have been recognized and corroborate the estimated total displacement derived from the Pinnacles/Neenach volcanics. The splitting of the volcanic field by the San Andreas is estimated to have begun around 16 Ma. Dividing 315 km (195 mi) by 16 m.y. yields an average slip rate of ~2 cm/yr (<1 in/yr), comparable to the current average slip rate along the creeping segment of the San Andreas. Fault creep at this constant rate of movement adds up to significant amounts of total displacement, given long spans of geologic time.

Origin of Landforms at Pinnacles

The volcanic rocks composing the Pinnacles landscape were likely buried and tilted as they migrated episodically on their northward journey over the last 16 m.y. They were eventually exhumed to the surface where erosion removed the uppermost rocks of the original volcanic edifice, leaving remnants resting above the granitic basement. Along the way, the rocks were stressed, creating vertical joints that penetrate through the rock (Fig. 28.4). Rainwater and morning dew work their way into the fractures, chemically breaking down the rock and creating narrow gaps along joint planes. With time, weathering and erosion have conspired to sculpt the bizarre array of rounded spires, crags, and fins within the jointed volcanic rock of Pinnacles.

Most caves form by the slow dissolution of limestone or marble deep beneath the surface, such as those in nearby Sequoia–Kings Canyon national parks. Distinctive caves at Pinnacles, however, are known as **talus caves** and are formed in a different way (Fig. 28.7). Over the years, countless flash floods and *debris flows* have incised deeply into vertical joints within the volcanic rock, creating narrow, steep-walled gorges. Later, huge, rounded boulders and angular blocks of rock tumbled downward from higher elevations, becoming wedged above the narrow,

Figure 28.7   A) Huge boulder near the entrance to Balconies Cave. B) Light streams through gaps in the boulder-crowned ceiling of Bear Gulch talus cave, Pinnacles National Park.

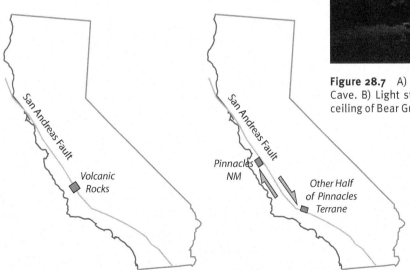

Figure 28.6   The Pinnacles volcanic field was split from its eastern half in Southern California by the San Andreas fault and offset 315 km (195 mi) to the northwest.

water-carved canyons. These massive rocks that form the roofs of talus caves at Pinnacles may have been toppled by groundshaking during earthquakes, crashing downward as *rockfalls* or *rockslides*. The newly formed "cave" is further excavated by flash floods that episodically rage through the network of passageways. The Balconies and Bear Gulch cave systems provide a cool, dark respite for visitors to Pinnacles and also serve as a haven for several species of bats.

If you happen to approach Pinnacles from the east along Route 25, be sure to look for signs on the landscape of the San Andreas fault, which trends subparallel with the highway. A sharp eye will notice dry sag ponds, *fault scarps*, and offset fence lines along either side of the road. You may even see aligned cracks in the road surface where the fault cuts across the highway, depending on how recently the road was repaved. Inside the park, be sure to hike the trail network, suitable for all abilities. A trek along the High Peaks Trail includes the surprise of steps carved directly into the volcanic rock. Explore the dark and quiet talus caves, and, if you're a climber, scale one of the hundreds of routes on the spires and cliff faces (or just watch others climb, if that suits you better). If you go after the winter and spring rains, anticipate a riot of color as wildflowers carpet the slopes. And if all that isn't enough to excite your senses, gaze upward to spy one of the resident condors soaring on a thermal updraft.

Key Terms

fault creep
pinning point

talus cave
volcanic breccia

Related Sources

Alt, D., and Hyndman, D. W. (2016). *Roadside Geology of Northern and Central California.* Missoula, MT: Mountain Press.

Johnson, E. R. and Cordone, R. P. (1984). *Pinnacles Guide: Pinnacles National Monument, San Benito County, California.* Stanwood, WA: Tillicum Press.

KellerLynn, K. (2008). *Geologic Resource Evaluation Scoping Summary: Pinnacles National Monument, California.* Denver, CO: Geologic Resources Division, National Park Service.

National Park Service—Pinnacles National Park faults and geologic formations. http://www.nps.gov/pinn/learn/nature/faults.htm (accessed February 2021) https://www.nps.gov/pinn/learn/nature/geologicformations.htm (accessed February 2021)

National Park Service—Pinnacles National Parks film. http://www.ksbw.com/news/central-california/salinas/new-video-of-pinnacles-national-park-captures-ancient-beauty/29926518 (accessed February 2021)

USGS Geology of National Parks, 3D and Photographic Tours, Pinnacles. http://3dparks.wr.usgs.gov/pinn/index.html (accessed February 2021)

Figure 29.1 Point Reyes National Seashore is bound on its eastern edge by the San Andreas fault zone, a strike-slip, transform plate boundary.

There aren't many places where you can stand on an active tectonic plate boundary, but it's possible when you visit Point Reyes National Seashore where the San Andreas fault slashes a linear path across the landscape (Fig. 29.1). The triangular-shaped peninsula is a fragment of continental rock moving to the northwest on the Pacific plate relative to the North American plate east of the fault. As you hike the trails of the linear Olema Valley or kayak the waters of Tomales Bay, you are directly on the narrow zone of weathered and eroded rock that defines the San Andreas transform plate boundary. During the 1906 San Francisco earthquake, the Point Reyes segment of the fault experienced the greatest surface rupture; traces of the event are still visible on the landscape today. In this chapter, we'll explore a variety of features and processes related to the transform plate boundary as well as a spectacular coastal landscape that justifies the Point Reyes Peninsula as a national seashore (Fig. 29.2).

The terrain of the peninsula includes a broad range of ecosystems that provide habitat for a diverse biota. Inverness Ridge parallels the San Andreas fault zone and forms the backbone of the park, reaching a peak elevation of 428 m (1407 ft) (Fig. 29.3). The ridge is covered with a forest of Douglas fir and Bishop pine, nourished by rain derived from Pacific storm fronts intercepted by the

Figure 29.2 View toward the northeast of the Point Reyes Headlands. The arcuate coastline in the distance marks the edge of Drakes Bay, while the linear coast to the left is part of Point Reyes Beach. The forested slopes of Inverness Ridge rise in the far background.

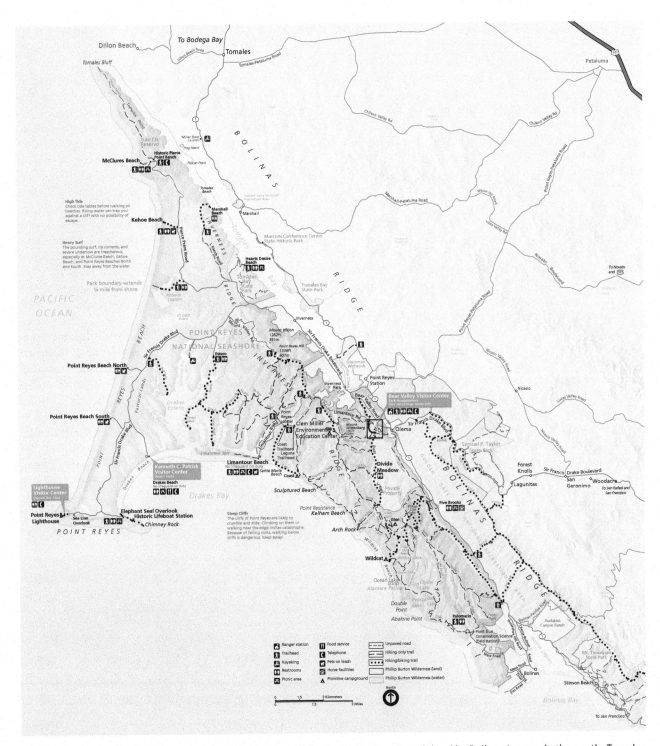

Figure 29.3 Point Reyes National Seashore is separated from the Marin County mainland by Bolinas Lagoon in the south, Tomales Bay in the north, and the narrow Olema Valley in between. The boundary of the park extends ~400 m (0.25 mi) seaward from the shore.

narrow highlands. The terrain west of Inverness Ridge is a rolling, open landscape of coastal grasslands and hardy scrub known as the pastoral zone. Don't be surprised by the grazing herds of cows. Ranching was established on the peninsula in the years after the Gold Rush and has a long and storied history. When the National Seashore was established in 1962, the ranchlands were bought and leased

back to the ranchers, who preserve the cultural heritage of dairy farming on the peninsula, as well as provide the Bay Area with locally produced food. Almost 25 percent of the land in the National Seashore is used for grazing and the related infrastructure.

In sharp contrast to the bovine population, the northernmost part of the park near Tomales Point is a reserve for a population of several hundred tule elk, separated from the rest of the park by a high fence. The progenitors of this herd were 10 tule elk that were relocated to the park in 1978. Native tule elk in California were threatened with extinction in the decades following the Gold Rush due to overhunting, conversion of elk habitat to agriculture, and competition with domesticated livestock. The trail leading through the reserve provides a splendid opportunity to see these magnificent ungulates in their native environment. The lush grasslands of the pastoral zone are also home to populations of black-tailed deer, coyote, fox, bobcat, skunk, weasel, raccoon, porcupine, and jackrabbit. Elusive brown bear and mountain lion are known to pass through the peninsula.

The coastline of the National Seashore is a diverse mosaic of sandy beaches, rocky headlands, estuaries, and marshlands. At the far southwestern tip of the peninsula is the Point Reyes Headlands, a rocky promontory surrounded by steep cliffs that plummet into the sea (Fig. 29.2). This array of coastal environments supports many species of birds, as well as marine mammals such as harbor seals, sea lions, and northern elephant seals that sprawl across pocket beaches nestled between rocky headlands. Point Reyes National Seashore is surrounded by the enormous Gulf of the Farallones National Marine Sanctuary, a biologically rich haven for marine life fueled by coastal upwelling of cold, nutrient-rich water. Gray whales can frequently be seen from seacliffs along the peninsula during their winter and early spring migration.

The ecology and landscapes of the Point Reyes Peninsula are ultimately related to the geologic framework of the park and the San Andreas fault that defines its eastern boundary. Be sure to review Chapter 26 on the active Pacific Margin to establish a foundation for understanding the origin and evolution of the San Andreas fault, the stick-slip mechanics of earthquakes, the uplift of the Coast Ranges, and the way active faults are expressed on the landscape.

Tracing the San Andreas Fault near Point Reyes

The San Andreas fault zone splits into several subparallel strands in the San Francisco Bay Area (Fig. 29.4). *Right-lateral, strike-slip faults* such as the San Gregorio, Hayward, Rodgers Creek, and Calaveras accommodate some of the tectonic strain between the Pacific and North American plates. Offshore from the peninsula, the Point Reyes *reverse fault* is also part of the San Andreas system. Recall from Chapter 26 that geologists place the plate boundary along the San Andreas

fault for convenience, simply because it's the longest and most dominant. In reality, the plate boundary is the entire broad zone of subparallel faults.

All of these faults in the Bay Area are seismically active, and many have experienced earthquakes greater than magnitude 6 over the past few centuries. Of course, the most catastrophic was the magnitude 7.8 San Francisco earthquake of 1906 that was centered along the San Andreas fault where it slices through the seafloor offshore from the Golden Gate. In 1868, the Hayward fault that runs beneath the East Bay ruptured with an estimated magnitude of 6.9. Most recently, the magnitude 6.9 Loma Prieta earthquake in 1989 rattled the Bay Area from its epicenter in the Santa Cruz Mountains. During each of these quakes, the Pacific plate lurched a few meters to the northwest along the transform plate boundary. Tens of thousands of major quakes over the last few million years add up to the average long-term *slip rate* of ~5 cm/yr (2 in/yr).

The San Andreas fault converges with the San Gregorio and Golden Gate faults near Bolinas Lagoon, then cuts northwest through Olema Valley and Tomales Bay before skirting the coastline on its way to Bodega Bay (Fig. 29.4). Countless earthquakes along the San Andreas have shattered the rock along the vertical fault plane, leaving it vulnerable to deepening by weathering and erosion. The linear valleys beneath Bolinas Lagoon and Tomales Bay are just deep enough for seawater to intrude, whereas the Olema Valley remains barely above sea level. The entire fault zone is only a kilometer or two wide as it cuts through the region. (Incidentally, many maps label this linear trough as the San Andreas "rift" zone. You should

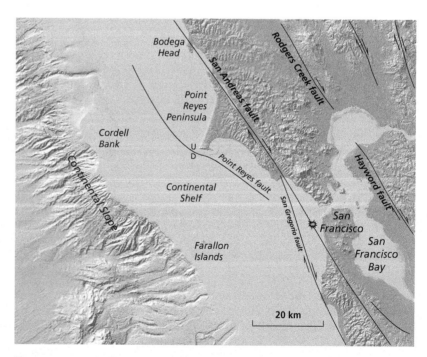

Figure 29.4 Major faults in the Bay Area and Point Reyes Peninsula. The orange symbol along the San Andreas fault seaward of the Golden Gate is the estimated epicenter of the 1906 San Francisco earthquake. "U" and "D" on either side of the Point Reyes fault indicate the upthrown and downthrown sides of the fault.

recall from Chapter 1 and Part 4 on parks of the Basin and Range that a rift valley is marked by *extensional* stress where blocks of rock on either side are moving away from one another, opening a narrow trough in between. In contrast, the San Andreas strike-slip fault is marked by *shear* stress where the blocks of rock on either side move laterally past one another, typically during episodic earthquakes. The linear valley is produced by erosion along the fault plane rather than extensional stress and is technically not a "rift" valley.)

Several places along the fault zone provide a feel for how the San Andreas transform plate boundary is expressed on the landscape. Driving Highway 1 from the Bolinas area northward through Olema Valley allows you to follow directly along the trace of the fault (Fig. 29.3). During the 1906 earthquake, this stretch of the San Andreas experienced the greatest amount of offset of anywhere along the 450 km (300 mi) length of rupture (Fig. 26.9). Maximum displacement of about 6 m (~20 ft) was recognized throughout the Olema Valley by offset fence lines, roads, buildings, and lines of trees. In the intervening years, humans have repaired most of the damage, and erosion has softened the linear scarps and offset features, but many have been restored by the National Park Service. You can see some of the effects of the 1906 quake along the short but well-described Earthquake Trail near the Bear Valley Visitor Center, as well as along the Rift Zone Trail that winds along the fault just inside the park boundary.

Another place to get a good feel for the San Andreas is located along the short stretch of Sir Francis Drake Boulevard where it enters the park perpendicular to the fault zone (Fig. 29.5). Turning off of Highway 1 toward the Bear Valley Visitor Center, we find that the road parallels Lagunitas Creek, imperceptibly crossing the strand of the San Andreas that ruptured in 1906. A gravel road to Olema Marsh leads to a small parking lot at the base of a low hill where you can wade through the grasses to the top for an overview. (As always, use Google Earth to plan your visit.) Figure 29.5A provides an oblique view of the transition from the Olema Valley northwest to Tomales Bay, with the location of the hill marked with a filled circle. The adjoining Figure 29.5B is a LiDAR image that partially overlaps with the photograph. (**LiDAR** is an imaging technique that uses pulses of light energy to create a high-resolution elevation map, permitting the identification of landscape features otherwise indiscernible in regular photographs.) The LiDAR image illustrates that the hill you can visit for an overview is the northern tip of a linear *pressure ridge* that formed along the eastern edge of the San Andreas. The ridge was moved laterally alongside the fault to its current location during countless earthquakes. Called a **shutter ridge** by specialists, the ridge deflects streams on its southwest side so that their channels make 90° turns to parallel the trace of the San Andreas. *Offset channels* are a key feature that can be used to identify the trace of faults across the landscape (Chapter 26).

Another overlook down onto the Olema Valley and Tomales Bay is located on the summit of Mount

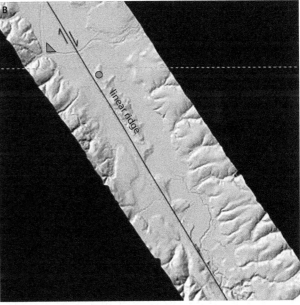

Figure 29.5 A) View toward the northwest from Olema Valley toward Tomales Bay along the trace of the San Andreas fault. Inverness Ridge on the left (west). B) LiDAR image of Olema Valley showing linear ridges that mark the location of the San Andreas fault. Triangle near sharp bend in Lagunitas Creek allows for comparison between images. White lines on photo are roads. Filled circles show the location of a viewpoint at the north tip of the linear ridge.

Wittenberg along the crest of Inverness Ridge. The peak is accessible via a solid hike from the Bear Valley Visitor Center, but it's well worth the effort. For easier access to the crest of Inverness Ridge, drive the road to the top of Mount Vision for equally incredible vistas. Inverness

Ridge itself is a gigantic linear landform (a shutter ridge, to use the jargon), moved into place by motion of the Pacific plate to the northwest along the San Andreas fault. As you gaze downward onto the linear gash of the fault zone, note the difference in vegetation between the grasslands on the east side of Tomales Bay with the forested slopes of Inverness Ridge on which you're standing (Fig. 29.5A). The contrast reflects the different rock types juxtaposed against one another by movement along the San Andreas fault.

The rolling hills to the east of Tomales Bay are composed of the *Franciscan assemblage*, a diverse array of rock types that accumulated within the *accretionary wedge* of the Farallon subduction zone (see Chapter 26). The oceanic rocks of the Franciscan are a chaotic mixture of *turbidites*, *chert*, *pillow basalt*, and *serpentinite*. These rocks readily weather to gently rounded hills and ridges and decompose to produce soils conducive to the growth of grasslands and scrub. In contrast, Inverness Ridge is composed of Mesozoic granite identical in age and composition to that of the Sierra Nevada. As you may surmise from previous chapters in Part 6, the granite of Inverness Ridge is part of the *Salinian Terrane*, a fragment of the *Sierran batholith* that was sliced off from somewhere in Southern California by the San Andreas and transported >300 km (~190 mi) to its current position. You are hiking on what is truly immigrant terrain, a piece of continental crust dislocated far from its place of origin over the past several million years. The soils derived from the decomposed granitic rock combined with rainfall to support the evergreen forest that blankets Inverness Ridge.

Origin of Rocks at Point Reyes

The Salinian granitic rocks that form Inverness Ridge are the same granitic rocks as those exposed in Pinnacles National Park to the southeast along the San Andreas (Chapter 28). On the Point Reyes Peninsula, these resistant plutonic rocks are also exposed along the narrow strip of Tomales Point, in the Point Reyes Headlands at the southwest end of the peninsula, and within seacliffs at McClures and Kehoe beaches along the northwest coastline (Fig. 29.6). These mid- to late Mesozoic age granitic rocks form the *basement beneath a thick veneer of Cenozoic*

sedimentary rocks, separated by an *unconformity* that spans a few tens of millions of years. (Unconformities are boundaries between disparate units of rock that represent episodes of nondeposition or prolonged erosion. They were discussed in many chapters of Part 1 on parks of the Colorado Plateau.)

A good place to examine the relationship between the granitic basement and the overlying sedimentary rocks is at the Point Reyes Headlands (Fig. 29.2). A short path from the parking lot below the lighthouse leads to the Sea Lion Overlook where you can see the granite in outcrop. Light-colored feldspars dominate the rock, with glassy quartz filling the space between crystals. Small, dark grains of blocky hornblende and platy mica stand out in contrast to the lighter feldspar and quartz. At this point in the book, you know that these granitic rocks were once molten masses deep beneath the surface that slowly cooled and solidified through time before being uplifted to the surface. The age when these granites crystallized is dated

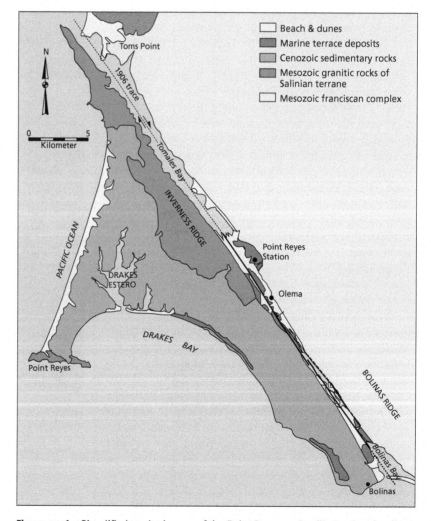

Figure 29.6 Simplified geologic map of the Point Reyes region illustrating the distribution of the major rock types and ages. The Franciscan assemblage dominates Marin County whereas Salinian granites and Cenozoic sedimentary rocks characterize the Point Reyes Peninsula. These disparate rock types were juxtaposed by motion along the San Andreas fault. Geologic maps illustrate the areal distribution of rock types beneath the surface veneer of soil and vegetation.

Figure 29.7 This conglomerate of early Cenozoic age rests unconformably on the granitic basement in the Point Reyes Headlands.

to about 80–100 Ma, comparable to the rest of the Sierran batholith from which the Point Reyes granites were derived.

From the Sea Lion Overlook, walk up the road to the Point Reyes Lighthouse visitor center to see outcrops of a clastic sedimentary rock called a *conglomerate* that rests directly on the granitic basement at the Headlands. (Review the origin of clastic sedimentary rocks, including conglomerates, in Chapter 1.) The conglomerate is composed of rounded pebbles, cobbles, and boulders bound together in a matrix of coarse sand (Fig. 29.7). The colorful, rounded clasts are made of pieces of the underlying granitic rock, purple and black volcanic rocks, metamorphic quartzite, and reddish chert. The rounding of the pebbles and cobbles indicates that they were repeatedly abraded and smoothed prior to their deposition, perhaps within a fast-flowing streambed or on a wave-washed beach. Rare fossils found within pebbles of siltstone within the conglomerate suggests an early Cenozoic age, perhaps 50–60 Ma.

This conglomerate is only found on the Point Reyes Headlands and is not present anywhere else on the peninsula. A conglomerate of identical age and composition, however, exists at Point Lobos near the Monterey peninsula, about 150 km (>90 mi) to the south. This southern conglomerate is composed of the same rounded pebbles and rests directly on Salinian granites, exactly like the conglomerate of the Point Reyes Headlands. Fossils in both conglomerates indicate deposition in deep water, and other sedimentary features suggest that the site of deposition was a submarine canyon.

How do we reconcile all of these pieces of the puzzle? The key is that the widely separated outcrops of this conglomerate lie on opposite sides of the San Gregorio fault (Fig. 29.4). During the early Cenozoic, the depositional environment of the Point Reyes Conglomerate was located near Monterey. Fragments eroded from nearby granites, volcanics, and other rocks were rounded as they were transported downstream, then perhaps abraded further as they were washed by waves and tides on a beach. Similar processes occur today on beaches at Point Reyes, where rounded pebbles and cobbles form gravel bars along the shoreline (Fig. 29.8). Changes in sea level and seaward-flowing currents may have enabled these rounded pebbles and cobbles to be flushed into deeper water by submarine debris flows that swept through canyons on the continental shelf and slope. The chaotic mixture of sand and rounded cobbles and pebbles settled out of the flow on top of granitic rocks on the early Cenozoic seafloor and were then buried by younger sediments.

Later in time, these deep-water conglomerates and their underlying crystalline basement were sliced by the right-lateral San Gregorio fault. The conglomerates now exposed on the Point Reyes Headlands were transported one earthquake at a time about 150 km (>90 mi) to the northwest, leaving behind their twin on the east side of the fault. This pair of conglomerates provides *pinning points* for calculating the amount of displacement along the San Gregorio fault, similar to the pinning points defined by the Pinnacles/Neenach volcanics along the San Andreas fault to the south (Chapter 28). The Point Reyes Conglomerate tells a strange story of deposition and displacement that required observation, analysis, and logic by inquisitive geologists over the past few decades.

Figure 29.8 Rounded cobbles and pebbles are washed by waves and tides on Palomarin Beach. Seacliffs in the background are mid- to late Cenozoic marine sedimentary rocks that have been gently tilted by slumping. Note the thin veneer of vegetation and soil draping the rocks.

The rolling pastoral hills of Point Reyes National Seashore other than Inverness Ridge and the Headlands are composed of over 4000 m (>13,000 ft) of mid- to late Cenozoic sedimentary rock that rests above the underlying granitic basement (Fig. 29.6). This thick pile of interbedded sandstone, siltstone, and shale was deposited in a spectrum of marine depositional environments whose water depths ranged from shallow to very deep. Interpretations of water depth are primarily based on the fossils found in each of the stacked layers of rock. Fragments of clamshells, barnacles, sand dollars, and other hard-shelled marine invertebrates indicate deposition in wave-washed, shallow waters. Fossils of microorganisms that are known to live on the deep seafloor suggest deposition in water hundreds to thousands of meters deep. Fossils of marine phytoplankton like diatoms indicate the suspension settling of the tiny shells of these microscopic plants to great depths in mid-Cenozoic oceans, comparable to where diatoms accumulate on today's ocean floor. Some of these sedimentary formations at Point Reyes contain the bones of whales and walrus, as well as shark teeth. These fossils of large-bodied animals provide deeper insight into the ecology of the ancient Pacific Ocean, but don't reveal much about water depth since they may have settled out anywhere on the seafloor after death.

Deposition of these clastic sedimentary rocks occurred when the crust beneath the Point Reyes region lay offshore from the Monterey area. This interpretation is based on similarities with comparable sedimentary rocks around Monterey Bay as well as by the pinning point along the San Gregorio fault defined by the Point Reyes Conglomerate discussed above. This thick pile of marine rocks, like the older granitic basement and conglomerates that lie beneath, was transported to its current location by episodic movement along the San Andreas, San Gregorio, and associated faults over the last several million years. Slight *transpressional stress* along the San Andreas system of faults lifted the sedimentary rocks of Point Reyes and their granitic basement above sea level during their northward journey. During uplift, the rigid block of rock was gently folded and faulted, features that are evident in many seacliff exposures.

You can get a close look at all of these sedimentary rocks along the seacliffs that line the beaches of the peninsula. (Be aware that the cliffs may spontaneously collapse and that it's dangerous to get too close to the base. If you notice a conical pile of debris at the bottom of a cliff, use it as a warning that the cliffs are unstable.) Remember also that these sedimentary rocks lie just beneath the thin veneer of soil and grazing cows of the pastoral zone throughout the park. In some places, the solid rocks crop out above the pasturelands, but the majority of good exposures occur along the bluffs behind beaches. You should also be aware that these rocks don't end at the shoreline, where they have been left exposed by coastal erosion. Offshore drilling indicates that each of the Cenozoic formations continue beneath the seafloor where they are buried beneath a thin cover of recent marine sediment. The Cenozoic rocks were once laterally continuous layers in the general region, but have been dissected and offset by faulting, erosion, and fluctuations in sea level. The shoreline is an ephemeral feature that migrates constantly as sea level changes through time, a concept that will be addressed below.

Uplift of the Point Reyes Peninsula

Point Reyes National Seashore is notable for its steep bluffs and headlands that commonly rise a hundred meters or more above the beach (Fig. 29.9). Even though wind, waves, and tides batter the shoreline constantly, doing the erosional work necessary to break down the seacliffs, the coastal landscape at Point Reyes remains elevated far above sea level in most places. The physical appearance of a coastline is determined by an ongoing battle between the agents of coastal erosion and tectonic stress. For instance, the *passive continental margin* that lines the Atlantic coast of the United States is tectonically quiet, so deposition has had plenty of time to create a gentle profile of broad beaches and coastal marshes (Chapter 26). Along the *active continental margin* that characterizes California, however, tectonic stress dominates, and thus coastal landscapes are elevated above the shoreline and relentlessly attacked by waves. Coastlines where the land is actively rising due to tectonic stress are known as **emergent coasts,** and the Point Reyes Peninsula is a prime example.

Some of the uplift along the tectonically active California coast is due to transpression, the combination of shear and compression that occurs where strike-slip faults bend (Chapter 26). Most of the mountains composing the Coast Ranges were uplifted due to transpression, and the Point Reyes Peninsula is no exception. Interaction between the San Andreas and its associated faults, like the San Gregorio, acts to squeeze rock along the fault planes and cause it to rise upward. A magnitude 5.0 earthquake in 1999 located where these faults merge near Bolinas attests to active transpressional uplift in the Point Reyes region.

The ongoing rise of the Point Reyes Peninsula is primarily related to the Point Reyes fault that cuts through the continental shelf seaward of the peninsula (Fig. 29.4). The Point Reyes fault is not a strike-slip fault like most of the others in the San Andreas system but rather is a *reverse fault*. You may recall from earlier chapters that reverse faults are those where a block of rock on one side of the fault is pushed upward and above the block of rock on the other side by compressional stress (Fig. 10.7). Along the Point Reyes fault, rock to the northeast, including the Point Reyes Peninsula, is being squeezed upward and above the rock underlying the continental shelf to the southwest. The plane of the Point Reyes fault exhibits a steep angle, so that the crustal rock composing the Point Reyes Peninsula is shoved upward above sea level.

You're likely wondering how geoscientists know that the Point Reyes fault exists. Isn't it hidden beneath the sea? They employ a battery of methods and technologies to map the contours of the seafloor and even to "see" deep into the crust. The seafloor off the California coast and the rocks beneath have been mapped in great detail as part of the California Seafloor Mapping Program, a consortium of federal and state agencies, universities, and private companies. The mapping program uses high-resolution sonar techniques and massive computer firepower to define the bathymetry of the seafloor. Additional detail is added by seafloor video and photography, as well as by physical sampling of the seafloor sediment. A geophysical technique called **seismic reflection profiling** uses powerful sound waves generated from a research vessel that penetrate through the seafloor into the underlying rocks before rebounding back up to the surface where the echoes are recorded by sensitive hydrophones then disentangled by computer software. The images produced by this seismic technique are a bit like an X-ray of the upper several kilometers of the crust, enabling researchers to identify the geometry of the rocks beneath the seafloor as well as to recognize faults such as the Point Reyes fault.

The Point Reyes fault comes closest to the land near the Point Reyes Headlands, the precipitous cliffs held up by the resistant Salinian granites and overlying conglomerates (Fig. 29.2). The fault plane is mostly hidden deep within the rocks beneath the sea, but it does create a step in the seafloor topography of a few tens of meters (evident as a subtle scarp southeast of the Headlands in the Google Earth image of Fig. 29.1). The seafloor a few kilometers offshore of the Headlands abruptly drops downward, marking the fault plane that separates the upthrown block to the north from the downthrown block to the south. The compressional Point Reyes reverse fault likely connects with the San Gregorio fault deep beneath the surface, making it part of the larger San Andreas network of faults in the region. The Point Reyes Peninsula and the surrounding continental shelf appear to be caught in a vise between the compressional stresses of the Point Reyes fault to the southwest and the San Andreas fault to the northeast.

The uplift rate on the Point Reyes fault is estimated to be very slow (~0.3 mm/yr; 0.01 in/yr), but there is discernible seismic activity on the fault. An earthquake centered a few kilometers offshore the Point Reyes Headlands on the Point Reyes fault occurred in 1970 with a magnitude of 2.9. Although small by human standards, this earthquake and other smaller quakes indicate that the Point Reyes reverse fault is indeed active and likely responsible for uplift of the Point Reyes Peninsula.

Coastal Landforms and Processes at Point Reyes

The Point Reyes Peninsula is surrounded by 130 km (~80 mi) of coastline that ranges from the protected shore along Tomales Bay to the crescent-shaped beach that lines Drakes Bay in the south to the dramatic seacliffs near Kehoe Beach and Point Reyes Headlands (Fig. 29.2). Estuaries and lagoons like Abbotts Lagoon, Drakes Estero, Estero de Limantour, and Bolinas Lagoon are quiet-water havens for wildlife. This section will focus on the dynamic array of processes that control the diversity of coastal landscapes where the land meets the sea along the Point Reyes Peninsula.

Emergent coasts are commonly marked by rocky headlands that protrude seaward, alternating with secluded pocket beaches backed by steep seacliffs. The west-facing coast of Kehoe Beach is a good example (Fig. 29.9). Waves along Kehoe Beach tend to be powerful and dangerous because they strike the coast head on, driven by the prevailing northwesterly winds. These waves focus their energy on the headlands and disperse it along the beaches due to **wave refraction**, the bending of waves as they encounter obstacles along the coast (Fig. 29.10). Waves off the Pacific meet the point of the protruding headlands first upon arrival at the shoreline. As the wave slows and

Figure 29.9 Oblique view looking east across the seacliffs, headlands, and pocket beaches near northern Kehoe Beach along the emergent northwest coast of Point Reyes National Seashore. The bluffs along this stretch rise over 120 m (400 ft) above the surf. Waves undercut the cliffs at their base, triggering landslides that maintain the steep cliff faces.

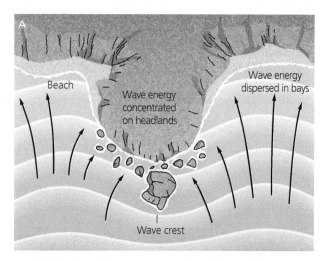
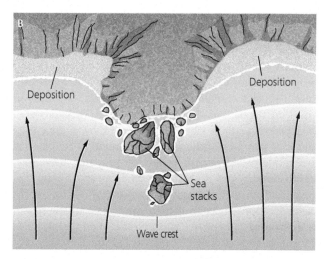

Figure 29.10   Wave energy tends to concentrate on headlands that protrude outward into the sea and also tends to spread out along adjoining beaches and bays due to wave refraction. Thus, wave erosion dominates at headlands, whereas deposition dominates on the beaches.

water piles up at the tip of the headland, the surrounding waves bend (refract) to wrap around the sides, attacking the rocky flanks of the headlands. By the time the rest of the wave reaches the adjoining beaches, the wave's angle of approach has spread out, resulting in a dispersal of wave energy. The end result is that the direct impact of wave energy acts to break down headlands to produce loose particles of sediment. Waves and tides redistribute these particles onto adjoining beaches where the lower wave energy permits deposition to occur. The scenic and diverse coastal landscape such as the one at Kehoe Beach is ever changing due to the continual breakdown of headlands and the transfer of the eroded debris onto nearby beaches in response to wave refraction.

Storm waves and high tides batter the base of seacliffs behind the beach, commonly eroding a slight notch. The overhanging cliff face eventually becomes unstable and will break away, perhaps along vertical joint planes, to collapse catastrophically onto the beach. The pile of debris that accumulates along the base of seacliffs is rapidly broken down by waves and tides and is redistributed along the beach. The relatively young and weak sedimentary rocks that compose most seacliffs circumscribing the Point Reyes Peninsula are particularly susceptible to wave erosion and collapse. Landslides are an inherent process along emergent coastlines such as the one at Point Reyes, and the ragged surfaces of most seacliffs are the scars left behind after collapse (Fig. 29.11). The precipitous seacliffs

Figure 29.11   South-facing granitic seacliffs, sea stacks, and pocket beach of Point Reyes Headlands. The precipitous cliffs fronting the headlands are relentlessly attacked by waves that undermine the cliffs, causing them to fail as landslides. Elephant seals sprawl on the sand at the base of cliffs in lower right.

and headlands along Point Reyes National Seashore are in a constant state of retreat due to the combined effects of wave erosion and landslides. Because the hazard posed by landslides and rockfalls at Point Reyes is ever present, visitors need to be aware of staying away from the base of seacliffs as well as avoiding the potentially unstable uppermost edges of bluffs and overlooks.

Other scenic coastal landforms created by wave erosion along headlands are sea caves, arches, and sea stacks (Fig. 29.12). **Sea caves** are rounded cavities at the base of seacliffs formed where wave refraction has focused its erosional energy. If the sea cave is developed along the side of an elongate headland, wave erosion may enlarge it to form a throughgoing **sea arch**. In time, the narrow ceiling of the arch will inevitably collapse. and the rubble will be broken down by the battering waves. The erosional remnant of the headland is left isolated out in the surf zone and is known as a **sea stack**.

If you think the process described above only happens on geological time scales, you are partially correct. In March 2015, the narrow neck of headland above Arch Rock near Kelham Beach suddenly gave way, collapsing catastrophically to the beach below (Fig. 29.13). Two people happened to be on the trail above the neck when it failed, sending them plummeting 20 m (~70 ft) onto the rocky debris on the beach below, killing one of them. The now-closed trail along the narrow fin of rock above Arch

A

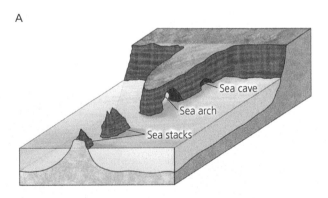

B

Figure 29.12 A) Wave erosion of headlands produces sea caves, sea arches, and sea stacks. B) Sea stacks near southeast tip of Point Reyes Headlands.

Figure 29.13 A) Arch Rock near Kelham Beach facing Drakes Bay, prior to collapse. B) Pile of debris left after the narrow arch suddenly failed in March 2015.

Rock led to a popular overlook, which today remains cut off as a result of the collapse. The National Park Service works hard to avoid tragedies like this and constantly assesses potential hazards throughout the park, restricting access as necessary. In addition, visitors to the National Seashore have to use common sense, anticipating dangers such as hazardous surf and unstable bluffs, and avoiding becoming trapped on pocket beaches by high tides.

Not all of the coastal landscapes around the Point Reyes Peninsula are marked by steep, rocky headlands. In places like Point Reyes Beach, located along the west-facing shoreline, the full force of Pacific waves pummels the coast, and the surf zone can be tumultuous (Fig. 29.14). Once-irregular headlands composed of weak Cenozoic sedimentary rock have retreated under the onslaught of wave attack to form a low, linear bluff behind the elongate beach, mantled by a field of sand dunes. The linearity of the beach along this stretch of shoreline is due to the systematic movement of sand along the beach by waves that approach dominantly from the northwest.

How do waves distribute sand along linear beaches? Waves are formed by frictional shear, with winds blowing just above the surface of the water. The two fluids interact along the air–sea interface, so that water will be driven in the direction of the wind. The prevailing northwest winds off the northern California coast force waves to strike

Figure 29.14   Linear strand of Point Reyes Beach as seen from the Lighthouse Visitors Center. A coastal dune field crowns the land behind the beach. The curving shoreline far to the north is the peninsula of Tomales Point.

narrow and expand as much as 30–60 m (100–200 ft) over the course of a single year. The long-term view of beaches like those at Point Reyes is that of a highly dynamic system, with longshore drift moving sand laterally along the beach at the same time that the sand is being redistributed on and off of the beach in tune with the seasons.

The beaches that line Drakes Bay along the south-facing shoreline of the peninsula are subject to less intense wave action than those along the northwest-facing coast (Fig. 29.3). The granitic rock of the Point Reyes Headlands protects Drakes Bay from the dominant waves approaching from the northwest. These waves refract around the headlands, eventually cycling back to approach the beaches around Drakes Bay from the south. The longshore drift produced by these refracted waves moves sand to the north and west, producing elongate extensions of beaches called **spits**, such as Limantour Spit. Another example is the spit that extends to the west from Stinson Beach where it forms a barrier near the mouth of Bolinas Lagoon. (During the 1906 San Francisco earthquake, the San Andreas fault offset this spit by a few meters.)

Where does the sand on beaches at Point Reyes come from? You know that some of the sand is derived from the erosion of rocky headlands and the collapse of seacliffs. The rocky debris from landslides is eventually broken

along Point Reyes Beach at a slightly oblique angle. As the waves first hit the beach, particles of sand are carried up the beach at an angle perpendicular to the wave crests. The sand within the backwash, however, flows back down the slope of the beach at a slightly different angle (Fig. 29.15). The end result is a zigzag pattern of sand movement laterally along the beach called **longshore drift**. The linearity of Point Reyes Beach is due to the inexorable longshore drift that redistributes sand toward the south where it piles up behind the western tip of the Point Reyes Headlands. Some of the sand on Point Reyes Beach is picked up by the unrelenting winds and transported a short distance inland where it forms a near-continuous belt of **coastal sand dunes**.

The appearance of sandy shorelines like that of Point Reyes Beach changes seasonally, with beaches becoming narrower during the winter and broader during the summer. Powerful, storm-driven waves during the winter months are capable of scouring sand from the beach and transporting it offshore where it is redeposited as elongate underwater bars parallel to the beach. Less intense summer waves act to rebuild and broaden the beach, transporting sand from the offshore bars back onto the face of the beach. Some beaches at Point Reyes may

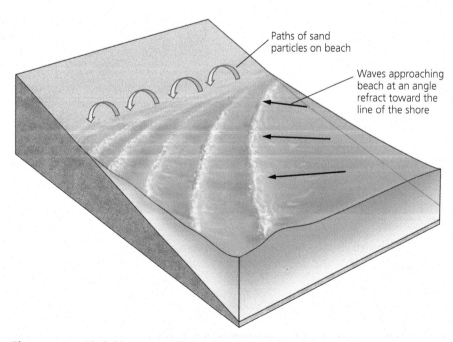

Paths of sand particles on beach

Waves approaching beach at an angle refract toward the line of the shore

Figure 29.15   Wind-driven waves that strike a beach at an oblique angle generate a longshore drift of sand that moves laterally along the shoreline, resulting in linear beaches.

down to smaller particles that contribute to the sediment supply. A close look at a handful of sand from any beach at Point Reyes reveals the nearby source rock. Sand grains composed of quartz and whitish feldspar likely were derived from the granitic basement. Sand composed of quartz and particles of sedimentary rock came from the erosion of the Cenozoic sedimentary rocks that dominate seacliffs on the peninsula. And some sand will have bits and pieces of shell fragments, derived from the breakage of invertebrate shells in the surf zone.

But many beaches at Point Reyes will include well-rounded sand grains composed of red chert, quartzite, volcanics, and metamorphic schist. These grain types are nowhere to be found in the rocks of the peninsula, but rather must have been derived from the Franciscan assemblage on the other side of the San Andreas fault to the east. The closest significant source from the mainland is the Russian River that drains the Franciscan terrain of Sonoma County, entering the Pacific north of Bodega Bay. It's possible that sediment discharged from the river was transported southward by longshore drift, ending up on the beaches of Point Reyes. But a more probable scenario is that the Russian River cut across the continental shelf during lowstands in sea level, perhaps during glacial phases of the Pleistocene Ice Ages, and transported particles of eroded Franciscan rock to the shelf near Point Reyes. Waves, currents, tides, and rising sea level would have redistributed these grains onto the beaches of Point Reyes. Thinking about all this while looking at a handful of grains from a Point Reyes beach can make you dizzy as you try to trace the life cycle of sand particles from their original source to your little patch of beach.

The southern part of the Point Reyes Peninsula near Bolinas is notable for its **marine terraces**—relatively flat or gently seaward-sloping surfaces formed along the shoreline by wave abrasion but now elevated above sea level (Fig. 29.16). The marine terrace near Bolinas rises 40–60 m (130–200 ft) above the shoreline. These terraces record the interplay of tectonic uplift and global sea-level change over the last half million years or so. They originate as **wave-cut platforms** formed by erosional beveling of a rocky coastline by wave action (Fig. 29.17). In the Bolinas area, the planar surface in the tidal zone that extends from Agate Beach south to Duxbury Point is an excellent modern example of a wave-cut platform.

Wave-cut platforms typically form during highstands in sea level as the surf zone is beveled to a flat surface by

Figure 29.16 Oblique view to the west–northwest across the southern tip of the peninsula near Bolinas. The darker area in the surf zone between Agate Beach and Duxbury Point is a wave-cut platform. The broad area of Bolinas Mesa and north is a marine terrace.

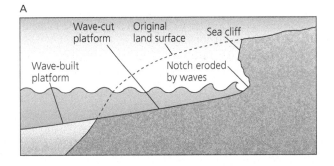

Figure 29.17 A) Wave erosion undercuts the base of seacliffs, causing them to retreat by repeated landslides. As the cliffs recede, a wave-cut platform forms along the base of wave action. B) Wave-cut platform near the southern end of Point Reyes Peninsula near Bolinas. These rocky surfaces are great for tide pooling when exposed during low tide.

waves and tides. As the adjacent seacliff retreats, the wave-cut surface grows landward. When sea level subsequently falls a few tens of thousands of years later, the wave-cut platform and the seacliffs are left behind as exposed land. Tectonic uplift related to motion along the network of faults continues during the lowstand in sea level, raising the wave-cut platforms far above the shoreline to become marine terraces. This process may repeat itself many times as sea level fluctuates coincident with the background of constant tectonic uplift. The result is a stepped series of marine terraces developed along the emergent coastline. Each terrace records a former sea-level highstand, now raised above the shoreline by episodic tectonic uplift. Marine terraces like those at Point Reyes are a common feature of Channel Islands National Park and many other places along the emergent coast of California.

Effects of Postglacial Sea-Level Rise at Point Reyes

During the Pleistocene Ice Ages of the last 2.6 m.y., sea level fluctuated in tune with the waxing and waning of continental ice sheets. As described earlier in Chapter 27 on Channel Islands National Park, sea level falls during glacial phases as water is locked up in gigantic sheets of ice on the continents. As the glaciers melt during warmer interglacial phases, the water is transferred back to the oceans and global sea level rises. During the Last Glacial Maximum about 20,000 years ago, sea level was ~120 m (~400 ft) lower, and the shoreline was about 40 km (25 mi) west of its current location. The future Point Reyes Peninsula was just a hilly part of the mainland, connected along the linear valley of the San Andreas fault. Streams drained the hills of Point Reyes, cutting channels and flowing far across the exposed continental shelf to the distant shoreline.

During this most recent sea-level lowstand, the land area of California extended as a broad coastal plain out to the very edge of the continental slope (Fig. 29.4). San Francisco Bay was a broad alluvial valley drained by streams that flowed north into the ancestral Sacramento River that cut through the exposed land of today's Golden Gate on its way to the Pacific off in the distance. Mammoth and bison almost certainly roamed the vegetated plains, likely pursued by the earliest human immigrants to America. The evidence of this lost world is now hidden beneath the waters covering the continental shelf, submerged as the vast ice sheets melted and sea level rose in response to climatic warming.

With the advent of the current interglacial phase about 15,000 years ago, the

shoreline transgressed inland in concert with rising sea levels (Fig. 29.18). The initial pace of sea-level rise was rapid, about 10 m/k.y. (33 ft/k.y.) but leveled off around 7000 years ago (~5000 BCE). This long-term rise in sea level and landward migration of the shoreline is known as the **postglacial marine transgression**, a global event that happened concurrent with a naturally warming planet and the rise of civilization. (For comparison, the current rate of sea level rise in our rapidly warming world is about 3 m/k.y. [10 ft/k.y.], a rate that is a rapid acceleration relative to the generally stable sea level of the past few thousand years.)

As the shoreline stabilized around the Point Reyes Peninsula a few thousand years ago, former stream systems were flooded by seawater, creating **estuaries** where saltwater mixes with freshwater. Drakes Estero is the most obvious example (*estero* means estuary in Spanish), with its finger-like appearance created as coastal stream valleys were slowly submerged (Fig. 29.19). Other estuaries at Point Reyes formed by the postglacial marine transgression include Estero de Limantour, Bolinas Lagoon, Abbotts Lagoon, and Tomales Bay (Fig. 29-3). The modern coastal landscape of Point Reyes National Seashore was created only over the last few thousand years as the shoreline stabilized near its current position. Tectonic uplift maintains the elevated topography of the peninsula even as waves and tides batter its margins.

The windswept beaches, stunning vistas, and diversity of wildlife make Point Reyes a treasure along the northern California coast. There aren't many places where you can see tule elk, elephant seals, and gray whales breaching just offshore all in one day (Fig. 29.20). The weather can be fickle though, with warm and sunny conditions

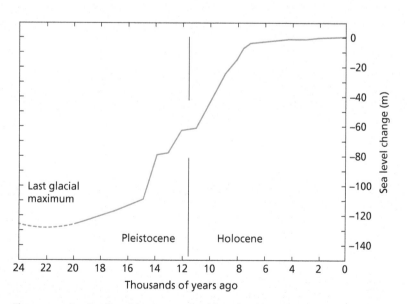

Figure 29.18 Sea level began to rise sharply about 15,000 years ago accompanied by the landward migration of the shoreline. The pace of sea-level rise was rapid from ~15,000 to ~7000 years ago but has since stabilized. The postglacial rise in sea level is ~120 m (~330 ft) higher than during the Last Glacial Maximum, inundating coastlines globally.

Figure 29.19 Drakes Estero, a former stream network now flooded by the Holocene rise in sea level. The estuary is a mosaic of eelgrass meadows, sandbars, and tidal channels that support a robust ecosystem.

Figure 29.20 Tule elk grazing above the waters of Tomales Bay.

suddenly replaced by howling wind and driving rain. Fog commonly envelopes the peninsula in the mornings, typically burning off as the temperature rises. But if you time it right on a day when both visitation and the tides are low, you'll revel in the solitude as well as the elemental sounds of wind and waves that define this national seashore.

Golden Gate National Recreation Area

The Olema Valley and much of the strip of land adjoining Point Reyes National Seashore along its eastern boundary are part of the Golden Gate National Recreation Area (NRA), administered by the National Park Service (Fig. 29.21). The NRA is a semicontinuous collection of special places that extends from Marin County through San Francisco and into San Mateo County. Because of its location within the urban corridor, it is widely visited, with most people not even recognizing that they're inside a unit of the National Park System. The NRA was established to preserve sites of aesthetic, cultural, and military significance, as well as to provide a haven for harried citizens of the densely populated region. In an incidental way, the Golden Gate NRA protects some hazardous segments of the San Andreas fault zone from development.

The Golden Gate NRA includes 45 km (28 mi) of coastline, including seacliffs and rocky headlands, beaches, estuaries, and wetlands. Outstanding state parks like Mount Tamalpais and Samuel P. Taylor border the NRA, as does the urban oasis of city-administered Golden Gate Park. The NRA's many geological attractions include the Marin Headlands, with its exceptional exposures of contorted Franciscan cherts, and Point Bonita, a narrow headland composed of pillow basalts that guards the entrance to the Golden Gate. The stately old-growth coastal redwoods of **Muir Woods National Monument** in Marin County are enclosed within the NRA, as is the history-rich military site of the Presidio on the northern tip of the San Francisco peninsula. For added measure, the exhilarating expanse of Ocean Beach along the west side of San Francisco and the calmer setting of Stinson Beach in Marin County provide refuge for those seeking a direct connection with the Pacific Ocean.

Key Terms

coastal sand dunes
emergent coast
estuary
LiDAR
longshore drift
marine terrace
postglacial marine transgression
sea arch
sea cave
sea stack
seismic reflection profiling
shutter ridge
spit
wave refraction
wave-cut platform

Related Sources

KellerLynn, Katie. (2008). *Geologic Resource Evaluation Scoping Summary: Point Reyes National Seashore, California.* Denver, CO: Geologic Resources Division, National Park Service.

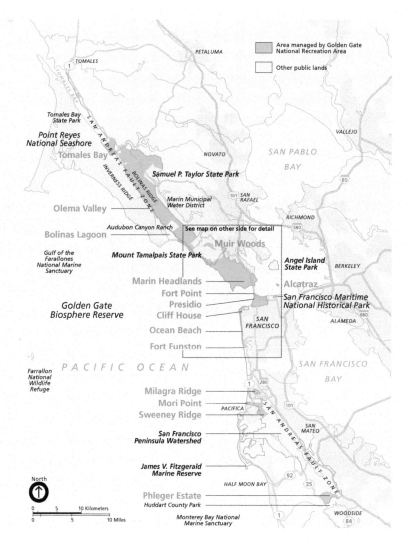

Figure 29.21   The Golden Gate National Recreation Area welcomes about 15 million visitors per year.

Konigsmark, Ted. (1998). *Geologic Trips: San Francisco and the Bay Area*. Gualala, CA: GeoPress.

Sloan, D. (2006). *Geology of the San Francisco Bay Region. California Natural History Guide Series 79*. Berkeley: University of California Press.

Stoffer, P. W. (2006). *Where's the San Andreas Fault? A Guidebook to Tracing the Fault on Public Lands in the San Francisco Bay Region*. General Information Product 16, U.S. Geological Survey.

National Park Service—animation of marine terrace formation. Accessed February 2021 at http://www.nps.gov/chis/photosmultimedia/models-of-change-geology.htm

National Park Service—Point Reyes National Seashore: Geologic Activity. Accessed February 2021 at https://www.nps.gov/pore/learn/nature/geologicactivity.htm

USGS Guidebook to the Geology at Point Reyes National Seashore. Accessed February 2021 at http://pubs.usgs.gov/of/2005/1127/chapter9.pdf

30 Olympic National Park

Figure 30.1 The Olympic Mountains from near the Hurricane Ridge Visitor Center.

Located in the northwest corner of Washington, Olympic National Park is one of the most dramatic and diverse in the entire national park system (Fig. 30.1). Serrated mountains with glacier-clad slopes; moss-draped rainforests with countless shades of green; storm-charged rivers raging through narrow valleys; and even sea stacks rising above a churning coastline—if that isn't enough scenic and ecologic variety to overload the senses, consider that Olympic's geologic underpinnings are deformed masses of oceanic rocks, lifted from the seafloor by the colossal forces at the leading edge of the Cascadia subduction zone.

The national park dominates the Olympic Peninsula, which is separated from the Cascades volcanic province to the east by the Puget Lowlands (Fig. 30.2). To the north lies the Strait of Juan de Fuca and to the west is the Pacific Ocean. Olympic National Park is actually two parks. One is centered above the Olympic Mountains, a roughly circular highland incised by alpine glaciers and fast-flowing rivers. The other is an elongate, west-facing coastline marked by rocky headlands and wave-battered beaches. Dense coastal woodlands of the Olympic National Forest separate the two parts of the park.

The highest peak in the Olympic Mountains is Mount Olympus at 2432 m (7980 ft), draped on all sides by alpine glaciers and snowfields. Dozens of glaciers crown the highest peaks of the Olympics, with the Hoh Glacier reaching

over 5 km (3 mi) in length. Glacial meltwaters, snowmelt, and copious rainfall feed a network of rivers and streams that radiate outward from the central highlands (Fig. 30.3). Water in all its forms dominates Olympic National Park because the mountains intercept storm fronts coming off the Pacific. Rainfall in the park is variable but averages well over 3 m/yr (~120 in/yr) along the coast and the western flank of the mountains, more than any other place in the contiguous United States. In the higher elevations, the moisture falls as snow, up to 9–20 m/yr (350–790 in/yr) on the tallest peaks. The jagged, muscular topography of the mountains is drained by the array of rivers and countless tributaries that diverge from the glacial summits. The rapidly flowing waters transport large amounts of sand, silt, and gravel from the highlands, effectively abrading deep canyons on their way downslope.

The ecosystems of Olympic National Park are as wide-ranging as its landscapes. The abundant rainfall and fog support a temperate rainforest along the coast and on the lower western slopes of the Olympic Mountains (Fig. 30.4). Sitka spruce is the dominant tree of the rainforest, forming a canopy above a dense groundcover of ferns and mosses. Deer and Roosevelt elk graze the lush

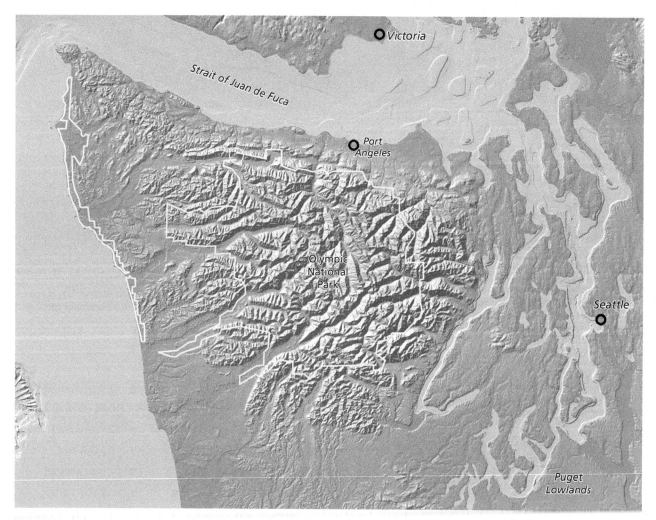

Figure 30.2   Olympic National Park is located on the Olympic Peninsula of northwest Washington, bound on three sides by water.

Figure 30.3 A radial pattern of rivers and streams drains the Olympic Mountains, with water flowing into the Pacific, Strait of Juan de Fuca, and Puget Sound.

along the convergent plate boundary (Fig. 26.16). As described in Chapter 12 on the Cascades volcanic province, the Pacific Northwest lies along a continental-oceanic convergent boundary known as the *Cascadia subduction zone* (Fig. 30.6). This plate boundary forms the northern part of the active Pacific Margin Geologic Province, with the San Andreas transform boundary in California forming the southern part. The transition occurs at the Mendocino *triple junction* in northern California where the North American, Pacific, and Gorda plates meet. Just offshore, the San Andreas fault turns westward to become the Mendocino fracture zone, the plate boundary between the microplates of the Cascadia convergent margin and the Pacific plate on the south.

This last chapter of Part 6 on the tectonically active Pacific Margin builds on several concepts introduced in earlier chapters such as the dynamics of the Cascadia subduction zone (Chapter 12), the tectonic setting of accretionary wedges (Chapter 26), the mechanics

green palette of vegetation. Elsewhere in the park, differences in rainfall and elevation create a mosaic of lowland, montane, and subalpine forests. The mountains generate a rain shadow along the northeast side of the peninsula, marked by an oak savanna ecosystem. Above the treeline at elevations around 1800 m (6000 ft), a rugged world of bare rock, glaciers, and snowfields crown the mountains.

Dramatic coastal ecosystems of the park encompass ~100 km (60 mi) of shoreline, marked by rocky seacliffs and sea stacks, sandy coves, and tidepools filled with invertebrates like sea stars and anemones. Offshore from the coastal strip is the Olympic Coast National Marine Sanctuary (Fig. 30.5), a haven for seals and sea lions, sea otters, whales, and dolphins. Hundreds of species of birds share in the bountiful resources of the park, and spawning Pacific salmon have been using the coastal rivers for millennia. All of this biodiversity is enabled by the combination of plentiful rainfall, rugged and generally inaccessible mountains, an isolated wilderness coast, and human foresight that created the park in 1938.

The diverse ecosystems and landscapes of Olympic National Park form a surficial cover above a complex geological foundation. The entire Olympic Peninsula provides insight into the exhumed interior of an *accretionary wedge*, that part of a subduction zone where rock formed on the deep seafloor is scraped off the top of a descending oceanic plate to accumulate as a thick stack of deformed crust

Figure 30.4 Moss-covered trees of the Hoh temperate rainforest.

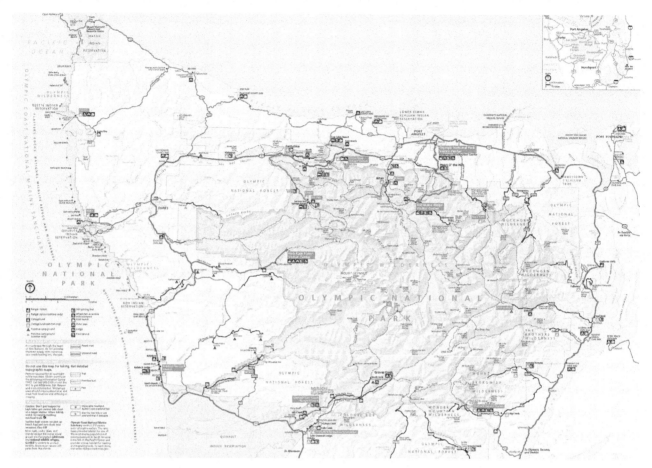

Figure 30.5 Map of Olympic National Park.

of alpine glaciers and their role in shaping mountain landscapes (Chapters 14, 17, 18), and the variety of processes that influence emergent coastlines (Chapter 29). We'll refer back to the relevant sections of these earlier chapters as necessary so that you can fill in the details as they relate to Olympic National Park.

Modern Tectonic Setting of Cascadia

Recall from earlier chapters that the oceanic Juan de Fuca, Gorda, and Explorer microplates are remnants of the much larger Farallon plate that subducted beneath western North America after the Mesozoic. Convergence of the Farallon plate is responsible for much of the mountainous topography of the American West. Its modern descendants along the Cascadia subduction zone are continuing that mountain-building process by supplying the magma that rises upward to feed the active Cascades volcanic arc (Fig. 30.6). The oceanic Juan de Fuca plate is descending beneath the North American plate at a net rate of about 3–4 cm/yr (1.2–1.6 in/yr). Even at this ultra-slow speed, the Juan de Fuca plate can move tens of kilometers over a few million years and hundreds of kilometers in several million years. As the Juan de Fuca plate migrated eastward

from its volcanic source along the mid-ocean ridge over time, sediment accumulated atop the volcanic foundation. Near the boundary with the North American plate deep beneath the sea, the sediment pile is ~2.4 km (1.5 mi) thick. In time, much of this sediment and underlying volcanic rock will be shoved down into the subduction zone, while the rest will be crushed against the leading edge of the North American plate to build the accretionary wedge.

The physical boundary on the seafloor between the subducting Juan de Fuca plate and the overlying North American plate occurs about 120 km (~75 mi) from the coastline at a water depth of around 3 km (1.9 mi). In other subduction zones, a *deep sea trench* forms along the plate boundary as the descending oceanic slab is flexed and bent downward. Some deep sea trenches may reach water depths of 10–11 km. But there is no deep trench along the Cascadia plate boundary, mostly because the wet climate of the Pacific Northwest enhances the weathering and erosion of rock, producing abundant sedimentary debris that is transported seaward by rivers such as the Columbia. Past lowstands in sea level exposed part of the continental shelf, permitting river systems to reach closer to the continental slope. There, they discharged their sediment load as turbidity currents into submarine canyons that funneled

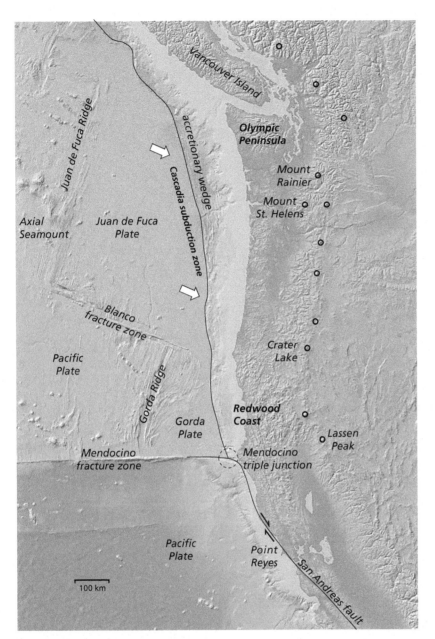

Figure 30.6 The transition from the San Andreas transform plate boundary to the Cascadia convergent boundary occurs at the Mendocino triple junction where the North American, Pacific, and Gorda plates converge. Aligned small circles mark the location of Cascades volcanoes.

depth of about 200 m (~660 ft), where it abruptly steepens onto the rugged continental slope, incised by submarine canyons. Near a depth of around 1500 m (5000 ft), the continental slope becomes less steep and is marked by a series of elongate ridges and valleys oriented parallel to the plate boundary. This rumpled lower slope is the actively deforming accretionary wedge, composed of semiconsolidated sediments, sedimentary rocks, and volcanic rocks scraped off the top of the slowly descending Juan de Fuca oceanic plate.

Seafloor mapping and *seismic reflection profiling* (Chapter 29) enable a view into the modern Cascadia accretionary wedge deep beneath the sea off the coast of Washington (Fig. 30.8). The interior of the accretionary wedge consists of a series of stacked sheets of actively deforming rock pushed along *thrust faults* so that they stack against one another. The thrust sheets are arranged so that older rocks are located near the landward portions of the wedge, with progressively younger rocks located in more seaward positions. You might envision the process that formed the anatomy of the wedge as younger slivers of marine rock being dragged *beneath* the older slivers. The elongate rumples on the lower continental slope are places where a thrust sheet has been pushed upward by underthrusting beneath. In essence, the accretionary wedge grows by the addition of material at the leading edge of the pile where the seafloor sediments are being squeezed into the subduction zone. At the same time that this compressional deformation is occurring beneath the surface, a veneer of sediment is being deposited on the continental slope by turbidity currents and submarine landslides derived from the adjacent continental shelf and upper slope. The complexity of the accretionary system is mind-bending.

The accretionary wedge is somewhat like a pile of paperback textbooks pushed up against each other at the end of the checkout conveyor belt at the university bookstore. As the books are shoved into the pile, they slip beneath the books already there, rotating and curling upward as the stack grows. Consider the surfaces between each book as a thrust fault where one book slides up and over an adjacent book. Now think of the books as layers of wet sediment, sandstones, shales, and volcanic basalt from the underlying crust, all bulldozed together as the conveyor belt of the

the mud, sand, and gravel downslope, filling the trench along the plate boundary to within a few kilometers below sea level.

The origin and depositional setting of the rocks composing Olympic National Park is intimately related to their tectonic setting along the Cascadia subduction zone. Today, depositional events occur contemporaneously with tectonic compression within the modern accretionary wedge about a hundred kilometers offshore (Fig. 30.7). This modern system is an ideal analog for comparable events that occurred earlier in the Cenozoic, now exhumed and exposed in Olympic National Park. A broad continental shelf about 40 km (25 mi) wide extends seaward to a

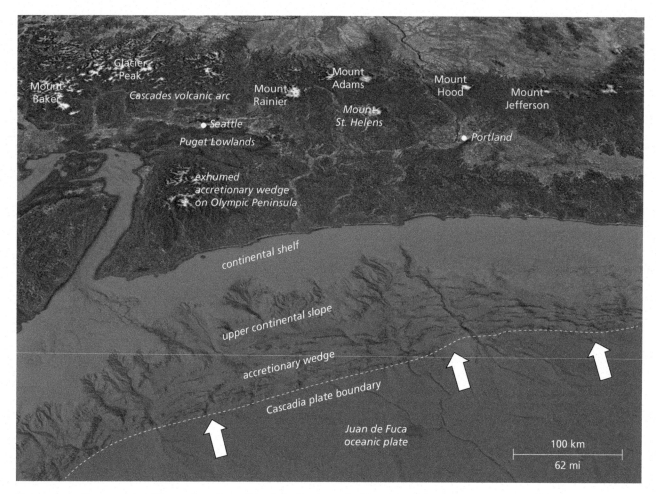

Figure 30.7   Components of the modern Cascadia subduction zone, looking toward the east.

rigid oceanic lithosphere slides relentlessly downward beneath the continent. Imagine the titanic forces, contorting and deforming the rocks within the vise of the subduction machine. In time, that messy pile of rock might be squeezed upward toward the surface to become the latest addition to the outer edge of a continent. This is the process of *accretion*, one of the fundamental ways that continents grow outward.

Origin of the Rocks of Olympic National Park

With an understanding of the modern tectonic setting of Cascadia and the dynamics of the accretionary wedge, we can more easily comprehend the origin of rocks composing Olympic National Park. The arrangement of rock bodies within the Olympic Mountains is exceedingly complex, as you might imagine for a stack of sedimentary and volcanic rock that accumulated in an accretionary wedge. Considering the wet climate, deep weathering, abundant vegetation, and steep terrain, deciphering the geology of the Olympic Mountains was a challenge for a generation or two of geologists. In this section, we'll deconstruct the

geologic architecture of the national park so that it becomes comprehensible to the nonspecialist. Remember to think in the three dimensions of space and the fourth dimension of time as you digest the following description.

The distribution of rocks composing Olympic National Park forms an arcuate pattern broadly shaped like a horseshoe open to the west (Fig. 30.9). (Geologic maps use colors and symbols to illustrate how bodies of rock are arranged across an area as if the surficial cover of soil and vegetation were removed. Standing at any one place, a geologic map would show the bedrock directly beneath the thin veneer of soil under your feet.) The rocks form two distinct groups that are both Cenozoic in age; a network of thrust faults separates the two groups. On the outside of the horseshoe along the southern, eastern, and northern side of the Olympic Peninsula is a very thick pile of volcanic basalt. On the inside of the horseshoe in the central Olympic Mountains and along the western side of the peninsula are variably metamorphosed sedimentary rocks. In general, the age of the rocks becomes progressively younger from the outside of the horseshoe toward the coastal exposures on the west. A scattering of very young glacial and alluvial sediments forms a thin covering over the older rocks.

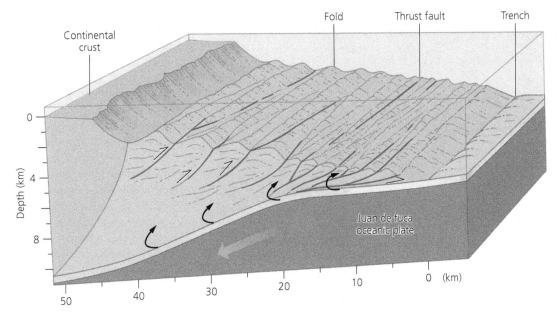

Figure 30.8 Generalized internal structure of the Cascadia accretionary wedge off the coast of the Pacific Northwest showing the stacking of thrust sheets of marine sediment and basaltic crust as they are scraped off the subducting Juan de Fuca plate.

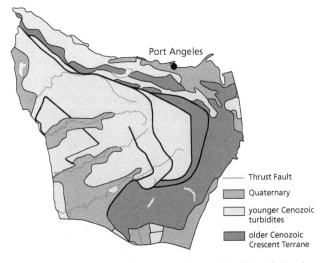

Figure 30.9 Simplified geologic map of the Olympic Peninsula showing the horseshoe-shaped geometry of the two major groups of rocks. Dark lines mark the trace of thrust faults where they breach the surface. Thrust faults separate the two major terranes. Slightly older Cenozoic pillow basalts form the outer margin, whereas slightly younger Cenozoic turbidites form the interior core.

The volcanic basalts on the outside of the horseshoe are called the Crescent Terrane. The tectonic term *terrane* (as noted earlier, not to be confused with the landscape term *terrain*) is a fault-bounded body of rock that originally formed elsewhere but that was tectonically transported and eventually accreted to a continental margin (Chapter 19). The volcanic rocks of the Crescent Terrane are mostly *pillow basalts*, formed underwater as lava solidified into bulbous mounds deep on the seafloor (Fig. 30.10). (The origin of pillow basalts was addressed in Chapter 26

in a discussion of the Franciscan assemblage of rocks comprising the Mesozoic accretionary wedge exposed in the Coast Ranges of California.) The age of the Crescent pillow basalts is early to mid-Cenozoic, ranging from 58 to 31 Ma. This is a massive stack of thrust sheets composed of volcanic rock that reaches a total thickness of about 16 km (10 mi) along the east side of the peninsula.

The lavas of the Crescent Terrane may have erupted on the seafloor within the Juan de Fuca mid-ocean ridge before being transported laterally and then accreted within the wedge. More likely, due to their enormous volume, they may have erupted from a submarine volcano, building up into a thick vertical pile. Volcanoes that don't breach the surface of the sea are called **seamounts.** A nearby modern example is Axial Seamount, located along the Juan de Fuca Ridge (Fig. 30.6). The Axial volcano resides over a highly active source of magma beneath the mid-ocean ridge and has grown into a huge edifice of pillow basalt. More complex tectonic interpretations of the Crescent Terrane are debated within the geologic community, but the origin of the basalts beneath the sea is indisputable.

The variably metamorphosed sedimentary rocks inside the Olympic horseshoe were originally deposited as couplets of sand and clay deep on the seafloor. After burial and cementation, these layers form alternating beds of sandstone and shale called *turbidites* (Fig. 30.11). The name is derived from turbid slurries that race downslope through submarine canyons, ultimately dropping their sedimentary load within a **submarine fan** on the deep seafloor (Fig. 26.13). As the turbidity current loses momentum, the coarser particles settle out first, forming a layer of sand. The finer particles of silt and clay remain in suspension as a milky cloud above the seafloor. In time, these finer sediments settle onto the sand layer, completing the

Figure 30.10   Pillow basalt of the Crescent Terrane.

deposition of the turbidite. The entire process may take anywhere from an hour to perhaps a day for the sand–clay couplet to be deposited.

Turbidity currents are a common mode of deposition on continental slopes and the adjacent deep seafloor. They may be set loose by severe storms that stir up sediment near the shelf–slope break, or they may result from rivers in flood, especially during sea-level lowstands when rivers extend out to the edge of the continental shelf. Earthquakes may destabilize steep underwater slopes of loose sediment, triggering turbidity currents that flush downward through submarine canyons. Thick layers of turbidites accumulate today on the continental slope and deep seafloor along the Cascadia plate boundary. In time, they may be deformed and incorporated within the accretionary wedge perched at the western margin of the North American plate.

Fossils within the thick pile of turbidites composing the interior of the Olympic horseshoe suggest ages ranging from mid- (~45 Ma) to late Cenozoic (~16 Ma). These dates partially overlap with those of the surrounding Crescent basalts, suggesting that during the mid-Cenozoic they formed contemporaneously with one another. An analogy can be made with today's Cascadia margin where seafloor volcanism is occurring in the same general area and over the same time scale as turbidite deposition. The thick stack of turbidites that form the interior of the Olympic horseshoe was incorporated into the mid- to late Cenozoic accretionary wedge where they were squeezed, sheared, rotated, and stacked into overlapping thrust sheets (Fig. 30.8).

During deformation within the accretionary wedge, some of the sedimentary rocks were dragged down along the subduction zone where heat and pressure caused the finer grained rocks to metamorphose into slate and phyllite (Fig. 1.30). Continued movement within the wedge brought these mildly metamorphosed rocks back toward the surface. In general, older rocks composing the core of the Olympic Mountains tend to show variable degrees

of metamorphism, whereas younger turbidites from the western coast retain their sedimentary characteristics (like those in Fig. 30.11). Collectively, the rocks within the interior of the Olympic horseshoe form stacked thrust sheets of distinct terranes that become progressively younger toward the west, separated from one another by thrust faults. This arrangement occurred when these rocks were deformed within the slow-motion chaos of the mid- to late Cenozoic accretionary wedge of the Cascadia margin.

How did this arcuate configuration of Cenozoic terranes arrive at their current location within the Olympic Peninsula? During the early to mid-Cenozoic, the Crescent basalts were accumulating along the seafloor atop the Juan de Fuca plate. As the plate and its thick mass of basalt migrated eastward toward the continent, turbidity currents deposited thick piles of sandstone and shale along the margins of the volcanic rocks (Fig. 30.12A). The slightly older Crescent basalt, perhaps projecting high above the seafloor as a seamount, was scraped off and incorporated into the accretionary wedge

Figure 30.11   Turbidites of the Hoh Terrane, Beach 4. The protruding light-colored beds are sandstones, whereas the thin, recessive layers are less resistant siltstones and shales. The originally horizontal layers were tilted sometime after they were deposited, buried, and solidified into rock,

by the mid-Cenozoic. With continued subduction, slices of younger turbidite deposits were jammed into the base of the accretionary wedge where they were thrust beneath the older, rigid Crescent Terrane (Fig. 30.12B). Through time, more slices of turbidites were pushed deep into the base of the accretionary pile-up where they were deformed, rotated, and smeared out along thrust faults, a process that continues today on the lower continental slope.

As new thrust sheets of turbidites were added to the base of the wedge, the overlying slabs of rock were raised so that they were elevated above sea level by mid- to late Cenozoic time. This phase marked the emergence of the Cascadia accretionary wedge as newly formed land of the Olympic Peninsula. With accretion and uplift, the shoreline shifted to the west and the North American continent became a bit bigger. Today, after millions of years of accretion and uplift, this huge stack of thrust sheets forms a collection of discrete terranes, rotated and tilted on end (Fig. 30.12C). The slightly older Crescent Terrane forms the highlands on the north, east, and south sides of the Olympic Mountains. Slightly younger metamorphosed turbidites form the central core of the Olympic Mountains. The seacliffs and headlands of the western coast are composed of relatively unmetamorphosed turbidites of the youngest terranes.

The *rock cycle* forms a tight loop between the Olympic Peninsula and the subduction zone just 120 km (~75 mi) to the west (Fig. 1.32). With uplift of the marine rocks to high elevations, weathering and erosion decompose the rocks, creating loose sedimentary debris that is transported back to the nearby ocean by rapidly flowing streams and rivers. Alpine glaciers carve deeply into the highlands, then carry the silt, sand, and gravel downslope where rivers do the rest of the work in getting the loose sediment to the coast. Along the shoreline itself, storm-driven waves batter the rocks, creating even more loose sediment to be transported seaward. During past phases of lower sea level when the continental shelf was partly exposed, rivers and currents transported the eroded particles to the edge of the continental slope where turbidity currents sweep them down onto the modern accretionary wedge. Some of the sediment will get dragged down into the subduction zone where it might transform into metamorphic rock, perhaps to be exhumed to the surface much later in time. Other sediment smeared along the subduction zone may be melted and incorporated into newly formed magma, possibly to erupt from an overlying volcano. The long-term recycling of Earth materials via the rock cycle requires energy, which is ultimately supplied by the Earth's internal heat and by the Sun. Few places on Earth express this timeless loop in such a compact system as well as the Olympic Peninsula.

Uplift and Seismicity at Olympic National Park

The modification of landscapes on the Olympic Peninsula by running water, glacial ice, and gravity occurs in response to uplift of the land surface. Cenozoic basalts and turbidites formed in marine settings cap the highest peaks in the mountains, providing clear evidence for recent uplift. The youngest terranes comprising the peninsula are about 16 m.y. old, suggesting that even younger rocks are jammed into the underlying accretionary wedge deep beneath the surface. The growth of the accretionary wedge begins along the lower continental slope near the interface with the subducting Juan de Fuca plate. As long as subduction continues, material will be added to the base of the accretionary wedge and compressional stress will translate inland, resulting in the progressive uplift of the Olympic Peninsula. As erosion removes material from the highest elevations, the land surface rises upward isostatically in response, maintaining the balance between uplift and erosion. Current uplift rates on the Olympic Peninsula are about 1 cm/100 yr (0.4 in/100 yr).

As expected for an active subduction zone, a constant drumbeat of low-magnitude seismicity occurs throughout

A   Early to mid-Cenozoic

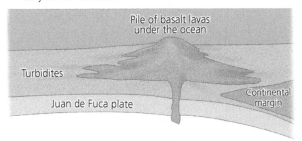

B   Mid- to late Cenozoic
Sedimentary rocks from continent deformed and stacked beneath lava pile

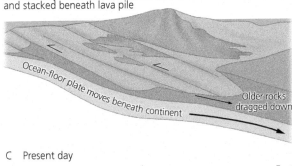

C   Present day

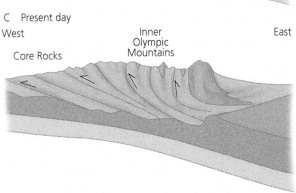

Figure 30.12   Sequence of events involved in the formation of the geologic foundation of Olympic National Park.

Cascadia, especially beneath the Olympic Peninsula. Most of the earthquakes are lower than magnitude 2.0, with occasional larger quakes randomly interspersed. The vast majority of the quakes beneath the Olympic Peninsula are focused about 5–40 km (3–25 mi) deep along the **megathrust** between the descending microplates and overriding North American plate. The low-magnitude quakes release a tiny amount of stored strain energy along the planar contact between the gigantic slabs of rock, but the megathrust today is considered to be **locked** and ready for a large-scale rupture. The threat of a large-magnitude Cascadia earthquake was addressed in Chapter 12 on the Cascades volcanic province, but we'll briefly build on that earlier discussion here.

The last major earthquake along the Cascadia subduction zone was in the year 1700, based on paleoseismic studies. Previous major quakes are estimated to have occurred around 900 CE, 700 CE, and 300 CE. Efforts to decipher the past ages of earthquakes on a fault enable estimates of the long-term average duration between major earthquakes, a value known as the **recurrence interval**. The recurrence interval on large quakes in Cascadia is around 300–500 years, so the abrupt unlocking of the megathrust beneath Cascadia and the Olympic Peninsula could occur at any time. The problem is that, even though everyone knows it's coming, no one knows when or exactly how large a future quake might be.

It's instructive to assess the effects of recent major earthquakes from comparable tectonic settings. The 2011 Tohoku megathrust earthquake centered on the subduction zone offshore from Japan was a magnitude 9.0, and groundshaking lasted for over 4 minutes. Along the plate boundary beneath the sea, the crust was abruptly shifted about 7 m (23 ft) to the east along a slab about 180 km (112 mi) wide, displacing a massive volume of seawater. The resulting **tsunami** reached heights of over 30 m (~100 ft) near the adjacent coast of Japan, and the water swept several kilometers inland. Almost 16,000 people died, most from drowning. The disruption to Japanese society was enormous, with the repercussions still being felt years later. Imagine a comparable scenario happening along the locked megathrust beneath the Olympic Peninsula. Tsunami warning signs are a fact of life along the coast from northern California to British Columbia, and public awareness is high in the region. The phenomenal natural beauty of the Pacific Northwest and the Olympic Peninsula in particular is a result of tens of thousands of major earthquakes that have occurred over the past tens of millions of years. There's no reason to think that the pattern of seismicity will end any time soon, so the best way to deal with the threat is to anticipate it with a concentration on emergency preparedness and education.

Glacial Landforms at Olympic National Park

The dramatic landscape of the Olympic Peninsula is a product of the interplay of continental ice sheets, alpine glaciers, downcutting by streams and rivers, a wet climate,

and a coastline exposed to the near-constant pounding of waves. These interacting factors are further influenced by differential erosion of the rocks, the ongoing uplift of the peninsula, and the postglacial rise in sea level. In this section, we'll focus on the role of glacial ice in sculpting the striking vistas that we enjoy in the park today.

The dynamics of the *Pleistocene Ice Ages* over the past 2.6 m.y. were addressed in many earlier chapters, and details of the advance and retreat of *continental ice sheets* near the Cascade Range were described in Chapter 14 on nearby Mount Rainier. The Cordilleran ice sheet, centered above the mountainous terrain of British Columbia (Fig. 22.12A), extended lobes southward into western Washington as many as six times during the Ice Ages. During the *Last Glacial Maximum* around 20,000 years ago, a projection of the ice sheet reached into northwestern Washington, splitting into two lobes when it encountered the northeastern corner of the Olympic Mountains (Fig. 30.13). One lobe flowed ponderously westward along the trough of the Strait of Juan de Fuca, while another lobe pushed southward through the Puget Lowlands. The ice was about 100 km (60 mi) wide and about 1 km (3500 ft) thick near Seattle. Both lobes of the ice sheet lapped onto the northern and eastern slopes of the Olympic Mountains where they merged with *alpine glaciers* flowing down from the highlands. Sea level was about 120 m (400 ft) lower during the Last Glacial Maximum, and the shoreline lay tens of kilometers to the west. Much of the exposed

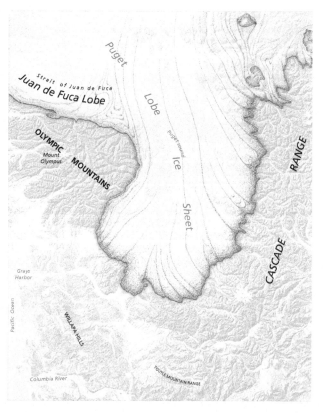

Figure 30.13 The Puget Lobe and Juan de Fuca Lobe of the Cordilleran ice sheet wrapped around the eastern and northern margins of the Olympic Mountains.

Figure 30.14 U-shaped valley left behind by a receding alpine glacier, Mount Carrie, Olympic Mountains.

continental shelf was a broad coastal plain covered in a green sheath of subalpine vegetation. The landscape at the time would have been magnificent, with the glacial crown of the mountains and surrounding ice sheets glistening in the sunshine.

By about 13,000 years ago, the planet was warming, and the Cordilleran ice sheet was in full retreat as it melted along its southern fringe. As the toe of the ice sheet migrated northward and the alpine glaciers receded into the highlands, a new landscape was revealed, with glacier-smoothed hills, *U-shaped valleys*, *arêtes*, and *cirques* emerging from beneath the ice (Fig. 30.14). The ebbing glaciers left behind elongate *moraines* of sedimentary debris as well as house-sized *glacial erratics* (Chapter 18). Surging meltwater streams bled from the receding toe of the ice, transporting huge volumes of sand and gravel that were eventually deposited in the surrounding lowlands as fields of *glacial outwash* (Chapter 22). The postglacial landscape of the Olympic Peninsula was a transformed world from its ice-covered appearance only a few thousand years earlier.

The alpine glaciers that chiseled the highlands of the Olympic Mountains during the Pleistocene were entirely melted as the warm interglacial climate of the Holocene progressed. The alpine glaciers we see today (Fig. 30.15) were born during the temporary cold climate of the Little Ice Age that gripped the Northern Hemisphere between

~1350 and ~1850 CE. Today, many of the glaciers are in rapid retreat due to a warming planet, with some isolated within north-facing cirques and some disappearing entirely. Not only are many glaciers in Olympic National Park receding, they are also becoming thinner due to an overall loss of volume. Moisture-laden fronts from the Pacific will continue to generate massive dumps of snow in the high country of the Olympic Mountains for the foreseeable future, but as the planet warms, some of that precipitation may begin to fall as rain rather than snow. You might want to do that glacier hike you've been planning sooner than later.

Coastal Landforms at Olympic National Park

The western coastal strip of Olympic National Park encompasses ~100 km (60 mi) of lonely beaches and rocky shoreline segmented by the mouths of rivers draining the mountains to the east. The coastal strip exhibits many of the same classic landforms expressed along other emergent coasts of the active Pacific Margin, such as Channel Islands National Park (Chapter 27) and Point Reyes National Seashore (Chapter 29). Much of the Olympic coast comprises narrow, linear beaches backed by vegetated bluffs about 30–90 m (100–300 ft) high. North of Hoh Head, the coastline becomes more rugged, with rocky headlands and hundreds of *sea stacks* littering the shallow surf (Fig. 30.16). Waves refract around the protruding tips of the headlands and focus their energy along the flanks (Fig. 29.10). The incessant battering of the headlands by waves creates picturesque coves, *sea caves*, and *sea arches*. When narrow remnants of headlands and arches eventually weaken and collapse, sea stacks are left isolated just offshore.

Beaches along the Olympic coast are composed of sand, pebbles, and cobbles with a variety of compositions. These sediments are derived from three main sources. Rivers draining the adjacent Olympic Mountains deposit their sediment load where they meet the sea. These sediments are redistributed onto beaches and *spits* by *longshore drift*, the lateral movement of sand along a coastline by prevailing waves (Fig. 29.15). The erosion of seacliffs creates a second source of beach sediment due to the near-constant assault by waves and tides. Many bluffs show evidence of recent landslides and rockfalls, owing to the relentless undercutting by wave erosion at their base. Upon collapse, the blocky debris is rapidly broken down into smaller fragments and then swept laterally by longshore drift. A third source of sediment accounts for the pebbles and cobbles of metamorphic gneiss and schist, as well as igneous granite. None of these rock types are found in the bedrock of the Olympic Peninsula. They are

Figure 30.15 Blue glacier on the north side of Mount Olympus, Olympic National Park.

Figure 30.16   Split Rock sea stack, Rialto Beach, Olympic National Park.

immigrants, transported by ice sheets from northern sources, then left behind as outwash. Scatterings of reddish sand on Ruby Beach are composed of rounded grains of garnet, a resistant mineral weathered from metamorphic rocks among the deposits of glacial debris.

The position and appearance of the modern landscape of the Olympic coastal strip is a product of the *postglacial marine transgression* of the past several thousand years (Fig. 29.18). Since the Last Glacial Maximum, sea level has risen globally about 120 m (400 ft) as the continental ice sheets melted and their water returned to the oceans. The vegetated coastal plain that once extended across the continental shelf offshore from the Olympic Peninsula was slowly submerged as the shoreline migrated landward.

Sea stacks located far offshore mark the former position of the Olympic western coastline, left behind as adjacent cliffs retreated under attack from

the rising sea (Fig. 30.17). The precipitous headlands and bluffs lining the coast are composed mostly of relatively erosion-resistant sandstones, but in the battle between the rocks and the waves, the waves eventually win. The rate of seacliff retreat can be rapid. Estimates of seacliff erosion rates using historical photographs suggest that some parts of the Olympic coast have retreated almost 70 m (~225 ft) in 60 years.

Another coastal feature related to rising sea levels and wave erosion along the Olympic coast are *wave-cut platforms*—partially submerged, planar surfaces created by erosional beveling from waves and tides (Fig. 29.17). Many parts of the Olympic coast, especially those marked by clusters of sea stacks, are underlain by wave-cut platforms. The ongoing tectonic uplift of the Olympic Peninsula has raised many ancient wave-cut platforms far above the waterline where they form relatively flat or gently seaward-sloping *marine terraces*.

All of the scenic coastal landforms we enjoy today along the western strip of Olympic National Park are characteristic of tectonically active *emergent coastlines*. But in the long term they are ephemeral features, subject to the vicissitudes of storm intensity, the occasional tsunami, sea-level change, and constant retreat in response to wave erosion.

The sheer diversity of landscapes and ecosystems at Olympic National Park provide something for everyone, whether you're looking for solitude or just wanting to take some can't-miss photographs. To fully enjoy the ragged mountains, the sinuous rivers, the verdant rain forests, and the dramatic coastline, you'll need a few days at least to take it all in. Like many of our best national parks, no roads cross through the heart of the Olympic Mountains, but a network of trails provide access for the intrepid backpacker. Many other trails, such as those leading from the Hurricane Ridge Visitor Center, provide reasonable

Figure 30.17   Sea stacks record former positions of the coastline. Giants Graveyard, Olympic National Park.

day hikes that lead to scenic highlands. If you plan to walk remote sections of the beaches, be sure to check the tide tables. The weather on the Olympic Peninsula can be unpredictable and sometimes intense, but just go with it—the rewards of the wilderness in Olympic National Park will always outweigh a bit of temporary discomfort.

Redwood National and State Parks

The Redwood Coast of northern California occupies a similar tectonic setting as the Olympic Peninsula (Fig. 30.6). The Gorda microplate is subducting beneath the North American plate along the Cascadia plate boundary a little over 80 km (50 mi) off the coast of Redwood National and State Parks. This stretch of the coastline and the nearby offshore area are among the most seismically active regions in the contiguous United States. The Redwood Coast is located just north of the Mendocino triple junction where the competing motions of three plates trigger countless small-to-moderate earthquakes.

The cumbersome name of the park reflects the cooperative management by the National Park Service and the California Department of Parks and Recreation. Redwood National Park shares boundaries with Jedediah Smith, Del Norte Coast, and Prairie Creek State Parks, collectively stretching along 80 km (50 mi) of the Redwood Coast (Fig. 30.18). The creation of the parks entailed a contentious series of legal battles to protect old-growth coastal redwoods from logging, which is still the main industry today in the adjacent forests outside the parks. The coastal redwood is a magnificent tree, impervious to decay, fire, and insects, but not to chainsaws. A recently discovered redwood in the park towers over 115 m (379 ft) above the forest floor, the world's tallest known tree. For comparison, the Statue of Liberty is 93 m (305 ft) high. (Their relatives, the giant sequoia of the Sierra Nevada, are larger in volume but are not as tall.)

The temperate climate, abundant rainfall, and summer fog provide the ideal conditions for the coastal redwoods along their narrow strip of Pacific Coast. Their current extent reaches from Santa Cruz to southern Oregon, a relatively small remnant from their pre-logging range. The

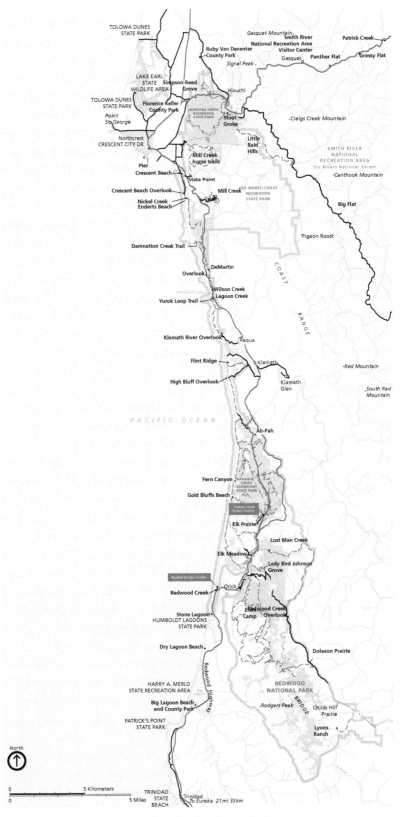

Figure 30.18 Map of Redwood National and State Parks.

lush ecosystems of Redwood National and State Parks support a wide variety of large mammals like elk, blacktail deer, and black bear. Game fish abound in the rivers and adjacent ocean, and sea lions and seals, whales and dolphins roam the shallow waters offshore.

Figure 30.19   View of Redwood Coast from High Bluff Overlook along Coastal Drive.

Redwood National and State Parks is located in the northern Coast Ranges, a rugged series of elongate ridges and valleys all trending slightly to the northwest. In the parks, the highest peak reaches over 940 m (3092 ft), with many of the mountains plunging steeply down to the coastline. The *Franciscan assemblage*, an exhumed accretionary wedge of Mesozoic age, underlies the soils and forests of the park. The best exposures of the Franciscan occur along the seacliffs and rocky headlands, accessible from the narrow beaches (Fig. 30.19). As in the rest of the California Coast Range (Chapter 26), sandstones and shales deposited by turbidity currents in the deep sea dominate the Franciscan assemblage of rocks. The sandstones are composed of poorly sorted, angular grains of sand mixed with clay, suggesting rapid deposition without time for *sorting* of grains to occur.

Many of the marine rocks of the Franciscan were dragged downward within the Mesozoic accretionary wedge where they were metamorphosed and intensely deformed. Some of the rocks have been rotated and sheared so severely that they form a unique style of rock called a **mélange**, composed of angular blocks of a variety of rock types chaotically distributed in a clay-rich matrix. This mixture of disparate rocks is a characteristic feature of many accretionary wedges and is particularly diagnostic of the Franciscan assemblage. Bodies of clay-rich mélange are inherently weak and susceptible to erosion and landslides. In Redwood National and State Parks, mélanges commonly form rounded, hummocky terrain, in contrast to the steep slopes developed above more resistant sandstones.

The rocks of the Franciscan traveled a circuitous path before arriving at their current location along the Redwood Coast. After accumulating on the deep ocean floor of the Farallon plate, they were bulldozed off the downgoing plate by the overriding North American plate and incorporated into the accretionary wedge. There they experienced millions of years of metamorphism and deformation within the crushing tectonic vise. The Franciscan rocks were exhumed to the surface only in the last 5 m.y. or so as the northern Coast Ranges were raised above sea level by *transpressional stress* along *strike-slip faults*. Ponder this incredible natural history as you wander the trails, gaze upward at the towering redwoods, and admire the sea stacks along the picturesque coastline.

Key Terms

locked fault
megathrust
mélange
recurrence interval
seamount
submarine fan
tsunami

Related Sources

Covington, S. (2004). *Geologic Resource Evaluation Scoping Summary: Redwood National and State Parks, California.* Denver, CO: Geologic Resources Division, National Park Service.

Orr, W. N., and Orr, E. L. (2019). *Geology of the Pacific Northwest*, 3rd ed. Long Grove, IL: Waveland Press.

Tabor, R. W. (1975). *Geologic Guide to Olympic National Park.* Seattle: University of Washington Press.

Tucker, Dave. (2015). *Geology Underfoot in Western Washington.* Missoula, MT: Mountain Press.

Washington State Department of Natural Resources: Olympic Mountains. Accessed February 2021 at https://www.dnr.wa.gov/programs-and-services/geology/explore-popular-geology/geologic-provinces-washington/olympic#geologic-history

National Parks Related to Hotspots

31 Yellowstone National Park

Figure 31.1 Excelsior Geyser and Grand Prismatic hot springs, Yellowstone National Park.

Mention the word "Yellowstone" to anyone who's ever visited the park and you'll hear a wide variety of responses. Most people immediately think of Old Faithful geyser and the plentiful steam vents and colorful hot springs (Fig. 31.1). For others, the bison, elk, moose, grizzly bear, wolves, and other wildlife that inhabit the park come to mind. Yet others wax rhapsodic about the trout-filled rivers, streams, and sparkling lakes, ideal habitat for those fixated on fishing as sport. Only incidentally does someone mention the singular most central feature of Yellowstone—the immense body of partially molten rock just a few kilometers beneath the surface. The tranquility of the rolling, forested terrain and the lack of any obvious conical volcanoes obscure the geologic reality that Yellowstone resides above the bowels of an active volcano—and not just any volcano, but rather a gigantic, dynamic "supervolcano" responsible for three of the planet's most cataclysmic eruptions of the past 2 million years.

Yellowstone National Park occupies an enormous region (larger than Rhode Island) in the northwestern corner of Wyoming, with edges lapping over into Montana and Idaho (Fig. 31.2). Grand Teton National Park lies just to the south and shares many of the same charismatic megafauna with Yellowstone. Both parks form the centerpiece

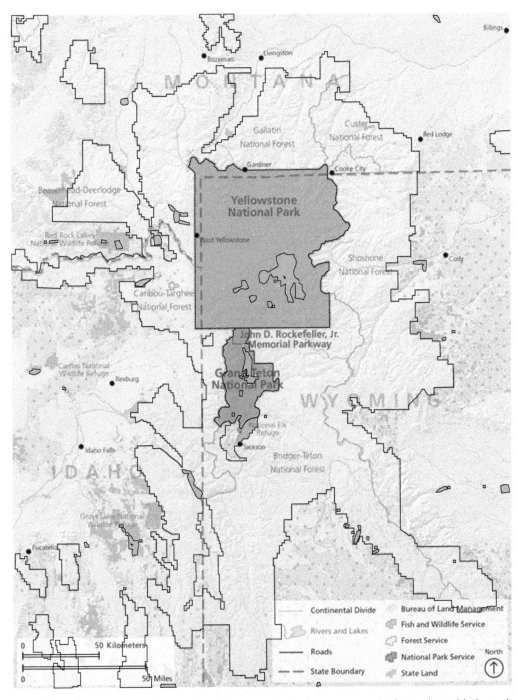

Figure 31.2   Yellowstone National Park is located in the northwestern corner of Wyoming, with the park boundaries extending into adjacent Montana and Idaho. Grand Teton National Park lies just to the south. The light green area on the map surrounding the parks marks public lands of the Greater Yellowstone Ecosystem.

of the 22.6 million acre **Greater Yellowstone Ecosystem,** the largest contiguous area of public wildlands in the continental United States.

Yellowstone sits at the nexus of three geologic provinces and shows characteristics of each (Fig. 20.1). The park is part of the Central Rocky Mountain physiographic province and exhibits some of the same *compressional* geologic features as Glacier National Park in Montana and Rocky Mountain National Park in Colorado. Yellowstone also expresses ample evidence of Basin and Range *extensional* geology, comparable to Grand Teton National Park and other nearby ranges. Yellowstone is dominated by youthful volcanism, however, and is intimately connected to the Snake River Plain that extends through southern Idaho to the southwest. Catastrophic eruptions at Yellowstone have occurred three times over the past 2 m.y. during the time period known as the Pleistocene. (Recall from Chapter 14 that the Pleistocene began 2.6 m.y. ago and ended about 12,000 years ago.) Each of these geologic connections to nearby parks and provinces will be explored in greater detail throughout this chapter.

The topography of the park is dominated by gently rolling terrain of the Yellowstone Plateau (Fig. 31.3). The plateau has an average elevation of 2400 m (~8000 ft) and slopes gently to the southwest toward the Snake River Plain. Subalpine forests, dominated by lodgepole pine, blanket the highland plateau. The hulking Absaroka Range flanks the plateau to the east, with several peaks reaching elevations over 3200 m (~10,500 ft); Eagle Peak in the Absarokas is the highest point in Yellowstone at 3462 m (11,358 ft). To the northeast lie the rugged Beartooth Mountains, while the Gallatin Range dominates the northwestern corner of the park.

The *continental divide* meanders in an irregular pattern across the southwestern corner of the park, separating the Snake River drainage to the south from the watersheds of the Yellowstone and Madison rivers to the north and east (Fig. 31.4). (Recall from earlier chapters that a *watershed* is the total area that contributes runoff from snowmelt and rainfall to a river and its tributary network of streams.) The Snake River flows southward into Jackson Hole in Grand Teton National Park before meandering through the Snake River Plain on its way to join with the Columbia River and then on to the Pacific Ocean.

The Yellowstone River originates in the Absaroka Range and drains into the southeastern corner of Yellowstone Lake before exiting at the north end. The river meanders for several kilometers before dropping over two large waterfalls into the narrow gorge of the Grand Canyon of the Yellowstone (Fig. 31.5). The river then flows north into Montana where it runs for over 1000 km (620 mi) before converging with the Missouri River. The Yellowstone is the longest undammed river in the contiguous United States. The name of the river is derived from a Native American pronunciation of yellow-colored sandstones far downstream in eastern Montana. Mountain men and the Lewis and Clark Expedition popularized the name of the river as Yellowstone, with the national park later being named after the river. It's a common misconception that the park is named for the yellow-colored rocks lining the Grand Canyon of the Yellowstone.

Yellowstone Lake forms the sparkling focal point of the national park. With an elevation of 2357 m (>7700 ft) and a maximum depth of 131 m (430 ft), Yellowstone Lake is one of the largest high-elevation lakes in North America. The amoeboid shape of the lake is due to a combination of glacial and volcanic processes that will be explored more fully below.

Yellowstone was established in 1872 as America's first

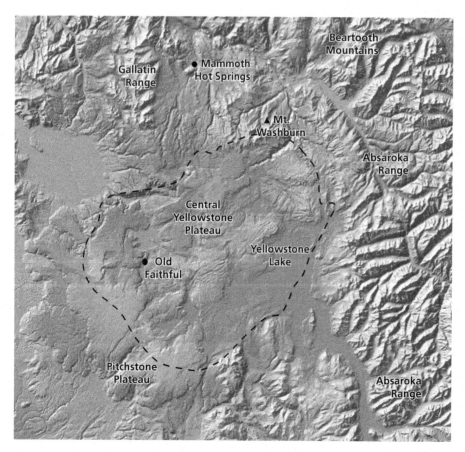

Figure 31.3 Satellite view of Yellowstone National Park, looking north. The dashed line marks the edges of the caldera.

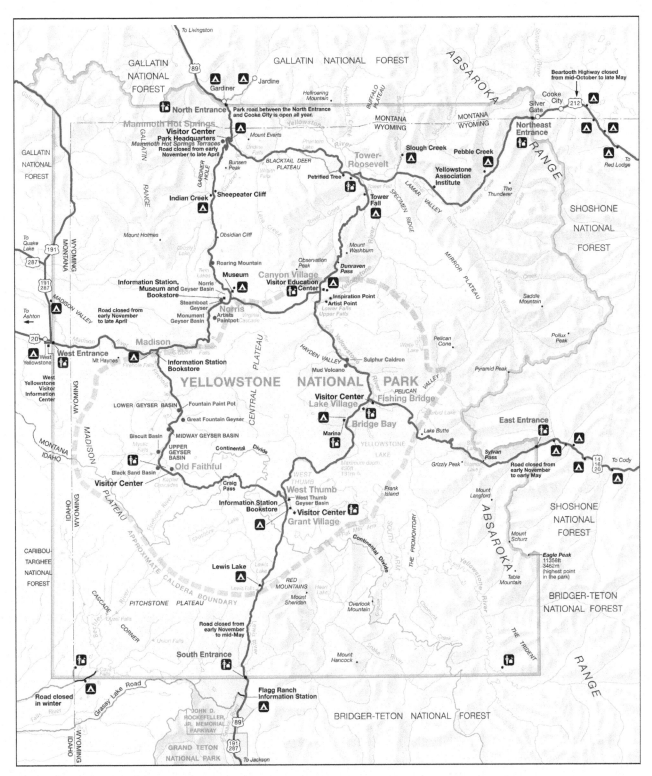

Figure 31.4 Map of Yellowstone National Park showing roads and main physiographic features. The continental divide is marked by the bold yellow line. The margins of the caldera are shown by the thick, dashed gray line.

national park and is generally accepted as the first national park in the world. At the time, the primary resource to be protected was the 10,000 or so *hydrothermal features*, including geysers, colorful hot springs, bubbling mud pots, and sulfurous steam vents, which represented the greatest concentration of hydrothermal features on Earth. The

hundreds of geysers in the park constitute more than half of all the geysers on the planet. The key to all this geologic dynamism is the massive magma chamber that resides just a few kilometers beneath the surface. In turn, the heat energy that creates the molten rock originates deep beneath Yellowstone in the Earth's mantle. The rest of this

Figure 31.5 Artistic perspective of Yellowstone National Park toward the Teton Range off to the southwest. The Yellowstone River flows through the Grand Canyon of the Yellowstone on its way north (toward the viewer).

chapter will address Yellowstone's history of "supereruptions," its unique tectonic setting, the factors that control its hydrothermal activity, and the climatic and volcanic events that shaped its modern landscape. You'll see how the geologic activity of this iconic national park controls the ecologic diversity, ranging from the largest bison to the most miniscule microbe.

Pre-Pleistocene History of Yellowstone

Most of the rocks and much of the landscape of Yellowstone National Park is geologically youthful, with relatively recent volcanism occurring contemporaneously with the terrain-modifying effects of Ice Age glaciation over the past 2 m.y. or so. But older rocks do exist in the Yellowstone region, recording the same ancient events that affected many of the other national parks discussed in previous chapters. This short section recapitulates some of the key episodes that affected the American West prior to the Pleistocene, as expressed in Yellowstone country.

Some of the oldest rocks in the National Park System are found just to the south of Yellowstone in Grand Teton National Park (Chapter 25), where metamorphic gneiss as

old as 2.7 b.y. is intruded by 2.4 b.y. granite. Even older Precambrian rock is found within the Beartooth Mountains bordering the northeastern corner of the park. There, crystalline rocks are 3.4–3.8 b.y. old and are among the oldest on Earth. Ancient metamorphic and igneous rocks of the Tetons and Beartooths were formed deep underground, perhaps 15 km (9 mi) or more within the roots of ancient Precambrian mountains that once towered above the Yellowstone region. After billions of years of erosion and burial, these rocks were uplifted to their current heights above 3700 m (>12,000 ft). It's notable that in the Yellowstone region some of the oldest rocks on Earth are juxtaposed against some of the youngest.

In the earlier chapters on national parks of the Colorado Plateau, you learned that Paleozoic and Mesozoic sedimentary rocks were deposited in response to the rise and fall of sea level through time that caused the shoreline to migrate back and forth across the western flank of the North American continent. As the coastal lands were submerged beneath tens of meters of seawater during sea-level rise, a variety of marine sediments were deposited. During falls in sea level, sediments were deposited in continental depositional systems such as rivers and their floodplains. The sedimentary rocks recording these *transgressions* and *regressions* of the shoreline range from southern Canada to northern Mexico and as far east as the Midcontinent. The region that would much later become the volcanic Yellowstone country was inundated as well, with the evidence preserved within scattered exposures of Paleozoic and Mesozoic sedimentary rock.

Yellowstone is located in the Central Rocky Mountain physiographic province where the uplift of mountains is related to the late Mesozoic through early Cenozoic Laramide orogeny. The compressional tectonism of the Laramide raised several mountain ranges in the greater Yellowstone region, including the thick crystalline block of the Beartooths where old Precambrian rocks were squeezed upward along high-angle *reverse faults*. This style of mountain-building is identical to that within Rocky Mountain National Park (Chapter 21) and many other mountain ranges within the Rocky Mountain Province.

The nearby Teton Range just to the south of Yellowstone is a *tilted fault-block* lifted upward along *normal faults* by extensional tectonic forces (Chapter 25). The Tetons are part of the northeastern portion of the Basin and Range Province, an enormous region of crustal stretching that began around 16 Ma and continues to this day. To the northwest of Yellowstone, the Gallatin and Madison Ranges are also tilted fault-blocks of the Basin and Range, and several ranges to the west in Idaho share the same extensional style of uplift. The rolling, volcanic terrain of the Yellowstone Plateau, bound on three sides by Basin and Range-style mountains, is itself cut by faults aligned north–south and northwest–southeast. It's likely that the faults are the remains of tilted fault-block mountains that once extended across Yellowstone but that were destroyed during the cataclysmic volcanism of the last 2 m.y. We'll expand on the extensional tectonism of Yellowstone country below within the context of its volcanic history.

A final pre-Pleistocene geologic event that profoundly impacted the Yellowstone region was early Cenozoic Absaroka volcanism. Today, the variety of rocks exposed within the Absaroka Range cover a large portion of the north and east sides of the park. These rocks, deposited between about 48 and 55 Ma, include not only *extrusive volcanic* rocks (Chapter 1), but also the *intrusive plutonic* rocks that were once the magma chambers deep beneath the Absaroka volcanoes, now exhumed by uplift and erosion. This prolonged phase of early Cenozoic volcanism is tectonically unrelated to the more recent volcanism at Yellowstone.

The most revealing rocks left behind by Absaroka volcanism are sedimentary *debris flow* deposits that surged down the steep flanks of *stratovolcanoes*. (Recall from earlier chapters that debris flows are turbulent slurries of sediment and water that flush though channels along mountainsides. Sediment in debris flows can range from tiny particles of clay to fist-sized cobbles to boulders the size of cars.) Debris flows ripping down the sides of Absaroka volcanoes during the early Cenozoic uprooted and buried countless trees on the densely forested slopes, entombing the wood as the debris flows came to a halt.

Figure 31.6 Petrified tree trunks with remnant cobbles and boulders left from debris flows that once entombed the trees, Specimen Ridge.

Silica-rich groundwater, derived from its passage through silica-rich volcanic rocks, permeated the cell tissue of the dead wood, replacing the organic material with silica at the molecular level and preserving the finest of details. This **petrified wood** is found in several places within the park and surrounding region and is notable for the wide diversity of preserved species of trees that provide paleobotanical evidence of the regional climate during the early Cenozoic (Fig. 31.6). A particularly good place to see these "fossil forests" is along Specimen Ridge above the Lamar River Valley.

Yellowstone Caldera

The "volcano" that dominates Yellowstone National Park is actually the barely perceptible outline of a massive *caldera* created during an enormous eruption 640,000 years ago (Fig. 31.3). Recall that a caldera is a large depression that forms by the catastrophic vertical collapse of an erupted volcano to form a deep basin (see Chapter 15 on Crater Lake National Park). Calderas are very different from the much smaller, near-circular *craters* on volcanoes from which eruptions occur. The dimensions of the Yellowstone

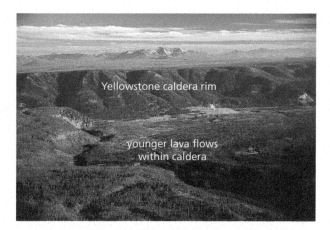

Figure 31.7 Low-angle oblique view of the northwest rim of the Yellowstone caldera near Madison Junction. Visualize the area in the foreground abruptly collapsing along the now-eroded fault plane of the caldera rim. Later lava flows smoothed the surface of the basin floor. Note the broad plateau above the caldera rim with the Gallatin Range rising in the distance.

Figure 31.8 Outcrop of Lava Creek Tuff above the Gibbon River composed of pyroclastic flows that were welded into rock soon after deposition. The Lava Creek Tuff was deposited 640,000 years ago during the most recent caldera-forming eruption.

caldera are about 50 x 75 km (~30 x 45 mi), centered on the Yellowstone Plateau. The steep walls of the actual caldera rim are difficult to discern because of hundreds of thousands of years of smoothing by glacial erosion, as well as the leveling of the land surface by lava flows that occurred after the caldera was created. Add a surface veneer of soil and a dense lodgepole pine forest and the caldera rim is all but invisible. It's no wonder that many visitors to the park have left without any idea of where the "volcano" actually is at Yellowstone. Geologists identified the boundaries of the caldera after careful mapping and the recognition of subtle topographic features. Remnants of the caldera rim can be seen from the Washburn Hot Springs overlook as well as near Madison Junction where the caldera wall rises 500 m (1600 ft) above the downdropped basin floor (Fig. 31.7).

The caldera-forming eruption 640,000 years ago deposited a huge volume of *rhyolitic tuff* in the Yellowstone region. The **Lava Creek Tuff** is composed of solidified particles of *volcanic ash* and fragments of *pumice* that reaches a maximum of almost 500 m (~1600 ft) within the park (Fig. 31.8).*

The Lava Creek *caldera eruption* was a natural catastrophe of unimaginable proportions, far beyond anything ever witnessed by modern humans. The story begins deep within the magma chamber that extends from 5 to 16 km (3 to 10 mi) beneath the Yellowstone Plateau. The magma chamber today is estimated to be about 5–15 percent molten liquid that fills pores and fissures within a

spongy, yet solid, matrix of rock. But 640,000 years ago, that proportion of liquid melt was at least 50 percent and likely much higher. The composition of the magma was rich in silica and dissolved gases, the ideal combination for an explosive eruption. The intense heat and a continual supply of magma from deeper sources caused the overlying crust to bulge upward, somewhat like a hot soufflé (Fig. 31.9A). It's unlikely that there was ever a conical volcano, but rather a broad, convex-upward swelling of the ground surface. As the rigid slab of overlying crust flexed higher, it developed a series of cylindrical fractures that originated near the surface but propagated downward over time toward the magma chamber. These *ring-vent fractures* would eventually be the conduits through which the magma would escape to the surface.

One very bad day, likely accompanied by a swarm of earthquakes, the fractures penetrated down into the magma chamber, triggering a cataclysmic release of gas-driven pressure. The magma in the upper chamber was violently injected into the concentric zone of ring vents where it was pulverized by the expanding gases and instantaneously crystallized to volcanic ash and pumice (Fig. 31.9B). The pyroclastic debris exploded into the atmosphere, accompanied by a deafening roar, forming a *Plinian eruption column* (Fig. 31.10). The tumescent column may not have been a single, broad pillar of ash and rock fragments, but rather a cluster of columns and smaller plumes that merged as they rose upward from individual ring vents. (We really don't know because no one has ever witnessed a caldera eruption of this magnitude in modern times.) As the rising column encountered dense layers of colder air in the atmosphere, it would have spread out laterally, forming horizontal bands of dense ash. Much of the immediate region around Yellowstone would have been enveloped in darkness as the broadening cloud of ash blocked sunlight.

Much of the volcanic ash was carried downwind before gravity took over, creating a *pyroclastic fall* of tephra that extended across an enormous area of the United States,

---

A primer on volcanism was provided in Part 2 on the national parks of the Cascades Geologic Province. Many concepts were introduced, including the role of silica in the viscosity and explosivity of eruptions, as well as the importance of gases within magma. The four main types of volcanic rocks were described, and the characteristics of the pyroclastic debris called tephra were defined. Furthermore, the mechanics of caldera eruptions were described in Chapter 15 on Crater Lake National Park, as well as in Chapter 23 on the ignimbrite flare-up. To avoid repetition in this chapter, we'll focus on those volcanic features and processes that are specific to Yellowstone, referring back to earlier chapters as necessary.

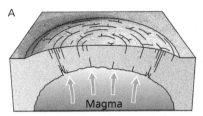

Swelling magma chamber arches
and fractures overlying rocks.

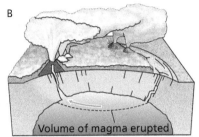

Fractures tap gas-charged rhyolite in
magma chamber releasing pressure
and triggering vast ash flows.

Crust collapses to form caldera.

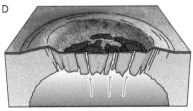

Later basalt lavas flood caldera floor.

Figure 31.9 Sequence of events related to the caldera eruptions at Yellowstone.

reaching all the way to the Gulf Coast (Fig. 31.11). The most distant tephra layer from the Lava Creek eruption is only a few centimeters thick but must have been devastating to the flora and fauna, with the thin blanket of powdered glass and fine particles of rock inhibiting photosynthesis for an extended duration of time. Closer to the caldera itself, pulse after pulse of incandescent *pyroclastic flows* raced across the nearby landscape, incinerating every plant and animal in their path. Pyroclastic flows travel at high speeds as a turbulent, ground-hugging cloud, eventually settling out as a hot blanket of ash and volcanic debris. The innumerable tiny shards of glass and particles of rock within the pyroclastic flows were still extremely hot when they eventually came to rest across the surface, welding together to form a dense tuff called an *ignimbrite* (Chapter 23).

The towering Plinian column and the devastating pyroclastic flows must have dissipated over a few days as the gas pressure diminished. At some point, the partially emptied magma chamber could no longer support the weight of the overlying slab of rock. The crust collapsed downward along the ring fractures, likely displacing any remaining gas-charged magma out onto the surface as a final pyroclastic burst (Fig. 31.9C). The formation of the oval-shaped Yellowstone caldera may have occurred in a few hours to days—we just don't know for sure. But after all the violence, what was left behind was a gaping hole in the ground, several hundred meters deep and shrouded in steam as remnant hot gases escaped to the surface.

The surrounding region must have been an eerily quiet moonscape, with only the steepest cliff faces and the occasional carbonized tree trunk visible above the sterile carpet of gray ash. The entire planet likely cooled for a few years as sulfur-rich gases from the eruption mixed with water in the atmosphere to form a diffuse layer of aerosol droplets that blocked solar energy from penetrating down to the surface. The total volume of tephra ejected during the Lava Creek eruption is estimated to be 1000 km^3 (240 mi^3). This quantity is about 20 times the amount of tephra erupted 7700 years ago at Crater Lake in Oregon and over a thousand times the amount of the 1980 eruption of Mount St. Helens in Washington.

Other Pleistocene Caldera Eruptions at Yellowstone

As colossal as the Lava Creek caldera eruption was, the earlier **Huckleberry Ridge eruption** 2.1 m.y. ago ejected 2.5 times the volume of material—likely the largest eruption in Yellowstone's geologic history. The eruption left behind the enormous Island Park caldera that stretches from eastern Idaho into Yellowstone National Park (Fig. 31.12). The northeastern part of the caldera was obliterated by the subsequent Lava Creek eruption discussed above. Volcanic ash, pumice, and rock fragments deposited by this immense eruption are known as the Huckleberry Ridge rhyolite tuff, which accumulated to about 185 m (~600 ft) in thickness within the Yellowstone region (Fig. 31.13). The lateral extent of the tephra deposited from the ash cloud spanned a huge area from Southern California to Iowa (Fig. 31.11). Consider the devastation that accompanied the eruption of Mount St. Helens in 1980 (Chapter 13) where about 1 km^3 (0.24 mi^3) of pyroclastic material was ejected. Now imagine the chaos during the Huckleberry Ridge eruption that released an estimated 2500 km^3 (600 mi^3) of material. It's been estimated that the volume of volcanic debris erupted during the eruption 2.1 m.y. ago would cover California in a layer 6 m (20 ft) thick. Contemplate an event of this magnitude occurring in today's modern world.

The Yellowstone region experienced a third catastrophic eruption about 1.3 m.y. ago called the **Mesa Falls eruption**, which left behind a 150 m (500 ft) thick deposit of tuff. The Henry's Fork caldera created by this eruption

Figure 31.10   1989 Plinian eruption column from Redoubt Volcano in Alaska. Caldera eruptions at Yellowstone ejected a much greater volume of pyroclastic debris and lasted much longer than this recent Alaskan eruption.

is located in eastern Idaho, nested inside the corner of the much larger Island Park caldera (Fig. 31.12). The volume of pyroclastic debris ejected during this second-oldest eruption was "only" 280 km³ (67 mi³), nine times less than the Huckleberry Ridge event, but more than five times larger than the caldera eruption at Crater Lake 7700 years ago.

The collective fury of the three major caldera eruptions at 2.1 m.y., 1.3 m.y., and 640,000 years ago completely altered the pre-eruption landscape at Yellowstone. The Huckleberry Ridge event probably destroyed entire mountains that cut across the region, likely tilted fault-block ranges that connected the Teton Range to the south with the Gallatin and Madison Ranges to the north. Evidence for these "lost"

mountains are north–south-oriented faults that end abruptly at the southern and northern margins of the Yellowstone caldera. Some of the mountains may have been completely demolished by the outward blast of the eruptions, while others were partially swallowed by the downward collapse of the caldera.

Post-Caldera Lava Flows at Yellowstone

After the Lava Creek eruption 640,000 years ago, the chaotic arrangement of blocks within the newly formed caldera was gradually smoothed over by the episodic release of tens of rhyolitic lava flows that escaped from the partially depleted magma chamber below (Fig. 31.14). Molten rock leaked upward through the highly fractured floor of the caldera, with the viscous, high-silica lava slowly flowing laterally from elongate vents. The rhyolite tuff and lava along the walls of the Grand Canyon of the Yellowstone River were laid down during this post-caldera volcanic phase, as were the overlapping flows beneath the caldera rim near Madison Junction (Fig. 31.7). The youngest rhyolitic lava flow in the park occurred about 72,000 years ago, forming part of the Pitchstone Plateau in the southwest corner of the park.

Less common are low-silica basaltic lava flows that poured out over older Lava Creek tuff and rhyolitic lava flows, smoothing the rough edges and leveling parts of the upper plateau surface. Some of the basaltic flows (and a few of the rhyolitic flows and tuffs) exhibit **columnar jointing**, which formed as ponds of hot lava cooled and contracted (Fig. 31.15). The systematic alignment of vertical joints form tightly packed polygonal columns of rock. Most columns are 6-sided hexagons, but others may have as few as 3 or as many as 12 sides, with contraction occurring uniformly around a central point within each column. Cooling begins at the upper surface of the flow where it is in contact with the atmosphere and along the lower surface where the lava is in contact with the ground. As the lava cools and solidifies, cracks propagate toward the interior of the cooling mass. Computer simulations suggest that during cooling, smaller cracks consolidate into larger ones, with the angles between cracks self-organizing toward 120°, the optimum angle that releases the most

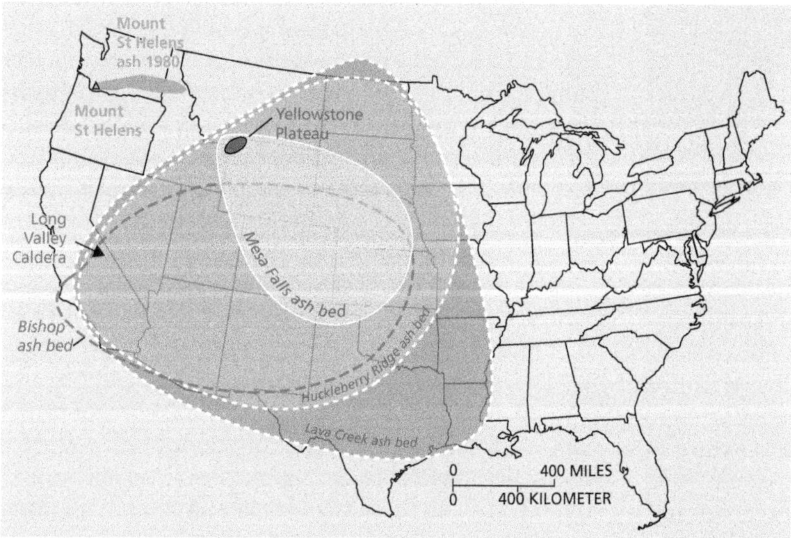

Figure 31.11   Distribution of ashfall deposits from the three most recent caldera-forming eruptions at Yellowstone National Park. The areal extent of tephra from the Long Valley caldera in eastern California and from the 1980 eruption of Mount St. Helens are shown for comparison.

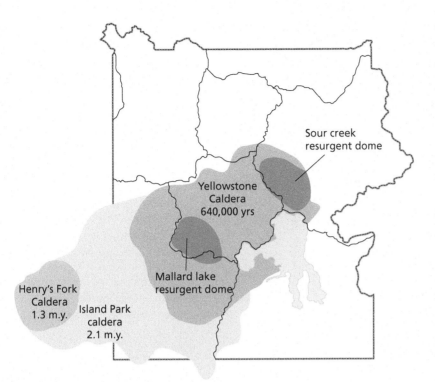

Figure. 31.12 Areal extent of three most recent calderas and two modern resurgent domes at Yellowstone National Park.

Figure. 31.13 Huckleberry Ridge rhyolite tuff a few kilometers south of Mammoth Hot Springs at Golden Gate. This exposure is about 20 km (12 mi) north of the Island Park caldera margin.

heat energy during contraction. (120° is the angle between two adjacent sides of a hexagon, the most common shape of columns.)

Together, these most recent high- and low-silica lava flows filled much of the caldera basin, with total volumes estimated at around 3000 km³ (720 mi³). This total is comparable to the ~3700 km³ (~900 mi³) of material ejected during the three major caldera eruptions, but the younger lava flows occurred much more episodically over hundreds

of thousands of years. A veneer of soil and lodgepole pine forest blankets the gently undulating, lava-covered surface of the Yellowstone Plateau, obscuring the violence of its volcanic past.

Scale of Yellowstone Caldera Eruptions

Volcanologists rank the explosivity of eruptions using a logarithmic scale known as the **Volcanic Explosivity Index (VEI)**. This relative scale uses estimates of the total amount of material erupted, the height of the eruption column, and other qualitative measures to rank ancient and historic volcanic events on a scale of 0 to 8. The lowest ranking eruptions, like Hawaiian volcanoes, have a VEI of 0 ("effusive") or 1 ("gentle"). Plinian eruptions, with towering pillars of ash, have VEIs of 6 ("colossal"), 7 ("super-colossal"), or 8 ("mega-colossal"). The Huckleberry Ridge and Lava Creek caldera eruptions at Yellowstone both have VEIs of 8, while the 1.3 m.y. old Mesa Falls event has a VEI of 7.

When plotted by volume of material erupted, the three events from Yellowstone are among the largest in recent Earth history (Fig. 31.16). Only the Toba eruption in Indonesia ~74,000 years ago was larger than the Huckleberry Ridge eruption. The Long Valley caldera eruption in eastern California that occurred 760,000 years ago was intermediate in scale between the Lava Creek and Mesa Falls events. Historic eruptions witnessed by modern humans are relatively puny in comparison. Novarupta Volcano in Alaska erupted in 1912 with a VEI of 6 and was the largest eruption of the 20th century, but it was 20 times smaller than the Mesa Falls event. The 1980 eruption of Mount St. Helens (Chapter 13) was powerful enough to kill 57 people and devastate the nearby landscape, but it was miniscule relative to past Yellowstone eruptions. The early 20th-century eruption of Mount Lassen (Chapter 16) was a mere burp in comparison.

The term **supervolcano** is a somewhat slippery concept, but the U.S. Geological Survey defines it as a volcano that erupted with a VEI of 8, emitting more than 1000 km³ (240 mi³) of material at any one point in time. Only 40 or

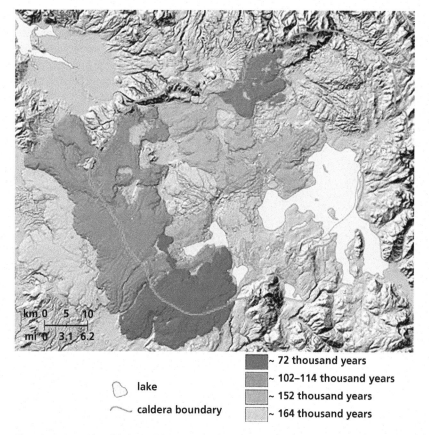

○ lake

◠ caldera boundary

■ ~ 72 thousand years
□ ~ 102–114 thousand years
□ ~ 152 thousand years
□ ~ 164 thousand years

Figure 31.14 Distribution of large lava flows that occurred after the Lava Creek caldera eruption. The eruption 72,000 years ago was a single event, whereas the others were multiple eruptions.

Figure 31.15 Sheepeater Cliff, just south of Mammoth Hot Springs, is a 500,000-year-old basaltic lava flow marked by spectacular columnar jointing. A closely spaced set of horizontal joints cuts the vertical columns, creating the appearance of adjoining piles of stacked poker chips.

so supereruptions have been identified since the mid-Cenozoic, with none occurring in the past 10,000 years. By this definition, both the 2.1 m.y. Huckleberry Ridge and 640,000 year Lava Creek events would qualify Yellowstone as a supervolcano. Based on the size of the magma

chamber lurking below the surface at Yellowstone, discussed in detail below, you'd be justified in saying that the park today is a "supervolcano-in-waiting."

Hotspot Volcanism at Yellowstone

Yellowstone National Park is located well within the North American tectonic plate, far from the nearest plate boundaries along the west coast. The ancestral *Farallon plate* is now buried within the mantle, far too deep to generate magma. So what is the source of the heat energy that generates magma beneath the park? Earth is continuously convecting heat energy through the body of the planet and eventually releasing it out into space, commonly by volcanism (Chapter 1). *Heat flow* is a measure of the heat emitted from a particular area on the surface of the Earth over a specified length of time (Chapter 23). The heat flow at Yellowstone is 30 to 40 times that of the average for North America and is concentrated in a bull's-eye pattern centered above the caldera. The proximal source of the heat is the gigantic magma chamber just beneath the caldera, but the ultimate source is an elongate **mantle plume** of hot, plastic rock that originates deep in the mantle and rises upward by *convection*. This stationary column of heat energy governs the volcanism at Yellowstone; the local region directly above the plume is called a **hotspot**. Tens of hotspots exist in today's world, some coincident with divergent plate boundaries (such as Iceland, along the Mid-Atlantic Ridge), but most occur in the interiors of plates. The Big Island of Hawaii manifests a hotspot in the middle of the oceanic Pacific plate (Chapter 32), while others, such as the Yellowstone hotspot, pierce continental plates.

How do geoscientists prove that hotspots are located directly above mantle plumes? We can tell a lot about the deep interior of the Earth by analyzing the echoes of earthquakes using a global network of seismometers to record the subtle vibrations of seismic waves. The fundamental premise is that seismic waves slow down as they pass through the lower density of hot rock and speed up as they move through higher density cooler rock. When certain types of seismic waves encounter a liquid, such as magma, their energy is absorbed, leaving a "shadow zone" of missing seismic waves on the far side. This sophisticated computational technique is called **seismic tomography**, an ingenious method of visualizing the interior of our planet, including the dimensions of mantle plumes. Much

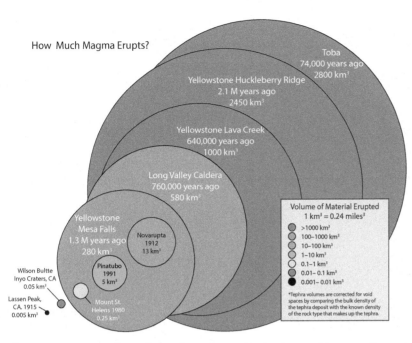

How Much Magma Erupts?

Toba
74,000 years ago
2800 km³

Yellowstone Huckleberry Ridge
2.1 M years ago
2450 km³

Yellowstone Lava Creek
640,000 years ago
1000 km³

Long Valley Caldera
760,000 years ago
580 km³

Yellowstone
Mesa Falls
1.3 M years ago
280 km³

Novarupta
1912
13 km³

Pinatubo
1991
5 km³

Wilson Bultte
Inyo Craters, CA
0.05 km³

Lassen Peak,
CA. 1915
0.005 km³

Mount St.
Helens 1980
0.25 km³

Volume of Material Erupted
1 km³ = 0.24 miles³
◯ >1000 km³
◯ 100–1000 km³
◯ 10–100 km³
◯ 1–10 km³
◯ 0.1–1 km³
◯ 0.01–0.1 km³
● 0.001–0.01 km³

*Tephra volumes are corrected for void
spaces by comparing the bulk density of
the tephra deposit with the known density
of the rock type that makes up the tephra.

Figure 31.16 Comparison of eruption volumes for the three caldera eruptions at Yellowstone compared with other ancient and historic eruptions.

of the following discussion is based on the interpretation of tomographic imagery.

The mantle plume beneath the Yellowstone hotspot is interpreted to be about 80 km (50 mi) wide and to originate at least 700 km (440 mi) beneath the surface in the upper mantle (Fig. 31.17). (Recall from Chapter 1 that the mantle reaches a depth of 2900 km (1800 mi) where it meets the outer core.) The plume is not molten, but rather is composed of very solid, very hot, low-silica rock that slowly flows as a dense, *plastic* material. The heat energy contained within the plume of rock makes it less dense than the surrounding rock of the mantle, allowing the plume to rise upward as a buoyant mass. When the plume contacts the base of the cooler, denser, North American lithosphere (composed of the crust and upper mantle), the plume stalls and spreads out laterally. At this level, the pressure from the weight of overlying rock is low enough so that the solid rock of the plume can begin to partially melt, producing molten, low-silica magma that buoyantly rises upward through fractures in the uppermost mantle. The magma accumulates to form vast chambers of molten rock a few kilometers beneath the surface of Yellowstone.

The magma chamber just beneath the Yellowstone caldera is not a balloon-shaped blob of molten rock,

sloshing around in a huge cavern. More recent tomographic studies reveal that Yellowstone is underlain by not one, but two distinct magma chambers (Fig. 31.18). The deeper chamber extends from 20 to 50 km (12 to 28 mi) in depth and contains only about 2 percent liquid melt of low-silica, basaltic composition, with the rest of the chamber being solid hot rock. The shallower chamber that resides 5 km (3 mi) beneath the Yellowstone Plateau extends downward to about 16 km (~10 mi). It has lateral dimensions of about 30 x 90 km (20 x 55 mi), so this upper chamber has a flattened shape, somewhat like a deflated football. The shallow chamber contains about 5–15 percent liquid melt of high-silica, rhyolitic composition and is the reservoir for the three caldera eruptions of the past 2 m.y. The discovery of the lower magma chamber, over four times larger than the shallower chamber, provides the connection between the mantle plume at depth and the upper chamber from which eruptions originate.

The composition of the low-silica basaltic magma of the lower chamber evolves as it interacts with the

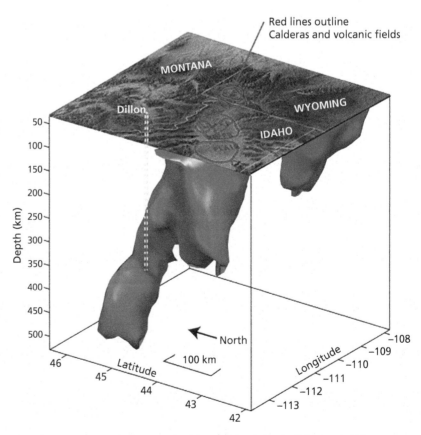

Red lines outline
Calderas and volcanic fields

MONTANA

Dillon

WYOMING

IDAHO

Depth (km)

50
100
150
200
250
300
350
400
450
500

North

100 km

Latitude
46
45
44
43
42

Longitude
−108
−109
−110
−111
−112
−113

Figure 31.17 Tomographic image of the mantle plume (red) beneath Yellowstone. The blue mass is cooler, denser rock.

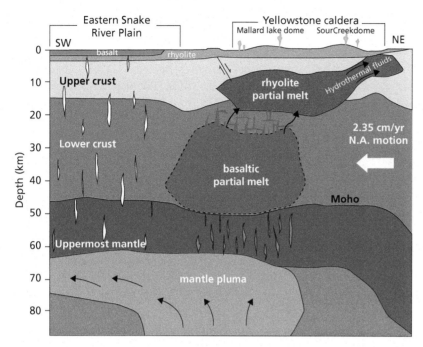

Figure 31.18 Simplified graphic based on seismic tomography showing the two stacked magma bodies in the crust beneath Yellowstone.

Yellowstone Hotspot Trail

Relief maps of the American West such as that shown in Figure 20.1 exhibit a huge banana-shaped swath across southern Idaho, extending from Yellowstone in the east to the Oregon border in the west. This is the **Snake River Plain**, a broad lowland with average elevations of ~1500 m (~5000 ft). Basin and Range mountains end abruptly along the southern margin of the Plain and reappear at the northern end. It's an anomalous flatland smack in the middle of otherwise mountainous terrain. The alluvial soils and farms of the Snake River Plain, however, conceal a geologic history of volcanic violence a kilometer or more below the surface. In places, the Snake River has carved canyons as it flows westward across the broad, flat surface (Fig. 25.5), providing a glimpse into the rocks that lie deep beneath the potato fields and sagebrush. Those rocks are pink and yellowish rhyolitic lavas and tuffs comparable to those at Yellowstone today. These same rocks are also encountered in deep exploration wells drilled into the Plain. What those rocks reveal is a record of caldera eruptions at the Yellowstone hotspot occurring over the past 16 million years.

Careful mapping and radiometric dating of the tuffs show them to be part of gigantic extinct calderas, some with discrete margins, others with overlapping edges (Fig. 31.19). They form an irregular chain aligned along the eastern Snake River Plain, extending about 800 km (~500 mi) in a gentle arc from Yellowstone to the border of northern Nevada and Oregon. The ages of the calderas are oldest at the southwestern end and become progressively younger toward the northeast, with the youngest being the 640,000-year-old Yellowstone caldera. This alignment is called the Yellowstone **hotspot trail** and is critical evidence for interpreting the long record of volcanism at Yellowstone.

In essence, each of the calderas along the hotspot trail is an "ancient Yellowstone" in that they erupted when they once occupied the geographic position where today's Yellowstone exists. In the preceding section, you learned that a stationary mantle plume beneath Yellowstone supplies heat energy for the overlying hotspot volcanism. But you also know that lithospheric plates are constantly on the move, slowly but surely sliding laterally above the underlying asthenosphere. Over the past 16 m.y. since the Yellowstone hotspot originated, the North American plate has slowly been migrating to the west–southwest over the stationary hotspot at an average rate of about 5 cm/yr (2 in/yr). In order to understand how the trail of "ancient Yellowstones" was created, you have to mentally drag the North American plate back to the east–northeast, repositioning each extinct caldera over Yellowstone's geographic location.

high-silica, granitic *country rock* of the lower crust. The heat within the lower chamber melts the surrounding silica-rich rock, assimilating the fluid and causing the melt within the lower chamber to evolve over time into the rhyolitic composition of the upper chamber. As more heat energy is supplied to the upper chamber over time, the silica-rich rocky matrix surrounding the liquid phase is itself melted, adding to the body of rhyolitic magma available for eruptions. The key to all this is that the explosive eruptions at Yellowstone are determined by the gas-rich, high-viscosity, rhyolitic magma of the upper chamber just beneath the surface. The images obtained by tomography are increasing in resolution over time, allowing for a more realistic view of the gigantic heat source that drives not only eruptions, but the variety of hydrothermal features in the park as well.

It's important to grasp the uncertainty associated with these interpretations of the volcanic "plumbing system" beneath Yellowstone and realize that our understanding of the relationship between the mantle plume and hotspot volcanism is a work in progress. Some geoscientists are skeptical about the fixed position of mantle plumes, suggesting that they may wander due to inherent turbulence within the mantle. Indeed, recent tomographic studies suggest that the plume beneath Yellowstone may originate at the core–mantle boundary beneath Mexico, extending as a tilted body toward the northeast where it reaches the crust. Others question the very existence of mantle plumes. As researchers generate more data, earlier models are challenged and are either improved, modified, or abandoned. This is the ongoing process of science and is part of its beauty as a systematic way of explaining the natural world.

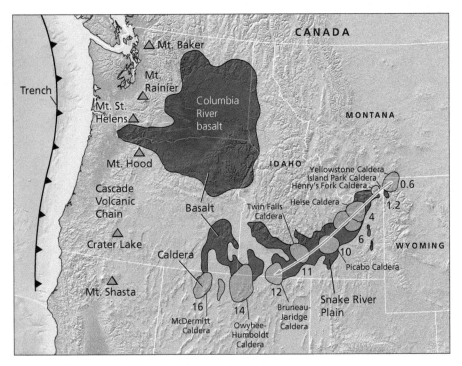

Figure 31.19 Track of the Yellowstone hotspot trail across the Snake River Plain. Orange splotches with numbers are ancient calderas and their ages. Note how they become progressively older away from Yellowstone. The red colors show the distribution of basalts, including the massive Columbia River basalt. Orange triangles are Cascades volcanoes. Dark line with triangular hachures marks the Cascadia convergent margin.

Sliding the North American plate to the east–northeast repositions the Heise caldera over the Yellowstone hotspot when it erupted in multiple events between 4 and 6 m.y. ago. Sliding the North American plate even further to the northeast places the Bruneau-Jarbidge Caldera, now located along the Idaho–Nevada border, above the hotspot where its eruptions occurred about 12 m.y. ago. Mentally pull the North American plate 800 km back so that the McDermitt Caldera, the oldest in the chain, is positioned above the Yellowstone hotspot. This was its location ~16 m.y. ago when its eruptions occurred, marking the arrival of the mantle plume at the surface and the origin of hotspot volcanism at Yellowstone.

Now run the process in reverse (Fig. 31.20). The North American plate slowly migrates laterally to the west–southwest above the stationary mantle plume beneath Yellowstone. The hotspot above the plume episodically blasts out magma during caldera eruptions every 2 to 3 m.y. as the North American plate slides above. As a caldera migrates away from the hotspot, it loses its source of magma and goes extinct. Eventually the magma chamber above the plume is replenished and a new caldera becomes active at the hotspot, before it too migrates

laterally to die out. The Yellowstone hotspot has been the focus of explosive volcanism continuously for the past 16 m.y. The calderas produced during each volcanic phase are progressively moved toward the west–southwest, away from the hotspot, leaving behind a trail of extinct calderas. It's important to reiterate that it's not the hotspot that's migrating across the Snake River Plain, but rather the overlying North American plate that's moving at a few centimeters per year.

Careful mapping and dating of volcanic deposits along the hotspot trail suggest that there were collectively about 100 individual eruptions from all of the calderas along the chain, implying that most of the calderas had multiple eruptions before moving off the hotspot. Estimates of the volumes of material deposited by some eruptions suggest that they were comparable in scale to the three most recent caldera eruptions at Yellowstone. An eruption about 8.1 m.y. ago, for instance, is estimated to have ejected more than 1900 km³ (~450 mi³) of material—not as much as the Huckleberry

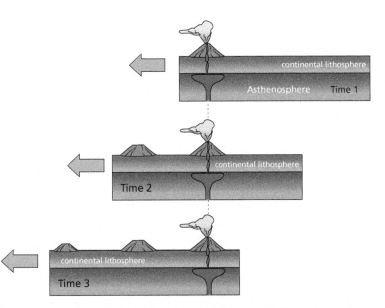

Figure 31.20 Formation of a continental hotspot trail: The mantle plume that is the source of the heat energy is stationary. Magma derived from the plume supplies the hotspot volcanism. The continental lithosphere continually moves laterally above the hotspot, carrying the caldera with it. The older caldera slowly goes extinct, with a new volcanic center developing above the mantle plume.

Ridge event, but more than the Lava Creek eruption. The Bruneau-Jarbidge caldera erupted with such force about 12 m.y. ago that an ashfall about a half-meter thick killed and buried a large number of animals in what is today Ashfall Fossil Beds State Historical Park in Nebraska almost 1600 km (1000 mi) away. The fossilized remains of extinct rhinoceros-like mammals, ancient horses, camels, and even birds indicate that they tried to seek relief from the heavy ashfall near a waterhole where they succumbed to lung failure from inhaling the powdered shards of glassy ash.

Why are the calderas spaced somewhat evenly along the hotspot trail? If the hotspot is stationary, why isn't there a continuous series of overlapping calderas? It may be that caldera eruptions follow a cycle that involves a sequence of stages between eruptions, including the resupply of magma into the chamber. As magma gradually replenishes the reservoir beneath Yellowstone, the North American plate slowly moves laterally, displacing the earlier caldera from the future caldera above the hotspot. The cycle may take several hundred thousand years to a few million years before the next major eruption, more than enough time for the older calderas to migrate to the west–southwest.

Why is the Snake River Plain such a featureless lowland? Why aren't the rocks of the calderas visible at the surface? When each caldera was positioned above the hotspot at Yellowstone, the tremendous heat from below buoyantly raised a broad bulge in the overlying crust of perhaps 600 m (2000 ft), just as it does today. Caldera eruptions occurred near the crest of the uplifted region, destroying any mountains that may have existed above the bulge. As the calderas moved laterally away from the hotspot, the rocks cooled and contracted, progressively subsiding by a half-kilometer or more within the eastern Snake River Plain.

Smoothing of the surface occurred over the past few million years as the sunken Plain and its chain of old rhyolitic calderas were blanketed by extensive basaltic lava flows. The fluid, runny lava poured from broad *shield volcanoes* and fissures, accumulating to thicknesses of over a kilometer. **Craters of the Moon National Monument** in southern Idaho is composed of extensive lava flows and small cinder cones ranging in age from 15,000 years to as young as 2000 years (Fig. 31.21). The stark, black terrain at Craters of the Moon looks more like a Hawaiian volcanic landscape than southern Idaho. At some point, the Snake River positioned itself on the lowland surface, meandering widely back and forth and depositing thick layers of alluvial sediment on its broad floodplain. These river sediments and a veneer of soil further buried the chain of "ancient Yellowstones."

Figure 31.21 Eastern Snake River Plain showing the location of Craters of the Moon National Monument relative to the Yellowstone Plateau. Note how Basin and Range-style mountains extend north and south from the Plain.

What event may have triggered the origin of hotspot volcanism at Yellowstone? Recall that the region around Yellowstone is part of the northeastern margin of the Basin and Range Geologic Province. Extensional tectonics began around 16 m.y. ago in the Basin and Range (Chapter 23), accompanied by a thinning of the crust and the formation of tilted fault-block mountains like the Grand Tetons just to the south of Yellowstone and the Gallatin and Madison ranges to the north. It may be more than coincidence that both crustal extension and the origin of the Yellowstone hotspot occurred about 16 m.y. ago. Perhaps the thinning of the crust and the concurrent development of faults and fractures permitted the upward movement of the mantle plume and its overlying magmatic system into the Yellowstone region.

Recent Uplift and Seismicity at Yellowstone

Careful surveys of surface elevations at Yellowstone reveal two **resurgent domes** rising a few hundred meters above the landscape (Fig. 31.12). The Mallard Lake dome northeast of Old Faithful is about 10 km (6 mi) across, while the Sour Creek dome north of Yellowstone Lake is slightly larger. Over the last few decades, these domes have been measured by the Global Positioning System (GPS) of satellites to rise upward by several centimeters over a few years, then just as rapidly subside before repeating the cycle. These vertical motions likely reflect the movement of hydrothermal fluids deep beneath the surface, suggesting that a network of fractures connects the underlying magma chamber with the surface beneath each dome. The rising and sinking behavior at each dome is regarded as typical for a landscape underlain by a massive body of hot, partially melted rock.

The upper magma chamber also controls the highly active seismicity at Yellowstone. The region is rattled by 1000 to 3000 earthquakes every year, most too small to be felt, but a few with magnitudes of 3 to 4 (Fig. 31.22). (The mechanics of earthquakes were described in several earlier chapters.) The 5-km-thick layer of rock above the magma body is brittle and readily breaks in response to the injection of hot fluids under pressure, forming a dense network of faults and fractures that extend to the surface. Seismicity at Yellowstone typically occurs as an **earthquake swarm**—clusters of earthquakes that happen over a relatively small region over a relatively short amount of time. A typical swarm may consist of hundreds of quakes over a few weeks. Earthquake swarms are typical phenomena near volcanically active regions because of the constant movement of hot fluids above the magma chamber. They are not considered to be precursors to eruptions.

Less common larger earthquakes at Yellowstone are related to Basin and Range extensional tectonics rather than magmatic activity. Many of the earthquake epicenters are aligned along north/south and northwest/southeast-oriented *normal fault* systems. This is the same fault trend seen in other parts of the Basin and Range, such as along the nearby Teton, Gallatin, and Madison Ranges. A particularly destructive example of a tectonically related earthquake was the magnitude 7.5 Hebgen Lake event in 1959 when two parallel normal faults ruptured with instantaneous displacement of up to 6.5 m (~22 ft). It may be that the huge offset was not only related to extensional stress along parallel normal faults, but was perhaps enhanced by flexural upwarping related to the magma chamber beneath the park. The horrific events associated with the Hebgen Lake earthquake were described in Chapter 25 on Grand Teton National Park.

Hydrothermal Features at Yellowstone

Over 10,000 hydrothermal features delight visitors to the park, who travel the Grand Loop Road from one highlight to the next (Fig. 31.23). At this point, you know that the heat

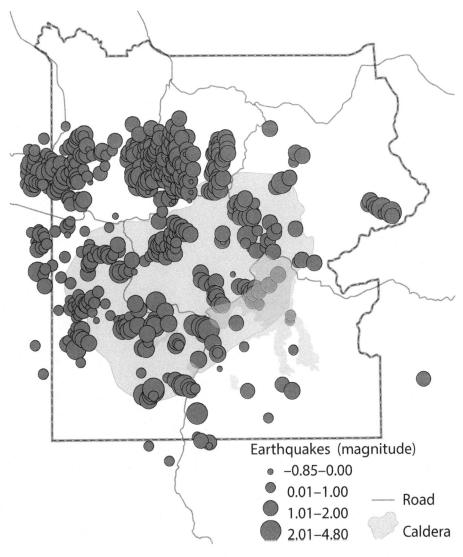

Figure 31.22 Seismicity map of Yellowstone region showing epicenters of ~2000 earthquakes during 2014.

Figure 31.23 Norris Geyser Basin, a dense cluster of geysers, milky blue hot springs, fumaroles, and a widespread surface crust of white siliceous sinter and orange bacterial mats.

source for the geysers, hot springs, mudpots, and steam vents is the gargantuan body of hot rock and magma about 5 km (3 mi) below the surface. Beyond heat energy, one of the other necessities is water, which is derived from the copious amounts of snowmelt and rainfall that drain through the region. Yet another critical factor for creating hydrothermal systems is a network of fractures and faults through which the fluids can travel, connecting the heat source at depth with the surface. These three components—hot rock and magma, water, and fractures and faults—compose the natural plumbing system that drives hydrothermal activity at Yellowstone.

The mechanics of hydrothermal systems were introduced in Chapter 16 on Lassen Volcanic National Park (Fig. 16.12), but the fundamentals will be reiterated and expanded upon here. Each of the four types of hydrothermal features—geysers, hot springs, mud pots, and fumaroles—are controlled by the same basic circulation system beneath the ground, with local variations near the surface determining the type of feature seen on the landscape. The process begins with cold snowmelt and rainfall percolating downward through faults and fractures to depths beyond a kilometer or two. There, the cold water mixes with hot water that's been heated by proximity to hot rock and magma of the underlying chamber. The water is **superheated** above the boiling point to temperatures over 200°C (~400°F) but doesn't convert to vapor because of the high pressures at those depths. (Remember that the freezing and boiling points of water are dependent not only on temperature but also on pressure.)

The superheated water expands, becoming buoyant because of its lower density, and it rises upward through a network of faults and fractures. The acidic, hot water is in constant contact with the silica-rich rhyolitic rocks of the upper crust. Some of the silica is dissolved into the water and transported upward. In places, the silica in solution will precipitate onto the walls of natural conduits like fractures, pipes, and cavities, somewhat like the scale that

builds up in water pipes in homes using "hard" water rich in dissolved elements. ("Precipitation," for those of you who have forgotten your high school chemistry, is the deposition of a solid from water oversaturated with certain dissolved elements. In this case, solid silica will precipitate from water charged with abundant dissolved silica.) This process is critical for hydrothermal systems because the silica-lined conduits become sealed against leakage, enhancing the efficiency and longevity of the system.

Many of the hydrothermal features in Yellowstone are clustered in **geyser basins**, low, flat areas commonly situated between the margins of lava flows. Rainwater and snowmelt percolate into the ground within these basins, providing a continual supply of water for geysers and hot springs. The other important factor that contributes to the concentration of hydrothermal features is the presence of a dense network of faults and fractures through which water can flow. Norris Geyser Basin is one of the most active in the park because it's located where three major faults intersect, focusing the flow of hot water beneath the basin. Scientific drilling in the Norris Geyser Basin revealed water temperatures of 238°C (460°F) a little over 300 m (~1000 ft) beneath the surface, the hottest measured hydrothermal temperature in the park.

HOT SPRINGS

Hot springs, the most common hydrothermal feature at Yellowstone, are pools of hot water supplied by a continuous source from below (Fig. 31.24). The key is an open plumbing system without constrictions that permits superheated water

Figure 31.24 Heart Spring in Upper Geyser Basin. The white crust surrounding the blue water of the hot spring and the white mound to the back right is siliceous sinter, deposited from the overflow of hot water enriched in dissolved silica. Castle Geyser erupts in the background.

Figure 31.25 Grand Prismatic Spring is the largest and most colorful hot spring in Yellowstone. The concentric bands of color are created by various heat-loving communities of microbes that live at different temperatures around the pool. The orange tendrils are overflow channels populated by heat-loving microbes.

microbial communities of archaea and bacteria color the water green, then yellow (~70°C; ~160°F), then a variety of orange and brownish hues along the margins and outflow channels (~60°C; 140°F). Fungi, algae, and some single-celled protozoans form distinct ecosystems with the microbes. Some of the thermophiles live on hydrogen or sulfur in the water, but most photosynthesize, using sunlight to fulfill their energy needs. Thermophilic archaea are considered to be relatives of the earliest life forms on Earth and are of considerable interest to astrobiologists and other life scientists.

A special type of hot spring occurs in the northern part of Yellowstone. Mammoth Hot Springs is notable for its spectacular terraces and "frozen" cascades of white **travertine** (Fig. 31.26). As opposed to most hydrothermal deposits composed of siliceous sinter, those at Mammoth are composed of the mineral *calcite*, which has the chemical formula of $CaCO_3$. Travertine is notable for its beautiful banded appearance when cut and polished as building stone. As hot groundwater rises through fractures from the depths beneath Mammoth, it picks up CO_2 emitted from the magma reservoir below, forming a weak carbonic acid. As the slightly acidic water rises toward the surface it passes through buried limestones deposited long ago by transgressions of the Paleozoic seas. The calcite composing the limestone is dissolved by the circulating groundwater and transported upward

to reach the surface. The water at the surface of the spring cools due to contact with the atmosphere, causing it to sink and be replaced by new superheated water from below. So the water in hot springs is continuously recirculated by convection. At times, the water may overflow from the pool and spread out across the surrounding ground. As the water either evaporates or runs off downslope into a nearby river, a white to gray deposit of **siliceous sinter** may precipitate, composed primarily of silica, SiO_2. This is the same silica-rich material that lines the conduits of the plumbing system beneath the surface. The silica is derived from dissolution of the rhyolitic rocks through which the superheated water passes on its way upward. Crusts of sinter take a long time to form, precipitating at rates of a centimeter or two per century.

The center of many hot springs, such as Grand Prismatic Spring, is the hottest and clearest part of the pool (Fig. 31.25). The water absorbs most wavelengths of sunlight, but the blue wavelengths are scattered back to our eyes as an intense shade of azure. The concentric bands of color surrounding the blue interior owe their vibrant colors to heat-loving microbes known as **thermophiles**. Countless bacteria and other primitive single-celled microbes called *archaea* (Chapter 16) not only thrive in each zone of scalding water, but require a specific narrow range of temperatures to survive. As the temperature of the water decreases outward from the blue center of a hot spring like Grand Prismatic (~90°C; ~190°F), various

Figure 31.26 Travertine terraces at Mammoth Hot Springs. The yellowish orange tint is caused by microbes living in the warm waters that flow over the terraces.

Figure 31.27   Bubbling mudpot, Artists' Paintpots.

Figure 31.28   Roaring Mountain fumarole, one of the hottest hydrothermal features in the park.

where it flows onto the surface at hot springs. Upon contact with the atmosphere, the CO_2 gas is released, changing the chemistry of the waters and triggering the precipitation of calcite. The water from the springs commonly pools along flat areas, creating terraces of travertine that cascade one into another downslope. The calcite that was once part of ancient hard-shelled organisms living in Paleozoic seas is currently being recycled through the Yellowstone plumbing system to form the ornate travertine terraces at Mammoth Hot Springs.

MUDPOTS

If the supply of water to a hot spring is limited, a thick slurry of grayish clay-rich water may form a bubbling pool called a *mudpot* (Fig. 31.27). The heat for mudpots is supplied by steam leaking upward from deeper conduits. Certain microbes living in the shallow plumbing system use hydrogen sulfide gas in the groundwater for energy, producing sulfuric acid as a by-product. The acidic water and steam react with the surrounding rhyolitic rock to convert it to clay, which combines with the pooled water to form the slurry. The bubbles within mudpots are not due to boiling, but rather to water vapor and other gases churning the surface on their way to the atmosphere. In places, iron oxides create a spectrum of pastel colors that tint the clay, giving rise to the alternative name "paintpots." These peculiar hydrothermal features can be

seen at Artists' Paintpots, Fountain Paint Pots in Lower Geyser Basin, West Thumb Geyser Basin, and at Mud Volcano and Dragon's Caldron in Hayden Valley.

FUMAROLES

Places where hot steam and other gases are emitted directly from the ground are known as steam vents or *fumaroles*, the hottest hydrothermal feature in the park (Fig. 31.28). Fumaroles occur because of a limited amount of water feeding into the underlying plumbing system. In water-rich systems, it's the weight of overlying water that keeps superheated water in the liquid phase. Less water means less weight, so the limited amount of water at depth beneath fumaroles converts more easily to the vapor phase, which rises through conduits to vent at the surface. The steam at fumaroles may be as hot as 138°C (280°F) and may be released with a whistling or roaring sound, like a massive teapot. Good examples are seen at Roaring Mountain and at Black Growler Steam Vent in Norris Geyser Basin.

Figure 31.29   Aerial view of Old Faithful geyser in mid-eruption, with its plume reaching heights of 32–56 m (106–184 ft). The white mound around the opening of the geyser is siliceous sinter.

GEYSERS

Geysers are defined by the intermittent spouting of water as a turbulent plume upward into the air. Fewer than a thousand geysers exist worldwide, with about half located in Yellowstone. Old Faithful is the iconic example (Fig. 31.29), but other geysers in the park reach higher in the air, others pulsate in a series of bursts, and yet others spray sideways, creating rainbows in their mist. The primary difference between a geyser and a hot spring is the existence of a narrow constriction within the uppermost plumbing system of geysers. The narrowing acts like a nozzle on a hose, causing the uprushing water to back up, increasing the pressure. The result is a jet of water that spurts out forcefully, rising meters to tens of meters above the surface.

The conduit beneath Old Faithful was carefully investigated with a mini-video camera in the early 1990s (Fig. 31.30). The vent at the surface of Old Faithful is an open crack about 0.6 x 1.5 m (2 x 5 ft) in area. At a depth of 7 m

(23 ft), however, the conduit narrows to a slot-like channel about 11 cm (~4 in) across. The constriction is lined with siliceous sinter and is underlain by a broader cavity that acts as a reservoir for water. During eruptions, superheated water and steam are forced through the narrow opening with tremendous force, shooting into the air for about 2 to 5 minutes before the pressure dissipates and the geyser begins the cycle anew. The duration between eruptions at Old Faithful varies, but the average interval in early 2017 was 90 minutes, with superheated water replenishing the reservoir beneath the constriction during that time.

The temperature of water ejected from Yellowstone's geysers averages 95°C (204°F), superheated well above the boiling temperature of water at Yellowstone's elevation. It cools significantly while airborne and is no longer scalding as it spatters back to the ground. Most geysers at Yellowstone are found in the Upper Geyser Basin, the site of Old Faithful, but others are widespread throughout the park. Geysers are transient features that are relatively unpredictable over time. For instance, Steamboat Geyser rapidly increased its activity in 2018, jetting water and steam to heights beyond 90 m (300 ft) and lasting from a few minutes to an hour. Earthquakes may open new fractures, or they may break up the silica deposits clogging older conduits, triggering a dormant geyser to erupt. Alternatively, earthquakes may seal conduits, shutting down a geyser that was behaving predictably. Other factors that influence geyser activity include changes in the rate at which water recharges the system (a function of weather and climate) and the slow closure of conduits by the precipitation of silica along fracture margins.

Glaciation and the Modern Yellowstone Landscape

During the multiple glacial-interglacial cycles of the 2.6 m.y. of the Pleistocene Ice Ages (Chapters 14, 17, 22), thick *ice caps* expanded and contracted over the Yellowstone Plateau and adjoining mountains. These glaciations happened contemporaneously with the three caldera eruptions and the hundreds of lava flows that occurred over that time. There must have been fiery interactions between active lava flows and glacial ice. Parts of the ice cap were likely obliterated during the gargantuan caldera eruptions. Hydrothermal energy almost certainly melted the base of the ice cap near geyser basins, creating subglacial streams of meltwater. But no evidence of those interactions exists today, having been erased by the erosive effects of subsequent glaciations, especially during the Pinedale glaciation of the Last Glacial Maximum (~18 to 20 ka).

The ice cap reached thicknesses of over 1 km (3500 ft) above the Yellowstone Plateau during the Last Glacial Maximum (Fig. 25.13), comparable in scale to those that crowned the Sierra Nevada and Colorado Rockies. The ice cap blanketed almost the entire area of the park, with only the highest summits of the Gallatin and Absaroka

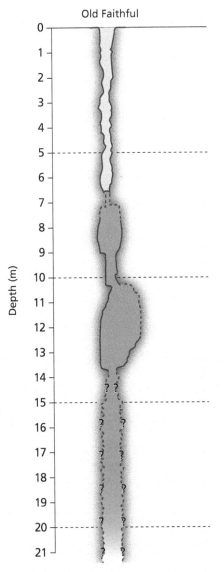

Figure 31.30 Schematic illustration of the conduit geometry beneath Old Faithful, derived from video imagery and in-hole measurements. Note the narrow constriction around 7 m (23 ft) deep.

Ranges peaking above the sea of ice as rocky nunataks. The Yellowstone ice cap was discussed extensively in Chapter 25 on Grand Teton National Park, including descriptions of a variety of glacial landscape features such as moraines, outwash plains, and glacial lakes. That chapter added to the extensive array of glacial processes and landforms discussed in earlier chapters, such as abrasion, plucking, cirques, horns, arêtes, tarns, striations, erratics, and U-shaped valleys. Here we'll focus on the effects of the Pinedale ice cap that ended around 14 ka and that created the modern landscape of the park.

Most prominent glacial landscape features occur along the periphery of the park in the higher elevations of the Beartooth, Gallatin and other surrounding mountains. In the national park itself, the sheer bulk of the ice cap acted like a gigantic rasp, abrading and plucking volcanic rock to sculpt the smooth, rolling terrain of the Yellowstone Plateau. Imagine standing on the crest of the ice cap 20,000 years ago above what is now Yellowstone Lake. The air is thin at an elevation of around 3350 m (11,000 ft), and a kilometer of ice lies between you and the future lake deep beneath your feet. A broad, white dome of frozen water extends in all directions, but it's far from stationary. Everywhere the ice is moving, flowing as a thick plastic downslope away from the crest of the ice cap.

At the very base of the ice cap where it was in contact with the terrain, the flow followed river valleys, rounding their V-shaped profiles into U-shaped troughs. The basal ice below the crest scoured two valleys that would in the future form the South Arm and Southeast Arm of Yellowstone Lake (Fig. 31.31). Other parts of the basal ice plucked rock as it ponderously moved downslope, deepening and modifying the main basins of Yellowstone Lake. The volcanic origins of the lake basin were obscured and reshaped by the power of glacial ice.

As the climate warmed over the last 15,000 years or so, the tremendous volume of ice melted and flowed outward from the margins of the ice cap. U-shaped glacial valleys were modified into V-shaped river valleys by the torrents of glacial meltwater. The Grand Canyon of the Yellowstone River

Figure 31.31 Bathymetric map of Yellowstone Lake obtained by multibeam sonar imaging and seismic mapping. The blue and violet colors show the relative depths of the lake floor. The muted colors of the land surrounding the lake show the surface geology.

Figure 31.32 Lower Falls, Grand Canyon of the Yellowstone River, from Artist's Point. Note the V-shape of the canyon, indicating incision by the river. The vibrant yellow, pink, and orange hues of the canyon walls result from intense hydrothermal alteration by hot, acidic fluids. The rhyolitic rock converts to weak clay minerals upon contact with corrosive gases and groundwaters.

was subjected to episodes of intense glacial flooding that cut into the soft, hydrothermally altered rock along the canyon walls (Fig. 31.32). The valleys of the Snake, Madison, Gallatin, Lamar, and other rivers draining the region

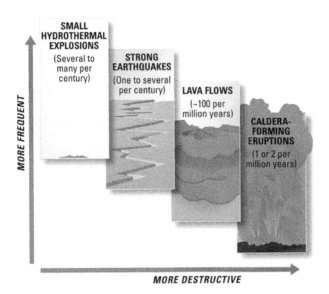

Figure 31.33 Probability of hazardous events at Yellowstone. In general, less hazardous events are relatively common, whereas catastrophic events are exceedingly rare.

were also reshaped by the higher meltwater flows and sediment loads triggered by the melting ice cap.

Future Eruptions at Yellowstone?

Visitors to Yellowstone are understandably curious about the possibility of future eruptions from the magma chamber beneath their feet. In terms of risk, the probability is exceedingly low that an event as catastrophic as a caldera eruption will happen in our lifetimes (Fig. 31.33). Volcanic events of that magnitude happen very rarely, on the order of one event every several hundred thousand years to every few million years. Adding to the small probability of a cataclysmic eruption is the low liquid content of the magma chambers beneath the park. Much more actual magma is necessary for a large eruption to occur.

Slightly more likely would be a lava flow. This event happens on the scale of tens of thousands of years, with the last lava flow at Yellowstone occurring 70,000 years ago. Even then, the eruption would most likely be a slow-moving rhyolite flow that would ooze across the landscape over months to years. The impact on the landscape would be significant, but the potential danger to humans would be negligible. Geologists, however, would likely be thrilled to observe a rhyolitic lava flow happening in real time.

The volcanic event with the highest probability of occurring would be **hydrothermal explosions**, which recur on a decadal to centennial scale. These events are triggered suddenly when a body of superheated water beneath the surface abruptly loses pressure, causing the water to flash to steam. The upward pressure from the rapid expansion of steam disintegrates overlying rock, sending an explosive blast of mud, rocky debris, and water into the air. The superheated groundwater may lose its confining pressure due to a lowering of the water table by drought or

the leakage of shallow groundwater into a newly formed fracture system. There would be a reasonable amount of danger to people and structures in the local area.

Large-magnitude earthquakes, such as the M7.5 Hebgen Lake event in 1959, pose the highest risk to the Yellowstone region. Extensional stresses related to Basin and Range tectonism as well as to flexural upwarping from the underlying body of partially melted magma are constant and ongoing. The Yellowstone Volcano Observatory, a consortium of federal and university expertise, carries out real-time monitoring using a network of seismic stations, GPS sensors, geochemical analysis of gases and water near hydrothermal areas, and satellite measurements to detect any changes in the geologic activity of the region.

The three most recent caldera eruptions over the last 2 m.y. are just the latest in the past 16 m.y. at Yellowstone, and there's no reason to think they're over. In time, the youngest calderas will slowly slide toward the west–southwest as part of the North American plate, with a supply of new magma reigniting the hotspot. Although some geologists think that the new continental crust currently moving over the Yellowstone hotspot is thicker and thus more difficult to penetrate with magma, it's inevitable that another caldera eruption will occur at Yellowstone sometime in the distant future. A supervolcanic eruption from Yellowstone poses an existential risk to humanity, a thought that nobody wants to contemplate.

Ecogeology of the Yellowstone Region

The dynamic geology of Yellowstone exerts a profound control not only on the landscape of the park, but also on the ecology. The high elevation of the park influences the regional climatic conditions, which in turn control the flora and fauna that live there. The elevated plateau is higher than it otherwise would be due to the underlying welt of magmatic energy that adds buoyancy to the crust. The geologic history of Yellowstone also determines the rock types spread across the surface, which in turn influence the distribution of forests, alpine meadows, sagebrush flats, and grasslands. The dominant rhyolitic bedrock forms a loose, granular soil that is inherently poor in nutrients. It tends to not retain water or organic matter due to its porous character—ideal conditions for the highly tolerant lodgepole pines that dominate the forests. In turn, the lack of floral diversity makes the forests particularly vulnerable to wildfire, such as the catastrophic event that burned a third of the park in the summer of 1988.

In contrast, the andesitic rock in the Absaroka Mountains is derived from early Cenozoic volcanism. The soils formed from andesite are richer in nutrients, enabling a more diverse flora and fauna to colonize the highlands and adjoining valleys. For instance, whitebark pine trees thrive in these soils and nourish grizzly bear with their pine nuts. Steep andesitic terrain along the slopes of Mount Washburn supports the growth of grasses and scrub, which provide forage for bighorn sheep. In places where old glacial lakes have been filled with clay-rich sediments to form wetlands and meadows, a diverse plant community thrives, attracting a variety of birds and larger animals like moose.

Soils derived from sediments deposited on broad river floodplains like the Hayden and Lamar valleys sustain grasslands, which in turn are habitat for mule deer, elk, pronghorn, and bison. These large herbivores attract carnivores like wolves, which were reintroduced into the park in 1995 and 1996 to restore ecological balance. The incredible array of wildlife that thrives in the Greater Yellowstone Ecosystem is directly connected to the vegetation, which in turn depends on the soil conditions, which in turn are dependent on the underlying composition of the rock. And the rock, of course, is the physical record of the volcanic and tectonic events that affected the region through time.

There's just so much to see and do in the natural wonderland of Yellowstone National Park. Marvel at the vibrant hot springs and pulsating geysers. Admire the abundant wildlife (from a distance). Gaze at the inspirational vistas and canoe the trout-filled lakes. But as you enjoy all the park has to offer, remember that it all exists because of the colossal source of heat energy just a few kilometers beneath your feet.

Key Terms

columnar jointing
earthquake swarm
geyser
geyser basin
hotspot
hotspot trail
hydrothermal explosion
mantle plume

petrified wood
resurgent dome
seismic tomography
siliceous sinter
superheated water
supervolcano
thermophiles
travertine
Volcanic Explosivity Index

Related Sources

Christiansen, R. L., et al. (1994). A Field-Trip Guide to Yellowstone National Park, Wyoming, Montana, and Idaho—Volcanic, Hydrothermal, and Glacial Activity in the Region. Denver, CO: *U.S. Geological Survey Bulletin* 2019.

Fritz, W. J., and Thomas, R. C. (2011). *Roadside Geology of Yellowstone Country*. Missoula, MT: Mountain Press.

Hendrix, M. S. (2011). *Geology Underfoot in Yellowstone Country*. Missoula, MT: Mountain Press.

Smith, R. S., and Siegel, L. J. (2000). *Windows into the Earth: The Geologic Story of Yellowstone and Grand Teton National Parks*. New York: Oxford University Press.

National Park Service—Yellowstone geologic history. http://www.nps.gov/yell/learn/nature/volcano.htm (accessed February 2021)

U.S. Geological Survey—Geology of Yellowstone. Accessed February 2021 at http://volcanoes.usgs.gov/volcanoes/yellowstone/yellowstone_geo_hist_52.html

U.S. Geological Survey Yellowstone Volcano Observatory. Accessed February 2021 at https://volcanoes.usgs.gov/observatories/yvo

Yellowstone hydrothermal features. Accessed February 2021 at https://www.nps.gov/yell/learn/nature/hydrothermal-systems.htm

To take a tour down the vent of Old Faithful from a mini-video camera in 1991, watch: https://www.youtube.com/watch?v=8luNCFUnvBw (accessed February 2021).

32 Hawai'i Volcanoes National Park

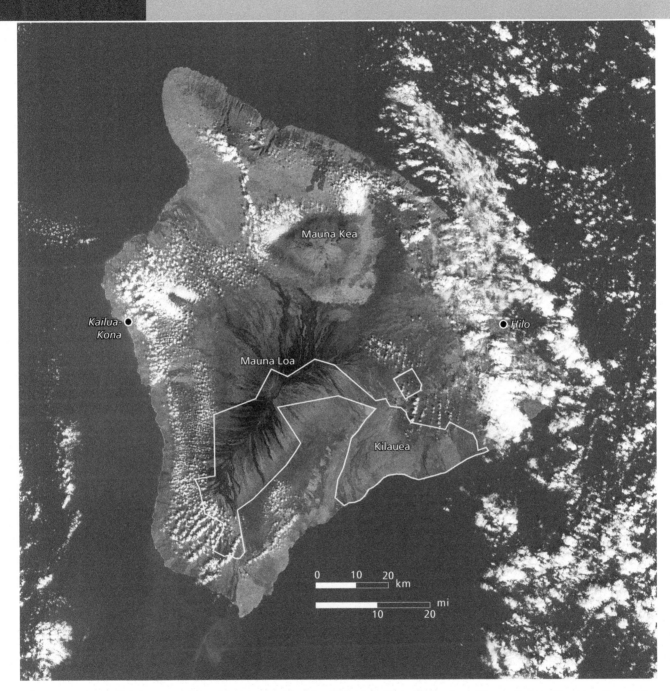

Figure 32.1 Hawai'i Volcanoes National Park is located on the Big Island of Hawai'i. Borders of the park are outlined in yellow. Note the contrast in colors between the eastern windward side and the rain shadow on the western side of the island.

No other national park displays as much real-time geologic activity as Hawai'i Volcanoes. Located on the Big Island of Hawai'i, the largest of the eight main islands in the Hawaiian chain (Fig. 32.1), active lava flows have been occurring within the park almost continuously since 1983. No other park has Earth's *largest* volcano adjacent to Earth's *most active* volcano. No other park has two gigantic calderas, both appearing as sunken black moonscapes of solidified lava. Inside one of the calderas is a deep crater with an active lava lake that episodically releases orange-hot fountains of molten rock. No other park has streams of lava flowing slowly toward the coast where they create a mesmerizing spectacle when they meet the ocean, accompanied by giant plumes of steam as the seawater immediately quenches the molten rock (Fig. 32.2).

All of these stark volcanic features are juxtaposed against a lush rainforest of ohi'a trees and extravagant ferns, creating a primeval environment that is constantly being reshaped. As if all this volcanic dynamism isn't enough, the Big Island of Hawai'i is one of the most seismically active locations in the United States. Hawai'i Volcanoes National Park is unlike any other in the National Park System due to its unparalleled display of geology in action.

The angular borders of Hawaii Volcanoes National Park occupy the southeastern part of the Big Island and encompass the crest of the historically active Mauna Loa Volcano and much of the body of currently active Kilauea Volcano (Fig. 32.3). The highest elevation in the park is 4170 m (13,680 ft) on the crest of Mauna Loa, while the lowest elevation is at sea level along the coast. The diversity of climates and ecozones within the park is influenced not only by the extremes of elevation, but also by the prevailing northeast trade winds that pick up moisture from the tropical Pacific. As the winds ride up the northeast-facing volcanic slopes, moisture in the air condenses and rains out, nourishing tropical rainforests and woodlands. Precipitation along the eastern windward part of the Big Island may reach 980 cm (386 in) per year. The western slopes downwind of the volcanic peaks occupy a *rain shadow* and are much drier, with scrubby vegetation. Rainfall along the western leeward slopes is only about 25 cm (10 in) per year. The highest volcanic summits commonly rise above a halo of midlevel clouds and are rocky landscapes of subalpine to alpine ecosystems.

The Hawaiian Islands are over 3500 km (~2200 mi) from North America, the nearest continent. The geographic isolation of the island chain in the central Pacific and the ecologic and climatic diversity within the national park foster an abundance of endemic species. Many of the plants and animals in the park are threatened or endangered, primarily due to competition from introduced non-native species. To preserve and protect its unique biodiversity, Hawai'i Volcanoes National Park was named an International Biosphere Reserve in 1980 and a World Heritage Site in 1987.

The primary purposes of this chapter are to explain the characteristics of modern Hawaiian volcanism, to identify volcanic features unique to Hawai'i Volcanoes National Park, and to describe the ongoing volcanic activity within the park. The broader goal is to understand the significant differences between hotspot volcanism at Hawai'i and hotspot volcanism at Yellowstone, discussed in Chapter 31. In the same vein, we'll explore the contrasting styles of volcanism between Hawai'i volcanoes and other volcanic national parks, such as those of the Cascades geologic province of Part 2. Be sure to refer back to the relevant pages on volcanic rocks and processes in Chapters 1 and 12 as necessary to fill in the details.

Big Island of Hawai'i

The Big Island is composed almost entirely of volcanic rock derived from five main volcanoes that overlap and coalesce to form the body of the island (Fig. 32.4). The island covers an area of about 130 x 150 km (80 x 90 mi) and is actively growing along its southeastern margin where lava flows from Kilauea Volcano meet the sea. All of the volcanoes that make up the Big Island are massive *shield volcanoes*, characterized

Figure 32.2 Lava flows from Kilauea Volcano enter the ocean along the south coast of the Big Island of Hawai'i, 2016.

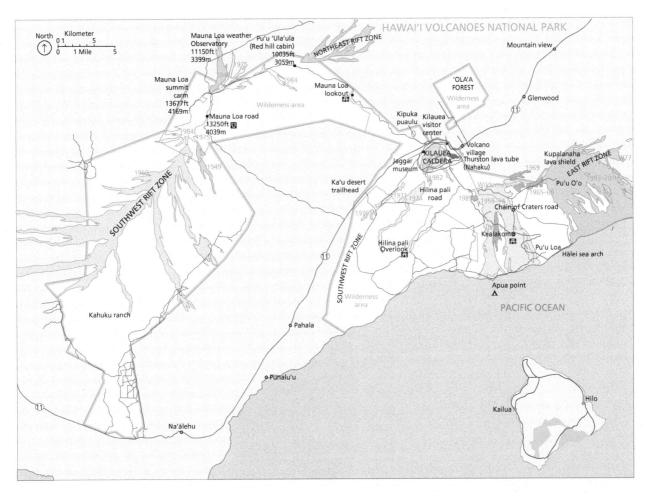

Figure 32.3 Map of Hawai'i Volcanoes National Park showing main roads, locations of two calderas, and recent lava flows.

by gentle slopes and a broad profile (Fig. 32.5). They are composed almost entirely of *basalt*, a dark, fine-grained rock with low silica content (Fig. 1.21). The magma that supplies Hawaiian volcanoes is derived directly from the underlying low-silica mantle. (Recall that molten rock beneath the ground is called *magma*, while the same fluid erupted on the surface is called *lava*.)

The low-silica content of the magma creates a low viscosity liquid, which in turn allows gases to escape from the magma chamber via fractures in the surrounding country rock. The low gas pressure within the magma beneath Hawaiian volcanoes tends to generate *effusive eruptions*, marked by the outpouring of very hot, very fluid lava that flows downslope within sinuous channels. The runny lava may spread laterally for tens of kilometers away from its source vent, filling in between older, solidified lava flows. Countless basaltic lava flows build up through time to create the broad profile of shield volcanoes. (The relationship between silica content, viscosity, gas pressure, and volcano architecture was addressed in Chapter 12 on the Cascades volcanic province.) The effusive eruptions occurring today on Kilauea Volcano in Hawai'i Volcanoes National Park are reasonably accessible; one of the most rewarding experiences in the park is a hike across the volcanic landscape to see active flows.

The *Volcanic Explosivity Index* of Kilauea eruptions ranges from 0 ("effusive") to 1 ("gentle"). (The Volcanic Explosivity Index was described in Chapter 31 on Yellowstone.) The relatively benign style of volcanism exhibited currently within Hawai'i Volcanoes National Park is distinctly different from the explosive style of volcanism expressed in national parks of the Cascades and Yellowstone. Cascades volcanoes typically erupt as pyroclastic columns or as thick, sticky lava flows, building upward to form steep-walled stratovolcanoes or lava domes, whereas Yellowstone is characterized by rare caldera eruptions. In contrast, the near-continuous eruptions of lava from Kilauea that have been occurring since 1983 are an entirely different style of volcanism than that expressed at other volcanic national parks. We'll address the tectonic controls on these differences later in this chapter.

Older Volcanoes Outside the National Park

Three shield volcanoes occupy the northern half of the Big Island and are considered to be in various states of repose (Fig. 32.4). The natural plumbing system beneath volcanoes evolves through time and every volcano behaves in its

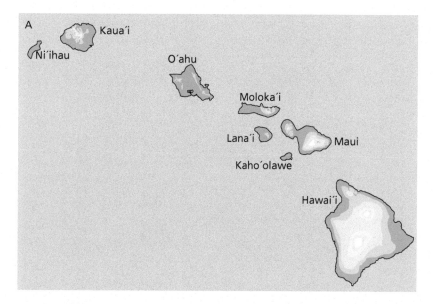

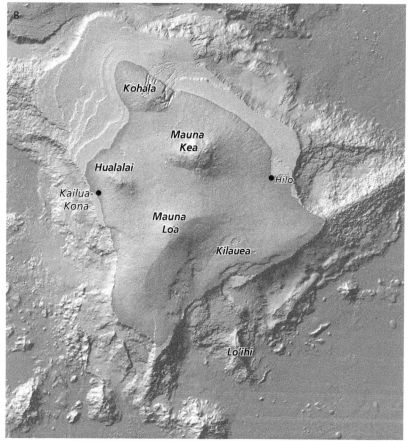

Figure 32.4 A) Eight main islands in the Hawaiian chain. B) Big Island of Hawai'i with surrounding seafloor bathymetry shown by shading.

at least once in historical time where humans were able to document the event. **Dormant** volcanoes are historically inactive but have the potential to erupt again in the future. As an example of a dormant volcano, the last eruption at Yellowstone was 72,000 years ago, long before humans were on the scene, but future eruptions are inevitable.

The **Kohala** shield volcano forms a broad peninsula on the northern tip of the island and is considered to be the oldest volcano, with an estimated age of over 1 million years. Its last eruptions occurred about 100,000 years ago, so subsidence and erosion have dominated since then. The northeast-facing, windward flank of the volcanic edifice is incised by deep, stream-cut gorges that provide habitat for a variety of endemic species.

The summit of **Mauna Kea** ("White Mountain" in the Native Hawaiian language) is the highest point on the Big Island at 4205 m (13,797 ft). The dormant volcano is high enough that it is occasionally crowned with snow during winter storms, hence its Hawaiian name. The volcano is estimated to be about 1 m.y. old, built upward by innumerable basaltic lava flows. The crest of Mauna Kea is marked by *cinder cones*, creating a lumpy summit topography. The youngest eruptions occurred between 4000 and 6000 years ago, old enough to make a new eruption unlikely, but not quite enough to consider the volcano to be extinct. It's hard to believe, but evidence for Ice Age glaciers abounds on the upper slopes of Mauna Kea. Imagine how cold the global temperatures must have been to support a crown of glaciers on the tropical Big Island of Hawai'i.

Hualalai is an active volcano located along the central Kona coast, having last erupted in 1800 and 1801 when lava flows reached the west coast. Much of its surface was covered by lava flows over the last 5000 years and the average time interval between eruptions is about every 300 years. An *earthquake swarm* in the early 20th century was likely triggered by the movement of magma or hot fluids beneath the volcano. The historical eruptions and seismicity suggest that Hualalai is a distinct threat to the Kona coast at some point in the future.

own way, so the definitions of the terms *extinct, dormant,* and *active* are somewhat arbitrary. As defined by the U.S. Geological Survey, **extinct** volcanoes are those that haven't erupted in the past 10,000 years and that are not expected to erupt in the future. **Active** volcanoes are defined either as those that are currently erupting or those that erupted

Figure 32.5 Broad profile of Mauna Loa, a shield volcano that is the largest active volcano on Earth. The crest reaches 4169 m (13,677') in elevation, about 3000 m (9800 ft) above the caldera crowning the nearby Kilauea Volcano (the oval near left center of photo). Hualālai Volcano looms in the distance to the northwest.

Active Volcanoes within the National Park: Mauna Loa

The park boundaries encompass the calderas of Mauna Loa and Kilauea, as well as volcanic rift zones lined with craters and jagged fields of solidified lava. The two shield volcanoes adjoin one another and collectively far exceed the size of typical stratovolcanoes, attesting to the voluminous amount of lava erupted over the last 300,000 years or so (Fig. 32.6).

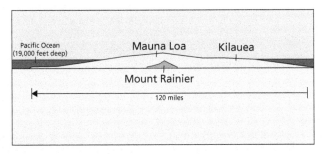

Figure 32.6 The combined size of the Mauna Loa and Kilauea shield volcanoes dwarf the size of the Mount Rainier stratovolcano.

Mauna Loa is the most massive shield volcano on the Big Island, about 35 m (117 ft) lower in elevation than nearby Mauna Kea but much larger in volume. It covers more than half the area of the Big Island and is almost as large as all the rest of the Hawaiian Islands combined. Having last erupted in 1984, Mauna Loa is considered to be the largest active volcano on Earth. The sheer mass of Mauna Loa depresses the oceanic crust deep beneath, somewhat like a bowling ball on a mattress (Fig. 32.7). The volcano rises 4170 m (13,680 ft) above sea level, but the edifice extends down another 5500 m (18,000 ft) to the seafloor, making the mountain almost 9700 m (~32,000 ft) high from base to summit. If Mauna Loa were magically transported to the Himalaya Mountains, it would tower almost a kilometer above Mount Everest.

The summit of Mauna Loa is marked by a *caldera*, a large, downdropped depression about 5 km (3 mi) long and 2.5 km (1.5 mi) wide (Fig. 32.8). Hawaiian calderas typically form through the collapse of the volcanic summit after magma in the underlying chamber drains laterally into fissures along the flanks of the volcano. The unsupported roof of the shield drops abruptly downward along *ring fractures*, creating the steep-walled caldera basin. The calderas atop Mauna Loa and a comparable one at the summit of Kilauea form in a slightly different way than the much larger calderas at Crater Lake and Yellowstone national parks. In those higher silica settings, the calderas collapsed downward after their magma chambers were emptied by colossal eruptions.

Calderas are created in a different manner than *craters*, which tend to be smaller and form by the outward expulsion of lava or pyroclastic debris during eruptions. The summit caldera of Mauna Loa is oriented northeast–southwest along the axis of the edifice and is about 180 m (~600 ft) deep. The floor of the caldera is formed from the solidified remains of several historic lava flows.

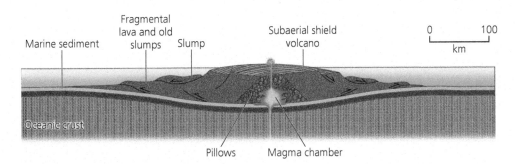

Figure 32.7 The mass of Mauna Loa depresses the underlying oceanic crust. When measured from the seafloor to the summit, the volcano is almost 9700 m (~32,000 ft) high. Island volcanoes begin their lives on the seabed as voluminous pillow lavas before growing high enough to breach the water surface as an island. Collapse and slumping of the volcano's margins counterbalance the upward growth.

Figure 32.8 A) Satellite image of Mauna Loa showing the summit caldera and two rift zones. The yellow line marks the border of the national park. B) Close-up of the caldera and recent lava flows.

Emanating outward from the Mauna Loa summit caldera are two rift zones—one to the northeast and one to the southwest. **Volcanic rift zones** are narrow areas on the flanks of volcanoes marked by highly fractured, planar zones of weakness. The near-vertical fractures are ideal

conduits for magma to reach the surface, resulting in **flank eruptions** that build the volcano laterally. Figure 32.8A shows abundant dark lava flows that erupted from vents along the length of the two rift zones. Many of the lava flows are tens of kilometers long, each adding mass to the surface of the elongate shield. Mauna Loa means "Long Mountain" in Native Hawaiian, which reflects its broad lateral extent along the two rift zones emanating outward from the caldera at the crest.

The oldest volcanic rocks on Mauna Loa are *pillow lavas* that erupted onto the seafloor about 600,000 to 1 m.y. ago. Since then, the volcano has gradually built upward, breaching sea level to become part of the Big Island about 300,000 years ago. Countless basaltic eruptions added to the mass of the volcano, including hundreds over the past 4000 years, which cover about 90 percent of the surface. Careful mapping and dating of the lava flows show that phases of eruptions from vents near the summit alternate with episodes of eruptions from fissures along the rift zones.

Mauna Loa is not currently erupting but is considered to be a highly active volcano that is almost certain to erupt in the future. Almost 40 eruptions have occurred since 1832, with the latest happening in 1984 when lava flows reached within 7 km (~4 mi) of Hilo along the eastern coast (Fig. 32.9). The eruption began near the

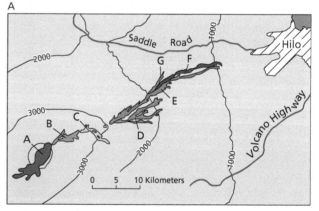

Figure 32.9 A) Map of seven distinct lava flows during the 1984 eruption of Mauna Loa. B) Fissure eruptions along the upper Northeast rift zone of Mauna Loa in 1984.

summit caldera, then migrated along the Northeast rift zone over just a few hours to days. Many of the lava flows originated as **fissure eruptions**, a type of flank eruption characterized by fountains of lava shooting upward from elongate cracks along the rift zone before spattering back to the surface. The chronology of events during this 20-day-long event suggests that magma beneath the summit drained laterally into fractures along the rift zone, which provided a pathway to the surface.

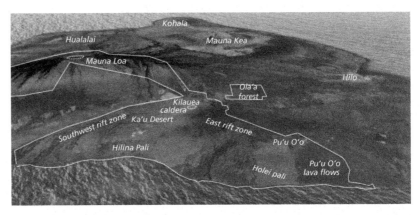

Figure 32.10 Low-angle oblique view looking northwest across the Big Island of Hawai'i. National park boundaries shown in yellow. The summit caldera of Kilauea is about 3000 m (~9800 ft) below the summit caldera of Mauna Loa.

Active Volcanoes within the National Park: Kilauea

Looking at a map of the Big Island of Hawai'i, you'd be hard-pressed to identify Kilauea as a shield volcano. It appears as sort of a shoulder on the southeastern margin of Mauna Loa, with the crest marked by an oval-shaped caldera (Fig. 32.10). Kilauea is not merely the active southeastern portion of Mauna Loa though but is rather a separate volcano with its own magma chamber. It rises about 1277 m (~4200 ft) above sea level, yet resides in the hulking shadow of Mauna Loa just to the north. The surface of the volcano slopes gently seaward from the crest.

The landscape of the Kilauea shield inside Hawai'i Volcanoes National Park is a diverse array of dark, youthful lava fields, craters, and elongate cliffs called *pali*. About 90 percent of the volcano's surface is blanketed with solidified lava flows less than 1000 years old, many of which are younger than 200 years. Along the eastern boundary of the park, slow-moving tongues of lava drain seaward from the Pu'u O'o vent. As the lava flows cool and solidify, they form the youngest rocks on the planet. The windward side of Kilauea is green and forested, while the leeward side on the west is arid, exemplified by the Ka'u Desert. Patches of ohi'a forest isolated by rugged lava fields are called **kipuka** and are havens for wildlife within the otherwise inhospitable terrain. Along the coast, lava flows have built broad benches above steep cliffs, which are constantly battered by waves.

Kilauea has a caldera at its summit, a little over 4 km long and 3 km wide (2.7 x 1.8 mi) and about 165 m (540 ft) deep (Fig. 32.11). Nestled within the caldera is the circular crater of Halema'uma'u, about a kilometer across and ~80 m (260 ft) deep. Inside the crater is an active **lava lake** whose surface fluctuates, occasionally spilling over the rim and spreading across the adjacent crater floor (Fig. 32.12).

Two rift zones extend outward from the caldera at the summit of Kilauea, broadly parallel to the two rift zones along Mauna Loa (Fig. 32.10). Global Positioning System (GPS) satellite measurements indicate that the rift zones on Kilauea are gigantic planes of weakness where the southeastern flank of Kilauea is sliding gravitationally toward the ocean. Extensional stress creates fractures in the narrow rift zones that are intruded by magma, further widening the zone of weakness within the volcanic edifice.

The Southwest rift zone cuts across the Ka'u Desert and has been relatively inactive in recent years. The East rift zone, however, is pocked by numerous craters, cinder cones, and narrow bands of fissures that have vented lava flows over the past several decades. One of these vents, Pu'u O'o, began erupting basaltic lava in early 1983 accompanied by fountains of lava over 100 m (330 ft) high (Fig. 32.13). **Lava fountains** are driven by the expansion of gas bubbles within the magma deep below. The cone surrounding the

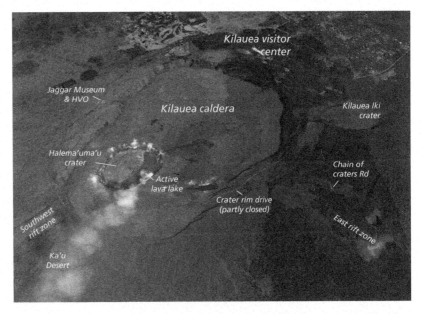

Figure 32.11 Oblique view of Kilauea caldera (~4 km long) and surrounding features. A plume of volcanic smog (vog) is blown downwind from the active lava lake in Halema'uma'u crater.

Figure 32.12 A lava lake ~250 m (800 ft) across filling the "Overlook crater" within the larger Halema'uma'u crater in 2018. A surficial crust floats above the molten lake, creating a mosaic pattern where it breaks apart. Gas-driven burps of lava spatter upward along the margins of the crustal skin. Sulfur dioxide gas mixes with steam to create a toxic volcanic smog (vog) that rises from the edge of the crater.

vent builds upward as the lava spatters back to the surface, accompanied by cinder, ash, and fragments of pumice (i.e., tephra). Currently, Pu'u O'o is an asymmetric **cinder-and-spatter cone** that is over 210 m (~700 ft) high. Over the years, eruption of lava shifted to other craters along the East rift zone but eventually moved back to Pu'u O'o where it continues today. A pond of lava commonly occupies the crater within the cinder cone, filling and draining as the magma supply within the fissures of the rift zone waxes and wanes.

The robust volcanic activity over the past 1000 years or so has covered most of the older flows that built Kilauea over the millennia. Some of the oldest rock is exposed in **pali**, the steep cliffs that cross the island somewhat parallel to the coastline (Fig. 32.10). In places, the pali may be >500 m (~1600 ft) high and partially draped with solidified lava that cascaded over the edge. Pali are the *fault scarps* of gigantic slumps where the volcanic edifice of Kilauea has slid toward the ocean in the past. The oldest basalts exposed in the pali erupted between 50,000 and 70,000 years ago. The very oldest rocks composing Kilauea, however, are found beneath the sea along the submarine flanks of the volcano. There, rocks are dated to have erupted between 210,000 and 280,000 years ago, marking the minimum age of Kilauea volcanism.

This brief summary of Kilauea's volcanic landscape provides just enough background to investigate the deeper controls on volcanism at Hawai'i Volcanoes National Park. The details of modern and historic eruptions from Kilauea will be addressed later in this chapter.

Hotspots and Mantle Plumes beneath the Big Island

The Hawaiian Islands are among the most isolated on Earth, with the nearest continents thousands of kilometers away. They are also isolated tectonically, with the nearest plate boundaries circumscribing the margins of the Pacific plate (Fig. 32.14). If you read the previous chapter on Yellowstone National Park, you understand the concept of *hotspots*, a stationary point of localized volcanism driven by *mantle plumes*, broad columns of hot, plastically flowing rock that originate deep in the mantle. Over 20 hotspots have been identified, many of which occur in the interior of tectonic plates. Yellowstone is a *continental hotspot*, with the magma interacting with silica-rich continental rock to produce explosive pyroclastic eruptions. In contrast, the Big Island of Hawai'i, specifically the region near Hawai'i Volcanoes National Park, is an *oceanic*

Figure 32.13 Lava fountains erupt from the vent of Pu'u O'o along the east rift zone on Kilauea, 1983.

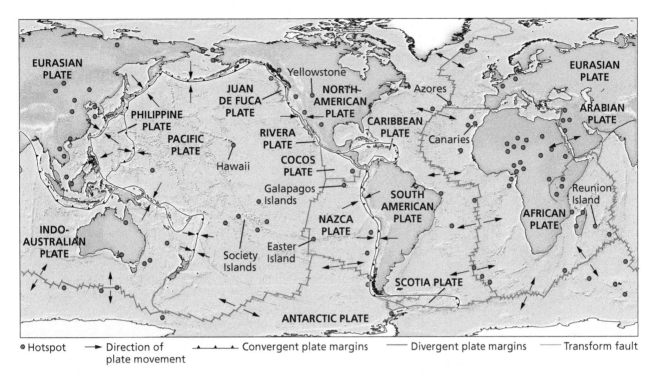

Figure 32.14 Global map of hotspots and plate boundaries. Note the isolation of the Hawaiian hotspot in the middle of the vast Pacific plate.

hotspot, with the basaltic magma derived directly from the low-silica mantle (Fig. 32.15). Eruptions from oceanic hotspots like those at Hawai'i tend to be effusive lava flows that produce broad shield volcanoes.

Mantle plumes are a physical expression of the movement of heat energy via convection as Earth's internal heat is transported toward the surface where it is released to space via volcanism (Chapter 1). You might imagine a plume as being comparable to a buoyant pillar of wax rising in a lava lamp, except on a planetary scale. The plume is not composed of magma but rather of very hot rock, moving at ultra-slow rates of cm/year as a dense plastic material. Evidence for a mantle plume directly beneath the southeastern portion of the Big Island of Hawai'i is provided by *seismic tomography*, a technique that analyzes the speed of seismic waves from earthquakes to create a visual image

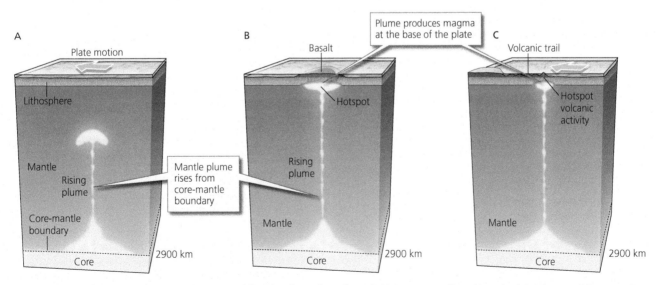

Figure 32.15 A) Mantle plumes are presumed to arise from the release of heat energy from the core into the overlying mantle. B) The immobile column of hot, plastic rock buoyantly moves upward through the mantle until it reaches the base of the lithosphere where it spreads out laterally. Here, the hot rock partially melts due to decompression to produce magma, which leaks upward to feed volcanism at the surface. C) Lateral motion of the lithospheric plate moves older volcanic islands away from the plume, allowing for the creation of a new volcanic island on the hotspot.

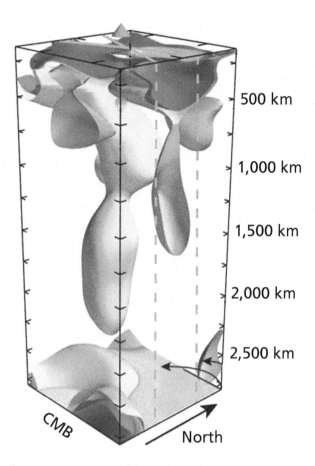

Figure 32.16 Tomographic image of the mantle plume beneath the Big Island of Hawai'i. The various colors represent bodies of rock with similar physical characteristics based on the speed of earthquake waves as they pass through the body of the Earth. The broad yellow column arises from near the core–mantle boundary (CMB) and supplies the heat energy for the Hawaiian hotspot.

of the deep Earth (Chapter 31). Recent tomographic imagery reveals that the mantle plume beneath Hawaii originates near the core–mantle boundary, 2900 km (1800 mi) beneath the surface (Fig. 32.16). The plume is calculated to be 650 km (~400 mi) across near its base but broadens as it rises upward.

As the plume nears the upper mantle and crust, the overlying weight of rock progressively decreases, causing the rock in the plume to decompress. The decrease in pressure allows the hot rock to partially melt, a style of magma generation called **decompression melting**. The magma rises upward through faults and fractures in the overlying crust to collect in mushy magma chambers beneath the active volcanoes within the national park (Fig. 32.17).

Evidence from tomographic studies and geochemical analysis of basaltic

rocks suggests that Kilauea is fed by a shallow reservoir of magma less than 2 km (1.2 mi) beneath Halema'uma'u crater. The lava lake within the crater is the surface expression of a conduit that connects the reservoir below with the surface. The level of the lake swells and recedes in tune with pressure from the underlying magma supply. Another, deeper chamber has been imaged about 2–4 km (1.2–2.4 mi) beneath the south rim of the Kilauea caldera. And another even deeper chamber of "magma mush" has been discovered at depths of 8–11 km (~5–7 mi) beneath the East rift zone of Kilauea. This is likely the magma supply that has been leaking upward through fractures to feed the near-continuous eruptions along the East rift zone since 1983. The magmatic plumbing system beneath Kilauea is clearly very complex, but it's likely that magma moves back and forth between the reservoirs underlying the caldera and reservoirs beneath the rift zones.

Hawaiian-Emperor Hotspot Trail

The hotspot model explains how the active volcanoes of Mauna Loa and Kilauea were formed. But how were the rest of the volcanic islands in the Hawaiian chain created? One clue comes from the ages of basalts on each of the islands (Fig. 32.18). On the Big Island of Hawai'i, the volcanoes on the northern part of the island are older than the active volcanoes of the national park to the southeast. Basalts on the next island in the chain, Maui, range in age from about 800 years to 1.3 m.y., slightly older than those composing the Big Island. The volcanic rocks of Moloka'i have average ages between 1.3 and 1.8 m.y, and the rocks of O'ahu are even older. On Kaua'i, at the northwestern end of the chain, the dates of basalts range from 3.8 to 5.6 m.y. So the ages of basaltic lavas on the main Hawaiian Islands increase with distance from the stationary hotspot.

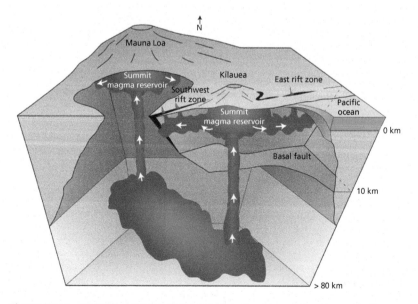

Figure 32.17 Simplified model of the magma plumbing system beneath Mauna Loa and Kilauea volcanoes. The magma consists of a "mush" of mostly rock infused with relatively small amounts of liquid melt.

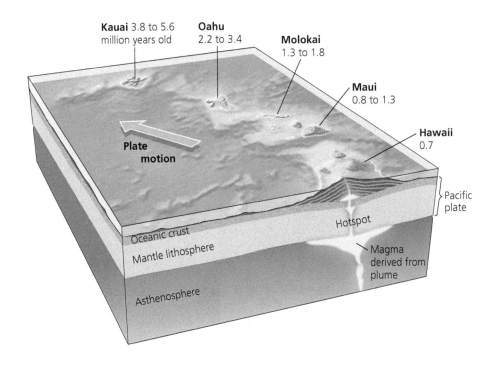

Figure 32.18 The age of basaltic lavas on the Hawaiian Islands increases with distance from the hotspot. Each island formed when they were located above the stationary hotspot. They were then transported to the northwest as part of the slowly moving Pacific lithospheric plate.

This age progression reflects the slow movement of the Pacific lithospheric plate toward the northwest, moving each island away from its origin at the fixed hotspot above the immobile mantle plume, somewhat like a conveyor belt on an assembly line. The hotspot generates a near-continuous magma supply, triggering volcanism which created the elongate submarine ridge that forms the foundation beneath the individual Hawaiian Islands. The Hawaiian Islands are the tail end of a much longer *hotspot trail* that extends across the northwest Pacific seafloor.

The chain of coalesced shield volcanoes that compose the main Hawaiian Islands continues to the northwest as an archipelago of small islands, seamounts, and atolls. A *seamount* is an underwater mountain that commonly forms as extinct volcanoes subside beneath the sea through time. An **atoll** is a ring-shaped array of low-lying islands that encircle a deep lagoon. They form from the upward growth of coral reefs that nucleate as a halo around the outer margins of subsiding volcanic seamounts. This northwest extension of the Hawaiian chain is part of **Papahanaumokuakea Marine National Monument**, established in 2006. The boundaries of the national monument extend for 1900 km (>1200 mi) from Kaua'i to Midway Atoll to the northwest.

As you might expect, the ages of basalts on each of the islands and seamounts of the northwest extension of the Hawaiian Islands increase progressively with distance from the hotspot beneath the Big Island (Fig. 32.19). The basalts that form the volcanic core of Midway Atoll are dated to be 28 m.y. old. Slide the Pacific plate back toward the southeast so that Midway was over the hotspot 28 m.y.

ago. At the time, it was likely a shield volcano comparable in size to the other Hawaiian Islands. Over its 28 m.y. of motion to the northwest, Midway's volcanoes became extinct, slowly eroding and subsiding until nothing was left above water except a ring of coralline islands.

The Hawaiian seamount chain continues to the northwest beyond Midway Atoll and then shifts orientation to a more northerly trend as the Emperor seamount chain. The entire length of the **Hawaiian-Emperor seamount chain** is over 6000 km (~3700 mi) and consists of about 130 islands, seamounts, and atolls. You might think of the Hawaiian-Emperor seamount chain as an alignment of "ancient Hawai'is" because each island and seamount once occupied the position where the Big Island of Hawai'i is today. The Meiji Seamount at the very northern end of the Emperor chain is dated to be 81 m.y. in age. Its eventual fate is to be consumed within the Aleutian subduction zone in the northwestern corner of the Pacific. Notably, 81 m.y. ago, the Meiji Seamount was an island composed of shield volcanoes positioned above the Hawaiian hotspot (now occupied by the Big Island). Any older islands that may have formed a hotspot trail northwest of Meiji are now located somewhere deep beneath the Aleutian Island chain, subducted as part of the northern Pacific plate.

The Hawaiian-Emperor hotspot trail marks the path of the Pacific plate over the last 81 m.y. The change in direction of plate motion occurred at the "elbow" about 43 m.y. ago when the movement of the Pacific plate changed from northerly (~43–81 m.y. ago) to northwesterly (the last 43 m.y. or so). This change in absolute motion of the Pacific plate around 43 Ma coincides in time with the initial convergence of the Indian plate with South Asia that culminated in the uplift of the Himalayas. This massive tectonic collision may have triggered a global reorganization of plate motions during the mid-Cenozoic.

The ages and distances of the islands and seamounts along the Hawaiian-Emperor hotspot trail are invaluable for calculating the absolute direction and rate of Pacific plate motion over the past 81 m.y. The current rate of motion of the Pacific plate as measured by the GPS of satellites is ~10 cm/yr (~4 in/yr). The long-term average over the past 81 m.y. or so, calculated from the length of the hotspot trail and age dating of volcanic rocks is 8.5

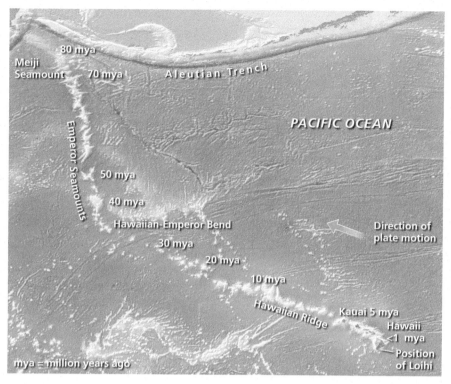

Figure 32.19 The Hawaiian-Emperor Seamount Chain extends as a hotspot trail for over 6000 km (~3700 mi) to the northwest. The ages of volcanic rock composing each seamount become progressively older with distance from the hotspot.

cm/yr (~3.3 in/yr). These are extremely slow annual rates, but they translate into enormous distances of travel over geologic time spans.

Volcanic islands evolve as they move off the hotspot as part of the Pacific lithospheric plate (Fig. 32.20). As the source of heat energy slowly recedes with plate motion, volcanism ceases and erosion begins to dominate. Islands may lose mass during catastrophic landslides when immense coastal slices of land collapse into the adjacent seas. The seafloor around the Hawaiian Islands is strewn with the debris from ancient landslides. Moreover, without the underlying heat energy to buoy the island upward, the island and the portion of the lithospheric plate on which it resides will cool and become denser. This causes the island to slowly subside with distance from the hotspot.

The main Hawaiian Islands show this progression, with each island to the northwest having a slightly lower profile and a greater degree of erosional degradation. The deeply incised canyons and steep, fluted cliffs that characterize the island of Kaua'i at the northwest tip of the chain exemplify this island-to-island landscape evolution. So not only do the Hawaiian Islands increase in age with distance from the hotspot, they also exhibit an increase in subsidence as well as in their degree of erosion. These

trends continue along the entirety of the Hawaiian-Emperor seamount chain.

Take a moment to contemplate the scale of the Hawaiian-Emperor chain of islands, seamounts, and atolls. Its 6000 km length is about the distance from San Francisco to Bogota or from New York to Paris. Imagine if the Pacific was somehow drained, exposing the linear array of old volcanoes to view. It would be the weirdest mountain range on Earth, with one huge peak dropping down to a valley followed by a climb to another huge peak, one after another. And the peaks would progressively decrease in elevation with distance from the gigantic mountain of Mauna Loa, adding to the strangeness.

What happens in the future when the Big Island of Hawai'i moves northwest away from the hotspot? Will a new island emerge from the seafloor? It's already happening about 35 km (22 mi) off the southeastern coast where an active submarine volcano called **Lo'ihi** is building upward (Figs. 32.4). The seamount rises about 3 km (1.9 mi) above the seafloor (taller than Mount St. Helens in the Cascades) and is within about a kilometer of sea level. It erupts basaltic *pillow lava*, which forms through the interaction of molten rock and seawater. At its current rate of growth, Lo'ihi will emerge as an island in several tens of thousands of years. Because of its proximity, it may

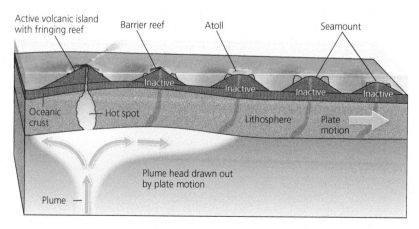

Figure 32.20 Evolution of volcanic islands formed above a hotspot. As the lithospheric plate moves laterally above the stationary hotspot, the islands progressively cool and contract. Erosion dominates, lowering the profile of the islands with distance from the hotspot. In time, the degraded islands subside beneath the waves, becoming seamounts.

merge with the edifice of Kilauea, adding to the overall size of the Big Island.

Historic and Modern Volcanism at Hawai'i Volcanoes National Park

Recent volcanic activity at the national park has been concentrated at two areas on Kilauea Volcano (Fig. 32.21). Since 1983, effusive eruptions have been occurring along the East rift zone, particularly near the Pu'u O'o cinder-and-spatter cone. Since 2008, activity has centered on the lava lake within Halema'uma'u crater, itself located within the Kilauea caldera. We'll use these two dynamic areas to illustrate some of the volcanic features and processes currently active in the national park. Later in this chapter, significant historic eruptions will be described that exemplify the range of volcanic activity at Kilauea. It's no wonder that the Native Hawaiian meaning of Kilauea is "much spewing."

LAVA FLOWS FROM THE EAST RIFT ZONE

The broad gray area in Figure. 32.21 marked "1983–2000s lava fields" is the region blanketed by eruptions from the East rift zone, primarily from Pu'u O'o. For scale, the Pu'u

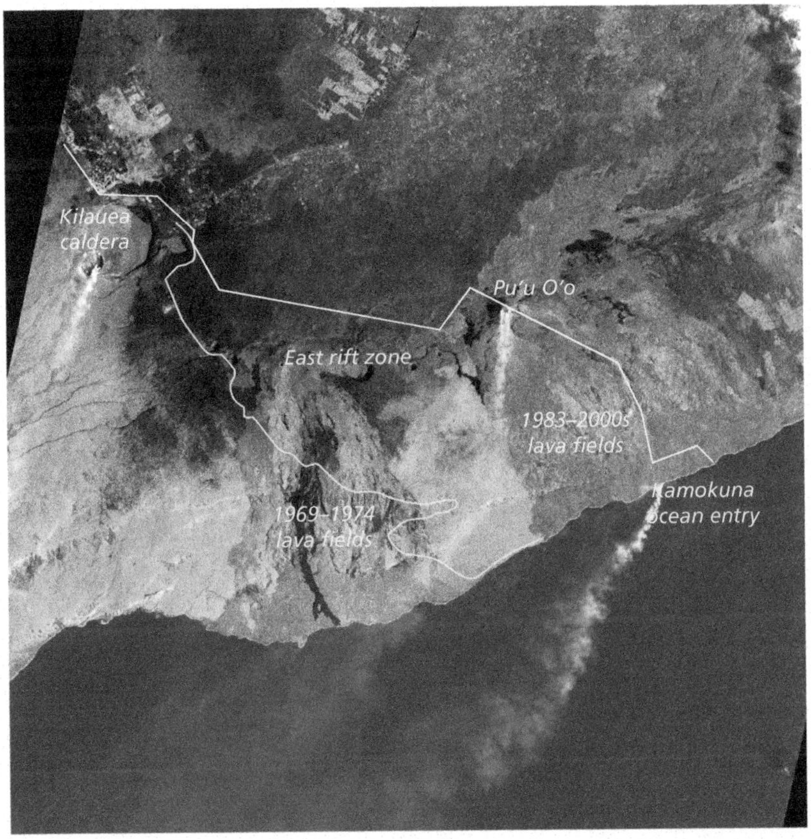

Figure 32.21 Natural color satellite image acquired November 2016 showing the active vent near Pu'u O'o as well as the lava lake within Kilauea caldera. Plumes of volcanic smog are blown downwind from each of the vents. A third plume marks the location of the Kamokuna ocean entry where active flows from Pu'u O'o meet the sea. Park boundary shown in yellow. Chain of Craters Road shown as the white line.

O'o vent is about 9.5 km (~6 mi) from the coast. Between 1983 and 2016, eruptions from the East rift zone emitted over 4 km³ (~1.1 mi³) of lava. A map of the most recent flows over the older solidified lava field is shown in Figure 32.22. It may come as a surprise to learn that most lava doesn't travel through channels on the surface of the lava field but rather flows just beneath the surface in lava tubes.

Lava tubes form as the upper surface of a flowing stream of lava loses heat due to contact with the much cooler air and then solidifies (Fig. 32.23). This surficial crust acts to insulate the fluid beneath, keeping it hot enough to continue streaming through the newly formed tube. Over time, the upper roof of the tube may grow thick enough so that lava flushes through the natural conduit for as long as the supply of lava continues. On occasion, a portion of the roof may collapse and fall into the lava beneath where it is either re-melted or carried along with the flow. The hole left behind is called a **skylight**, a window into the flowing lava below (Fig. 32.24). When the lava supply is diverted from the tube, or when the lava supply runs out, the lava in the tube drains downslope, leaving the tube as a tunnel-like cave (Fig. 32.25).

Most lava flows on Kilauea occur just beneath the surface within lava tubes rather than as open channels cutting across the lava fields of older flows. Lava tubes are very efficient systems for transporting lava over long distances and contribute to the lateral growth of shield volcanoes. Their insulating effect permits lava to remain as a liquid much longer than it otherwise would if flowing across the surface, exposed to the atmosphere. The extensive cave system of lava tubes on Kilauea exhibits exotic ornamentation such as "lavacicles," formed as remnant lava dribbles from a ceiling, cooling and solidifying mid-drip to form a downward-hanging pendant. Many lava tubes provide habitat for particular plants and animals, as well as odd species of microbes. Ancient Hawaiians used the caves as part of their culture, leaving behind burial chambers, rock art carved into walls, and various artifacts. To many modern Native Hawaiians, those lava tubes used by ancestors have a sacred significance.

The solidified basaltic lava erupted from the East rift zone (and from Kilauea as a whole) takes two forms that primarily reflect the temperature and thus the viscosity of the flow. Each type has the same chemical composition but differs in appearance. **Pahoehoe** (pronounced "*pa-hoy-hoy*") lava has a smooth, hummocky, or ropy surface and is the dominant type of flow

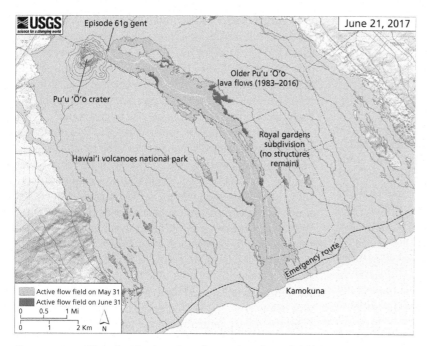

Figure 32.22 Map showing location of recent lava flows (pink) from Pu'u O'o to the ocean at Kamokuna. The gray region represents >30 years of older flows, while the blue lines are descent paths of flows to the ocean. The yellow line (within the pink) is the trace of an active lava tube through which much of the lava flows.

Figure 32.23 Lava tubes begin as the upper surface of lava flows forms a thin crust while in contact with the much cooler atmosphere. The crust keeps heat from escaping the interior of the flow so that the lava remains a runny liquid, moving within a tube just beneath the surface.

from the East rift zone (Fig. 32.26). It is the hotter of the two types, flowing at temperatures of ~1100 °C (~2000°F). The extreme heat energy contained within pahoehoe flows lowers its viscosity, creating a very runny fluid that may travel for tens of kilometers before spreading laterally as a thin lobe. Pahoehoe flows commonly form lava tubes that extend their distance of travel downslope, building

the shield outward. Much of the young, upper surface of Kilauea is covered by solidified pahoehoe lava that flowed through lava tubes.

The hot upper surface of many pahoehoe flows may rapidly chill while in contact with the much cooler air, creating a paper-thin shell of iridescent glass. Another very common feature of pahoehoe lava fields are **tumuli**, elliptical or domal mounds that rise a meter or more above the surface that mark places where the brittle crust of basalt was inflated upward by the motion of underlying gaseous lava (Fig. 32.27). These rounded blisters often have a broad fracture running along the long axis of the mound, comparable to the elongate notch along the top of a loaf of baked bread.

The second type of basaltic lava at Kilauea is called **a'a** (pronounced "*ah-ah*"). A'a flows have an extremely rough, rubbly surface of angular blocks of solidified lava carried along by a pasty interior (Fig. 32.28). Flowing at temperatures of ~1000°C (~1800°F), slightly lower than pahoehoe, a'a flows have a higher viscosity and thus tend to flow for shorter distances. A pahoehoe flow may evolve into an a'a flow as the lava loses heat energy and internal gases, increasing its viscosity. The jagged, irregular surface of a'a flows makes for difficult hiking. Contrary to what your intuition tells you, the derivation of the term *a'a* does not necessarily reflect the sound someone might make while walking barefoot across its surface; rather, it has many different meanings in Native Hawaiian, including "stony" and "fire."

Basalt flows like those from the East rift zone can travel up to 10 km/hr (6 mph) where the slopes are steep, but they slow to about 1 km/hr (about a foot per second) where the slopes are gentle. Some flows have been clocked at >30 km/hr (~19 mph) where they are confined within channels or lava tubes. (If you're wondering, the average healthy human can run 24 km/hr (~15 mph) for short distances. We also have the awareness that we must not get caught at the front of an advancing lava flow!)

Recent flows from Pu'u O'o have reached the coast at Kamokuna where the convergence of hot lava and cold seawater creates a hypnotic spectacle called an "ocean

Figure 32.24 A skylight is created when a portion of the roof of a lava tube collapses, allowing a glimpse into the streaming lava beneath.

Figure 32.26 Pahoehoe lava flow forming the characteristic ropy texture as the hot fluid cools and solidifies.

entry" (Fig. 32.29). The lava makes a hissing sound as it is instantaneously quenched by the water, sending up billowing clouds of gas and steam as the lava turns into rock. This violent interaction between water and lava is how volcanic islands like the Big Island build outward. Since the modern eruptions began in 1983, several hundred acres of new land have been added along the southeast coast. Over the much longer term, the entire Big Island formed through this process.

This amazing display of geology in action naturally draws visitors to the area, from both the land and the sea. The National Park Service strongly advises caution, however, because of the numerous hazards associated with ocean entries. Hot debris can fly off in any direction due to the explosive interaction of hot lava with seawater. The coastal bench formed above ocean entries can be unstable as it is built above unconsolidated fragments of solidified

Figure 32.27 Tumulus formed on a pahoehoe lava flow on Kilauea. The axial fracture forms as the brittle crust flexes upward from the pressure of underlying lava. In this example, lava has oozed up within the crack.

lava. People have died when the superficially stable coastal bench on which they were standing suddenly collapsed. Another hazard is the acidic plume of gas and steam, charged with fine particles of glassy ash, given off as the lava is abruptly quenched.

BIRTH OF THE LAVA LAKE IN HALEMA'UMA'U CRATER

As of 2018, the active volcanism near the summit of Kilauea occurs within the oval-shaped lava lake tucked inside Halema'uma'u crater (Fig. 32.12). Only a few lava lakes exist on Earth, and the one at Kilauea is among the largest; today it is about 250 m (~800 ft) across and several tens of meters deep. U.S. Geological Survey volcanologists at the Hawaiian Volcano Observatory (HVO) perched on the rim of Kilauea caldera were able to document the birth of the

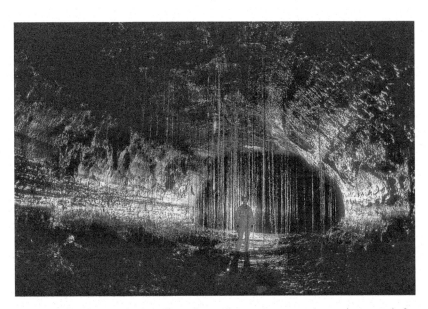

Figure 32.25 Formed by lava flows from Kilauea, Kazumura lava tube extends for over 65 km (40 mi), the world's longest mapped lava tube. The grooves along the walls were scoured into the older basalt by lava flowing through the tube about 600 years ago. Tree roots hang from the ceiling.

Figure 32.28 Rubbly a'a lava flowing across a surface of older pahoehoe. Note how the interior of the a'a flow is a glowing fluid that carries along the blocky front and upper surface.

Figure 32.29 Pahoehoe lava streams from lava tubes and surface flows into the ocean at Kamokuna, sending up plumes of steam. The lava will immediately solidify, creating new land along the coastline.

lava lake as it happened. This is a rare event and seldom occurs in view of humans.

The lake formed in 2008 when a *fumarole* began emitting steam and gases near the south wall of Halema'uma'u crater. The plume of gases was high in sulfur dioxide and posed a danger to tourists, forcing the closure of the western portion of Crater Rim Drive (Fig. 32.3). Within a month, a small explosive eruption accompanied the collapse of the fumarolic area of the crater floor, blowing out angular blocks of old rock. What was not known at the time was that a conduit was opening beneath the crater through which magma was rising toward the surface from the chamber about 2 km (1.2 mi) below. The explosive eruption of old rock was the gas-driven expression of the conduit "clearing its throat." Although mostly obscured by a dense plume of gas and steam, a bubbling pond of lava developed over the next few months. The only real evidence of the birth of the lava pond was the orange glow from the area at night.

Through 2009, the level of the lava pond rose and fell, occasionally spilling over the rim to spread out across the floor of Halema'uma'u crater. At times, the entire pond would drain back into its underlying conduit, leaving a steaming, barren crater floor. By 2010, the lava pond had grown into a full-blown lava lake, with a surficial skin of black, floating crusts of basalt. Gas-driven burps of lava spattered along the incandescent margins of the crust. The crater containing the lava lake grew larger as portions of the overhanging rim would collapse downward into the lake, sending up surges of molten rock. Over the years, the level of the lake inflates and deflates in response to changes in gas pressure within its underlying plumbing system. During occasional inflationary events, the lava overtops the rim of its crater and spills out across the floor of Halema'uma'u, laying down a thin sheet of fresh basalt.

A visit to the overlook at the Jaggar Museum at dusk is a must for visitors to the park to view the eerie orange glow against the darkening sky. A closer overlook above the lava lake off Crater Rim Drive remains closed. If you can't get to the park itself, you can view the lake activity on your computer via a USGS Hawaiian Volcano Observatory webcam.

HISTORIC EXPLOSIVE VOLCANISM AT KILAUEA CALDERA

Even though the ongoing effusive volcanism at Kilauea consists primarily of lava flows emanating from vents along the East rift zone and from the lava lake within Halema'uma'u crater, past episodes of activity at Kilauea have been explosive. They're not energetically equivalent to the pyroclastic eruptions that characterize the Cascades or Yellowstone, but they're dangerous nevertheless. The record of explosive eruptions is revealed in layers of tephra that crop out in various locations on Kilauea (Fig. 32.30). Age dating of mineral crystals within the volcanic ash, pumice, and rock fragments that compose the tephra establish the chronology of explosive events. Below are two accounts of these episodes from the historical recent past that reveal the magnitude and danger associated with explosive eruptions at Kilauea.

The modern caldera at the summit of Kilauea likely formed around 1470 CE after a phase of large effusive eruptions emptied the magma chamber, triggering collapse of the summit into the underlying void. Native Hawaiian oral histories substantiate what must have been a truly dramatic event. Over the next few decades, the depleted column of magma

Explosive eruptions can occur when

1) Magma column drops below water table
2) Groundwater interacts with hot rock
3) Steam pressure builds then explodes.

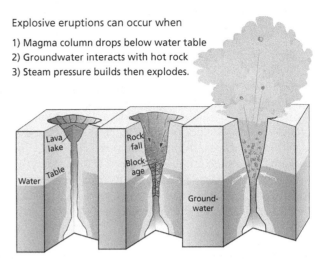

Figure 32.31 The volatile mixture of groundwater and magma triggers phreatic explosions of steam and pyroclastic debris.

Figure 32.30 Layers of tephra from a prolonged phase of explosive eruptions from Kilauea between ~1500 and 1800 CE. Multiple individual events accumulated to create this 11 m (36 ft) thick deposit.

Figure 32.32 Visitors pose in front of the eruption plume from Kilauea caldera in 1924 before being warned of the potential danger.

beneath the caldera floor fell low enough so that groundwater leaked into the rubble-filled conduit (Fig. 32.31). In time, the water mixed with the 1200°C magma and flashed to steam, which provided the pressure for explosive *phreatic eruptions* of ash, pumice, fragments of rock, and steam. Over the next few centuries, phreatic eruptions from the caldera laid down an 11 m (36 ft) thick layer of tephra (Fig. 32.30). In 1790, one of these explosive eruptions released a ground-hugging *pyroclastic flow* down the western flank of Kilauea that scorched and suffocated a large number of Native Hawaiians who were in the Ka'u Desert when the turbulent cloud enveloped them. Footprints preserved in the solidified ash from this eruption likely record that terrifying day.

A second series of explosive eruptions occurred in 1924 (Fig. 32.32). The event began with the sudden draining of a lava lake from Halema'uma'u crater. Soon after, an intense earthquake swarm far to the east suggests that the magma beneath the crater flowed tens of kilometers laterally through fractures along the East rift zone. The crater floor of Halema'uma'u collapsed into the depleted magma chamber over the next few days, with the upper level of the magma dropping below the water table. Prior to collapse, outward pressure from the magma-filled conduit kept groundwater from leaking in. But with the conduit now filled with the debris from the collapsed crater floor, water infiltrated into the conduit and mixed with magma. The rapid accumulation of

steam pressure provided the energy for a series of phreatic eruptions that lasted 17 days. Some of the explosion clouds reached an estimated 9 km (5.6 mi) in altitude, and wet ash fell across the southern part of the Big Island. During some of the explosions, blocks of rock weighing as much as 14 tons were thrown from the crater. One person was killed by falling debris during the event. After the explosions dissipated, Halema'uma'u crater had more than doubled in size.

It's inevitable that another explosive event of this magnitude will occur near the Kilauea caldera in the future. The spark for the event would occur if the conduit that connects the current lava lake to the underlying magma chamber begins to drain. Earthquake swarms along the flanks of Kilauea would indicate that magma is migrating laterally away from the summit reservoirs. The next step would be for groundwater to seep into the depleted conduit, triggering the volatile interaction of water and magma. These warning signs will hopefully occur with enough time for the National Park Service, in coordination with the U.S. Geological Survey and local decision makers, to prepare for the explosive event.

The current effusive style of volcanism at Kilauea began about 200 years ago, driven by a high supply of magma from the underlying reservoirs, which in turn were derived from the hotspot deep below. Prior to this phase, Kilauea experienced an explosive interval that lasted about 300 years (~1500–~1800 CE), as recorded in the tephra deposits along the margins of the caldera. The pyroclastic eruption of 1790 occurred near the end of this phase. At some point in the future, as the magma supply beneath the volcano diminishes in tune with the underlying hotspot, the volcano will return to an explosive phase and Kilauea will take on a more dangerous personality, significantly different from its current designation as a "safe" volcano.

HISTORIC EFFUSIVE VOLCANISM AT KILAUEA IKI

Kilauea Iki ("Little Kilauea") is a well-defined crater located to the east of the main Kilauea caldera (Fig. 32.11). The view from the overlook at Kilauea Iki today reveals the steep walls of the crater rising 120 m (400 ft) above the black basaltic floor of the crater (Fig. 32.33A). A reddish spatter-and-cinder cone named Pu'u Pua'i rises at the far side of the crater. This placid scene was very different in 1959, when a magnificent and terrifying eruption occurred within the crater.

The eruption was preceded by a series of earthquakes that progressively increased in frequency and intensity over a few months, indicating the upward movement of magma beneath the crater. The event began in November 1959, when a fissure eruption opened along the south edge of the crater. Streams of lava flowed down the wall as a molten cascade, ponding on the crater floor. Over a few days, activity migrated to a lava fountain near what is today the base of Pu'u Pua'i ("gushing hill"). The height of the fountain fluctuated over the month-long duration of the event, reaching a maximum of 580 m (1900 ft). Lava spatter and cinders from the fountain built upward to form Pu'u Pua'i, while a lava lake grew to a maximum depth of 126 m (413 ft) within the crater (Fig. 32.33B). Over the years, the lava lake partially drained back into the vent, with the remainder solidifying to form the basalt floor of the crater. As you hike the moderate-to-challenging Kilauea Iki trail across what was once a lake of lava, look around and try to imagine the hellish scene of a few decades earlier.

Natural Hazards at Hawai'i Volcanoes National Park

A variety of *volcanic hazards* exist within the confines of Hawai'i Volcanoes National Park, including the inundation of park structures and nearby communities by lava flows, explosive phreatic and pyroclastic eruptions, collapse of thin crusts above lava tubes, collapse of coastal benches, forest fires set by encroaching lava, and airborne hazards such as ashfalls and volcanic smog. Between 1983

A

B

Figure 32.33   A) View into Kilauea Iki today, with the gas plume from Halema'uma'u's lava lake in the background. The reddish cone on the middle left is Pu'u Pua'i, built from the spatter and cinder erupted from the vent at the lower right during the 1959 eruption. Note people on the trail for scale. B) A lava lake filled the crater of Kilauea Iki in 1959, solidifying to become the solid floor of the crater today.

and 2017, lava flows destroyed almost 200 houses just outside the boundaries of the park, as well as a visitor center in the park. Roads have been covered in lava, including the coastal stretch of Chain of Craters Road. In early 2017, a total of 22 acres of a newly formed shelf composed of freshly solidified lava collapsed into the ocean without warning. (Five visitors who disregarded warning signs and rope barriers to visit the shelf fortuitously left about 15 minutes before the collapse.)

A serious volcanic hazard is the noxious plume given off by the lava lake within Halema'uma'u crater. Sulfur dioxide-rich gases in the plume react with the atmosphere and sunlight to produce volcanic smog known as **vog** (Fig. 32.12). The sulfur dioxide originates as a dissolved gas within the magma chamber, along with several other gases. As the magma rises upward, the pressure decreases, causing the gases to form bubbles that escape to the atmosphere from the surface of the lava lake. The plumes emanating from the lake are primarily composed of water vapor, carbon dioxide, and sulfur dioxide, as well as small amounts of elements like mercury and lead. The sulfur dioxide reacts with water vapor to form tiny droplets of sulfuric acid that in turn combine with minute solid particles to form vog. The near-constant emission of vog creates a variety of deleterious effects on humans and the environment. The U.S. Geological Survey, the Park Service, and health professionals work to educate and protect the public from this airborne natural hazard.

Hawai'i Volcanoes National Park is subject to many other hazards such as earthquakes, landslides, and *tsunami*. The movement of magma through the natural plumbing system beneath Mauna Loa and Kilauea triggers thousands of low-magnitude earthquakes every year. They don't cause much damage, but they are invaluable for tracking the location of magmatic activity beneath the surface, which aids in predicting volcanic events. Other earthquakes near the park are tectonic in origin, occurring along faults that cut through the volcanic edifice of both active volcanoes. Two of the largest earthquakes in recent history were the 1868 magnitude 7.9 quake and the 1975 magnitude 7.4 quake.

Groundshaking from the 1868 quake was severe, and the southern flanks of Mauna Loa and Kilauea slumped seaward during the event. Large parts of the coastline of Kilauea sank into the sea, displacing seawater and generating a tsunami. Wave heights from the tsunami were reported to be as high as 12–15 m (>40 ft) along the west coast of the Big Island. Coastal villages were inundated, causing 46 deaths. The enormous quake also triggered widespread landslides, one of which killed 31 people along the flanks of Mauna Loa.

The 1975 earthquake occurred along a 60 km (40 mi) section of the Hilina fault system that stretches along the southeastern flank of Kilauea. The *normal faults* composing the system are expressed on the landscape as the Hilina and Holei pali, steep cliffs along the southern flank of Kilauea that face seaward (Figs. 32.10, 32.34). The pali are the exposed fault scarps of a gigantic slump block that reaches as deep as 9 km (~6 mi) into the volcano and outward for 40 km (25 mi), emerging along the submarine

Figure 32.34 The Holei Pali is a fault scarp up to 400 m (~1300 ft) high. It is part of the Hilina fault system that marks the landward edge of an enormous slump composing much of the southeastern margin of Kilauea. During earthquakes, the lower portion on the right may drop several meters. Elongate fingers of pahoehoe lava drape the escarpment.

flanks of Kilauea. It is the largest active slump block of any oceanic island on Earth and GPS measurements indicate that the entire mass is sliding seaward about 10 cm/yr (4 in/yr).

During the 1975 quake, a large volume of the slump slid about 3 m (11 ft) downward and about 8 m (26 ft) seaward. The tsunami generated by the displacement of seawater by the slump block reached a maximum height of 14 m (47 ft). Miraculously, only two people lost their lives. The possibility exists that the entire southern flank of Kilauea will catastrophically detach along the Hilina fault system and collapse into the adjacent seafloor. All of the Hawaiian Islands show evidence of these types of *debris avalanches* in their history, and there's no reason to think that the actively growing southern flank of Kilauea won't break away at some point in the future. The event would trigger a "mega-tsunami" that would devastate the entire Pacific Margin, let alone the coastline of the Hawaiian Islands. The *recurrence interval* between these types of disasters is on the order of hundreds of thousands of years.

The U.S. Geological Survey maintains the Hawaiian Volcano Observatory, which monitors the active volcanoes in the national park with a wide array of instrumentation, including seismometers that record the tiniest of tremors, tiltmeters that measure deformation of the land surface, spectrometers that calculate gas concentrations, and an armada of other devices and techniques. Satellites are used to assess changes in elevation of the land surface, thermal signatures of heat flow from the volcanoes, and visual changes in the landscape. Volcanologists at HVO also perform the fundamental fieldwork required to ground-truth the measurements from all the sophisticated technology.

Differences between Hawai'i Volcanoes and Other National Parks

Most people tend to think of geologic processes as happening at extremely slow rates, taking millions of years to affect any significant change. That's certainly true in most of our national parks, and it's the long periods of geologic time that are required for their stunning landscapes. But at Hawai'i Volcanoes National Park, geology happens at the human scale of decades to years. Go to the other parks to stir your imagination about former worlds preserved in the rocks and the landscape, but go to Hawai'i Volcanoes to witness the "geology of today."

Everything about Hawai'i Volcanoes National Park is young. The caldera at Kilauea's summit that looks so timeless and ancient was only formed in 1470 CE. The horrific pyroclastic eruption that killed several Native Hawaiians occurred in 1790, while a less explosive eruption sent a column of steam and ash 9 km (5.6 mi) skyward in 1924. The moonscape that is the floor of Kilauea Iki was formed in 1959 by lava from a fissure eruption and an enormous lava fountain. Lava flows from Mauna Loa threatened Hilo in 1984. Massive earthquakes rocked the park in 1868 and 1975. All of these historic events occurred "yesterday" in geologic time. In addition to all these recent geologic events are the ongoing lava flows from Pu'u O'o and the surreal lava lake in Halema'uma'u crater that looks like something out of *Jurassic Park*. A visit to Hawai'i Volcanoes National Park is truly a unique geologic experience relative to what you see and do at other national parks.

Visiting Hawai'i Volcanoes National Park

The Big Island of Hawai'i is a true paradise, with perfect weather, stunning scenery, and a relaxed culture. Visiting the national park should be on everyone's bucket list. Drive the 30 km (19 mi) long Chain of Craters Road to its end where lava flows have buried it. You'll drop more than 1100 m (~3700 ft) from near the summit to the windy coastline, and you'll see several craters, cinder cones, a'a, and pahoehoe lava fields from the 1969–1974 eruptions, tracts of kipuka forest, and steep pali covered with bizarre forms of lava. Drive the open portion of Crater Rim Drive to the Jaggar Museum to witness the orange glow from the lava lake in Halema'uma'u crater. Along the way, stop at the Kilauea Iki overlook, sniff the acrid gases from Sulfur Banks, and explore the easy path through Thurston lava tube.

The best way to experience Hawai'i Volcanoes National Park, however, is to hike the trails to fully appreciate the landscape. If you decide to cross lava fields, do so judiciously with adequate footwear (no flip-flops!), a hat and sunscreen, more than enough water, and an awareness of your own fitness. If you hike at night to see the ocean entry, be sure to bring along a flashlight or headlamp.

Even though Hawaiian lava flows are considered to be among the safest to visit in the world, lava is, for lack of a better word, hot. Lava viewing opportunities in the park are constantly changing, so check with the Visitor Center for information and plan your visit with sufficient forethought. Be sure to respect all warning signs and rope barriers—they're there for a reason. All that said, make plans to visit the planet's most active volcano.

Postscript: Renewed Eruptions at Hawai'i Volcanoes National Park

In early May 2018, a fissure eruption occurred in a residential subdivision in the lower Puna district, located within the southeast corner of the Big Island outside of the national park (Fig. 32.3). The event was preceded a week earlier by increased seismicity along the East rift zone and the overflow of the lava lake onto the floor of Halema'uma'u crater near the summit. The level of the lava lake then began to drop, suggesting that the magma chamber was draining laterally through the East rift system. By the end of the month, 24 fissures were active within neighborhoods, accompanied by fountaining, spatter, lava bombs, lava flows, vog, and earthquakes, with magnitudes up to 6.9. Over 2000 residents were evacuated.

By midsummer 2018, about 35 km² (14 mi²) were covered with lava flows, while about 3 km² (>1 mi²) of new land had been added to the coastline. Two communities were almost completely buried beneath the flows, with over 700 houses destroyed and many more threatened. The lava lake within the crater near the summit of Kilauea sank over 450 m (~1500 ft) from its pre-eruption level. A series of phreatic eruptions larger than those of 1924 widened the crater, with massive sections of the weakened crater walls slumping downward. The eruptions came to an end in August 2018, with damage estimated at $800 million and only a few tens of injuries reported.

Haleakala National Park

Haleakala is one of two shield volcanoes that compose the island of Maui, located 45 km (26 mi) across the Maui Channel from the northern tip of the Big Island (Fig. 32.35). West Maui volcano is probably extinct, but Haleakala last erupted only about 500 years ago. Haleakala is in the declining stages of its lifespan, but it is very likely to erupt again in the future, although probably with limited intensity.

The east-facing slopes of Haleakala are covered in dense rainforest nourished by the moisture-laden trade winds. Rainfall may reach a meter per year (400 in/yr). Erosion is intense, with deep valleys cut into the slopes, and waterfalls dropping from the edges of resistant basalt flows. The western slopes in the rain shadow are considerably drier but may get up to 150 cm/yr of rain (60 in/yr), supporting a cover of dry forest and scrub. The highest elevations

Figure 32.35 The island of Maui is composed of two volcanoes separated by an isthmus. Haleakala National Park is part of the much larger volcanic edifice. The park boundaries are shown in yellow, and the winding entrance road to the summit is shown in black.

are relatively barren subalpine and alpine ecosystems that harbor a sparse, highly adapted flora and fauna.

The summit of Haleakala Volcano reaches 3055 m (10,023 ft) and is located just inside the western boundary of the national park, near the Visitor Center. A steep-walled basin filled with youthful lava flows and cinder cones spreads out to the east of the summit and is the most visited area of the national park. The elongate crest of the volcano to the southwest is a rift zone, pocked with cinder cones along its length. Another rift zone extends eastward from the summit toward the remote community of Hana along the easternmost point of the island. The east rift zone continues beneath the sea as the Haleakala Ridge, which extends for several tens of kilometers. The vast majority of the volcanic edifice of Haleakala is beneath the sea surface, reaching down almost 5500 m (~18,000 ft) to the adjacent seafloor. From base to peak, Haleakala is over 8500 m (28,000 ft) high. During its prime, the volcano was likely a kilometer or more higher than it is today.

Evidence for the oldest basaltic lava flows on Haleakala is buried far below the surface of the volcano. The oldest exposed flows are about 1.1 m.y. old, but the volcano is estimated to have been born on the seafloor about 2 m.y. ago when it resided above the hotspot where the southern

Big Island is today. During its long history, the shield grew upward and laterally, with basaltic flows reaching all the way to the base of the West Maui Volcano, forming the foundation of the isthmus that connects the two parts of the island. Over the years, Haleakala has drifted over 220 km (140 mi) to the northwest (Fig. 32.18). With time and distance from the hotspot, both the frequency and vigor of volcanism at Haleakala has decreased, with erosion taking over as the dominant process.

Even though Haleakala is no longer fed by magma from the hotspot, there is still remnant magma beneath the surface. Over the past 30,000 years, when Maui was much closer to its current position than to the hotspot, volcanism has occurred mostly along the southwest and east rift zones. Moving even nearer in time, at least 10 eruptions have been mapped and dated as occurring in the last 1000 years at Haleakala. The very youngest lava flow occurred along the lower southwest rift zone outside the park sometime between 1420 and 1600 CE, only ~500 years ago. Clearly, enough magma still occupies reservoirs within the volcano to keep it active.

The waning stage of an older Hawaiian volcano's life, like that of Haleakala, is marked by a final "rejuvenation" phase before the magmatic tap is completely shut off. This

Figure 32.36 Alignment of cinder cones along east rift zone near crest of Haleakala.

final phase is characterized by a change in the chemistry of the magma, typified by an enrichment in the element sodium. It's still basalt, but it comes in names that only a geologist would love and recognize, such as hawaiite and ankaramite. Lava flows still dominate, but cinder cones have become more common. The rejuvenated rift zones along the east–west long axis of Haleakala are marked by an alignment of cinder cones, creating a blistered terrain (Fig. 32-36).

The magma that feeds the cinder cones uses the fractures along the rift zones as conduits to reach the surface. Cinder cones form by *Strombolian eruptions* (Fig. 16.8), short-lived but energetic eruptions of droplets of incandescent lava that cool while in contact with the atmosphere, falling as innumerable pebble-sized cinders known as *scoria*. The particles build vertically around the vent, tumbling down along the flanks to form the conical shape around the central crater. The pile of scoria composing the

cone is structurally weak, so the final, gas-depleted dregs of magma may burst through the base of the cone as a lava flow.

Haleakala "Crater"

Haleakala means "House of the Sun" in Native Hawaiian; many visitors to the national park drive the winding, 43 km (27 mi) road to the summit specifically for the sublime sunrises and sunsets. The steep-walled, elongate basin near the summit of the volcano where most visitors go is called Haleakala crater (Fig. 32.37), but it's not really a crater or a caldera. Rather, it's a deep erosional valley sculpted by landslides and flowing water. Streams draining the crest of the volcano created the Ko'olau Gap to the north and the Kaupo Gap to the southeast, whereas landslides formed the steep, scalloped walls of the summit basin. An enormous debris avalanche that rumbled down the south flank of Haleakala in the distant past further modified the Kaupo Gap. All evidence of the summit crater that must have existed somewhere nearby has been eroded away or buried by young volcanic deposits.

The entire summit basin and the Ko'olau and Kaupo gaps have been filled by very young lava flows and a chain of cinder cones. The oldest basaltic flows are only 3000–5000 years old, whereas the youngest cinder cones and lava flows are only about 800 years old. A cluster of youthful cinder cones (called "pu'u") extends across the summit basin as part of the east rift zone (Fig. 32.38). The soft ocher hues of the cinder cones were created as steam and gases percolated through the hot pile of scoria after the eruption ended. Iron in the scoria particles was oxidized to a variety of reddish shades, imparting an artistic tint to the landscape. Some of the larger blobs of lava ejected during Strombolian eruptions cool as they traveled through the air, forming streamlined **lava bombs**. Older rock clogging the central conduit of a cinder cone may be blasted outward during the eruption, fragmenting into angular blocks that crash back to the surface. Lava bombs and blocks are common around the margins of cinder cones. Don't be tempted to drop a lava bomb into your backpack; it's illegal to remove rocks from a national park without a permit.

The best way to experience the summit basin and its surreal landscape is to hike either the Sliding Sands or Halemau'u trails. You'll see the endangered silversword plant and perhaps the protected Hawaiian geese known as nene. You may even be lucky enough to glimpse a rare honeycreeper, a highly adapted family of birds that has found sanctuary within the confines of the

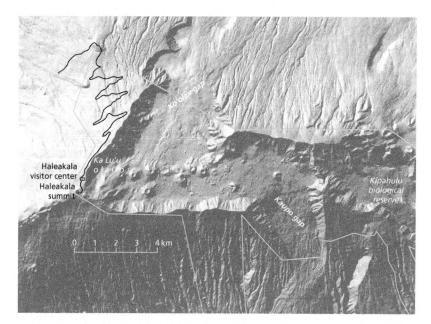

Figure 32.37 Haleakala crater is an erosional basin at the summit of the volcano. Ko'olau and Kaupo gaps are stream-cut valleys. The zigzag dark line marks the switchbacks of the entrance road to the park.

Figure 32.38 Youthful cinder cones within the summit basin of Haleakala "crater." The silversword plants in the foreground are endemic to Haleakala National Park. They flower only once in their lifetimes and die soon after.

park. Be prepared for your adventure, though. The elevation is high, the Sun can be intense, and the weather can change quickly.

For now, Haleakala is a volcano in repose. But active magma bodies reside beneath the ground, waiting for the right conditions to send magma rising to the surface. The next eruption from Haleakala may not occur in the next few centuries, but then again, it may occur next year.

Key Terms

a'a lava
active volcano
atoll
cinder-and-spatter cone
decompression melting
dormant volcano
extinct volcano
fissure eruption
flank eruption

kipuka
lava bomb
lava fountain
lava lake
lava tube
pahoehoe lava
pali
rift zone volcanism
skylight
tumuli
vog

Related Sources

Carey, R. J., Cayol, V., Poland, M. P., and Weis, D. (2015). *Hawaiian Volcanoes: From Source to Surface. Geophysical Monograph Series.*

Decker, R., and Decker, B. (2006). *Volcanoes*, 4th ed. New York: W. H. Freeman.

Hazlett, R. W. (2014). *Explore the Geology of Kilauea Volcano.* Hawai'i Pacific Parks Association, 145 p.

Hazlett, R. W., and Hyndman, D. W. (1996). *Roadside Geology of Hawai'i.* Missoula, MT: Mountain Press.

Thornberry-Ehrlich, T. (2009). *Hawai'i Volcanoes National Park Geologic Resources Inventory Report.* Natural Resource Report NPS/NRPC/GRD/NRR—2009/163. Denver, CO: National Park Service.

Hawai'i Volcanoes National Park. Accessed February 2021 at https://www.nps.gov/havo/index.htm

Hawaiian Volcano Observatory: Mauna Loa and Kilauea. Accessed February 2021 at https://volcanoes.usgs.gov/observatories/hvo

Hawaiian Volcano Observatory: Haleakala. Accessed February 2021 at https://volcanoes.usgs.gov/volcanoes/haleakala

The Hawaiian Volcano Observatory maintains a series of webcams that transmit from various angles around Kilauea caldera, the active vent at Pu'u O'o, and on Mauna Loa. Find them at: Accessed February 2021 at https://hvo.wr.usgs.gov/cams

A variety of videos of the volcanic activity in the park can be seen at https://www.nps.gov/havo/planyourvisit/lava2.htm (accessed February 2021)

Evolution of the American Cordillera

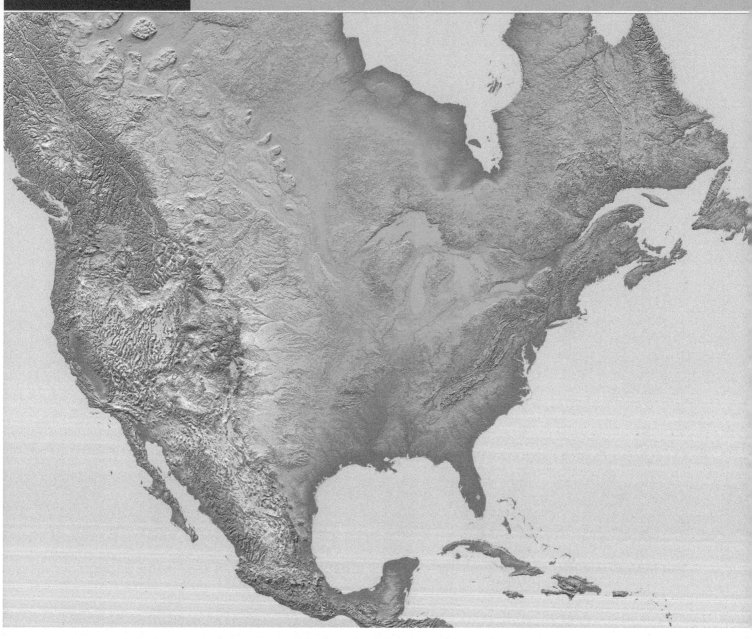

Figure 33.1   North American relief map generated with data from the Space Shuttle Radar Topography Mission. The lowest elevations are in green, yellows and tans are higher, and the white to gray tones are the highest.

North America exhibits stark contrasts in physiography between the American West and the rest of the continent (Fig. 33.1). In the east, elongate wrinkles extend from the American South to northeast Canada, hinting at the ridges, valleys, and plateaus of the Appalachian Mountains. Lowlands along the Mississippi corridor give way westward to the High Plains, incised by eastward-draining river systems. In the west, a broad belt of mountainous terrain stretches north-to-south from Canada to southern Mexico and west-to-east from the Pacific Coast to Colorado. This is the *North American Cordillera*—the focus of this chapter.

The part of the Cordillera within the coterminous United States includes all of the geologic provinces discussed in previous chapters—the Sierra Nevada, Coast Ranges, and Transverse Ranges of California, the Cascades and Olympic Mountains of the Pacific Northwest, the hundreds of basins and ranges of Nevada and western Utah, the Colorado Plateau, and the Rocky Mountains along the eastern flank (Fig. 33.2). At its widest, the U.S. Cordillera is about 1600 km (1000 mi) across. The highest peak is Mount Whitney in the Sierra Nevada at 4421 m (14,505 ft). Seventy-nine mountains rise above 4000 m

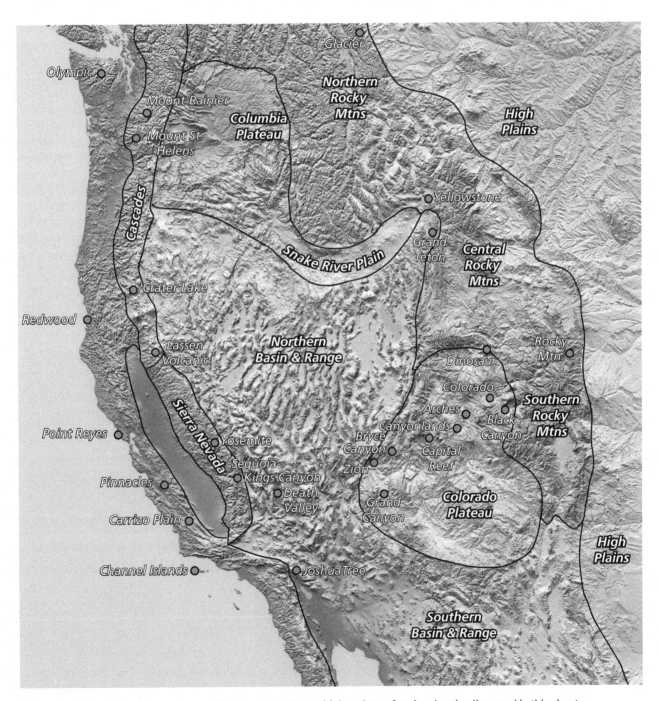

Figure 33.2 Physiographic provinces of the American West, with locations of national parks discussed in this chapter.

(13,123 ft) in elevation, mostly in Colorado, California, and Wyoming.

The north–south texture of the U.S. Cordillera reflects the east–west *compressional* tectonic events that began around 200 m.y. ago. The topographic fabric was enhanced by east–west *extensional* tectonism within the Basin and Range over the last 16 m.y. Along the Pacific Margin, the north–south trend is controlled by tectonic *shear* related to the development of the San Andreas fault system over the past 30 m.y. Earlier parts of the book addressed the origin and characteristics of individual geologic provinces and their national parks, but this synthesis chapter is intended to assemble the disparate pieces of the geologic jigsaw into a single narrative of the origin of the U.S. Cordillera.

This story of the evolution of the U.S. Cordillera will use the national parks of the American West as "touchstones" to synthesize the chronology of events because each park is a compact representative of its province. In turn, the provinces tell the collective story of the assembly of the American West. Since this narrative is told from the perspective of the national parks, not every important geologic event involved in the origin of the U.S. Cordillera is discussed. Geologic purists may say, for example, "Where are the Sonoma and Sevier orogenies?" These important episodes are minimized in the text because evidence for these events is not recognized in the national parks discussed earlier. The intent here is to tell the story in as straightforward a manner as possible to link all of the parks together in a continuous chronicle of geologic events.

To best comprehend this synthesis chapter, you'll need to integrate all the information you've acquired earlier in the book. You'll need to recall the three main types of rocks, as well as the various types of faults. You'll need to incorporate your understanding of deep time and the fundamental principles of relative and numerical age dating. And you'll need to remember the array of tectonic settings that you've read about, primarily convergent and transform plate boundaries. Refer back to Chapter 1 and specific chapters on individual parks as necessary for details on concepts and principles that may be unclear.

It's human nature to view the Earth through the prism of our lives, which change on the scale of a few seconds to a few decades. We tend to see the landscape around us as static and everlasting, varying only by the cycle of the seasons. But through this book, you've learned to push your imagination to think in terms of geologic time scales, which span a wide range from

thousands to billions of years. For this chapter, be sure to use your newfound ability of thinking in the fourth dimension of time to fully comprehend the narrative of the origin and evolution of the American West. To keep you oriented, a series of timelines are interspersed through the text.

Precambrian Assembly of North America

The oldest rocks exposed in the national parks of the U.S. Cordillera are 2.7 b.y. old metamorphic rocks composing the highest peaks of Grand Teton National Park. Nearby, rocks as old as 3.4 to 3.8 b.y. are found in the Beartooth Range along the northeastern border of Yellowstone. Both sets of rocks were formed during a part of the Precambrian known as the *Archean*, which ranges from 4.0 to 2.5 Ga, about 45 percent of Earth's history (Fig. 33.3). To comprehend what Earth may have looked like that long ago requires a vivid imagination. Life on Earth consisted of microbes like *archaea* and bacteria that were just beginning to develop photosynthesis as a process of gaining energy from sunlight. The oxygen content of the atmosphere was negligible compared to what it is today; it was nowhere near enough to support multicellular forms of life. Plants hadn't yet evolved, so continents were barren expanses of rocky land drained by rivers choked with sediment. For 1.5 billion years, the Archean Earth was a barren, inhospitable planet other than for the splotches of colorful microbes inhabiting the shallow margins of oceans and lakes.

Around 2.5 Ga, North America consisted of the central portions of Canada, much of Greenland beneath today's

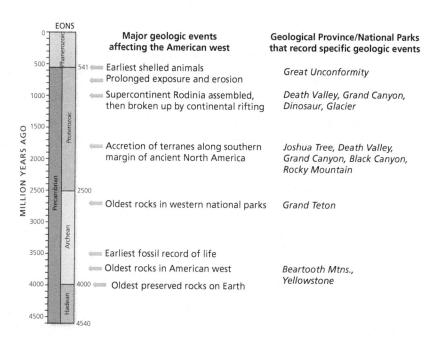

Figure 33.3 Relative chronology of major geologic events that affected the provinces and national parks of the American West through Precambrian time. Dates based on the Geological Society of America 2012 time scale.

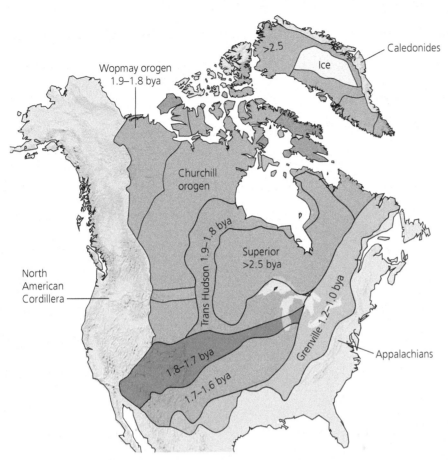

Figure 33.4 Simplified map of ages of major crustal provinces of North America. Continents build outward from a Precambrian nucleus by tectonic accretion. The oldest rocks in North America are exposed in Canada where glaciation has scoured away overlying rocks. The youngest parts of North America are the yellow shades around the margins of the continent.

The intense heat and pressure associated with tectonic accretion metamorphose the rock within the core of the growing mountain range. Granitic magma commonly rises upward from the subduction zone to intrude into the chaotic mass of deformed, metamorphosed rock composing the terrane. The rocks within individual terranes have unique characteristics that distinguish them from the main continental landmass as well as from adjacent terranes.

North America grew significantly around 1.8 Ga when a huge chain of volcanic island arcs was accreted to its southern boundary (Fig. 33.6). This event is recorded by 1.7–1.8 b.y. old metamorphic rocks exposed today in Joshua Tree, Death Valley, Grand Canyon, Black Canyon of the Gunnison, and Rocky Mountain national parks and Colorado National Monument. The common age reflects the phase when the original rock of the terrane was metamorphosed, likely coincident with accretion. All of the gneisses and schists in each of the parks were intruded by granitic magma around 1.4 b.y.

mantle of ice, and part of the northcentral United States. (Fig. 33.4). The rocks of the Tetons and Beartooths were part of the southwestern corner of the young Archean continent. Through the rest of Precambrian time, belts of crustal rock were attached to the margins of North America by tectonic *accretion* (Fig. 33.5). This fundamental process of continental growth occurs when a volcanic island arc, continental sliver, or larger continental landmass is tectonically sutured to the margins of a continent during subduction. As the oceanic lithosphere between the two converging landmasses is consumed into the mantle, the intervening seaway is lost forever. When the buoyant, lower density island arc or continental fragment finally encounters the larger landmass, it resists descending. Instead, it detaches from the downgoing plate and is heaved up and over the adjoining continent in a colossal, slow-motion collision.

These compressional tectonic episodes build an elongate coastal mountain belt along the convergent margin where the newly arrived mass of rock (called an *accreted terrane*) is attached to the larger continental mass. Rocks within the terrane are severely folded and faulted as they are sutured against the margins of the growing continent.

ago, during the peak phase of mountain growth along the former subduction zone. This thick belt of rock extends from northwest Mexico to Michigan (Fig. 33.4), mostly hidden beneath a cover of younger rocks, but exposed in deep canyons like those of the Colorado Plateau, or in mountain ranges like those in Death Valley and the Colorado Rockies. In the American Midwest, drill holes encounter these Precambrian rocks a few hundred meters to a few kilometers beneath the surface.

The crystalline metamorphic and igneous rocks formed during Precambrian continental assembly constitute the crustal foundation of all continents. These rocks of the *basement* compose by far the greatest volume of continental crust, with all younger rocks creating a relatively thin veneer (Fig. 3.15). Continental crust averages about 35 km (~22 mi) in thickness, all composed of rock with a slightly lower density than the rock of the underlying mantle. The difference in density permits continental crust to *isostatically* "float" above the mantle, somewhat like an iceberg in seawater. You might think of the crust beneath continents as the roots of long-lost mountain ranges, tangled together one at a time by tectonic convergence through the Precambrian and infused with massive volumes of

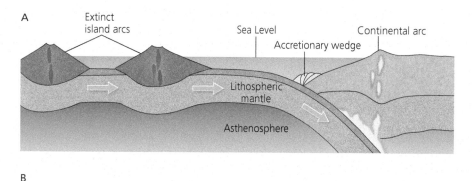

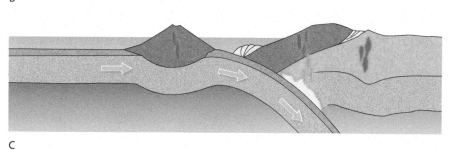

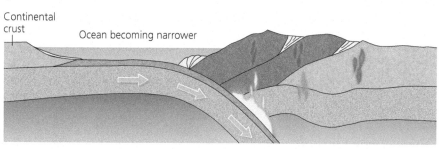

Figure 33.5 Process of continental accretion. A) In this example, two extinct volcanic island arcs ride passively as part of an oceanic lithospheric plate, which is consumed within the subduction zone. B) The buoyant, lower density island arcs collide, detach from the downgoing plate, and accrete to the margin of the continent, building it outward and raising a coastal mountain belt. C) High temperatures and pressures accompany accretion and mountain-building, metamorphosing rock. Granitic magma rising from the subduction zone commonly intrudes the growing mountain range.

granite. In much of Canada, Precambrian rocks of the crustal basement were exposed as continental ice sheets scraped away the surficial layers of younger rock. In lower latitudes, like those of the national parks in the U.S. Cordillera, these deep-seated rocks are visible only in places where they were lifted upward during mountain-building events, then exposed by erosion of overlying rock or by river incision.

Supercontinent Breakup and the Birth of Western North America

During the later *Proterozoic*, the vast span of Precambrian time from 2.5 Ga to about 540 Ma, continental accretion continued, building ancient North America into a larger landmass one terrane at a time. By about 1 Ga, the cores of today's continents had assembled into a single supercontinent known to geologists as *Rodinia*

(Fig. 33.7). Laurentia, the name given to ancestral North America and Greenland, formed the nucleus for the larger supercontinent. The western margin of Laurentia was likely sutured to East Antarctica and Australia, a bizarre arrangement considering their current geographic positions. In eastern and southern Laurentia, collisions with parts of ancestral South America and Africa raised a welt of mountains known today as the Grenville belt. The remnant roots of the Grenville mountains extend from northern Mexico through Texas and Arkansas, then northeast all the way through the Appalachians to the Labrador coast of Canada (Fig. 33.4). The Grenville event will be discussed in greater detail in Chapter 35 on the origin of the Appalachians.

Around 800 Ma, Rodinia began a prolonged phase of *continental rifting* that eventually isolated ancestral North America. Plates and their floating continents slowly disassembled, pulling away from one another in a stately dance with narrow ocean basins opening between the rifting continental margins. Along the nascent western margin of North America, rifting was recorded by the deposition of thick accumulations of sediment within narrow *rift basins* oriented perpendicular to the continental edge (Fig. 33.8). Four national parks preserve these rift–basin sediments, whose ages reflect the breakup of Rodinia around 800 Ma. In Death Valley National Park, rocks of the Pahrump Group suggest deposition in a variety of environmental settings, whereas contemporaneous rocks in the Grand Canyon are composed not only of sedimentary rocks but rift-related volcanic rocks as well. In Dinosaur National Monument, the Uinta Mountain Group records deposition in sediment-choked streams and rivers. To the north in Glacier National Park, rocks of the contemporaneous Belt Supergroup accumulated in shallow-water environments that were at times ankle-deep. Along the rifted eastern margin of North America, the Grenville Terrane remained attached, adding a massive slice of real estate to the continent. Proterozoic rocks within Great Smoky Mountains and Shenandoah national parks record the breakup of Rodinia where it separated from ancestral South America (Fig. 33.7).

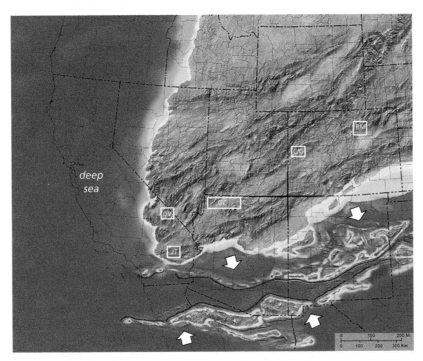

Figure 33.6 Speculative reconstruction of the southwestern United States during the accretion of Proterozoic rock to the young North American continent (~1.7 Ga). Note Four Corners region on the map as a point of reference. Rectangles show locations of future Joshua Tree (JT), Death Valley (DV), Grand Canyon (GC), Colorado Monument and Black Canyon of the Gunnison (C/G), and Rocky Mountain (RM) national parks.

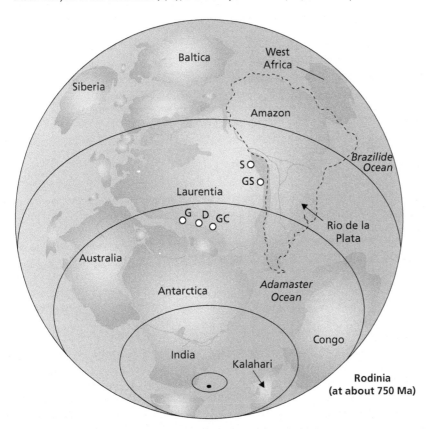

Figure 33.7 One interpretation of the Rodinian supercontinent around 750 Ma during the earliest phases of continental rifting. Compare with version of Rodinia at 1 Ga, Fig. 22.8. Positions of continents composing Rodinia are a subject of vigorous debate. National parks with exposures of rocks deposited in late Proterozoic rift basins shown in their approximate locations. GC-Grand Canyon; D-Dinosaur; G-Glacier; GS-Great Smoky Mountains; S-Shenandoah

Late Proterozoic rocks in the four western national parks reach thicknesses of several kilometers but range over a limited areal extent, suggestive of deposition in rapidly subsiding, fault-bounded rift basins. The only visible life forms in these depositional settings were *stromatolites*, built by sticky mats of cyanobacteria. You might envision these rift basins as narrow embayments of the sea that extended inland along the continental margin, somewhat comparable to the Great Rift Valley of Ethiopia or the Salton Trough inland extension of the Gulf of California. The rifting continued until seafloor spreading created an elongate ocean basin between western North America and the conjoined mass of Antarctica and Australia (Fig. 1.11). This ocean basin was the distant ancestor of today's Pacific.

The final few hundred million years of the Proterozoic in western North America were marked by a prolonged phase of exposure and erosion. The western margin of North America was tectonically quiet as it passively drifted farther away from its mirror image on the margins of East Antarctica and Australia. During this time, the upper surface of Precambrian rock was erosionally beveled down to a broad, relatively smooth coastal plain. It wasn't until the great marine transgressions of the early Paleozoic that the rocky surface was again buried by sediment. This erosion surface is the *Great Unconformity*, exposed in many national parks and monuments such as Grand Canyon (Fig. 3.22), Colorado National Monument (Fig. 10.4), Black Canyon of the Gunnison (Fig. 11.5), and Grand Teton (Fig. 25.8). Depending on the locality, the depth of erosional removal of underlying rock, and the age of overlying rock, the Great Unconformity spans a few hundred million years to as many as 2.2 billion years. This is an inconceivably long period of time preserved along a single surface separating Precambrian crystalline basement below from younger sedimentary rock above. The upper surface of crystalline rock just below the unconformity marks the barren coastal plain of western North America prior to the rise of sea level at the beginning of the Paleozoic, the topic of the next section

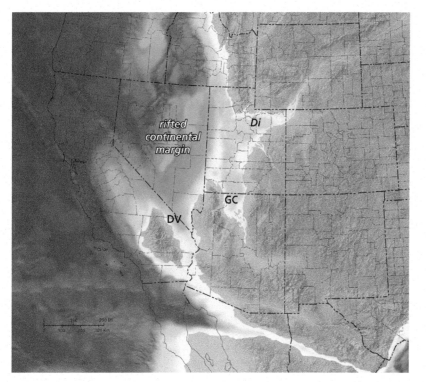

Figure 33.8 Paleogeographic map of the southwestern margin of North America about 730 Ma during continental breakup. As North America broke away from Antarctica and Australia, narrow rift basins opened along the continental margin. The sediments that accumulated in these basins are preserved in the rocks of Death Valley (DV), Grand Canyon (GC), and Dinosaur National Monument (Di).

Early to Mid-Paleozoic Passive Margin

Sea level began a long-term rise at the beginning of the Paleozoic, forcing the shoreline to transgress far onto the gently seaward-sloping margin of North America at that time (Fig. 3.10). As the shoreline migrated landward, a broad sheet of beach sands accumulated above the exposed Precambrian basement. These earliest transgressive beach deposits have widespread lateral continuity across the American West and are exposed as time-equivalent sandstone layers within Death Valley, Grand Canyon, Grand Teton, and several other national parks (Fig. 33.9). The North American continent at that time was a relatively low-elevation terrain, and as a result, the earliest Paleozoic sea-level rise flooded much of the landmass, reaching as far inland as Wisconsin and beyond (Fig. 3.11). Most of the world's continents at the time, many geographically isolated as large islands following the breakup of Rodinia, were extensively submerged beneath shallow seas.

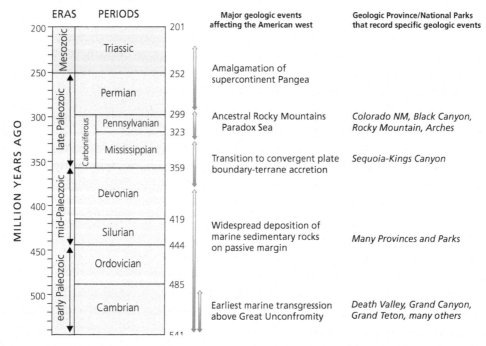

Figure 33.9 Relative chronology of major geologic events that affected the provinces and national parks of the American West through the Paleozoic Era. Dates based on the Geological Society of America 2012 time scale.

A

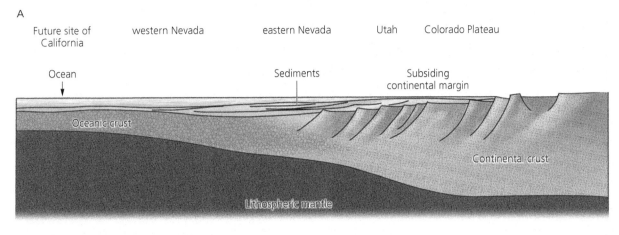

B

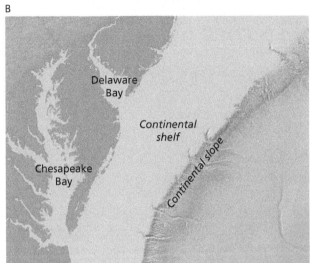

Figure 33.10 A) Cross section of the Paleozoic continental shelf of western North America. Sedimentary rocks accumulated on the passive margin shelf during transgressions and regressions of the sea above the rifted Precambrian basement. Continental shelves far removed from a tectonic plate boundary are called passive margins. B) Modern passive margin continental shelf off the U.S. east coast.

All of the continental margins surrounding North America were tectonically quiet during the early Paleozoic. The broad continental shelf that extended seaward on all sides was repeatedly flooded and exposed as sea level rose and fell through time. With each transgression and regression of the shoreline, vast layers of sand, silt, clay, and limy sediment accumulated within an array of beaches, lagoons, deltas, and shallow seas. These sediments would eventually be buried and solidified into tabular beds of sandstone, siltstone, shale, and limestone. This type of tectonically quiet continental shelf is called a *passive margin* because these shelves drift passively within the interior of a tectonic plate (Fig. 33.10).

The modern continental shelf of the east coast of North America is a passive margin because it is thousands of kilometers from the nearest plate boundary, the Mid-Atlantic Ridge. Water depths above modern continental shelves may reach about 140 m (460 ft) before they deepen abruptly along the continental slope. (Considering that the average depths of the oceans are about 4000 m (>13,000 ft), the continental shelves are very shallow extensions of the continents.) The width of the modern continental shelf

off the U.S. east coast is 80–200 km (50–125 mi). During the earliest Paleozoic, the water depths on the submerged continental shelves of the western United States were comparably shallow, but the shelves were much wider, extending from the Dakotas to a continental slope cutting across Idaho, central Nevada, and southeastern California (Fig. 3.11).

As sediment accumulates on the continental shelf, the underlying crust subsides due to the weight of the sediment load, creating more space beneath sea level for sediment to fill. Commonly, a balance develops between the rate of *subsidence* and the rate of sediment filling the available space between the seabed and the sea surface. This acts to maintain a diversity of coastal and shallow marine depositional environments over long periods of time and explains why most of the Paleozoic sedimentary layers composing the rocks of many national parks exhibit evidence for long-term deposition in very shallow water. The strata of the upper Grand Canyon, deposited in a variety of coastal and shallow marine environments, provide a good example of this long-term balance between the rates of deposition and subsidence.

Figure 33.11 Sedimentary strata deposited on the Early Paleozoic passive margin. The originally horizontal layers of seafloor sediment were buried and solidified. Much later in time, the rocks were uplifted and tilted by block faulting, then eroded to expose the succession of layers. Last Chance Range, Death Valley National Park.

Late Paleozoic Mountains and Basins

The long trend of tectonic quiescence in the American West would come to a close during the late Paleozoic when a phase of prolonged mountain-building ensued. Chains of volcanic islands and elongate fragments of continents collided with the western continental margin, slowly transforming what was a passive margin into a convergent plate boundary (Fig. 33.13A). These island chains and narrow ribbons of continental rock accreted to the western continental edge, adding material to what is today western Nevada and

The accumulation of a seaward-thickening prism of sedimentary rock along passive margin shelves is another way that continents build outward. Much later in time, these sedimentary strata may be deformed and lifted upward during mountain-building events, exposing stacks of layered rock along the flanks of mountains or on the walls of incised canyons (Fig. 33.11). In national parks of the Basin and Range and Colorado Plateau, many of the rocks composing the ranges, plateaus, and canyon walls are the tangible record of this early Paleozoic passive margin.

The vast coastal systems and shallow seas of the early Paleozoic created a variety of oxygenated environments that catalyzed the rapid diversification of marine invertebrate life (Fig. 33.12A). Complex animals such as trilobites, mollusks, echinoderms, and corals developed hard outer skeletons, ideal for fossilization in the muddy seafloors on which they lived. An "evolutionary arms race" developed in those early Paleozoic seas, with new innovations emerging in response to predator–prey relationships. It was also in those early Paleozoic shallow seas that the first fish arose and diversified. These were the first vertebrates, the subphylum that would lead to all other animals with a vertebral column running through the length of the organism, including us.

The pattern of quiet sediment accumulation on passive continental margins continued into the mid-Paleozoic (Fig. 33.12B). The shoreline reached even further inland, creating an intricate system of bays, estuaries, and wetlands along the coast. Plants began to colonize the land, first along the margins and then further inland until large swaths of the continents were covered in forests by about 350 Ma. The color of the continents was transformed from gray-brown to green. The invasion of the land included the first known insects, as well as the first amphibians. Pulses of mass extinction occasionally decimated the biosphere, but the empty ecospace was eventually refilled by a radiation of new plants and animals, both on land and in the sea.

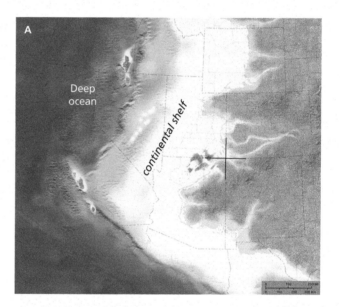

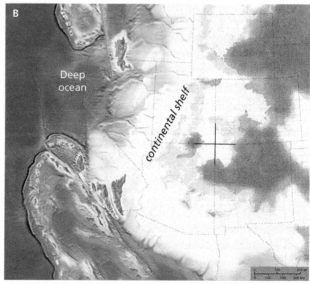

Figure 33.12 A) Paleogeographic maps showing the changes in land and sea of the American West at 515 Ma and B) 370 Ma. Chains of volcanic islands were converging from the west, eventually to accrete along the continental margin. Four Corners used as a point of reference.

northern California. An arm of the ocean extended from future Southern California up through eastern Nevada and western Utah into Idaho and Wyoming, geographically isolating the western terranes as elongate islands. Some of the metamorphosed *roof pendants* in the Sierra Nevada at Sequoia–Kings Canyon national parks are remnants of these accreted terranes. This pattern of continental growth by the accretion of terranes along the western margin of North America would continue into Mesozoic and Cenozoic times, creating much of the crustal foundation of California, Nevada, and the Pacific Northwest. To the north of the U.S. Cordillera, accretionary tectonics would build much of western Canada and Alaska. The North American Cordillera was beginning to take shape.

In Colorado, the *Ancestral Rocky Mountains* rose upward, with mountain-building reaching its peak between 315 and 300 Ma (Fig. 33.13A). These mountains consisted of two elongate highlands, the Uncompahgre on the west and the Front Range uplift on the east. The core of the Ancestral Rockies was composed of Precambrian crystalline basement, raised along steep *reverse faults* (Fig. 10.7). The remnant roots of these mountains are exposed within Colorado National Monument, Black Canyon of the Gunnison National Park, and Rocky Mountain National Park. Early to mid-Paleozoic sedimentary rocks once formed a thick cap above the basement and were lifted upward to become the highest peaks of the Ancestral Rockies. By the late Paleozoic, however, these sedimentary rocks were eroded back to loose sediment and redeposited by running water and wind into nearby lowland basins.

Examples of these late Paleozoic redeposited rocks include the redbeds of the Supai Group in the upper Grand Canyon (Fig. 3.13), the colorful Cutler Group in Canyonlands National Park (Fig. 8.6), and the imposing white cliffs of the Weber Sandstone in Dinosaur National Monument (Fig. 9.5). *Alluvial fans* along the flanks of the Uncompahgre and Front Range highlands are preserved as resistant outcrops of red sandstones and conglomerates of the late Paleozoic Fountain Formation, exposed along the margins of Rocky Mountain National Park (Fig. 21.9). The Ancestral Rockies were beveled down to sea level through the rest of the Paleozoic and early Mesozoic. They would rise once again late in Mesozoic time as the roots of today's Colorado Rockies.

A crustal sag southwest of the late Paleozoic Ancestral Rockies was low enough that seawater flooded in through a narrow connection with the open ocean (Fig. 33.13A). This was the *Paradox Sea*, a restricted basin comparable to the modern Caspian Sea. It was located in the rain shadow of the Uncompahgre highlands at about 10°N latitude, which created an arid, evaporative environment. Thick layers of salt precipitated within the semi-enclosed basin during phases of lower sea level that would cut off the outlet with the open ocean to the southeast. In time, the salt layers would be buried by kilometers of younger rock eroded from the adjoining highlands. The salt responded to the overburden pressure by flowing upward as a thick plastic, forming ridges and pillars that buckled and fractured the

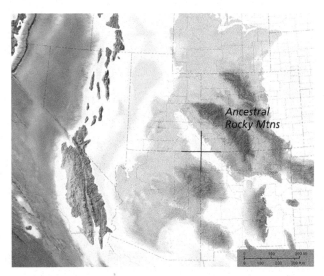

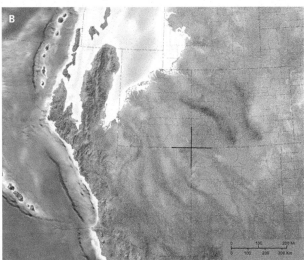

Figure 33.13 A) Paleogeographic maps showing the changes in land and sea of the American West at 300 Ma and B) 240 Ma. Four Corners used as a point of reference. Note the transition from dominantly marine environments during the late Paleozoic to dominantly terrestrial environments during the early Mesozoic.

overlying rock. Today, the Paradox salt deposits underlie both Canyonlands and Arches national parks, influencing their landscapes and controlling the location of arches.

Mountain-building was widespread during the late Paleozoic in response to the amalgamation of plates to form the supercontinent of *Pangea* (Fig. 33.14). Many ocean basins were closed by subduction as continents approached one another during supercontinent assembly. When continents collided, the buoyant landmasses merged to create colossal mountain ranges. North America was fused to Africa, Eurasia, and South America, creating the equatorial, Himalayan-scale Appalachian Mountains. By the late Paleozoic, Pangea had grown so large that the interiors became arid wastelands because they were so distant from weather fronts bringing moisture in from the oceans. Mountain ranges created rain shadows and distorted wind patterns, further contributing to the desiccating climatic conditions in the centers of continents. Huge expanses of wind-blown sand became the Coconino Sandstone in

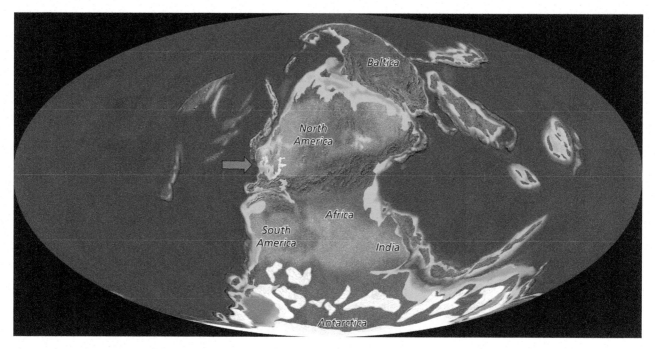

Figure 33.14   Paleogeographic map of Pangean supercontinent around 280 Ma during the latest Paleozoic. Arrow points to western U.S. margin of North America.

the Grand Canyon and the Weber Sandstone in Dinosaur National Monument. The supercontinent configuration and associated arid climates would continue into the early Mesozoic.

Adding to the dynamism that was the late Paleozoic was a mosaic of continental ice sheets centered over Antarctica, southern Africa, southern South America, India, and Australia. As the glacial ice expanded and contracted in tune with climatic rhythms, sea level would fluctuate rapidly by several tens of meters. When the planet cooled, water would be drawn from the oceans via the hydrologic cycle to feed the growing continental ice sheets. Sea level would fall in response and coastlines would migrate seaward. When the planet warmed, glacial meltwater would drain back into the world ocean, causing sea level to rise and coastlines to flood the margins of continents. These glacial cycles and associated sea-level fluctuations influenced depositional patterns worldwide, creating a cyclic arrangement of late Paleozoic rocks that is recognized globally.

As if the widespread burst of mountain-building, arid climates, and glacial fluctuations weren't enough, the Paleozoic ended around 250 m.y. ago with the greatest mass extinction ever on Earth. Between 80 and 85 percent of all species died, affecting plants and animals in the oceans as well as on continents. Although the ultimate cause is still debated, evidence points toward a global environmental catastrophe triggered by profuse and long-lived eruptions of lava in Siberia coincident with the extinction. The abrupt change recognized globally in the fossil record at the extinction horizon defines the boundary between the Paleozoic and Mesozoic eras.

Early to Mid-Mesozoic Convergent Tectonics

The American West at the start of the Mesozoic was strikingly different from that of today (Fig. 33.13B). Elongate embayments of the ocean extended southward deep into northern Utah and Nevada. Large islands of newly accreted terranes lay off to the west in what is today western Nevada and northern California. To the east and south of the seaway, continental deposition was dominant, with rivers and streams draining north and west toward the embayment. The future Colorado Plateau was located near sea level along the coastal plain, and the erosional remnants of the Ancestral Rockies formed low, elongate hills. Sluggish streams lined with marshes and lakes drained the region that would become the Colorado Plateau, depositing the multicolored layers of clay and silt of the Chinle Formation (Fig. 6.4), exposed in several parks of the Colorado Plateau. As opposed to the dominantly marine deposition of the Paleozoic, sediments accumulated in mostly terrestrial environments during the early to mid-Mesozoic (Fig. 33.15). Dinosaurs roamed the coastal plain of the future Colorado Plateau region, leaving bones and trackways in the sediments as evidence of their existence.

Many of the scenery-forming redrock layers of the Colorado Plateau were deposited at this time, including the windblown sands of the Wingate Sandstone (e.g., Capitol Reef, Arches, Canyonlands, and Colorado National Monument), the river-laid silts of the Kayenta Formation (e.g., Zion, Colorado NM), and the cross-bedded sands of the Navajo Sandstone (e.g., Zion, Capitol Reef, Arches,

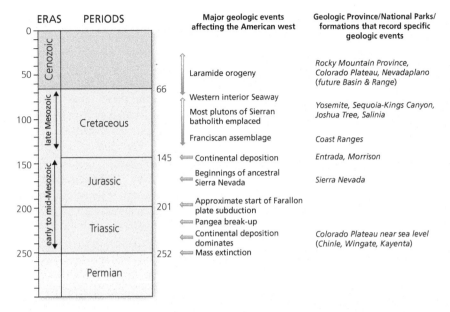

Figure 33.15 Relative chronology of major geologic events that affected the provinces and national parks of the American West through the Mesozoic Era. Dates based on the Geological Society of America 2012 time scale.

Canyonlands, Dinosaur National Monument). The Wingate and Navajo sand seas were about the size of the modern Taklamakan Desert in western China and just as forbidding.

Recall from Chapter 4 on Zion National Park that many of the quartz grains composing the Navajo were originally derived from the eastern side of North America. Geologic sleuthing determined that the quartz grains had to have been eroded from 1 b.y. old rock of the Grenville belt that forms part of the Appalachians. Long-vanished rivers transported these grains westward across the continent where they were deposited along the coast of the seaway in Utah and eastern Nevada. From there, the resistant quartz grains were blown inland by the prevailing northwesterly winds to accumulate on the vast Navajo coastal desert (Fig. 4.8). Today we see these quartz grains as part of the cross-bedded Navajo Sandstone that forms prominent cliffs throughout the Colorado Plateau. What a long, strange trip it's been for those sand grains!

Roughly coincident with the spread of the Wingate and Navajo deserts around 200 Ma, Pangea was in the early stages of breaking apart. A continental rift opened between eastern North America and northwestern Africa that in time would evolve into a linear seaway. As seafloor spreading created fresh oceanic lithosphere along the newly formed Mid-Atlantic Ridge, the North Atlantic Ocean grew wider. The North American plate began moving toward the west, starting a chain of events that culminated in the formation of the North American Cordillera. That westward motion continues today.

The eastern edge of the North American continent became a passive margin as it migrated away from the newly formed plate boundary in the Mid-Atlantic (Fig. 33.16). The once-mighty Appalachians started their

inexorable decline as erosion began to dominate over uplift. The eroded sediments accumulated along both sides of the decaying mountains, building outward along the east as a broad coastal plain that extended seaward as a continental shelf. The east coast of North America remains a passive margin to this day as seafloor spreading actively widens the North Atlantic.

On the western side of the North American plate, the accretionary tectonics that began in the late Paleozoic evolved into a full-blown subduction zone around 200 Ma as the oceanic *Farallon plate* began to descend eastward beneath the west coast. This subduction system would dominate tectonism in the U.S. Cordillera through the mid-Cenozoic. The *ancestral Sierra Nevada* began to rise in response to the compressional stress and intrusion of granitic magma (Fig. 17.8). The oldest granitic plutons of the *Sierran batholith* record this initial emplacement of magma, exposed around the periphery of Yosemite National Park. The molten rock rising from the descending Farallon plate intruded into accreted terranes that were attached during the late Paleozoic through mid-Mesozoic. Metamorphism of roof pendants at Sequoia–Kings Canyon national parks likely records the high temperatures associated with Farallon convergence and magmatism. The ancestral Sierra Nevada, capped by a linear chain of volcanoes, formed the westernmost land of mid-Mesozoic North America. Tides lapped against the lowest flanks of the mountains, somewhat similar to the Andean margin of South America today. Further inland, a wave of compressional mountain-building would sweep across Nevada, Utah, and Idaho through the rest of Mesozoic time.

Global sea level began to rise around 200 Ma, perhaps in response to seafloor spreading that increased the volume of the Mid-Atlantic Ridge, displacing seawater up onto the margins of continents. By about 170 Ma, the shallow Sundance Sea covered large parts of the American West from Utah to the Dakotas (Fig. 33.16). It was along the arid margins of this seaway that the Entrada Formation was deposited, which today is prominently seen in Arches National Park and Colorado National Monument. In the vicinity of Black Canyon of the Gunnison National Park, the waters of the Sundance Sea lapped onto Precambrian basement, depositing sands and silts of the Entrada on the unconformable surface (Fig. 11.5). The Sundance Sea retreated from the region by about 150 Ma, leaving a broad interior plain drained by rivers flowing to the northeast. The mosaic of floodplains, wetlands, and lakes provided an inviting environment for an array of dinosaurs and

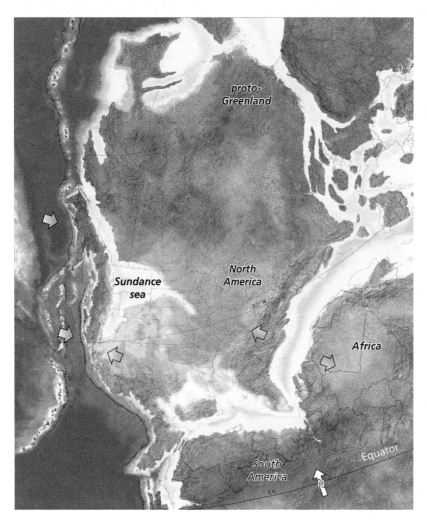

Figure 33.16 Paleogeographic map showing Sundance Sea extending southward across western Canada and flooding much of the western United States around 170 Ma, the largest marine incursion since the late Paleozoic. Volcanic islands and peninsulas lay to the west. The opening of the Atlantic and Gulf of Mexico progressed as Africa and South America pulled away from the margin of eastern North America.

between 140 and 80 Ma. The dramatic landscapes of Yosemite and Sequoia–Kings Canyon national parks are carved into these granites. The monoliths and *inselbergs* of Joshua Tree National Park are formed from these same granitic rocks. And as strange as it may seem, these granites crop out as fault-bounded slivers within the *Salinian Terrane*, exposed at Pinnacles National Park and Point Reyes National Seashore (Fig. 26.17). During the emplacement of all this granitic magma, a chain of volcanoes along the crest of the ancestral Sierra Nevada exploded in towering pyroclastic eruptions that spread across the late Mesozoic landscape.

Much of the rock that forms the mass of California accumulated during the late Mesozoic (Cretaceous). A large slice of oceanic crust detached from the descending Farallon slab around 140 Ma and was accreted to the western flank of the ancestral Sierra Nevada. This terrane would become the foundation of today's Central Valley, now buried beneath several kilometers of sediment washed off the flanks of the ancestral Sierra Nevada. To the west, seafloor sediment and volcanic crust were scraped off the subducting Farallon plate to accumulate as an *accretionary wedge* (Fig. 26.16). These highly deformed turbidites, cherts, pillow basalts, and serpentinites of the *Franciscan assemblage* were lifted up from the sea during the late Cenozoic to form the bulk of the California Coast Ranges. The Franciscan is nicely exposed in coastal bluffs at Redwood National and State Parks and also forms the oldest rocks of Channel Islands National Park. Modern California came into being during the Cenozoic, but the rocks that were uplifted to create the state are primarily Mesozoic in origin.

other Mesozoic plants and animals. Many of these organisms were preserved within the colorful mudstones, sandstones, and conglomerates of the famous Morrison Formation, the fossil-rich unit of Dinosaur National Monument.

Late Mesozoic Mountains and Seaways

Subduction of the Farallon plate dominated the late Mesozoic in the American West, with the wave of compressional mountain-building migrating ever further toward the east. A *fold-and-thrust belt* of mountains extended from Utah northward into western Wyoming and then across Montana into Canada (although evidence for these mountains is not well expressed in any U.S. national parks). The mountainous terrain of the U.S. Cordillera was growing wider.

Along the California margin, the greatest volume of magma was emplaced beneath the ancestral Sierra Nevada

Widespread seafloor spreading drove the dispersion of continents away from their Pangean configuration (Fig. 33.17). The expansion of mid-ocean ridges by seafloor spreading increased their volume, causing the seas to spill over onto the continents. Sea level was thus very high, and late Mesozoic shallow marine deposits are widespread across all continents. In the western United States, the fold-and-thrust belt of mountains created a load on the underlying lithosphere that depressed the adjacent crust to the east, forming the *Western Interior Seaway* that connected waters of the Gulf of Mexico with the Arctic Sea (Fig. 17.9). This elongate *foreland basin* split North America in two, separating a broad island on the east (known to

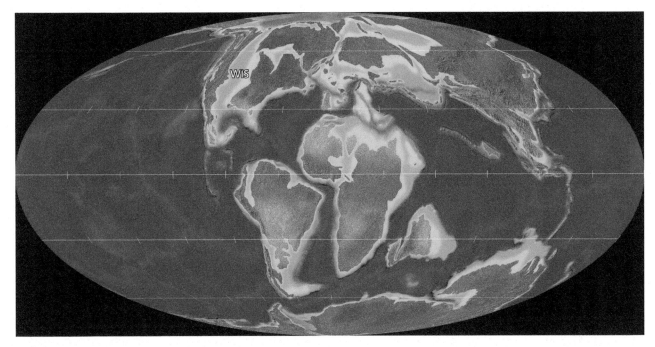

Figure 33.17 By 90 Ma, the continents had almost entirely separated from their Pangean arrangement. Australia was detaching from Antarctica, and India was moving northward across the ancestral Indian Ocean. The Atlantic had widened considerably over the 110 m.y. since breakup began. Sea level was very high during the late Mesozoic, flooding the continents with shallow seas. In North America, the Western Interior Seaway (WIS) connected the Gulf of Mexico with the Arctic Sea

geologists as Appalachia) from an elongate island on the west (known as Laramidia).

At its maximum, the Western Interior Seaway was over 750 m (~2500 ft) deep and extended laterally from central Utah to western Iowa. Texas, New Mexico, Colorado, Wyoming, and much of Montana and the Dakotas were submerged beneath the seaway. Marine reptiles, long-necked and toothy, hunted for prey in what is today Kansas and Nebraska, based on fossil evidence. Dinosaurs migrated along the coastlines, leaving trackways imprinted in the mud, which were later exposed by uplift and erosion. Flying reptiles soared high above, swooping down to skim the water for fish. The limestones and clastics that record the existence of the seaway are widespread but are not particularly prominent in any of the national parks discussed in this book. Limited exposures of shales and mudstones of the Seaway are present in Bryce Canyon and Capitol Reef national parks, but much more complete outcrops occur at Mesa Verde National Park in southwestern Colorado.

As the Laramide orogeny began around 80 Ma, much of the crust beneath the seaway was slowly heaved upward by the growth of mountains, causing the waters to drain north and south back into the larger ocean. By the end of the Mesozoic around 66 Ma, the seaway was gone, and marine waters would never again cover the American West. Today, the highlands of the Colorado Plateau, Rocky Mountains, and High Plains occupy the former site of the Western Interior Seaway in the United States.

Late Mesozoic to Mid-Cenozoic Laramide Orogeny

Magmatism within the ancestral Sierra Nevada ended around 80 Ma. This suggests that the underlying Farallon plate was no longer deep enough for rock to melt along the upper surface of the descending plate. The angle of the subducting slab was flattening, generating the compressional tectonic stress of the Laramide orogeny due to the frictional interaction of the Farallon plate grinding eastward directly beneath the North American plate (Fig. 33.18). A wave of mountain-building swept eastward, deforming rock and raising mountains and plateaus first in Nevada and then in central Utah (Fig. 33.19A). These highlands are called the *Nevadaplano* and may have looked like the modern Altiplano of South America, an elevated plateau that extends eastward from the volcanic crest of the Andes Mountains. Laramide mountain growth reached Colorado by about 67 Ma when the Rocky Mountains were lifted upward, far inland from the subduction zone beneath future California. The sediment that had been shed eastward from the Colorado Rockies filled the final remnants of the Western Interior Seaway. From this point onward, continental depositional environments would dominate the American West.

The eastward migration of mountain-building was accompanied by a sharp decrease in volcanism throughout the California-to-Colorado corridor, supporting the model of *flat-slab subduction* as the controlling mechanism for the broad extent of the Laramide orogeny. Volcanism did

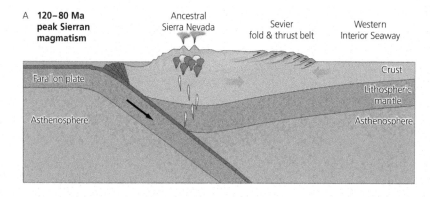

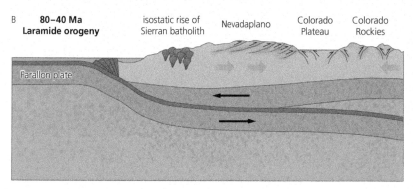

Figure 33.18 A) Late Mesozoic subduction of the Farallon oceanic plate, producing the ancestral Sierra Nevada volcanic arc. B) Model of flat-slab subduction during the Laramide orogeny. Arrows denote the orientation of compressional stress within the upper plate.

continue to the south in Southern California and Arizona, and in a swath to the north in Oregon, Idaho, and the Yellowstone region, represented by the early Cenozoic Absaroka volcanics. These two zones of continued volcanism suggest that the angle of subduction in those regions remained steep. You might imagine the Farallon plate beneath North America at this time as having a broad convex-upward warp beneath the California-to-Colorado latitudes.

What is the evidence for the Laramide orogeny in each of the geologic provinces and their national parks? We'll begin in the west and move eastward, tracking the wave of orogenic uplift. In Yosemite and Sequoia–Kings Canyon, the lack of a magma supply shut off volcanism, and the balance tilted toward erosion in the ancestral Sierra Nevada. During the late Mesozoic and early Cenozoic, flowing water and landslides slowly decapitated the volcanic crest, with the eroded particles accumulating along the eastern and western flanks of the range. Much of today's Central Valley of California is filled with kilometers of sediment shed from the ancestral Sierra Nevada. As the weight of the mountain range decreased due to erosion, the crustal root rose upward isostatically in response, supported by flow within the underlying mantle (Fig. 17.10). By the mid-Cenozoic, the once deeply buried granitic plutons of the Sierran batholith were exhumed and exposed at the surface,

At the end of the Laramide around 40 Ma, the Sierra were reduced to a westward-sloping series of rolling hills underlain by granite and metamorphic country rock. To the east, the Nevadaplano highlands supplied sand and gravel to gravel-choked, braided rivers that flowed westward down the ramp of the once-mighty ancestral Sierra Nevada. The granitic rocks of Yosemite and Sequoia–Kings Canyon were closer to the surface, but the landscape of the parks was yet to be created.

To the north along the Farallon subduction zone near Washington, volcanism was active along the ancestral Cascade chain. The roots of these volcanic mountains form the foundation for younger Cascades volcanoes. Seaward of the ancestral Cascades, a massive volume of pillow lava was erupting from a volcanic *seamount*. Through time, these early to mid-Cenozoic basalts and interfingering turbidites would be transported laterally until they were scraped off to become part of the accretionary wedge (Fig. 30.12). There, they were intensely deformed and stacked into overlapping thrust sheets, eventually to emerge above sea level by the mid- to late Cenozoic as the Crescent Terrane of Olympic National Park.

The Colorado Plateau was lifted upward as a stable, rigid block during the Laramide orogeny as old Precambrian faults along the margins were reactivated by the compressional stress. The entire crust underlying the Colorado Plateau was raised from below sea level (when it was submerged by the Western Interior Seaway ~70 Ma) to a kilometer or more in elevation, based on evidence from fossil plants. Geologists aren't entirely sure why the Colorado Plateau remained relatively undeformed during uplift. It may be that upward pressure was exerted from hot mantle rock below to cause the Plateau to rise like a buoyant plug.

Even though the Colorado Plateau was now a highland, it still formed a broad intermontane basin between the Nevadaplano to the west and the Colorado Rockies to the east (Fig. 33.19A). By the early Cenozoic, rivers drained the surrounding mountains, depositing a broad layer of sand and gravel along floodplains and within lakes on the Plateau (Fig. 33.19B). The hoodoos of Bryce Canyon National Park are sculpted from the Claron Formation, laid down within an extensive lake during the early Cenozoic phase of the Laramide orogeny. Contemporaneous Absaroka volcanism near Yellowstone may have supplied pulses of ash that settled to the lake floor, forming discrete layers within the Claron.

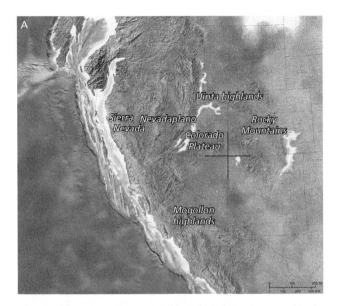

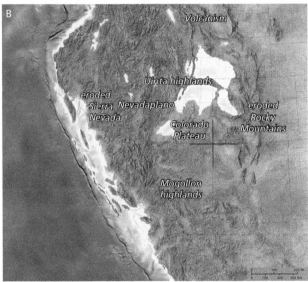

Figure 33.19 A) Paleogeographic map of the American West during the peak of the Laramide Orogeny ~65 Ma. Note how the American West is beginning to look more familiar. Rivers drained highlands of the Nevadaplano, central Arizona, and Colorado, depositing their sedimentary load in rivers, floodplains, and lakes occupying what is today the Colorado Plateau. B) Western United States at 50 Ma when the Colorado Plateau was buried in its own erosional debris. Large intermontane lakes were separated by the east–west Uinta Uplift. The Claron Formation exposed in Bryce Canyon National Park was deposited in a related lake in southern Utah.

In general, Laramide deformation was much less intense in the Colorado Plateau than in surrounding areas of the American West. High-angle reverse faults lifted Precambrian crystalline rocks and their overlying veneer of Paleozoic and Mesozoic sedimentary rocks upward without deforming them very much. Because most of the reactivated reverse faults didn't rupture the surface, the overlying sedimentary layers draped over the buried faults

like a carpet over a step, forming elongate *monoclinal folds* (Fig. 6.7). The north-trending Waterpocket monocline of Capitol Reef National Park is an iconic example of how a horizontally layered stack of rocks abruptly flexes downward to the east at a steep angle before returning to horizontality a kilometer or two away. A monocline is also beautifully expressed in Colorado National Monument where layers of Mesozoic redrock step downward above a reverse fault (Fig. 10.5). At nearby Black Canyon of the Gunnison National Park, Precambrian rock and its overlying carapace of sedimentary rock were raised along reverse faults as the Gunnison uplift.

In the Grand Canyon region, Precambrian crystalline rocks exposed today in the Inner Gorge were uplifted during Laramide compression as part of the Kaibab upwarp, a north-trending arch of rock that crosses the eastern side of the park (Figs. 3.4, 5.10). A monocline marks the east side of the upwarp and drapes downward to the east over a buried, north-trending fault, raising the Kaibab Plateau almost a kilometer above the adjacent plateau to the east. The rise of the Kaibab upwarp lofted its overlying layers of Mesozoic sedimentary rock to high elevations, making them more susceptible to weathering. Erosion has removed the blanket of Mesozoic and early Cenozoic rocks from the crest of the Kaibab upwarp. Remnants of these younger rocks are exposed in erosional escarpments in the Grand Staircase north of Grand Canyon (Fig. 5.10). (The Kaibab upwarp plays an important role in the formation of the Grand Canyon, as described in the next chapter.)

Throughout the entire Colorado Plateau, several other monoclines, upwarps, and "swells" were created during Laramide compression, typically with north–south orientations, perpendicular to the compressional stress. Near Dinosaur National Monument, however, the Uinta Uplift trends east–west, raising Proterozoic rocks of the Uinta Mountain Group as a broad, convex-upward *structural arch* (Fig. 9.9). These 800 m.y. old rocks are exposed in the Canyon of Lodore along the Green River. The east–west orientation of the Uinta Uplift likely reflects reactivation of old Precambrian faults that bounded the rift basin in which the Uinta Mountain Group was deposited.

In the Northern Rocky Mountains near Glacier National Park, Laramide compression created a fold-and-thrust belt that pushed Proterozoic rocks over and above late Mesozoic rocks (Fig. 22.10). The master thrust in Glacier is the Lewis Thrust, which laterally displaced a thick slab of rock 80 km (50 mi) to the east. The fold-and-thrust belt continues north as the Canadian Rockies.

An entirely different style of mountain-building occurred in the Wyoming and Colorado Rockies, however, where old faults were reactivated as high-angle reverse faults to form *basement-cored uplifts*. The *Archean* rocks of the Beartooth Mountains, along the northern border of Yellowstone National Park, were raised by Laramide compression. The Proterozoic rocks of the Front Ranges in Rocky Mountain National Park were lifted skyward along reverse faults as well (Fig. 21.8). In both regions, younger Paleozoic and Mesozoic rocks that blanketed the basement rocks were stripped away, exposing the older crystalline rocks. The U.S. Cordillera was growing wider with the rise of the Rocky Mountain Province.

The Laramide uplift of the Colorado Rockies was the third time that the crystalline basement rocks of the region would be part of a mountain chain. The first was during the Proterozoic when rocks were metamorphosed within the roots of mountains formed during accretion along the outer margin of ancient North America. The second was during the late Paleozoic as part of the ancestral Rocky Mountains. By the mid-Cenozoic, the Laramide version of the Colorado Rockies would bury themselves in their own erosional debris, just as the earlier versions did. The modern Front Ranges of Rocky Mountain National Park would once again rise in the latest Cenozoic, discussed later in this chapter.

By the end of the Laramide orogeny in the mid-Cenozoic, the American West was an elevated region, but without the sharp, jagged peaks of the earliest phases of uplift (Fig. 33.20A). The Sierra Nevada were reduced to low rolling hills draining the broad, upland surface of the Nevadaplano to the east. The Colorado Plateau was buried beneath a thick blanket of erosional debris from the surrounding highlands. And the Colorado Rockies had been beveled to an extensive upland plateau that sloped eastward toward the High Plains. The general dimensions of the U.S. Cordillera were set by the end of Laramide time, but the modern, rugged topography of the American West was yet to be created. The long phase of compressional tectonism that built the foundation of the U.S. Cordillera was coming to a close, soon to be replaced by the tectonic stress of extension and shear through the mid- to late Cenozoic.

One other important event that occurred during the Laramide Orogeny that influenced the landscapes of many national parks of the American West was the initial formation of *joint systems*. Brittle rocks of the upper crust tend to fracture when subjected to the flexural stress of uplift. Many of the fractures formed parallel sets of near-vertical joints, commonly at sharp angles to one another. These ubiquitous planes of weakness within the bodies of rock composing the American West would be exploited by weathering and erosion later in the Cenozoic to create the various landscapes we see today in the national parks.

Around 66 Ma, coincident with the Laramide uplift of the American West, a meteor about 10 km (6 mi) across streaked through the atmosphere, crashing into the northern margin of Mexico's Yucatan Peninsula. The environmental devastation triggered by the impact had dire consequences for the biosphere. The dinosaurs vanished, along with marine reptiles and several other groups of plants and animals, both on land and in the sea. This mass extinction marks the end of the Mesozoic and the beginning of the Cenozoic. Recovery of the biosphere took a few

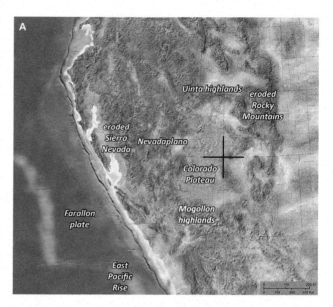

Figure 33.20 A) Paleogeographic map of the American West just after the Laramide orogeny ~35 Ma. The terrain was elevated, but generally subdued, as erosion and deposition dominated over uplift. The ignimbrite flare-up was active, covering the region in a thick blanket of ash. Off the west coast, the East Pacific mid-ocean ridge was getting closer to the subduction zone. Note how close the Colorado Plateau is to the west coast, prior to Basin and Range extension. Dark line just off the west coast is the Farallon subduction plate boundary. B) Basin and Range extension began about 16 Ma. The modern Colorado River watershed had not yet developed; streams drained into interior basins. The youthful San Andreas fault system was beginning to translate continental fragments to the northwest. A large inland sea covered today's Central Valley west of the southern Sierra Nevada. The Colorado Plateau and Rocky Mountains were blanketed by sediments and volcanic ash.

million years, but eventually opportunistic species took advantage of the newly available ecospace to repopulate the planet. Mammals and birds were the main beneficiaries of the catastrophe, including one particularly clever group of primates, the four-limbed, large-brained lineage that eventually led to humans by the latest Cenozoic.

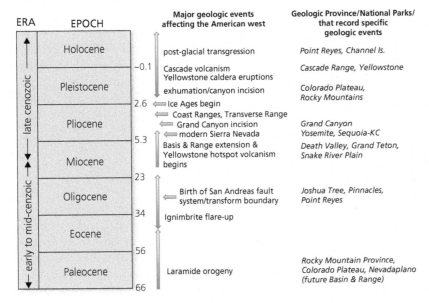

Figure 33.21 Relative chronology of major geologic events that affected the provinces and national parks of the American West through the Cenozoic Era. Many of these events overlap in time, so the relative order on the chart is not exact. Dates based on the Geological Society of America 2012 time scale.

Mid- to Late Cenozoic Tectonic Transition: Basin and Range

The Farallon slab began to steepen around 40 Ma, bringing an end to the Laramide orogeny (Fig. 33.21). The increased angle of subduction triggered the renewed production of magma and the initiation of the *ignimbrite flare-up* throughout the American West (Fig. 33.22A). These massive *caldera eruptions* spewed pyroclastic debris across a huge area ranging from the Nevadaplano of Nevada and Utah to the beveled remnants of the Laramide Rockies in Colorado. This prolonged phase of volcanism began around 43 Ma and ended around 21 Ma, marking the tectonic transition from east–west Laramide compression to east–west extension in the region. By the end of the flare-up, a thick veneer of volcanic tuff covered a widespread region. The rocks of the flare-up are recognized in Rocky Mountain National Park, the region around Black Canyon of the Gunnison National Park, and Death Valley National Park.

At this point, the narrative shifts to the west coast where the *East Pacific Rise*, the mid-ocean spreading center that created the oceanic crust for both the Farallon and Pacific plates by seafloor spreading, was approaching the North American plate (Fig. 26.10B). It was at the plate junction that the critical transition from the Farallon convergent boundary to the San Andreas transform boundary happened. Compressional forces had been occurring along this plate boundary since 200 Ma, so the transformation to shear stress beginning around 30 Ma profoundly changed the stress regime in the American West. The origin of extension in the Basin and Range Province is inextricably linked with the birth of the San Andreas transform plate boundary. The following

account is a synthesized version of the transformation of the plate boundary after 30 Ma previously described in Chapter 23 on the Basin and Range and Chapter 26 on the Pacific Margin.

The oceanic Farallon plate had been descending into the mantle beneath North America for about 170 m.y. when the system stalled as the East Pacific Rise encountered the subduction zone. The rate of subduction along the Farallon–North American tectonic boundary was higher than the rate of seafloor spreading along the East Pacific Rise, so the Rise and its system of transform faults was slowly dragged toward the subduction zone, coming into contact with the North American plate around 28–30 Ma. The hot and buoyant rock within the Rise resisted descending, and in time the Farallon plate is presumed to have separated from the adjoining Pacific plate and continued its slow-motion plunge into the mantle, pulled downward by the sheer weight of the slab (Fig. 33.22B). The Farallon plate that drove uplift of the American West for so long is today deep in the mantle and far to the east. Tomographic images show that the leading edge of the Farallon plate is approaching the core–mantle boundary beneath the eastern Atlantic Ocean (Fig. 33.23).

As the distance increased between the descending Farallon plate and the stalled Pacific plate, plastic rock of the asthenosphere flowed upward into the gap, eventually rising beneath the Nevadaplano (Fig. 33.22D). The buoyant plume of hot rock caused the overlying crust to flex upward, stretching it laterally. By about 16 Ma, the reduced compressional stress from the plate boundary to the west combined with the upward flexure from the hot rock below to initiate crustal extension beneath the Nevadaplano. These large-scale forces triggered brittle normal faulting in the upper crust and ductile flow in the lower crust, beginning the formation of *tilted fault-block* ranges and basins (Fig. 23.8). The once-elevated Nevadaplano slowly collapsed due to east–west crustal stretching and thinning to become the northern Basin and Range Province. In sum, even though active block-faulting didn't begin till 16 Ma, the tectonic transition from compression to extension in the American West began around 30 Ma with the origin of the San Andreas transform plate boundary.

The region between the east side of the Sierra Nevada and the west side of the Colorado Plateau doubled in width over the past 16 m.y. of crustal extension (Fig. 33.20B). The surface of the northern Basin and Range became a network of internally drained streams—the enormous, closed watershed of the Great Basin. The high heat flow, thin crust, and high elevations that characterize the modern Basin

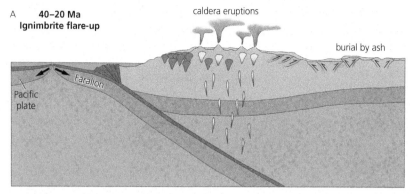

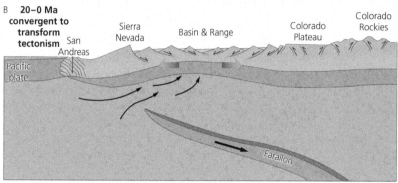

Figure 33.22 A) By about 30 Ma, the mid-ocean ridge separating the Farallon plate from the Pacific plate was approaching the subduction zone. Massive pyroclastic eruptions buried much of the American West in tephra. B) By about 16 Ma, the Farallon plate had separated and continued its downward journey beneath North America. Hot rock of the asthenosphere rose into the gap and flexed the Nevadaplano upward, eventually to form the Basin and Range.

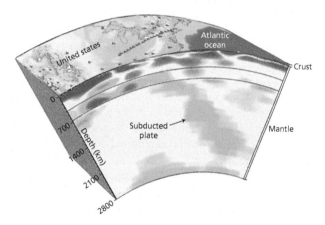

Figure 33.23 Seismic tomographic image of the Farallon plate beneath North America. Blue and green colors are relatively cooler, rigid rock. Reds and yellows are hotter regions of more ductile rock. The inclined slab of cool, stiff rock is interpreted to be the Farallon oceanic plate. Today, the leading edge of the plate is over 2500 km (>1500 mi) deep under the eastern Atlantic Ocean.

and Range are attributed to the welt of hot asthenospheric rock that underlies the region. The Basin and Range is an active continental rift system, although it is much broader than other modern continental rifts.

What are the effects of extensional tectonics in the geologic provinces and their national parks of the American

West? Block faulting began in the area near Death Valley National Park by 16 Ma and continues today. Movement along the Teton normal fault in Grand Teton National Park began about 9–10 Ma, with thousands of earthquakes resulting in the extreme vertical relief between the eastern escarpment of the Teton Range and adjacent Jackson Hole. Along the western side of the Basin and Range, the tilted fault-block of the Sierra Nevada began to rise about 5 m.y. ago, released from its topographic connection with the slowly downdropping Nevadaplano to the east. On the eastern side of the Basin and Range, normal faulting triggered a renewed phase of uplift of the adjacent Colorado Plateau around 5–6 Ma. (The rise of the modern Sierra Nevada and Colorado Plateau will be addressed in greater detail below.)

The beginning of extensional stress in the Basin and Range around 16 Ma was accompanied, perhaps not coincidentally, by the very first eruptions at the Yellowstone hotspot. The oldest caldera in the Yellowstone hotspot chain is dated to 16.5 Ma and is today located several hundred kilometers to the west–southwest of the hotspot (Fig. 31.19). The connection may be that the mantle plume rose through the gap created by the detached Farallon plate. As the plume rose ever higher, crustal thinning and the associated growth of faults and factures permitted magma to more easily work its way through the continental crust, generating the explosive eruptions at the hotspot.

Mid- to Late Cenozoic Tectonic Transition: Pacific Margin

The previous section related the transition from compressional tectonics to extensional tectonics in the Basin and Range. In this section, the transition from compression to shear along the San Andreas transform plate boundary will be addressed, emphasizing the impact on national parks of the active Pacific Margin. Most of California and much of the Cascadia Margin came into existence during the mid- to late Cenozoic.

Let's return to 28–30 Ma when subduction stalled as the East Pacific Rise met the North American plate (Fig. 26.10B). Transform faults bisecting the Rise began to accommodate the tectonic strain along the boundary, gradually shifting the stress field from one dominated by compression to one dominated by shear. The North

American plate was moving slowly toward the west–southwest, whereas the Pacific plate was moving more rapidly toward the northwest along the transform faults of the East Pacific Rise. It was this system of transform faults that commandeered the relative motion between the Pacific and North American plates, converting what was once a convergent boundary into a transform boundary. The network of strike-slip transform faults would eventually coalesce into the dominant San Andreas fault and its subparallel system of faults (Fig. 26.4).

The San Andreas fault grew longer as *triple points* at both ends migrated north and south as subduction continued along the remnant Farallon plate (Fig. 26.10C). Geologists call the main remnant of the Farallon plate north of the transform boundary the Juan de Fuca microplate. Today that convergent margin drives volcanism along the *Cascadia subduction zone*. Youthful explosive volcanism at national parks and monuments of the Cascades Range are all supplied with magma derived from the trailing remnants of the Farallon plate. Seaward of the Cascades volcanic arc, the accretionary wedge at the base of the continental slope progressively grew through the Cenozoic as turbidites and pillow basalts of the seafloor were scraped off the downgoing plate (Fig. 30.12). As the chaotic pile of rock accumulated, compressive forces elevated it above sea level, with the prism of rock eventually accreting to the edge of the North American plate. Today, these accreted terranes make up the rock of Olympic National Park.

Right-lateral, strike-slip motion along the San Andreas fault has displaced the sliver of continental rock to the west of the fault over 315 km (195 mi) to the northwest over the past 16 m.y. (Fig. 26.11). A key piece of evidence for the distance is the offset of 23 m.y. old volcanic rocks from their origin in the west Mojave Desert to Pinnacles National Park southeast of Monterey Bay. Motion along the San Andreas also displaced granitic rocks of the southern Sierran batholith to the northwest as the *Salinian Terrane* (Fig. 26.17). These Mesozoic granites were eventually dissected into slices by faulting and interweaved with Franciscan terranes of the Coast Ranges. A large mass of Salinian granite forms the topographic backbone of Point Reyes National Seashore. The mid- to late Cenozoic sedimentary rock that composes the gently rolling terrain of the Point Reyes Peninsula was translated into place along the San Andreas system over the last several million years as well.

In Southern California, *transpression* along the Big Bend of the San Andreas fault controlled the clockwise rotation and uplift of the Transverse Ranges through the mid- to late Cenozoic (Fig. 27.9). The Proterozoic metamorphic rocks and granites of Joshua Tree National Park were exposed at the surface as part of the uplift of the Transverse Ranges. At the western end of the Range, mid- to late Cenozoic sedimentary rocks and pillow lavas dominate Channel Islands National Park. This suite of marine rocks accumulated contemporaneously

with crustal *transtension* of the Continental Borderland, the submerged continental fragment marked by fault-bounded banks and ridges bordered by deep basins. In the last 5 m.y., a few of these banks and ridges were tectonically raised above sea level, creating the eight Channel Islands.

Latest Cenozoic Formation of the American West

Certain geologic events of the last 5–6 m.y. deserve special consideration because they created many of the modern landscapes we enjoy today in our western national parks. These contemporaneous events will be described in a west-to-east fashion, with the focus on how individual national parks achieved their current appearance. The final section of this chapter will address the effects of Ice Age climates on the landscapes of national parks of the American West.

The Gulf of California linear sea began to open around 5–6 Ma as the sliver of continental rock of Baja California was split off as part of the Pacific plate (Fig. 33.24). The Baja Peninsula is gradually separating from mainland Mexico as the Pacific plate moves to the northwest along a series of parallel transform faults beneath the waters of the Gulf. Seafloor spreading is occurring within short segments of the northern extension of the East Pacific Rise that cuts through the Gulf, with the result being the oblique opening of the *linear sea*. Rifting continues northward from the Gulf through the Salton Trough of southeastern California, an active continental rift basin partially filled with sediments from the Colorado River. This connection between the opening of the Gulf of California and the Colorado River will be more fully explored in Chapter 34 on the evolution of the Grand Canyon.

Transpression along the San Andreas fault during the northwesterly movement of the Baja-to-Point Reyes sliver of continental rock raised the Transverse Ranges and Coast Ranges, a process that is ongoing today (Fig. 33.25). All the national parks along the San Andreas fault system were progressively exposed to weathering and erosion as these mountains grew, typically in abrupt spasms during countless earthquakes. Parks located in arid climates, like Joshua Tree and Pinnacles, developed desert landforms. Parks located along coastlines, like Channel Islands and Point Reyes, were modified by waves, tides, and changes in sea level.

To the east, the modern Sierra Nevada tilted fault-block began its upward rise around 5 Ma. The system of normal faults lining the eastern base of the Sierra lifted the range one earthquake at a time, raising the granitic rocks of the Sierran batholith ever higher (Fig. 17.4). With uplift, the granitic foundation of Yosemite and Sequoia–Kings Canyon national parks was progressively exposed to the elements. As erosion removed the weight of overlying

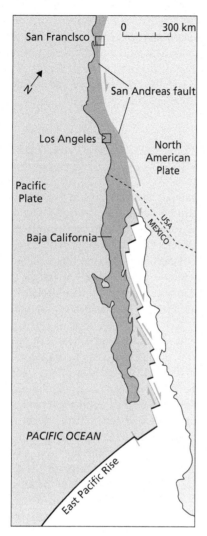

Figure 33.24 The sliver of continental rock comprising the Baja Peninsula and the western part of California is moving to the northwest as part of the dominantly oceanic Pacific plate. The Gulf of California is growing as a linear sea due to seafloor spreading and movement along transform faults of the East Pacific Rise.

rock, *exfoliation joints* developed, creating the rounded domes of Yosemite and Sequoia–Kings Canyon. Snowmelt and runoff from rainfall penetrated the rock along the joint systems cutting the plutons. Rockfalls and rockslides occurred along weakened joint planes, steepening the topography. The network of rivers draining the Sierran national parks, such as the Tuolumne, Merced, and Kaweah, flowed westward down the broad slope of the Sierra into the Central Valley. The sediments transported from the growing Sierran highlands accumulated within the Central Valley, transforming it from an inland sea to the fertile continental basin of today.

Normal faulting in the Basin and Range enhanced the relief between tilted fault-block ranges and downdropped basins, such as those in Death Valley National Park. The Badwater *pull-apart basin* sank ever lower as strike-slip motion initiated normal fault activity, deepening the basin

and raising the adjacent Black Mountains ever higher (Fig. 24.9). Today the Badwater Basin contains the lowest point in North America at –86 m (–282 ft), only 140 km (90 mi) from the highest point atop Mount Whitney in Sequoia National Park.

At the northeast corner of the Basin and Range, the Teton Range had been rising along the Teton fault since 9 Ma. Much of the 9 km (~30,000 ft) of total offset between the highest peaks of the range and the deepest sediments within the adjoining Jackson Hole basin occurred over the last few million years during thousands of large-magnitude earthquakes. The entire northern Basin and Range is highly active today, marked by youthful fault scarps along the base of ranges, high rates of seismicity, recent volcanism, and thousands of hot springs.

To explain the origin of the modern landscape of the Colorado Plateau, we need to back up a bit. Prior to 16 Ma, the Nevadaplano highlands lay to the west of the Colorado Plateau, while the subdued highlands of the Rocky Mountains lay to the east (Fig. 33.20A). Rolling mountains crossing central Arizona bounded the Colorado Plateau on the south, and the Uinta highlands lay to the north. None of these regions were particularly high in elevation, but they were higher than the intermontane basins on the Colorado Plateau. Stream networks flowed from these surrounding uplands into the plateau, depositing a veneer of sediment within large lake basins. The data are sparse, but it is likely that at the time the Colorado Plateau had *internal drainage*, similar to the modern Great Basin. If there was an outlet from the plateau to the ocean, evidence for it has yet to be discovered.

Several rivers in the modern world empty into large interior basins, never reaching the open sea. The Okavango River drains into a large inland delta in Botswana, Africa. The Tarim River in western China ends in a salt-crusted basin with the Taklamakan Desert. And in Chapter 23 you read how the Humboldt River crosses the Great Basin of Nevada to evaporate within the Humboldt Sink. These rivers provide modern analogs for visualizing the high-elevation basin that was the Colorado Plateau during the early to mid-Cenozoic.

Prior to ~16 Ma, the precursors to the Colorado and Green river drainage systems had not yet organized into the southwest-flowing network seen today. Instead, rivers and streams developed meandering patterns on the low gradient carpet of loose sediment that buried the older rocks and Laramide structures below. The stage was set for the formation of the deeply entrenched, meandering canyons that form the centerpiece of many of the national parks on the plateau.

Normal faulting along the boundary between the Basin and Range and the Colorado Plateau began around 16 Ma but appears to have accelerated around 5–6 Ma (Fig. 33.26). The Basin and Range to the west and south of the Colorado Plateau episodically dropped downward along normal faults, increasing the relief between the two provinces. It is likely that, as the Basin and Range

Figure 33.25 The American West was beginning to look more modern by 5 Ma. Basin and Range extension widened the western United States by about 10–5 Ma. The California Coast Ranges and Transverse Ranges had begun to rise upward as they moved into place along the San Andreas fault. The modern Sierra Nevada was tilted upward by faulting along its eastern flank. The Gulf of California had evolved into a linear sea, opened by a combination of shear and extension. The trace of the ancestral San Andreas fault is shown by the red dashed line.

collapsed, the Colorado Plateau was simultaneously lifted upward as an intact crustal block, perhaps because it was detached from the thinned crust of the Basin and Range by faulting. The sedimentary rocks of the Colorado Plateau maintained their relatively horizontal orientation during uplift, whereas the sedimentary rocks within the Basin and Range were rotated and tilted within fault blocks. Another generation of jointing may have formed on the Colorado Plateau during this latest Cenozoic phase of uplift and gentle tilting.

The increased difference in elevation between the collapsing Basin and Range and the rising Colorado Plateau amplified the gradient of streams on the southwest margin of the Plateau. The meandering streams and rivers on the plateau responded to the higher gradient with an increase in velocity and thus erosive power. They rapidly cut downward through the cover of loose sediments until encountering the flat-lying, sedimentary rock beneath (Fig. 33.27). These *superposed streams* maintained their meandering pattern as they incised into the underlying

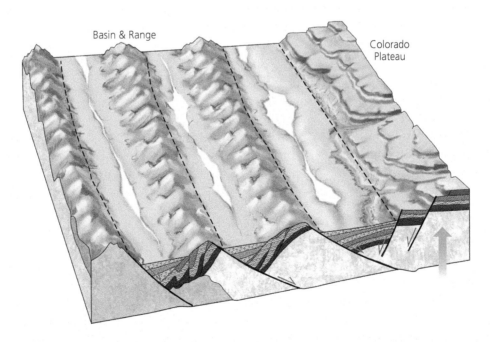

Figure 33.26 Lowering of the Basin and Range Province beginning around 16 Ma increased the relief along the boundary with the adjoining Colorado Plateau. Stream gradients increased, with the stream response being an increase in the rate of incision on the Plateau.

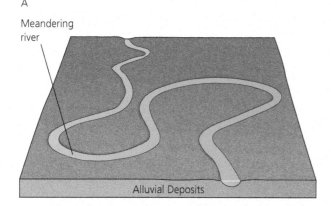

A

Meandering
river

Alluvial Deposits

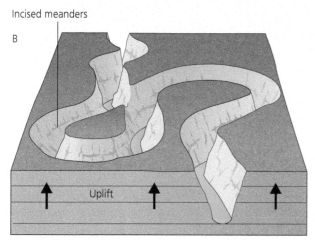

Incised meanders

B

Uplift

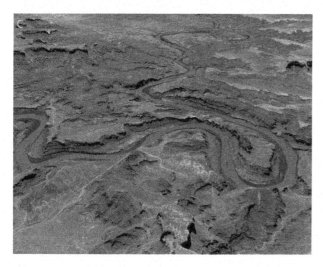

Figure 33.27 A) Streams develop meanders as they flow across low-gradient landscapes. B) As the region is uplifted, the rivers carve through the cover of loose sediment until reaching the solid bedrock beneath. With continued uplift, the superposed rivers maintain their original pattern, carving the meandering, entrenched canyons that characterize the Colorado Plateau. C) Incised meander of the Colorado River, Canyonlands National Park.

rock, carving the looping, entrenched canyons that are so ubiquitous on the Colorado Plateau. During this enhanced phase of river incision, the tributaries of the incipient Colorado River system reorganized to flow toward

the southwest. The Colorado River probably attained its modern form around 5–6 Ma, originating in the highest elevations of the Colorado Rockies, flushing through the adjacent Colorado Plateau, then descending through the lowest elevations of the Basin and Range on its way to sea level at the Gulf of California.

What was the effect of superposed stream incision in the national parks of the Colorado Plateau? The Colorado and Green rivers and their tributaries cut deep canyons near Arches National Park and directly through Canyonlands National Park (Fig. 33.27C). The Virgin River cut downward to create Zion Canyon, aided by *cliff retreat* and *headward erosion*. Tributaries of the Paria River system dissected the eastern margin of the Paunsagunt Plateau, sculpting the ornate architecture of Bryce Canyon National Park (Fig. 5.12). Rivers and streams stripped away the cover of loose sediments that obscured the underlying older rocks and Laramide structures, exhuming features like the Waterpocket fold in Capitol Reef National Park. The rejuvenated Green and Yampa rivers within Dinosaur National Monument uncovered the buried Uinta structural arch, carving deep gorges into resistant rock. The entire erosional landscape of the Colorado Plateau rapidly evolved as the rate of stream incision increased in response to this late Cenozoic phase of uplift. Over the last 5–6 m.y. of erosional modification by running water, all the stunning features of the Colorado Plateau were formed: long escarpments of horizontally layered sedimentary rock, deeply incised sinuous canyons, and a variety of joint-controlled mesas, buttes, and pinnacles (Fig. 33.28).

The Grand Canyon we see today was cut by a fully developed Colorado River during the late Cenozoic uplift of the Colorado Plateau, but the story there is more complicated than just stream superposition. To comprehend the origin of the Grand Canyon, it must be viewed as an evolving system, requiring a sequence of somewhat disparate events to explain its ancestry. That's the topic of the next chapter.

By the latest Cenozoic, the Laramide Colorado Rockies had buried themselves in their own erosional debris. Based on plant fossils, the upper surface was around 1800 m (~6000 ft) in elevation, about half the rarified elevation of the mountains today. The topography was a subdued erosional surface and nothing like the modern chiseled terrain. Rivers flowed eastward on the blanket of sand and gravel covering the roots of the Laramide mountains beneath, redistributing sediment as an enormous eastward-thinning wedge. The Front Ranges near Rocky Mountain National Park would have merged imperceptibly into the broad ramp of the High Plains to the east.

Around 5 Ma, the buried Laramide Rockies were exhumed as superposed rivers incised downward through the thick carapace of sand and gravel. One of two mechanisms was responsible for the exhumation of the long-hidden mountains, or perhaps a combination of the two

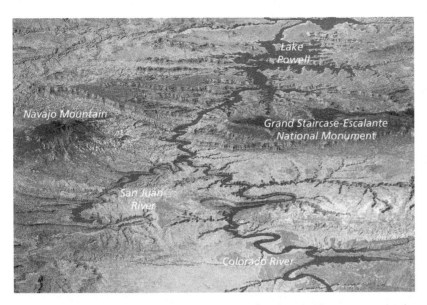

Figure 33.28 View southwest across the redrock Colorado Plateau toward Lake Powell. Higher elevations are green in this International Space Station photo. Meandering path of the Colorado River and San Juan River in the foreground.

created the magnificent landscape we see today. One possibility is that the crust was raised upward into the rivers, with the drainage system responding by cutting downward through the veneer of loose sediment to expose the underlying rock and Laramide folds and faults. Geophysical measurements reveal a welt of hot, mobile asthenosphere beneath the region that may have pushed up the overlying crust into the river systems. The other possibility was that the erosional power of the rivers was strengthened, enhancing the river's ability to wear away the blanket of loose sediment and eventually exposing the rock and Laramide structures beneath. The power of the rivers may have been intensified by wetter and cooler climates that affected the American West over the last 5 m.y. or so. Whatever the catalyst for the exhumation of the Colorado Rockies, the end result in Rocky Mountain National Park was the widespread extent of Precambrian crystalline rock at high elevations and the dissection of the highlands by fast-flowing, superposed rivers.

The final nonglacial geologic events during the latest Cenozoic in the Rocky Mountains were the stupendous caldera eruptions at the Yellowstone hotspot. As the Middle Rocky Mountains were being rejuvenated in the Yellowstone region, the supply of magma was continually building beneath the surface. Explosive thresholds were reached 2.1, 1.3, and 0.64 m.y. ago, with the catastrophic eruptions blanketing large parts of the American West in volcanic ash (Fig. 31.11). Later eruptions smoothed the surface of the Yellowstone Plateau and Snake River Plain. The U.S. Cordillera was complete, from Glacier National Park in the north to Joshua Tree National Park in the south, and from Point Reyes National Seashore in the west to Rocky Mountain National Park in the east. The final artistic touches on the

landscape of the American West were climatic in origin and sculpted by glaciers.

Icing on the Cake: Pleistocene Landscape Evolution in the American West

The serrated mountains and smoothly rounded valleys that decorate many of our high-elevation national parks are remnants of the Pleistocene Ice Ages of the last 2.6 m.y. Many lower-latitude, lower-elevation national parks were affected as well, but the sculptor was not glacial ice, but rather the meltwater runoff created during interglacial warm phases. Glacial modification and erosion by running water acted in concert to create the scenic vistas we enjoy today in our western national parks. Three types of glaciers either directly or indirectly impacted the landscapes of almost all national parks in the American West: continental ice sheets, ice caps, and alpine glaciers.

Large areas of the American Midwest and New England were intensely modified by the enormous continental ice sheets that crept southward from Canada (Fig. 33.29). But the continental ice sheets reached only a short distance into the American West during the Last Glacial Maximum around 20,000 years ago, and their effects are indirect on the national parks in the region. In the Montana Front Range, a lobe of the Cordilleran ice sheet blanketed the region just west of Glacier National Park (Fig. 22.12). In the Pacific Northwest, thick lobes of ice at the southern end of the Cordilleran ice sheet flowed westward into the Strait of Juan de Fuca, while another lobe extended south across what is today the Puget Lowlands east of the Cascades (Fig. 30.13). These kilometer-thick tongues of ice merged with alpine glaciers descending from the adjacent Olympic Mountains and Cascade Range. You would need to visit the modern coastlines of Greenland or Antarctica to see comparable landscapes.

Far more effective in their effects on sculpting the landscape were the ice caps that crowned the Cascades, the Sierra Nevada, the Colorado Rockies, and the Yellowstone Plateau (Fig. 33.29). Alpine glaciers extended downslope from the ice caps, rounding and smoothing valleys created earlier by stream erosion. The periodic alternation of glacial cold phases and interglacial warm phases during the 2.6 m.y. of the Ice Ages caused ice caps and their fringing alpine glaciers to expand then contract 20 to 30 times. The back-and-forth movement of glacial ice acted like a gigantic scraper on the landscape of many national parks,

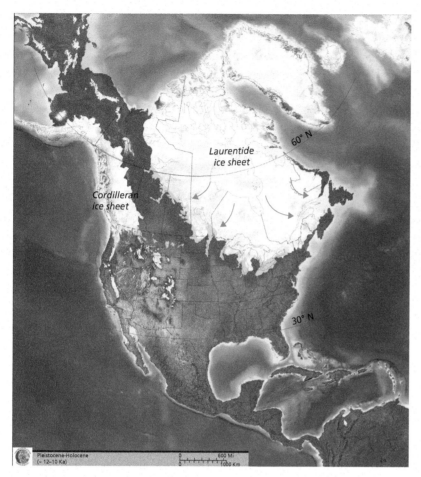

Figure 33.29 Extent of continental ice sheets across North America between 10 and 12 ka, a time of glacial retreat after the Last Glacial Maximum around 20 ka. Note the large pluvial lakes in the Great Basin.

continually modifying the terrain. The most apparent effects of glaciation in our national parks were created during the Last Glacial Maximum that peaked around 20 ka. The advance of ice caps and alpine glaciers during this event erased many of the effects of earlier glaciations and left behind its own set of glacial landforms.

Erosional landforms created as glaciers advanced include cirques, arêtes, cols, horns, hanging valleys, U-shaped valleys, roche moutonnee, striations, glacial polish, and a variety of glacial lakes. Depositional features left behind as glaciers retreated include lateral and terminal moraines composed of unsorted glacial till, gravelly outwash plains, erratic boulders, and the silty glacial flour that turns many glacial lakes a turquoise blue. All of these remnant Ice Age glacial features are expressed to a variable degree in the national parks and monuments of the American West.

During warmer interglacial phases of the Ice Ages, meltwater streams surged from the receding toes of glaciers, transporting huge amounts of sedimentary outwash within their torrential flows. As meltwater streams charged with sand, gravel, and boulders merged to form

larger rivers, the erosive power must have been far beyond anything witnessed today. Massive floods coursing through drainage systems of the American West battered the rocky walls of canyons. Many river valleys, canyons, and gorges were dramatically deepened and widened by countless floods during times of glacial melting.

Consider just the Colorado River system and its effects on the watershed as the Last Glacial Maximum came to a close around 15,000 years ago (Fig. 33.30). Meltwater runoff from the ice cap crowning Rocky Mountain National Park fed into the head of the Colorado River. The high-volume, high-velocity flows raged across the western Colorado Rockies, incising deep into bedrock. The *discharge* of the Colorado River increased as it flowed westward onto the Colorado Plateau, augmented by hundreds of swollen tributaries. One of the main tributaries, the Gunnison River, transported huge volumes of rocky debris picked up as it rampaged through the narrow gorge of the Black Canyon. The Colorado River grew even more fearsome near its confluence with its largest tributary, the Green River, in what is today Canyonlands National Park. The Green River, itself charged with meltwater and glacial debris from its headwaters in the Wind River Range of Wyoming, barreled through the Uinta Mountains, deepening the Canyon of Lodore in Dinosaur National Park.

As the ever-growing Colorado River surged through the Grand Canyon, its load of sand, cobbles, and house-sized boulders bashed against the bedrock of the canyon walls. And as the Grand Canyon was cut deeper, tributaries tried to match the pace of incision, flushing torrents of watery *debris flows* through side canyons and widening the overall canyon system. When the Colorado River reached its mouth near the Gulf of California, it deposited a huge volume of eroded sediment, building a delta outward into the linear sea. Now multiply that meltwater scenario of 15,000 years ago about 20 to 30 times to account for each interglacial phase during the Pleistocene. In concert with late Cenozoic uplift and the associated increase in stream gradients, Ice Age swings in climate played a profound role in the exhumation and landscape evolution of the Colorado Rockies and Colorado Plateau, expressed so elegantly in many of our national parks.

Even in the normally arid Basin and Range, Pleistocene climate fluctuations allowed for the accumulation of freshwater within closed-basin lakes, sourced by snowmelt and rainfall runoff in adjoining ranges. Individual lakes connected through low passes to create a linked network of closed *pluvial lakes* (Fig. 23.14). Glacial meltwater and snowmelt from the Sierra Nevada drained through a series of pluvial lakes and connecting streams to create the 180 m (~600 ft) deep Lake Manly in Death Valley National Park. Narrow wave-cut ledges notched into the surrounding mountains mark the ancient shorelines. The salt-encrusted surface of the Badwater Basin is all that remains of the long-vanished lake.

Two other significant effects of the Pleistocene Ice Ages influence the appearance of many national parks of the American West to this day. One consequence of the glacial-interglacial cycles is that modern river systems established their current pathways in response to glaciation. One example among many is the network of rivers draining the western slopes of the Sierra Nevada (Fig. 33.30). The positions of the sparkling rivers cutting through Yosemite and Sequoia-Kings Canyon national parks were

set as the Sierra were tilted westward during the latest Cenozoic. The stream network was deepened by meltwater runoff from the ice cap that crowned the mountains through the Pleistocene. The populace of the American West owes much of their current water supply and many of their water-based recreation to the aftereffects of Ice Age glaciation.

A second consequence of the Pleistocene Ice Ages is the change in sea level that accompanied the growth and decay of continental ice sheets. As sea level fell from the buildup of glaciers during phases of planetary cooling, broad continental shelves were exposed by the receding shoreline. Point Reyes National Seashore was tens of kilometers inland, and the four main islands of Channel Islands National Park were combined into a single huge island. As sea level rose during warm interglacial phases, the shoreline migrated landward. Continental shelves were submerged, and wave erosion carved the seacliffs and sea stacks along the emergent coasts of Point Reyes National Seashore and Channel Islands and Olympic

Southwestern United States: Topography and Rivers

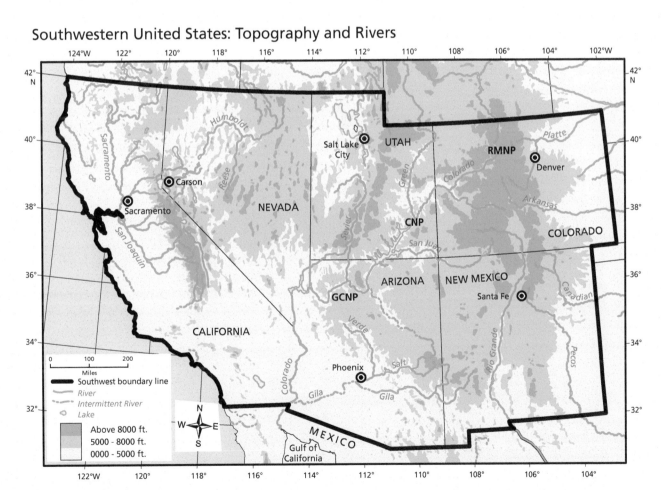

Figure 33.30 Major river systems of the southwestern United States. RMNP = Rocky Mountain National Park; CNP = Canyonlands National Park; GCNP = Grand Canyon National Park.

national parks. Over the past 15,000 years or so, the *post-glacial marine transgression* of the shoreline (Fig. 29.18) has set the natural boundaries of coastal national parks ranging from Channel Islands in the south to Olympic far to the north.

It may be that the growth and decay of ice caps and alpine glaciers exerted the single most important influence on the landscapes of national parks in the American West. If we include the erosional impact of streams and rivers charged with glacial meltwater and sedimentary debris, then there isn't much of an argument that Ice Age climate change put the finishing touches on the evolution of landscapes in our western national parks.

Related Sources

Baldridge, W. S. (2004). *Geology of the American Southwest.* Cambridge: Cambridge University Press.

Blakey, R. C., and Ranney, W. D. (2008). *Ancient Landscapes of the Colorado Plateau.* Grand Canyon Association.

Blakey, R. C., and Ranney, W. D. (2018). *Ancient Landscapes of Western North America: A Geologic History with Paleogeographic Maps.* New York: Springer.

Dickinson, W. (2004). Evolution of the North American Cordillera. *Annual Review of Earth and Planetary Sciences* 32: 13–45.

Meldahl, K. H. (2013). *Rough-Hewn Land: A Geologic Journey from California to the Rocky Mountains.* Berkeley, University of California Press.

34 Origin of the Grand Canyon

Figure 34.1 Colorado River within the Marble Canyon section of the Grand Canyon. View from the Nankoweap granaries.

The simplest way to understand the origin of the Grand Canyon is to just imagine the Colorado River and its tributaries carving inexorably downward into the uplifted southwestern corner of the Colorado Plateau for millions and millions of years. For many casual visitors to the Grand Canyon, that's enough to know. But as you might guess, the actual origins of the Grand Canyon are much more complex than just the progressive incision by the Colorado River system. The narrative involves long-vanished ancestral rivers that carved segments of the Grand Canyon, a chain of lakes that spilled one into the other, and the "piracy" of one river's flow by another. You'll come to see that the Colorado River we glimpse today deep within the Grand Canyon was finally assembled about 5 to 6 m.y. ago by a series of extraordinary events that span 70 m.y. of Earth history.

The age and origin of the Grand Canyon are among the great unanswered questions in the geosciences. Everyone agrees that the most recent incision of the canyon was done by the Colorado River, which you can see deep within the serpentine bends of the Canyon (Fig. 34.1). It's also well understood that canyon incision was triggered by uplift of the Colorado Plateau, as well as by changes in climate through the late Cenozoic. But the exact timing and mechanics of the origin of the Grand Canyon, as well as the exact role of the Colorado River, remain a focus of debate within the geoscience community. Many models have been proposed after a century of careful boots-on-the-ground fieldwork, geophysical measurements of the crust and mantle beneath the Colorado Plateau, and painstaking geochemical analysis of rocks and minerals in the region, but geologists are hardly in agreement on a single, conclusive explanation.

One of the recent geochemical techniques used to address this vexing question is called **thermochronology**, a method that dates the temperature history of specific minerals and thus the rock in which the mineral is found. The radiometric measurements permit determination of how long the mineral was buried beneath the surface and the rate at which canyon incision removed the overlying rock. In essence, the technique provides a way to reconstruct the elevations of past landscapes that have long since been eroded away. The methods of thermochronology are reliable, but the tricky part lies in interpretation of the data.

Arguments regarding the age of the Grand Canyon can be distilled into two broad categories. The "old Grand Canyon" camp has accumulated evidence that the majority of the incision of the Grand Canyon occurred around 70 m.y. ago during the latest Mesozoic in response to Laramide uplift. In this view, substantial portions of the Grand Canyon were incised by ancestral precursors to the Colorado River when dinosaurs were still around. Only in the latest Cenozoic did the Colorado River as we know it opportunistically occupy the old canyons to cut the final few hundred meters downward into bedrock on its way to the Gulf of California. The "young Grand Canyon" adherents, based on equally robust evidence, assert that the main phase of canyon incision occurred over the last 5 to 6 m.y. during the latest Cenozoic, primarily by the Colorado River. It may be that parts of both models are correct. Recent results indicate that different parts of the canyon formed at different times, requiring an evolutionary model for the age and origin of the Grand Canyon.

The earlier chapters of this book, especially Chapter 33, should give you a full command of the geologic processes and features that create the landscapes of our western national parks. In the present synthesis chapter, we can delve much deeper into the evolution of Grand Canyon National Park, far beyond our original discussion in Chapter 3. This chapter will begin with a brief review of the physiography of the Colorado Plateau and Grand Canyon, adding details beyond that of earlier chapters. Then the known sequence of events that most geologists agree on is described. Finally, a simplified model for the age and origin of the Grand Canyon is presented, along with the inherent uncertainties.

Geologic Framework of the Colorado Plateau and Grand Canyon

The Southern Rockies border the redrock country of the Colorado Plateau on the east, while the lower elevation Basin and Range Province surrounds the plateau on the west and south (Fig. 34.2). The Colorado River steps downward in elevation across all three provinces on its circuitous path to sea level at the Gulf of California (Fig. 2.4). Thousands of *superposed* tributaries of the Colorado River incised deep canyons into the level surface as the plateau was lifted upward during the late Cenozoic (Chapter 33). Canyons range in size from the deeply incised gorges of Canyonlands National Park to narrow slot canyons no wider than the width of your outstretched arms. The Grand Canyon is exceptional, not only because of its immensity, but because it experienced a more complicated evolution than other incised canyons in the watershed. The southwesterly flow of the Colorado River from its headwaters to its mouth has likely been in place for the past 5–6 m.y., but ancestral versions of the river system didn't always follow the same drainage channels as today because river systems evolve through time in response to geologic and climatic events. This is an important concept to grasp before the models for the age and origin of the Grand Canyon are presented.

The Colorado Plateau is not a single flat-topped surface, but rather a series of smaller plateaus separated by north–south-trending faults and *monoclinal folds*. Elevations of plateaus range between 1500 and 3000 m (~5000–10,000 ft). The plateaus where Zion and Bryce Canyon national parks are located are good examples of this fault-block topography where linear valleys, commonly occupied by rivers and streams, separate

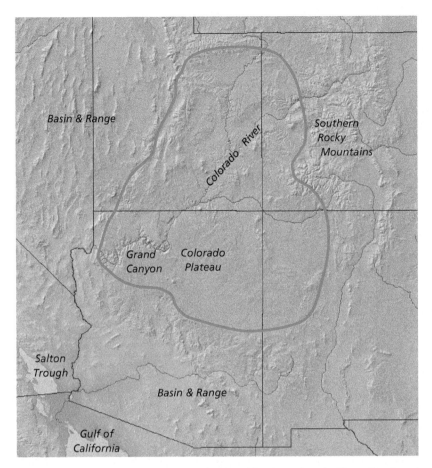

Figure 34.2 The Colorado Plateau is bisected by the Colorado River. The Grand Canyon is located in the southwestern corner of the Colorado Plateau and can be identified by its two great southward-looping bends.

cliff (Fig. 34.3). You may recall from Chapter 5 that this thick pile of Mesozoic and Cenozoic rocks once extended across the area of the Grand Canyon but was subsequently stripped away prior to the final incision of the Canyon 5–6 m.y. ago. The horizontal texture of reddish Mesozoic and Cenozoic sedimentary rock dominates the landscape of the Colorado Plateau, except along the walls of the Grand Canyon where multicolored Paleozoic sedimentary rocks lie above crystalline basement.

One of the defining features of the Grand Canyon region is the extraordinarily planar upper surface of the plateaus bordering the deep chasm of the Canyon (Fig. 34.4). The surrounding plateaus seem to extend endlessly in all directions, with their margins near the Grand Canyon notched by *headward erosion* of tributaries of the Colorado River. The late Paleozoic Kaibab Limestone (Chapter 3) usually forms the upper surface of these plateaus, with younger rocks almost entirely eroded away. The highest of the plateaus is the north–south-oriented Kaibab Plateau, elevated almost a kilometer above the Marble Platform to the east. The *topographic* Kaibab Plateau is the physical expression of the *geologic* Kaibab upwarp, an elongate highland formed during Laramide compression (Fig. 34.5). The East Kaibab monocline parallels the upwarp along its east side, draped over a partially buried fault.

The upper Colorado River flows generally southwest from its headwaters in Colorado to the Utah–Arizona border. There, it bends southward through Marble Canyon,

high-elevation plateaus (Fig. 5.12). Differences in the amount of offset on adjoining faults and monoclinal folds are responsible for the variable elevations and gentle tilt of most plateaus.

The linear escarpments of the Grand Staircase lie to the northeast of Grand Canyon, with each colorful cliff face marking the erosional limit of the rocks composing the

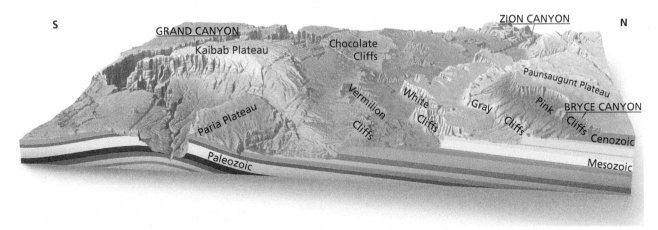

Figure 34.3 Linear escarpments of the Grand Staircase are the erosional edges of Mesozoic and Cenozoic rocks that once extended above the Paleozoic and Precambrian rocks of the Grand Canyon. Uplift of the Kaibab Plateau elevated the younger rocks and exposed them to greater amounts of weathering and erosion.

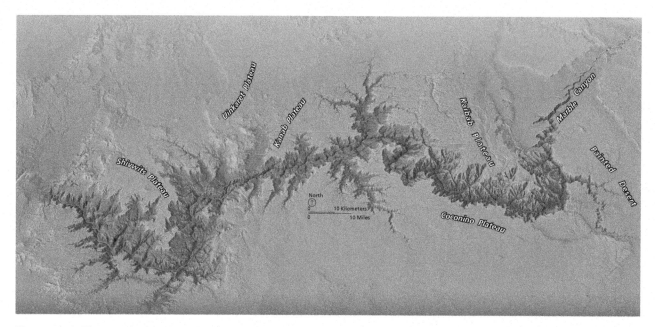

Figure 34.4 The Grand Canyon and its network of tributary streams incised downward into the level upper surfaces of smaller plateaus composing the larger Colorado Plateau in the region.

controlled by the north–south trend of the East Kaibab monocline (Fig. 34.6). A few kilometers south of its confluence with the Little Colorado River, however, the Colorado River makes an odd change in its direction of flow. It inexplicably takes a westward turn directly across the axis of the Kaibab upwarp. In Marble Canyon, the river flows through erodible Paleozoic sedimentary rocks, but as it abruptly turns to cross the Kaibab upwarp, it is forced to incise through the highly resistant crystalline rocks in the core of the arch, carving the Upper Granite Gorge below Grand Canyon Village (Fig. 34.7). This abrupt westward bend is one of the most perplexing features of the Colorado

River in the Grand Canyon and is an important aspect of all models for the origin of the Canyon. We'll return to this peculiar bend in the river later in this chapter.

The Coconino Plateau forms the lower elevation, southern extension of the Kaibab Plateau on the south rim of the Canyon (Fig. 34.6). To the west, the Kanab Plateau is separated from the Uinkaret Plateau by the Toroweap *normal fault*. Further to the west, the Uinkaret Plateau ends along the Hurricane normal fault which drops the rocks of the broad Shivwits Plateau into lower elevations. The Hualapai Plateau on the southwest rim of the Canyon is the southern extension of the Shivwits Plateau. The western edge of the

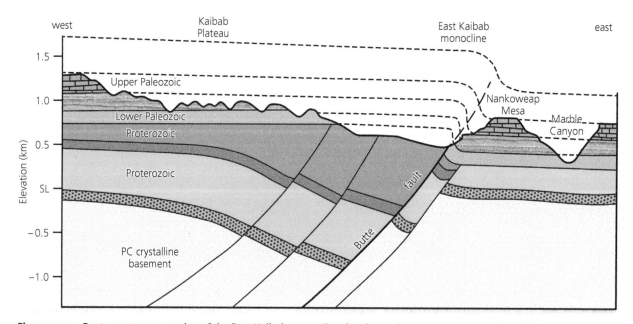

Figure 34.5 East–west cross section of the East Kaibab monocline that forms the eastern margin of the Kaibab upwarp.

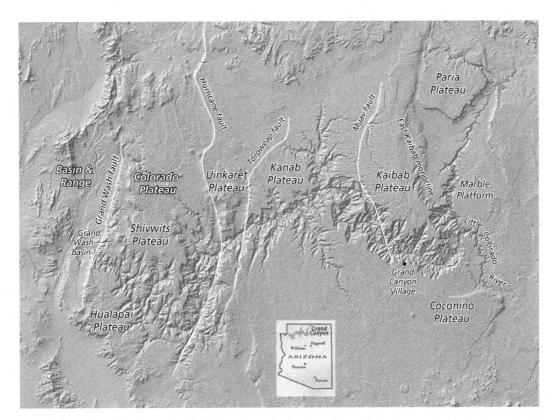

Figure 34.6 Several plateaus bound the Grand Canyon to the north and south, separated by faults or mono-clinal folds. The north rim of the Canyon has an average elevation of about 2500 m (8200'), whereas the south rim stands at 2100 m (~7000 ft).

Figure 34.7 Glimpse of the Colorado River deep within the Upper Granite Gorge of the Grand Canyon, as viewed from the South Rim.

offset, and collectively they form the stepped boundary with the lower elevation Basin and Range to the west. Think of each of these plateaus as a remnant surface where a few kilometers of Mesozoic and Cenozoic sedimentary rock have been stripped off by erosion. We see these younger rocks today in the northward-retreating escarpments of the Grand Staircase, including at Zion and Bryce Canyon national parks.

The vertical perspective of the Grand Canyon dramatically contrasts with the horizontal sightlines of the adjoining plateaus. This is why the first-time visitor to the south rim of the Grand Canyon is so awestruck. You become accustomed to the broad, level surface of the Coconino Plateau as you drive north toward Grand Canyon Village. But upon reaching the overlooks on the south rim of the Canyon, your eyes are immediately drawn downward toward the sublime spectacle spread out below. The Colorado River and its hundreds of tributaries have carved out a chasm about 1.6 km (1 mi) deep, and on average about 16 km (10 mi) wide. From Lees

Colorado Plateau in the Grand Canyon region is marked by the Grand Wash normal fault. Each of these three major north–south-oriented normal faults (Toroweap, Hurricane, and Grand Wash) exhibits down-to-the-west

Ferry, a few kilometers downstream from Glen Canyon Dam, the Colorado River traces a sinuous path for 446 km (277 mi) to its exit at the Grand Wash Cliffs at the very edge of the Basin and Range. Over that distance, the Colorado River maintains a consistent *gradient* of 1.3 m/km (7 ft/mi), dropping from an elevation of 950 m (~3100 ft) near Lees Ferry to 360 m (~1200 ft) at Lake Mead. (Recall that the gradient of a river is the change in elevation over a given distance, more simply known as the slope of the river.)

The origin of the Grand Canyon is directly related to the uplift history of the Colorado Plateau because uplift is the key to increasing the gradients of rivers and streams. As the gradient of a river increases, its velocity increases due to the higher slope. Rivers with higher velocities can transport greater amounts of sedimentary debris, which is why rivers in flood can move huge volumes of not only sand and silt, but cobbles and boulders as well. It stands to reason that fast-flowing rivers carrying large amounts of sediment will have tremendous erosive power due to physical abrasion on the bedrock walls of the channel. Thus, the greater the amount of uplift near the headwaters of a river, the greater the river's ability to incise into the underlying surface. This also works in reverse; if the elevation at the downstream end of a river system decreases, the overall gradient increases. Rivers will adjust to the increased gradient by cutting downward. The dynamics of river systems in response to episodes of uplift or subsidence exerts a powerful control on landscape evolution. The other primary control on the dynamics of river systems is the natural variability of climate, a factor addressed later in this chapter.

Known Chronology of Events on the Colorado Plateau

WESTERN INTERIOR SEAWAY

The Precambrian basement and overlying Paleozoic and Mesozoic sedimentary rocks of the Grand Canyon and the rest of the Colorado Plateau were submerged below sea level during the late Mesozoic around 90 Ma when the Western Interior Seaway covered the region (Fig. 34.8A). Sediments poured into the Seaway from highlands to the west, blanketing the site of the future Grand Canyon. Erosional remnants of Late Mesozoic mudstones are exposed near Bryce Canyon National Park, recording the infilling of the Seaway in the region. Clearly, there could not have been a Grand Canyon at that time. Since the late Mesozoic,

the Colorado Plateau has been lifted upward in at least two discrete stages to its current average elevation of over 2 km (1.2 mi). The Laramide orogeny was the first of the mountain-building events.

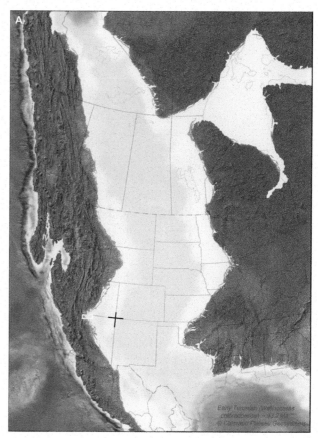

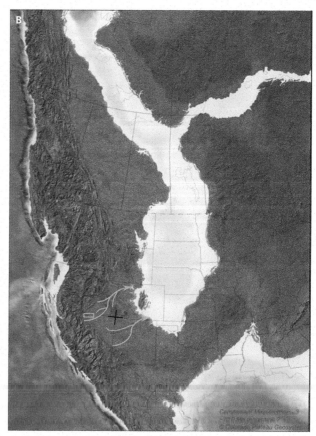

Figure 34.8 A) Extent of the Western Interior Seaway at 93 Ma. Note how the future Colorado Plateau is completely submerged. B) By ~70 Ma, Laramide mountain-building had forced the shoreline to the east, creating a broad coastal plain above the future Colorado Plateau. Northeast flowing rivers likely crossed the region as they drained into the seaway (blue arrows). Black cross marks Four Corners and white rectangle marks future Grand Canyon.

LARAMIDE UPLIFT

With the advent of the Laramide phase of mountain-building around 80 Ma, the future Colorado Plateau was lifted above sea level, forcing the shoreline to migrate eastward. By 70 Ma, the shoreline extended through Colorado, and rivers drained to the northeast across a broad coastal plain into the shrinking Seaway (Fig. 34.8B). The headwaters of these rivers were the ancestral Sierra Nevada, the Nevadaplano, and highlands in central Arizona. These ancient river systems flowed in the exact opposite direction to the modern Colorado River system and would maintain this orientation until at least 30 Ma. Some researchers have proposed that the Grand Canyon was carved to within a few hundred meters of its current depth by what they call the "California River" around 70 Ma, based on thermochronologic data. We'll return to this provocative idea in a later section, but for now we'll stay with those geologic events that are more broadly accepted.

Compressional tectonism during the Laramide reactivated old reverse faults around the margins of the Colorado Plateau, lifting it upward by a kilometer or more. For reasons that are not entirely understood, the plateau rose as a single, stable block of crust, with local flexures creating structural arches bounded by faults and monoclinal folds. The Kaibab upwarp in the eastern Grand Canyon was one of these broad north–south arches.

The actual amount of Laramide uplift is debatable, but elevations were high enough in the Grand Canyon region that erosion began to remove Mesozoic-age rock along the crest of the Kaibab upwarp. Other evidence in the region suggests that canyons near the modern Grand Canyon were incised perhaps a kilometer deep during the Laramide phase of uplift. One of these Laramide-age rivers flowed northward from the area of the present-day Hualapai Plateau toward the Shivwits Plateau, based on the distribution and orientation of river-deposited gravels of that age. It may be that a portion of the modern Grand Canyon inherited its orientation from this Hualapai **paleocanyon**, a possibility that will be addressed in a later section.

EARLY TO MID-CENOZOIC INTERIOR DRAINAGE

By the end of the Laramide orogeny, the Colorado Plateau was an elevated intermontane basin, perched between the surrounding highlands

of the Rockies to the east and north, the Nevadaplano to the west, and highlands cutting across central Arizona on the south (Fig. 34.9). In response, nearby rivers were forced to drain internally into large, closed basins, such as the enormous lake where the sediments of the Claron Formation accumulated, today exposed in Bryce Canyon National Park and many other places on the plateau.

As Laramide tectonism came to an end around 40 Ma, the record becomes spotty due to a lack of exposures of this age in the region. Ignimbrite volcanism was occurring elsewhere in the American West (Chapter 33), but ash deposits are uncommon on the plateau, likely because of removal by later erosion. During this Mid-Cenozoic phase of internal drainage, a thick blanket of sand and gravel was deposited across the Colorado Plateau, and river networks presumably meandered across the low-gradient landscape. By 25–30 Ma, the large lakes had mostly filled in, and a broad, sandy desert covered the Four Corners region, marking a phase of aridity. Streams on the Plateau may have become ephemeral due to the lack of precipitation in the region.

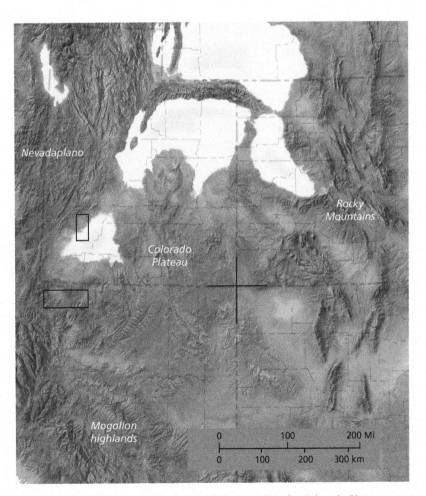

Figure 34.9 Rivers draining the highlands surrounding the Colorado Plateau. ~50 Ma flowed into a set of large intermontane lakes extending across central Utah into western Colorado. The Laramide-age Uinta Arch separated the lakes from an equally large lake to the north in Wyoming. The Colorado Plateau at the time was likely internally drained. Two rectangles show locations of future Bryce Canyon and Grand Canyon national parks.

BASIN AND RANGE EXTENSION

Distant events along the Pacific plate boundary generated extensional stress in the region south and west of the stable Colorado Plateau beginning around 16 Ma (Chapter 33). Normal faults along the eastern margin of the Basin and Range where it abuts the western edge of the Colorado Plateau caused the region to collapse episodically during countless earthquakes (Fig. 33.26). This spasmodic downdropping of the eastern and southern Basin and Range increased the relief with the adjacent highlands of the Colorado Plateau. The once meandering streams and rivers on the plateau now had a distinct gradient to follow, resulting in the realignment of the Colorado River system toward the southwest corner of the plateau.

This **drainage reversal** was intact by about 10 Ma, with the Colorado River flowing southwest from its headwaters in the Colorado Rockies toward the lowlands of the Basin and Range. The increase in gradient triggered an increase in the velocity of streams and their ability to carry a larger sediment load. Thus, many superposed streams and rivers on the plateau began to incise downward into the veneer of sand and gravel that buried older rock and Laramide structures. The modern Colorado River system was coming into focus.

A small, closed basin to the west of the Grand Wash fault (Fig. 34.6) accumulated sediments that provide important constraints on the timing of the formation of the modern Grand Canyon. The Grand Wash normal fault originated around 13 Ma, marking the birth of the boundary between the Basin and Range and the Colorado Plateau in the area of the Grand Canyon. By about 10 Ma, total slip along the Grand Wash fault reached ~3000 m (10,000 ft), with a deep basin forming along the downthrown, western side of the fault. Sediments of the Muddy Creek Formation accumulated in lakes and rivers within the Grand Wash basin between 13 and 6 Ma, attaining a thickness of over 500 m (>1600 ft). The sediments were derived from nearby fault-block mountains, as well as from an ancestral precursor of the Virgin River to the north. The key is that none of the sediments were derived from the Colorado River, with the implication being that the modern Colorado River in the Grand Canyon didn't exist prior to 6 Ma. This is the critical date around which all explanations of the modern Grand Canyon must pivot.

LATE CENOZOIC AGE FOR THE MODERN GRAND CANYON

The second main phase of uplift of the Colorado Plateau appears to have begun around 5–6 Ma. Several mechanisms driving surface uplift of the Colorado Plateau at this time have been proposed, most involving the flow of hot rock in the upper mantle that caused the western margin of the plateau to buoyantly rise upward as an intact crustal block. Recent images from *seismic tomography* reveal a mass of hot mantle rock currently positioned beneath the southwestern edge of the Colorado Plateau, which may have been the driver for late Cenozoic uplift.

This late Cenozoic phase of uplift triggered a variety of dynamic responses on the plateau. Gradients of streams and rivers were increased by uplift, with their amplified velocity starting a renewed phase of erosional denudation of the Plateau by running water. Older Laramide structures (like the Kaibab upwarp) were exhumed, with the erosional retreat of layered rock overlying the upwarp expressed as the vibrantly colored escarpments of the Grand Staircase. The rejuvenated Colorado River network of streams and rivers, previously constrained to sinuous meanders across the low-gradient, mid-Cenozoic surface, rapidly cut downward to carve deeply entrenched canyons (Fig. 33.27). The modern Grand Canyon was cut by the superposed Colorado River and its tributaries at this time, but the Grand Canyon has multiple segments and odd twists and turns that require a more complicated explanation, detailed here.

In the Grand Canyon region, motion along the north–south-oriented Toroweap and Hurricane faults (Fig. 34.6) occurred in the last 2–4 m.y., perhaps triggered by the flexure of the rigid crust above the rising hot rock of the mantle deep below. Total offset on these two down-to-the-west normal faults is ~580 m (1900 ft), occurring episodically during thousands of earthquakes. The main consequence of the fault activity was the downdropping of the western Grand Canyon block and thus an increase in the gradient of the Colorado River through the Grand Canyon. It's no coincidence that the Grand Canyon's main phase of incision occurred during the last 6 m.y. as the region was uplifted and faulted, forcing the river to cut downward as it dropped off the edge of the Colorado Plateau on its way to the lower elevations of the Basin and Range and, ultimately, to the Gulf of California.

Evidence for this late Cenozoic age for the formation of the *modern* Grand Canyon by the *modern* Colorado River is the appearance of sediments derived from the Colorado watershed in the Lake Mead area after 6 Ma. These younger river gravels cut through the older sediments of the Muddy Creek Formation in the Grand Wash Basin, indicating that the Colorado River was actively exiting the Colorado Plateau through the Grand Wash cliffs at this time (Fig. 34.6). Once it broke through the edge of the Colorado Plateau, where did the Colorado River go? Recall from the previous chapter that the Gulf of California began to open by 5 to 6 Ma (Fig. 33.25). The Colorado River, now streaming through the Grand Wash cliffs onto the Basin and Range, finally had a destination.

The Colorado migrated southward toward sea level by sequentially draining into a series of closed lake basins in a process known as **basin spillover** (Fig. 34.10). The sedimentary rocks that record this stepwise fill-and-spill sequence within the Colorado River corridor are known as the Bouse Formation. The closed basins formed a series of topographic steps in the extended and broken terrain that reached southward toward sea level at the Gulf of California. The sequence began with the filling of the northernmost basin near modern-day Las Vegas with Colorado River sediment and water. As the rapidly filling basin reached its brim, the water spilled over into the next closed basin to the south. The pattern of basin filling followed by

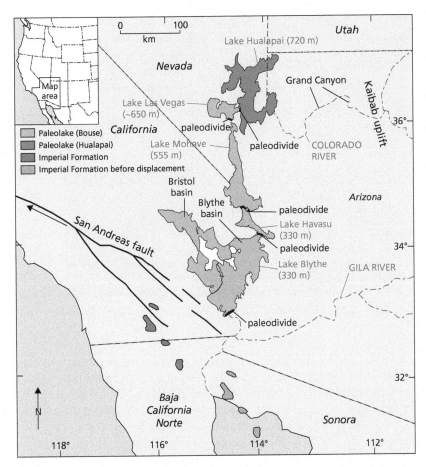

Figure 34.10 As the modern Colorado River exited the Colorado Plateau onto the lower elevation Basin and Range, it migrated southward by sequentially spilling over through a series of closed lake basins. This map shows the maximum extent of each of the basins during the late Cenozoic, inferred from exposures of the Bouse Formation.

cases building large *cinder cones* like Vulcan's Throne (Fig. 34.12). Basaltic lava poured over the rim of the canyon as **lava falls**, cooling in place to create a dark drape of volcanic rock. Several lava flows were voluminous enough to continue downcanyon, in one instance extending over 125 km (78 mi) from its vent on the north rim.

During his 1869 exploration of the Grand Canyon by boat, geologist John Wesley Powell waxed rhapsodic about the lava flows, writing: "What a conflict of water and fire there must have been! Imagine a river of molten rock, running down into a river of melted snow. What a seething and boiling of the waters; what clouds of steam rolled into the heavens!" This remarkable interaction of lava and water within the deep, narrow gorge would truly have been an astonishing sight to behold, and Powell's exuberance is justifiable.

In places, the lava accumulated to thicknesses of over 700 m (2300 ft) within the canyon, creating a **lava dam** that temporarily backed up the Colorado River into a huge, elongate lake that must have reached far upstream. Downstream from the lava dam, the mass of basalt tapered into a wedge-shaped body, based on the thickness of erosional remnants of basalt flows lining the river (Fig. 34.13). Lava dams formed at least 17 times during the several hundred thousand years of active volcanism from the Uinkaret Plateau.

Most of the lava dams were probably unstable, due to their weak foundation of river gravels and shattered fragments of rapidly quenched basalt. The dams were eventually undercut by the river working away at the base, causing the dams to collapse catastrophically. Alternatively, the dams may have been breached along their upper surface by the overflow of water from the backed-up lake. No one is really sure how long the lava dams lasted; perhaps a few decades to several centuries passed before they were sufficiently weakened by the force of the river against the dam's upstream face. What is revealing are deposits of huge, angular blocks of basalt as big as a train car far downstream, carried along by enormous **outburst floods** marking the catastrophic collapse of the lava dams. These mega-floods must have been as spectacular as the lava flows that formed the dams in the first place.

The significance of these lava flows is that the Grand Canyon had to have been almost fully incised by ~800 ka when the lava poured into the river. Post-lava flow downcutting by the Colorado River is measurable, but the majority of Canyon incision must have occurred between ~6 Ma and ~800 ka.

spillover continued sequentially southward, forming a descending chain of elongate lakes.

As each of the lake basins filled with river sediment, the Colorado River incised downward through the loose silt, sand, and gravel, cascading southward until finally reaching the Gulf of California by 4.8 Ma. In a little over a million years, the Colorado River had carved its channel southward for ~450 km (280 mi) from the Las Vegas area to the Gulf of California. Finally, snowmelt from the Colorado Rockies had reached sea level, and the modern Colorado River was established. The stage was set for the final incision of the modern Grand Canyon.

LAVA FLOWS AND THE MODERN GRAND CANYON

Before the model for the fully integrated Colorado River and the origin of the Grand Canyon is described, one final, startling geologic event adds an important time constraint. Incredible as it seems, basaltic lava flows cascaded over the rim of the Grand Canyon from vents on the Uinkaret Plateau during four volcanic phases between 800,000 and 100,000 years ago (Fig. 34.11). One hundred and fifty distinct lava flows have been mapped. The magma migrated upward along the network of faults associated with the Hurricane and Toroweap main faults, in some

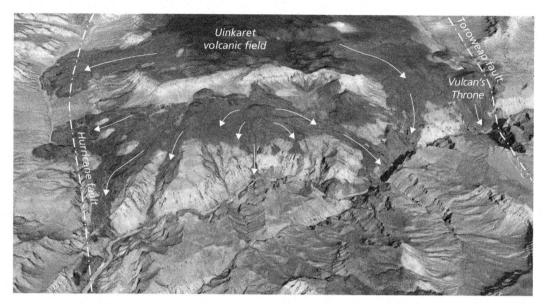

Figure 34.11 Dark basaltic lava flows of the Uinkaret volcanic field drape the lighter Paleozoic rock beneath. North is to the top of this oblique view.

As a side note, the famous Lava Falls rapid on the Colorado River occurs at the base of the lava flow beneath Vulcan's Throne on the north rim (Fig. 34.12). This exhilarating rapid with its huge waves is arguably the scariest half-minute on the entire river. You might assume from the name that the Lava Falls rapid was created by remnants of the ancient lava flows from the north rim, but instead it was formed by debris flows emanating from Prospect Canyon cutting the south rim. Turbulent slurries of sediment, boulders, and water episodically flush through the canyon, coming to rest within the main channel of the Colorado. The huge **debris fan** at the mouth of Prospect Canyon constricts the width of the main channel, increasing the flow velocity. The debris also creates an uneven riverbed, increasing the turbulence. Repeated debris flows from Prospect Creek create ever-changing conditions at Lava Falls rapid, one of many highlights experienced during raft trips through the Grand Canyon.

Model for a Fully Integrated Colorado River

A unified model for the age and origin of the Grand Canyon must take into account the generally accepted chronology of events described in the preceding section. This includes evidence for at least two main phases of uplift that influenced the gradient of streams and

rivers traversing the Colorado Plateau, as well as a complete reversal of the drainage direction of the Colorado River system. A unified model also requires that the full course of the Colorado River be set by about 5 Ma when it reached the Gulf of California. By this point in time, the Grand Canyon was not incised to its full depths, but it is generally accepted that the modern Colorado River flowed

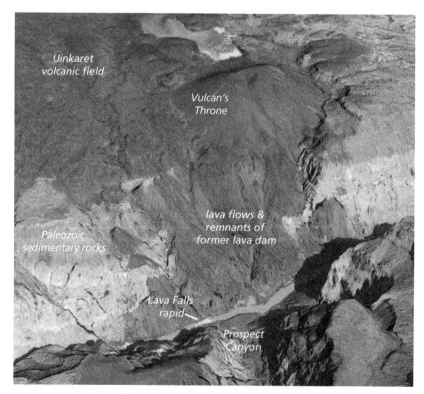

Figure 34.12 Vulcan's Throne is a cinder cone over a kilometer in diameter, perched on the north rim of the Grand Canyon. Lava Falls rapid is located near the mouth of the partially submerged Prospect Canyon debris fan.

Figure 34.13 Erosional remnant of basaltic lava flow near river level. The flows originally extended across the entire river channel but were subsequently incised by the sediment-choked river.

through the Canyon at some level. The simplified model presented here is full of uncertainties and is a work in progress. But it pulls together much of the evidence and ideas that have been developed over the last century or so by hundreds of geologists.

A key question is whether the Colorado River carved the Grand Canyon entirely by itself or whether ancient rivers cut major segments of the Canyon that the Colorado simply inherited through time. The "young Grand Canyon" school of thought leans toward the southwest-flowing Colorado River as the main driver of Canyon incision, with the vast majority of the Canyon cut in the last 5–6 m.y. The "old Grand Canyon" supporters view the modern Colorado River as an opportunist, merely occupying an ancient Grand Canyon that was deeply incised long ago during the late Mesozoic by an ancestral river flowing to the northeast. Reconciling these disparate models has been a contentious and ongoing debate in the geosciences. You would think that geologists would have already figured out how one of the most iconic landscapes on the planet was formed, but you'd be only partly right.

Recent syntheses of all the available geologic and thermochronologic data suggest that both models appear to be somewhat correct and that the Colorado River in the Grand Canyon was stitched together from a combination of ancient paleocanyons and more recently cut segments. A key element of the synthesis that most geologists agree on is that the Colorado River became fully integrated between 5 and 6 Ma. In this context, "integration" means that both old and young segments of the Colorado River in the Grand Canyon were completely connected as a single, throughgoing system. The nexus where integration was achieved is believed to be near the 90-degree bend where the Colorado River abruptly swings westward from Marble Canyon directly across the nose of the Kaibab upwarp (Fig. 34.14). How this might have happened is described below.

Recall that the geochemical technique of thermochronology enables the reconstruction of ancient landscapes and the ages of canyon incision. This method was applied to segments of the Grand Canyon, revealing that parts of the Canyon are very old and thus inherited by the modern Colorado River, while other parts of the Canyon are relatively young. In particular, the oldest part of the Canyon is the "Hurricane" segment, dated to between 50 and 70 m.y. in age (Fig. 34.14). The next oldest is the "eastern Grand Canyon" segment, dated as being formed 15—25 m.y. ago. The youngest segments are the "westernmost Grand Canyon" and "Marble Canyon," dated to the past 5–6 m.y. What does all this mean? How can one continuous canyon have different ages along its length?

The oldest Hurricane segment may have been originally carved during Laramide uplift by the north-flowing Hualapai paleoriver, defined by river-deposited gravels of early Cenozoic age located in now-dry Peach Springs Canyon (Fig. 34.4). The Hurricane segment is near-continuous with Peach Springs Canyon, suggesting that they are part of the same Hualapai paleocanyon. The Hualapai paleoriver may very well have been part of the proposed "California River" that drained northeast from highlands in the southwest toward the Western Interior Seaway around 70–80 Ma (Fig. 34.8B). Whatever the case, the paleocanyon may have been cut to perhaps half of the current depth of the Hurricane segment of the Canyon, creating a deep notch in the landscape that was later inherited by the modern Colorado River.

A different, long-vanished river carved the next-oldest eastern Grand Canyon segment to a depth of several hundred meters during the mid-Cenozoic. The ancient river that cut this segment flowed northwest across the crest of the Kaibab upwarp, superimposing a partially incised paleocanyon across the Kaibab by 15–25 Ma. It's unclear where this paleoriver originated and where it ended due to later modification of the landscape by uplift and erosion. The paleoriver could not have been connected to Marble Canyon because this younger segment hadn't yet been cut.

The two outermost segments of the Grand Canyon were incised by the modern Colorado River after 6 Ma. It was these two younger segments that began the process of connecting with the two older paleocanyons in between to fully integrate the Colorado River to complete the incision of the modern Grand Canyon. What river processes may have been instrumental in linking together the younger segments with the older?

One way that rivers and streams connect through time is by means of the related processes of *headward erosion* and *stream capture* (aka *stream piracy*) (Fig. 34.15). Two rivers may flow close to one another on either side of a natural divide. The stream with the higher gradient (and thus faster velocity) may erode headward through the divide, lengthening its channel until it reaches the lower gradient stream. The flow of the lower gradient stream will eventually be captured by the flow of the higher gradient stream, creating a single main channel and tributary channels.

Figure 34.14 Grand Canyon from Lake Powell in the northeast to Lake Mead and Colorado River corridor to the southwest. Dated segments of the Grand Canyon shown by yellow labels.

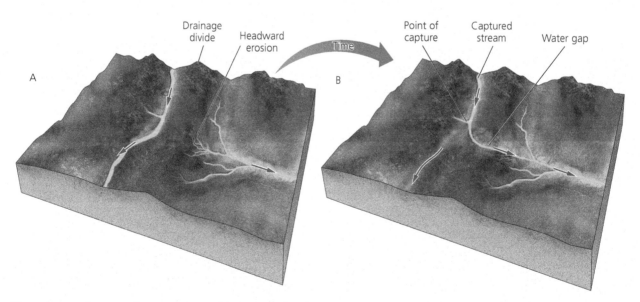

Figure 34.15 Concept of stream capture (aka stream piracy). A fast-flowing stream may capture the flow of a slower-flowing stream. Headward erosion by the higher velocity stream eventually cuts through the divide separating the streams, capturing the flow of the slower-flowing stream.

You may have read about this process in Chapter 10 on Dinosaur National Monument where the Green River became fully integrated by headward erosion and stream capture.

In the Grand Canyon between 5 and 6 Ma, the west-flowing drainage in the eastern Grand Canyon segment may have incised headward across the drainage divide of the Kaibab upwarp until it encountered the lower gradient ancestral Colorado River in Marble Canyon (Fig. 34.16). Flow through Marble Canyon may at one time have continued southeasterly through the channel of the Little Colorado River, emptying into a lake basin in northeastern Arizona. As the eastern Grand Canyon segment captured the flow of the Colorado River in Marble Canyon, a single main channel was created. This combination of headward erosion and stream capture may account for the strange, 90-degree bend in the Colorado River across the nose of the Kaibab Plateau. The Little Colorado River was forced to flow in the opposite direction to the north as a tributary of the main Colorado. This scenario may also explain the oblique angle of confluence between the Little Colorado and the main channel of the Colorado. (Most tributary streams merge with the main channel at acute angles in the direction of flow, like the tailfeathers of an arrow.)

A similar sequence of events may have occurred farther downstream where drainage in the westernmost segment, flushing down the steep gradient along the edge of the Colorado Plateau, eroded headward through the Shivwits Plateau to capture the older Hurricane segment (Fig. 34.14). In time, the entire network of young canyons and older paleocanyons were connected as a single throughgoing Colorado River in the Grand Canyon. The scenarios above are somewhat speculative, but they provide a way to visualize the integration of the Colorado River over the last 5–6 m.y.

The ultimate driver for the full integration of the modern Colorado River in the Grand Canyon was late Cenozoic uplift of the southwestern corner of the Colorado Plateau and associated downdropping along the Toroweap, Hurricane, and Grand Wash faults. This increased the gradients of the various segments of the Colorado River in the region and enhanced headward erosion, leading to the formation of a single river system.

The process of basin spillover downriver from the Grand Wash fault accomplished the final step in the integration of the modern Colorado River. As described earlier, the segment of the Colorado River entirely within the southern Basin and Range Province progressively migrated southward to the Gulf of California

after 6 Ma by the stepwise process of basin filling, followed by spillover into a lower closed basin. The connection with the westernmost segment of the Grand Canyon may have occurred by headward erosion through the Grand Wash cliffs. The process continued sequentially toward the south until the Colorado River incised a single channel into the sediment-filled chain of basins. By 4.8 Ma, the modern Colorado River was fully integrated, with all segments of the river within the Grand Canyon connected before flowing through the Basin and Range to its rendezvous with the open ocean.

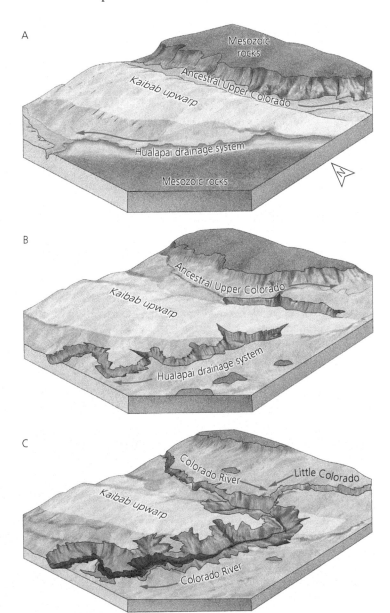

Figure 34.16 A) The west-flowing drainage in the Eastern Grand Canyon segment of the ancestral Colorado River was separated from the south-flowing Marble Canyon segment by the Kaibab upwarp. B) The paleoriver occupying the Eastern Grand Canyon segment eroded headward through the Kaibab upwarp, eventually capturing and redirecting the flow of the Marble Canyon segment after 6 Ma. C) The Little Colorado River, once the southern continuation of the Marble Canyon segment, was forced to flow in the opposite direction as a tributary.

This brief section has left many related processes (e.g., karst collapse) out of the discussion for the sake of simplicity and brevity. Even so, the model as described is anything but simple, and many unanswered questions remain. Maybe you'd prefer to just revert back to visualizing the Colorado River as eternal and stationary, cutting downward slowly but surely into the southwestern corner of the Colorado Plateau to create the magnificent landscape of the Grand Canyon. That's fine if you'd prefer to just let your eyes soak in the vistas without thinking too deeply about how it all came to be.

Final Landscape Evolution of the Grand Canyon

The modern appearance of the Grand Canyon developed over the last 5 m.y. or so, initiated by the full integration of the Colorado River within the Canyon. Since then, the Canyon has widened and deepened considerably, with multiple tributaries carving side canyons from the edge of the plateaus down to river level (Fig. 34.17). Landscape evolution within the Canyon was controlled by the two large-scale factors of mantle-driven uplift of the Colorado Plateau and late Cenozoic climate change. In turn, the combined forces of uplift and climate influenced the flow of the Colorado River and its tributaries, as well as many smaller scale processes that still occur within the Grand Canyon to this day. Those finer-scale, landscape-forming processes are the focus of this brief section, which builds on the earlier account presented in Chapter 3 on Grand Canyon National Park.

Recall that the alternating cliffs and slopes of the upper Grand Canyon are controlled by *differential erosion* of the Paleozoic sedimentary rocks. In the semiarid climate, resistant sandstones and limestones tend to erode into cliffs, while less resistant siltstones and shales tend to erode into slopes. As the softer rocks on slopes slowly decompose near their contact with overlying harder rocks, the cliff above becomes destabilized as it loses support from below. Huge slabs of rock may abruptly detach along weakened *joint planes*, collapsing downward as *rockfalls* and *rock avalanches*. The vertical faces of sandstone and limestone cliffs in the upper Grand Canyon are the topographic expression of failure along joint planes. Through time, the slow weathering of softer rocks and the rapid erosion of harder rocks cause *cliff retreat*, resulting in the progressive widening of canyons like that of the upper Grand Canyon. The combined effects of differential erosion and cliff retreat result in the ubiquitous "stairstep" topography of sedimentary rocks, not only within the upper Grand Canyon, but throughout the Colorado Plateau as well.

(I recall camping along a low terrace above the Colorado River during a raft trip and being startled awake by the clattering of falling rock against the walls of the canyon. It was raining that night, which likely triggered a mass of rock to detach from someplace high above. It was a disconcerting feeling for me as a vulnerable human, but that rockfall was just one of innumerable events like it that have been widening the Grand Canyon for millions of years.)

The crystalline rocks of the Precambrian basement are by far the most resistant to erosion and tend to form narrow, V-shaped gorges. The tight chasm of the Inner Gorge, with the sinuous Colorado River at the bottom, contrasts sharply with the broad upper Grand Canyon spread out along the tributaries (Fig. 34.17). The Inner Gorge was incised episodically during powerful floods when the churning waters of the Colorado transported huge boulders and other debris that bashed against the bedrock lining the channel. These floods don't happen today because the Glen Canyon Dam controls the flow of the Colorado, but they raged through the Grand Canyon countless times in the past 5 to 6 m.y., especially during wetter phases of climate.

The Colorado River has carved the Grand Canyon to an average *depth* of about 1.6 km (1 mi), but the river itself has probably always been about 90–120 m (300–400 ft) from bank to bank. The vast *width* of the Grand Canyon is created by the relentless downcutting of tributary streams.

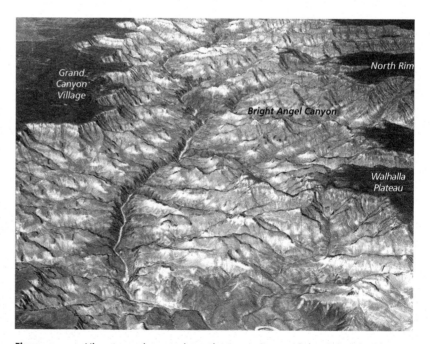

Figure 34.17 View toward west along the Inner Gorge of the Colorado River as it crosses the Kaibab Plateau. Note how the V-shaped canyon of the Inner Gorge is superimposed within the broader network of tributaries cut into the overlying Paleozoic rocks.

As the main channel of the Colorado inexorably deepens through time, the gradients of side streams incrementally increase, with the tributaries cutting downward in response. To a certain extent, side canyons deepen gradually by weathering and erosion from the runoff of snowmelt and rainfall. Much more effective, however, are episodic *debris flows* that may remove huge volumes of sediment during a single event. These dense, chaotic slurries may be triggered by intense summer storms on adjacent plateaus. As the runoff flushes through side canyons, the rapidly flowing water entrains loose, weathered sediment or remnant landslide debris. The water-saturated debris flow itself has a high erosional capacity, scouring the walls of the side canyon and gaining mass as it surges downslope. Depending on the intensity of the flow, the entire turbulent body may end up as a debris fan within the main channel of the Colorado River (Fig. 34.18). The side canyon is left just a little deeper and a little wider after excavation by the debris flow. The boulder-strewn deposits of debris flows create almost all of the 161 named rapids on the Colorado River in the Grand Canyon.

Tributaries on both sides of the Colorado River grow longer by headward erosion at *knickpoints* along their drainages. Recall from Chapter 8 that a knickpoint is simply an abrupt increase in slope where the acceleration of runoff is focused. In the Grand Canyon, knickpoints are commonly formed where a resistant cliff underlies a gentler slope. Runoff flowing through rivulets and gullies on slopes (like the Tonto platform shown in Figure 34.19) tends to converge at the knickpoint, commonly positioned along a joint plane. The velocity of the runoff increases sharply as the water flushes over the cliff face, incrementally cutting a notch ever deeper into the rock. The cumulative effect over long periods of time is the upslope migration of the knickpoint by headward erosion. In this manner, tributary streams gradually lengthen upslope toward their head. The edges of the plateaus lining the Colorado River have been dissected into intricate patterns over the last 5–6 m.y. by the incessant headward erosion of tributary streams.

Visitors to the south rim of the Grand Canyon are treated to dramatic views, including glimpses of the Colorado River far below. Visitors to the north rim witness equally magnificent vistas but are typically surprised by how far away the Colorado River is from their vantage point. The reason is that the gradient of streams coursing through tributary canyons is higher on the north side of the Colorado River than on the south side. This is due to the north rim being about 370 m (1200 ft) higher than the south rim, on average. The higher gradients allow for higher velocities of flow within side channels draining the north rim, which increases their erosive power and thus their ability to erode headward. Consequently, side channels on the north side of the Colorado River are much longer than those draining the south side, creating the asymmetry of the Grand Canyon (Fig. 34.17).

Figure 34.18 A) Tributary channel connecting South Rim with Colorado River. Distance from head of channel to river level is about 6.5 km (~4 mi). Arrowed lines show path of debris flows. B) Debris fan from side canyon, forming Hance Rapid in Colorado River. The bouldery debris constricts the channel and creates turbulent flow.

Along with uplift of the surface of the Colorado Plateau, the other major control on the incision of the Colorado River and its innumerable tributaries within the Grand Canyon is climate change, specifically during the glacial-interglacial cycles of the past 2.6 m.y. During glacial phases, water was locked up within the continental ice sheets, ice caps, and alpine glaciers worldwide. Meltwater

Tonto platform

Figure 34.19 View from south rim toward Inner Gorge of Grand Canyon. Thin dashed line marks Great Unconformity of Precambrian basement rocks below the cliff-forming Tapeats Sandstone. The Tonto platform is developed on the slope-forming Bright Angel Shale. Arrows mark knickpoints at the tips of side canyons where active headward erosion is occurring.

b.y. age of the rocks deep within the Inner Gorge, then they do record the geologic history of a truly ancient Earth. But if you're referring to the Canyon itself, then you're looking at a mere 5–6 m.y. of Earth history, an exceedingly short period of time considering the 4.6-billion-year age of the planet. Given enough time, nature can create wondrous landscapes such as those that delight the eye at Grand Canyon National Park.

"Leave it as it is. You cannot improve on it. The ages have been at work on it, and man can only mar it."

Theodore Roosevelt

Key Terms

basin spillover
debris fan
drainage reversal
lava dam
lava falls
outburst flood
paleocanyon
thermochronology

Related Sources

Beus, S. S., and Morales, M. (Eds.). (2003). *Grand Canyon Geology*, 2nd ed. New York: Oxford University Press.

Blakey, R., and Ranney, W. (2008). *Ancient Landscapes of the Colorado Plateau*, 1st ed. Grand Canyon Association.

Karlstrom, K. E., et al. (2014). Formation of the Grand Canyon 5 to 6 Million Years Ago through Integration of Older Palaeocanyons. *Nature Geoscience* 7: 239–244.

Powell, J. L. (2005). *Grand Canyon—Solving Earth's Grandest Puzzle*. New York: Pi Press.

Ranney, W. (2012). *Carving Grand Canyon: Evidence, Theories, and Mystery*, 2nd ed., Grand Canyon Association.

Timmons, J. M., and Karlstrom, K. E. (Eds.). (2012). *Grand Canyon Geology: Two Billion Years of Earth's History*. Geological Society of America Special Paper 489. Boulder, Colorado.

was continually released from the margins of glaciers, but the supply was relatively low and rivers like the Colorado probably had a meager discharge. But during episodes of interglacial warming, enormous amounts of ice melted, sending torrents of runoff through the river systems draining the glaciers. When the ice cap crowning the Colorado Rockies partially melted during warmer stages, the Colorado River through the Grand Canyon must have been unimaginably powerful. The constant impact of entrained rocky debris against the bedrock of the canyon walls was a highly efficient erosional force. Much of the incision of the Grand Canyon likely occurred during the multiple interglacial phases of the Pleistocene Ice Ages.

Let's place the 5–6 m.y. age of the Grand Canyon in perspective. When you visit, the Canyon looks eternal, as if you were staring back into the abyss of time. If you're referring to the 1.8

Select National Parks of the Appalachian Geologic Province

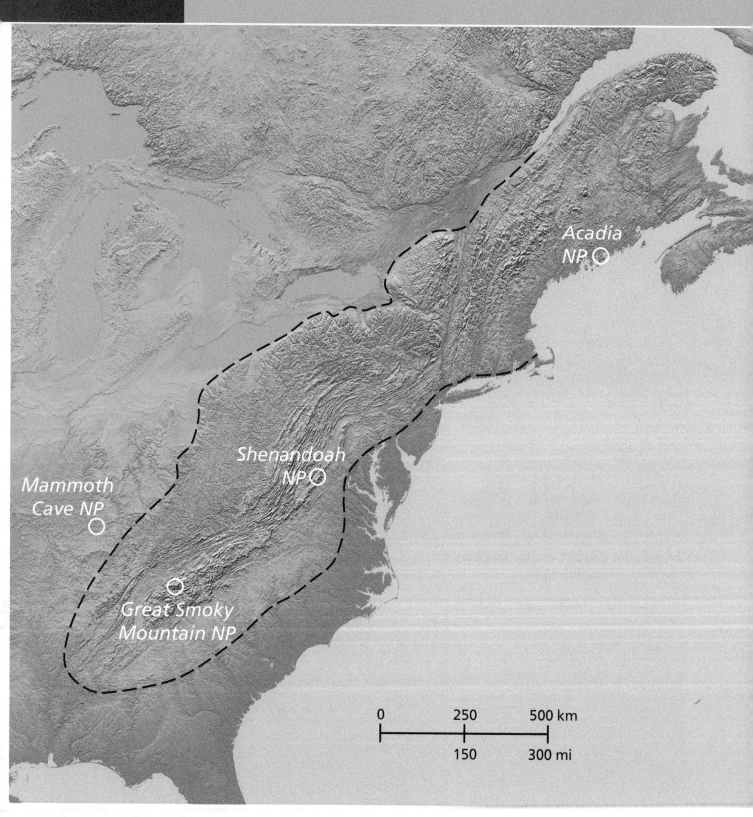

Figure 35.1 Relief map of the Appalachian Geologic Province in the eastern United States showing the locations of the four national parks discussed in Part 9. Dashed line marks the outer limits of the five main subprovinces that compose the Appalachians.

The Appalachian Geologic Province comprises the Appalachian Mountains, adjacent lowlands to the southeast, and a broad, dissected plateau to the northwest (Fig. 35.1). The province loosely parallels the eastern flank of North America, extending ~2600 km (1600 mi) from northern Alabama to Newfoundland. Landscapes of the Appalachian Province are highly variable, including clusters of parallel ridges, dissected plateaus, rolling lowlands, and elongate valleys (Fig. 35.2). The Appalachian Mountains form the central axis of the province, averaging about 900 m (~3000 ft) in elevation. Mount Mitchell in North Carolina reaches 2037 m (6684 ft), the highest in the chain.

Three national parks exhibit a diverse array of geologic processes and landforms of the Appalachian Province. **Great Smoky Mountains** and **Shenandoah** national parks, located within the smooth contours of the Southern Appalachians, preserve the roots of ancient continental collisions beneath their densely forested terrain. The two parks are connected by the winding **Blue Ridge Parkway** that stretches for 755 km (469 mi) along the crest of the

Blue Ridge Mountains, a distinct range within the larger Appalachian chain. In the Northern Appalachians, the rounded islands and peninsulas of **Acadia National Park** are granitic landscapes sculpted by continental ice sheets and coastal processes. **Mammoth Cave National Park** is not technically part of the Appalachian Geologic Province but rather is located in the Interior Low Plateaus Province. It is included within Part 9 based on its geological connections to the Appalachians, regardless of its provincial location.

Five distinct subprovinces compose the Appalachian Province, each defined by their unique geology and terrain (Fig. 35.3). The **Appalachian Plateau** forms the westernmost subprovince, extending from New York to northern Alabama. Flat-lying to gently tilted sedimentary rocks compose the plateau, creating the relatively level upper surfaces that define the subprovince. In northern Pennsylvania and southern New York, glaciation has smoothed and rounded the contours of the plateau, creating a rolling, hilly topography. The rest of the Plateau country is deeply dissected by streams and rivers, creating a rugged, irregular landscape of broad, forested summits separated by narrow valleys. The Paleozoic rocks that make up the Appalachian Plateau are only mildly deformed and thus are geologically distinct from the highly deformed rocks of adjacent subprovinces to the east.

The **Valley and Ridge Subprovince** snakes northward as a series of sinuous forested ridges and intervening valleys (Fig. 35.4). The geologic foundation of the Valley

Figure 35.2 View of the Southern Appalachians in North Carolina from the Blue Ridge Parkway.

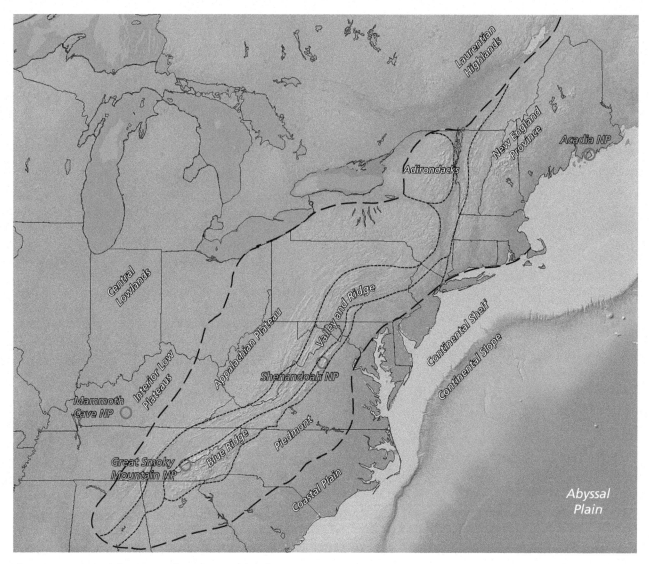

Figure 35.3 Map of the major subprovinces within the Appalachian Geologic Province in the eastern United States. Dotted lines mark the five main subprovinces within the larger Appalachian Province.

and Ridge is a *fold-and-thrust belt* of Paleozoic sedimentary rocks. (Fold-and-thrust belts were introduced in Chapter 22.) Compressional mountain-building during the late Paleozoic squeezed the layered rocks into broad, subparallel *anticlinal* and *synclinal folds* (Fig. 35.5). In places, the tectonic compression was intense enough to break thick sequences of rock along *thrust faults*, slowly pushing sheets of folded rock over and above adjacent sheets. Through time, the indefatigable effects of weathering and erosion have stripped away overlying layers, exposing the elongate patterns of folds. The sedimentary rocks erode differentially in the humid climate, with resistant sandstones forming narrow ridges and less resistant shales and limestones forming linear valleys. Thus, "valley and ridge" refers to the surface landscape, whereas "fold-and-thrust belt" refers to the underlying structure of the rocks. The "Great Appalachian Valley" is a continuous trough that marks the eastern edge of the Valley and Ridge where it abuts the Blue Ridge Subprovince. It extends from

the Champlain Valley on the New York–Vermont border all the way south to Alabama. The Valley and Ridge will be addressed in greater detail in the next chapter on Shenandoah National Park.

The Paleozoic sedimentary rocks of the Valley and Ridge and Appalachian Plateau subprovinces are sometimes referred to as the **Sedimentary Appalachians**, or the **Appalachian Basin** (Fig. 35.6). It can be confusing when uplifted sedimentary rocks in highlands are called a "basin," but that terminology refers to the original depositional area where the sediments accumulated through the Paleozoic. This distinction will become clearer later in the text as the origin of these rocks is addressed.

The **Blue Ridge Subprovince** is a narrow belt of rugged mountains extending from northern Georgia to southern Pennsylvania (Fig. 35.3). It forms the central axis of the highest elevations of the Southern Appalachians, with almost 40 peaks reaching heights above 1800 m (~6000 ft). Both Great Smoky Mountains and Shenandoah national

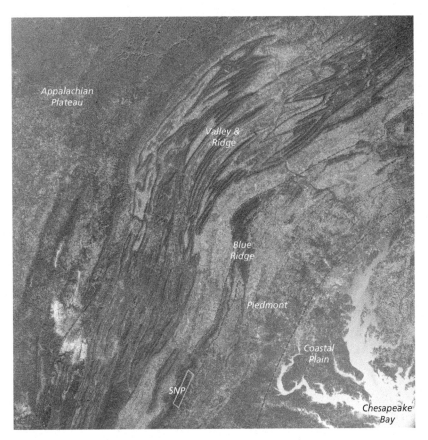

Figure 35.4 Satellite image of subprovinces within the central Appalachian Province. Note the different texture of the flat-lying, undeformed rocks of the Appalachian Plateau relative to the sinuous topography of the adjacent Valley and Ridge. Image centered around Pennsylvania, Maryland, and Virginia. Shenandoah National Park (SNP) is located in the Blue Ridge subprovince.

parks are located within the Blue Ridge Mountains. The reason for the high elevations and steep topography is that the Blue Ridge Subprovince is dominantly composed of highly resistant *crystalline rocks*, the general term for complex assemblages of igneous and metamorphic rocks. These rocks are mostly late Precambrian in age, having formed deep within the bowels of ancestral precursors of the Appalachian Mountains. During a late Paleozoic mountain-building event, the crystalline rocks of the Blue Ridge were thrust westward over and above younger Paleozoic sedimentary rocks (Fig. 35.6), forming an immense mountain range. Those late Paleozoic mountains have slowly worn away over the past 250 m.y., exposing their deep roots in the modern Blue Ridge Mountains.

East of the Blue Ridge is the broad **Piedmont Subprovince**, a low-elevation region of rolling lowland hills cut by deeply incised rivers flowing toward the Atlantic (Fig. 35.3). The bedrock beneath the thick soils of the Piedmont is composed of highly deformed crystalline rocks that preserve the effects of several phases of mountain-building. The Piedmont grades laterally into the **Atlantic Coastal Plain Subprovince**, a gently sloping landscape crossed by meandering river systems. The two subprovinces are separated by the **fall line**, a low escarpment that marks the boundary between the resistant crystalline rock of the Piedmont

and the feather-edge of unconsolidated Cenozoic sediments of the coastal plain (Fig. 35.7). Rivers crossing the fall line form waterfalls and rapids due to *differential erosion* between the strong rocks of the Piedmont and the weaker, less consolidated rocks of the coastal plain. The Atlantic fall line historically marked the inland limit of navigability and several major cities are located along it, originating as inland ports.

The fifth main subprovince of the Appalachian Province is the **New England Subprovince**, which is somewhat similar in composition to the Blue Ridge and Piedmont subprovinces (Fig. 35.3). The bedrock of the New England subprovince is highly variable, however, including both crystalline and sedimentary rocks of late Precambrian and Paleozoic age. The landscape of the New England Subprovince is a product of the *Pleistocene Ice Ages* when kilometer-thick continental ice sheets covered the region. The Adirondack Mountains west of the New England Subprovince have a greater geologic affinity to the Laurentian highlands of Canada to the north but are commonly considered to be part of the Appalachian Geologic Province due to their proximity and comparable topography.

The New England, Blue Ridge, and Piedmont subprovinces are sometimes referred to as the **Crystalline Appalachians**, reflecting their dominantly metamorphic and igneous composition (Fig. 35.6). These subprovinces are distinct in origin from the adjacent Sedimentary Appalachians to the west, composed of the Paleozoic rocks of the Appalachian Plateau and Valley and Ridge subprovinces. The Sedimentary and Crystalline belts of the Appalachian Geologic Province are useful entities for understanding the geologic history of the Appalachians, described below.

Rivers and streams draining the Appalachian Geologic Province flow into three main *watersheds*, which in turn drain into different parts of the Atlantic Ocean (Figs. 23.6, 35.8). Recall that a watershed is simply a topographic basin where surface runoff from rainfall and snowmelt drain through countless rivulets and creeks into larger tributary streams that in turn merge into a single main river. In eastern North America, the main drainage divide between the major watersheds follows a circuitous path northward along the Blue Ridge Mountains before crossing over into the Appalachian Plateau. Water to the west of the main divide flows into the Mississippi River watershed, which empties into the Gulf of Mexico. Runoff east of the divide flows through rivers that drain to the southeast across the Piedmont and coastal plain into the Atlantic Ocean.

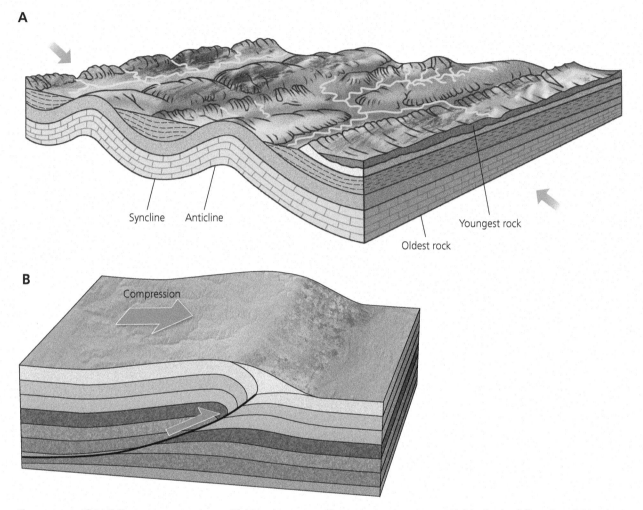

Figure 35.5 A) Anticlines are convex-upward folds, whereas synclines are concave-upward folds. Rocks deform into folds when subjected to tectonic compression (arrows). B) Low-angle thrust faults commonly form as layered rocks break during folding.

A secondary divide north of the main divide trends northeast, parallel to Lake Erie, then crossing back and forth along the northern edge of the New England Subprovince. North and northwest of this divide, runoff flows into the eastern Great Lakes and Saint Lawrence River, which flows northeast into the North Atlantic. Southeast of this divide in New England, streams and rivers drain into the eastern Atlantic.

Passive versus Active Continental Margins

The gently sloping coastal plain that borders the Appalachian Geologic Province along its eastern margin south of New York City continues seaward as the continental shelf (Fig. 35.3). (Part of the New England coastline is an *emergent coast* due to postglacial adjustments. The details will be addressed in Chapter 38 on Acadia National Park.) The seaward-thickening wedge of sediments beneath the coastal plain and continental shelf was derived from erosion of the ancestral Appalachians during the Mesozoic

and Cenozoic (Fig. 35.9). The pile of sediments reaches several kilometers in thickness beneath the continental shelf and slope.

The sedimentary rocks at the bottom of the wedge were the first to be eroded from the Appalachians during the Mesozoic, while the loose sediments at the top of the wedge were washed down from the Appalachians during the latest Cenozoic. So in a sense, the sediments beneath the Atlantic continental shelf are the inverted remains of the ancestral Appalachians. Drillholes through this thick pile of sediment and rock, made in the search for oil and gas, reveal the progressive degradation of the Appalachians through time. The smooth contours of the Appalachian Mountains that we see today are remnants left behind after massive volumes of overlying rock were eroded and transported through streams and rivers to be deposited on the coastal plain and continental shelf.

Recall from earlier chapters that the modern Atlantic edge of North America is a *passive continental margin* because the nearest plate boundary is over 3000 km (2000 mi) to the east along the Mid-Atlantic Ridge (Fig. 35.10). The Atlantic continental margin is thus located within

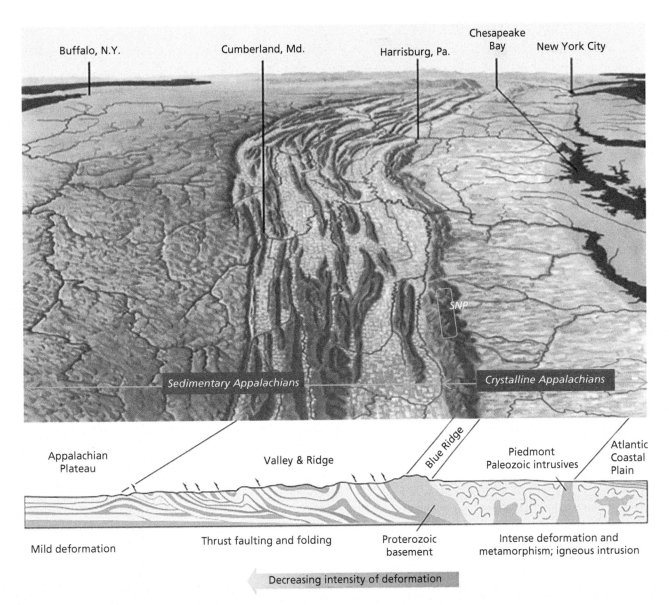

Figure 35.6 Paired topographic sketch of an oblique view north across the central Appalachians and idealized cross section of five subprovinces of the central Appalachians. SNP marks the location of Shenandoah National Park.

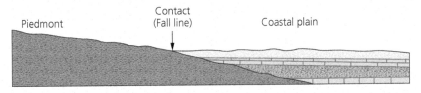

Figure 35.7 Simplified cross section showing the relationship between the crystalline rocks of the Piedmont and the Atlantic coastal plain. The fall line marks the contact where unconsolidated sediments of the coastal plain lap onto the bedrock of the Piedmont.

the interior of the North American plate, with the coastal plain and continental shelf drifting passively along as the plate migrates westward. The Atlantic continental margin has no volcanism and minimal seismicity. The geomorphic response to this tectonic calm along the Atlantic shoreline

is an array of coastal wetlands, broad beaches, and barrier islands that confine lagoons and bays.

These characteristics are in direct contrast to the *active continental margin* along the Pacific edge of North America where the physical margin of the continent coincides with the San Andreas and Cascadia tectonic boundaries (Fig. 26.2). Active seismicity and volcanism attest to the geologic dynamism of the Pacific Margin. This ongoing tectonism results in coastal mountain belts, steep seacliffs, and elevated marine terraces strung out along the Pacific Coast.

During the early Paleozoic, the eastern edge of North America was a passive margin, similar to the modern Atlantic

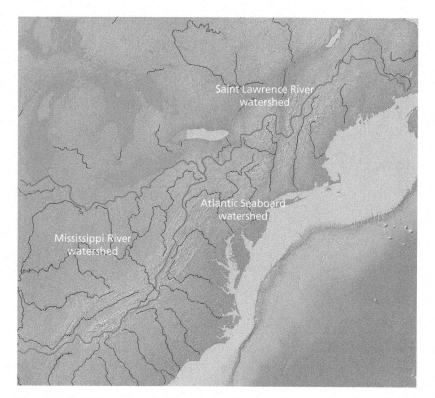

Figure 35.8 Rivers, watersheds, and main drainage divides (orange lines) of eastern North America.

Similarities and Differences between the Appalachians and the Cordillera

The processes, ancient landscapes, and tectonic history of geologic provinces of the U.S. Cordillera were described in Chapter 33 (Fig. 33.1). The purpose of that synthesis chapter was to integrate the various provinces and national parks into a single, coherent story. This chapter lays out the physiography and geologic history of the Appalachian Geologic Province to provide context for later chapters in Part 9. As you'll see, the Appalachians and the Cordillera share many common processes, but there are distinct differences in the origin and age of the two mountain belts that collectively tell a larger narrative of the assembly of the North American continent. The pivotal point occurred around 200 Ma during the Mesozoic when northwestern Africa and eastern North America began to split away from one another, leading to the birth of the Atlantic Ocean. As the North American continent slowly migrated westward after 200 Ma, a series of orogenic episodes began throughout the American West, building the mountains and basins that comprise the broad U.S. Cordillera. So, as the active tectonism of the Appalachians waned in the early Mesozoic, the action shifted to the other side of the continent with the accelerated pace of events in the American West.

continental shelf. But the rest of Paleozoic time in eastern North America was marked by a continuum of orogenic events spanning a few hundred million years. The multiple phases of uplift of the Appalachian Mountains and its ancestral precursors occurred along plate boundaries that were every bit as geologically active as those of today's Pacific Coast. A solid understanding of the differences between active and passive continental margins is essential for grasping the geologic history of the Appalachians and its national parks.

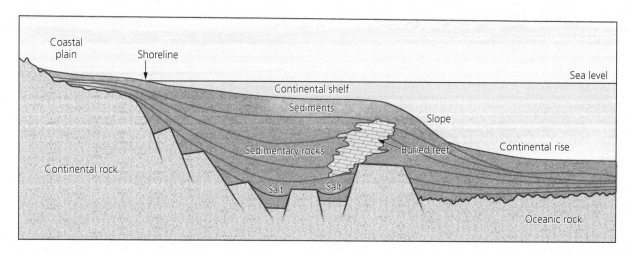

Figure 35.9 Passive continental margin of the Atlantic coast. Note how the seaward-thickening wedge of sediments rests on a foundation of faulted continental rock, formed during the rift phase of continental breakup. The transition to oceanic rock occurs near the continental slope. Water depths above the Atlantic continental shelf reach about 200 m (~660 ft) before abruptly deepening across the continental slope. The average depth of the world ocean is 4 km (2.5 mi), so in comparison, continental shelves are quite shallow.

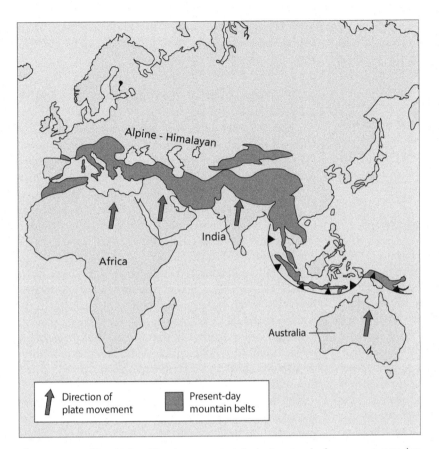

Figure 35.10 The Alpine-Himalayan mountain belt extends for over 10,000 km (6000 mi) from southern Europe, across the Middle East, through the Himalaya and Tibetan Plateau, and all the way to Southeast Asia. Elevations reach over 8 km (5 mi) in the Himalaya. This mountain belt is a modern analog for the Appalachians of the late Paleozoic.

What are some of the similarities between the Appalachians and the Cordillera? (1) Both orogens border the Precambrian nucleus of the North American continent along its eastern and western margins (Fig. 33.4). The Cordillera is oriented generally north–south, whereas the Appalachians trend slightly northeast–southwest. Both orientations are perpendicular to the main directions of tectonic stress that lifted the mountains upward. (2) The geologic history of each mountain belt is equally complicated, with long phases of *accretionary tectonism* contributing to the lateral growth of North America on each side. (3) Both the Appalachians and Cordillera preserve the rock record of the assembly of the late Precambrian supercontinent of *Rodinia*, as well as its eventual breakup. (4) Thick sequences of sedimentary rock accumulated along passive margins and inland seas near the Appalachian and Cordilleran continental edges at various times in the past. These massive volumes of layered rock were later uplifted and deformed during tectonic events affecting each margin, adding more real estate to the outer edges of the continent.

There are also a few physical similarities between subprovinces of the Appalachians and provinces of the Cordillera. For example, the horizontally layered sedimentary rocks of the Appalachian Plateau are comparable to those of the Colorado Plateau, specifically to the Paleozoic rocks of the upper Grand Canyon. Another example is that the fold-and-thrust belt of the Valley and Ridge Subprovince is identical in structural style to the fold-and-thrust belt of the Northern Rockies near Glacier National Park, although the topographic expression of each is radically different.

What are some of the differences between the Appalachians and the Cordillera? (1) Mountain-building in the Cordillera began during the late Paleozoic and accelerated through the Mesozoic and Cenozoic. Conversely, the uplift of the ancestral Appalachians occurred entirely during the Paleozoic. (2) Uplift of the Cordillera involved tectonic events driven by the *subduction* of the Farallon oceanic plate along its outer margin for about 170 m.y. from the early Mesozoic to the mid-Cenozoic (Chapter 33). The Appalachians also experienced orogenic phases related to subduction, but the ultimate episode of uplift involved *continental collision* with parts of Africa, South America, and Eurasia. If we could have somehow seen the Appalachian Mountains during their late Paleozoic climax, we would have been dumbfounded. They were the size and extent of the modern Alpine-Himalayan chain (Fig. 35.10), and they were located within the interior of the Pangean supercontinent along equatorial latitudes (Fig. 33.14). Elevations certainly reached 6000 m and perhaps 8000 m (~20,000–26,000 ft).

(3) The modern mountains of the Cordillera are steep and jagged, reflecting their relatively recent uplift. In contrast, the Appalachians are much lower in elevation, and their rounded contours echo their ancient origins and the dominance of erosion over the last 200 m.y. (4) The Cordillera is characterized by a wide diversity of climate conditions, including large regions of aridity in the rain shadows of mountain ranges. The Appalachians, broadly speaking, have a temperate and humid climate driven by proximity to the Atlantic Ocean. Ample precipitation contributes to a thick veneer of soil supporting vast tracts of hardwood forest.

With these basic similarities and differences in mind, you are ready for the geologic history of the Appalachian Geologic Province, with special emphasis on those regions near the four national parks: Shenandoah, Great Smoky Mountains, Acadia, and Mammoth Cave. Graphical timelines are interspersed through the following narrative to more easily visualize the chronologic sequence of events.

Concepts explored in the following sections include styles of mountain-building, accretionary tectonics and continental growth, metamorphism within the roots of mountains, and continental rifting. If you've already read the earlier chapters, most of these concepts are familiar to you. But you'll become reacquainted with them in the context of the Appalachians, and a few new concepts will be introduced, such as mountain growth by continental collision and the effects of continental ice sheets on eastern landscapes.

Geologic History of the Appalachian Geologic Province

Rocks exposed in the national parks and other regions of the Appalachian Geologic Province record a geologic history that began in the late Precambrian. Careful mapping, age dating, and geomagnetic analyses over the last century tell an incredible, billion-year-long story of the tectonic assembly of two distinct supercontinents, as well as their subsequent breakup into smaller landmasses. Embedded within this vast history is evidence for the slow-motion collision of continents, the rise and decay of mountain belts, and the opening and closing of entire ocean basins. In the process, chains of volcanic islands, slices of ancient oceanic crust, and fragments of continents were attached along the eastern continental margin of North America as *accreted terranes*.

You may recall from earlier chapters that accreted terranes (aka "exotic" terranes) are discrete bodies of rock that originally formed far away from their current location (Fig. 33.5). These terranes were tectonically transported across great distances as parts of lithospheric plates to physically attach (i.e., accrete) along a continental margin. Rather than descend into the subduction zone, these landmasses were scraped off the downgoing oceanic plate, then plastered to the continental edge. The boundaries between adjacent terranes are typically networks of sub-parallel faults that represent the *suture zone* of one terrane with the next. Accretionary tectonics is a fundamental process of continental growth and is one of the primary ways that both the Cordillera and the Appalachians were assembled. In fact, most of the continental crust of the eastern margin of North America is composed of a collage of accreted terranes, stitched together like a poorly made quilt. Fragments of these accreted terranes are exposed at the surface in Shenandoah, Great Smoky Mountains, and Acadia national parks, but the greatest volume of the terranes is hidden beneath younger rocks and the surficial cover of soil and vegetation.

The geologic variability along the length and breadth of the Appalachian Province reflects a diversity of ages and origins of the rock composing the region. As you'll see, different parts of the Appalachians were deformed and uplifted at different times, but this complexity can be distilled into a linear chronology of events. With this

brief background in mind, the great narrative of the origin and evolution of the Appalachian Geologic Province can be told.

LATE PRECAMBRIAN ASSEMBLY AND BREAKUP OF RODINIA

Around 1 billion years ago, a supercontinent was constructed by the collision and accretion of drifting plates. Geologists have named this long-vanished landmass *Rodinia*, with the ancestral North American continent at its core (Figs. 22.8, 33.7). The Precambrian nucleus of North America was conjoined with Greenland in a single landmass called *Laurentia*. Tectonic reconstructions of Rodinia reveal a strange geography, with Antarctica and Australia sutured to the ancient western margin of Laurentia, and fragments of South America attached to the eastern margin. Part of the growth of Rodinia involved the collision of eastern Laurentia with ancestral South America around 1.1 b.y. ago during a prolonged phase of mountain building known as the **Grenville Orogeny**. The rocks caught in the vice of colliding continents were intensely deformed and metamorphosed within the welt of mountainous highlands (Fig. 35.11A).

Supercontinents have an inherent weakness that causes them to eventually break apart. Their enormous crustal mass traps hotter rock in the underlying mantle, allowing heat to build up, similar to how a blanket traps your body heat as you sleep. The accumulation of hot rock and magma causes the overlying supercontinent to flex upward to accommodate the buoyant mantle beneath. The rigid crust fractures in response to the flexure, providing pathways for the intrusion of magma into the base of the supercontinent. In time, the supercontinent begins to break apart along the upward bulge, initiating the process of *continental rifting* (Fig. 1.11). This scenario of continental breakup was described in greater detail in Chapter 1, with the example of the East African rift system used as a modern analog for the rifting of continents and the birth of ocean basins.

Rodinia first began to break apart along the western boundary of Laurentia as the combined continents of Australia and Antarctica pulled away. Thick rift–basin deposits exposed in several national parks of the American West record this phase of continental rifting beginning around 800 Ma (Chapter 33). The supercontinent began to break apart along the eastern boundary of Laurentia around 750 Ma, as the ancestral South American continent rifted away. A huge slab of crystalline continental rock called the **Grenville Terrane** was left behind after rifting that today forms the crustal foundation of a huge portion of eastern North America, reaching as far inland as Ohio and Kentucky (Fig. 33.4).

During the late Precambrian, the Grenville Terrane formed enormous mountains, but today all that's left are the deeply eroded metamorphic roots. Grenville crystalline rocks are exposed in a broad swath of the Laurentian Highlands of eastern Canada and the Adirondacks of New

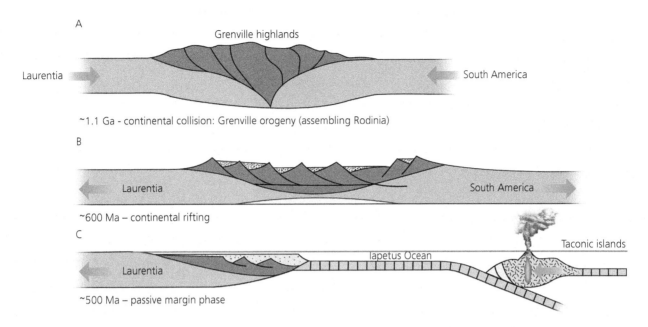

A

Grenville highlands

Laurentia   South America

~1.1 Ga - continental collision: Grenville orogeny (assembling Rodinia)

B

Laurentia   South America

~600 Ma – continental rifting

C   Taconic islands

Laurentia   Iapetus Ocean

~500 Ma – passive margin phase

Figure 35.11   Sequence of events along the eastern continental margin of North America during the late Precambrian to early Paleozoic.

York but are mostly buried beneath the U.S. Appalachians. Drilling reveals that they form the *basement* beneath the Paleozoic sedimentary rocks of the Appalachian Plateau and Valley and Ridge. Fragments of these Grenville rocks are exposed in the Blue Ridge Mountains where they've been lifted upward by later mountain-building events. You can touch the rocky core of the ancient Grenville Mountains in Shenandoah National Park and along the Blue Ridge Parkway.

Recall from earlier chapters that the Grenville highlands supplied quartz sand to continent-crossing rivers, eventually to end up as windblown sand grains on Mesozoic deserts of the American West. Some of the quartz sand composing sandstones of the Colorado Plateau, like the Navajo and Entrada formations, was derived from exposures of the Grenville Terrane on the eastern side of the continent. It's somewhat surreal to recognize the crustal foundation of the Appalachians within grains of sand in Mesozoic rocks of western national parks.

As eastern Laurentia broke away from ancestral South America during the latest Precambrian, sediment accumulated in deep continental rift basins comparable in appearance to the modern East African rift (Fig. 35.11B). In Great Smoky Mountains National Park, rift–basin sediments reach thicknesses of ~15 km (9 mi), reflecting rapid subsidence and deposition as these narrow, deep troughs were pulled apart by tectonic extension. In the region that would become Shenandoah National Park and the Blue Ridge Mountains, magma intruded the fractured and faulted Grenville rocks, erupting along fissures to form thick accumulations of *basaltic* lava that pooled within the rift basins. In places, the pile of volcanic rocks reaches 700 m (~2300 ft) in thickness, attesting to the dynamic stretching of crust during continental rifting. The volcanic rocks were mildly metamorphosed during later

mountain-building events in the Appalachians, and they form prominent outcrops along the Blue Ridge Mountains. The chronology of late Precambrian events in the Appalachians is summarized in Figure 35.12.

Continental rifting ended by the early Paleozoic, with a broad passive margin developing along the eastern margin of Laurentia (Fig. 35.11C). The low remnants of the Grenville Mountains supplied sediment that washed out onto the coastal plain and continental shelf, perfectly analogous to sedimentation on the modern Atlantic continental margin. The newly born ocean basin that opened between the receding continents is called the **Iapetus Ocean**, a long-gone ancestor of the Atlantic Ocean. (In Greek mythology, Iapetus was the father of Atlas, for whom the Atlantic Ocean was named.) The tectonic transition from continental rifting to passive margin development is commonly called the **rift-to-drift phase**, marking the time when continental rifting ends and seafloor spreading begins (Fig. 1.11). It may be valuable to review the process of the birth and evolution of ocean basins in Chapter 1.

Clastic sediments were deposited across the southward-facing early Paleozoic passive margin within coastal plain environments (e.g., rivers, deltas, bays, wetlands), as well as on beaches, tidal flats, and shallow-water environments on the continental shelf (Fig. 35.13A). These clays, silts, sands, and gravels would, after burial and cementation, become thick layers of shale, siltstone, sandstone, and conglomerate. The low-latitude position of the eastern Laurentian passive margin at the time was conducive to the growth of calcite-secreting marine organisms. Their skeletal debris would be buried and solidified to become layers of limestone, remnants of tropical seas that transgressed inland across eastern North America. These early Paleozoic passive margin sediments reach 8 km (5 mi) in thickness and compose the lower part of the stack of

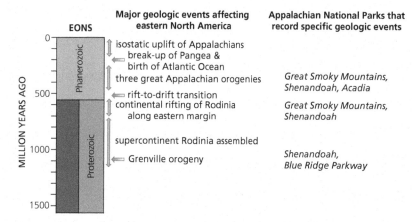

Figure 35.12 Relative chronology of major geologic events that affected national parks of the Appalachians through time. Dates based on the Geological Society of America 2012 time scale.

sedimentary rocks in the Sedimentary Appalachians. Compressive stress during later Paleozoic **orogenesis** (i.e., mountain-building) would elevate the layered sedimentary rocks to form the Appalachian Plateau and Valley and Ridge subprovinces.

EARLY PALEOZOIC TACONIC OROGENY

The tectonic quiescence of the early Paleozoic passive margin would come to a close as a chain of volcanic islands slowly approached from the Iapetus Ocean (Fig. 35.13A). Oceanic lithosphere descended beneath the so-called Taconic volcanic islands, with magma supplied from partial melting along the subducting plate (Fig. 35.14A). As oceanic lithosphere was progressively consumed beneath the encroaching islands, the Iapetus Ocean gradually became smaller. When the volcanic islands were located near the continental slope of North America, they shed muddy sediment eastward onto limestones of the former passive margin, marking the transition to an active continental margin. Volcanism along the island chain blasted massive ash falls onto the North American continent, indicating the proximity of the Taconic Islands. Two ash layers from these eruptions are among the most voluminous of the past 500 m.y. in North America and are recognized as far west as Iowa. Their radiometric ages provide important time constraints on the accretionary history of the **Taconic terrane**.

By about 450 Ma, the oceanic lithosphere between the continental margin of eastern North America and the Taconic volcanic islands was completely subducted, resulting in the collision of the two landmasses in an event known as the **Taconic orogeny** (Figs. 35.13B, 35.14B). The slow-motion convergence thrust the volcanic islands up and over the adjacent continent, suturing the Taconic terrane to eastern North America. Sedimentary rocks of the earlier passive margin were pushed westward as thick thrust sheets by the tectonic stress. The orogenic collision thickened the crust and formed a deep root beneath the newly formed Taconic highlands. The sheer weight of the thickened crust depressed the adjacent crust to the west,

creating a deep inland basin that captured sediment shed from the Taconic highlands. Recall that inland basins formed by flexure of the lithosphere due to the load of nearby mountain belts are called *foreland basins* (Chapter 21).

As the Taconic orogeny came to a close around 440 Ma, eastern North America once again returned to tectonic quiescence. Clastic sediments derived from the eroding Taconic Mountains filled the foreland basin to the west, forming a westward-thinning wedge. Today these mid-Paleozoic sedimentary rocks compose a thick portion of the Sedimentary Appalachians exposed in the Appalachian Plateau and Valley and Ridge subprovinces.

The Taconic highlands were most prominent in New England and eastern Canada, with lesser mountains located near the modern Piedmont Subprovince of the Southern Appalachians. As erosion stripped away the uppermost rocks of the Taconic highlands, the metamorphic roots were isostatically exhumed toward the surface. These rocks of the Taconic terrane are exposed today in a discontinuous belt spanning New York to northern Maine, and they locally crop out above the soil and vegetation of the Piedmont as well. Evidence for these ancestral Taconic Mountains is expressed within deformed and metamorphosed rocks in Great Smoky Mountains National Park, but this evidence is not apparent in the other three national parks discussed in Part 9. But the Taconic Orogeny is clearly a critical event in the evolution of eastern North America. During this episode of mountain-building, the Iapetus Ocean basin decreased in size due to subduction, a volcanic island arc was accreted to the eastern continental margin, and massive volumes of sediment were shed into the Appalachian foreland basin west of the Taconic highlands. North America once again grew larger by the addition of continental rock along its eastern edge.

Roughly coincident with the end of the Taconic orogeny around 440 Ma, Earth experienced a short-lived ice age as well as the second-largest mass extinction of all time. The global glaciation was related to the position of a supercontinent called **Gondwana** above the South Pole. This huge landmass was a remnant of the earlier Rodinian supercontinent and consisted of the conjoined continents of Africa, South America, Antarctica, Australia, and India. Widespread glacial features are preserved in rocks of northwest Africa, which occupied the coldest polar location around 445 Ma. (Yes, what is today the Sahara was once positioned above the South Pole!) Sea level fell by perhaps 80 m (~260 ft) in response to the expansion of continental ice sheets, decreasing the amount of shallow marine ecospace and likely contributing to the extinction of ~85 percent of marine invertebrates. (Plants and animals had not invaded the continents to any appreciable degree by the time of this mass extinction event.) Marine biodiversity rebounded as the glacial conditions waned, sea level rose, and shorelines transgressed onto the continents.

MID-PALEOZOIC ACADIAN OROGENY

While the Taconic orogeny was building eastern North America outward, an elongate fragment of Gondwana had broken loose and was slowly drifting northward toward a rendezvous with the growing continent (Fig. 35.15A).* Geologists call this chain of microcontinental islands **Avalonia**, and its eventual collision and accretion with eastern North America is the second great mountain-building event of the Paleozoic Appalachians, the **Acadian orogeny** (Fig. 35.16). Recall that a continent is not merely land that rises above the ocean but rather is a body of low-density crustal rock that is able to buoyantly float above the denser mantle beneath. Islands of continental crust like Avalonia are just a little too small to be called a continent, and so they are instead called microcontinents. The islands of Indonesia (Sumatra, Java, Bali, and the rest) are a good modern analog for what Avalonia may have looked like.

As the Avalonian Islands moved toward eastern Laurentia around 420 Ma, the intervening oceanic plate was consumed via subduction (Fig. 35.14C). Partial melting provided magma for the formation of granitic plutons and explosive volcanoes within Avalonia. At the same time, weathering and erosion patiently broke down the Avalonian rocks into particles of sediment, which were shed into the adjacent deep ocean trench where they accumulated as an *accretionary wedge*. All of these continental rocks of Avalonia formed as the islands drifted across thousands of kilometers, eventually becoming part of eastern North America in the mid-Paleozoic. Touch the granitic rocks of Acadia National Park and you touch the microcontinental islands of Avalonia.

The docking of Avalonia with eastern Laurentia began in the north, then progressively migrated southward as the remnants of the Iapetus oceanic lithosphere were consumed via subduction, somewhat like a closing zipper

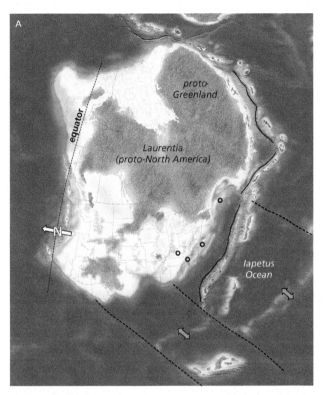

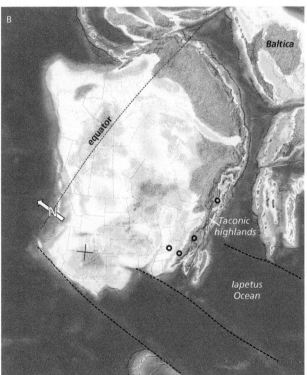

The ancient geographic locations of landmasses are determined by measuring the magnetic properties of certain rocks, which in turn preserve their latitudinal position at the time the rock was created. Using specialized instruments in a lab, we can measure the magnetic signatures of rocks collected from the field, marking the changing latitude of a continent as it migrates across the planet. Paleogeographic maps like those of Figures 35.13 and 35.15 are based partly on numerous magnetic measurements of rocks from Avalonia, dated to specific geologic ages. The maps also rely on the distribution and composition of sedimentary rocks, as well as their assemblage of fossils. Visualizations of the ancient geographic appearance of a landmass are based on enormous amounts of data, mixed with a healthy dose of speculation.

Figure 35.13 A) Paleogeography of Laurentia at 500 Ma. Passive margins of the continent were covered with shallow seas (light blue) where sediment accumulated. The Taconic volcanic arc and microcontinental islands (bordered by dark lines marking the deep ocean trench along the subduction zone) were approaching from the south, and the Iapetus Ocean was becoming smaller. Double arrows mark divergent plate boundaries along mid-ocean ridges. B) Paleogeography of Laurentia at 450 Ma as the Taconic volcanic islands collided with eastern North America. Sea level rose through the early Paleozoic, submerging much of the continent in shallow tropical seas conducive to the accumulation of limestone. Circles mark locations of the four national parks of the Appalachians.

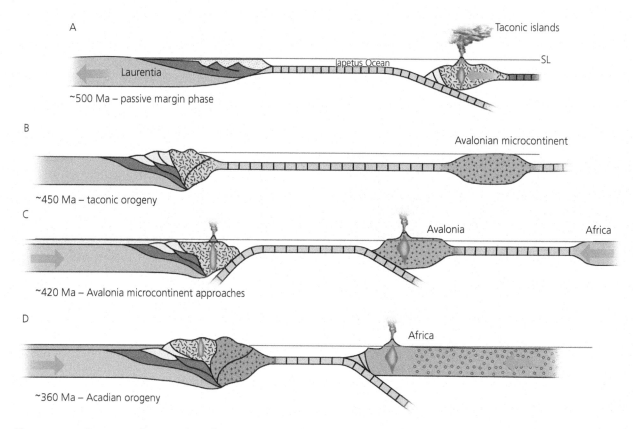

A

Taconic islands

Laurentia   Iapetus Ocean   SL

~500 Ma – passive margin phase

B

Avalonian microcontinent

~450 Ma – taconic orogeny

C

Avalonia   Africa

~420 Ma – Avalonia microcontinent approaches

D

Africa

~360 Ma – Acadian orogeny

Figure 35.14   Sequence of events along the eastern continental margin of North America during the early Paleozoic Taconic orogeny and mid-Paleozoic Acadian orogeny. Horizontal blue line represents sea level (SL).

(Fig. 35.15). Eventually, the entire Iapetus ocean basin would vanish, completely subducted beneath the converging continents and integrated into the mantle far below. Acadian orogenesis occurred at different times along the eastern Laurentian continental margin, ranging between about 400 and 360 Ma.

The Acadian orogeny mostly affected the northern Appalachians. (Avalonia was named for the Avalon Peninsula of Newfoundland, Canada.) Metamorphosed pieces of the **Avalon terrane** are exposed in a belt from Newfoundland through eastern New England. Igneous rocks compose part of the Avalon terrane along the coast of Maine. Acadia National Park is dominated by granitic plutons emplaced around 420 Ma, as the final remnants of the Iapetus oceanic lithosphere were subducted beneath Avalonia (Figs. 35.14C, 35.15A). This Avalonian granitic crust was later accreted to the northeastern margin of Laurentia during Acadian orogenesis.

Evidence for the Acadian orogeny is somewhat cryptic in the southern Appalachians. Fragments of the Avalon terrane compose part of the eastern Piedmont Subprovince, while other slices are buried beneath the sediments of the Atlantic coastal plain. None of the national parks of the Southern Appalachians preserve direct evidence of the Acadian orogeny. Where exposed

in the Piedmont, accreted fragments of the Avalon terrane contain metamorphosed fossils of organisms that were exotic to ancestral North America, reflecting their distant origins on the other side of the closing Iapetus Ocean.

As the Avalon terrane docked sequentially from north to south, its rocks were thrust over the remnant roots of the earlier Taconic terrane, forming a second version of the ancestral Appalachians (Fig. 35.14D). Like the Taconic highlands before them, these Acadian highlands exerted a load on the underlying crust and mantle, creating an elongate depression to the west (Fig. 35.17). This rejuvenated Appalachian foreland basin extended from New York to southern Virginia and received massive amounts of sediment shed from the Acadian highlands to the east. Geologists refer to this westward-thinning prism of sedimentary rocks deposited within the foreland sea as the Catskill **clastic wedge**. Depositional environments preserved in these mid-Paleozoic rocks range from deep marine to coastal deltas to meandering rivers. As the Acadian Mountains slowly eroded, the deep metamorphic core was isostatically exhumed toward the surface. These orogenic roots of the ancient Acadian highlands are seen today wherever the Avalon terrane is exposed within the Appalachian Province.

LATE PALEOZOIC ALLEGHENIAN OROGENY

The end of Acadian orogenesis by ~360 Ma was marked by rising sea levels that flooded the interior of the equatorial continent in extensive shallow tropical seas. The Appalachian foreland basin was separated from the Illinois Basin to the west by a low peninsula called the Cincinnati arch (Fig. 35.18A). Both of these basins were embayments that extended inland from the open ocean. The subtropical

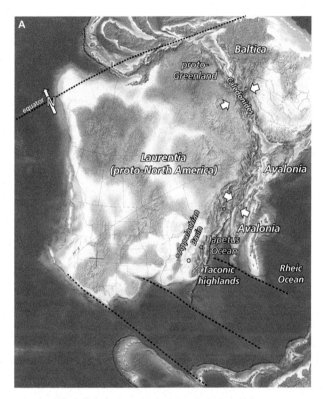

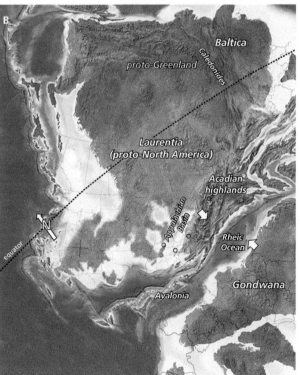

setting was conducive to the expansion of coral reefs and meadows of sea lilies that spread across the shallow marine waters. (Sea lilies are animals known as crinoids that anchor themselves to the seafloor, filtering particles of food that float by. Crinoids dominated the early late Paleozoic shallow seas, with their calcite skeletal debris building up through time to form thick beds of limestone.) The limestones of Mammoth Cave National Park accumulated at this time along the southern margin of the Illinois Basin.

The tectonic quiescence would soon end, however, as the African portion of the supercontinent of Gondwana slowly approached from the south, closing the intervening Rheic Ocean via subduction (Fig. 35.19). (Rhea was the sister of Iapetus in Greek mythology, and, like her older brother ocean, the Rheic Ocean would eventually vanish as the two continents converged.) By about 330 Ma, the part of Gondwana that is now northwest Africa began to physically converge with the eastern margin of Laurentia, marking the start of the prolonged **Alleghenian orogeny** (aka Appalachian orogeny). The two buoyant continental landmasses crushed against each other in ultra-slow motion, squeezing and deforming rocks within the tectonic vice. This type of continent-continent orogenesis is called *continental collision* (Fig. 1.17). The compressional stress pushed rock downward to form a deep metamorphic root while simultaneously raising other rock into highlands (Fig. 35.19B). As the collision progressed, reaching a peak around 300–280 Ma, this third version of the Appalachians would attain the scale of the modern Alpine-Himalayan mountain belt (Fig. 35.18B).

The Alleghenian orogeny involved the collision of enormous continental landmasses, building tectonic stress beyond that of the two earlier Paleozoic orogenies. Most of the physical deformation occurred in the southern and central Appalachians, with the New England Subprovince experiencing widespread metamorphism. In response to the immense pressure, the rocks of the Piedmont, Blue Ridge, and Valley and Ridge subprovinces were shoved inland along low-angle thrust faults. Thick slabs of rock, bound above and below by thrusts, stacked on top of each other like a poorly made shingled roof (Fig. 35.19B). The overlapping thrust sheets laterally compressed the crust

Figure 35.15 A) Paleogeography of Laurentia at 420 Ma during the mid-Paleozoic. The remnants of the Taconic highlands were eroding, with sediments shed into the Appalachian basin to the west. An irregular chain of microcontinental islands (Avalonia) was approaching from the southeast. South of the paleo-equator, the continent of Baltica was converging with Greenland, raising the Caledonide Mountains (preserved today along the coasts of Greenland, Norway, and Scotland). B) Paleogeography of Laurentia at 360 Ma during the Acadian orogeny. A thick wedge of sediment was being deposited in the Appalachian Basin to the west of the Acadian highlands. Note the approach of Gondwana from the south. Thick arrows show the relative sense of plate movement. Circles mark locations of the four national parks of the Appalachians.

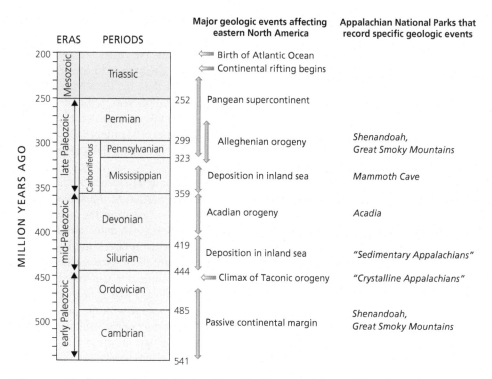

Figure 35.16 Relative chronology of major geologic events that affected national parks of the Appalachians through the Paleozoic. Dates based on the Geological Society of America 2012 time scale.

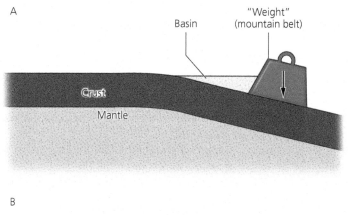

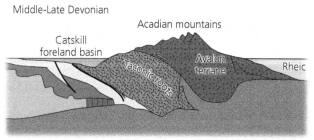

Figure 35.17 A) Mountain belts create a load on the underlying crust and mantle. As the lithosphere flexes in response, a basin forms in the "foreland" of the mountain belt. These basins accumulate sediment shed from the adjacent highlands. B) Foreland basin created by the tectonic load of the Taconic and Avalon terranes on the crust.

within the mountains by perhaps 320 km (200 mi), while raising the rocks to elevations of 6–8 km (~3.5–5 mi). When you stand on mountaintops in Shenandoah or Great Smoky Mountains national parks, the rocks beneath your feet were originally far to the east prior to Alleghenian thrusting.

Some of the highest peaks in the late Paleozoic Appalachians may have been African in origin because it is likely that part of the colliding Gondwanan plate was thrust over and above the Laurentian plate. Direct evidence is lacking, however, due to the erosional removal of those rocks through the Mesozoic and Cenozoic. They now exist as particles of sand and clay composing the deeper portions of the sedimentary prism of the Atlantic continental shelf.

The tectonic stress of the Alleghenian continental collision moved as a wave across the southern and central Appalachians, with variable expressions in each subprovince. Rocks of the Piedmont, composed of older Taconic and Avalon crystalline terranes, experienced renewed metamorphism and deformation because they were the first to absorb the gradual impact. The entire Piedmont Subprovince was displaced westward along a system of near-horizontal thrust faults (Fig. 35.19B).

Grenville-age crystalline rocks of the Blue Ridge Subprovince were thrust more than 160 km (100 mi) westward, pushed over the Paleozoic rocks of the adjacent Sedimentary Appalachians in the future Valley and Ridge Subprovince. During their slow lateral transport deep beneath the surface of the growing Appalachian Mountains, the rocks of the future Blue Ridge were rotated and folded. The details of this deformation will be addressed in Chapters 36 and 37, respectively, on Shenandoah and Great Smoky Mountains national parks.

Compressional stress during the Alleghenian orogeny reached inland to the Sedimentary Appalachians, forming the fold-and-thrust belt of the future Valley and Ridge Subprovince. The Paleozoic sedimentary rocks and a thick slice of their underlying Grenville-age crystalline basement were thrust westward (Fig. 35.19B). During movement as subhorizontal thrust sheets, the layered rocks were crumpled into folds oriented perpendicular to the tectonic stress (Fig. 35.5). The corrugated topography of the Valley and Ridge (Fig. 35.4) reflects the orientation of anticlinal and synclinal folds, modified by erosion. The "hinges" of the folds in the Valley and Ridge are tilted downward as if plunging into the ground (Fig. 35.20A). (Take a paperback textbook, squeeze it to create a fold, then tilt it downward to get a rough idea of plunging folds.) The folds and thrusts of the Valley and Ridge were created deep underground during Alleghenian compressional stress. Through time, overlying rocks were removed by erosion. Resistant sandstone layers weathered to form sinuous ridges that create zig-zag shapes along the axis of folds (Fig. 35.20B). Less resistant layers of shale and limestone weathered to form the intervening valleys.

The effect of Alleghenian tectonic compression dissipated at the eastern edge of the Appalachian Plateau, with the Paleozoic rocks left horizontal or only gently deformed into broad folds. The entire Plateau Subprovince was lifted upward, however, elevating the sedimentary rocks above sea level to create an extensive upland. Further westward along the late Paleozoic equator, Alleghenian stresses were evidently great enough to reactivate long-dormant faults in the American West, raising the Ancestral Rocky Mountains (Fig. 35.18B).

THE PANGEAN WORLD

The Alleghenian orogeny was one of a cluster of mountain-building episodes during the late Paleozoic that resulted in the formation of the supercontinent of Pangea (Fig. 35.21). The Gondwanan supercontinent formed the greatest mass of Pangea, with large continents and smaller continental fragments sutured along the margins by multiple collisions. The "other side" of the Appalachians in northwest Africa is the Atlas Mountains of Morocco, now

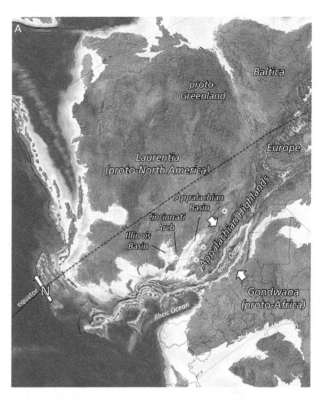

Figure 35.18 A) Paleogeography of Laurentia at 325 Ma near the start of the Alleghenian orogeny. B) Paleogeography of Laurentia at 300 Ma during the peak of the Alleghenian orogeny. Thick arrows show the relative sense of plate movement. White circles mark locations of the four national parks of the Appalachians.

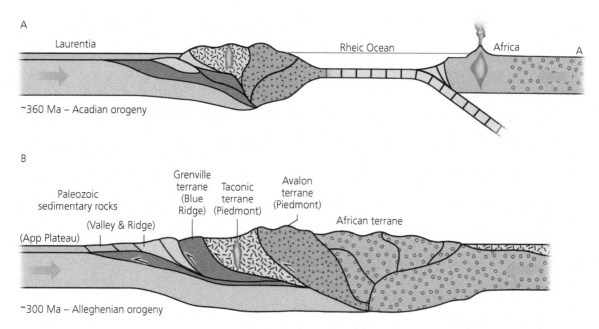

Figure 35.19   Schematic cross sections of mid-Paleozoic Acadian orogeny and late Paleozoic Alleghenian orogeny.

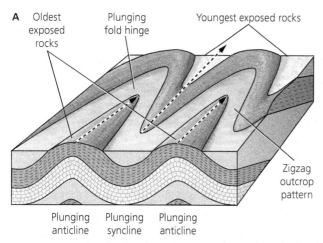

Zigzag outcrop pattern of plunging folds

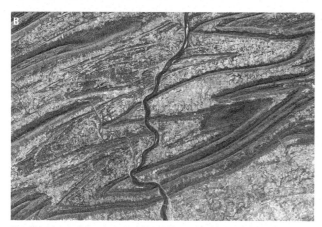

Satellite view of the Appalachian Valley and Ridge Province

Figure 35.20   A) Synclinal and anticlinal folds are not always horizontal but may "plunge" at angles along their hinge. B) After weathering and erosion, the plunging folds may be expressed as a zig zag pattern of valleys and ridges, such as these from central Pennsylvania

separated by the Atlantic Ocean. Northern South America had collided with the southeastern margin of North America during the Alleghenian, raising the Ouachita Mountains of Arkansas and Oklahoma. Southern Europe was welded to northern Africa, while Scandinavia and the British Isles were sutured to Greenland. Russia and far eastern Asia were attached to Europe along the Ural Mountain chain. It's almost as if the continents were gravitationally attracted to one another during the late Paleozoic, aligning themselves along a pole-to-pole axis. An intrepid late Paleozoic reptile could theoretically have walked from Australia to Antarctica, then to South America and Africa before traversing North America, Greenland, and Eurasia. Isolated fragments of East Asia were chains of islands, while a single, vast ocean covered the remainder of the planet. Pangea would last for over 100 m.y.

The newly formed Appalachian Mountains and their equatorial location profoundly influenced the late Paleozoic climate of North America (Fig. 35.21). The American West was located about 30–35°N, prime latitudes for aridity and windswept deserts. This was the time of the Coconino Desert, now exposed in the upper Grand Canyon, as well as the wind-deposited Cedar Mesa sandstones of Canyonlands National Park. The evaporitic Paradox Sea subsided along the flank of the Ancestral Rocky Mountains, precipitating vast amounts of salt that would eventually rise upward beneath the Colorado Plateau.

The climate was much wetter closer to the equatorial latitudes north of the late Paleozoic Appalachians. Lowlands in the American Midcontinent and Appalachian Basin (paleogeographically just north of the ancient Appalachian highlands) were actively accumulating sediment in a variety of subtropical environments. Dense forests of hardwood trees grew in swampy wetlands, much like those of southern Louisiana today. The

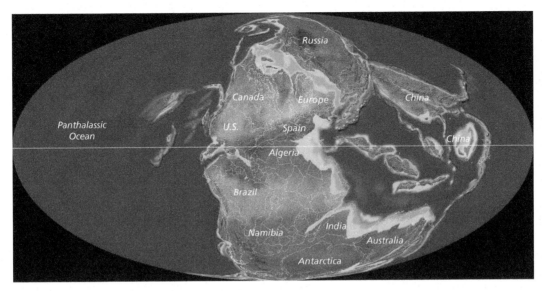

Figure 35.21   Paleogeographic map of Pangean supercontinent (~250 Ma) showing the political boundaries of today's countries.

profuse growth and death of wetland forests resulted in the accumulation of plant debris in the muck of the wetland forest floor. So much plant material was produced that the bacterial decomposers couldn't keep up, permitting the preservation of thick layers of organic matter. Over time, this carbon-dense plant debris would be buried and compacted by kilometers of sediment shed from the adjacent Appalachians, eventually transforming into extensive beds of **coal**. Later uplift and exposure of the coal made it available to humans, fueling the Industrial Revolution. It's not too much of a stretch to say that our modern civilization was triggered by the late Paleozoic Alleghenian orogeny and its impact on global climate and life.

The southern portion of Pangea was located over the South Pole during the late Paleozoic, activating a prolonged and intense ice age (Fig. 33.14). Glacial features are widespread in late Paleozoic rocks in South Africa, southern South America, India, and Australia, attesting to the growth and decay of continental ice sheets. As described in Chapter 33 and elsewhere in the book, sea level fluctuated in tune with phases of glaciation and deglaciation. Cyclic patterns of sedimentary deposits worldwide record the repeated transgressions and regressions of the shoreline. In the American Midcontinent and Appalachian Basin, continental deposits (including the remains of ancient coal swamps) are interbedded with marine limestones deposited in shallow tropical seas, reflecting geologically rapid oscillations in sea level.

By about 270 Ma, the Alleghenian orogeny was waning, replaced by relentless degradation of rock driven by tropical weathering and erosion. Rocks along the crest of the mountains were reduced to small particles of sediment that were shed westward into the Appalachian Basin through the rest of Paleozoic time. By the early Mesozoic, the elevation of the Appalachians was significantly diminished.

Pangea would last several tens of millions of years longer, however, until the inevitable process of continental rifting began. The next stage in the long narrative of the Appalachian Mountains involves the breakup of Pangea and the origin of the modern configuration of continents. The physical settings of the four major national parks of the Appalachian Province were beginning to come into focus.

MESOZOIC RIFTING AND THE BIRTH OF THE ATLANTIC OCEAN

Through geologic time, supercontinents inevitably break apart, and Pangea was no exception. Similar to its late Precambrian ancestor, Rodinia, heat energy accumulated in the mantle beneath the insulating blanket of Pangean continental rock. During the early Mesozoic, the lithospheric bulge and its underlying thermal welt were located close to the axis between the decaying Appalachian Mountains and their African counterpart. The extensional stress within the flexing lithosphere triggered the *reactivation* of older faults, transforming former thrust faults into normal faults. In places, faults and fractures acted as conduits for the upward movement of magma. By about 220 Ma, continental rifting, extensional tectonism, and the fragmentation of Pangea had begun (Fig. 35.22).

A series of elongate rift basins developed within the conjoined continents along the axis of greatest extension between eastern Laurentia and African Gondwana (Fig. 35.23A). Offset along normal faults, accompanied by earthquakes, opened deep, narrow depressions in which sediment accumulated. Tropical weathering and erosion produced copious amounts of sediment that rapidly filled the margins of the growing rift basins with conglomeratic debris building up as alluvial fans. Rivers and streams emanating from remnant Appalachian hills transported pebbles and sand into the basins. Intense rainfall in the

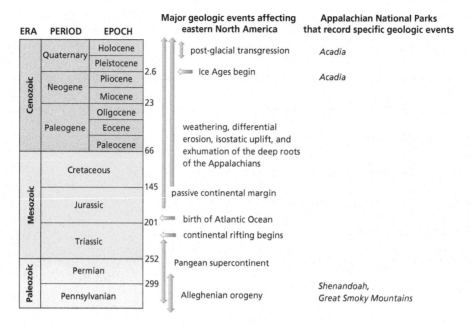

Figure 35.22 Relative chronology of major geologic events that affected national parks of the Appalachians through the Mesozoic and Cenozoic. Dates based on the Geological Society of America 2012 time scale.

equatorial latitudes drained into deep rift lakes, which may have been comparable in scale to the great lakes of the modern East African Rift system (Fig. 1.12). In places, basaltic magma welled up along faults and fractures, spilling out across the floor of the rift basins. These sedimentary and volcanic rift deposits reach 6 km (4 mi) in thickness in some regions. The remnants of these narrow rift basins are preserved within the New England, Piedmont, and Atlantic Coastal Plain subprovinces, ranging from Newfoundland to Alabama.

By 200 Ma, a series of rift basins amalgamated to become the dominant locale of extension between Laurentia and African Gondwana, widening to become a *linear sea* (Fig. 33.17). *Seafloor spreading* began within the linear sea, marking the birth of the Atlantic Ocean basin. As the young ocean widened through time, it connected at both ends with linear seas separating Europe from North Africa and the nascent Gulf of Mexico between South America and southern North America (Fig. 35.23B). In time, a fully divergent plate boundary developed along the young Mid-Atlantic Ridge, with North America and Africa going their separate ways. The Atlantic grew via seafloor spreading at a rate of about 2.5 cm/yr (1 in/yr), opening to more than 1600 km (1000 mi) wide by the mid-Mesozoic. In time, the entire Pangean supercontinent would split into distinct smaller continents, resulting in today's modern global geography (Fig. 35.24).

The parting of ways between eastern North America and northwest Africa did not go smoothly, however. A large fragment of Africa, known today as the Florida

Peninsula, remained attached to the southeastern corner of the continent after the rest of Africa split away. Other large masses of the Gondwanan continent were left behind after continental rifting, hidden today beneath the sediments of the Atlantic coastal plain and continental shelf. Drilling in the search for oil and gas, together with seismic imaging and other geophysical techniques, have produced evidence of these accreted African terranes deep beneath the surface.

With the opening of the Atlantic Ocean around 200 Ma, the North American plate began to drift westward, transferring the locus of mountain-building to the left side of the continent. The great Paleozoic orogenies of eastern North America were over, replaced by continental rifting in the early Mesozoic and subsequent tectonic quiescence through the remainder of the Mesozoic and Cenozoic. The North American Cordillera, however, was just beginning its dynamic history of mountain-building, coincident with the Appalachian slide into senescence. Around 200 Ma, the vast Wingate and Navajo sand seas extended across the American West (Fig. 4.8). During the acme of Pangea, huge rivers transported quartz sand grains from the exposed Grenville basement of the Appalachians across North America, where they were redistributed by wind onto the Navajo and Entrada deserts. Volcanic island arcs were being accreted to the western margin of North America, conveyed to the active continental boundary by the Farallon oceanic plate. Soon, by geological standards, the ancestral Sierra Nevada would rise, followed by a wave of orogenesis that swept eastward across the Cordillera.

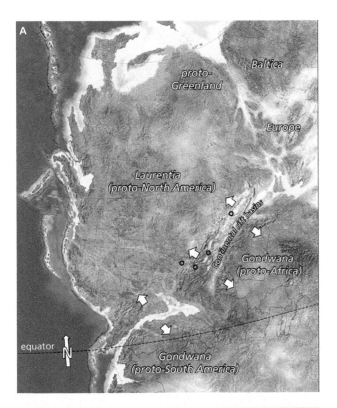

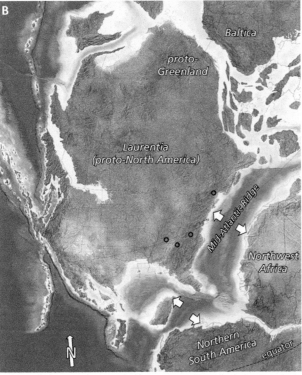

Figure 35.23 A) Paleogeography of Laurentia and Gondwana at 220 Ma near the start of supercontinental breakup. Elongate rift basins opened along the axis of divergence. B) Paleogeography of Laurentia and Gondwana at 150 Ma after seafloor spreading has fully developed in the young Atlantic Ocean. For both maps, thick arrows show the relative sense of plate movement and circles mark locations of the four national parks of the Appalachians.

MESOZOIC-CENOZOIC EXHUMATION AND THE MODERN APPALACHIANS

As continental rifting gave way to seafloor spreading around 200 Ma, the continental margins of eastern North America and northwestern Africa began a long history of passive drift away from the newly forming oceanic lithosphere along the Mid-Atlantic Ridge. This tectonic rift-to-drift transition is physically marked by the burial of narrow rift basins by broad sedimentary layers of the Atlantic passive continental shelf (Figs. 35.9, 35.25). The thick wedge of clastic sediments beneath the coastal plain and continental shelf was derived from the erosion of the Appalachians, beginning in the early Mesozoic and continuing through today. As more and more sediment was added to the seafloor of the continental shelf, the underlying crust subsided in response, creating more space for the continued accumulation of sediment.

The modern Appalachian Mountains are the exhumed roots of the colossal late Paleozoic Appalachians (Fig. 35.26). As the mountains were weathered and eroded through the Mesozoic and Cenozoic, the deformed crust deep beneath rose upward in response. In many places, rocks that were metamorphosed 15–20 km deep in the crust are exposed today at the surface. This *isostatic uplift* occurred episodically, as recorded by the sediments of the Atlantic continental shelf. (The concept of isostasy was introduced in Chapter 17.) During phases of rapid weathering and erosion, perhaps coincident with warmer, wetter climates or a lower sea level, large volumes of sediment were shed eastward onto the coastal plain and continental shelf. Consequently, deeper portions of the Appalachians bobbed upward in response to the removal of overlying weight. During phases of slower weathering and erosion, perhaps coincident with cooler, drier climates or higher sea level, lesser volumes of sediment were removed and the smaller the isostatic response.

In essence, the more rock erosionally removed from the mountains, the greater the isostatic uplift. Episodes of uplift triggered the rejuvenation of streams and rivers that cut ever deeper into the mass of rock. Evidence from the sedimentary rocks of the coastal plain and continental shelf suggests that the Appalachians experienced three or more major episodes of erosional "unroofing" and isostatic uplift through the Mesozoic and Cenozoic, culminating in the rounded contours of today's Appalachian Province.

The modern Appalachian topography is a product of differential erosion in a humid climate, with resistant crystalline rocks in the Blue Ridge and tightly cemented sandstones in the Valley and Ridge forming elevated ridges. In contrast, the elongate valleys of the Valley and Ridge are underlain by easily weathered shale and limestone, while the old rocks of the Piedmont are mostly buried beneath a veneer of soil and vegetation. The Paleozoic rocks of the Appalachian Plateau, centered over West Virginia, have been deeply incised by streams and rivers, creating the rugged relief of steep-sided ravines dissecting flat-topped highlands. In the New England Subprovince, continental

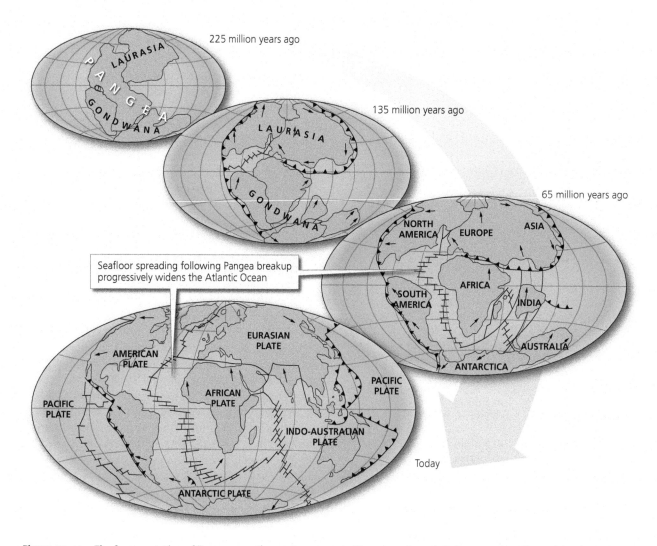

Figure 35.24   The fragmentation of Pangea over the past 200 m.y. led to our modern global geography of large island continents.

glaciation stripped away enormous amounts of rock, with the underlying crust rebounding in response.

PLEISTOCENE CLIMATE CHANGE

The *continental ice sheets* that flowed southward from Arctic Canada multiple times over the past 2.6 m.y. had profound effects on the New England Subprovince of the Appalachians. The Laurentide ice sheet reached as far south as central Pennsylvania and New Jersey, covering New England in a sheath of ice a kilometer or more thick (Fig. 35.27). During the *Last Glacial Maximum* about 20,000 years ago, sea level was ~120 m (~400 ft) lower than today, with shorelines regressing by 100 km (60 mi) or more, exposing the broad Atlantic continental shelf. Coastal Maine and Acadia National Park were far inland but entombed in a mantle of glacial ice. The modern landscape of Acadia National Park is a product of multiple episodes of glacial advance and retreat, with widespread evidence for glacial erosion and deposition across the terrain. Several earlier chapters described the modifying effects of alpine glaciers and ice caps on the appearance of

national parks and provinces in the American West. Only near Glacier and Olympic national parks did lobes of continental ice sheets peripherally influence the formation of glacial features. Chapter 38 on Acadia National Park will provide an opportunity to contrast the effects of continental ice sheets with those of smaller alpine glaciers on the landscapes of national parks.

The impact of glacial maxima on national parks and subprovinces of the southern Appalachians was primarily related to dramatic shifts in climate belts that accompanied the expansion of the ice sheets (Fig. 35.28). In Shenandoah and Great Smoky Mountains national parks, the elevations were high enough and cold enough that the highest ridges supported tundra ecosystems. Sediment cores taken from ponds and bogs preserve a pollen record that reveals the hardy grasses and dwarf vegetation indicative of tundra flora. In the surrounding lower-elevation latitudes, dense forests prevailed, marked by conifers like pine, spruce, and fir. The midlatitudes of Virginia, Tennessee, Kentucky, and the Carolinas would have resembled the vast boreal forest that covers much of Canada today.

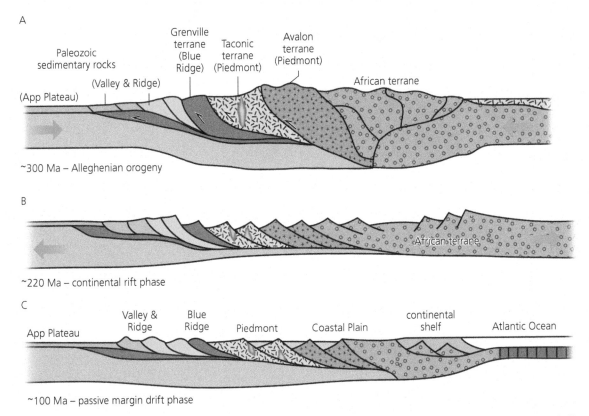

A

Paleozoic
sedimentary rocks

(Valley & Ridge)

(App Plateau)

Grenville
terrane
(Blue
Ridge)

Taconic
terrane
(Piedmont)

Avalon
terrane
(Piedmont)

African terrane

~300 Ma – Alleghenian orogeny

B

African terrane

~220 Ma – continental rift phase

C

App Plateau

Valley &
Ridge

Blue
Ridge

Piedmont

Coastal Plain

continental
shelf

Atlantic Ocean

~100 Ma – passive margin drift phase

Figure 35.25 Schematic cross sections from the late Paleozoic Alleghenian orogeny, early Mesozoic continental rifting, and the late Mesozoic passive margin phase.

In a swath along the Gulf Coast, hardwood forests of oak, hickory, and chestnut dominated during glacial maxima.

Streams and rivers draining the terminus of the continental ice sheet flowed milky white from glacial silt released from the melting southern margin of the ice. During glacial maxima, low sea level allowed for the expansion of coastal plain ecosystems across the exposed Atlantic continental shelf. A thick-coated megafauna of mammoths and mastodons, wooly rhinos and musk oxen, cave bears and saber-toothed cats roamed the forested terrain of the shelf. There were no humans yet occupying the Americas during the Last Glacial Maximum, as far as we can tell, but small bands of nomadic hunters would soon arrive by about 15–13 ka, stalking the large-bodied sources of meat.

With climatic warming during the last 12,000 years of the *Holocene*, the climate belts migrated northward,

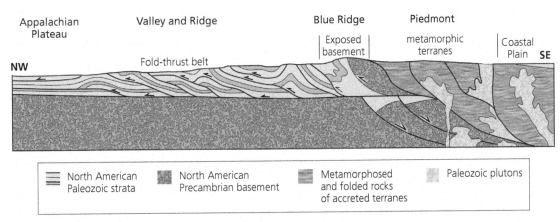

Appalachian
Plateau

Valley and Ridge

Blue Ridge

Piedmont

Exposed
basement

metamorphic
terranes

Coastal
Plain

NW

Fold-thrust belt

SE

| | North American Paleozoic strata | | North American Precambrian basement | | Metamorphosed and folded rocks of accreted terranes | | Paleozoic plutons |

Figure 35.26 Cross section from the Appalachian Plateau to the Atlantic coastal plain, illustrating the deep roots of the late Paleozoic Appalachians exposed today. Huge amounts of overlying rock were removed by erosion, causing the underlying rock to rise upward as the isostatic response to the decrease in confining pressure.

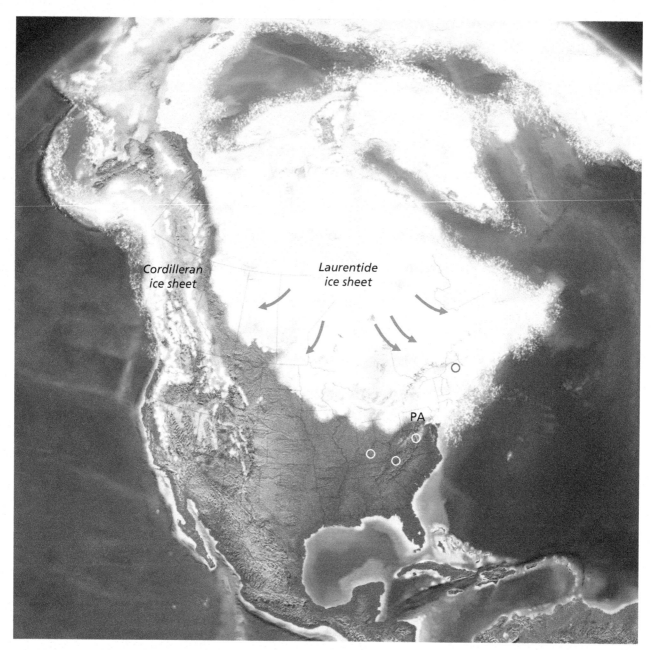

Figure 35.27 Extent of continental ice sheets across North America 12,000–15,000 years ago. The Laurentide ice sheet covered New England (including Acadia National Park, red circle), with the maximum extent reaching across central Pennsylvania (PA). The other three national parks (white circles) experienced the peripheral effects of colder climates and changing ecosystems.

tracking the retreat of the continental ice sheets. The tundra and boreal forest belts drifted toward Canada, while hardwood forests occupied much of the Appalachian latitudes. The melting ice sheets left behind elongate ridges of glacial debris within *terminal* and *recessional moraines*, seen today on Cape Cod, Long Island, and elsewhere along the New England coast. Some moraines acted as natural dams that ponded glacial meltwater into large lakes. A prime example is the Great Lakes, which occupy elongate depressions gouged by the continental ice sheets, with dams along their southern margins created by morainal ridges.

A *postglacial marine transgression* of the shoreline accompanied Holocene warming as glacial meltwaters poured into the world ocean (Fig. 29.18). Deep river valleys that cut across the formerly exposed continental shelf were drowned by rising sea level, creating great estuaries such as the Chesapeake and Delaware. The shoreline stabilized near its current position by about 7000 years ago (~5000 BCE), leaving coastal national parks like Acadia susceptible to the constant pounding of waves and tides. The modern Atlantic coastline was set, separating the meandering rivers and streams of the coastal plain from the shallow marine environments of the continental shelf.

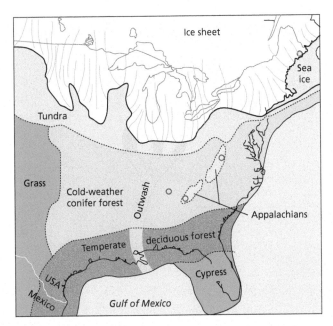

Figure 35.28 Climate belts migrated in concert with the advances and retreats of continental ice sheets. During glacial maxima, as shown here, tundra conditions prevailed along the ice front as well as on the higher elevations of the southern Appalachians. Boreal conifer forests expanded southward across much of eastern North America, comparable to the modern conifer forests of Canada. Circles mark the locations of the four national parks in Part 9.

Further inland, drainage systems realigned themselves in response to the retreat of glacial ice and associated changes in precipitation. The final phases of landscape modification by running water in the Appalachian Province (and specifically in our eastern national parks) occurred during the last few thousand years of the Holocene.

Summary of the Origin and Evolution of the Appalachian Province

This synthesis chapter has laid the foundation for understanding the four national parks discussed in the rest of Part 9. It should be clear that the preceding narrative on the origin and evolution of the Appalachians and the eastern margin of North America is every bit as complicated as that of the American West, described in previous chapters. The story of the American East involves the assembly and deconstruction of two supercontinents, as well as the growth and demise of several mountain ranges. Entire ocean basins opened, closed, and reopened as the continental margin alternated between passive and active settings. North America expanded outward by over 800 km (~500 mi) along its eastern margin by accretionary tectonics, continental collision, and sediment accumulation. The final phases of landscape evolution throughout the Appalachian Province occurred during the last few thousand years of a billion-year-long story.

Details about the origin and evolution of the Appalachians are far from complete, and controversies exist today among researchers. Most broadly agree about the version in this chapter, but it's almost certain that the details will evolve in light of new technologies and further discoveries. The transparency of the scientific method is one of its greatest strengths, and inevitably, the great narrative of the Appalachians will be fine-tuned in the future.

Key Terms

clastic wedge
coal
fall line
orogenesis
rift-to-drift transition

Related Sources

Graham, J. (2010). *Acadia National Park: Geologic Resources Inventory Report.* Natural Resource Report NPS/NRPC/GRD/NRR—2010/232. Fort Collins, CO: National Park Service.

Hatcher, R. D., Jr., Thomas, W. A., and Viele, G. W. (Eds.). (1989). *The Appalachian-Ouachita Orogen in the United States.* Boulder, CO: Geological Society of America.

Thornberry-Ehrlich, T. L. (2014). *Shenandoah National Park: Geologic Resources Inventory Report.* Natural Resource Report NPS/NRSS/GRD/NRR—2014/767. Fort Collins, CO: National Park Service.

Tollo, R. P., Bartholomew, M. J., Hibbard, J. P., and Karabinos, P. M. (Eds.). (2010). *From Rodinia to Pangea: The Lithotectonic Record of the Appalachian Region.* Boulder, CO: Geological Society of America Memoir 206.

36 Shenandoah National Park

Figure 36.1 View across the Blue Ridge Mountains of Shenandoah National Park from the Big Run Overlook.

Most visitors to Shenandoah National Park savor the dense forests of hardwood trees that blanket the mountainsides, particularly as they turn on their autumn colors (Fig. 36.1). But hidden beneath the leafy veneer is a rich geological history recorded in the rocks that compose the rumpled contours of the mountains. More than enough outcrops exist to piece together a fascinating story of supercontinent assembly and breakup, the inundation of the region by shallow seas, and the growth of a titanic ancestor to the modern Appalachians. We'll begin with a description of the physical attributes of the park, move on to the origin and uplift of its rocks, and finish with the evolution of the rolling landscape.

Shenandoah National Park is located in the Blue Ridge Mountains of north-central Virginia, flanked on the west by the elongate textures of the Valley and Ridge Subprovince and on the east by the subdued topography of the Piedmont Subprovince (Figs. 36.2, 35.4). The mountains rise 900 m (~3000 ft) above the adjacent Shenandoah Valley, just outside the park to the west, with Hawksbill Mountain reaching the highest elevation in the park at 1230 m (4050 ft). The park is traversed by Skyline Drive, a scenic byway that stretches for 169 km (105 mi) from one tip of the park to the other (Fig. 36.3). The famous **Appalachian National Scenic Trail** extends for 162 km (101 mi) along the crest of the park.

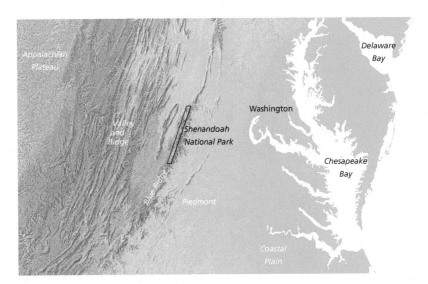

Figure 36.2 Color-coded relief map of the region surrounding Shenandoah National Park. The park is located along the narrow ridgetops of the Blue Ridge Mountains due west of Washington D.C.

The Blue Ridge Mountains are a subprovince of the larger Appalachian Geologic Province, described in Chapter 35. In north-central Virginia, the Blue Ridge form a narrow belt of resistant crystalline rocks that rise above the less resistant rocks of the adjoining subprovinces. On humid days, the mountains sometimes take on a bluish tinge due to the release of organic compounds from the dense cover of trees, giving rise to the name "Blue Ridge." In Shenandoah National Park, the crest of the Blue Ridge marks the drainage divide for three watersheds. To the northwest of the divide, streams and rivulets fed by rainfall and snowmelt drain into the South Fork of the Shenandoah River, which in turn flows northward into the Potomac River. Runoff to the southeast of the crest of the Blue Ridge in Shenandoah flows into either the James or Rappahannock watersheds. All of the drainage from the national park eventually finds its way through these rivers to Chesapeake Bay.

Before describing the rocks and the array of geologic events that they represent, it's useful to reiterate a fundamental geologic tenet first introduced in Chapter 3 on the Grand Canyon. *The modern landscape that we see around us (e.g., elongate ridges, narrow valleys, rounded peaks) is much younger than the rocks beneath the ground.* This may seem intuitive, but the rocks had to have already been present before the terrain we hike on could develop on the surface. Think of the rocks as analogous to the raw lump of clay that a potter begins with to create a pot. The potter is analogous to the landscape-modifying effects of weathering, erosion, and time. Both the potter and the natural processes of weathering and erosion are artists, slowly sculpting their raw materials into the pot/landscape that

we enjoy today. This simple relationship between the age of the rocks and the age of the landscape is based on the principle of *cross-cutting relations* (Chapter 1).

Origin and Age of Rocks at Shenandoah National Park

If you read the previous chapter on the geologic history of the Appalachian Geologic Province, then you have a good background for understanding the origin of the rocks at Shenandoah National Park, as well as their history of uplift as part of the Appalachian Mountains. The metamorphic, igneous, and sedimentary rocks of Shenandoah can be broadly divided into three distinct groups, with each group representing a specific geologic phase of Earth history in the southern Appalachians (Fig. 36.4). Precambrian rocks of the *crystalline basement* are the roots of the 1-billion-year-old Grenville Mountains that were created as the Rodinian supercontinent was assembled. A thick pile of metamorphosed lava flows overlie the basement rocks and represent the late Precambrian breakup of Rodinia. Early Paleozoic sedimentary rocks are the youngest in the park and manifest a transgressing shoreline as the passive margin of eastern North America was inundated by rising sea level. A timeline of these events in the Appalachians is shown in Figure 35.12.

These three sets of rocks compose the bedrock that underlies the forested ridges of Shenandoah National Park. Geologists have defined the relative arrangement of the rock units and their ages by methodically piecing together the information from isolated outcrops and roadcuts, combined with radiometric age dating of certain units. Collectively, the rocks of Shenandoah tell an extraordinary story of the earliest geologic history of the east-central United States.

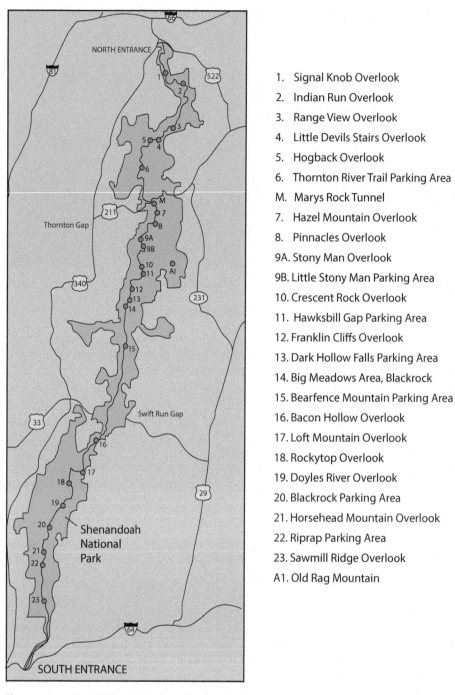

1. Signal Knob Overlook
2. Indian Run Overlook
3. Range View Overlook
4. Little Devils Stairs Overlook
5. Hogback Overlook
6. Thornton River Trail Parking Area
M. Marys Rock Tunnel
7. Hazel Mountain Overlook
8. Pinnacles Overlook
9A. Stony Man Overlook
9B. Little Stony Man Parking Area
10. Crescent Rock Overlook
11. Hawksbill Gap Parking Area
12. Franklin Cliffs Overlook
13. Dark Hollow Falls Parking Area
14. Big Meadows Area, Blackrock
15. Bearfence Mountain Parking Area
16. Bacon Hollow Overlook
17. Loft Mountain Overlook
18. Rockytop Overlook
19. Doyles River Overlook
20. Blackrock Parking Area
21. Horsehead Mountain Overlook
22. Riprap Parking Area
23. Sawmill Ridge Overlook
A1. Old Rag Mountain

Figure 36.3 Map of Shenandoah National Park. Skyline Drive runs along the axis of the park, providing access to trails, overlooks, and other points of interest.

PRECAMBRIAN GROWTH OF EASTERN NORTH AMERICA

The crystalline basement of Shenandoah National Park is composed of *granite* and *granitic gneiss* (Fig. 36.5). Granitic magma was emplaced within the core of the Grenville Mountains about 1.2–1 billion years ago during the construction of the Rodinian supercontinent (Chapter 35). As the magma slowly cooled beneath the surface, releasing its thermal energy to the surrounding rocks,

minerals crystallized out of the fluid. In time, the magma solidified into the interlocking mosaic of mineral crystals that define granite (Fig. 1.20). (See Chapter 1 for a more thorough explanation of the formation of granite from magma.)

Older granites within the Grenville Mountains were subjected to intense pressures and temperatures during *orogenesis*, perhaps at depths of 15–20 km (9-12 mi) within the deep roots of the mountain range. Under the extreme

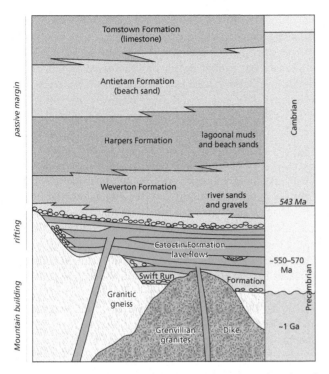

Figure 36.4 Precambrian and early Paleozoic rock units of Shenandoah National Park. The tectonic and depositional significance of each unit is noted along the left side of the diagram.

Figure 36.5 A) Granite of Old Rag Mountain, Shenandoah National Park, marked by an interlocking mosaic of blocky white feldspar crystals, darker hornblende or biotite crystals, and bluish-gray quartz. Width of view is ~15 cm (6 in). B) Granitic gneiss showing a foliated texture of minerals arranged in a banded pattern, Shenandoah National Park. Width of view is ~30 cm (12 in).

conditions of metamorphism, the minerals within the granite recrystallized into granitic gneisses. The banded texture, called *foliation*, is created as newly metamorphosed minerals align in orientations perpendicular to the main direction of tectonic stress.

Today these late Precambrian igneous and metamorphic rocks are nicely exposed near the Hogback and Hazel Mountain overlooks along Skyline Drive, as well as atop Old Rag Mountain, a separate peak east of the main Blue Ridge. As you look closely at these rocks, remember that they are the exhumed roots of the ancient Grenville Mountains, originally formed several kilometers deep beneath the surface. The rocks of the Grenville terrane were originally located about 160 km (100 mi) to the east, but thick slices were subsequently thrust to the northwest during the *Alleghenian orogeny* of the late Paleozoic when the Appalachians were in their prime (Fig. 35.26). Erosion and *isostatic uplift* over the last 200 m.y. or so have raised these resistant rocks to their current heights in Shenandoah National Park and throughout the Blue Ridge Mountains.

Thinking at a broader scale, these crystalline rocks lie several kilometers below the stacked thrust sheets of sedimentary rock of the Valley and Ridge subprovince to the west. And at an even larger scale, the granites and gneisses of the Grenville Terrane form the crustal basement of an enormous region of eastern North America (Fig. 33.4). Those lichen-covered rocks may look somewhat benign as you take in the nearby scenery, but they record a multifaceted story of Earth's deep past.

LATE PRECAMBRIAN CONTINENTAL RIFTING

The second major rock unit composing Shenandoah National Park is a thick stack of metamorphosed *basalts* called the **Catoctin Formation** (Fig. 36.4). The layered basalts originated as lava flows that poured out across the late Precambrian landscape around 570–550 m.y. ago. Basaltic lavas are low in silica (SiO_2), which causes them to be very fluid and runny (Chapter 12). Consequently, they tend to spread for great lateral distances. The Catoctin basalts can be recognized across a large region extending northward along the Blue Ridge into southern Pennsylvania. Age-equivalent basalts are recognized throughout New England and northeastern Canada, suggesting that the volcanism was a widespread regional event. You might imagine the Catoctin basaltic lava flows as nonexplosive *effusive* eruptions, like those occurring today at Hawaii Volcanoes National Park (Chapter 32).

Basalts are typically black, but the Catoctin basalts have been metamorphosed to shades of green (Fig. 36.6).

Figure 36.6 Catoctin basalt, metamorphosed to greenstone. The green tint of the rock is due to the minerals chlorite and epidote.

These "metabasalts" are commonly called **greenstones**, reflecting the abundance of chlorite and epidote—green minerals that form during metamorphism of basalt. In places, irregularly shaped holes pit the surface of exposures. Called *vesicles*, these holes record the moment when gas bubbles in the lava were trapped as the lava solidified. Vesicles are commonly filled with minerals such as quartz or epidote that crystallized within the voids from percolating groundwater long after the lava stopped flowing. These mineral-filled vesicles are called **amygdules** and

may stand out in relief against the greenstone matrix due to their greater resistance to weathering.

Catoctin greenstones are widespread in Shenandoah National Park, extending discontinuously along the entire length of the park. Some of the highest peaks, including Hawksbill, Hightop and Stony Man, are composed of stacked lava flows of Catoctin greenstone (Fig. 36.7). They are particularly prominent in outcrops along the northern third of the park from the Front Royal to Thornton Gap entrances, as well as at several other places along Skyline Drive. At Indian Run Overlook, the metabasalts exhibit *columnar jointing*, composed of the systematic alignment of vertical joints that create an angular bundle of five- and six-sided columns of rock. As lava flows cool and solidify along their upper and lower surfaces, the rock contracts. Hexagonal joint planes propagate inward from the outer surfaces toward the interior of the still-cooling flow. The end result is a cluster of tightly packed prisms of basalt, oriented perpendicular to the layering. The example at Indian Run is the best one along Skyline Drive, but other spectacular outcrops of columnar basalts can be seen at Compton Peak and along the Limberlost Trail.

The Catoctin lava flows are interpreted to be **flood basalts**, voluminous outpourings of lava that erupt from fissures cutting the crust. The fluid lava spreads out laterally for great distances due to its low viscosity, smothering the landscape and burying older flows. The Catoctin lava flows likely occurred episodically, with each individual flow lasting perhaps a few months and reaching thicknesses of ~10–30 m (30–100 ft). Collectively, these flows lasted about 20 m.y. from 570 to 550 Ma. The end result was a thick stack of basaltic lava flows with broad lateral extent, lacking any obvious volcanic vent.

Flood basalts are commonly associated with the process of continental rifting, which during the latest Precambrian involved the breakup of the Rodinian supercontinent (Fig. 33.7). As eastern *Laurentia* separated from ancestral South America, extensional tectonic stress created faults and fissures with northeast–southwest orientations, perpendicular to the main direction of rifting. These planes of weakness within the preexisting crystalline rocks of the

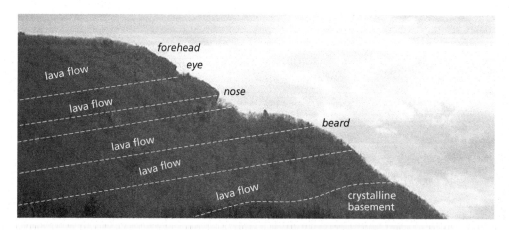

Figure 36.7 Stony Man Mountain with contacts between stacked Catoctin lava flows superimposed.

Grenville basement acted as conduits for magma to rise toward the surface, eventually erupting across the landscape as the Catoctin flood basalts.

The volcanic fury of these eruptions is hard to imagine, as no comparable volume of flood basalt has been erupted in historic times. Try to imagine the Earth splintering along a series of fractures that extend from Virginia to Newfoundland, with curtains of lava bursting upward along fissures before spreading out laterally across hundreds of square kilometers. The environmental devastation during each individual event over the 20 m.y. duration of eruptions would have been catastrophic. The modern East African Rift system and its relatively youthful volcanism (Chapter 1) provide a snapshot of what the breakup of Rodinia may have looked like during the latest Precambrian.

Much of the magma rising upward along planar faults and fractures became trapped, eventually cooling and solidifying to form *dikes*. Over a hundred of these dikes are exposed by erosion in Shenandoah, providing a glimpse into the natural magmatic plumbing network within the late Precambrian rift system. A metabasalt dike cuts near-vertically through the crystalline basement near Marys Rock Tunnel along Skyline Drive, and a more accessible outcrop can be seen at Little Devils Stairs Overlook. The summit trail on Old Rag Mountain exhibits narrow gaps where metabasalt dikes were preferentially eroded, leaving steep walls of resistant Old Rag granite along each side (Fig. 36.8). Weathering and erosion along columnar joints oriented perpendicular to the sides of the dike created natural steps along the trail through the gap. Remember as you look at these dikes that they were once a kilometer or two beneath the surface, churning with magma flowing upward through the granitic country rock to eventually erupt onto the late Precambrian landscape.

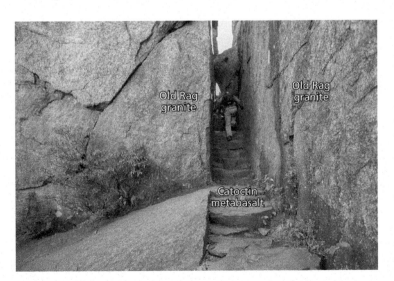

Figure 36.8 Narrow gap between granitic rocks of the basement complex on the summit trail to Old Rag Mountain. The gap was created as a dike of Catoctin metabasalt was preferentially eroded, leaving the adjacent, more resistant crystalline rocks intact.

EARLY PALEOZOIC PASSIVE MARGIN

As continental breakup continued and the nascent *Iapetus Ocean* opened wider, the rift basins filled with Catoctin flood basalts were left behind as part of the newly formed continental margin of eastern North America (Fig. 35.11). The heat supply for magma formation dissipated as continents diverged further away from each other, with the Catoctin volcanics cooling and subsiding over time. Rivers and streams draining the continental interior meandered across the volcanic surface on their way to the newly developing Iapetus Seaway to the east. The rivers deposited sands and gravels in the low-elevation regions along the continental margin, effectively smoothing the topography into a broad coastal plain. These coarse-grained sediments would become the sandstones and conglomerates of the Weverton Formation, which overlies the Catoctin volcanics (Fig. 36.4).

With continued subsidence, the shoreline gradually migrated inland across the continental margin near the future location of Shenandoah National Park. A variety of coastal depositional environments tracked the westward transgressing shoreline, including tidal flats, lagoons, and wave-washed beaches. *Clastic sediments* like clays, silts, and sands deposited in these environments constitute the Harpers and Antietam formations (Fig. 36.9).

The Weverton, Harpers, and Antietam formations compose the **Chilhowee Group**, which manifests sediment accumulation during a marine *transgression* onto the early Paleozoic *passive margin*. The tectonic significance of the rift-related Catoctin volcanics and the overlying Chilhowee sedimentary rocks is interpreted as the *rift-to-drift transition* of the youthful eastern continental margin of North America. As the Chilhowee sediments were deposited, Laurentia was drifting passively to the north through lower latitudes during the earliest Paleozoic (Fig. 35.13A). As a more equatorial position was reached, shallow tropical seas occupied the continental shelf, with limestones of the Tomstown Formation accumulating above the Chilhowee Group (Fig. 36.4). The Tomstown is not exposed in Shenandoah but overlies the Antietam in the Valley and Ridge subprovince to the northwest.

The clastic sedimentary rocks of the Chilhowee Group were subjected to metamorphism during later Paleozoic tectonic events. Clay-rich shales would convert to mica-rich *phyllites*, while sandstones would recrystallize to *quartzites*. The Chilhowee "metasedimentary" rocks are exposed primarily in the southern part of Shenandoah near Rockytop, Doyles River, and Sawmill Ridge overlooks, as well as atop the knob of Blackrock (Fig. 36.9). An outcrop of the Antietam Formation at Calvary Rocks exhibits the oldest fossilized traces of animal life in the park—vertical tubes in the quartzite known as *Skolithos*. The cylindrical

Figure 36.9 Lichen-covered quartzite of the Harpers Formation at the crest of Blackrock, southern Shenandoah National Park. The original quartz sand grains were deposited on an early Paleozoic beach and were much later metamorphosed to resistant quartzite.

Skolithos tubes are relatively common features of well-washed beach sands of early Paleozoic age and are interpreted to be fossilized burrows made by a soft-bodied worm. *Skolithos* burrows are **trace fossils** that record the biologic activity of ancient life rather than the actual hard parts of the organism itself. Complex behavior among invertebrates was just beginning to emerge in the shallow seas that bordered the continents during the early Paleozoic, and some of the evidence is found in Shenandoah National Park.

Metamorphism, Uplift, and Deformation of Rocks at Shenandoah National Park

The three main groups of rocks composing the bedrock of Shenandoah originally formed between about 1 billion and 500 million years ago, eventually becoming buried by younger sedimentary rocks. They would have to wait until about 300 m.y. ago before they were uplifted into the ancestral Appalachian Mountains. Over the most recent 200 m.y. or so, the rocks of Shenandoah would rise isostatically at extremely slow rates as overlying rocks were weathered and eroded, eventually revealing the deep roots of the ancestral Appalachians that we see in the park today.

All of the three mountain-building episodes of the Paleozoic were tectonically capable of inducing the metamorphism of Catoctin basalts to greenstone and the Chilhowee sedimentary rocks to phyllites and quartzites. Terranes of both Taconic and Acadian origin are recognized in the Piedmont subprovince to the east (Fig. 35.25A), suggesting that their tectonic effects may have been near enough to trigger low-grade metamorphism of the older Catoctin

and Chilhowee rocks. Whether there was one main event or whether there were overlapping metamorphic events is still being determined. What is agreed upon is that all of the rocks of Shenandoah were deformed and lifted upward within the ancestral Appalachians during the Alleghenian orogeny that crested around 300 Ma.

The late Paleozoic Alleghenian *continental collision* of the northwest African part of Gondwana with eastern North America generated colossal tectonic stress (Chapter 35). The Appalachians were raised to heights approaching the modern Alpine-Himalayan mountain chain, with crustal thickening occurring by the compressional folding of rock and the stacking of slabs of rock along low-angle *thrust faults*. The deeply buried rocks of the Grenville basement, the Catoctin flood basalts, and the Chilhowee Group were detached from their original positions and pushed westward about 160 km (100 mi) toward the continental interior. Folding and thrust faulting of the Shenandoah rocks occurred deep beneath the surface within the roots of the late Paleozoic Appalachians.

Alleghenian compressional stress pushed the detached mass of Grenville basement and its overlying carapace of Catoctin and Chilhowee rocks over and above the adjacent *fold-and-thrust belt* of sedimentary rocks now exposed in the Valley and Ridge to the west. During thrusting, the rocks composing Shenandoah were squeezed into a convex-upward *anticline* that "rolled over" along the surface of the underlying thrust (Fig. 35.5B). You might imagine a throw rug becoming snagged on an old wooden floor beneath, crumpling up on itself as it gets pushed farther along. The end result is the "Blue Ridge overturned anticline," the main structural feature within Shenandoah National Park (Fig. 36.10). The massive crystalline rocks of the Grenville basement form the core of the anticline, while the layered rocks of the Catoctin and Chilhowee wrap around the top. Both of these upper units have been eroded from the crest of the overturned fold, exposing the granites and gneisses of the basement at the surface. Shenandoah National Park and the ridgeline of the Blue Ridge Mountains are centered over the "hinge" of the fold along the western edge of the structure.

The network of thrust faults beneath the overturned anticline is called the **Blue Ridge fault system**. The leading edge of the fault system where it cuts the land surface is located along the western base of the Blue Ridge near its border with the Valley and Ridge subprovince in the Shenandoah Valley. The trace of the fault system across the vegetated landscape is difficult to discern but is commonly marked by a topographic change from the steep western

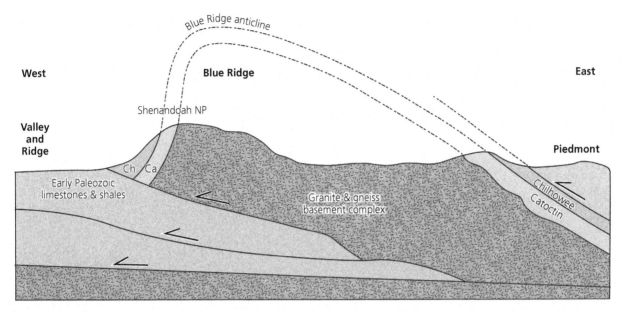

Figure 36.10 Generalized cross section of the Blue Ridge overturned anticline. Half arrows indicate major thrust faults. The dashed outline shows the hinge of the fold that was subsequently removed by erosion.

slopes of the Blue Ridge to the gentle floodplain of the Shenandoah Valley.

Many overlooks along Skyline Drive afford a view of Massanutten Mountain off to the west, providing an opportunity to understand the formation of the distinctive topography of the adjacent Valley and Ridge subprovince. Similar to the Blue Ridge, the fold-and-thrust belt of the Valley and Ridge formed during the Alleghenian orogeny as well. Massanutten Mountain comprises a series of prominent linear ridges that, from above, form what looks like a giant eye of a needle (Fig. 36.11). Visualized in cross section (Fig. 36.12), the structure beneath the ridgelines is that of a complex series of folds, flexed into a larger, convex-downward fold called a **synclinorium**.

The core of the structure is composed of middle Paleozoic shale that weathers easily, forming the elongate central valley in the eye of the topographic needle. Underlying and surrounding the shale is an older middle Paleozoic sandstone that forms the limbs of the synclinorium. The sandstone is much more resistant to weathering and erosion than the overlying shale, so it forms the linear ridgelines of Massanutten where the sandstone breaches the surface. Older early Paleozoic limestones and shales within the synclinorium underlie the surrounding flatlands of the Shenandoah Valley. The limestones readily dissolve in the presence of water, whereas the shales break down into particles of silt and clay. The broad floodplain of the meandering Shenandoah River flows along the lowland topography above the weathered limestones and shales. The river overtopped its banks during countless floods, laying down a veneer of sand, silt and clay particles. When combined with organic matter from plants and microbes, a fertile **alluvial soil** developed on the floodplain.

Landscape Evolution at Shenandoah National Park

The last 200 m.y. or so in eastern North America have been marked by tectonic quiescence and the progressive degradation of the ancestral Appalachians. Several kilometers of rock were eroded from the mountains, broken down into particles of sediment then transported by rivers and streams toward the Atlantic Ocean where they accumulated as the sedimentary wedge of the coastal plain and continental shelf. As the weight of the rock at the highest elevations was erosionally removed, the rock beneath rose isostatically, exhuming the deep metamorphic roots of the ancestral Appalachians.

The connected processes of erosion and uplift continue today, with downcutting by flowing water performing the bulk of the work. Estimates of the modern rate of erosion in the Blue Ridge range from 4 to 12 m (13 to 40 ft) per million years, a rate that translates to a thickness of about four sheets of paper every century.

The modern topography of Shenandoah National Park is a consequence of weathering and erosion acting differentially on the main rock types in the park. Broadly speaking, greenstones and quartzites are the most resistant rocks in the park and thus tend to form the highest peaks, including Hawksbill Mountain. Many of the park's picturesque waterfalls cascade over ledges formed by stacks of Catoctin lava flows (Fig. 36.13). Horizontal layers of Catoctin greenstone underlie Big Meadows, a broad, high-elevation expanse near the center of the park. The overlying Chilhowee metasediments were removed by erosion, allowing for the development of thick soils and wetlands directly on the flat-lying greenstone beds at Big Meadows.

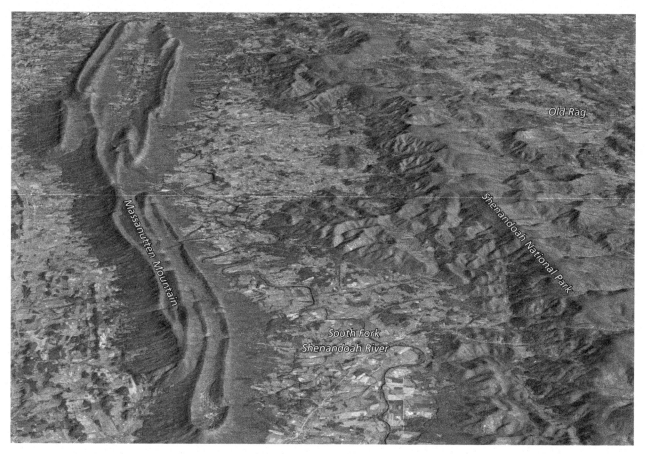

Figure 36.11   Landsat image draped over a digital elevation model of the Blue Ridge Mountains, with the distinctive Massanutten Mountain off to the northwest. The series of linear ridgelines are surrounded by the broad Shenandoah Valley and the widely meandering North and South forks of the Shenandoah River.

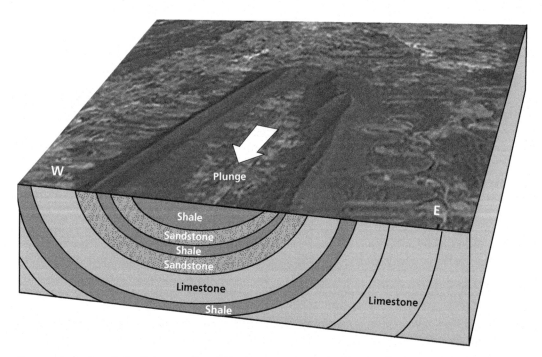

Figure 36.12   Google Earth perspective of Massanutten Mountain, showing the disposition of units within the synclinorium. The synclinorium plunges toward the lower part of the figure (southwest). Resistant layers of sandstone form the ridges, whereas less resistant shales and limestones form the valleys. Note the widely meandering North Fork (left) and South Fork (right) of the Shenandoah River.

Figure 36.13 The cascades of Dark Hollow Falls are developed on ledges and benches formed by layers of Catoctin lava flows.

Figure 36.14 Spheroidal weathering along intersecting joint planes creates the stack of rounded granite boulders of Old Rag Mountain.

Old Rag Mountain, a popular hiking destination east of the main Blue Ridge, stands high above the adjacent valleys because the Old Rag Granite that composes the mountain is more resistant to breakdown than the rocks that surround it. The granite is cut by an array of vertical joints and near-horizontal *exfoliation joints* that dissect the massive rock into angular blocks (Fig. 36.14). *Frost wedging* physically pries apart the rock as water alternately freezes and thaws within the joint planes. Chemical weathering also widens joints as water infiltrates along the planar surface, decomposing feldspar and mica grains into loose clay minerals that readily wash away. Over time, the combination of physical and chemical weathering causes the granite blocks to become rounded into a precariously stacked arrangement of boulders. This process of *spheroidal weathering* is responsible for the characteristic appearance of the highest elevations of Old Rag Mountain. Many of the spheroidal boulders detach from the main mass and tumble into "hollows"—small valleys cut by running water along the flanks of the mountain, where they accumulate as **boulder streams**.

Quartzites of the Antietam and Harpers formations crop out along the southwestern side of the Blue Ridge in Shenandoah where they form cliffs, steep ridges, and rugged **block fields** (Fig. 36.15). Beds of phyllite beneath the quartzite readily decompose by weathering, leaving the rigid beds of quartzite unsupported. The quartzite erodes along joints and bedding planes to form angular blocks that accumulate downslope as fields of talus. The mass of blocks likely migrates slowly downslope over time as underlying soils alternately freeze and thaw, incrementally heaving the overlying debris downward.

Continental ice sheets never reached farther south than central Pennsylvania during the last 2.6 m.y. of the Pleistocene Ice Ages (Figure 35.27). But the Blue Ridge near Shenandoah experienced much colder **periglacial** conditions during glacial phases. ("Periglacial" refers to areas in close proximity to the margins of ice sheets and alpine glaciers.) The frigid periglacial climate profoundly affected the flora of the region, with much of the crest of the mountains converting to windswept tundra. The surrounding lower elevations were swathed in a dense forest of conifers (Figure 35.28). The bitterly cold conditions also enhanced the physical weathering of exposed rock above *treeline*, mainly by the activity of frost-wedging that physically broke rock apart along joint planes. Periglacial soils contracted when they were frozen and expanded when they warmed, incrementally moving overlying masses of loose rock in boulder streams and block fields downslope.

Figure 36.15 Block field of angular fragments of Antietam quartzite along the southwestern flank of the Blue Ridge below the Rockytop overlook.

It's likely that these accumulations of rocky debris are relict features of Pleistocene periglacial processes, with only minimal activity during the current Holocene interglacial phase.

PATTERNS AND PROCESSES OF STREAMS AND RIVERS

During the three orogenic events of the Paleozoic, most rivers and streams flowed westward from the highlands toward the Appalachian Basin. The *clastic wedges* of sediment that we see today in the Appalachian Plateau and Valley and Ridge subprovinces are evidence for the westward direction of drainage systems. As the ancestral Appalachians were erosionally denuded during the Mesozoic, they were likely buried in their own debris, with a broad, eastward-sloping alluvial surface blanketing the deformed roots of the mountains beneath. At some point, regional drainage systems reversed their flow to discharge eastward toward the Atlantic across the gently sloping ancestral coastal plain. As the overlying veneer of sand and gravel was slowly stripped away by the river system, the rivers eventually encountered the resistant rocks of the Blue Ridge beneath. The rate of downcutting was rapid enough that the rivers incised canyons across the elongate mass of crystalline rock, forming **water gaps** transverse to the trend of the now-exposed ridgelines.

The Potomac River cuts eastward across the Blue Ridge north of Shenandoah National Park through a water gap, as does the James River south of the park. These are *superposed streams*, comparable to those of the Colorado Plateau described in earlier chapters (Fig. 6.9). Other rivers once incised downward across the Blue Ridge as well, but have since been confined to either the east or west flank. As evidence for the former existence of these ridge-crossing

rivers there are **wind gaps**, such as Thornton Gap, Swift Run Gap, and Rockfish Gap. These passes across the Blue Ridge were formerly occupied by rivers or streams but were subsequently abandoned. Today, the wind gaps provide the easiest routes for the main roads and entrances into the park.

Why did these drainages abandon their incised channels across the Blue Ridge? As the upstream ends of the east-flowing rivers eroded *headward* toward the west, they encountered the northward-flowing Shenandoah River, which today meanders across the limestone terrain of the Shenandoah Valley. In time, the flow of these ridge-crossing rivers was diverted into the larger Shenandoah River by the process of *stream capture* (Fig. 34.15). Their water gaps were converted to wind gaps, with the crest of the Blue Ridge transformed into the major drainage divide that exists today, separating streams flowing west from those flowing east.

MODERN LANDSCAPE OF SHENANDOAH NATIONAL PARK

Most of the bedrock of the Blue Ridge is hidden beneath a veneer of residual soil and young surface deposits. *Residual soils* originate in place by the slow disintegration of underlying bedrock. The ample rainfall at Shenandoah penetrates deeply into the rock, primarily along joint planes, chemically decomposing it to form smaller particles. Vegetation takes hold where water collects in the pores between particles, eventually decaying to add organic matter to the incipient soil. Given enough time, the mixture of unconsolidated sediment and organic matter evolves into residual soils, which in Shenandoah National Park supports widespread hardwood forests.

Surface deposits in Shenandoah include those loose sediments that accumulated within *alluvial fans* and *debris flows*. Alluvial fans form a gently sloping apron along the base of the west side of the Blue Ridge where the mountains merge with the Shenandoah Valley. The sediments composing the fans were transported downslope as surface runoff in creeks and rivulets, then deposited as broad lobes at the base of the slope. The unconsolidated fan deposits were eventually stabilized by vegetation. Debris flows are chaotic masses of rock, soil, and vegetation that mix with stormwater to form thick slurries that race downslope through hollows and stream channels. They are usually triggered by intense rainfall events and occur on average about every 10–15 years over the past several decades.

Go to Shenandoah to take in the vistas from the overlooks along Skyline Drive, to catch a glimpse of the abundant wildlife, to relish the autumn colors, and to hike the wide array of trails. But don't forget to look closely at the exposures of rock that crop out along roadcuts, waterfalls, and near the crests of ridges. The outcrops of rock may appear isolated and random but remember that they underlie all of the forested ridgelines and extend deep into the ground beneath your feet. The exposures of rock are windows into the crust that forms the very foundation of eastern North America. Moreover, appreciate that the rocks of Shenandoah tell a remarkable story of colliding continents, the birth and closure of ocean basins, and the rise and fall of immense mountains.

Key Terms

alluvial soil
amygdules
block fields
boulder streams
flood basalt
greenstones
periglacial
Skolithos
synclinorium
trace fossil
water gap
wind gap

Related Sources

Badger, R. L. (2012). *Geology along Skyline Drive: Shenandoah National Park, Virginia.* Helena, MT: Falcon Publishing.

Hackley, P. C. (2000). *A Hiker's Guide to the Geology of Old Rag Mountain, Shenandoah National Park, Virginia.* Open-File Report 00-0263. Reston, VA: U.S. Geological Survey. Accessed February 2021 at http://pubs.er.usgs.gov/publication/ofr00263.

Southworth, S., et al. (2009). *Geology of the Shenandoah National Park Region.* Reston, VA: 39th Annual Virginia Geological Field Conference. Accessed February 2021 at http://csmgeo.csm.jmu.edu/geollab/eaton/web/eaton_files/Publications/vgfc2009.pdf.

Thornberry-Ehrlich, T. L. (2014). *Shenandoah National Park: Geologic Resources InventoryRreport.* Natural Resource Report NPS/NRSS/GRD/NRR—2014/767. Fort Collins, CO: National Park Service. Accessed February 2021 at www.nps.gov/articles/nps-geodiversity-atlas-shenandoah-national-park-virginia.htm#gri.

37 Great Smoky Mountains National Park

Figure 37.1 View across Great Smoky Mountains National Park from Clingmans Dome.

Great Smoky Mountains National Park is consistently the most popular national park in America (Fig. 37.1). In 2019, over 12.5 million people visited the park, 6 million more than visited Grand Canyon, the second most popular of the 63 national parks. The park straddles the state line separating Tennessee from North Carolina, which extends northeast–southwest along the crest of the Great Smoky Mountains (Fig. 37.2). This location places it within easy driving distance of several major cities in the American South, making it accessible to those living in the region as well as those driving to and from Florida. Tourists flock to Great Smoky Mountains National Park to

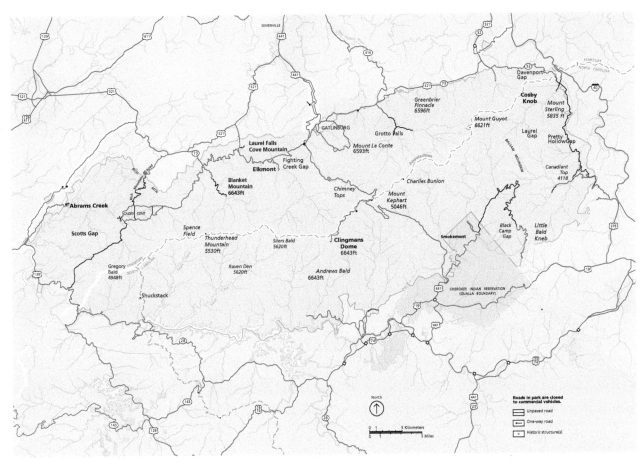

Figure 37.2 Great Smoky Mountains National Park encompasses a huge area of the Southern Appalachians straddling the Tennessee–North Carolina state line. The Newfound Gap Road (Route 441) cuts across the park, providing access to other roads, trails, overlooks, and cultural points of interest.

enjoy the lush forests blanketing the steep ridgelines, rushing streams and waterfalls, wide diversity of plants and animals, and historical richness that preserves a vanishing mountain culture. The lack of an entrance fee adds to the high visitation.

The national park is situated within the broad welt of the Great Smoky Mountains (aka Great Smokies), a dense cluster of ridges and peaks that forms the southern end of the Blue Ridge subprovince of the Appalachian Mountains (Fig. 35.3). The rolling terrain of the Piedmont lies to the southeast, whereas the sinuous contours of the Valley and Ridge bound the Great Smoky Mountains to the northwest. The topography within the national park has a rounded character to it due to millions of years of weathering and erosion in the rainy, temperate climate, but the slopes of ridges can be very steep.

The mountainous landscape in the park is the highest and most rugged in the Appalachians, with more than a dozen peaks rising above 1830 m (6000 ft). The highest elevation in the park is Clingmans Dome at 2025 m (6643 ft), accessible

by car as well as via the Appalachian Trail that winds along the spine of the park (Fig. 37.2). Mount LeConte is only the third highest summit in the park but measures 1615 m (5300 ft) from its base, placing it among the highest single mountains in the eastern United States. For comparison, the Blue Ridge Mountains in Shenandoah National Park to the north are about 600 m (2000 ft) lower in elevation.

Access to the interior of Great Smoky Mountains National Park is provided by U.S. Route 441, which cuts directly across the roughly east–west axis of the park (Fig. 37.3). Newfound Gap marks the highest point on the highway (1539 m/5048 ft) and provides sweeping vistas of surrounding ridgelines and summits. The drive along Route 441 from park entrances on either the north or south passes through a diversity of microclimates and forest ecosystems, rising 900 m (~3000 ft) to the cooler temperatures and spruce–fir forest at Newfound Gap. Other paved roads wind through lower elevations of the park, providing access to trailheads and cultural attractions such as Cades Cove. The portion of the Foothills Parkway that traverses along the ridge of Chilhowee Mountain near the northwestern border of the park provides spectacular views of the western Great Smokies as well as the Valley and Ridge subprovince to the west.

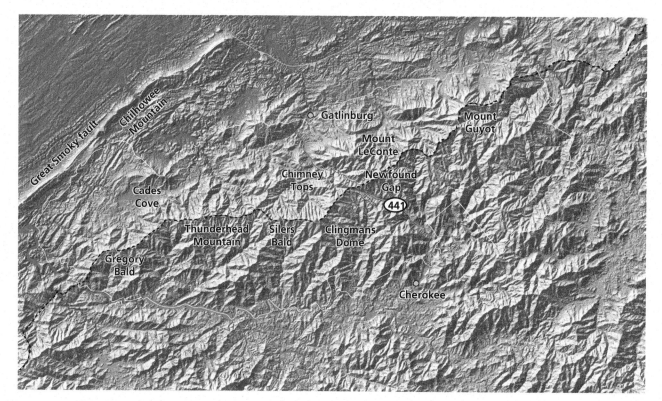

Figure 37.3 Color-coded elevation map of the region near Great Smoky Mountains National Park. Yellow line marks the park boundaries. Black dashed line denotes the Tennessee–North Carolina state line. Route 441 across the park is shown in orange. Some of the key topographic features discussed in the text are shown as red triangles. Great Smoky fault is shown as a tan line with barbs on the upper thrust sheet.

The mountains of the park tend to capture moist air masses from both the Atlantic and Gulf coasts, resulting in high annual amounts of precipitation. The 130 cm (50 in) of rainfall in the lower elevations and the 200 cm (85 in) in the highlands support dense woodlands that blanket the slopes. Most of the trees were logged in the 19th and 20th centuries, but about a third of the original trees remain, forming one of the largest intact old-growth forests in the eastern United States. The vegetation releases organic compounds that combine with water vapor in the air to form the misty blue "smoke" that gives the Great Smokies their name.

The combination of copious rainfall, a wide range of elevations, and a variety of forest habitats support a diverse array of about 10,000 species of plants and animals. High-elevation ecosystems in the park sustain species typically found in higher latitudes like those of southern Canada. These species likely migrated along the axis of the Appalachians to the lower latitude ridgelines and peaks in the Great Smokies during glacial maxima when the expansion of continental ice sheets forced climate and vegetation belts to shift southward (Chapter 35). Great Smoky Mountains National Park was named an International Biosphere Reserve in 1976 due to its remarkable biodiversity.

The dense vegetation makes life complicated for those hardy geologists who have worked to decipher the geologic framework of the Great Smoky Mountains. Exposures of rock tend to occur as isolated masses on ridgetops and summits, in road cuts, along some streambeds, and beneath waterfalls.

Furthermore, the geologic relationships are a challenge to untangle due to the tremendous thicknesses of relatively monotonous bodies of rock, the lack of fossils, the variable degrees of metamorphism, and the dissection of the rocks by faulting. Nevertheless, a great deal of meticulous fieldwork, microscopy, and radiometric age dating by geologists from the U.S. Geological Survey and various academic institutions reveals a reliable geologic history of the rocks composing the park.

If you read the two previous chapters, Chapters 35 and 36, in Part 9 on the origin of the Appalachian Geologic Province and Shenandoah National Park, then you have a good background for understanding the history of the rocks at Great Smoky Mountains National Park. There are many similarities with the rocks composing Shenandoah National Park, but those at Great Smoky Mountains have been deformed and metamorphosed to a greater degree. Furthermore, they have been sliced and displaced by many more *thrust faults* than at Shenandoah, creating a stacked series of thrust sheets that have been weathered and eroded into the corrugated landscape we see today.

Origin and Age of Rocks at Great Smoky Mountains National Park

The vast majority of the rocks composing the landscape of Great Smoky Mountains National Park are late Precambrian in age. The three main sets of rocks at the park represent a sequence of tectonic settings in the assembly and breakup of the Rodinian supercontinent and are directly comparable to the three sets of rocks at Shenandoah National Park to the north along the Blue Ridge Mountains. The three sets consist of (1) metamorphic rocks of the *crystalline basement*, (2) a thick assemblage of metamorphosed sedimentary rocks deposited in *rift basins*, and (3) less deformed sedimentary rocks that accumulated on the early Paleozoic *passive margin* of eastern North America. The following narrative on the rocks composing Great Smoky Mountains National Park weaves together the composition of the rocks with their tectonic and depositional settings through time.

The oldest rocks are 1.0 Ga *gneisses* and *schists* composing the crystalline basement upon which all other rocks were deposited. Recall from the previous two chapters that these rocks are the exhumed roots of the ancient Grenville Mountains that formed by convergent tectonism along the margin of Rodinia (Fig. 33.7). These highly deformed metamorphic rocks were originally sedimentary and igneous rocks that were later transformed by extreme pressures and temperatures during Grenville mountain-building. The original minerals composing the rocks slowly recrystallized into new minerals that aligned along thin bands, forming the foliated texture in the gneisses and schists we see today (Fig. 37.4). (If necessary, revisit Chapter 1 to review metamorphism and metamorphic rocks.)

Rocks of the Grenville basement are exposed in the southeast portions of the park, particularly well in a roadcut along Route 441 near the Oconaluftee Ranger Station. Even though we can only catch a small glimpse of these rocks, remember that they form an enormous volume of the crustal foundation deep beneath the surface of the Great Smoky Mountains.

LATE PRECAMBRIAN CONTINENTAL RIFTING

The crystalline rocks of the Grenville Mountains experienced a prolonged phase of erosion prior to the disassembly of Rodinia (Fig. 35.12). Perhaps beginning around 750 Ma,

Figure 37.4 Billion-year-old gneiss showing foliated texture (vertical alignment of mineral grains). This rock forms the crystalline basement upon which all younger rocks of the park were deposited. Mingus Mill roadcut, Great Smoky Mountains. Field of view is about 10 cm (4 in) across.

ancestral South America began to separate from what is today eastern North America. Over the next 200 million years, a series of narrow and deep rift basins opened along the northeast–southwest axis of maximum extension between the two diverging continents (Fig. 35.11). The fault-bounded highlands surrounding the rift basins were composed of Grenville-age rocks, which supplied the abundant pebbles, sand, silt, and clay that filled the adjacent lowland basins. In the region that would become the Southern Appalachians, these unconsolidated sediments were eventually buried and cemented to become the clastic sedimentary rocks of the **Ocoee Supergroup**, the dominant rocks composing Great Smoky Mountains National Park. This massive pile of rocks reaches an estimated thickness of ~15 km (9 mi), recording the progressive subsidence and infilling of the Ocoee rift basin through the latest Precambrian.

The clastic rocks of the Ocoee Supergroup were later metamorphosed to variable degrees, so that what were originally sandstones are now "metasandstones" and quartzites. What were once conglomerates composed of pebbles and sand grains are now "metaconglomerates." And what were once shales composed of clay particles are now slates and phyllites. Thick slabs of Ocoee rocks were later thrust over and above each other during late Paleozoic mountain-building, a process addressed in greater detail below. Today, this huge volume of late Precambrian rock underlies the thin veneer of soil and woodland that carpet the undulating topography of the Great Smokies (Fig. 37.5).

The metasedimentary rocks of the Ocoee form a **supergroup**, which is simply a term of convenience to organize a large volume of rock. The Ocoee Supergroup is divided into *groups*, which in turn are divided into several formations. The rocks of the Ocoee have a widespread distribution beyond the national park, ranging from the western Carolinas to northern Georgia, recording the regional extent of the Ocoee rift basin during the latest Precambrian. The rocks of the Ocoee Supergroup are estimated to have accumulated between ~750 and 550 Ma, although the exact dates are very uncertain due to the lack of fossils as well as the absence of ash beds or other volcanic layers that could be dated radiometrically. The youngest rocks of the Ocoee were deposited at the same time as the Catoctin flood basalts (570–550 Ma), which flowed across the surface of a separate rift basin now exposed in Shenandoah National Park to the north.

Clastic sediments of the Ocoee Supergroup accumulated in a wide variety of depositional environments ranging from rivers and deltas to deep-water settings. Rift basins subside as *normal faults* bounding their margins are abruptly offset during earthquakes, providing space for sediment to fill. The modern rift basins within Death Valley National Park as well as the entire Basin and Range (Chapters 23, 24) are analogs for parts of the Ocoee rocks that were deposited in continental settings. Much of the Ocoee was deposited in deep-water environments,

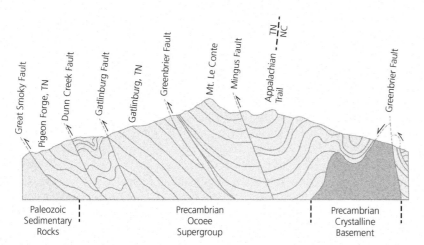

Figure 37.5 Highly generalized cross section across Great Smoky Mountains National Park (essentially along Route 441) showing the three main sets of rocks and a few of the thrust faults that stack sheets of rock against one another. This cross section has a very large vertical exaggeration, meaning that the vertical axis has been stretched relative to the horizontal axis to emphasize the topography. The thrust faults are much closer to horizontal than shown here.

however, so a good modern analog for those rocks may be deep lakes like Lake Tanganyika along the East African rift system (Fig. 1.12A).

The oldest metasedimentary rocks of the Ocoee (Snowbird Group) accumulated in broad stream channels, deltas, and beaches directly above the exposed crystalline rock of the Grenville basement. The sediments of these earliest deposits are composed of sand, silt, and clay grains derived from the weathering and erosion of granites and gneisses in nearby highlands. You can see these oldest rocks of the Ocoee along the Roaring Fork Motor Trail, but be sure to get out of your car to take a closer look (Fig. 37.6). You can also view blocks of *cross-bedded* metasandstone of the Snowbird Group in the walls of the Sugarlands Visitor Center and the Oconaluftee Ranger Station. Remember that the original sand grains composing these quarried blocks of rock were once washed back and forth along a late Precambrian shoreline.

Overlying rocks of the Ocoee (Great Smoky Group) were laid down as *turbidites* and underwater landslides within a deep-water setting. In the absence of any fossil evidence, it's difficult to tell whether the turbidites were deposited in freshwater lakes or perhaps an extension of the ocean that filled the Ocoee rift basin. Recall from Chapter 26 that turbidites formed as sediment-laden currents swept downslope before spreading out across the deep floor of the basin, laying down a lower layer of sand and a finer upper layer of silt and clay (Fig. 26.13). If the turbidity current originated close to the mouth of a river, then the resulting turbidite may consist of pebbles within a sandy matrix. These turbidite metasandstones, metasiltstones, and metaconglomerates are interbedded with finer grained slates and phyllites, which are interpreted to have originally formed as clay particles settled out of suspension within deep-water, oxygen-poor settings. Visualize the

deepest part of the Ocoee rift basin, perhaps covered by several hundred meters of water, slowly blanketed by layers of mud over long expanses of time. Then, possibly triggered by earthquakes along the margins of the basin, turbid currents of sand and pebbles flushed across the basin floor in relatively short bursts. This depositional mode lasted for several tens of millions of years.

The rocks of the Ocoee Supergroup in Great Smoky Mountains National Park record the rifting of the Rodinian supercontinent along its eastern margin toward the end of the Precambrian. In a tectonic sense, the Ocoee rocks are related to late Precambrian rift–basin deposits in national parks of the American West, such as the Grand Canyon Supergroup (Chapter 3), the Uinta Mountain Supergroup (Chapter 9), the Belt Supergroup (Chapter 22), and the Pahrump Group (Chapter 24). Each of these four bodies of rock records the breakup of Rodinia along its western margin, the only difference being that rifting occurred earlier there (~800 Ma) than along the eastern margin.

Figure 37.6 Exposure of lower Ocoee metasandstones along the Roaring Fork Road at Place of a Thousand Drips, Great Smoky Mountains National Park

WHERE TO SEE OCOEE ROCKS IN GREAT SMOKY MOUNTAINS NATIONAL PARK

Coarse-grained turbidites of the Great Smoky Group are the thickest and most widespread rocks of the Ocoee Supergroup in the national park. Resistant exposures of these rocks crop out along the crests of ridgelines and peaks as well as in numerous roadcuts along Route 441. The two formations of the Great Smoky Group that compose the majority of the crest of the Great Smokies in the park are the Thunderhead and Anakeesta.

The massive layers of conglomeratic metasandstone composing the Thunderhead Sandstone are highly resistant to weathering and stand out as light-colored cliffs and ledges on Clingmans Dome, Mount LeConte, and Thunderhead Mountain (Fig. 37.7A–C). Many outcrops of the Thunderhead occur in roadcuts along Route 441 as it traverses through the park. The Thunderhead also underlies some of the park's waterfalls and streams, such as Ramsay Cascades, Laurel Falls, and The Sinks along the Little River. As you look closely at these rocks, today lifted upward as part of the Great Smokies, think about how they were originally deposited by bottom-hugging turbidity currents in a deep basin several hundred million years ago.

The other resistant formation of the Great Smoky Group exposed along steep slopes and jagged ridgelines is the Anakeesta. These metasiltstones, phyllites, and slates are commonly stained shades of rusty orange due to oxidation of the mineral pyrite upon exposure to the atmosphere. The presence of pyrite, an iron sulfide, indicates low-oxygen conditions during deposition. The pyritic, fine-grained sediments of the Anakeesta were likely deposited in the lowest energy, deepest water environments within the Ocoee rift basin. The Anakeesta Formation crowns Mount LeConte, holds up the pinnacles of Chimney Tops (Fig. 37.8), forms large outcrops along the Alum Cave Bluffs trail, and is well exposed in roadcuts near Newfound Gap.

EARLY PALEOZOIC PASSIVE MARGIN

By the beginning of the Paleozoic Era ~540 Ma, the rifting of the eastern margin of Rodinia was complete, with the newly formed Iapetus Ocean growing wider by seafloor spreading. The rifted continental margin began to subside below sea level, forming a passive margin continental shelf. Layers of sand, silt, and clay derived from the exposed continent buried the Ocoee rift basin beneath a seaward-thickening wedge (Fig. 37.9). This *rift-to-drift transition* is recorded in rocks of the Chilhowee Group, described in the previous chapter on Shenandoah National Park.

The quartzite, sandstone, siltstone, and shale of the Chilhowee Group can be traced along the entire western flank of the Blue Ridge subprovince, from southern Pennsylvania to northern Georgia. In Great Smoky Mountains National Park, the Chilhowee Group is not exposed within the main park boundaries but crops out along the Foothills Parkway where it forms the crest of Chilhowee

Figure 37.7 A) Inclined beds of Thunderhead Sandstone on Cliff Tops, Mount LeConte. B) Massive beds of Thunderhead Sandstone exposed near the parking lot to the Clingmans Dome trail. C) Close-up of quartz pebble conglomerate of Thunderhead Sandstone exposed near the parking lot to the Clingmans Dome trail.

Mountain (Fig. 37.3). Exposures of Chilhowee sandstone reveal vertical tubes of the early Paleozoic trace fossil *Skolithos*, similar to those in Shenandoah National Park. Rocks younger than the Chilhowee are generally absent in Great Smoky Mountains National Park, other than in isolated "windows" eroded through thrust sheets, discussed below. Younger Paleozoic rocks dominate the Valley and Ridge and Appalachian Plateau to the west of the park.

Figure 37.8   A steeply tilted spine of Anakeesta metasandstone forms the jagged crest of Chimney Tops, northwest of Newfound Gap. Weathering of rock along intersecting joint planes in the Anakeesta has left behind the angular blocks that jut as much as 15 m (50 ft) above the adjacent slopes.

Recall from Chapter 35 that those two provinces constitute the "Sedimentary Appalachians."

Uplift and Deformation of Rocks at Great Smoky Mountains National Park

The crystalline basement was once part of the deep roots of the billion-year-old Grenville Mountains, and the rocks of the Ocoee Supergroup were deposited in active rift basins. The earliest Paleozoic Chilhowee Group was once part of a vast continental shelf below sea level. How did these rocks become the rugged highlands of the national park we see today?

The late Precambrian and earliest Paleozoic rocks of the Great Smokies were initially deformed and metamorphosed during the first of the three great orogenic episodes that created the ancient Appalachians (Fig. 35.16). The docking of the *Taconic Terrane* around 450 Ma pushed

rocks of the basement, the Ocoee Supergroup, and early Paleozoic sedimentary rocks to the northwest, stacking thrust sheets one atop another and generating enough heat and pressure to trigger metamorphism.

The most significant uplift and deformation in the Great Smokies, however, took place during the *continental collision* of the *Alleghenian orogeny* that peaked around 300 Ma. As the northwest African part of the Gondwanan supercontinent slowly crushed against eastern North America, continental crust in the region that is today the Great Smoky Mountains was intensely deformed and re-metamorphosed deep beneath the surface. With continued compression, slivers of the crust detached along multiple thrust faults and were pushed over 100 km (60 mi) toward the northwest to their current location. This long displacement of the rocks didn't occur gradually and continuously, but rather it's likely that movement occurred just a few meters at a time during innumerable major earthquakes when the faults would abruptly rupture. Over millions of years and countless large earthquakes, the emplacement of overlapping thrust sheets was complete, and the ancestral Great Smokies towered far above the adjacent lowlands as part of the larger Appalachian Mountains of the late Paleozoic.

The primary thrust created during Alleghenian orogenesis is called the **Great Smoky fault**, a very low angle surface that juxtaposes the rocks of the Ocoee Supergroup and Chilhowee Group over and above early Paleozoic sedimentary rocks. You may recall from Chapter 22 on Glacier National Park that tectonic compression first causes layered sedimentary rocks to deform by folding before a stack of rocks breaks along a thrust plane (Fig. 22.9). With continued compressional stress, thick sheets of older rock are juxtaposed above sheets of younger rock along near- horizontal thrust faults. This stacking of

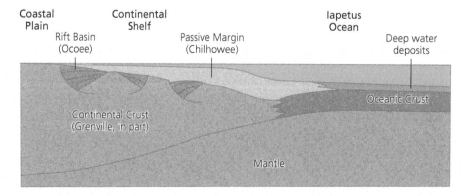

Figure 37.9   Early Cambrian tectonic settings of the three main sets of rocks composing Great Smoky Mountains National Park.

thrust sheets creates a thickening of the crust, forming topographic highlands as well as a deep root of deformed rocks beneath the surface. The trace of the Great Smoky fault where it intersects the vegetated landscape is difficult to recognize, but it can be identified by the abrupt juxtaposition of metamorphosed rocks of the Great Smoky Mountains against unmetamorphosed rocks of the Valley and Ridge *fold-and-thrust belt* to the west (Fig. 37.3). At a larger scale, the Great Smoky fault is one of a network of thrust faults that extend along the western margin of the Blue Ridge subprovince. You can consider it to be a cousin of the Blue Ridge fault that bounds Shenandoah National Park to the north.

Within the boundaries of the national park, the trace of the Great Smoky fault can be inferred where it is breached by erosion, exposing the underlying younger rock in **tectonic windows** (Fig. 37.10A). In places such as Cades Cove, a scenic open valley surrounded by high ridges, erosion has cut down through the Ocoee rocks above the plane of the Great Smoky fault, revealing the Paleozoic rocks beneath (Fig. 37.10B). The contact between the Paleozoic limestones within the gently rolling terrain of the cove and the late Precambrian rocks of the encircling ridges generally marks the location of the surface of the Great Smoky fault. The window of Paleozoic limestone just beneath the surface within Cades Cove readily weathers to produce a fertile soil, prized by early settlers.

Other tectonic windows through the Great Smoky thrust in the national park include Wear Cove and Tuckaleechee Cove, where a small segment of the fault is exposed behind a waterfall (Fig. 37.11). To fully appreciate the broad extent of the Great Smoky fault, use your imagination to project the tiny exposure of the fault plane at Tuckaleechee Cove beneath the landscape across the entire Great Smokies and part of the adjoining Piedmont. The actual fault plane is an enormous, complicated surface, hidden almost entirely out of sight deep beneath the terrain.

The modern Great Smoky Mountains we see today are only partly due to compressional uplift during the late Paleozoic Alleghenian orogeny. That event deformed the rocks and stacked them into thrust sheets, raising the ancestral Appalachians. Those highlands were eventually weathered and eroded, and then were slowly split apart as Africa separated from eastern North America during the Mesozoic birth of the Atlantic Ocean (Chapter 35). The remnants of the ancient Appalachians were reduced to particles of gravel, sand, and clay, and transported by rivers and streams onto the surrounding coastal plains and continental shelves of the Atlantic and Gulf of Mexico.

As more and more rock was removed, the deeper roots of the ancient Appalachians *isostatically* rebounded upward toward the surface, somewhat analogous to what happens to the mattress when you get out of bed in the morning, only much much more slowly. Those deep roots are the rocks of the crystalline basement and the Ocoee exposed today along the crest of the Great Smokies. It's been calculated that about 15–20 km (9–12 mi) of rock were erosionally removed over the past 200 m.y. to reveal the rocks we see today in the national park. The mountains were never that high, of course, but erosion continually removes the highest elevation rock, allowing deeper rock within the roots of the mountains to rise toward the surface. Erosion and isostatic uplift are still occurring in the region, driven by rivers and streams that slowly slice their way downward through the rock.

Landscape Evolution at Great Smoky Mountains National Park

The related processes of physical and chemical weathering act on the rocks of the national park to different degrees. In general, rainfall, snowmelt, running water, and the intrusive roots of plants have less of an effect

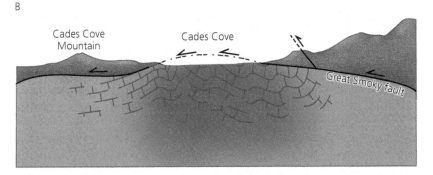

Figure 37.10 A) A tectonic window is an erosional opening through the rocks of a thrust sheet, revealing the rocks below the plane of the thrust. B) Simplified cross section of the tectonic window of Cades Cove showing the low-angle Great Smoky thrust fault separating older Ocoee rocks above from younger Paleozoic rocks below.

Figure 37.11 The Great Smoky thrust fault is difficult to see on the landscape, but at this locality in Great Smoky Mountains National Park, a small segment of the fault is expressed as the horizontal surface (thin dashed line) separating late Precambrian Ocoee rocks above from early Paleozoic limestones below. White Oak Sink, Tuckaleechee Cove window.

the linear Oconaluftee River Valley is developed above crushed and decomposed rock along the trace of the Oconaluftee fault.

The dense woodland forest that covers most of the rock in Great Smoky Mountains National Park is developed on a thick layer of *residual soil* that may be 2–10 m (6–32 ft) thick. Recall from Chapter 36 on Shenandoah National Park that residual soils are those that form in place from the weathering of underlying bedrock. Decaying vegetation and microbes add organic matter to the decomposed particles of sediment, creating a fertile mix. These soils are different from those that form from sediment that has been transported from somewhere else by running water, such as *alluvial soils* developed along the floodplains of rivers. The thick residual soils and their entangling mat of rooted vegetation act to stabilize slopes in most places in the park.

Rivers and streams have cut deeply into the rocks of the park, creating very steep slopes that are prone to rapid collapse as *landslides* and *debris flows*. Bedding planes and the network of joints that cut through the bodies of Ocoee rock permit water to penetrate into the interior, decomposing the rock and making it susceptible to failure. The foliated texture of the slates and phyllites of the Anakeesta Formation makes it particularly predisposed to landsliding; scars of past landslides are common along steep slopes of Anakeesta rock near Newfound Gap (Fig. 37.12). Planes of weakness may also develop near the intersection of joint planes where seeping water has decomposed the rock. Heavy rains can trigger landslides of the weakened material along the intersecting joint planes, leaving behind a long, V-shaped ravine as an unvegetated scar on steep slopes.

Episodes of intense rainfall can also activate debris flows, thick slurries of boulders, soil, and vegetation that flush through chutes and streamcuts along steep slopes, especially in areas underlain by rocks of the Anakeesta. Swaths of bare rock along the flanks of vegetated ridges mark the sites where debris flows have stripped away the soil and vegetation. A large pile of boulders may mark the spot where the flow came to an end. Severe debris flows have occurred in the park during torrential rains over the last several decades. The scars of these events are still

on the resistant rocks of the Thunderhead and Anakeesta formations than they do on less resistant rocks of the rest of the Ocoee Supergroup. Thus, those two bodies of rock form the highest ridgelines and summits in the park, whereas the weaker rocks tend to form valleys and lowlands. But *differential erosion* is only one of many factors that influence the modern landscape of the national park. For instance, some valleys in the park are formed above the splintered and deeply weathered rock where thrust faults breach the surface. (The rock along fault zones tends to be crushed and fragmented by repeated movement during earthquakes, making them vulnerable to weathering and erosion.) The Cosby Valley in the northern part of the park is formed on weakened rock where several thrust faults, including the Great Smoky fault, converge onto the landscape. And

Figure 37.12 Landslide scar developed in phyllites of the Anakeesta Formation. The foliation of the phyllite is oriented in the same direction as the slope, enhancing the probability of failure. Anakeesta Ridge, viewed from Newfound Gap Road.

evident today due to the lack of slope-stabilizing vegetation in the absence of soil.

During the Last Glacial Maximum around 20,000 years ago, the Great Smoky Mountains were subjected to intensely cold *periglacial* temperatures. Tundra vegetation replaced the hardwood forests at higher elevations, based on evidence from pollen. It wasn't cold enough for alpine glaciers to form, but the rocks were exposed to severe *frost wedging* (Chapter 5) that split layers of rock along vertical joint planes and horizontal bedding planes. The rigid quartzites and metasandstones of the Thunderhead Formation were particularly susceptible to this process, forming *block fields* of angular talus. Over time, these boulders were slowly transported downslope as underlying soils contracted when they froze and expanded as they thawed, inching the boulders downward. Today, block fields of Thunderhead rock are common features along the base of steep slopes in the park.

Most rivers and streams in the Great Smokies are part of the Tennessee River watershed that eventually drains into the Ohio River in western Kentucky. From there, the Ohio flows a short distance to the west where it merges with the Mississippi River. You may recall from earlier chapters on Glacier, Rocky Mountain, and Yellowstone national parks that snowmelt from those parks eventually finds its way into the Missouri River, which meanders its way eastward to the confluence with the Mississippi

at St. Louis. So the end result is that waters from Great Smoky Mountains National Park mingle with waters from three parks of the American West, creating a fluid connection between distant national parks at opposite ends of the country.

Air Quality in Great Smoky Mountains National Park

Great Smoky Mountains National Park is consistently among the parks with the worst air pollution and lowest visibility. The rugged topography constrains traffic to only a few mountain roads, with the daily exhaust from thousands of cars becoming trapped within narrow valleys. Add to the mix air pollution from outside the park, generated by the combustion of fossil fuels in power plants and automobiles in the surrounding region. The elevated topography of the park captures currents of air and its transported pollutants, degrading visibility and visitor enjoyment. Furthermore, compounds of nitrogen and sulfur within air pollution reach the surface through rainfall and fog, producing elevated acidity that affects plants, soils, and aquatic animals.

Particulates within polluted air masses create a whitish haze in the Great Smokies that impairs visibility and dilutes the natural blue "smoke" for which the park is named. Episodes of poor air quality during the summer tourist season can reduce visibility on the summits by as much as 80 percent. As of 2019, average annual visibility at Great Smoky Mountains National Park was 40 km (25 mi), compared to 150 km (90 mi) under natural conditions. On particularly bad days, visibility may be less than a kilometer, a disheartening concern for visitors to the park. But conditions are getting better, with measurements indicating that the skies around the national park are becoming clearer due to regulations related to the Clean Air Act of 1970 and amendments added in 1977 and 1990. Power plants are emitting less sulfur dioxide and nitrogen oxides than in decades past, reducing the acidity of rainfall and fog. The trends are toward a broad improvement in air quality and visibility in Great Smoky Mountains National Park, a pattern that is a literal breath of fresh air for visitors to the park.

Key Terms

supergroup
tectonic window

Related Sources

Clark, S. H. B. (2001). *Birth of the Mountains: The Geologic Story of the Southern Appalachian Mountains*. Reston, VA: U.S. Geological Survey.

Merschat, A. J., Hatcher, R. D., Jr., Thigpen, J. R., and McClellan, E. A. (2018). *Blue Ridge—Inner Piedmont geotraverse from the Great Smoky fault to the Inner Piedmont.* In A. S. Engel and R. D. Hatcher, Jr. (Eds.), *Geology at Every Scale: Field Excursions for the 2018 GSA Southeastern Section Meeting in Knoxville, Tennessee.* Boulder, CO: Geological Society of America Field Guide 50, pp. 141–209. Accessed February 2021 at https://doi.org/10.1130/2018.0050(09).

Moore, Harry. (1988). *A Roadside Guide to the Geology of the Great Smoky Mountains National Park.* Knoxville: University of Tennessee Press.

Southworth, S., Schultz, A., Aleinikoff, J. N, and Merschat, A. J. (2012)., *Geologic Map of the Great Smoky Mountains National Park Region, Tennessee and North Carolina. U.S.* Geological Survey Scientific Investigations Map 2997. Accessed February 2021 at https://pubs.er.usgs.gov/publication/sim2997.

Thornberry-Ehrlich, T. (2008). *Great Smoky Mountains National Park Geologic Resource Evaluation Report.* Natural Resource Report NPS/NRPC/GRD/NRR—2008/048. Denver, CO: National Park Service.

Williams, D. D. (2013). *The Rocks of Great Smoky Mountains National Park.* Athens, GA: Possum Publications.

38 Acadia National Park

Figure 38.1 View from Cadillac Mountain summit across the granitic bedrock and forested slopes of Mount Desert Island, Acadia National Park.

The smoothly rounded contours and wave-battered coastline of Acadia National Park draw almost three million visitors a year to enjoy the stunning beauty and expansive views (Fig. 38.1). Hidden beneath the bucolic landscape, however, is a geologic history of molten rock and explosive volcanic fury, followed much later in time by a massive sheet of ice over a kilometer thick that sculpted Acadia into the rounded, undulating terrain we see today. Even stranger to imagine is the fact that some of the rocks in the park were formed thousands of kilometers away in an ocean basin that doesn't exist anymore. Add the surreal idea that much of the underlying crustal rock of Maine was originally part of northwestern South America and your view of Acadia National Park as a serene, coastal landscape will be forever changed. The narrative of each of these events forms the body of this chapter.

Acadia National Park is located within the New England Subprovince of the Appalachians (Fig. 35.3). Even though the underlying crust is highly variable in composition, the New England Subprovince is considered to be part of the *Crystalline Appalachians*, comparable to the Blue Ridge and Piedmont subprovinces of the southern Appalachians. The boundaries of the national park are highly fragmented, spread out over several islands and a mainland peninsula along the central coast of Maine (Fig. 38.2). The core of Acadia National Park is located on Mount Desert Island and most of the discussion in this chapter will refer to features there. But important pieces of the geologic puzzle are found within park boundaries on the Schoodic Peninsula about 8 km (5 mi) to the east of Mount Desert Island and Isle au Haut, about 25 km (15 mi) to the southwest. Other significant geologic features addressed in this chapter are located on the nearby Cranberry Islands outside the park boundaries.

Mount Desert Island is the largest of the thousands of islands strung out along the coast of Maine. The highest point, Cadillac Mountain, is 466 m (1530 ft) above sea level and is the highest peak along the eastern seaboard of the United States. The rounded summit is composed of *granite*, split by fractures and partly covered by splotches of

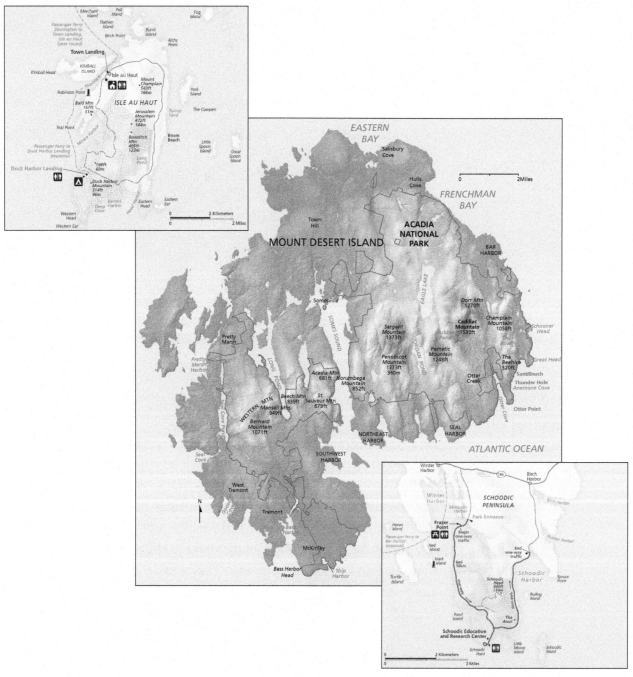

Figure 38.2 Most of Acadia National Park is centered on Mount Desert Island, but large portions are found on the Schoodic peninsula, Isle au Haut, and a few smaller islands.

lichen. The surrounding slopes are blanketed in a patchy mosaic of spruce–fir woodlands and hardwood forest. Several elongate lakes, oriented slightly northwest–southeast, separate the streamlined ridges in between.

The irregular coast of Mount Desert Island is dominated by rocky headlands that jut into the sea, marking the boundaries of bays and inlets (Fig. 38.3). Somes Sound cuts deeply into the midsection of the island. A large portion of the island consists of both saltwater and freshwater wetlands, and tide pools support a robust intertidal ecosystem. The narrow beaches at the base of rocky cliffs are typically composed of fist-sized cobbles or large boulders, whereas sandy beaches are rare. The variety of environments on Mount Desert Island attracts a diverse array of plants and animals as well as hordes of summer visitors.

As part of the northern Appalachian Geologic Province, Acadia National Park shares a common history with Shenandoah and Great Smoky Mountains national parks to the south. But Acadia also has similarities to Yosemite and Sequoia/Kings Canyon national parks, sharing landscapes of dominantly granitic bedrock sculpted by ice. There are parallels with Crater Lake and Yellowstone national parks as well because the rocks of Acadia record colossal caldera eruptions of the distant past, comparable to its younger brethren. The effects of continental ice sheets

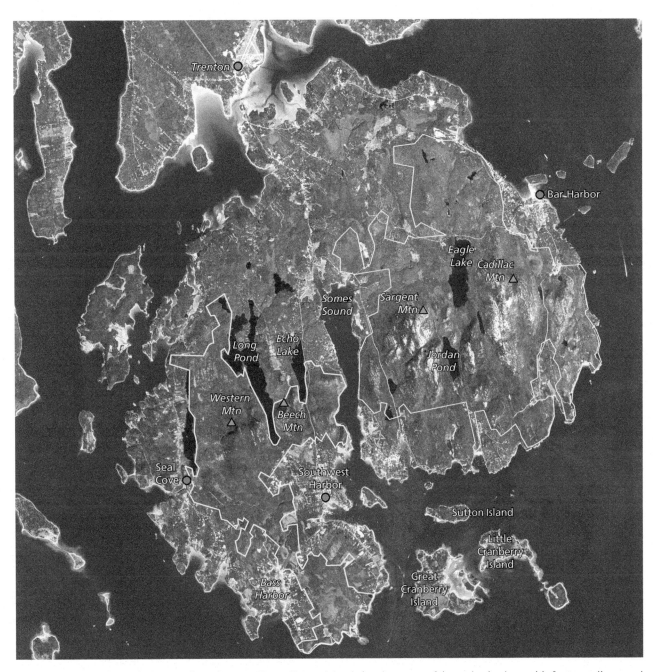

Figure 38.3 Google Earth view northward across Mount Desert Island showing some of the main physiographic features discussed in the text. The largest parts of Acadia National Park on the island are shown in yellow.

on the Acadia landscape bring to mind the influence of glaciers on national parks of the Rocky Mountains, Sierra Nevada, and Cascade Range on the opposite side of the continent. And the rocky, elevated coastline of Acadia is reminiscent of that of Point Reyes National Seashore along the Pacific Margin. In light of these similarities, many of the geologic concepts introduced in earlier chapters will be readdressed within the context of Acadia National Park.

Origin and Age of Rocks at Acadia National Park

The bedrock underlying Acadia National Park records events that occurred in the northern Appalachians during the Paleozoic Era, whereas the surficial landscape of the park was sculpted much more recently by Ice Age glaciation and shoreline processes. You might think of the rocks of Acadia as analogous to your body, while the landscape is comparable to your personality. This section deals with the origin and evolution of the rocks composing the foundation of the park.

The dominant rock by far on Mount Desert Island is *granite*, but the less abundant metamorphic, sedimentary, and volcanic rocks are equally important for understanding the full story of the geologic origins of Acadia National Park. The following sections on the main sets of rocks in the park combine simple descriptions of the rocks with their depositional and tectonic settings. A timeline will help you organize the sequence of events affecting Acadia National Park. A later section will integrate all of the main sets of rocks into a chronologic history of the geologic evolution of Maine and the northern Appalachians, referring liberally to the paleogeographic maps and cross sections of Chapter 35 as necessary.

EARLY TO MID-PALEOZOIC METAMORPHIC AND SEDIMENTARY ROCKS

The oldest rock on Mount Desert Island is the **Ellsworth Schist**, dated to be early Paleozoic in age (~510 Ma). The Ellsworth in the area of the park shows characteristics of both a *schist* and a *gneiss*, indicating a relatively high degree of metamorphism of the original rock (Fig. 1.30). Typical outcrops of the Ellsworth are greenish-gray, with bands of lighter colored quartz and feldspar alternating with darker layers of the green mineral *chlorite* (Fig. 38.4A). The Ellsworth is commonly contorted into tight folds. The easiest place to see outcrops of the Ellsworth Schist is on Thompson Island along the bridge connecting the mainland to Mount Desert Island, with other exposures occurring along the north and west margins of the island.

When looked at under a microscope, there is evidence that the precursor rock of the Ellsworth prior to metamorphism was volcanic in origin. The original lava is interpreted to have erupted along an underwater rift system similar to that within the modern Red Sea or Gulf of California. But this volcanism didn't happen anywhere near its current location along the coast of Maine. Instead, it

occurred thousands of kilometers away along the margins of *Gondwana* where a slice of continental crust was breaking away (Fig. 35.13A). That continental fragment rifted apart from what today we would today call northern South America, which at the time was sutured to Africa, Antarctica, and Australia as part of the Gondwanan supercontinent. It would take many millions of years for that Japan-sized sliver of continental crust to migrate across

Figure 38.4 A) Ellsworth Schist showing the foliated texture of a gneiss within the formation. Lighter bands are composed of the minerals quartz and feldspar, whereas the darker bands are mostly greenish chlorite. B) Sedimentary layering of turbidites in the Bar Harbor Formation.

the intervening ocean basin before it collided with eastern Laurentia as an *accreted terrane*.

Recall from several earlier chapters that a terrane (again, not to be confused with the landscape term *terrain*) is a unique body of rock that contrasts sharply with adjacent bodies of rock. Terranes may travel for great distances as part of larger plates before becoming tectonically accreted to the outer edge of a larger continental mass. The boundaries between terranes are commonly abrupt and marked by fault zones that represent the eroded remains of an ancient subduction zone where accretion took place. The larger body of rock on which the Ellsworth volcanic rocks traveled was one of the terranes that today compose a large part of the crust of Maine. At some point during its long migration toward its rendezvous with Laurentia, the volcanic rocks of the Ellsworth were metamorphosed and deformed into the folded schists and gneisses seen today.

The second-oldest rock unit in Acadia National Park was deposited directly on top of the Ellsworth Schist while the terrane was drifting across what was then the *Iapetus Ocean* (Chapter 35). The continental terrane on which the Ellsworth rocks were traveling eroded to form sediments that were shed into the adjacent ocean by turbidity currents, laying down thin beds of sand and silt. These *turbidites*, along with interbedded layers of volcanic ash, would eventually become buried to become beds of sandstone and siltstone of the **Bar Harbor Formation** (Fig. 38.4B). Good exposures of these rocks can be seen along the shore path in the town of Bar Harbor along the eastern edge of Mount Desert Island.

Deposition of the Bar Harbor Formation took place about 465 Ma, prior to the accretion of the terrane that would form the crust beneath Acadia National Park (Fig. 38.5). So,

like the underlying Ellsworth Schist, these rocks originally formed far away from their current location. Both formations were part of a larger terrane that would "dock" against the flank of eastern North America during the prolonged Acadian orogeny around 400–360 Ma. Recall that the dominant rock on Mount Desert Island is granite and that *plutonic igneous rocks* like granite intrude as a molten fluid into previously existing solid rock. The Ellsworth and Bar Harbor formations are the *country rock* that the granitic magmas intruded into around 420 Ma.

MID-PALEOZOIC PLUTONIC AND VOLCANIC ROCKS

The greatest mass of rock underlying Mount Desert Island, the Schoodic Peninsula, and Isle au Haut is the **Cadillac Mountain Intrusive Complex** (CMIC), which comprises a variety of plutonic igneous rocks. These types of rocks form a few kilometers beneath the surface within magma chambers or fractures as molten rock slowly crystallizes into solid rock (Chapter 1). The slow rate of cooling and solidification within the magma chamber allows mineral crystals to grow, and plutonic igneous rocks like those at Acadia National Park typically are composed of relatively large crystals visible to the naked eye. The majority of plutonic rocks within the CMIC are granites, rich in silica and composed of light-colored minerals such as quartz and feldspar (Fig. 1.20). Intrusion of the granitic magma into the older country rock took place during the mid-Paleozoic around 420 Ma (Fig. 38.5).

Most of the bedrock near the surface of Mount Desert Island is composed of **Cadillac Mountain granite**, which forms the greatest volume of the CMIC. Large crystals of

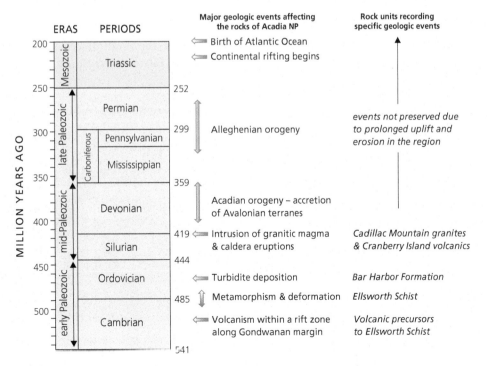

Figure 38.5 Timeline of major geologic events that created the rocks of Acadia National Park

feldspar impart a pink cast to the granite, which contrasts with the adjacent glassy crystals of quartz. The Cadillac Mountain granite differs from the other granitic rocks of the CMIC because of the large size of its constituent mineral crystals and the abundance of hornblende, a darker mineral that forms rod-shaped crystals within the mosaic of quartz and feldspar. The CMIC includes three other bodies of granite that exhibit smaller crystal sizes and a slightly different chemical and mineral composition from that of the Cadillac Mountain granite.

All of the plutonic rocks of the CMIC are cross-cut by *dikes*, vertical sheets of dark-colored rock with a basaltic composition (Fig. 38.6). Their chemistry suggests that they were derived from silica-poor magmas and injected into cracks within younger, solidified rocks beneath the surface. The cracks were widened by the force of the injection combined with melting of the adjacent rock by the heat of the magma. Dikes cutting the plutonic rocks of the CMIC may be a meter or more thick and typically form parallel to one another in a north–south orientation, likely following a pattern of preexisting fractures. The basaltic dikes are particularly well exposed on the headlands of Schoodic Point across the bay from Mount Desert Island.

Each of the granitic bodies and the darker plutonic rocks represents distinct magmas that intruded at different times during the overall emplacement of the CMIC in the region that would become Acadia National Park. The CMIC is part of a much larger mass of plutonic igneous rock known as the Coastal Maine Magmatic Province, which comprises over 100 individual plutons that were emplaced slightly before and during the Acadian orogeny

between about 420 and 360 Ma. The alignment of magma chambers within the Coastal Maine Magmatic Province fed molten rock into a linear array of volcanoes on the surface, analogous to the modern Cascades Range of the Pacific Northwest (Chapter 12).

The overlapping bodies of molten rock of the CMIC, located today within Acadia National Park, supplied magma to a volcano whose remnants are exposed on Mount Desert Island and surrounding islands. The volcanic rocks erupted from the magma chambers of the CMIC are called the **Cranberry Island series**, named for thick exposures on the nearby Cranberry Islands and Isle au Haut, and along the south side of Mount Desert Island near Bass Harbor (Fig. 38.3). (A "series" of rocks simply means that the kilometers-thick mass of rock is composed of a variety of rock types.) The Cranberry Island volcanics are dated to ~420 Ma, supporting the interpretation that they are the extrusive equivalent to the contemporaneous intrusive rocks of the CMIC. The rocks of the Cranberry Island series are primarily *rhyolite tuffs*, which form from *pyroclastic eruptions* of volcanic ash derived from magma with high-silica, granitic compositions. (It might be useful to review Chapter 12 to become better acquainted with the fundamentals of volcanism.)

Another common rock of the Cranberry Island series is *volcanic breccia*, composed of angular chunks of rhyolite encased within a matrix of solidified volcanic ash (Fig. 38.7). They form during explosive eruptions when fragments of older volcanic rock from the throat of the volcano are ejected within the plume of volcanic ash, with the larger debris and the ash falling out onto the flanks of the volcano then fusing together in place. On Isle au Haut, the rocks of the Cranberry Island series are over 3 km (~2 mi) thick and record a long-lived series of violent eruptions ~420 m.y. ago, supplied by the gas-charged magma of the CMIC below.

Many of the rocks described above—the metamorphics of the Ellsworth Schist, sedimentary rocks of the Bar

Figure 38.6 Dark-colored dike of basaltic composition cutting across pinkish granite. Acadia National Park.

Figure 38.7 Volcanic breccia consisting of dark, angular fragments of volcanic rock surrounded by a matrix of finer grained volcanic ash. Isle au Haut, Acadia National Park.

Figure 38.8 Rock of the shatter zone consists of angular blocks of country rock encased within light-colored Cadillac Mountain granite.

Harbor Formation, plutonic rocks of the CMIC, and basaltic rocks of the dikes—are in places exposed as angular fragments surrounded by a matrix of Cadillac Mountain granite (Fig. 38.8). Individual pieces of debris may be as small as a book or as large as a building. This chaotic assemblage of rock rubble encased in granite forms a band about 300 m (1000 ft) to 1.6 km (1 mi) wide that bounds the Cadillac Mountain granite along the east and south sides of Mount Desert Island. Called the **shatter zone**, these rocks are interpreted to represent the catastrophic collapse of the country rock composing the roof of a magma chamber into the underlying body of molten rock. The floating debris was trapped in place as the magma quickly cooled and solidified into granite, enclosing the fragments of rock. The rocks of the shatter zone are exposed in many places on Mount Desert Island, but particularly well at Little Hunters Beach and the east end of Sand Beach.

These descriptions of the disparate rocks and their geologic origins can be confusing and difficult to organize in your mind. Be patient. The chronology of events connecting all of the rocks of Acadia National Park will be assembled in the following section detailing the remarkable tectonic and volcanic history of the region. For now, think of the wide diversity of rocks at Acadia as a peek into the roots of a Paleozoic magma chamber, including glimpses of the surrounding country rock as well as remnants of the overlying volcano. There aren't many places in the National Park System where all of these elements come together to reveal such an exceptional sequence of events.

Tectonic and Uplift History of Rocks at Acadia National Park

The terranes that compose the crust of Maine and Acadia National Park in particular were originally created during the early Paleozoic along the rifted margin of Gondwana,

several thousand kilometers to the south of the ancestral North American landmass (Laurentia). At the time, the continental margin was near what would become the Canadian border; Maine and much of New England didn't yet exist. Through the Paleozoic, several terranes would detach from the northern part of Gondwana, migrate across long-vanished ocean basins, and then sequentially accrete onto Laurentia to form a large part of the New England Subprovince of the Appalachians.

The oldest rocks at Acadia National Park originated ~510 Ma within an underwater rift system as a slice of northern Gondwana peeled away from the rest of the supercontinent and migrated toward Laurentia. During its northward drift, these rift volcanics would be metamorphosed and deformed into the Ellsworth Schist. This narrow piece of continental crust that included the Ellsworth rocks is called the **Gander terrane**, which for our purposes can be considered to be part of *Avalonia*, a chain of microcontinental islands that would accrete to Laurentia during the Acadian Orogeny (Chapter 35). By 465 Ma, the drifting Gander microcontinent would shed sediment into the adjoining deep ocean via turbidity currents, with the resulting turbidite sediments of the Bar Harbor Formation accumulating directly above the eroded surface of the Ellsworth Schist.

As the Gander terrane and related fragments of Avalonia slowly migrated toward Laurentia, older rocks of the Taconic chain of volcanic islands would accrete around 450 Ma during the *Taconic orogeny*, forming the northwestern part of Maine (Fig. 35.13). The Gander terrane and the rest of the Avalonian Islands were far behind, approaching from the south as the intervening Iapetus Ocean became narrower due to subduction (Fig. 35.14). By about 420 Ma, the subducting oceanic plate beneath the Gander terrane supplied molten rock to a chain of more than 100 magma chambers and overlying volcanoes that would later become the Coastal Maine Magmatic Province (Fig. 35.15A). The Cadillac Mountain intrusive complex beneath Acadia National Park was part of the larger magmatic province. At this time, Avalonia and its mass of granitic and volcanic rock were still far offshore in the Iapetus Ocean, hundreds of kilometers and 20 million years from docking onto eastern Laurentia.

The repeated intrusion of granitic magma created an enormous body of molten rock that fed an overlying volcano that erupted a succession of violent *pyroclastic falls*, *pyroclastic flows*, and lava flows. These rocks today form the Cranberry Island series. The great thickness of the volcanics suggests that the Cadillac Mountain intrusive complex fed a *supervolcano*, defined as a volcano that erupted more than 1000 km³ (240 mi³) of material at any one point in time (Chapter 31).

An immense *rhyolitic tuff* about a kilometer thick composes the uppermost part of the Cranberry Island series that may represent a single catastrophic *caldera eruption*, like those that formed at Crater Lake and Yellowstone national parks (Figs. 15.10, 31.9). The final eruption from the Cadillac Island magma chamber was powered by pent-up gas pressure that drove a cataclysmic release of pyroclastic debris. During the course of the eruption, the magma chamber partially emptied, leaving the overlying rock of the volcano unsupported. The crest of the volcano abruptly collapsed into the remaining magma, which cooled and solidified around the fragmental debris, creating the rocks of the shatter zone seen today on Mount Desert Island. The caldera formed after collapse is estimated to have been about 20 km (12 mi) across, much larger than the Crater Lake caldera and perhaps a third the size of Yellowstone. This colossal eruption happened on the Gander microcontinent ~20 m.y. prior to its accretion onto Laurentia.

The Gander terrane and the remnants of the Cadillac Mountain caldera docked with Laurentia ~400 Ma during the early stages of the Acadian orogeny, followed over the next few tens of millions of years by other microcontinental fragments of Avalonia (Fig. 35.15B). The accretion of these terranes during the Acadian orogeny would form the majority of the crust of Maine, including the bedrock foundation of Mount Desert Island and Acadia National Park.

By about 330 Ma, *continental collision* between the eastern margin of Laurentia and the part of Gondwana that is now northwest Africa initiated the *Alleghenian orogeny* (Fig. 35.18). The region around Acadia National Park was uplifted and tilted during the prolonged phase of late Paleozoic mountain-building. It's possible that the network of vertical fractures (i.e., *joint sets*) that cut across Mount Desert Island were formed at this time.

Through the Mesozoic and Cenozoic, significant geologic events affected the eastern continental margin of North America, including the rifting of the Pangean supercontinent and the opening of the Atlantic Ocean (Chapter 35). During this time, the region around Acadia National Park was dominated by an extended phase of weathering and stream erosion that reduced the ancient highlands to a rolling, hilly terrain. About 3 km (2 mi) of rock was removed, with the deep roots of the Cadillac Mountain caldera complex isostatically exhumed toward the surface. Evidence for the once mighty Paleozoic

supervolcano was reduced to a few remnant exposures of volcanic rock and a narrow rim of splintered rock of the shatter zone. By the latest Cenozoic, the bedrock foundation of Acadia National Park was ready for the arrival of the continental ice sheets descending from the north.

Modern Landscape at Acadia National Park

The landscape we enjoy today in Acadia National Park is a product of events that occurred during the Pleistocene Ice Ages that span the last 2.6 million years of Earth history. The Pleistocene is characterized by the advance and retreat of continental ice sheets across the high northern latitudes. The Earth entered the Holocene Epoch about 12,000 years ago as the planet began a natural interglacial phase of warming. The last few thousand years at Acadia National Park were dominated by shoreline processes that modified the margins of Mount Desert Island via the constant attack from waves and tides, combined with changes in the position of the shoreline.

Landscape evolution and the impact of erosion by streams, the ocean, and glaciers at Acadia National Park was strongly influenced by the network of joint sets that cut across Mount Desert Island (Fig. 38.9). Joint planes permit the infiltration of water into the rock, decomposing the minerals and widening the space between adjacent blocks. The *freeze–thaw cycling* of water within joints incrementally wedges the rock apart. Plants send roots into

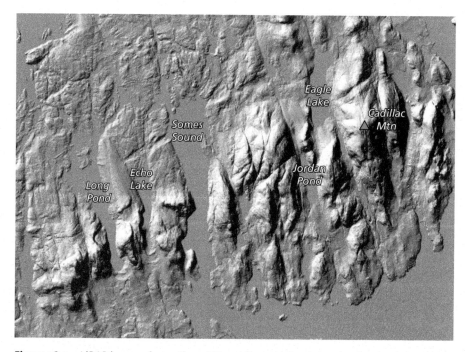

Figure 38.9   LiDAR image of a portion of Mount Desert Island showing the dominant northwest–southeast alignment of linear joints cutting across the landscape. A subtle northeast–southwest joint orientation is evident as well. LiDAR imagery permits the removal of vegetation and surface structures so that high-resolution details of the terrain are revealed. North is at the top of the image.

the planar space along the joints, physically prying apart the rock. Lichens, composed of algae and fungi living symbiotically, add colorful stains to the pinkish granites but slowly decompose the rock along joints with their organic acids. The influence of jointing on the topography and appearance of a landscape like that at Acadia National Park cannot be underemphasized.

The predominant orientation of vertical joints on Mount Desert Island is slightly northwest–southeast, but other sets of vertical joints trend northeast–southwest, cutting the granite and other rocks into angular blocks. Long phases of erosion removed the weight of overlying rock, causing the underlying granite to expand and develop near-horizontal *exfoliation joints* (Fig. 18.12). Weathering of rock along the intersecting network of joints focused the flow of streams that cut V-shaped valleys. Pleistocene glaciers occupied those linear valleys, modifying them into *U-shaped valleys* that separate elongate, smoothly rounded ridges (Fig. 38.10). The near-parallel orientation of many streams, lakes and ponds, bays and inlets, Somes Sound, and linear ridges on Mount Desert Island is almost entirely a product of erosion by streams, wave action, and glaciers superimposed on the underlying network of joints.

GLACIAL MODIFICATION OF THE ACADIA LANDSCAPE

The terrain of Acadia National Park is the cumulative result of multiple advances and retreats of continental ice sheets that flowed southward across Mount Desert Island over the past 2.6 m.y. Glacial advances would entomb the region with well over a kilometer of ice and extend seaward for several hundred kilometers across the exposed continental shelf. Each glacial advance would erase the effects of earlier glacial episodes, further streamlining ridges and deepening and straightening the valleys. During warmer interglacial phases, sea level would rise due to the global influx of glacial meltwater; rejuvenated

streams charged with *glacial outwash* would incise into the underlying rock.

The current landscape at Acadia National Park reflects the Last Glacial Maximum that peaked in the region around 21,000 years ago. A wide variety of both erosional and depositional landforms were left behind on Mount Desert Island, comparable to those addressed in earlier chapters on the Cascades, Sierra Nevada, and Rocky Mountain geologic provinces. The difference is that Acadia National Park is the only park whose glacial landscape was created by continental ice sheets rather than by smaller *alpine glaciers*. The same mechanical processes apply, however, with both *abrasion* and *glacial plucking* performing the erosional work.

Glacial abrasion is the smoothing and rounding of underlying rock by the grinding of gritty sediment at the base of a moving glacier. This process creates the linear *striations* and *glacial polish* common on granite surfaces at Acadia National Park. Plucking occurs as meltwater along the lower part of a glacier infiltrates into the underlying rock along joint planes, then freezes and expands to slowly wedge blocks of rock away from the bedrock (Chapter 18). Repeated freezing and thawing eventually detaches the angular block, with the fragment incorporated into the base of the moving ice sheet. Plucking is especially effective where joint planes are closely spaced.

The combined effects of abrasion and plucking are nicely expressed on *roche moutonee*, asymmetric hills such as The Bubbles (Fig. 38.10), the Beehive, and Acadia Mountain. The gradual north slopes of these landforms were smoothed by abrasion as the southward-flowing ice piled up behind the granitic knob, increasing pressure along the base of the glacier and enhancing the erosive effectiveness of the entrained silt and sand (Fig. 18.19). As the ice sheet overtopped the crest of the knob, the pressure along the ice–rock contact decreased. Meltwater at the base of the ice sheet penetrated downward along vertical joint planes and froze, eventually plucking slabs of rock away from the bedrock on the downflow, south side of the knob and leaving behind steep, angular cliffs. The coastal headlands of the Otter Cliffs on the southeastern edge of Mount Desert Island express another good example of glacial plucking.

Other common landforms created by the erosional effects of flowing ice sheets across Acadia National Park are U-shaped valleys, typically oriented along a slightly northwest-southeast trend (Fig. 38.9). Plucking of jointed rock along the base of the ice sheet deepened the valleys as the glaciers advanced. When the ice sheet receded, meltwater would pond within the troughs, forming elongate bodies of water. Long Pond, Echo Lake, Jordan

Figure 38.10 Computer-generated image showing inclined joints cutting the granite into layers. Continental ice sheets took advantage of the horizontal and vertical jointing to create the smoothly rounded topography of Acadia National Park. View looking north.

Pond, and Eagle Lake are all aligned along the same general trend, reflecting their origin as elongate basins carved by the massive ice sheet. A related erosional feature is Somes Sound, a glacially scoured embayment flooded by the sea. In high-latitude places like Norway, intense glacial erosion has scored steep and deep notches into the coastline, creating the famous fiords, hundreds of meters deep. Somes Sound is superficially similar but is only about 40 m (130 ft) deep, reflecting its origin as a U-shaped valley that has been flooded by the postglacial rise in sea level. Somes Sound is a **fjard**, a Scandinavian term that describes its broad, shallow dimensions and distinction from true fiords.

During the multiple glacial phases of the Ice Ages, *erosional* processes that occurred along the base of the advancing ice sheet created the glacial landforms just discussed. During the warmer interglacial phases, however, the retreating ice sheets left behind a variety of remnant *depositional* features in Acadia National Park. Almost all of the unconsolidated depositional landforms were stripped away by later glaciations, but those seen today in the park are very young, formed over the past few thousand years since the Last Glacial Maximum. Among the most obvious depositional remnants are *glacial erratics*—huge boulders left behind as they dropped out of the melting front of the ice sheet (Fig. 38.11). Bubble Rock on South Bubble Mountain and Balance Rock on the shore path near Bar Harbor are two notable examples. An erratic composed of gray granite near the top of Cadillac Mountain contrasts sharply with the pink granite of the underlying bedrock. The nearest source for the gray boulder is about 30 km (19 mi) to the north on the mainland, suggesting that the boulder was transported at least that distance within the glacial ice.

Much of Mount Desert Island is blanketed by a thin layer of *glacial till*, an unsorted mixture of sediments ranging from clays and silts to pebbles and cobbles. Some of the till was deposited as sheets of outwash from meltwater streams choked with sedimentary debris. In other places where the receding front of the ice sheet

paused for an extended period of time, glacial till accumulated as *moraines*, elongate ridges a few tens of meters high. Moraines commonly act as natural dams that impound meltwater in U-shaped valleys. On Mount Desert Island, moraines are found at the southern ends of Long Pond, Echo Lake, and Jordan Pond. Several moraines left behind as the ice sheet stalled in its retreat after the Last Glacial Maximum form the narrow mouth of Somes Sound.

The effects of continental ice sheets at Acadia National Park are different from those of alpine glaciers at parks of the American West, which we addressed in several earlier chapters. The smoothly rounded contours of Acadia lack the sharper glacial landforms of western national parks such as *cirques*, *arêtes*, and *horns*. Part of the reason is that mountain-building in the Appalachians ended by the late Paleozoic, with only slow, isostatic uplift rejuvenating the topography. Weathering and erosion have been dominant in the northern Appalachians for the last 200 million years. In contrast, *orogenesis* is much younger in the American West, with uplift still occurring today in many places. The cold temperatures of the high elevations promoted the growth of alpine glaciers, enhancing the intensity of Pleistocene glacial incision in the West. The other major factor in the softer appearance of glacial landforms in Acadia National Park is that the sheer mass of continental ice sheets is much greater than that of ice caps or alpine glaciers. The repeated advance and retreat of the enormous ice sheets across New England have ground down the sharp edges, culminating in the low-relief, streamlined terrain seen today.

POSTGLACIAL SEA-LEVEL CHANGE AT ACADIA NATIONAL PARK

During glacial maxima, water evaporates from the oceans to fall as snow on land, eventually building up to form vast continental ice sheets. The enormous transfer of water causes sea level to fall and coastlines to migrate seaward, exposing the continental shelves. Conversely, during the warmer glacial minima, the melting of ice sheets returns the water to the oceans, causing sea level to rise, flooding the shallow continental shelves once again. These changes in sea level are global in extent, affecting continental margins worldwide. With the advent of interglacial warming over the last 15,000 years or so, coastlines everywhere are experiencing a *postglacial marine transgression*. The early pace of global sea-level rise was rapid but tapered to a slower rate over the last 7000 years of the Holocene (Fig. 29.18).

But the postglacial rise in sea level hasn't been uniform at Acadia National Park because the position of the coastline is affected not only by global sea-level change, but also by contemporaneous regional changes in the level of the land surface. The land surface subsides in response to the load emplaced by advancing ice sheets, with the underlying *plastic* rock of the mantle migrating laterally to accommodate the sinking crust. As the ice sheets melt and

Figure 38.11 Bubble Rock, a glacial erratic left behind as the most recent ice sheet retreated from Mount Desert Island.

the weight of the overlying ice is removed, the land surface isostatically rebounds upward, with the underlying crust and mantle returning to equilibrium—a process called **glacial rebound**. Global changes in sea level interact with glacial rebound of the land surface to create **relative sea level** change and a complicated history of the position of the shoreline on Mount Desert Island.

During the Last Glacial Maximum at Mount Desert Island around 20,000 years ago, the land surface beneath the ice was depressed a few hundred meters. During the succeeding interglacial warming, the land surface has rebounded a comparable amount. But the difference between the *rate* of global sea-level rise and the *rate* of change in the height of the land surface along the coast of Maine has produced a complicated postglacial history at Acadia National Park. The fluctuating location of the shoreline over the past several thousand years can be summarized in three broad phases.

1. As the ice sheet progressively retreated from the region after the Last Glacial Maximum, the rate of global sea-level rise was rapid, flooding Mount Desert Island to a height of about 64 m (~210 ft) above the present sea level about 16,000 years ago. Evidence for this ancient highstand in sea level is preserved as boulder beach deposits, seacliffs, *wave-cut platforms* (Fig. 29.17), a *sea stack* (Fig. 29.12), and a *sea cave* perched 64 m above today's shoreline. A layer of greenish clay with fossil shells of clams, mussels, and other marine invertebrates was deposited across the submerged margins of Mount Desert Island at this time. Today, the marine clay is exposed in streambeds and along the margins of coves.

2. By about 12,000 years ago, the rate of isostatic rebound of the land surface of Mount Desert Island outpaced the rate of global sea-level rise. Relative sea level fell as the land rose rapidly, reaching 55 m (180 ft) *below* the current shoreline. The land that is today an island was a peninsula of low, rolling hills attached to the mainland; Native Americans began to occupy the region by about 9000 years ago.

3. By about 7000 years ago, the rebound of the island had slowed significantly, allowing global sea-level rise to dominate once again, reaching a height just below the present coastline and submerging Somes Sound. Since then, the shoreline has transgressed slowly around the margins of Mount Desert Island, with waves and storms relentlessly battering and modifying the coast.

During the postglacial retreat of the ice sheet from Mount Desert Island, the modern landscape of Acadia National Park came into view. The diverse array of erosional glacial landforms was gradually exposed. Depositional features such as glacial moraines marked the retreat of the ice front, erratic boulders were left scattered about, and sheets of glacial outwash covered the lowlands. In time, freshwater draining from the surrounding hills filled U-shaped valleys dammed at their southern ends by moraines, evolving into the elongate lakes and ponds nestled between the elongate ridges. Forests colonized the

soils formed in glacial outwash along lower slopes, while wetlands rimmed the lakes and lowlands.

SHORELINE FEATURES AT ACADIA NATIONAL PARK

The photogenic coast of Acadia National Park is dominated by rocky headlands that jut outward into the surrounding surf, protecting small inlets and harbors. The rocky cliffs that dominate the *emergent coastline* of Mount Desert Island are a product isostatic glacial rebound of the region. The slow global rise in sea level is gradually returning the Acadia shore to a *submergent coastline*, but the highly resistant rocks composing the island are able to endure the constant attack from storm-driven waves and tides (Fig. 38.12).

Piles of wave-washed boulders and cobbles dominate small, narrow beaches along the Acadia shoreline. The large, rounded blocks were derived from the erosional breakdown of jointed granitic seacliffs, or they were washed out of glacial till by the methodical sorting of waves and tides. The lone beach composed of sand-sized particles is the appropriately named Sand Beach, which occupies the head of Newport Cove. The beach is composed of sand-sized shell fragments derived from the calcitic skeletons of marine invertebrates that were washed into the head of the inlet and trapped within the shelter of the adjacent rocky cliffs.

Headlands that protrude seaward, like those at Otter Cliff, Bass Harbor Head, Great Head, and Schooner Head, are composed of granite that resists wave erosion. In other places where the coastal rock is cut by closely spaced joints

Figure 38.12 Rocky cliffs lining the inlet to Sand Beach, southeastern corner of Mount Desert Island. Note the intersecting joints that cut through the resistant Cadillac Mountain granite. Sand Beach forms the narrow strand at the distant head of the inlet.

or where it is composed of less resistant basaltic dikes, waves have methodically dislodged angular blocks, cutting notches and inlets into the coast. Thunder Hole is a good example, where the constant pounding by waves has preferentially eroded a narrow zone of fractured basaltic rock, creating a vertical slot that focuses the energy of waves and tides (Fig. 38.13). Wave erosion along the low-tide line has created a sea cave at the head of the slender inlet. Large waves compress air within the cave that then creates a thunderous boom as it is expelled outward by the surging water.

Other features of emergent coastlines such as sea stacks, sea arches, sea caves, and wave-cut platforms are nicely expressed along the margins of Mount Desert Island. At Monument Cove along the southeastern corner of the island, a sea stack about 6 m (20 ft) high has been detached from the adjacent seacliff by the focusing of wave action along a vertical fracture (Fig. 38.14). The granite of the sea stack is partitioned into angular blocks by the intersection of vertical joints and horizontal exfoliation joints, leaving it vulnerable to breakdown. The boulders lining the beach at the base of

the sea cliff accumulated from the long-term weathering and erosion of the highly jointed seacliff and then were rounded into their current shapes by abrasion as they were jostled against one another by storm-driven waves.

Sea caves are found along the east side of Newport Cove, at Anemone Cave near Schooner Head, and at The Ovens outside the national park on the northeastern side of Mount Desert Island. A sea arch created by weathering along a vertical joint exists near Sand Point. A wave-cut platform is visible just below the surf at the base of Otter Cliff along the southeastern coast. Each of these characteristic features of emergent coasts formed over the past few thousand years by the relentless assault of waves and tides along the western North Atlantic.

Final Thoughts on Acadia National Park

The picturesque charm of Acadia National Park is a product of ancient geologic events that formed the bedrock foundation, massive ice sheets that carved the landscape, and constant battering by the surrounding ocean. Visitation has surged in recent years, as people converge on the park to enjoy its hiking trails, carriage roads, stone bridges, and magnificent views. Traffic gridlock and limited parking detract from the pastoral calm the park is supposed to induce. You can help to diminish overcrowding and enhance your experience by considering a few simple suggestions, without losing contact with everything the park has to offer.

Begin by using the free Island Explorer bus to enter the park and to access many of the trailheads. Visit during the early morning or late afternoon when the crowds are thinner, or even better, visit during the winter to cross-country ski on the carriage roads. See the park from a boat in Frenchman Bay or by skirting the coast in a sea kayak. Tour the less crowded Schoodic Peninsula part of the park, or take the ferry to Isle au Haut (pronounced I-la-HO). The nearby Cranberry Isles aren't in the park but are well worth a visit for their relative calm. Overcrowding is a serious threat to all of our national parks, but especially to those smaller parks constrained by a rocky coast and rugged surf such as at Acadia.

Whatever your plans, remember that as you explore the park, think about how you're actually standing on the deeply eroded roots of what used to be a 1000°C (1800°F) body of magma a few kilometers beneath an enormous volcano. Try to imagine the catastrophic eruptions and eventual collapse of the volcano that happened long ago and far offshore. Then visualize yourself beneath the crushing pressure of a kilometer-thick mass of ice that once occupied the island. These geologic musings will instill a deep and unique perspective on the idyllic landscape of Acadia National Park.

Figure 38.13 The narrow slot of Thunder Hole formed where tightly spaced joints and a basaltic dike were preferentially eroded by waves and tides, leaving steep, resistant walls of granite on either side. The sea cave at the low tide line at the head of the slot is the source of the thunderous boom as air is rapidly compressed, then expelled by large waves.

Figure 38.14 The highly fractured pinnacle at Monument Cove is a sea stack, raised above sea level by isostatic glacial rebound. The seacliff to the left is about 10 m (33 ft) high. Wave erosion wearing away at a vertical joint has detached the sea stack from the seacliff. Rounded boulders compose the narrow beach.

Key Terms

fjard
glacial rebound
intrusive complex
relative sea level
shatter zone

Related Sources

Braun, D., and Braun, R. (2016). *Guide to the Geology of Mount Desert Island and Acadia National Park*. Berkeley, CA: North Atlantic Books.

Gilman, Richard A., C. A. Chapman, T. V. Lowell, and H. W. Borns, Jr. (1988). *The Geology of Mount Desert Island: A Visitor's Guide to the Geology of Acadia National Park*. Augusta, ME: Maine Geological Survey. Accessed February 2021 at https://digitalmaine.com/cgi/viewcontent.cgi?referer=https://www.google.com/&httpsredir=1&article=1035&context=mgs_publications

Graham, J. (2010). *Acadia National Park: Geologic Resources Inventory Report*. Natural Resource Report NPS/NRPC/GRD/NRR—2010/232. Ft. Collins, CO: National Park Service. Accessed February 2021 at https://www.nps.gov/articles/nps-geodiversity-atlas-acadia-national-park-maine.htm#gri

Maine Geological Survey. (2018). *Maine Geologic Facts and Localities*. Accessed February 2021 at https://www.maine.gov/dacf/mgs/explore/explore_map.shtml.

National Park Service. *Written in Acadia's Rocks*. Accessed February 2021 at https://www.us-parks.com/acadia-national-park/geology.html

Figure 39.1 The passage called Broadway (or "Main Cave") is an abandoned underwater stream channel carved through layers of limestone. This part of Mammoth Cave National Park is among the most visited.

Many national parks are described by superlatives, but only Mammoth Cave National Park in south-central Kentucky can lay claim to the title of "longest known cave system in the world." The Mammoth Cave network of interconnected passageways is currently surveyed at 652 km (405 mi) and will undoubtedly be lengthened in the future as cave explorers discover new connections and extensions. The next longest cave, Mexico's Sac Actun, is only half the length and is almost entirely underwater. The underground labyrinth of Mammoth Cave consists of multiple levels of cave passages and vertical shafts, with several kilometers accessible by guided tour (Fig. 39.1)

Mammoth Cave National Park is located in the Interior Low Plateaus Province west of the Southern Appalachians, a region of forested rolling hills and rock-capped ridges dissected by deep river valleys (Fig. 35.3). The province shares some of the characteristics with the nearby Appalachian Plateau but was unaffected by Appalachian folding and faulting during growth of the mountain chain, with the widespread sedimentary rocks maintaining a near-horizontal layering.

Mammoth Cave National Park is situated along the southeastern margin of the Chester Uplands (aka Mammoth Cave Plateau), a hilly terrain marked by sandstone-capped ridges and limestone-floored valleys (Fig. 39.2). The Green River meanders westward through the park and plays an important role in the origin and evolution of the cave network. The uplands rise about 60 m (200 ft) above the Pennyroyal Plateau to the southeast, a large physiographic region that spans much of south-central Kentucky. The Chester Escarpment marks the boundary between the Chester Uplands and the Pennyroyal Plateau. The entire region is notable for its *karst landscape*, an irregular topography of low hills and ridges pocked by thousands of *sinkholes*, *springs*, and "disappearing streams" that relate to dissolution of limestone beds and cave formation just beneath the surface (Fig. 39.3). The karst landscape of the region continually evolves due

to the abundant annual rainfall of about 130 cm (~50 in) that rapidly seeps into the ground through sinkholes and fractures in the rock.

The Mammoth Cave system comprises four smaller cave networks that were explored separately but were subsequently determined to be connected as a single large system. About a third of the cave system extends outside of the park boundaries to the southeast (Fig. 39.2). Over 300 smaller caves occur elsewhere in the park, unconnected to the larger cave system. Like most caves, the Mammoth Cave system was created as limestone rock was dissolved by the movement of slightly acidic groundwater over the last few million years. The countless passageways mark the former channels of underground streams, and some of the deeper caves have active streams flowing today.

Native Americans explored Mammoth Cave as long as 5000 years ago, using the wide entrances for shelter and mining the deep recesses for materials such as gypsum and chert. They left behind many artifacts, including torches, footwear, baskets, and even desiccated mummies. In the early 19th century, deposits of nitrates in bat guano were mined for making gunpowder. Black slaves carried out some of the earliest mapping of the caves through the mid-1800s, including giving the first guided tours. Exploration, mapping, and tourism continued sporadically over the intervening years, culminating in the establishment of

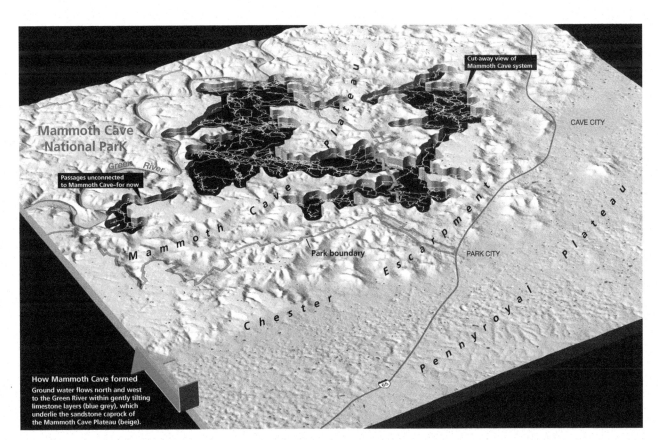

Figure 39.2 Perspective view of Mammoth Cave National Park (green outline), showing the distribution of the cave system beneath the surface. Part of the cave system extends outside the park boundaries. The groundwater that sculpted the network of caves originated in the sinkhole plain of the Pennyroyal Plateau, then flowed northwest through the rock of the Mammoth Cave Plateau.

Figure 39.3 Sinkhole plain of the Pennyroyal Plateau, Kentucky. This is a characteristic feature of the karst landscape in the region.

the national park in 1941. Over a half-million people visit the park each year, with most descending beneath the surface on one of the many guided tours.

Mammoth Cave has a surprising biological diversity of fish, snails, insects, mussels, mammals, birds, crustaceans, and reptiles. Twelve species of bats make the caves their home, although they are not particularly abundant and pose no harm to visitors. Some animals use the caves as shelter or to hunt, but others have adapted to live exclusively in the absolute darkness. This ecological richness earned Mammoth Cave recognition as part of an International Biosphere Reserve in 1990.

This chapter focuses on the origin and evolution of Mammoth Cave but builds on an earlier discussion of the basics of cave formation at Sequoia–Kings Canyon national parks (Chapter 19). There, relatively small caves occur within pods of marble (metamorphosed limestone), decorated with a variety of ornate flowstone deposits. Moreover, a large area of Sequoia National Park exhibits a prominent karst landscape comparable to that of the region surrounding Mammoth Cave. But as you'll see, the Mammoth Cave network of karst and caverns is much more extensive and complex than its western cousin.

Origin and Age of Rocks at Mammoth Cave

The Mammoth Cave system is developed within middle Paleozoic limestones that were deposited about 330–340 m.y. ago (Mississippian Period) during a phase of high sea level. *Transgression* of the shoreline reached far inland, with shallow seas flooding much of the

continental interior. In the American West, the thick Redwall and Madison limestones addressed in earlier chapters were deposited at about the same time. The subtropical position of the region around Mammoth Cave supported a highly productive marine ecosystem of calcite-secreting organisms such as crinoids, corals, algae, and many others. As these organisms died, fragments of their hard outer skeletons accumulated on the seafloor, building up over long periods of time to form thick layers of squishy lime sediment. In time, rivers and deltas would fill in the shallow sea with layers of sand, silt, and clay. All of these layers would eventually be buried by younger sediments, then solidified to form limestone rock that gradually passes upward into layers of sandstone and shale (Chapter 1).

The limestones and clastic rocks in the Mammoth Cave region were deposited along the southeastern margin of the **Illinois Basin**, a shallow embayment of the open ocean (Fig. 35.18). The Illinois Basin was separated from the Appalachian *foreland basin* by a low-elevation peninsula called the Cincinnati arch. The limestones and overlying siliciclastic rocks in the area around Mammoth Cave National Park were relatively unaffected by the late Paleozoic orogenic events of the Appalachians. They were left as widespread, near-horizontal layers, with only a very gentle tilt of much less than one degree downward to the northwest toward the center of the Illinois Basin. This simple arrangement of the limestone beds and overlying sandstones plays an important role in the evolution and architecture of the Mammoth Cave system.

The multiple cave levels of the national park are developed within three limestone formations that total about 120 m (~390 ft) in thickness (Fig. 39.4). The oldest and lowest formation is the St. Louis Limestone, which contains minor amounts of dolomite (Mg-rich limestone), shale, and

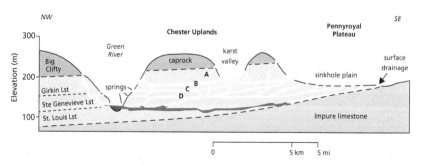

Figure 39.4 Simplified cross section from the sinkhole plain of the Pennyroyal Plateau across the Chester Upland to the Green River showing the generalized hydrologic system at Mammoth Cave National Park. At the southeast edge of the sinkhole plain, surface streams "disappear" into the labyrinth of underground cavities. The final destination for infiltrating groundwater is the Green River. A–D are the four main cave levels discussed in the text.

layered *chert* nodules. The overlying formation is the Ste. Genevieve Limestone, a relatively pure layer where most of the cave system is developed. The uppermost limestone is the Girkin Formation, which includes a few shaly beds that become more common upward in the unit. A resistant layer of sandstone known as the Big Clifty Formation overlies the three limestone units. The sandstone represents the encroachment of river and delta systems over the shallow tropical seas within the Illinois Basin during the middle Paleozoic. Younger rocks buried the limestones and sandstones later in Paleozoic time, but since then the region has been dominated by weathering and erosion.

Karst Landscape

The unique hydrologic framework that controls the formation of the Mammoth Cave system is related to the variable way that the two major rock types erode. The limestone units readily dissolve with prolonged contact with mildly acidic water, whereas the overlying impermeable sandstones are more resistant to weathering. The very gentle northwest tilt of the bedding near Mammoth Cave National Park is a second major factor in the evolution of the current karst landscape. Southeast of the park on the widespread Pennyroyal Plateau, the sandstone layers of the Big Clifty and overlying units have been completely eroded (Fig. 39.4). Runoff from rainfall and snowmelt has interacted directly with the underlying limestone units for millions of years, lowering the level of the land through dissolution and creating the thousands of circular depressions that compose the **sinkhole plain**. No surface streams exist on the sinkhole plain because small rivulets and gullies of runoff rapidly disappear into soil-filled sinkholes that funnel the water into underlying fractures and cave passages (Fig. 19.10).

The Chester Uplands to the northwest maintain a higher elevation than the deeply eroded sinkhole plain due to the caprock of relatively resistant Big Clifty sandstone that protects the underlying limestones. The Visitor Center on Mammoth Cave Ridge is built atop the Big Clifty caprock. In places, streams have incised through the upper sandstone, allowing surface water to come into direct contact with the limestones, forming steep-walled karst valleys. Most of the karst valleys have no streams because surface runoff flushes directly into sinkholes that pierce the limestone of the valley floors. The exception is the Green River, which occupies a deep valley entrenched about 90 m (300 ft) into the Chester Uplands (Fig. 39.4). Together, the Pennyroyal Plateau and the Chester Uplands compose the karst landscape of the Mammoth Cave region.

Groundwater Flow Paths and Cave Formation

The flow of water that formed the intricate passageways of the Mammoth Cave system has been traced using harmless fluorescent dyes. Of course, water flows from higher elevations to lower elevations due to gravity, and that applies to the underground flow of water as well. The water begins as rainfall and snowmelt on the Pennyroyal Plateau to the southeast and ends its circuitous journey over 30 km (18 mi) away at the Green River on the Chester Uplands to the northwest (Fig. 39.4). This subterranean drainage system is an underground version of an aboveground *watershed*, defined in earlier chapters as a region of land drained by an interconnected network of streams that converge and empty into a river, lake, or ocean.

Most of the water that formed the Mammoth Cave system originated in the sinkhole plain of the Pennyroyal Plateau, far outside the boundaries of the national park. At the southeastern edge of the low plateau, surface streams disappear as the water infiltrates into sinkholes and fractures. Sinkholes may open at their base directly into underground passageways, but most are filled with soil and fragmented rock that impede the flow of water. Rainwater is always mildly acidic due to interaction with carbon dioxide in the atmosphere, producing a weak carbonic acid. As the water passes through the soils within the sinkholes, it becomes even more acidic as it incorporates carbon dioxide from microbial decay of organic matter and other sources.

As the water comes in contact with the Paleozoic limestones beneath the surface of the sinkhole plain, it focuses its dissolving power within narrow partings along horizontal bedding planes. The calcite ($CaCO_3$) composing the limestone along the bedding planes is dissolved, and its constituent elements become ions (Ca^{2+} and HCO_3^-) that are carried away within the water. Because of the gentle northwest tilt of the limestone beds, the water and its dissolved ions flow in that direction. As the water becomes more and more saturated with dissolved calcite, it becomes less and less acidic. Thus, a continual replenishment of new water from the surface is required to create openings in the rock wide enough to efficiently move groundwater. As you might imagine, the process of *dissolution* and the creation of conduits for water flow takes many tens of thousands of years for passageways to open and many millions of years for an intricate cave system like that at Mammoth Cave National Park to form.

At this point in the process, the caves are just empty passageways devoid of the elaborate *speleothems* (cave deposits like stalagmites and stalactites) that most people associate with caves. The openings in the limestone have to form first by the dissolution of calcite before they can be decorated by the reverse process of calcite *precipitation*. As long as the waters are charged with carbon dioxide and thus are chemically balanced toward dissolving calcite, openings and passageways will develop. The abundant rainfall and relatively thick soils of the Mammoth Cave region provide ideal conditions for the flow of acidic groundwater and the continued enlargement of caves.

The groundwater flowing northwest from the Pennyroyal Plateau combines with acidic rainwater and snowmelt draining off the caprock of the Chester Uplands, as well as with water infiltrating through karst valleys (Fig. 39.4). Rainwater flowing through the soil at the tops of sandstone-capped ridges drops off the edges until it encounters the underlying limestone. The acidic waters may penetrate into the body of the limestone through vertical fractures, slowly enlarging them to form cylindrical tubes known as **vertical shafts**. The largest known vertical shaft in Mammoth Cave is Mammoth Dome, about 60 m (200 ft) long (Fig. 39.5). As shafts open wider, they focus the flow of acidic water into the underlying limestone where it drains laterally through horizontal passages developed along bedding planes. Blocks of sandstone may lie scattered at the base of shafts, dropped into the void by streams draining off the edges of the caprock.

Underground streams flow through the Mammoth Cave system toward the northwest, guided by the almost imperceptible tilt of the beds and protected from collapse by the overlying sandstone caprock. The water emerges into the entrenched valley of the Green River through a series of springs. From there, the dissolved ions of the Paleozoic limestone are carried first into the Ohio River and then into the Mississippi River where they mingle

with water draining from national parks of the American West. Their final destination is the Gulf of Mexico where the dissolved ions contribute to the overall chemistry of the world ocean.

This subterranean plumbing system of infiltrating groundwater and underground streams is still active today, and cave formation is an ongoing process at Mammoth Cave National Park. The karst region of south-central Kentucky has all the right ingredients for the longest cave system in the world. (1) Plentiful rainfall maintains a reliable supply of water and contributes to the development of soils that enhance the acidity of runoff. (2) The beds of readily dissolvable limestones are separated by numerous horizontal bedding planes that act as conduits for lateral flow of groundwater. (3) The widespread layers of limestone are overlain by a caprock of impermeable sandstone that prevents cave collapse. (4) The beds of rock are tilted ever so slightly to the northwest, providing a hydraulic gradient for the flow of groundwater along bedding planes. (5) The final piece of the cave-forming puzzle that makes Mammoth Cave so unique is the millions of years of time needed for the continuous dissolution of calcite and the removal of its constituent elements by a river system.

Origin and Age of Cave Levels of the Mammoth Cave System

The Mammoth Cave system consists of four cave levels, now mostly inactive, and one modern level where passageways are currently being created by underground stream flow (Fig. 39.6). A **cave level** is a series of interconnected horizontal passageways that occupy a relatively narrow range of depths beneath the surface. Each level represents a stage of cave development that relates to former positions of the *water table*, the boundary beneath the ground that marks the uppermost surface of water-filled cavities in the rock. The **vadose zone** is the region of rock above the water table where cavities are filled with air, whereas the **phreatic zone** is the region of water-filled cavities beneath the water table. The water table will fluctuate with the seasons, long-term changes in climate, and vertical movements of the land surface such as uplift or subsidence.

During the multimillion-year history of cave development at Mammoth Cave, the position of the water table mirrored that of the Green River, the **base level** for groundwater flow in the immediate region. Base level is defined as the lowest elevation to which a landscape can erode and is theoretically equivalent to sea level. In the context of a drainage system like that of the Mammoth Cave region, however, the Green River marks the deepest level of erosion and the final destination for all groundwater flow through the cave network. The gentle northwesterly tilt of bedding planes in the Paleozoic limestone units determines the direction of groundwater flow and thus the orientations of cave levels.

Figure 39.5 Staircase tower in Mammoth Dome, a vertical shaft that connects a sinkhole at the surface with underlying cave levels. It cuts through the entire Ste. Genevieve Limestone and the upper part of the St. Louis Limestone.

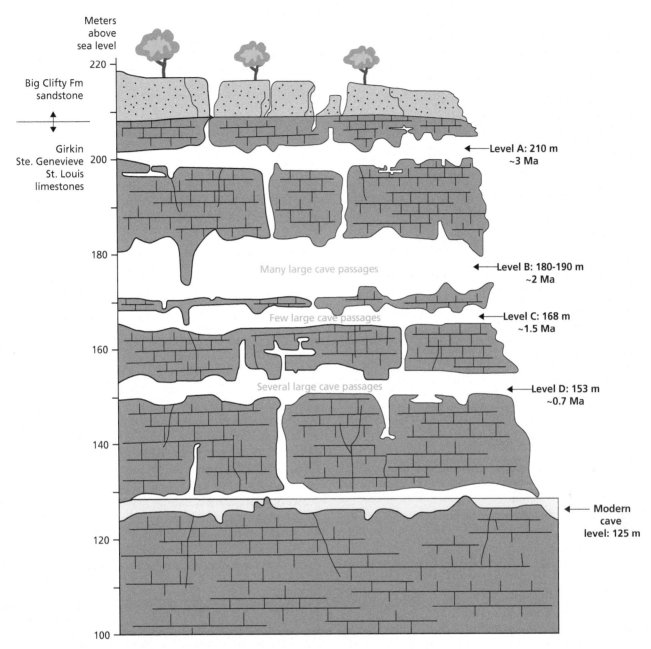

Figure 39.6 Four main abandoned cave levels of the Mammoth Cave system and their estimated ages of formation. Modern cave level is at very bottom.

The upper cave level is the oldest, with lower-elevation layers becoming progressively younger. (Determination of ages for the cave levels will be explained below.) The decreasing age of cave levels with increasing depth beneath the surface reflects the stepwise entrenchment of the Green River within its valley. As the Green River incised its channel deeper through time, the water table and thus the level of cave formation deepened as well. Downcutting by the Green River was not a continuous, gradual process, but rather happened in pulses that may relate to *Pleistocene Ice Age* changes in climate. The connections are somewhat speculative, but the overlap in time between the ages of

cave levels and the geologically rapid climatic events of the Pleistocene are difficult to ignore.

Two distinct types of caves formed during the sequential incision of the Green River and the associated increase in depth of the water table – canyon passages and tubular passages. Each type corresponds to the rate of base level lowering through time. During phases when the river rapidly incised its channel, the water table tracked the abrupt lowering of the base level. Groundwater dissolved limestone vertically downward, forming narrow caves known as **canyon passages** (Fig. 39.7). Dissolution occurred along shallow streams that flowed down the tilt of bedding planes

Figure 39.7 Canyon passage formed in the vadose zone as an underground stream progressively dissolved the floor of the cave as the water table dropped.

at the floor of the canyons. As the water table continued to fall, the canyon passageways were deepened, stranding the canyon within the vadose zone.

The second type of passage formed when the rate of incision of the Green River stalled for an extended duration of time. The water table stabilized as well, with dissolution focused laterally along bedding planes. This resulted in the formation of oval, **tubular passages** that are wider than they are tall (Fig. 39.8). These passages formed primarily at or below the water table within the phreatic zone. Vertical shafts connect the various horizontal cave levels of canyon passages and tubular passages.

Careful mapping of the transitions between narrow canyon passages and the wider tubular passages revealed the history of episodic lowering of the vadose-to-phreatic position of the water table. The four abandoned cave levels were recognized as distinct phases of cave evolution that marked pauses in the stepwise lowering of base level as the Green River episodically incised downward through time. The transitions in passage type for each of the four cave levels occur at elevations of 210 m (690 ft), 180–190 m (590-620 ft), 168 m

(550 ft), and 153 m (500 ft) above sea level (A through D on Figures 39.4 and 39.6). Today, all four of the abandoned cave levels occupy the vadose zone, whereas the currently active cave network at the base of the system lies within the phreatic zone.

The Green River didn't always erosionally cut downward, but would on occasion fill its valley with sediment. This caused the level of both the river and the water table to rise. Loose sand and gravel were transported laterally by flowing water into the adjacent cave level and deposited as a horizontal layer of sediment on the cave floor. This seemingly incidental detail turns out to be extremely fortuitous because grains composed of quartz (SiO_2) can be radiometrically dated to estimate the time of deposition and thus the age of cave development. Called **surface exposure dating**, the technique measures the radioactive signature of specific elements within the surface layers of quartz grains to determine how long the grains have been buried. The dates determined from the sediment fill in each of the main cave levels are interpreted to represent the latest phases of cave development for each level (Fig. 39.6).

The floor of the uppermost and oldest cave level A at 210 m (690 ft) elevation was covered with sediment about 3 m.y. ago. This suggests that cave enlargement had begun within the Girkin Formation several million years earlier, assuming typical rates of dissolution and cave formation. The next lowest cave level B was abandoned about 2 m.y. ago, and cave level C was left stranded about 1.5 m.y. ago. The youngest abandoned cave level D at 153 m (500 ft) in elevation was formed by 1.2 m.y. ago but experienced renewed phases of enlargement as recently as 700,000 years ago.

The trend of decreasing numerical age with deeper cave levels supports the model of stepped pulses of incision by the Green River as the controlling factor. Episodic shifts in the rate of river downcutting governed changes in local

Figure 39.8 Cleaveland Avenue is a tubular passage formed within the phreatic zone when the water table was stable for an extended period of time. The elliptical shape evolved as dissolution was focused along horizontal bedding planes.

base level, the water table, and cave development over the past several million years. Changes in the rate of Green River incision may relate to abrupt shifts in the regional climate driven by the advance and retreat of continental ice sheets during the Pleistocene Ice Ages of the last 2.6 m.y. The farthest advance reached about 100 km (60 mi) to the north of Mammoth Cave National Park, but the proximity of the ice front certainly affected the regional climate and the rate of river downcutting.

The current water table coincides with the modern level of cave formation in the park at 125 m (410 ft) in elevation. Active streams flow through these passages and include the Logsdon River, which discharges into the Green River about 23 km (14 mi) from its source on the sinkhole plain. It is one of the world's longest underground rivers that can be traced along its entire path. The upper four cave levels in the modern vadose zone are still slowly enlarging today as acidic runoff drains off the caprock and then drops through vertical shafts into the underlying cave network.

Mammoth Cave Speleothems

Much of the narrative above focused on the opening of cave passages related to the dissolution of limestone rock. Those massive underground canyons and tubular caves are the dominant features for which the national park is known. But Mammoth Cave also has a selective array of ornate speleothems as well, although they are not particularly abundant (Fig. 39.9). These cave decorations form by the precipitation of calcite from groundwater, the reverse chemical process of dissolution.

As groundwater charged with carbon dioxide trickles through fractures and small cavities in the vadose zone, it dissolves some of the calcite composing the limestone rock. When the water enters an open passage, its store of carbon dioxide gas is released to the atmosphere of the cave, just as it escapes when you open a bottle of carbonated soda. The acidity of the water rapidly decreases, instantaneously triggering the precipitation of minute crystals of calcite.

The continual seepage of groundwater at that site through time adds to the growth of *stalactites* hanging like icicles from the ceiling. As the pathway for groundwater becomes plugged with calcite, the water may trickle laterally along a fracture, resulting in a linear array of stalactites.

Drops of water falling from the tip of a stalactite will rapidly expel carbon dioxide as the drips hit the floor of the cave, resulting in the precipitation of calcite that builds vertically through time to create a *stalagmite*. If a downward-growing stalactite meets an upward-growing stalagmite, they will combine to form a *cave column*. Groundwater saturated with dissolved calcite may flow across the surface of a wall of limestone, release its carbon dioxide, and precipitate an undulating sheet of calcite called *flowstone*. Each of these types of speleothems is exhibited on the larger feature called Frozen Niagara (Fig. 39.9).

In select caves in the Mammoth Cave system that are considered to be dry, without any percolating groundwater, speleothems form from the sulfate mineral **gypsum** ($CaSO_4$). The dominant process in these dry caves is evaporation, which draws tiny amounts of moisture from within the rock outward toward the open void of the cave. The surrounding limestone in these caves contains crystals of pyrite, composed of iron and sulfur. Calcium dissolved from the limestone combines with sulfur dissolved from the pyrite to precipitate crusts and delicate "flowers" and "snowballs" of gypsum along the cave ceiling and walls (Fig. 39.10). Gypsum speleothems are common in dry caves directly beneath the sandstone caprock as well as deeper in the system in Cleaveland Avenue.

You can access many of the cave passages and decorations by taking one of the various tours offered by the National Park Service that explore the 23 km (14 mi) of underground trails. Some are easy, such as the Frozen Niagara tour, which takes you to the speleothems described above. Others are moderate in difficulty, such as the Domes and Dripstones tour, which traverses 500 descending steps

Figure 39.9 Frozen Niagara exhibits a variety of calcite speleothems, such as stalactites, stalagmites, columns, and flowstone.

Figure 39.10 Gypsum flower speleothem in Cleaveland Avenue, a tubular passage in level C. Many of these features were detached or vandalized by earlier visitors to the cave.

deep into the cave system. The Wild Cave tour is for those intrepid folks who enjoy scrambling and crawling through some of the tighter reaches of the cave. It's not often that humans are completely removed from the sounds and light of the Earth's surface, so take advantage of the opportunity to immerse yourself in the dark world beneath the ground at Mammoth Cave National Park.

Key Terms

base level
canyon passage
cave level
gypsum
phreatic zone
sinkhole plain
surface exposure dating
tubular passage
vadose zone
vertical shaft

Related Sources

Palmer, A. N. (2017). *Geologic History of Mammoth Cave.* In H. H. Hobbs III et al. (Eds.), *Mammoth Cave: A Human and Natural History,* pp. 110–121. New York: Springer.

Palmer, A. N. (2002). *A Geological Guide to Mammoth Cave National Park.* Teaneck, NJ: Zephyrus Press.

Palmer, A. N. (2017). Geology of Mammoth Cave. In H. H. Hobbs III et al. (Eds.), *Mammoth Cave: A Human and Natural History,* pp. 97–109. New York: Springer.

Thornberry-Ehrlich, T. (2011). *Mammoth Cave National Park: Geologic Resources Inventory Report.* Natural Resource Report NPS/NRSS/GRD/NRR—2011/448. Fort Collins, CO: National Park Service.

White, W. B., and White, E. L. (2017). Hydrology and Hydrogeology of Mammoth Cave. In H. H. Hobbs III et al. (Eds.), *Mammoth Cave: A Human and Natural History,* pp. 123–143. New York: Springer.

National Park Service. Caves and Karst. Accessed February 2021 at https://www.nps.gov/subjects/caves/park-resources.htm

Epilogue: Environmental Threats to Our National Parks

Figure 40.1 Traffic is a serious problem within our National Park System.

Our national parks are wonderlands of geology, ecology, and cultural history to the more than 327 million Americans and international tourists who visited the 423 national park units in 2019. A common perception of national parks is that they are pristine, unadulterated wilderness, separate from the overdeveloped world outside their boundaries. Nothing could be further from the truth, as our national parks are beset by a spectrum of ongoing stresses that threaten the tenuous balance between wilderness preservation and access for people. Overcrowding and the attendant issues of traffic, crime, and air pollution are an enduring headache for National Park Service administrators at many parks (Fig. 40.1). Invasive species are changing the ecological character of our parks at an alarming rate. Encroaching development by mining interests, oil and gas companies, and the tourism industry are continually gnawing at the margins of the parks. And the looming specter of climate change casts its shadow over our national parks, just as it does for the rest of the planet. Beyond these environmental threats, the national park system needs to address the skewed distribution of visitors to the parks, which leans heavily toward older white Americans. The purpose of this short epilogue is not to add to the sense of angst we all have about environmental deterioration, but rather to bring the issues to light so that we can devise long-term and sustainable solutions.

Let's set the stage with a little background about our National Park System. By the way, it's important to remember that you own it. As public lands, your tax dollars go toward maintaining the national parks as communal property, available to anyone who wants to visit. It's been said that our national parks are the most democratic and egalitarian places in the country, owned not by the wealthy elite, but rather by every American citizen. Though managed by the U.S. government, you own 47 percent of all land in the American West, including our national parks, monuments, forests, wilderness areas, and seashores. The total drops to just 4 percent east of the Mississippi River, where private ownership dominates for historical and political reasons.

All of the land in the National Park System is administered by the National Park Service (NPS), which was established in 1916 by the **Organic Act**. This law states that the role of the Park Service "is to conserve the scenery and the natural and historic objects and the wild life therein and to provide for the enjoyment of the same in such manner and by such means as will leave them unimpaired for the enjoyment of future generations." This wordy and vague directive has an inherent conflict built into it, which has led to a variety of interpretations through the years. On the one hand, the act states that the role of the NPS is to conserve wild places, leading some to infer that parklands should remain as undeveloped and "unimpaired" as possible. On the other hand, the act states that the charge of the NPS is to enhance the people's "enjoyment" of the parks. This tension between preservation and making the parks welcome to visitors lies at the crux of many of the environmental issues confronting our parks.

The NPS employed on average around 20,000 people per year between 2009 and 2020. This is about as many employees as CBS, the television network, had in 2019.

For comparison, in 2019 Whole Foods Market employed about 91,000 and Starbucks employed about 185,000. The NPS manages 423 units, including 63 national parks, 130 national monuments, and hundreds of national seashores, preserves, historic sites, battlefields, and recreation areas.

The budget for the National Park Service (an agency of the U.S. Department of the Interior) was $3.37 billion in federal year 2020. This total was 8 percent lower than it was in federal year 2009, when adjusted for inflation. The NPS budget is about 1/15th of 1 percent of the total federal budget and costs the average American taxpayer the equivalent of a cup of coffee per year. This is quite a bargain, considering that national parks like Grand Canyon and Yellowstone are among the most iconic places in America, rivaling the Washington Monument and the Statue of Liberty in popularity. Over 3.3 billion visits to the parks were recorded between 2009 and 2019, resulting in the generation of hundreds of thousands of private-sector jobs and billions of dollars in economic activity. The value of the National Park System is beyond reproach.

Yet the Park Service carried a burden of nearly $12 billion in deferred maintenance costs in federal year 2019. Even though many of the national parks charge entry fees, the revenue generated is not nearly enough to fully fund the system. Private philanthropy helps to flesh out the NPS budget but is not sufficient to fill the fiscal void. The budget crunch at our national parks is an ongoing concern, exacerbated by aging infrastructure, increasing visitation, and swings in the political climate. Many of the environmental threats described below are a consequence of the continual funding deficit. Part of the problem is perspective. NPS administrators and private-sector advocates for the parks take a long-term view, as the Organic Act intended. Conserving the natural resources of each park while also providing for the enjoyment of each visitor takes money and a vision deep into the future. This runs headlong into the reality of annual appropriations from a fractious Congress and the unpredictability of the four-year presidential cycle.

Surprisingly, about $9.5 billion over five years was appropriated in late 2020 by the passage of the **Great American Outdoors Act**. The unanticipated bipartisan support from Congress was triggered by the COVID-19 pandemic of 2020-21, which devastated tourism-based businesses in western states where many of the large and popular parks are located. The infusion of funds was intended to generate jobs and revitalize the economy, while lowering the deferred maintenance backlog that had accumulated over the years. The Great American Outdoors Act has been called the most significant conservation legislation in several decades.

Beyond the issue of chronic underfunding, an inherent vulnerability of our national parks is the artificial nature of their boundaries. Many eastern national parks were carved out of private land, while most western national park boundaries were established to protect unique and exceptional landscapes. Take a look at the park maps in this book: You'll see that in many places the borders are an angular assortment of straight lines that trace no natural feature, like a river or drainage divide. But national parks are open systems, with cars and RVs and helicopters and bears and plants and water and air all flowing across the porous membrane of park boundaries. In the sections below, you'll see how this intrinsic and unavoidable feature of our national parks contributes to a host of environmental threats.

This epilogue may be a challenge to read, but it's important that you take the time to understand its implications simply to counterbalance our mistaken image of national parks as unspoiled wilderness. The idea here is not to dampen your enthusiasm for learning about the geology and landscapes of our national parks, but rather to spur a proactive approach to dealing with the spectrum of environmental risks to our park system. None of the problems described in this chapter are intractable; the solutions will take concerted action by committed individuals and networks of volunteers, political will at both the state and federal levels, and a long-term perspective. College students, like many of you reading this book, come to mind as just the right people to help resolve the array of problems.

Visitation and Overcrowding

In 2019, over 327 million people visited at least one national park unit, the third highest year ever after 2016 and 2017 (Fig. 40.2). Of course, many people visited more than one park, so they may be counted multiple times in the total. But just considering the number as a whole, total visitation was comparable to the current U.S. population

of about 330 million. These huge numbers reflect the ever-increasing popularity of the national parks. In the years following World War II, visitation exploded, coincident with the expansion of America's highway system and growing prosperity. As the Baby Boom generation grew in concert with the environmental movement and the increasing popularity of outdoor recreation, visits to the national parks rose above 250 million per year by the mid-1980s and have accelerated since 2014. The skyrocketing visitation is happening at a time when staffing at the National Park Service is decreasing and as the number of underfunded park budgets is increasing.

The two most popular national park units in 2019 were the Golden Gate National Recreation Area and the Blue Ridge Parkway, both with around 15 million visits. Each of these park units is easily accessible to lots of people and neither has gates or entry fees. Among the national parks proper, Great Smoky Mountains is by far the most popular, with over 12.5 million visits in 2019, followed by Grand Canyon National Park with almost 6 million visitors. Rounding out the top ten in 2019 are Rocky Mountain, Zion, Yosemite, Yellowstone, Acadia, Grand Teton, Olympic, and Glacier national parks, each with annual visitation of 3 to 4 million. As you might imagine, the unprecedented number of people and their vehicles is placing a strain on NPS staff and the infrastructure of roads, buildings, and trails. It's been said many times that Americans are "loving their parks to death." The NPS at each park is caught in a quandary because of its dual mandate to not only conserve and protect the wilderness within the park, but also to provide for the enjoyment of visitors. How is the NPS supposed to accommodate the rising number of

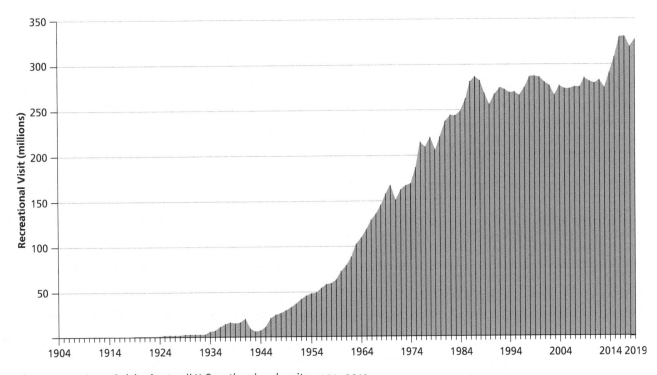

Figure 40.2   Annual visitation to all U.S. national park units, 1904–2019.

visitors and yet maintain the natural integrity of the terrain and its fauna and flora?

Among the many impacts of the surge of visitors at the most popular parks is a decline in the quality of the visitor experience and a degradation of the native environment. Traffic congestion and the attendant deterioration of air quality affect every popular park. Parking lots fill quickly, and long lines form at shuttle stops, restrooms, and visitor facilities. The effects of overcrowding are perhaps most acute at Zion National Park, the fourth most popular in the system in 2019. Most visitors, including hordes of tourists arriving on tour buses from nearby Las Vegas and Los Angeles, converge on the narrow main canyon and its dramatic redrock cliffs. Zion had over 450,000 more visitors than Yellowstone in 2019 but has only a fifteenth the area. To deal with the crush, officials at Zion put into operation a mandatory shuttle service in the year 2000, but visitation has doubled since then, overwhelming the system. Today, park managers are actively considering requiring reservations for entry into the park, a first for the National Park System.

Other popular national parks face comparable problems related to overcrowding. At Great Smoky Mountains National Park, visitation has increased by 25 percent over the past decade, while the number of full-time staff has declined by 23 percent. Complicating the situation is a deferred maintenance deficit of over $215 million, with most of that necessary for the road system. At Yellowstone National Park, visitation increased by 33 percent over the ten years prior to 2019. A budget deficit of about $516 million delays upkeep of the roads, boardwalks around thermal features, wastewater treatment plants, and the trail system. NPS personnel at Yellowstone are spending time dealing with people who stray from boardwalk paths and make their own trails across sensitive areas, those who toss coins into thermal pools, and visitors who get dangerously close to bison, elk, and moose.

What can be done about overcrowding in our most popular national parks? Widening roads, paving more parking lots, and building more accommodations would exacerbate the problem by facilitating the arrival of even more vehicles as well as further diluting the mandate to leave the parklands "unimpaired." Raising entry fees has been suggested, but only 118 of the 423 total units in the National Park System currently collect fees. The affordability of visiting our national parks is part of the ethos of the system, encouraging people of all socioeconomic classes to vacation in their parks. A recent suggestion by the Department of the Interior to nearly triple entry fees at the most popular parks was met with a furious backlash and was quietly scaled back. Some have suggested allowing corporate America to play a greater role, but public opinion is decidedly against encroaching commercialization in our national parks. Over 27,000 people recently signed a petition opposing a Starbucks in Yosemite, reflecting a clear opinion about mixing our public lands with private enterprise. (The Starbucks opened, regardless.)

Two policies are becoming more popular regarding the intertwined problems of overcrowding and underfunding. First, some parks are using shuttle systems to alleviate congestion. Yosemite National Park has a countywide bus system as well as a shuttle network that reduces the number of personal vehicles inside the park. At Acadia National Park, the Island Explorer bus service reduces congestion while also connecting visitors to amenities outside the park. If traffic gridlock continues to be a problem, many parks will be forced to create a mandatory shuttle system inside park boundaries. A second potential resolution that is gaining more traction is a reservation system, not only for entry into the parks, but also for use of the most popular trails. This is a touchy topic because it may be perceived as counter to the egalitarian ethic of the National Park System. The most direct solution would simply be for the federal government to acknowledge the tremendous economic and psychological benefits the National Park System provides and fund the Park Service appropriately on a continuing basis. This is a highly unlikely scenario unless a champion of the parks emerges who can engage political support.

Air Quality

Let's begin with the good news: Since the Clean Air Act was passed in 1970, air quality in our national parks has improved significantly, just as it has in most major cities throughout America. Reductions in vehicle emissions and the installation of pollution control technology in factories and power plants have resulted in clearer skies and healthier air. But anyone who's visited a national park recently knows that the hazy skies and muted colors of landscapes hint at deeper issues with the quality of the air (Fig. 40.3).

Atmospheric monitoring indicates that air pollution lowers visibility at every one of the major national parks and that 75 percent of these parks have unhealthy air quality during the summer months when visitation is at its peak. You may think that since your visit is only a few days long that air quality isn't that much of an issue. But many people may visit Grand Canyon or Yosemite only once in their lives, so the fuzzy views that they see are what they'll remember, both in their mind and in their photos. Park rangers and other personnel at national parks with seasonally unhealthy air have no recourse but to endure it. And why should we accept poor air quality in the first place? We have the technology and the public consensus to make improvements, so what barriers stand in the way of crisp views and healthy air? Let's begin by defining the components of air pollution and their sources.

The Environmental Protection Agency identifies the four major pollutants in air as fine particles, sulfur dioxide, nitrogen oxides, and ground-level ozone. Particulate matter consists of micron-sized specks of carbon soot, smoke, dust, and liquid droplets that are small enough to remain suspended in the atmosphere. These tiny particles are derived from natural sources such as wildfires and

Figure 40.3 View to the west into Shenandoah Valley from Dickey Ridge. Left image is a good-visibility day, whereas the right image was taken on a poor-visibility day.

windblown dust as well as from the combustion of fossil fuels like coal, oil, and natural gas. Burning fossil fuels in power plants, factories, and vehicles also releases gaseous compounds of sulfur and nitrogen. Agricultural sources such as crop fertilizer and animal waste further contribute nitrogen-based gases. The fourth major component of air pollution, **ozone**, occupies two levels in our atmosphere. Small amounts of ozone high in the stratosphere enable life on Earth to survive by filtering harmful ultraviolet radiation from the Sun. But ozone, consisting of three oxygen atoms, also forms at ground level when gaseous vehicle emissions react with sunlight on hot summer days. Each of the four constituents composing air pollution contributes to reduced visibility and detrimental effects to human health.

The most obvious effect of poor air quality in our national parks is on visibility, the sharpness and distance of views we take in from scenic overlooks and mountaintops. The fuzziness of the air is due to **haze**, created by the interaction of tiny atmospheric particles with sunlight. Fine particulates tend to scatter and absorb light, so the amount of particulate matter in the air and the warmth of that air determine the distance and clarity of your view. Haze can be especially thick where tiny droplets of water mix with sulfur dioxide released from industrial sources, producing sulfate aerosols that linger in the air. Haze pollution in our national parks diminishes the vibrancy of colors and the contrast of textures across a landscape. Lower visibility also lessens the experience of the visitor who may have driven a long distance to that overlook above Yosemite Valley or to the peak of Clingmans Dome in Great Smoky Mountains National Park.

On hot days during the summer tourist season, haze casts a yellowish tinge to the air at Yosemite and nearby Sequoia–Kings Canyon national parks. Pollution from the nearby San Joaquin Valley and San Francisco Bay Area is blown by the westerly winds up into the Sierra Nevada, where the tiny particulates settle into valleys and canyons. To the south, pollution from vehicles and industry in the Los Angeles Basin wafts inland toward Joshua Tree National Park, obscuring the views. Near Grand Canyon National Park, haze pollution reduces the natural visibility by over 30 percent, on average. Vehicle exhaust is one source of fine particulates near Grand Canyon, as is the presence of the largest coal-burning power plant in the American West a few tens of kilometers northeast of the park (Fig. 40.4).

Reduced visibility is also a problem in our eastern national parks. Haze pollution subtracts about 100 km (60 mi) from the views at Mammoth Cave National Park, located downwind of power plants, industries, and cities in Kentucky and Tennessee. In Shenandoah and Great Smoky Mountains national parks, the natural bluish haze produced by the release of organic compounds from trees is turned yellowish gray by haze from human-made sources. The average visibility at Great Smoky is 150 km (90 mi), but it currently stands at 40 km (25 mi). This is an improvement over the poor visibility of the 1990s due to the closure and retrofitting of certain power plants in the region. Even the visibility of the coastal setting of Acadia National Park in Maine is diminished by haze pollution derived from inland sources to the south and west.

A more insidious effect of air pollution in our national parks is its impact on human health. Ground-level ozone is a colorless and odorless gas that has a wide variety of negative effects on humans. It exacerbates asthma and other breathing problems, increases the risk of heart attacks, and reduces immune function. Everyone who works and plays outdoors in the national parks, visitors and NPS

Figure 40.4 Navajo Generating Plant, a coal-fired power plant near Grand Canyon National Park. The color of the sky in the background is unretouched.

staff alike, are susceptible to prolonged exposure to ozone, especially during hot summer days. Air-quality monitoring of over 30 of our largest national parks from 1990 to 2014 revealed ozone concentrations comparable to those of the 20 largest cities in the United States. In Sequoia National Park, ozone levels exceeded those in Los Angeles, the city with the highest ozone pollution in the country, in all but two years since 1996. From 2001 to 2014, Joshua Tree National Park had an average of 101 unhealthy ozone levels per year, far beyond the average of 78 for New York City. Yosemite National Park has the fourth-highest ozone levels of all national parks, right behind Sequoia, Kings Canyon, and Joshua Tree.

Ozone is harmful to plants as well, especially Jeffrey and ponderosa pines like those that dominate forests in many western national parks. Needles turn dry and yellow, enhancing their susceptibility to fire. The forests at Great Smoky Mountains National Park are also significantly affected by ozone pollution, with more than 30 species of plants exhibiting damage.

Air pollution is a problem without borders that requires a broad set of laws and regulations. In 1999, the U.S. Congress passed the **Regional Haze Rule**, which was specifically intended to reduce air pollution in the national parks, the lands that should have the cleanest air in the country. The Interagency Monitoring for Protected Visual Environments (IMPROVE) program at the University of California, Davis, my home institution, samples and analyses air quality at all of the large national parks and provides the data to state agencies to comply with the Regional Haze Rule. The goal of the rule was to aim for pollution-free air in the parks by the year 2064, but directives to states were vague about how to plan for reductions in pollutants. The rule needs to be strengthened to enhance state accountability and to reduce the influence of special interest groups and energy-industry lobbyists. Most Americans would agree that clean air and high visibility for future generations of visitors to our national parks should take precedence over the short-term profits of corporations.

Encroaching Development

The sanctity of our national parks and monuments is under constant assault from political and corporate interests who covet the economic potential of America's public lands. It's not the land inside the park boundaries that are vulnerable to resource exploitation, but rather the public lands just outside the parks that are most susceptible to mining, oil and gas drilling, and logging. The Bureau of Land Management is responsible for hundreds of millions of acres of public land that by law is available to companies for resource extraction. The potential profit of mineral and energy resources, combined with the excessive power of politicians, threatens the water, air, and landscapes along the porous margins of our national parks and monuments.

One immediate threat involves the downsizing of more than two dozen national monuments, including Grand

Staircase-Escalante and Bears Ears in Utah, Grand Canyon-Parashant in Arizona, and Katahdin Woods and Waters in Maine. National monuments are created by presidential proclamation under the *Antiquities Act* of 1906, "to protect significant natural, cultural, or scientific features." The Act has been used more than a hundred times and, until recently, no president has reversed earlier decisions since 1915. The overt reason for reducing the size of national monuments is to free up protected lands for mining and oil and gas drilling, as well as to push back against the regulations for habitat protection for endangered plants and animals. The downsizing is also part of an aggressive effort to support states' rights in the ideological battle against federal power regarding ownership of public lands. Predictably, the political attempts to reduce the size of national monuments have been met with litigation by conservation advocates, the outdoor tourism industry, and progressive lawmakers. Partisan efforts to weaken the Antiquities Act are likely to be an ongoing issue over the next few decades, with the outcome potentially determined by the intensity of public indignation.

The unprecedented attacks on the once-inviolate boundaries of national monuments belie the reality that the margins of many of our national parks and monuments are already subject to mineral extraction and oil drilling. In fact, there are hundreds of active oil and gas wells *inside* several national parks, permitted by arcane laws that allow private companies to own mineral rights beneath the surface while the federal government owns the land above. Current law does allow the NPS to charge fees for commercial use, which helps to pay for the restoration of land damaged by the drilling process.

Public lands surrounding several national parks and monuments are currently slated for possible energy development, such as near Dinosaur National Monument and Canyonlands, Capitol Reef, Zion, Mesa Verde, and Carlsbad Caverns national parks. Although legal, the infrastructure necessary for resource exploitation damages landscapes, disrupts scenic views, fragments habitats, generates noise and light pollution, potentially contaminates local water supplies, and is discordant with the visitor experience. In general, the Bureau of Land Management, which administers the largest acreage of public land, has operated with a scientifically driven set of decisions, but unforeseen factors may challenge their cautious stewardship. One scenario that would drive increased exploitation of energy resources near park boundaries would be a dramatic increase in the global price of oil combined with a political push for "energy independence."

Remarkably, about 200 units in the National Park System have existing claims that permit mining of metals and coal within park boundaries, mostly because the rights to mine were already in place when the park unit was established. Other federal lands along the margins of national park units are fair game for existing and new mining claims; many of these claims are currently active on the fringes of our national parks. Only 10 km (6 mi) south of Grand Canyon Village along the south rim, a private company has reopened a mine to extract high-grade ores of uranium and copper. The Havasupai Tribe, environmental groups, and county officials opposed the mine, fearing that local groundwater resources could be contaminated. In reality, no one really understands the complexity of the underground plumbing system for water flow in the region, but it's certain that the ultimate destination for groundwater is the Colorado River in the depths of Grand Canyon. The mining company's public reports promise "extensive controls" to protect natural resources in the area, such as sealing the 400 m (~1400 ft) production shaft to prevent leakage into aquifers, but it may be many years before any contamination is detected.

The U.S. Geological Survey has identified several springs and wells inside Grand Canyon park boundaries that have water that is unsafe to drink due to infiltration of uranium-rich fluids from older mines in the region. Two springs directly linked to the abandoned Lost Orphan uranium mine inside the park have elevated uranium concentrations. Both springs empty into the Colorado River within the Canyon, the drinking supply for about 25 million people downstream. Uranium mining during the Cold War years has left a toxic legacy across the American West, including areas within and very close to several national parks. The mandate of the NPS to maintain an "unimpaired" environment is incongruous with mining operations, especially near the permeable boundaries of national parks. Exploitation of mineral and energy resources benefits a relatively small number of people at the expense of the many concerned Americans who cherish the communal inheritance of our national parks.

Noise Pollution

Surveys show that most park visitors, day-trippers and backpackers alike, visit national parks to hear the sounds of nature and to experience the personal contentment that it brings. Whether it's the sound of water rushing across rocks in a stream in Shenandoah, the wind whistling through a narrow slot in Canyonlands, or the bugling of an elk at Yellowstone, the sounds of nature help people feel a deep connection with the natural world. Other visitors wander trails in the backcountry, searching for solitude and the profound silence that wild places can provide. The NPS supports a research program that measures the **soundscape** within national parks, the combination of both natural and human-made sounds that determine the acoustic environment within a given area.

The NPS considers the soundscape of national parks to be one of the resources they are obligated to protect as part of the Organic Act. To that end, specialists combine long-term measurements from listening stations in national parks with the background sounds from other parts of the country, urban and rural alike. Computer software filters variables such as weather, topography, human activity, season, and time of day to generate a soundscape map of the United States (Fig. 40.5). Without the high intensity of noise created by cities, towns, roads, air traffic, and other human influences, the map predicts the natural noise level of a typical summer day across the entire country. The most obvious pattern is that eastern America is louder than the American West. The reason is due to the presence of water and the attendant fertility that it promotes. Flowing water, wind blowing through dense forests, and the sounds of animals like birds and frogs animate the wetter eastern side of the country. The drier climates west of the Great Plains are ecologically less productive, so the sounds are muted, especially in the arid regions of the Great Basin, Colorado Plateau, and southern deserts. Many of our national parks echo this broad-scale soundscape, providing visitors with the sense of remoteness and well-being that only wilderness can provide.

That's why it's so disconcerting to hear the annoying whup-whup-whup of helicopters ferrying tourists up and down the corridor of the Colorado River in Grand Canyon National Park. Air tours are legal and very popular in the park, with hundreds of flights per day during the busiest summer months. Their high-decibel sounds reverberate within the canyon walls, reaching into even the most remote places in the park. In some parts of Grand Canyon, noise from air tours can be heard about 80 percent of daytime. The air-tour industry has a bank of lobbyists that influence politicians who seem impervious to the experience of the vast majority of visitors who view the Canyon from overlooks along the south rim and who hike the trails down toward river level.

The Grand Canyon is one of several national park units that permit air tours. Over 90 percent of the tours occur in nine parks, including the National Parks of New York Harbor, Hawai'i Volcanoes, Haleakala, and Mount Rushmore national parks. The air tours provide a unique aerial perspective for visitors, but the trade-off is a barrage of noise pollution for those tourists at ground level who may have hoped for a quieter, more tranquil national park experience. Furthermore, several studies have shown that noise from aircraft disrupts the behavior of birds and large mammals. The National Park Service walks a tightrope between protecting the natural soundscape of its wilderness areas versus policies that permit the discordant sounds of aircraft in the skies above. To their credit, the NPS is partnering with private interests to develop an electric plane that could reduce the noise from air tours.

Climate Change

Of all of the environmental threats to our national parks, none is more immediate or challenging than climate change. Evidence for long-term shifts in climate is particularly well documented in our national parks because they provide some of the most intact natural environments in the world. In addition, a wealth of archival data has been captured over the past several decades that permit scientists to compare changes over long time periods.

L50 SPL,
dBA re 20μPa

- 38 – 40
- 37 – 37
- 37 – 36
- 36 – 36
- 35 – 35
- 34 – 34
- 32 – 33
- 31 – 31
- 30 – 30
- 29 – 29
- 28 – 28
- 27 – 27
- 25 – 26
- 23 – 24
- 20 – 22
- < 20

Figure 40.5   Soundscape map of natural conditions on a typical summer day across the United States. Human-made noise was filtered out to produce this map. In general, wetter areas of the country are louder than drier areas.

Shrinking glaciers are one of the most apparent trends that reflect rising temperatures, but the parks also record an increasing intensity of drought, enhanced coastal erosion due to sea-level rise, increased frequency of wildfires, shifts in the ranges of plants and animals, and earlier dates for snowmelt. These indicators of a warming climate have happened many times in the geologic past, but never have they occurred at the accelerated rate seen today. The NPS acknowledges the reality of climate change and is actively engaging the public about the impact on our national parks and what measures are being taken to mitigate and adapt to the problem.

Tiny amounts of invisible greenhouse gases suffuse our atmosphere. The primary greenhouse gases—water vapor, carbon dioxide, methane, ozone, and nitrous oxide—have the ability to absorb heat energy in the atmosphere and reradiate it back down toward the surface. As the amount of these gases increases, the energy balance on our planet changes, resulting in measurable increases in global temperatures. Water vapor is the most powerful greenhouse gas, but it moves through the hydrologic cycle in a matter of days, limiting its long-term effect. In contrast, carbon dioxide stays in the atmosphere for centuries and is thus the primary driver for a warming planet. Carbon dioxide constantly cycles between plants and animals, but over the last century or two it has been increasing in the

atmosphere from the combustion of fossil fuels in vehicles and industry, as well as from deforestation.

Ancient concentrations of carbon dioxide have been measured from air bubbles trapped in ice cores from the Greenland and Antarctic ice sheets, providing a long-term record prior to human influence. Modern amounts of atmospheric carbon dioxide have been measured at the top of Mauna Loa on the Big Island of Hawai'i for several decades and are currently at their highest levels in the last 800,000 years. Ominously, the rate of increase of carbon dioxide concentrations in the atmosphere is orders of magnitude faster than rates measured over the pre-industrial geologic past. Global temperatures are directly linked to atmospheric carbon dioxide concentrations and are thus increasing rapidly as well.

One important consideration in this brief primer on climate change is the difference between weather and climate, two concepts that are frequently confused with one another. Weather is the hour-to-hour and day-to-day changes in temperature, precipitation, cloudiness, wind, and other variables that we experience constantly. In contrast, climate is the average of weather for a particular region and time period, usually taken over 30-year intervals. So climate change might be best understood as shifts in the long-term averages in daily weather. It's been said that climate is like your personality, but weather is like your mood.

What are some of the specific effects of climate change in our national parks? When Glacier National Park was established in 1910, it had over 150 named glaciers. Today it has 25 remaining glaciers that are rapidly thinning and retreating. You can hike to Grinnell glacier, but you may want to do it sooner than later because it is likely to be gone by 2030, along with all the other remaining glaciers in the park. In Mount Rainier National Park, the alpine glaciers that radiate downward from its crest have on average decreased in thickness by about 7 m (23 ft) since accurate measurements began in 1970. In Yosemite National Park, the Lyell and Maclure glaciers are projected to disappear in the next few decades (Fig. 40.6). The five-year drought that California experienced between 2012 and 2017 measurably accelerated their melting. These glacial examples are the most visible effects of the global increase in temperature of 0.85°C (1.5°F) since 1880, with most of the warming occurring since the mid-1970s.

The increasing heat in the Earth system promotes evaporation that raises the amount of water vapor in the atmosphere. Decades of recordkeeping show marked changes in temperature and precipitation at many national parks. Acadia National Park is experiencing wetter and warmer conditions, accompanied by more extreme storms. At Crater Lake National Park, snowpack has decreased 37 percent since the 1930s as more and more precipitation falls as rain. At Yellowstone, there are about 30 fewer days per year with snow on the ground, relative to the 1960s. Temperatures are higher overall in the park, with 80 more days above freezing per year, triggering an earlier spring snowmelt, a reduced snowpack, and an increased probability of more frequent and intense wildfires.

Other parks are experiencing rapid changes related to longer and more intense droughts. The iconic Joshua trees for which the national park is named are showing changes related to increasing temperatures, prolonged drought, and wildfire. Older Joshua trees are dying and are not being replaced with younger trees because the cool winters and freezing temperatures necessary for flowering aren't happening like they once did. One pessimistic prediction is that 90 percent of the park's namesake trees will be gone by the end of the century. Increasing temperatures and decreasing precipitation in the Sierra Nevada lower water availability and intensify the risk of wildfire for the majestic trees that give Sequoia National Park its name. The desert southwest and its array of parks like Death Valley and Grand Canyon will in the future be subject to even more extreme temperatures and sustained drought conditions, increasing stress on their populations of plants and animals. The rapid rate of change may inhibit the ability of plants and animals to migrate upslope or northward to more amenable environments.

Coastal national parks and seashores are feeling the effects of a steady rise in sea level of about 20 cm (8 in) since 1880, with projections of over a meter more by the year 2100. Sea level is rising partly due to the input of glacial meltwater into the world ocean but primarily because of the expansion of seawater as it absorbs excess heat from the atmosphere. This rapid pulse in sea-level rise is a recent phenomenon, superimposed on the longer-term *postglacial transgression* described in earlier chapters. The encroaching shoreline increases the odds of seawater flooding and coastal erosion on national seashores and their estimated $40 billion in infrastructure. Assateague Island National Seashore off the coast of Maryland and Virginia is currently migrating westward toward the mainland in response to rising sea level, forcing park administrators to move structures landward in tune with the shifting sands beneath. Everglades National Park on the tip of the Florida Peninsula is shrinking in size as the liquid boundary between land and sea encroaches

Figure 40.6   The Lyell glacier in Yosemite National Park has shrunk dramatically since the late 19th century.

further inland. National parks along the west coast, such as Golden Gate NRA, Channel Islands, Point Reyes, Redwoods, and Olympic, are all vulnerable to coastal erosion and submerged wetlands.

Climate change is also altering ecosystems in our coastal national parks. The oceans absorb about a quarter of the excess carbon dioxide emitted by humans, turning seawater ever more acidic. At coastal parks like Olympic and Point Reyes, efforts are being made to measure the changing ocean chemistry and its effect on marine organisms that have a harder time making their calcitic shells as seawater acidity increases. At Everglades National Park, saltwater is intruding into the park's freshwater marshes, making them brackish and rapidly changing the ecological balance.

In mountainous national parks, ecosystems are showing a clear response to warmer temperatures and changing precipitation. At Yellowstone, the earlier spring snowmelt changes water levels in streams, rivers, and wetlands, affecting the growth of plants and the migratory habits of animals. In Rocky Mountain National Park, cold winter temperatures usually kill the eggs and larvae of bark beetles, keeping their numbers in check. But winters have been warmer than average over the past decade or so, allowing the beetles to survive into the spring when their increased numbers ravage forests of lodgepole and ponderosa pine, spruce, and fir. When combined with periods of low precipitation, the ghostly array of dead and dying trees raises the risk of wildfire.

The National Park Service is actively dealing with climate change in each park with its "Green Parks Plan," developed in 2012 to reduce park emissions of greenhouse gases by integrating alternative forms of energy into the mix and by modifying their transportation systems to make them more efficient. One initiative of the larger plan is the "Climate Friendly Parks" program, designed to help individual parks find actionable ways to become more sustainable and to directly engage the public about climate change.

The physical effects of climate change on our national parks almost seem manageable when taken at face value. But remember that these ecological and landscape responses to changing climates are just what have been measured over the recent past. The more important point of view is to project these changes into the coming decades and centuries. Think on a longer time scale, that of your children or your future grandchildren, in order to comprehend the slow-moving trainwreck that is climate change. It's scary to contemplate, but the problem is tractable if we act in a timely and strategic manner.

Future of Our National Parks

None of the environmental threats we have addressed occurs in isolation, separated from the rest of the planet by the imaginary lines that circumscribe our national parks

Solutions require an approach that takes the larger scale of the greater ecosystem, the watershed, the country, and even the entire planet into account. Solutions also require a change in the behavior of you and me. Consider the interrelated problems of air quality and climate change. Remember that each of us is the trigger for air pollutants and greenhouse gases every time we start our cars, charge our phones, or heat our homes. The upstream provider may be a natural gas-fired power plant or oil refinery, but we choose to use the electricity or gasoline as sources of energy to power our lives. You may live hundreds of miles from a national park, but your input of gases affects the quality of their air and the global climate system as the gases mix and circulate in the atmosphere. Of course, we need a global consensus to transition away from fossil fuels so that we can decarbonize our world, but individuals acting collectively can profoundly contribute to the solutions as well.

How about the chronic problem of underfunded park budgets? The funding provided by the Great American Outdoors Act of 2020 will infuse money into the national park system for five years, but the funding deficits will return if park budgets remain flat while visitation progressively increases. Part of the answer lies in the creation of laws that transfer funds from revenue that the government earns for the commercial use of public lands directly into the NPS budget. For instance, energy and mining companies that exploit resources along the margins of national parks could be forced to funnel some of their profits directly toward the National Park System. Developers who take advantage of the steady stream of visitors to major national parks by turning nearby gateway communities into tourist traps could pay into a fund for the adjacent park. Air-tour operators might pay an amount proportional to the decibel level of their aircraft and the duration of each flight. The fees for ranchers who graze their cattle on public lands are obscenely low; our government could raise them just a small amount, with the extra funding directed toward our public lands. Or politicians could grow a backbone and pay less attention to the lobbyists and more to their constituents, who, according to multiple surveys, want to see their national parks properly funded and protected.

Philanthropy and volunteerism play a critical role in the well-being of our national parks. Nonprofit organizations like the National Parks Conservation Association and the National Parks Foundation contribute tens of millions of dollars per year to the park system and advocate for the parks within the halls of Congress. Park-specific groups like the Yosemite Conservancy provide grants to support projects such as providing outdoor learning opportunities for the public, improving the extensive trail system, and restoring habitat. In 2017, 2800 volunteers in Great Smoky Mountains National Park helped with trail maintenance, removed invasive plants, and provided information to visitors, alleviating the strain on the park staff.

The National Park Service recognizes that most people who visit the parks are older and whiter than the overall demographics of America. This is an unsustainable trend if our national parks are to maintain their vitality and importance in the future. The U.S. Census Bureau projects that people of color will compose the majority of the American population by 2044. If our national parks are perceived as irrelevant to the younger generation and unwelcoming to people of color, then the long-term support our parks require will diminish. Minorities compose 42 percent of the U.S. population but make up only 23 percent of visitors to the national parks, according to a recent NPS study. Moreover, less than 20 percent of the 20,000 National Park Service employees are minorities.

The NPS, acknowledging a diversity problem, has established marketing programs to reach out to underrepresented groups, including the newly created Office of Relevancy, Diversity and Inclusion. Part of their charge is to train NPS staff on racial sensitivity and to diversify the park workforce. To attract younger visitors to the parks, the NPS has implemented the "Every Kid in a Park" program that enables fourth graders and their families to visit hundreds of national park sites for free; at least two million people have benefited from the program thus far (Fig. 40.7). In the private sector, volunteer organizations such as Outdoor Afro and Latino Outdoors are promoting the national park experience to communities of color. If the national parks are to retain their role as our most democratic places in the country, the people who visit should reflect the age distribution and ethnicity of the entire American population.

The one thing that people who care about our national parks should *not* do is take them for granted. We may think of our national parks as inviolate assets of our American system, but they will always be targets for private interests and the politicians they can influence. Ongoing attempts to dilute the Antiquities Act and downsize some national monuments are potentially only the opening salvo in an attack on the sanctity of the system. Other protected areas may be downgraded, meaning that more human activities (e.g., mining, drilling, grazing) may become permissible within their borders. Yet, other protected areas in the National Park System may be targeted for **degazettement**—a legal term that describes the complete loss of protection for an entire national park. We may think of degazettement as an unlikely possibility, but the economic interests of a few wealthy individuals or corporations can challenge the majority who may not have the focused resources to push back.

Concerned citizens can take action by making their opinions known. The recent (2017) attempt to hike fees at 17 of our most popular national parks was met with a fierce public outcry. The increased fees were seen as discriminatory to low-income families who view the parks as inexpensive destinations, as well as to some people of color who saw the policy as just another way to keep the parks white. The near unanimous public uproar was enough to cause the Secretary of the Interior to rescind the plan and replace it with a more modest proposal. Activism works and needs to be exercised by everyone who cares about their national parks.

Throughout this book, you've gained insight into the distant past of our national parks, learning about their geologic origins and the evolution of their landscapes. Along the way, you've read about the value of wilderness and how it provides a refuge from the rigors of daily life, calms the mind, and rejuvenates the soul. With this last chapter, I hope you've learned to think about the future of our national parks. Now that you're more aware of the environmental threats to these phenomenal places that we all love, I hope you take action to keep them as healthy and vibrant as possible. By being proactive, we can not only preserve the landscapes, ecology, and natural wonders of our national parks, but also maintain the spirit of what our national parks mean to all Americans.

"National parks are the best idea we ever had. Absolutely American, absolutely democratic, they reflect us at our best rather than at our worst."

Wallace Stegner

Figure 40.7 Kids and national parks are a good mix and important for the future of both.

Key Terms

degazettement
Great American Outdoors Act
haze
Organic Act
ozone
Regional Haze Rule
soundscape

Related Sources

NASA Earth Observatory. (2016). *Natural Beauty at Risk: Preparing for Climate Change in National Parks.* Accessed February 2021 at: https://earthobservatory.nasa.gov/Features/NationalParksClimate.

Nash, Stephen. (2017). *Grand Canyon for Sale: Public Lands versus Private Interests in the Era of Climate Change.* Oakland: University of California Press.

National Park Service. (2019). *Annual Visitation Highlights.* Accessed February 2021 at https://www.nps.gov/subjects/socialscience/annual-visitation-highlights.htm

National Park Service. (2016). *Climate Change and Your National Parks.* Accessed February 2021 at https://www.nps.gov/subjects/climatechange/index.htm

National Park Service. (2016). *Green Parks Plan: Advancing Our Mission through Sustainable Operations.* Accessed February 2021 at https://web.archive.org/web/20171004031207/https://www.nps.gov/subjects/sustainability/green-parks.htm

National Park Service. (2018). *National Park Service Appropriations: Ten-Year Trends,* Congressional Research Service. Accessed February 2021 at https://fas.org/sgp/crs/misc/R42757.pdf

National Park Service. (2018). *A Symphony of Sounds.* Accessed February 2021 at https://www.nps.gov/subjects/sound/index.htm.

Glossary

Note: The number at the end of each definition indicates the chapter in which the term first appears.

a'a lava Hawaiian term for lava flows that have a rough surface composed of broken blocks of solidified lava carried along by a thick, pasty lava matrix. (32)

Acadian Orogeny A major phase of mountain-building and continental accretion along the ancestral Appalachian continental margin that ranged from ~430 Ma to ~360 Ma; resulted in the closure of the Iapetus Ocean and the suturing of the Avalonian Terrane. (35)

accreted terrane A crustal block sutured to the edge of a continent along zones of oceanic-continental convergence. (19)

accretion The process that involves the tectonic detachment of a crustal block from a subducting plate and its subsequent suturing to the edge of a continent; a fundamental process of continental growth. (19)

accretionary wedge A prism-shaped mass of seafloor sediment and rock that accumulates at the boundary between a subducting oceanic plate and a continental plate; the upper part of the subducting slab is scraped off as it is forced beneath the adjacent continent, forming the wedge. (26)

active continental margin Continental margins that occur close to tectonic plate boundaries; marked by earthquakes, volcanoes, mountain-building, and emergent coastlines. (26)

active volcano A volcano that is either currently erupting or that has erupted at least once in historical time where humans were able to document the event. (32)

actualism Principle that states that geologic processes and events of the past can be explained by actual processes and events occurring today; displaces the older principle of uniformitarianism in that it takes into account the wide range of rates at which geologic events can occur through time. (1)

alcove A shallow recess at the base of a cliff face created by the weathering and erosion of underlying rock, causing the now-unsupported overlying rock to break away as rockfalls; typically occur along the contact between a less resistant layer of rock below and a more resistant layer of rock above. (4)

Alleghenian Orogeny A major phase of mountain-building and continental collision between the Appalachian continental margin and the northwest African portion of Gondwana that ranged from ~330 Ma to ~270 Ma; resulted in the closure of the Rheic Ocean and the peak uplift of the ancient Appalachian Mountains. (35)

alluvial fan Fan-shaped accumulations of sediment that form near the mouths of stream channels where they meet a relatively flat valley; sediment is commonly derived from episodic debris flows; particularly well exposed in arid regions. (24)

alluvial soil Type of soil composed of water-deposited particles of sediment mixed with organic matter; common on river floodplains and deltas. (36)

alluvium Unconsolidated deposits of stream-laid sediments, such as gravel, sand, silt, and clay. (1)

alpine glacier (aka *valley glacier*) A glacier that is confined by surrounding mountain terrain; typically follows the route of preexisting stream-carved valleys. (14)

amygdules Mineral-filled vesicles or pockets within a larger body of volcanic rock; minerals may be quartz, calcite, epidote, chlorite, or many others. (36)

anastomosing stream A stream characterized by multiple smaller channels that interweave with one another within a main channel; differs from a braided stream in that it is more stable due to banks composed of silt and clay; typically develops on low gradients. (25)

Ancestral Rocky Mountain Orogeny The late Paleozoic uplift of two elongate mountain ranges located close to their modern counterparts in Colorado; involved the rise of basement blocks along reverse faults due to compressional tectonics. (10)

andesite Grayish volcanic rock with an intermediate silica (SiO_2) content. Andesite magma may erupt as either thick lava flows or explosive eruption columns. (12)

angular unconformity An erosion surface that separates horizontally layered strata above from tilted or folded strata below. (3)

anticline A fold in rock strata with a convex upward shape. (22)

Archaea A domain of single-celled microbes lacking cell nuclei; genetically distinct from bacteria; often tolerant of extreme environments such as hot springs, saline lakes, and submarine hydrothermal vents, but also recognized in almost all modern habitats. (16)

Archean An eon of the Precambrian beginning 4.0 Ga and ending 2.5 Ga; begins with the oldest rock formations on Earth and ends at the appearance of oxygen in Earth's atmosphere. (22)

arête A narrow, jagged ridge that separates—or that once separated—two adjacent glaciers; formed by glacial erosion; resembles the serrated edge of knives. (14)

aridification Long-term regional change toward progressively more arid climatic conditions. (27)

ash-fall tuff Volcanic rock composed of tephra deposited by fallout from an eruption column; may cover a very broad area downwind of the source; typically thicker near the source and thinner at the distal edge. (15)

ash-flow tuff (aka *welded tuff* or *ignimbrite*) Volcanic rock formed by widespread deposition and solidification of material in a pyroclastic flow; heat from the deposit welds the particles together. (15)

asthenosphere Hot and weak portion of the upper mantle composed mostly of solid rock with a small percentage of partially melted rock that allows for slow, ductile flow; upper boundary is defined by a temperature of ~1300°C, about 100 km (60 mi) beneath the oceans and about 150 km (90 mi) beneath the continents. (1)

astrobiologist A scientist concerned with the origins, early evolution, and possible habitats of life in the universe. (16)

atmosphere Earth's gaseous envelope composed mostly of nitrogen (N_2) and oxygen (O_2). (1)

atoll A ring-shaped, shallow island or set of islands encircling a lagoon; formed as coral reefs surrounding the rim of an extinct seamount or volcanic island, which eventually erodes below sea level. (32)

Avalonian Terrane Microcontinental fragments rifted away from Gondwana that were eventually accreted to eastern North America during the Acadian Orogeny; part of modern coastal New England, the eastern Piedmont, northeastern Canada, and parts of coastal Europe. (35)

axial rift valley A deep, narrow trough extending along the axis of a mid-ocean ridge; site of fissure volcanism and seafloor spreading. (1)

badlands A stark erosional terrain composed of soft, easily eroded sedimentary or volcanic rocks; typically lack soil cover and have minimal vegetation; often highly colorful, with numerous short channels scarring the surface. (24)

bajada An apron of rocky debris formed by the coalescing of alluvial fans along the base of a mountain; partially buries the range front in its own sediment. (24)

basalt Black, fine-grained volcanic rock with a low-silica (SiO_2) content; typically formed from effusive lava flows. Basaltic magma is derived directly from the mantle; composes most of the ocean floor and is thus the most common rock type of the Earth's crust. (1)

base level The lower limit of erosion by a stream, with the ultimate base level being sea level. Base-level fall will result in the seaward progradation of rivers and their deltas, whereas base-level rise will result in landward retreat of rivers accompanied by sediment aggradation within the channel. (39)

basement A complex assemblage of metamorphic and igneous rocks, usually but not always Precambrian in age, that forms the thick crustal foundation beneath an overlying blanket of sedimentary rocks. (3)

basement-cored uplift Style of mountain growth where Precambrian crystalline rocks composing the basement were pushed upward to high elevations along reverse faults; common during the Laramide orogenic event. (21)

basin spillover A process sometimes involved in the growth of a river channel where a flowing river fills a basin until it spills over a natural dam, draining downslope into another basin; the process repeats itself in a stepwise fill-and-spill fashion until the river meets sea level. (34)

batholith A large mass of intrusive igneous rock formed by multiple overlapping plutons; defined as having more than 100 km^2 (40 mi^2) of surface exposure and no known floor. (17)

bed A layer of sedimentary rock with distinct characteristics from layers above and below; distinguished by the composition, shape, size, and sorting of its constituent particles. (1)

bedrock The uppermost solid rock beneath any soil or surficial sediment; may be exposed at the surface.

biochemical sedimentary rock Sedimentary rocks composed of the fossil remains of organisms or of their hard skeletons; common examples include limestone, chert, chalk, and coal. (1)

biosphere The realm of life on Earth, including plants, animals, and microbes that are either living or in the process of decay. (1)

block fields Areas covered by angular blocks of resistant rock, commonly quartzite, that accumulate as talus along slopes;

may move slowly downslope by the heaving of soils that alternately freeze and thaw. (36)

boulder streams Accumulations of rounded blocks of granitic rock produced by spheroidal weathering that tumble into small stream valleys at the base of mountain slopes. (36)

braided stream A stream whose main channel is marked by multiple smaller channels that constantly diverge and remerge to form an interlacing pattern; common in streams choked with a large amount of coarse sediment; a dynamic system that may constantly change form with changing flow velocity. (25)

brittle behavior Reaction that occurs in rock or ice that breaks or fractures when placed under stress; results in the formation of faults and joints in the upper crust. (1)

brittle-ductile transition The boundary within the crust that separates brittle rock above from ductile rock below; located about 13–18 km (8–11 mi) beneath the surface. In the Basin and Range, the transition marks the base of listric normal faults. (23)

burial Fundamental process involved in the conversion of unconsolidated sediments into sedimentary rock; involves associated processes of compaction and cementation. (1)

butte An isolated erosional landform with steep sides and a flat top commonly composed of a cap of resistant sedimentary rock; derived from the erosion of a larger mesa. (8)

calcite A common mineral ($CaCO_3$) that is the main constituent of limestone and marble; commonly formed biologically as the shells of marine organisms. (1)

caldera A large basin-shaped volcanic depression with a diameter of 2–>70 km (1–45 mi). Commonly forms through the collapse of the volcanic edifice into the void left after a magma chamber has partially emptied, either by lateral underground flow or by a large explosive eruption. (15)

caldera eruption Colossal explosive volcanic event that emits large volumes of magma resulting in the collapse of the volcanic crown to form a caldera basin; occurs infrequently at scales of tens to hundreds of thousands of years. (15)

canyon passage Narrow, canyon-like caves within the modern vadose zone formed by the dissolution of limestone as the water table fell rapidly through time. (39)

cave column A type of speleothem shaped like a pillar formed by the merging of a downward-growing stalactite and an upward-growing stalagmite. (19)

cave level A series of interconnected horizontal passageways that occupy a relatively narrow range of depths beneath the surface. Each level represents a stage of cave development that relates to former positions of the water table. (39)

cementation The crystallization of minerals within the pore spaces between grains of sediment from groundwater saturated with ions; binds loose sediment into a solid mass. (1)

Cenozoic Most recent era of geologic time, beginning 66 Ma; separated from the previous Mesozoic era by a mass extinction. (1)

chemical weathering In-place decomposition of a rock or its constituent minerals by chemical reactions, usually involving water. (1)

chert Very fine-grained sedimentary rock composed of silica (SiO_2); commonly biologic in origin, consisting of the skeletal remains of diatoms and radiolarians deposited on the deep seafloor as a siliceous ooze. (26)

chlorite Silicate mineral commonly found in low- and medium-temperature metamorphic rocks, creating a green tint to the rock. (22)

cinder cone A steep conical hill formed by the accumulation of solidified fragments of pebble-sized cinders (scoria); commonly produced by Strombolian eruptions. (15)

cinder-and-spatter cone Conical-shaped volcanic vent produced as lava fountains spatter back to the surface, accompanied by cinder, ash, and particles of pumice. Pu'u O'o on Kilauea Volcano in Hawaii is a good example. (32)

cirque An amphitheater-shaped depression carved into mountainsides by alpine glaciers near their uppermost elevations; the zone of accumulation typically occurs above cirque basins. (14)

clastic sedimentary rock Sedimentary rock produced by the weathering, erosion, transportation, deposition, and burial of inorganic particles of sediment; common examples include shale, sandstone, and conglomerate. (1)

clastic wedge Areally extensive, prism-shaped package of sediment shed from a mountain range into an adjoining foreland basin. (35)

clay The finest of sedimentary particles, being less than 2 microns in size; flake-shaped grains typically derived from prolonged chemical weathering of silicate minerals; primary component of shales and mudstones. (1)

cleaver Local term to describe dark, rib-like artes that rise above and between adjacent alpine glaciers; commonly used at Mount Rainier National Park. (14)

cliff retreat An erosional process that occurs where an underlying layer of less resistant sedimentary rock is overlain by a more resistant layer; as the soft layer erodes into a slope, it leaves the base of the overlying layer unsupported, causing rockfalls. The upper cliff retreats through time via this process. (8)

coal A carbon-rich, combustible sedimentary rock formed from the burial and compaction of the organic remains of land plants. (35)

coastal sand dunes A linear belt of sand dunes located just landward of a coastline, formed by offshore winds that redistribute beach sand inland. (29)

cobble An angular or rounded chunk of loose rock with a diameter between 64 mm (2.5 in) and 256 mm (10 in). (1)

col Low gaps along the scalloped edges of artes; formed where two adjacent glaciers eroded through the rock before retreating and exposing the saddle along the ridgeline. (21)

columnar jointing Polygonal pattern of jointing that forms by the cooling and contraction of lava flows or within dikes and sills. (31)

compaction The physical compression of particles of buried sediment that forces grains to pack closer together, accompanied by the expulsion of water from pore spaces. (1)

compressional stress Force that squeezes bodies of rock, causing it to deform by folding or to break along faults, resulting in the horizontal "shortening" and the vertical "thickening" of the mass of rock; dominant force near convergent plate boundaries. (1)

confluence The location where two or more streams merge. (8)

conglomerate A clastic sedimentary rock composed of rounded pebbles and/or cobbles, commonly with a sandy matrix. (1)

continent-continent convergent boundary Plate interaction involving the compression of one continental mass against another, resulting in the uplift of rock as mountains; marked by the process of continental collision. (1)

continental collision The tectonic process of mountain growth involving the convergence of one continental mass against another, without subduction or volcanism. (1)

continental crust The part of the crust composed of granitic rock and other rock types of comparable density; rises higher above the underlying mantle relative to oceanic crust, allowing continents to "float" above oceanic crust; ranges in thickness from 30 to 70 km (20 to 45 mi). (1)

continental divide The topographic boundary that runs along the spine of the North American Cordillera separating drainage networks that flow into the Atlantic Ocean from those that flow into the Pacific. (21)

continental ice sheets A massive glacier that forms in a continental interior and then slowly spreads laterally across vast areas; exist today in Greenland and Antarctica where the ice thickness reaches 4 km (2.5 mi). (14)

continental rift A type of divergent plate boundary that occurs when continental crust is exposed to extensional tectonic stress; expressed by fault-bounded, linear rift valleys such as the East African rift. (1)

continental shelf The seaward continuation of the coastal plain of continents that may extend outward for tens to hundreds of kilometers. Shelves slope gently seaward until water depths of ~140 m (~460 ft) are reached at the edge of the continental slope; underlain by continental crust. (24)

continental slope The seaward margin of continental shelves where water depths abruptly increase; incised by submarine canyons; marks the outer edge of continental crust where it transitions into oceanic crust. (24)

continental-oceanic convergent boundary Plate interaction involving the compression of a continental lithospheric plate against an oceanic plate, resulting in the subduction of the denser oceanic plate; results in the destruction of oceanic lithosphere by descent into the mantle. (1)

continental volcanic arc A linear chain of volcanic mountains formed along the margins of continents roughly parallel to a deep-sea trench and an associated subduction zone. (1)

convection currents The dominant mechanism of heat transfer within Earth involving the absorption of heat by rock in the mantle, causing it to become less dense and slowly flow upward toward the surface; with the release of heat near Earth's surface, the cooler, now denser current sinks back down into the mantle, presumably near subduction zones. (1)

convergent plate boundary Plate interaction involving the tectonic compression of one plate against another; oceanic lithosphere is destroyed along convergent plate margins by subduction into the mantle. (1)

Cordillera That part of western North America characterized by a broad array of mountain ranges oriented in a generally north–south direction; extends from Alaska to Central America and from the Pacific Coast to central Colorado. (20)

core Earth's innermost sphere consisting of an inner solid metallic iron sphere surrounded by an outer liquid metallic iron sphere. (1)

country rock (aka *wall rock*) Preexisting rock into which magma intrudes; may consist of any rock type; forms the walls and roofs of magma chambers. (12)

crater Conical vent formed as a result of the eruption of lava or pyroclastic debris from a volcano. (15)

cross-bedding Layers of sedimentary rock composed of smaller internal layers oriented at an angle to horizontal bedding planes; formed during deposition on the inclined surface of ripples or dunes by flowing water or wind. (4)

cross-cutting relations Principle of relative age dating that states that the geologic feature which cuts across another is the younger of the two features; common examples include faults cutting through layers of rock or igneous dikes cutting across a body of rock. (1)

crust The outermost layer of the solid Earth, ranging from about 7 to 70 km (4 to 40 mi) in thickness; oceanic crust is denser than continental crust and therefore sinks deeper into the underlying mantle, creating the ocean basins. (1)

cryosphere The part of the Earth system composed of frozen water; includes continental ice sheets, ice caps, ice shelves, sea ice, alpine glaciers, snow cover, and frozen ground; expands and contracts with the rhythm of glacial-interglacial cycles. (1)

crystalline rocks General term for the interlocking mosaic of crystals within igneous and metamorphic rock; commonly used to describe the basement. (3)

cyanobacteria Group of photosynthetic bacteria whose colonies may form filaments and sheets that contribute to the growth of stromatolites. (22)

dacite Light gray volcanic rock with an intermediate silica (SiO_2) content higher than andesite but lower than rhyolite. Dacitic magma may erupt as either thick blocky lava flows, forming volcanic domes, or as explosive eruption columns. (12)

daughter element The final, stable nonradioactive element that remains after a parent element has fully broken down through radioactive decay. (1)

debris avalanche Rapidly moving masses of rock, soil, and snow that occur when the flank of a volcano collapses and slides chaotically downslope. They may travel for several kilometers before transforming into more water-rich lahars. (13)

debris fan A lobe of rubbly debris formed by deposition of a debris flow; commonly forms rapids where debris flows from tributaries empty into larger channels. (34)

debris flow Water-laden slurries of rock fragments and soil that flush rapidly through stream channels, entraining trees and other debris in their flow; rock and sediment may constitute 40–50 percent of the volume of the flow. (3)

decompression melting The formation of magma from solid rock triggered by the decrease in pressure from the weight of overlying rock as a current of plastically moving mantle rock moves higher toward the surface. (32)

deep ocean trench Elongate troughs in the seafloor that mark the downward flexure of an oceanic plate as it begins to subduct beneath an adjacent plate; convergent plate boundaries are commonly located along the axis of the trench. (1)

deep time A colloquial term for the vast depths of geologic time. (1)

deformation The physical process by which a body of rock is reshaped or moved in response to the application of tectonic stress; folds and faults are common forms of deformation. (23)

degazettement A legal term that describes the complete loss of protection for a national park. (40)

deposition The physical settling of loose particles of sediment within a low area on Earth's surface; commonly occurs underwater but may occur on land as well; includes the precipitation of chemical sediments from mineral-rich waters. (1)

depositional environment The physical setting where sediment is deposited; examples include low-energy settings such as a river floodplain, delta, lake, or ocean basin or higher energy settings like stream channels, windblown deserts, or beaches. (1)

desert pavement Land surface covered with tightly packed pebbles and cobbles forming an interlocking mosaic pattern. Thought to form as finer particles are removed by wind or by the upward movement of larger fragments by wetting and drying of the surface, frost heaving, microseismic vibrations, or the shrinking and swelling of clay particles. (24)

desert varnish Dark, micron-thick coating of silica, clay, and iron and manganese oxides that forms a veneer on stable rock surfaces in arid environments. Thought to form through the interaction of wind, dew, and microbial activity on the exposed surface. (6)

differential erosion Erosion that occurs at differing rates caused by differences in the resistance and hardness of rocks exposed to the atmosphere; softer and weaker rocks erode rapidly, commonly forming slopes; harder and more tightly cemented rocks erode slowly, commonly forming cliff faces or ledges. (3)

dikes Tabular or sheet-like bodies of solidified magma that cut through and across the layering of country rock, commonly emanating from a magma chamber; form by forcible injection into existing fractures or by creating their own before cooling and solidifying. (11)

disconformity An unconformity within parallel layers of sedimentary rock, representing a phase of erosion or nondeposition; may be marked by a gap in the fossil record, incised stream channels, paleosols, or other features of subaerial exposure. (3)

dissolution In a geochemical sense, the breakdown of a solid mineral or rock into its component ions by acidic groundwater; the ions are carried away in the groundwater solution leaving a void. (19)

divergent plate boundary Locations where two plates move apart from one another due to extensional forces; occurs by seafloor spreading along mid-ocean ridges or by normal faulting along continental rift zones. (1)

dolomite A mineral [$CaMg(CO_3)_2$] similar to calcite ($CaCO_3$) but with a significant amount of magnesium; typically forms as magnesium-rich waters interact with limy sediment or limestone, forming a sedimentary rock also called dolomite. (24)

dome A convex-upward landform commonly formed by exfoliation jointing within granitic rock. (18)

dormant volcano A volcano that has been inactive in historic time but that has the potential to erupt again in the future. (32)

drainage basin (aka *watershed*) The total land area where surface runoff from rainfall and snowmelt drain through countless rivulets and creeks into larger tributary streams that in turn merge into a single main river. (2)

drainage reversal The reversal of flow direction by a river due to either stream capture or a change in gradient due to tectonic activity within the watershed. (34)

ductile behavior A style of deformation where rock will bend or flow without breaking; folds in rock are common examples of a ductile response to tectonic stress; plastic flow of rock under high temperatures and pressures is another example of ductile behavior. (23)

earthquake The sudden release of pent-up energy along a fault resulting in the radiation of seismic waves outward from the point of rupture; commonly expressed as groundshaking on Earth's surface. (1)

earthquake epicenter The location on the surface directly above the focus of an earthquake.

earthquake focus The location along a fault plane at depth where rupture begins, triggering an earthquake. (26)

earthquake swarm A succession of low- to moderate magnitude earthquakes that occur within a local area over a relatively short period of time, lasting days, months, or years; common near volcanic and hydrothermally active regions. (31)

effusive eruption Volcanic eruption dominated by the outpouring of streams of lava onto the surface from a vent. (12)

element Fundamental substance composed of atoms with the same number of protons in their atomic nucleus; 94 naturally occurring elements exist on Earth. (1)

emergent coast A coastline that has been tectonically uplifted above sea level or that has been relatively elevated by a fall in sea level. (29)

end-Permian mass extinction (aka *Permian-Triassic extinction* or *P-T extinction*) Earth's most severe mass extinction, with up to 96 percent of all marine species and 70 percent of all terrestrial vertebrate species becoming extinct; marks the boundary between the Paleozoic and Mesozoic eras. (3)

eon Longest formal division of the geologic time scale; the four eons are the Hadean, Archean, Proterozoic, and Phanerozoic. (1)

ephemeral stream A stream that flows for a short interval of time in response to a burst of rainfall or rapid snowmelt in the region; flow may be minimal or as intense as a flash flood. (10)

epoch Unit of geologic time within a longer period; commonly used in the subdivision of Cenozoic time due to a higher preservation and diversity of fossils; we currently live in the Holocene Epoch. (1)

era Unit of geologic time shorter than an eon but longer than a period; for example, the Phanerozoic Eon comprises the Paleozoic, Mesozoic, and Cenozoic eras. (1)

erosion The physical removal of a weathered particle away from its source, perhaps by running water, wind, or gravity. (1)

eruption column The towering, rapidly rising pillar of pyroclastic debris that surges upward above an exploding volcano. Higher in the atmosphere, an eruption cloud of fine ash may drift for thousands of kilometers downwind, often encircling the Earth within days. (13)

escarpment An elongate cliff or steep slope that forms in response to faulting or erosion; may separate two relatively level areas or may express the fault scarp of a tilted fault-block mountain range. (2)

estuary Former stream valleys now flooded by the postglacial rise in sea level; marked by the mixing of freshwater from land and saline water brought inland by tides. (29)

exfoliation jointing (aka *onion-skin weathering*) A style of fracturing caused by the release of confining pressure from overlying masses of rock by erosion that causes underlying rock to expand and fracture parallel to the surface; commonly forms granitic domes. (18)

exhumation The process by which rocks buried within the crust are raised up toward the Earth's surface; may occur during phases of mountain-building or may occur isostatically in response to erosion. (1)

explosive eruption Violent expulsion of pyroclastic debris into the atmosphere from a volcano; typically driven by high gas pressure within the magma. (12)

extensional stress Force that stretches bodies of rock laterally, causing it to deform by breaking along normal faults, resulting in the horizontal "widening" and the vertical "thinning" of the mass of rock; dominant force near divergent plate boundaries. (1)

extinct volcano A volcano that hasn't erupted in the past 10,000 years or so and that is not expected to erupt in the future. (32)

fall line A low escarpment that marks the geologic boundary between the resistant crystalline rocks of the Piedmont and the feather edge of unconsolidated Cenozoic sediments of the Atlantic coastal plain; characterized by low waterfalls, rapids, and the location of several major cities. (35)

Farallon tectonic plate An ancient oceanic plate that subducted beneath western North America beginning about 200 Ma; primary tectonic source of uplift of the North American Cordillera. Modern remnants of the Farallon plate include the Juan de Fuca and Cocos minor plates. (17)

fault Planar fracture in a body of rock along which movement has occurred. (1)

fault creep The slow, but near constant, movement of rock on either side of a fault plane; may occur along fault planes with minimal frictional strength due to the presence of fibrous minerals. (28)

fault reactivation Renewed motion along a fault plane after a prolonged phase of quiescence. Motion along the fault may exhibit the same sense of offset at a later time, or an old fault may express itself with a different sense of motion if the tectonic stress regime changes. (10)

fault scarp A topographic step or offset on the ground surface that is the exposed physical expression of a fault plane that has moved vertically during earthquakes. (23)

faunal succession Principle of relative age dating that states that in an undeformed succession of sedimentary rock layers, the fossil record will be an orderly, nonrepetitive succession of fossils from one layer to the next; basis for correlation of fossil assemblages between distant localities. (1)

fissure eruption The release of lava from elongate fractures on the surface of a volcano or within a rift valley; sourced from magma moving through fractures that connect with the underlying magma chamber. (32)

fjard A U-shaped glacial valley along a coastline that has been flooded by the postglacial rise in sea level; differs from fjords in that it is broad and shallow rather than deep and steep-walled. (36)

flank eruption The release of lava or pyroclastic debris from the margins of a volcano rather than from a crater at the apex; may occur from fissures or from isolated vents. (32)

flash flood In the American West, a short-lived, turbulent flow of water through a narrowly confined channel like a slot canyon; may be triggered by intense rainfall far upslope in the drainage basin. (4)

flat-slab subduction A model for widespread mountain-building that involves a decrease in the angle of descent of an oceanic plate beneath a continent; commonly invoked as the tectonic geometry for the compressional Laramide Orogeny of the American West. (20)

flood basalt Eruptions of fluid basaltic lava from networks of fissures that may accumulate to great volumes and broad lateral extents; may occur over geological brief periods of time. (36)

flood stage The level of a river or stream when the water surface has risen high enough to overtop its banks, inundating its floodplain; for rivers cut into bedrock, flood stage refers to the highest volume flows when the velocity of flow is much higher than normal. (3)

flowstone A sheet-like speleothem formed as saturated groundwater flows over the surface of a cave wall; may also form downward-hanging curtains from groundwater dripping from a crack in a cave ceiling. (19)

fold A bend in layered rock formed beneath the Earth's surface in response to compressional stress. (6)

fold-and-thrust belt A zone of deformed rock characterized by folds and thrust faults that form elongate mountains, typically adjacent to and inland from a larger mountain belt. (22)

foliation A planar fabric in some metamorphic rocks created by the alignment of mineral crystals perpendicular to the primary orientation of stress. (21)

foreland basin An elongate depression that develops adjacent and parallel to a mountain belt; formed by the flexural downwarping of the lithosphere in response to the load emplaced by the nearby mountains; commonly filled by sedimentary debris from the adjacent highlands. (33)

formation A layer of sedimentary rock with distinct properties that enable it to be recognized and mapped over a broad region. (3)

fossil The preserved remains, trace, or imprint of an ancient form of life. (1)

fossilization The process by which evidence of an ancient life form is preserved in the sediment of a depositional environment where the organism once lived; usually requires rapid postmortem deposition to preserve hard physical remains. (9)

frost wedging A process of physical weathering whereby water infiltrates into rock along joints, freezing and expanding at night and thus wedging the rock apart; may repeat daily or seasonally, widening the joint plane until large masses of rock fragment into smaller masses. (5)

fumarole Vents on volcanoes or recently active lava flows or pyroclastic deposits from which volcanic gases escape. (14)

Ga Giga-annum or billions of years; used to designate specific *points* in geologic time in contrast to "b.y.," which signifies a *duration* of billions of years. (1)

geologic province A physiographic region characterized by its distinct geology, topography, and landscapes. (2)

geologic time The vast span of time beginning with the formation of the earliest Earth ~4.56 Ga through today. (1)

geologic time scale A chronological ordering of Earth history into blocks of time such as eons, eras, periods, and epochs; uses a combination of relative and numerical age-dating techniques. (1)

geyser A hot spring that intermittently erupts jets of steam and hot water; created by the heating of groundwater in contact with hot rock surrounding a magma body. (31)

geyser basin Broad, flat areas notable for multiple hydrothermal features and a surficial coating of light-colored sinter; common in Yellowstone National Park. (31)

glacial abrasion Process of glacial erosion whereby particles of sand and silt entrained in the basal ice grind at the underlying bedrock as the glacier advances, smoothing the surface and removing surficial rock. (18)

glacial erratic A boulder transported by glacial ice that is left behind during melting and glacial retreat; used to trace glaciers back to their source terrain. (18)

glacial flour Silt-sized sediment produced by glacial abrasion; commonly remains suspended in meltwater ponds, turning the water a milky turquoise color as the fine particles scatter sunlight. (22)

glacial plucking (aka *quarrying*) Process of glacial erosion whereby large blocks of bedrock are removed by the moving ice as water infiltrates into underlying joints and then freezes to progressively wedge the rock away; the block is eventually incorporated into the body of the glacier, then transported downflow. (18)

glacial polish Smooth, shiny surfaces on glaciated bedrock formed by repeated abrasion of fine particles of sediment entrained within the basal ice. (18)

glacial rebound The slow, upward movement of the land surface after a mass of ice has receded from the region; an isostatic response to the removal of the ice load. (38)

glacial striations Subparallel scratches on glaciated bedrock formed by the gouging action of particles of rock and sand within the basal ice; used to determine the direction of glacial movement. (18)

glacial till Chaotic, unsorted assemblage of sediment deposited by a receding glacier, ranging from fine muds to large boulders; commonly mounded to form moraines. (18)

Global Positioning System (GPS) Satellite-based system that provides precise locations and elevations of points on Earth where there is an unobstructed line of sight to four or more GPS satellites. Current resolution is to within 30 cm (12 in) of the exact location. (29)

gneiss A coarse-grained, foliated metamorphic rock composed of light bands of quartz and feldspar alternating with darker bands of micas and other dark-colored minerals. (1)

Gondwana Ancient supercontinent consisting of the conjoined landmasses of Africa, South America, Antarctica, Australia, and India; fully formed ~550 Ma and remained intact until rifting began ~180 Ma. (35)

gooseneck Informal term describing the shape of a highly sinuous river channel, commonly incised into bedrock. (8)

granite A coarse-grained plutonic igneous rock dominated by quartz and feldspar; the main rock type of the continental crust. (1)

Great American Outdoors Act A landmark law passed by Congress in 2020 that provides permanent funding for federal, state and local governments to acquire land and water for conservation and 5 years of funding to ease the maintenance backlog at national parks. (40)

Great Unconformity In western North America, a widespread surface separating Precambrian crystalline rocks from overlying sedimentary rocks of Paleozoic or Mesozoic age; depending on location, may represent a gap in the rock record of >2 billion years to hundreds of millions of years. (3)

greenstones Basaltic rocks that have been metamorphosed to a greenish shade by the formation of chlorite and epidote. (36)

Grenville Orogeny Prolonged phase of mountain-building and continental accretion that occurred ~1.1 Ga along what is today the Atlantic continental margin; with continental rifting ~750 Ma, a massive volume of continental rock called the Grenville Terrane remained attached to eastern North America. (35)

groundshaking Physical motion of the land surface in response to seismic waves produced during earthquakes. (23)

grus Granular sediment derived from the weathering and erosion of granitic rock, commonly in semiarid and arid climates. (27)

gypsum A common mineral ($CaSO_4 2H_2O$) formed by the evaporation of saline water from enclosed lakes and semi-enclosed seas; often forms sedimentary layers. May also precipitate from hot springs and saturated groundwater. (39)

half-graben basin A wedge-shaped, asymmetric depression located along the downrotated margin of a tilted fault-block; fills with sediment shed from the nearby range. (23)

half-life Length of time for the number of parent atoms of a radioactive element to decay to exactly half that number of daughter atoms; a critical value used in radiometric age dating. (1)

halite (aka *rock salt*) A common mineral (NaCl) formed by the evaporation of water from enclosed lakes and semi-enclosed seas; leaves behind a broad sedimentary layer of salt that, after repeated evaporative events, may reach thicknesses of over a kilometer. Upon burial, masses of salt may flow laterally and upward due to the overburden pressure. (7)

hanging valley A tributary to a U-shaped glacial valley which, instead of joining at the same level as the main valley, enters at a higher elevation; formed due to a slower rate of glacial erosion than the larger glacier that once filled the main valley, commonly has waterfalls dropping from its lip. (19)

harmonic tremor Continuous release of seismic energy due to the underground movement of magma or the motion of volcanic gases through the natural plumbing system beneath volcanoes. (13)

haze A form of air pollution produced when fine particulate matter (such as sulfate aerosols) interacts with sunlight to scatter light and diminish visual clarity. (40)

head The upper portion of a stream channel where water collects and begins to flow downslope. (5)

headward erosion Process by which a stream increases its length by eroding in the upslope direction near the head of the stream; occurs where the velocity of runoff abruptly increases, increasing the erosive power of the water. (5)

heat flow A measure of the heat emitted from a particular area on the surface of the Earth over a specified length of time. (23)

hematite A common mineral composed of iron oxide (Fe_2O_3) that forms in a wide variety of chemical environments; commonly forms a surface coating on sediment grains, turning the rock various shades of red. (22)

hogback ridge A long, narrow ridge with a narrow crest and steep slopes formed by the differential erosion of steeply tilted sedimentary strata. (9)

Holocene The current epoch of geologic time, beginning 11,700 years ago; marked by a globally equable climate and glacial minima. (14)

hoodoo Informal term used to describe a tall, thin spire of rock with layers of variable thickness, giving it the appearance of a totem pole; an erosional remnant of a larger body of rock. (5)

horn Sharp, pyramidal peaks formed at the intersection of three or more glaciers; steep flanks are the remnants of the headwalls of glacial cirques. (21)

horst-and-graben A landscape of elongate valleys and ridges formed by movement along parallel normal faults. The upraised blocks of rock composing ridges are called *horsts*, whereas the downdropped blocks of rock composing valleys are called *graben*. (8)

hot spring A pool of hot water underlain by a natural plumbing system that permits heated water to reach the surface as a continual supply. (16)

hotspot A localized region marked by high heat flow and active volcanism thought to be fed by magma from the underlying mantle; magma may be derived from mantle plumes or from the passive rise of magma through extensional fault networks. (31)

hotspot trail A linear track of extinct and deeply eroded volcanic edifices that extend from the stationary hotspot; the volcanic materials become older with distance from the hotspot. Formed by the motion of a lithospheric plate over the hotspot. Hotspot trails are indicators of the absolute direction and speed of a plate through time. (31)

hydrosphere All of Earth's water, including liquid water in oceans, lakes, rivers, and groundwater, frozen water in glaciers and ice sheets, and atmospheric water vapor. (1)

hydrothermal alteration The interaction of acidic hydrothermal fluids with volcanic rock that chemically changes certain minerals within the rock to clay; the alteration weakens the rock, contributing to landslides or even catastrophic collapse of the flank of a volcano. (14)

hydrothermal explosion An event that occurs when superheated water within a volcano's hydrothermal system flashes to steam, exploding upward, shattering rock, and blasting the debris into the air through a vent. (31)

Iapetus Ocean Ancient precursor ocean to the Atlantic that existed from ~600 Ma to ~400 Ma; the Iapetus oceanic lithosphere was consumed by the time of the Acadian Orogeny. (35)

ice cap A mass of glacial ice that covers less than 50,000 km² (20,000 mi²) of land area, typically crowning the crest of mountains or high-latitude regions; common Ice Age feature of the Sierra Nevada, Rocky Mountains, and Yellowstone Plateau. (17)

igneous rocks Formed by the cooling and crystallization of molten rock; includes volcanic rocks deposited on the Earth's surface and plutonic rocks that solidify beneath the surface. (1)

ignimbrite (aka *welded tuff* or *ash-flow tuff*) Volcanic rock formed by widespread deposition and solidification of material in a pyroclastic flow; heat from the deposit welds the particles together. (23)

impact crater Near-circular depressions on the surface of Earth or other bodies of the solar system formed by the high-velocity impact of an extraterrestrial object such as an asteroid or a comet. Most impact craters on Earth have been destroyed by tectonism or erosion. (8)

incised meander (aka *entrenched meander*) Looping meanders of a stream or river that have cut deeply into bedrock; requires a fall in base level, perhaps caused by a fall in sea level, isostatic uplift, or tectonic uplift. (8)

inselberg Isolated monolith of rock and large boulders that rises above a broad plain of alluvial sediment; remnant erosional feature of arid climates. (27)

intermontane basin A broad region surrounded by mountain ranges, partly filled by alluvium derived from the nearby highlands; may be high in elevation. (5)

internal drainage A drainage system where water flow ends at enclosed lakes within a watershed, without any connection to the open ocean; surface water either seeps into the ground or evaporates. (23)

intrusive complex A heterogeneous assemblage of plutonic rock types that compose a large region of continental crust. (38)

iron meteorites Fragments of the metallic cores of asteroids that were broken away by collisions that serendipitously intersected Earth's orbital path and survived the intense heating as they passed though Earth's atmosphere to land on the surface. (1)

isostasy A condition of gravitational equilibrium, similar to floating, in which a mass of lighter lithospheric rock is buoyantly supported from below by denser rock of the asthenosphere. The elevation of the crust depends on its thickness and density contrast with the asthenosphere. (17)

isostatic uplift The slow rise of continental crust in response to the removal of an overlying load. For example, continual erosion of rock from the crest of mountains allows for the uplift of the deeper crustal roots of the mountains; the melting and retreat of a continental ice sheet permit the gradual uplift of underlying crust. (17)

joint A fracture within a body of rock commonly occurring in sets of parallel, evenly spaced fractures; ubiquitous feature that acts to focus the agents of weathering and erosion within bodies of rock. (4)

ka Kilo-annum or thousands of years; used to designate specific *points* in geologic time in contrast to "k.y.," which signifies a *duration* of thousands of years. (1)

karst landscape A pitted and irregular topography formed from the dissolution of limestone, dolomite, or marble near the land surface; characterized by sinkholes, caves, and "disappearing" rivers that drain into voids in the underlying rock. (19)

kipuka Areas of forest woodland surrounded entirely by lava fields on the Hawaiian Islands. (32)

knickpoint An abrupt change in slope along a drainage channel that acts as the focus of headward erosion by a rivulet, stream, or river; the sharp increase in flow velocity across the knickpoint enhances localized erosion and channel incision. (8)

lahar (aka *volcanic mudflow*) Slurry of water mixed with volcanic debris that moves rapidly down the flanks of volcanoes, commonly through stream channels; may be triggered by rapid melting of snow and ice by pyroclastic eruptions, intense rainfall on loose volcanic ash deposits, failure of a lake dammed by volcanic deposits, or as a consequence of debris avalanches. (13)

landslide A general term for a variety of mass wasting events, including rockfalls, rockslides, slumps, mudflows, and debris flows; driven by gravity acting on unstable cliff faces or slopes. (25)

Laramide Orogeny A phase of mountain-building in the American West that began around 80 Ma and ended around 40 Ma, although the exact timing varies regionally; widespread compression likely occurred in response to flat-slab subduction of the Farallon oceanic plate beneath western North America. (6)

Last Glacial Maximum The most recent phase of peak glaciation of the longer Pleistocene Ice Ages, peaking between 18,000 and 21,000 years ago. (14)

lateral blast Violent sideways explosion from the flanks of a volcano; may rapidly devastate broad areas of the surrounding countryside by impact, burial, and heat. (13)

lateral continuity A principle of relative age dating that states that layers of sedimentary rock were deposited as laterally continuous sheets or wedges; if the layers were subsequently dissected by erosion or deformed during uplift, one can assume that they were once originally continuous. (9)

lateral moraine Elongate, sharp-crested ridges of glacial debris that accumulate along the flanks of alpine glaciers. (21)

lava General term for molten rock that flows across the surface of volcanoes from effusive eruptions. (1)

lava bomb Elongate masses of solidified lava that acquire their aerodynamic shape during travel through the air after expulsion from a volcano. (32)

lava dam Outpourings of lava into a river or stream that impede the flow of water; the dam may create a lake or pond, or it may eventually fail from weathering and erosion to produce outburst floods. (34)

lava dome A rounded, steep-sided mass of volcanic rock that forms by the accumulation of viscous, gas-depleted, dacitic, or rhyolitic lava near the vent of a volcano. If located within the crater of a volcano, domes may act as a plug to trap gases within the underlying magma chamber. May form entire volcanic edifices such as Lassen Peak. (13)

lava falls Falls that occur when lava flows drain over the edge of a cliff or deep river canyon. (34)

lava fountain Jets of lava that shoot upward from vents or along fissures, driven by the expansion and escape of gases from the magma. (32)

lava lake Roiling pools of molten rock that form within the crater of some volcanoes; typically basaltic in composition. (32)

lava tube Elongate caves or passageways left behind as lava drains downslope beneath an insulating crust of solidified lava. (32)

left-lateral slip Sense of motion on a strike-slip fault in which the fault block on the far side of the fault line has moved to the left relative to the block on the near side. (24)

LiDAR Light Detection and Ranging; a surveying method that employs pulses of light energy to measure the distance to a target; used to make high-resolution maps. (29)

limestone A biochemical sedimentary rock formed by the accumulation of hard, shelly remains of invertebrates on the seafloor (or lakebed, in some cases); with burial and cementation, the limy sediment solidifies into limestone. (1)

linear sea A narrow seaway that commonly forms as an intermediate phase between continental rifting and a broad, open ocean basin; examples include the Red Sea and the Gulf of California. (1)

listric normal fault A type of rotational normal fault where the fault plane curves and flattens with depth, commonly along the brittle-ductile transition within the crust. (23)

lithosphere The rigid outer surface of Earth composed of the crust and upper mantle; about 100–150 km (60–90 mi) thick; forms tectonic plates. (1)

Little Ice Age A phase of natural climatic warming that primarily affected the Northern Hemisphere between ~1350 and ~1850 CE. (14)

locked fault An active fault held in place by the frictional strength of rock along the fault plane; the long-term "stick" phase of the stick-slip earthquake cycle when tectonic stress accumulates along the fault. (30)

longshore drift A zigzag pattern of sand movement laterally along a beach driven by waves striking a coastline at a slightly oblique angle. (29)

Ma Mega-annum or millions of years; used to designate specific *points* in geologic time in contrast to "m.y.," which signifies a *duration* of millions of years. (1)

magma Molten or partially molten rock found beneath Earth's surface. (1)

magma chamber A large reservoir of molten rock, typically located within the upper 20 km (12 mi) of the crust. Magma may rise to the surface where it is erupted via volcanism, or it may slowly cool and solidify within the chamber to become a pluton. (1)

mantle The hot but solid layer of rock beneath Earth's crust but above its outer metallic core. The rock of the mantle can slowly flow convectively in response to intense heat and pressure; comprises about 84 percent of Earth's volume. (1)

mantle plume A convectively driven current of hot rock that buoyantly moves upward through the mantle by plastic deformation; produces magma by decompression melting as the plume moves higher into the crust, feeding volcanism at the surface, commonly at hotspots. (31)

marble A nonfoliated rock formed from the metamorphism of limestone or dolomite. (1)

marine terrace Relatively flat or gently seaward-sloping surfaces originally formed along the shoreline by wave abrasion but now elevated tens of meters above sea level; they record the interaction between tectonic uplift and sea-level change. (29)

meandering stream A channel morphology typically developed on low gradient slopes where the river forms a sinuous series of loops and bends. Meanders form as fast-moving water on the outside of a slight bend erodes sediment from the outer bank, redepositing it downstream on an inner bank to form a point bar. Meanders grow in sinuosity as this process continues through time. (8)

megathrust An informal term that describes the massive thrust-type architecture of a subduction zone; the Cascadia subduction zone is commonly referred to as a megathrust where the North American plate is being thrust over the descending Juan de Fuca and smaller microplates. (30)

mélange A chaotic mass of rock characterized by large angular fragments of rock "floating" in a fine-grained matrix; marked by a lack of obvious bedding, serpentinite fragments, and a greenish tinge; forms within the accretionary wedge and is later exhumed to the surface along thrust faults. (30)

melt The liquid phase of magma; other components of magma are solid crystals formed earlier from the magma that float in the melt and gases dissolved in the melt. (12)

mesa A broad, isolated, flat-topped landform surrounded on all sides by steep escarpments; formed by the erosional dissection of a larger plateau composed of sedimentary strata. (8)

Mesozoic Era of geologic time beginning ~252 Ma and ending 66 Ma; bracketed with a mass extinction at its start and its end. (1)

metamorphic rock Major group of rock formed by the solid-state transformation of preexisting rock under conditions of high pressure and temperature beneath Earth's surface; may also form from interaction with hot fluids. (1)

metasedimentary Sedimentary rocks that have been mildly metamorphosed, retaining their bedding and most original depositional characteristics. (22)

microplate Tectonic plates that are much smaller than major plates (like the Pacific) and even smaller than minor plates (like the Nazca); commonly become sutured to larger plates as an accreted terrane. (12)

mid-ocean ridge system A 60,000 km (37,000 mi) long, low-relief mountain chain that marks divergent plate boundaries, sites of submarine volcanism, and seafloor spreading; elevated above the adjacent seafloor by heat from the underlying asthenosphere. (1)

migmatite Metamorphic rock composed of tightly folded, looping bands of minerals reflecting high temperatures and pressures; texture of the rock created by plastic flow. (1)

mineral The building block of rocks; a naturally occurring, inorganic, crystalline solid. (1)

monocline A step-like fold in sedimentary strata where horizontally bedded rock abruptly bends downward at a steep angle before returning to horizontality a short distance away; commonly form as a drape above buried reverse faults. (6)

moraine A ridge of unconsolidated and unsorted till, usually deposited during the melting and retreat of a glacier. Terminal moraines mark the maximum advance of the glacier prior to receding. Recessional moraines are smaller ridges of till left as a glacier pauses during its retreat. (14)

moraine lake A body of freshwater impounded behind a morainal ridge; commonly forms within U-shaped valleys during glacial retreat. (25)

mouth The part of a stream or river where it empties into another river, a lake, or an ocean; opposite of the head of a stream. (5)

mudcracks Shallow, sediment-filled cracks within fine-grained sedimentary rocks formed by the exposure and desiccation of muddy tidal flats or lakebeds during deposition; may form polygonal patterns on bedding plane outcrops. (22)

mud pot Bubbling pools of gray, clay-rich water supplied by an underlying source of steam. The release of gases from within the slurry produces the bubbling surface of the pool. (16)

mudstone A fine-grained sedimentary rock composed of clay-sized particles. (9)

national monument A region set aside to preserve a single unique feature of either cultural or natural significance;

designated by presidential approval through the power of the Antiquities Act of 1906. (5)

national park Region of exceptional scenic and natural beauty set aside for purposes of conservation and public enjoyment; requires congressional approval for designation. (5)

natural arch An erosional landform created by weathering along a set of joints, commonly in sandstone or limestone. (7)

nonconformity A type of unconformity separating crystalline rocks below from sedimentary rocks above; represents a phase of subaerial exposure followed by the deposition of sediments on the exposed surface. (3)

normal fault A type of inclined fault formed by extensional stress; the block of rock above the fault plane moves downward relative to the block of rock beneath the fault plane. (5)

numerical age dating Technique of placing absolute numerical values on a rock unit using the methods of radiometric dating; when integrated with relative age-dating techniques, permits the refinement of the geologic time scale. (1)

nunatak Rocky island surrounded by flowing glacial ice. (18)

oceanic crust The part of the crust composed of dense basalt and gabbro; sinks deeper into the underlying mantle relative to continental crust, forming basins that hold the waters of the world ocean; ranges in thickness from 7 to 10 km (4 to 6 mi). (1)

oceanic-oceanic convergent boundary Plate interaction involving the compression of an oceanic lithospheric plate against another oceanic plate, resulting in the subduction of the older and thus denser oceanic plate; results in the destruction of oceanic lithosphere by descent into the mantle. (1)

offset (aka *displacement* or *slip*) The relative movement of geologic or topographic features along either side of a fault plane; commonly occurs during individual earthquakes. Offset during any one earthquake may range from less than a meter to several meters. The long-term cumulative offset along a major fault may be a few hundred kilometers. (22)

offset channel A stream channel that crosses a fault and that has been displaced laterally by episodic earthquakes; the angular pattern of the channel provides a distinctive topographic feature that marks the trace of faults. (26)

Organic Act A federal law that established the National Park Service in 1916 to protect and preserve U.S. national parks while also making them available for the enjoyment of the public. (40)

original horizontality A principle of relative age dating that states that layers of sediment are originally deposited horizontally due to the influence of gravity on the settling of grains. (1)

orogenesis The process of mountain-building by tectonic forces, resulting in the deformation, magmatism, metamorphism, and uplift of bodies of rock. (35)

orogeny A distinct phase of mountain-building due to tectonic activity. (6)

outburst flood Catastrophic flow of a large volume of water released from a lake or dammed river; may occur in response to the failure of a lava dam, a glacial ice dam, or a glacial moraine. (34)

outwash plain Broad, generally flat landscape composed of glacial sediment that occupies the region in front of melting glaciers; deposited by meltwater streams that drain outward from the toe of the receding glacier. (14)

ozone A colorless and odorless gas consisting of three oxygen atoms bonded together; ozone at ground level has a wide variety of deleterious effects on human health as well as on other animals and plants. (40)

pahoehoe lava Hawaiian term for a lava flow that forms a surface texture like a coiled rope; typically forms from low viscosity, basaltic lavas. (32)

paleocanyon An ancient canyon preserved by river-deposited sediments, such as rounded gravels. Paleocanyons record the path of old river systems and thus relict landscapes. The age of the sediment permits determination of the age of river incision of the paleocanyon.

paleogeographic map A visualization that shows a speculative ancient geography of continents, ocean basins, offshore islands, mountain ranges, and other physiographic features at a specific point in the geologic past; based on the distribution and composition of age-appropriate rocks, assemblages of fossils, the magnetic properties of rocks, and other geologic data. (3)

paleoseismology Technique for estimating the occurrence and frequency of past earthquakes; involves digging a trench perpendicular to the fault plane, recognizing indicators of past offset, and radiocarbon dating of organic materials to determine the age of the event. (25)

paleosol Ancient soils preserved by burial beneath sediments or volcanic deposits; typically solidified into rock. (5)

Paleozoic Era of geologic time beginning ~541 Ma and ending ~252 Ma; began with an abrupt radiation of life forms and ended with a mass extinction. (1)

pali Elongate escarpments that cross the Big Island of Hawaii subparallel to the coastline; formed as fault scarps of ancient gigantic slumps of the volcanic edifice toward the ocean. (32)

Pangea A supercontinent that lasted over 100 m.y. from ~300 Ma to ~200 Ma; the breakup of Pangea led to the dispersal of today's island continents. (6)

parent element The original unstable element that breaks down through radioactive decay to a daughter element. (1)

partial melting The incomplete melting of a rock where those minerals with the lowest melting temperatures become liquid, but those with higher melting temperatures remain solid. (1)

passive continental margin Continental margins that occur distant from tectonic plate boundaries; marked by slow subsidence, deposition of sediment, and lack of volcanism or seismicity. (24)

paternoster lakes Chains of small lakes that occupy formerly glaciated valleys; form from the accumulation of meltwater within small bedrock basins scoured by glacial erosion; often connected by small streams. (22)

pebble A rounded particle of sediment that is between 4 and 64 mm (0.075–2.5 in) in size; larger than sand grains but smaller than cobbles. (1)

pegmatite A type of intrusive igneous rock exhibiting large, interlocking crystals; commonly expressed as dikes cutting across older rock. (11)

periglacial The region in close proximity to the margins of alpine glaciers, ice caps, and continental ice sheets; marked by distinct ecological conditions and environments. (36)

period Time interval in the geologic time scale lasting tens of millions of years that is a subdivision of an era; typically defined on the basis of the fossil record. (1)

petrified wood Product of fossilization where the organic materials composing the cell walls of a tree are slowly replaced at the molecular scale by atoms of silica (SiO_2) transported by groundwater, commonly preserving the finest details. (31)

Phanerozoic Eon Unit of geologic time comprising the Paleozoic, Mesozoic, and Cenozoic eras. (1)

phreatic eruption Steam-driven explosions that occur when water beneath the surface or on the surface is suddenly heated by contact with magma or lava; the water will boil and flash to steam, generating an eruption of steam, water, ash, and blocks of rock. (13)

phreatic zone The zone of saturation below the water table where pores and cavities within rock are filled with groundwater. (39)

phyllite Type of foliated metamorphic rock that forms from the progressive metamorphism of slate; flakes of fine-grained mica minerals align in an orientation perpendicular to the orientation of stress. (1)

physical weathering In-place decomposition of a rock or its constituent minerals by the mechanical activity of ice that wedges blocks apart along fractures by freeze–thaw cycling, physical abrasion by water, ice and wind, the infiltration of plant roots along fractures, and thermal stress. (1)

pillow basalt Elongate masses of volcanic rock formed by the underwater eruption of lava; composes the majority of the basaltic seafloor. (26)

pinnacle An individual column of rock formed by the complete erosion of surrounding rock, leaving an isolated remnant. (8)

pinning point (aka *piercing point*) A distinct geologic or topographic feature on opposite sides of a fault that can be used to calculate the offset and slip rate along a fault. (28)

plastic behavior (aka *ductile behavior*) In rock, occurs when mineral grains change shape and volume in response to a stress and do not return to their original shape when the stress is removed; rock may flow as a solid plastic in response to high temperatures and pressures. In ice, occurs within the body of glaciers when the weight of overlying ice causes underlying crystals of ice to deform and slowly flow. (1)

plate Tabular slabs of Earth's lithosphere that are in constant slow motion relative to one another; may be thousands of kilometers in width but only 50–150 km (30–100 mi) thick. (1)

plate tectonics Unifying theory of the geosciences that explains the opening and closing of ocean basins, the uplift of mountains, and the occurrence of earthquakes and volcanoes. A modern synthesis derived from the precursor concept of continental drift. (1)

plateau A broad highland marked by a relatively level upper surface. (2)

playa A dry lakebed commonly crusted with evaporitic salts or mud; common in regions marked by internal drainage. (23)

Pleistocene Ice Ages The epoch of geologic time that began around 2.6 m.y. ago and that ended 11,700 years ago; marked by 20–30 glacial-interglacial cycles, each lasting tens of thousands of years. (14)

Plinian eruption Explosive pyroclastic eruptions that form dark pillars of tephra and gas that reach high into the atmosphere before broadening laterally into diffuse clouds due to prevailing winds. (15)

pluton A body of intrusive igneous rock ranging in size from tens to meters to hundreds of kilometers across that forms when a magma chamber cools and solidifies. Multiple overlapping plutons compose larger masses called batholiths. (1)

plutonic igneous rocks Coarse-grained rock that forms by the slow cooling and crystallization of magma beneath the surface. Common examples include granite, granodiorite, and gabbro. (1)

pluvial lakes Large lakes formed during phases of cooler temperatures and higher precipitation that accompany glacial-interglacial cycles, particularly in otherwise semiarid regions. (23)

postglacial marine transgression The long-term rise in sea level and the associated landward migration of the shoreline that happens in response to the melting of continental ice sheets during interglacial warm phases. (29)

Precambrian A "supereon" of geologic time, spanning 88 percent of Earth's history, from 4.6 Ga to 541 Ma; comprises the Hadean, Archean, and Proterozoic eons. (1)

precipitation In a geochemical sense, the formation of a solid substance from a liquid saturated with ions in solution. (19)

pressure ridge Elongate, narrow ridges that parallel the trace of strike-slip faults; forms by localized areas of compression and uplift along the fault plane. (26)

Proterozoic An eon of the Precambrian beginning 2.5 Ga and ending 541 Ma; begins at the appearance of oxygen in Earth's atmosphere and ends at the transition to the Phanerozoic Eon. (22)

pull-apart basin A long, narrow depression formed where two parallel strike-slip faults (or a bend in a single fault) cause the crust to stretch and subside; generated by transtensional stress. (24)

pumice Porous, glassy rock formed from the froth at the top of a mass of magma; the froth is created as rapidly escaping gases bubble up through the rising body of magma. Fragments of pumice are ejected as a component of tephra during explosive eruptions. (12)

pygmy mammoth A dwarf form of mammoth created by genetic isolation on islands that were detached from the mainland by the postglacial rise in sea level; went extinct about 13,000 years ago. (27)

pyroclastic debris Fragments of volcanic ash, pumice, and blocks of rock blasted skyward during explosive eruptions of intermediate- to high-silica magma. (12)

pyroclastic eruption An explosive ejection of pyroclastic debris from a volcano; driven by gas pressure within the underlying magma chamber. (12)

pyroclastic fall Ash, pumice, and rock fragments erupted explosively eventually fall back to Earth's surface, forming a broad blanket that smothers the landscape; thins downwind with distance from the volcanic source. (12)

pyroclastic flow A hot (often >800°C), chaotic mixture of rock fragments, ash, pumice, and gases that travel rapidly downslope as a ground-hugging density current. Commonly forms from the vertical collapse of eruption columns. (13)

quartz veins Thin, tabular layers of quartz that cut though larger masses of granitic rock; formed from silica-rich fluids injected into fractures beneath the surface. (27)

radial drainage Spoke-like pattern of streams that radiate out from a mountainous peak such as a volcano. (13)

radiocarbon dating (aka *carbon dating* or *carbon-14 dating*) A type of radiometric dating that involves determination of the age of an object containing organic material; the half-life of the radioactive form of carbon in organic material is relatively short, limiting the technique to objects younger than about 50,000 years. (1)

radiometric dating Method that uses the natural radioactive decay of certain elements within minerals in combination with the unstable element's half-life to determine a numerical age of the mineral and thus the rock in which it is found. (1)

rain shadow An arid region located on the downwind side of a mountain range; air masses drop their moisture in the highlands, leaving the region in the shadow of the range devoid of precipitation. (23)

recurrence interval The long-term average duration between natural events; for earthquakes, the estimate is usually based on paleoseismic studies along specific faults. (30)

Regional Haze Rule A federal law passed in 1999 intended to reduce air pollution in U.S. national parks; implementation of the rule is spotty due to arbitrary wording and a lack of state accountability. (40)

regression The seaward migration of the shoreline and associated coastal and marine environments due to a fall in relative sea level; recognized by an upward shallowing of apparent water depths in a vertical succession of sedimentary rock. (3)

relative age dating A qualitative method of determining the age of a body of rock by following a set of principles, such as superposition; typically uses phrases such as "younger than" or "older than." When integrated with numerical age dates, it allows for the refinement of the geologic time scale. (1)

relative sea level The level of the sea surface with respect to the shoreline at any one location; changes are driven by increases or decreases in the volume of water in the oceans and by changes on land such as subsidence, uplift, or isostatic rebound. (3)

residual soil Soil that develops from the in-place weathering of the bedrock directly beneath it. (36)

resurgent dome Topographic mounds that rise a few hundred meters above the surrounding landscape due to the upward movement of magma through natural conduits toward the surface. Two resurgent domes exist above the Yellowstone magma bodies. (31)

reverse fault A type of inclined fault formed by compressional stress; the block of rock above the fault plane moves upward relative to the block of rock beneath the fault plane; differs from a thrust fault in that the angle of the fault plane is steep. (10)

rhyolite Light-colored volcanic rock with a high-silica (SiO_2) content; may erupt explosively to produce ash and pumice or may erupt effusively to form thick, blocky lava flows. (12)

rift basin An elongate, steep-walled trough formed by extensional faulting during the breakup of a continent. (9)

rift-to-drift transition The tectonic evolution from the end of continental rifting to the beginning of seafloor spreading and the development of passive continental margins; marked by the transition from rift–basin sedimentation to passive margin sedimentation. (35)

rift zone Narrow regions on the flanks of volcanoes marked by a dense concentration of subparallel fractures that act as natural conduits for the upward flow of magma to the surface. Flank eruptions commonly occur along rift zones of Hawaiian volcanoes. (32)

right-lateral slip Sense of motion on a strike-slip fault in which the fault-block on the far side of the fault line has moved to the right relative to the block on the near side. (24)

ring vents A concentric array of fractures that develop on the volcanic edifice prior to caldera eruptions; act as conduits for the explosive eruption of magma toward the surface. (15)

ripplemarks Linear or gently curving low ridges on bedding planes of sedimentary rock; formed as flowing water- or wind-organized grains of sand into ripples; commonly preserved by the deposition of finer grained muds on the rippled surface. (22)

river gradient The slope of a river measured by the ratio of the drop in elevation per unit horizontal distance; usually expressed as meters per kilometer or foot per mile. High-gradient rivers commonly have steep, narrow V-shaped valleys, whereas low-gradient rivers tend to have wider and less rugged valleys, often with the channel developing a meandering pattern. (8)

roche moutonnee An asymmetric mass of bedrock with a broad, smooth side formed by glacial abrasion and an abrupt, angular downflow side formed by glacial plucking. (18)

rock A naturally occurring solid mass composed of an aggregate of minerals; the three main types are igneous, metamorphic, and sedimentary. (1)

rock avalanche A type of rockslide where the rock fragments travel for long distances above a cushion of trapped air; may evolve from large rockfalls or rockslides. (4)

rock cycle The continual transfer and recycling of matter within the Earth and on its surface through time; a set of processes that convert each of the three main rock types into one another. (1)

rockfall A sudden downward movement of rock from a cliff or steep slope; masses of rock commonly separate from the main mass along joint surfaces; steep cliff faces are maintained by countless rockfalls occurring through time. (3)

rockslide A type of mass movement in which rock debris detaches and slides down a slope under the influence of gravity; may occur along an inclined bedding plane, fault, or joint surface. (18)

Rodinia A late Proterozoic supercontinent that formed about 1 Ga, eventually breaking apart by about 750 Ma. The assembly and breakup of Rodinia is recorded in the rocks of several U.S. national parks. (22)

roof pendant Erosional remnants of preexisting country rock that formed the roofs and walls of magma chambers; typically composed of metamorphic rocks named for the way they may have hung, pendant-like, downward into the magma body. (19)

rounding A feature of particles of sediment produced when angular fragments of rock are made less angular by physical abrasion during transport. (4)

rhyolite Light gray volcanic rock with a high silica content; extrusive equivalent to granite. Rhyolitic magma is highly viscous and may erupt as either thick blocky lava flows, forming volcanic domes, or as explosive pyroclastic events. (12)

sag pond A narrow pond or lake formed by transtension along a gently curving fault plane. (26)

salt dome The broad domal crest of a vertical column of salt that flowed upward from a tabular layer under the influence of overburden pressure; may cause a topographic bulge on the overlying landscape. (8)

salt flat A broad expanse of dry lakebed covered by salt and other evaporite minerals. (24)

salt wall An elongate mass of salt with the shape of a broad wall formed by the slow migration of a mass of salt under pressure; dissolution by fresh groundwater along its upper surface may result in collapse of overlying rock along normal faults, forming linear "salt valleys." (7)

sandstone A clastic sedimentary rock composed of naturally cemented particles of sand; typically dominated by quartz grains; an abundant and weather-resistant mineral. (1)

schist Type of foliated metamorphic rock that forms from the progressive metamorphism of phyllite; flakes of platy mica minerals align in an orientation perpendicular to the orientation of stress, creating a "scaly" texture. (1)

scoria Pebble-sized particles of magma ejected during Strombolian eruptions; similar to but denser than particles of pumice; commonly compose cinder cones. (16)

scree Veneer of angular rock fragments that accumulate along the steep slopes of volcanoes or steep escarpments; differs slightly from talus deposits in that the debris forming scree is usually of comparable size. (15)

sea arch An opening in a headland overlain by a thin bridge of rock; created by the merger of two sea caves. (29)

sea cave An alcove carved into the base of a sea cliff by erosion from waves and tides. (29)

sea level Commonly measured as the midpoint between mean low and mean high tide at a particular location; the datum for elevations on land and water depths in the ocean. The position of sea level has changed by hundreds of meters in the geologic past. (3)

sea stack A solitary mass of rock located in the surf zone of emergent coastlines; formed by the collapse of the bridge above a sea arch, leaving the headland isolated just offshore. (29)

seafloor spreading The process by which new oceanic lithosphere is produced at mid-ocean ridge divergent plate boundaries. All ocean basins are formed by this process. (1)

seamount An underwater mountain that commonly forms as extinct volcanoes subside beneath the sea through time. (32)

sedimentary basin A large-scale topographic depression in which layers of clastic and biochemical sediment accumulate through time; commonly consists of a wide variety of depositional environments. Space is continually created within the basin by subsidence. (1)

seismic tomography A computer imaging technique that utilizes differences in the velocity of earthquake waves to generate a visualization of Earth's interior; commonly used to image mantle plumes and subducting plates. (31)

seismicity General term that describes the distribution and frequency of earthquakes. (1)

serpentinite A greenish metamorphic rock derived from the chemical interaction of superheated fluids with peridotite, an igneous rock that composes a large volume of the mantle; typically accumulates within the accretionary wedge of subduction zones, ultimately to be uplifted and exposed at the surface. (26)

shale A fine-grained clastic sedimentary rock composed of clay minerals and silt-sized grains of other minerals; characterized by thin laminations along which the rock will break into plates; the most common sedimentary rock. (1)

shatter zone A semicontinuous band of rock near Acadia National Park composed of a wide variety of angular rocks surrounded by a matrix of granite; interpreted to record the collapse of the roof of a magma chamber into an underlying magma chamber during caldera formation. (38)

shear stress A set of opposing forces acting on opposite sides of a fault or larger-scale plate boundary; dominant stress orientation along strike-slip faults and transform plate boundaries. (1)

shield volcano A broad, gently sloping volcanic edifice built by countless, mostly effusive eruptions of basaltic lava. (32)

shutter ridge A linear ridge that was moved laterally to its current location by countless earthquakes along a strike-slip fault. (29)

silica Silicon dioxide (SiO_2), the most abundant rock-forming compound on Earth and the main molecular component of plutonic and volcanic igneous rocks; critical to the viscosity, gas content, and explosivity of magma. (12)

silicate minerals Minerals in which a variety of elements are bonded with silicon and oxygen. (1)

silicate rocks The most voluminous rocks within the body of the planet, due to the abundance of silicon and oxygen in Earth's crust and mantle. (1)

siliceous sinter A white-to-gray surficial deposit of silica (SiO_2) precipitated from silica-rich fluids overflowing near hot springs and geysers; may also line the conduits of fractures and pipes beneath hydrothermal features. (31)

siltstone A clastic sedimentary rock composed of naturally cemented particles of silt, whose size is larger than clay but smaller than sand. (1)

sink A general term for closed-basin lakes that collect water from internally drained streams and rivers. (23)

sinkhole plain A karst landscape characterized by abundant circular depressions created by dissolution of limestone near the surface. (39)

Skolithos Trace fossils in sandy sedimentary rocks that are interpreted to be vertical, cylindrical burrows dug by a marine invertebrate. (36)

skylight Opening in the roof of a lava tube created by the collapse of the surficial crust above the flow. (32)

slate Type of foliated metamorphic rock that forms from the metamorphism of shale or volcanic ash; very fine-grained with thin laminations oriented in planes perpendicular to the direction of metamorphic compression. (1)

sliding stone Small boulders resting at the end of long furrows gouged into the muddy surface of a playa lakebed in Death Valley National Park; thought to form by wind pushing thin panes of ice against the boulders. (24)

slip rate The long-term average speed of movement along a fault; determined by dividing the amount of offset along a fault by the age of the offset feature. (24)

slot canyon A narrow channel bordered by steep, high rock walls; subject to flash floods in semiarid climates, posing a danger to hikers. (6)

soil Small particles of sediment mixed with organic matter; most common types include alluvial and residual. (1)

sorting A feature of sedimentary rocks produced when particles of the same size, shape, or density are naturally selected and separated from dissimilar particles. (4)

soundscape The combination of natural and human-made sounds that determine the acoustic environment in a given area. (40)

speleothem Cave deposits, usually of calcium carbonate ($CaCO_3$) or calcium sulfate ($CaSO_4$) formed by chemical reactions of groundwater interacting with limestone, dolomite, or marble. (19)

spheroidal weathering Process whereby masses of granite cut by multiple joint sets are broken down and rounded by weathering along the angular edges; produces piles of rounded boulders in desert climates. (27)

spit Elongate tongues of sand extending from a beach by longshore drift; commonly forms barriers near the mouths of lagoons and estuaries. (29)

spring A natural outlet of groundwater onto Earth's surface. (4)

stalactite A downward-pointing speleothem hanging from the ceiling of a cave. (19)

stalagmite An upward-pointing speleothem formed from the precipitation of calcium carbonate derived from the dripping of saturated groundwater from the ceiling of a cave. (19)

stick-slip The behavior along a fault plane that describes the mechanics of earthquakes. A fault plane may "stick" for long periods of time, held together by the frictional strength of rock on either side of a fault. At the threshold where the accumulated tectonic stress exceeds the frictional strength of the rocks, the blocks of rock on either side of the fault will "slip," resulting in the release of the pent-up energy as an earthquake. (1)

strata A term that refers to an individual layer or stack of layered sedimentary rocks. (1)

stratovolcano Steep-sided, conical edifices created by the accumulation of overlapping layers of pyroclastic debris and viscous lava flows. (12)

stream capture (aka *stream piracy*) A process where the drainage from a stream is diverted into the channel of a neighboring stream; may occur due to headward erosion, local tectonic activity, or damming of the stream by a landslide or ice sheet. (9)

stream discharge The volume of water in a flowing stream that passes a given location in a unit of time; frequently expressed as cubic feet per second or cubic meters per second; dependent on the cross-sectional area of the channel and the average velocity of the stream. (4)

stream terrace Broad, flat benches that step upward above the active channel of a stream; each terrace marks the location of a former floodplain and thus a previous position of the river. (25)

strike-slip fault Type of fault where the block of rock on one side slips laterally relative to the block on the other side, with minimal vertical motion; form in response to shear stress. (24)

stromatolite Mounded or columnar structures with internal laminations formed by the photosynthetic activity of cyanobacteria; typically preserved as limestone or dolomite, recording shallow water environments. (3)

Strombolian eruption Type of volcanic eruption characterized by fountains of lava jetting upward from a central crater; droplets of lava commonly form scoria particles that build to create cinder cones. (16)

structural arch A convex-upward flexure of sedimentary strata, commonly bounded on both sides by reverse faults or thrust faults, and underlain by crystalline basement; formed by compressional tectonics. (9)

subduction The process at convergent plate boundaries where a denser plate descends beneath a less dense plate to be consumed back into the mantle. (1)

subduction zone The region involved in subduction, including a volcanic arc, deep-sea trench, explosive volcanism, and large-magnitude earthquakes. (1)

sublimation The transition of solid ice directly to water vapor without an intermediate liquid phase; an important process on the zone of ablation of a glacier. (14)

submarine fan A fan-shaped, convex-upward depositional environment located near the mouths of submarine canyons along the base of continental slopes; turbidity currents transport sediment into the deep ocean through submarine canyons, then dump their load within submarine fans. (30)

subsidence Vertical sagging of the crust under the weight of an overlying load, such as a mass of sediment. (1)

supergroup A thick assemblage of sedimentary strata that combines formations into groups, which in turn may compose a supergroup. (37)

superheated water Under high pressures, water may be heated to beyond its usual boiling point of 100°C (212°F); contributes to the hydrothermal activity of hot springs and geysers. (31)

superposed stream A stream or river that establishes its channel on an unconsolidated layer of alluvium that overlies harder rock. With regional uplift, the stream easily erodes through the alluvium. Upon contact with the underlying harder rock, the stream incises downward while maintaining the same channel morphology, often carving deep and narrow incised meanders. (6)

superposition Principle of relative age dating that states that in an undeformed succession of sedimentary rock layers, the oldest layers will be at the bottom and the youngest at the top. (1)

supervolcano A somewhat arbitrary term that describes a volcano that has erupted more than 1000 km² (240 mi²) of material at any one point in time. Also pertains to volcanoes like Yellowstone that may erupt this volume of material in the future. (31)

surface exposure dating A set of radiometric dating techniques for estimating the length of time that a rock or mineral grain has been exposed at Earth's surface. (39)

suture zone Linear tectonic boundary between two adjacent accreted terranes, commonly expressed as a laterally continuous fault zone; interpreted to be the remnant of a former subduction zone and thus the scar of an ancient ocean basin, now consumed beneath the continent. (21)

syncline A fold in rock strata with a concave upward shape. (22)

synclinorium A large downwarp of sedimentary rock with superimposed smaller folds. (36)

Taconic Orogeny The first of the three great mountain-building episodes along the North American continental margin during the Paleozoic; involved the accretion of a series of volcanic islands, ending ~440 Ma; rocks of the Taconic terranes are exposed today in New England, the Piedmont, and Great Smoky National Park. (35)

talus An accumulation of angular rocky debris at the base of a cliff; commonly derived from rockfalls and other mass wasting events. Fragmental debris may range in size from pebbles to house-sized blocks. (4)

talus cave Large void within a body of rock initially formed by erosion from flash floods and debris flows, then covered by large boulders that tumble down from higher elevations. (28)

tarn Lake or pond that occupies the deepest part of a cirque basin; often dammed by a moraine at one end. (18)

tectonic window An erosional feature that reveals the rock beneath a thrust fault; the window itself is surrounded on the landscape by the rock overlying the thrust fault. (37)

tectonics The large-scale, long-term motion and deformation of Earth's lithospheric plates. (1)

tephra A general term describing fragments of pyroclastic debris ranging from tiny particles of ash to boulder-sized blocks. (12)

terrane A fault-bounded mass of crustal rock with unique characteristics relative to adjacent masses of rock; commonly form by tectonic accretion along an active continental margin. (19)

thermochronology An analytical method that involves the radiometric dating of the temperature history of specific minerals to determine how long the mineral was buried before it was exposed at the surface; provides a way to reconstruct the elevations of past landscapes. (34)

thermophiles Heat-loving archaea and bacteria that thrive in the high-temperature environments of hot springs and deep-sea hydrothermal vents. (31)

thrust fault A type of low-angle reverse fault formed by compressional stress; the block of rock above the fault plane is pushed up and laterally over the block of rock beneath the fault plane; differs from a high-angle reverse fault in that the angle of the thrust plane must be 45° or less; the angle of most thrust faults is more commonly closer to 15°. (22)

tilted fault-block A style of mountain-building caused by extensional tectonism and the formation of a series of normal faults; rotation of a fault-block through time results in the uplift of one corner of the block, forming an elongate mountain range, and the downdropping of the opposite corner of the block, forming an adjacent basin. (17)

topographic bulge A domal growth on a volcanic edifice indicating the intrusion of magma into the body of a volcano, deforming its outer surface; may presage a volcanic eruption. (13)

topographic relief The difference in elevation within a topographic region; commonly used to refer to the difference between the highest and lowest points in a given area. (1)

trace fossil A remnant trace of former biological activity without any preserved body fossils of the organism; may consist of impressions along the bedding planes of sedimentary rock, burrows, trackways, plant root cavities, and many others. (36)

transform fault A type of fault, usually strike-slip, that characterizes transform plate boundaries; motion along transform faults is mainly horizontal. (1)

transform plate boundary A type of plate margin dominated by tectonic shear stress, resulting in plates moving laterally in opposite directions, often during earthquakes. Transform boundaries, commonly expressed as a zone of strike-slip faulting, bisect mid-ocean ridge segments or cut through continental crust where they connect with divergent or convergent boundaries. (1)

transgression The landward migration of the shoreline and associated coastal and marine environments due to a rise in relative sea level; recognized by an upward deepening of apparent water depths in a vertical succession of sedimentary rock. (3)

transportation The movement of loose particles of sediment downslope under the force of gravity, usually by rivers, wind, glacial ice, and landslides. (1)

transpression The hybrid combination of compression and shear stresses that occurs along many faults; generates small features like pressure ridges or huge mountains like the Transverse Ranges of Southern California. (26)

transtension The hybrid combination of extension and shear stresses that occurs along many faults; generates local features such as sag ponds, or large pull-apart basins such as Death Valley. (26)

travertine A light-colored deposit of calcium carbonate formed near hot springs or within limestone caves by precipitation from calcite-saturated fluids. (31)

tributary stream Stream or river that flows into a larger stream or river; tributaries of all lengths direct flow from all corners of a watershed into the main river. (3)

trimline The near-horizontal horizon along the margins of highlands that marks the highest level of an alpine glacier;

distinguished by smoothed rock below and angular rock above. (17)

triple divide A mountain peak that separates drainage into three distinct watersheds. (22)

triple junction (aka *triple point*) The location where three lithospheric plates meet. (26)

tsunami A series of waves in oceans or lakes created by the physical displacement of water by underwater earthquakes, volcanic eruptions, or underwater landslides. As the mass of displaced water moves into shallower water near the coastline, it rises upward in height before washing inland for a kilometer or more, depending on the slope of the coastal plain. (30)

tubular passage Oval-shaped caves formed within the phreatic zone during long-term stability in the position of the water table. (39)

tuff A general term for all solidified deposits of pyroclastic debris. (15)

tumuli Elliptical or domal mounds that rise a meter or more above the surface and mark places where the brittle crust of lava flows was inflated upward by the flow of underlying gaseous lava. (32)

turbidites Layered couplets of sandstone and shale deposited at the mouths of submarine canyons by turbidity currents, bottom-hugging slurries of sediment-laden water; the larger sand grains fall out first, followed by the slow settling of finer clays from suspension. (26)

U-shaped valley Typical profile of valleys formed by glacial erosion, as opposed to the V-shape created by stream-carved valleys; alpine glaciers remove bedrock protrusions extending from the walls of the valley, leaving behind a smoothed valley shape. (18)

unconformity A buried surface that separates two rock masses of significantly different ages, indicating a break in deposition of some duration; represents a period of time dominated by the erosional removal of rock or by a phase of nondeposition. (3)

uniformitarianism Eighteenth-century principle that proposed that geological processes and events occur at the same rate today as they did in the geologic past; commonly simplified to the phrase "the present is the key to the past;" replaced by the modern principle of actualism. (1)

uplift A general term that describes the rise of crustal rock to higher elevations; commonly a regional response to larger-scale tectonic processes. (23)

vadose zone The unsaturated zone of rock and soil above the water table where pores and cavities are filled with air. (39)

valley glacier (aka *alpine glacier*) A glacier that is confined by surrounding mountain terrain; typically follows the route of preexisting stream-carved valleys. (14)

vent Any opening at the Earth's surface through which magma erupts or volcanic gases are emitted; craters and fissures are common examples. (12)

vertical shaft A cylindrical, vertical cave formed as acidic groundwaters penetrate into a body of limestone along vertical fractures; connects horizontal cave levels within the cave system. (39)

vesicles Rounded holes within volcanic rock that mark the locations where bubbles of gas were released during cooling. (16)

viscosity The resistance of a fluid to flow; fluids with a high viscosity resist flow, whereas fluids with a low viscosity flow freely. The viscosity of a magma is dictated by the amount of silica. (12)

vog A volcanic smog produced by the interaction of sulfur-rich gases, water vapor, and sunlight; poses a significant hazard to humans and other animals. (32)

volatile gases Reactive gases that remain in the gas phase at relatively low surface temperatures; examples include water, carbon dioxide, methane, and sulfur dioxide. (12)

volcanic ash Dust- to sand-sized particles of volcanic rock and tiny shards of glass less than 2 mm in size; formed by the explosive shattering of magma during an eruption and by magma being ejected as a fine spray. (12)

volcanic breccia Rock composed of angular, cobble-sized fragments embedded within a finer grained matrix; commonly form from rockslides along the flanks of volcanoes or from eruptions of pieces of rock from the throat of a volcano that eventually fuse within a matrix of volcanic ash. (28)

volcanic dome Steep-sided mass of viscous and often blocky lava extruded from a vent; typically has a rounded top and covers a roughly circular area; commonly silica-rich in composition. (13)

volcanic edifice The main body of a volcano, produced by effusive or explosive eruptions of lava or tephra. (15)

Volcanic Explosivity Index A relative measure of the explosiveness of eruptions based mostly on the volume of erupted material; the logarithmic scale ranges from 0 (nonexplosive) to 8 (mega-colossal). (31)

volcanic igneous rocks Fine-grained rock that forms by the rapid cooling and crystallization of lava or tephra on Earth's surface. Common examples include basalt, andesite, and rhyolite. (1)

volcanic island arc A chain of volcanic islands formed along subduction zones created by the convergence of an oceanic plate against another oceanic plate; examples include the Japanese, Indonesian, and Philippine islands. (1)

water gap A channel or canyon cut across a narrow mountain range or ridge through which water still flows; may form through headward erosion of a stream. (36)

water table The upper limit of groundwater that roughly parallels the surface topography; pores and cavities above the water table are filled with air, whereas pores and cavities below the water table are filled with water. (19)

watershed (aka *drainage basin*) The total land area where surface runoff from rainfall and snowmelt drain through countless rivulets and creeks into larger tributary streams that in turn merge into a single main river. (2)

wave refraction The bending of ocean waves as they encounter protruding headlands along a coastline. (29)

wave-cut bench Narrow, horizontal ledges cut into the flanks of mountains by wave erosion; mark the former position of lake level. (24)

wave-cut platform Flat or gently seaward-sloping rocky surface formed by wave erosion; may eventually be uplifted to form marine terraces. (29)

weathering The in-place breakdown of minerals in rock via chemical and physical activity; commonly followed by erosion, which is the physical removal of weathered grains away from their source. (1)

welded tuff (aka *ash-flow tuff* or *ignimbrite*) Volcanic rock formed by widespread deposition and solidification of material in a pyroclastic flow; heat from the deposit welds the particles together. (15)

wind gap A valley or channel carved by running water but now left dry as a result of stream capture; often used as passes for roadways and railroads. (36)

xenolith A distinct fragment of rock surrounded by plutonic rock; commonly forms by the breakage of rock along the roof or wall of a magma chamber and its incomplete melting within the adjoining body of magma. (18)

zircon A silicate mineral ($ZrSiO_4$) that crystallizes from magma; exceptionally resistant to heat and erosion, enabling zircon grains to be found in metamorphic and sedimentary rocks. Zircons incorporate uranium and thorium into their lattices during crystallization, making them invaluable for dating by a variety of radiometric techniques. (1)

zone of ablation The area near the toe of a glacier where the processes of melting, sublimation, evaporation, and calving of icebergs dominate; the toes of glaciers retreat upslope when the rate of ablation exceeds the rate of accumulation at the head. (14)

zone of accumulation The area near the head of a glacier where snow accumulates, becomes buried and transforms to ice, and then begins to flow downslope under the force of gravity; the toes of alpine glaciers advance downslope when the rate of accumulation exceeds the rate of ablation. On continental ice sheets, ice flows laterally outward away from the zone of accumulation. (14)

Index

Figures are indicated by "*f*" following the page number.